Frontiers in Probability and the Statistical Sciences

The "Frontiers" is a new series of books (edited volumes and monographs) in probability and statistics designed to capture exciting trends in current research as they develop. Some emphasis will be given on novel methodologies that may have interdisciplinary applications in scientific fields such as biology, ecology, economics, environmental sciences, finance, genetics, material sciences, medicine, omics studies and public health.

Narayanaswamy Balakrishnan • Tomasz Rychlik

Sharp Inequalities for Ordered Random Variables in Statistics and Reliability

Volume I: Standard Order Statistics

Springer

Narayanaswamy Balakrishnan
Department of Mathematics and Statistics
McMaster University
Hamilton, ON, Canada

Tomasz Rychlik
Department of Probability Theory
Polish Academy of Sciences, Institute of Mathematics
Warsaw, Poland

ISSN 2624-9987 ISSN 2624-9995 (electronic)
Frontiers in Probability and the Statistical Sciences
ISBN 978-3-031-61346-3 ISBN 978-3-031-61347-0 (eBook)
https://doi.org/10.1007/978-3-031-61347-0

Mathematics Subject Classification: 62N05, 60E15, 62G30, 62G32, 62E10, 62G70

This Springer imprint is published by the registered company Springer Nature Switzerland AG
The registered company address is: Gewerbestrasse 11, 6330 Cham, Switzerland

To my precious granddaughter,
Maya Gabriel Ewing Balakrishnan

NB

To Mariola, Marzena, Janek and Michał

TR

Preface

Order statistics and associated properties and applications have been studied rather extensively in the literature. The elaborate discussions on many different aspects pertaining to order statistics, as presented in the following books, provide an ample testimony to the intense activity seen in this area of research:

H.A. David and H.N. Nagaraja (2003). *Order Statistics*, Third edition, John Wiley & Sons, Hoboken, New Jersey;
B.C. Arnold, N. Balakrishnan and H.N. Nagaraja (1992). *A First Course in Order Statistics*, John Wiley & Sons, New York;
N. Balakrishnan and C.R. Rao (Eds.) (1998a). *Handbook of Statistics–16: Order Statistics–I: Theory and Methods*, North-Holland, Amsterdam, The Netherlands;
N. Balakrishnan and C.R. Rao (Eds.) (1998b). *Handbook of Statistics–17: Order Statistics–II: Applications*, North-Holland, Amsterdam, The Netherlands.

Even though many different aspects of order statistics and their applications have all been discussed in great detail in all these volumes, one topic that has seen considerable new developments during the last two to three decades is "Inequalities for Ordered Random Variables and Their Applications". In fact, the following two books synthesized all the past developments on this topic:

B.C. Arnold and N. Balakrishnan (1989). *Relations, Bounds and Approximations for Order Statistics*, Lecture Notes in Statistics–No. 53, Springer-Verlag, New York;
T. Rychlik (2001). *Projecting Statistical Functionals*, Lecture Notes in Statistics–No. 160, Springer-Verlag, New York.

In addition, an elaborate review article by T. Rychlik on this topic has been published in the volume by N. Balakrishnan and C.R. Rao (1998a). All these discussions and reviews of the literature are, however, more than two decades old. During the last 20 years, many new results on bounds for order statistics have been established, several new methods of obtaining bounds have been introduced, and some novel applications of these bounds to a wide range of applied problems have also been discussed in the literature. This is what provided an impetus to prepare this book in order to provide an up-to-date and thorough review of all the developments on "Inequalities for Ordered Random Variables and Their Applications".

But, while going through the extensively published literature on this topic, we realized that this one volume will not suffice to review all the results. Hence, we plan to cover all the results concerning only standard order statistics in this volume. Moreover, most of the bounds detailed here are sharp in the sense that they get achieved for some specific distribution and the corresponding discussions are also added at respective places. The second volume will then focus on inequalities for other forms of ordered data and their applications; specifically, these would cover inequalities for record values, progressively censored order statistics and generalized order statistics, and their diverse applications. This second volume is expected to be published next year!

It is our sincere hope that the researchers engaged in the area of order statistics and also those who intend to take on this research area to make their own contributions will find the present volume and the second one to be published next to be valuable resources! We welcome the readers to inform us of any mistakes or errors they find and also of any additional comments and insights for any of the results presented here, and we will certainly appreciate that!

Hamilton, ON, Canada N. Balakrishnan
Warsaw, Poland T. Rychlik
December 2023

Acknowledgements

Our sincere thanks go to all the researchers who have made fine contributions over the years to the area of "Inequalities for Ordered Random Variables". It is these works, which consolidated this area of statistical research, that formed the core for this volume.

Our thanks go to Dr. Eva Hiripi (Senior Editor, Statistics Books Springer) and Mr. Daniel Ignatius Jagadisan (Project Coordinator, Books Springer) for taking a keen interest on this book project and for providing constant support, starting from the proposal stage all the way to the final production stage. We sincerely appreciate the outstanding technical support and production assistance provided by the entire production team of Elsevier. Without their kind help and cooperation, the book would not have come to this fine final form!

We thank Ms. Debbie Iscoe (McMaster University, Hamilton, Canada) for skillfully transforming the manuscript into a nice and elegant manuscript suitable for submission for publication. We are grateful to Dr. Magdalena Szymkowiak for carefully performing a substantial part of the numerical computations presented in this volume for the purpose of illustrating the usefulness of the associated results. The first author also thanks the Natural Sciences and Engineering Research Council of Canada for the financial support provided through an individual discovery grant that assisted in the preparation of the manuscript.

Finally, we express our sincere gratitude to all our family members for bearing with us when we worked on this volume, and for providing us with constant support and great encouragement!

Hamilton, ON, Canada
Warsaw, Poland
December 2023

N. Balakrishnan
T. Rychlik

Contents

Chapter 1
Introduction and Notation

1.1 Preamble

Mathematical inequalities, seeking to set bounds for a mathematical expression, have a long and rich history. Though many of them were explored initially as a mathematical curiosity, they started to be studied subsequently with a rigorous mathematical purpose and intent, and sometimes even motivated by some applied practical problems. Then, the publication of the book *Inequalities* in 1934, by Hardy, Littlewood and Pólya (see Hardy et al., 1934), resulted in making the study of inequalities into a formal and active area of mathematical research. In the field of Probability and Statistics, various stochastic inequalities were also established, including the celebrated Chebyshev, Markov and Jensen inequalities.

In the context of ordered statistical data, discussions on inequalities and bounds seem to have started in mid 1930s. The first work in this direction appears to be in the work of Pearson and Chandra Sekar (1936) while proposing a criterion for testing for the presence of outlying observations. A formal mathematical proof of the inequality considered by Pearson and Chandra Sekar, along with a detailed discussion on the attainment (i.e., the sharpness) of the corresponding bound, was given by Scott (1936) in an Appendix to the paper of Pearson and Chandra Sekar (1936). Subsequently, some significant contributions to inequalities and bounds for order statistics were made by Plackett (1947), Moriguti (1951, 1953), Gumbel (1954), Hartley and David (1954), and Sugiura (1962, 1964). They all introduced different tools and techniques to the literature such as inequalities through calculus of variation, Cauchy-Schwarz type inequalities, greatest convex minorant inequalities, and inequalities and bounds through orthogonal inverse expansions. Two additional significant developments came through the works of Samuelson (1968) and Lai and Robbins (1976, 1978), with the latter particularly focusing on joint distributions that form as extreme cases for distributions of

N. Balakrishnan, T. Rychlik, *Sharp Inequalities for Ordered Random Variables in Statistics and Reliability*, Frontiers in Probability and the Statistical Sciences,
https://doi.org/10.1007/978-3-031-61347-0_1

possibly dependent maxima. These works were all followed by a flurry of papers dealing with a plethora of variations, extensions and generalizations.

Extensive reviews of all these historic as well as subsequent developments have been furnished in the books by Arnold and Balakrishnan (1989) and Rychlik (2001), and also in the elaborate survey article by Rychlik (1998). Since then, many new results on inequalities and bounds for order statistics have been developed, several new methods of deriving bounds have been introduced, and also some interesting applications of these bounds to problems of applied nature have been discussed in great detail. These, in fact, form the basis for this book aiming to provide an up-to-date review of all the work on inequalities and bounds for order statistics!

1.2 Scope of the Book

As the whole book deals with inequalities for order statistics, Chap. 2 provides a detailed overview of various analytic inequalities and some other related notions and tools. Specifically, it first introduces the basic definitions of function spaces including linear spaces, Banach spaces, Hilbert spaces and orthonormal systems in Hilbert spaces. Next, several inequalities for monotone sequences and functions are described and these include Chebyshev inequality, Abel inequality, inequalities based on stochastic orderings and Moriguti inequality. This is followed by a description of different norm inequalities including Schwarz inequality, projection inequalities, inequalities based on orthogonal inverse expansions, Hölder inequality, Minkowski inequality, Young inequality and inequalities obtained by norm maximization. Then, Steffensen inequality is discussed in detail. Next, as convex functions play a key role in many statistical problems, inequalities for convex functions are described and these include Jensen inequality, Jensen-Steffensen inequality and Bernoulli inequality. Finally, Descartes rule of signs and the notion of transform orders, including convex transform order and star order, are introduced. All the inequalities and related notions reviewed in this Chapter form the foundation for the development of various bounds and inequalities for ordered random variables and different statistics involving them in the subsequent Chapters.

Chapter 3 provides an elaborate discussion on deterministic bounds for order statistics and L-statistics (i.e., linear functions of order statistics). Specifically, Schwarz bounds are derived for these quantities as well as lower and upper bounds based on Moriguti-Schwarz inequality and bounds based on Moriguti-Hölder inequality. The case of nonnegative samples is also discussed in these cases. Moreover, negative upper bounds and positive lower bounds and lower inequality for scale estimators are developed based on norm maximization bounds. Finally, some generalizations of the presented results are detailed.

In Chap. 4, the focus is made on the cases of independent identically distributed random samples arising from arbitrery, symmetric and life distributions. To begin with, positive upper and negative lower bounds are derived based on Schwarz inequality for largest order statistic, sample range statistic and some other L-

statistics. Then, based on Schwarz-Moriguti inequality, positive upper and negative lower bounds are presented for individual order statistics, differences of order statistics and general L-statistics. Next, positive upper and negative lower bounds are discussed based on Hölder-Moriguti inequality in terms of different units. Also, the results for nonpositive upper and nonnegative lower bounds are presented. Results for the case of symmetric distributions are then detailed. When the order statistics arise from life distributions, bounds on their expectations as well as other moments are presented. The results on upper bounds in terms of Gini mean difference units are then given for individual order statistics, differences of order statistics, general L-statistics as well as linear estimators of location and scale parameters of population distributions. Then, based on mean residual life inequalities, lower and upper bounds are presented. The orthogonal inverse expansion method is then used to develop bounds for moments of order statistics from an arbitrary distribution as well as from a symmetric distribution and also for the case of distributions for which the mean of the distribution does not exist. Based on this method, improved imputation-type bounds are also presented by utilizing some moments of order statistics that are available exactly for some sample sizes. Finally, some variance bounds are described for individual order statistics, spacings and general L-statistics for an arbitrary distribution and also for a symmetric distribution.

Bounds for order statistics arising from samples from finite populations are described next in Chap. 5. Specifically, upper bounds for individual order statistics and differences of order statistics are presented. Further, positive upper and negative lower bounds as well as negative upper and positive lower bounds for general L-statistics are discussed. Additionally, bounds for variances of order statistics are given. Then, bounds for order statistics are presented under with replacement model from general distributions in the cases of fixed drawing distribution and arbitrary drawing distribution. Next, the case of order statistics under without replacement model is discussed in detail. The upper bounds for individual order statistics and differences of order statistics are given. Moreover, positive upper and negative lower bounds as well as negative upper and positive lower bounds for general L-statistics are presented. Also, bounds for variances of order statistics are given.

Finally, in Chap. 6, bounds for order statistics arising from independent and identically distributed random variables from some shape restricted families are described. To start with, upper bounds for individual order statistics and spacings are presented for decreasing density, decreasing failure rate, and decreasing reversed failure rate distributions based on projection inequalities. Also, nonnegative upper bounds for individual order statistics are given. Moreover, both positive upper bounds and nonpositive upper bounds for individual order statistics are presented for symmetric unimodal distributions. Next, the corresponding results for increasing density, increasing failure rate and increasing reversed failure rate distributions are presented. Specifically, positive upper bounds for the case of extreme order statistics for increasing density and increasing failure rate distributions are given, after which positive upper bounds for nonextreme order statistics and spacings are presented for increasing density and increasing failure rate distributions. Also, positive upper bounds for order statistics are given for increasing reversed failure rate distributions.

Then, for the cases of decreasing reversed failure rate and increasing reversed failure rate distributions, bounds for spacings are presented. Next, in a similar manner, results are developed for decreasing density on average, increasing density on average, decreasing failure rate on average and increasing failure rate on average families of distributions. Specifically, positive upper bounds are provided for order statistics and spacings for the cases of decreasing density on average and decreasing failure rate on average families of distributions while nonpositive upper bounds are given for individual order statistics in these cases. For increasing density on average and increasing failure rate on average families of distributions, upper bounds for general L-statistics are presented. Finally, some quantile inequalities are described for individual order statistics and convex functions of order statistics for arbitrary distributions and also some results for the special case of symmetric distributions.

Through out the book, many numerical evaluations and comparisons of bounds are made for different distributions and varying sample sizes. Moreover, most of the bounds detailed here are sharp in the sense that they get achieved for some specific distribution and the corresponding discussions are also added at respective places. These numerical evaluations demonstrate not only how efficient and accurate the various bounds are, but also how they can be readily computed for the required assessments!

1.3 Abbreviations and Notations

In order to facilitate an easy and smooth reading of all the materials presented here, we have adopted an unified set of abbreviations and notations as follows:

iid	independent and identically distributed
iff	if and only if
LHS	left-hand side
RHS	right-hand side
A	arbitrarily shaped (distributions)
S	symmetric
SU	symmetric unimodal
DD	decreasing density
ID	increasing density
DFR	decreasing failure rate
IFR	increasing failure rate
DRFR	decreasing reversed failure rate
IRFR	increasing reversed failure rate
α-DGFR	α-generalized decreasing failure rate
α-IGFR	α-generalized increasing failure rate
α-DGRFR	α-generalized decreasing reversed failure rate

α-IGRFR	α-generalized increasing reversed failure rate
DDA	decreasing density on average
IDA	increasing density on average
DFRA	decreasing failure rate on average
IFRA	increasing failure rate on average
VDP	variation diminishing property
x_+	$= \max\{x, 0\}$—nonnegative part of a number
x_-	$= -\min\{x, 0\}$—nonpositive part of a number
$\lfloor x \rfloor$	$= \max\{k \le x : k \text{ integer}\}$—floor of a number
$\lceil x \rceil$	$= \min\{k \ge x : k \text{ integer}\}$—ceiling of a number
$\mathbb{R}$	$= (-\infty, +\infty)$—set of real numbers
$\mathbb{R}_+$	$= [0, +\infty)$—set of nonnegative real numbers
$\mathbb{Z}$	$= \{0, 1, -1, \ldots\}$—set of integers
$\mathbb{N}$	$= \{1, 2, \ldots\}$—set of positive integers
$\mathbb{S}^n$	$= \{\tilde{s} = (s_1, \ldots, s_n) \in \mathbb{R}^n,\ s_i \ge 0,\ \sum_{i=1}^n s_i = 1\}$— simplex in $\mathbb{R}^n$
$\mathbf{1}(x)$	constant function equal to 1
$\tilde{1}$	$= (1, \ldots, 1) \in \mathbb{R}^n$ constant sequence of length n, with all coordinates equal to 1
$\mathbf{1}_A(x)$	indicator function of set A (=1 for $x \in A$ and 0 otherwise)
$\binom{n}{r}$	$= \begin{cases} \frac{n!}{r!(n-r)!}, & \text{for } r \le n, \\ 0, & \text{for } r > n, \end{cases} \quad r, n \in \mathbb{N}$
$\Gamma(p)$	$= \int_0^\infty x^{p-1} e^{-x} dx$—gamma function
$(\mathcal{B}, \|\cdot\|)$	Banach space with norm $\|g\|$, $g \in \mathcal{B}$
$(\mathcal{H}, \langle\cdot,\cdot\rangle)$	Hilbert space with inner product $\langle g, h\rangle \in \mathbb{R}$, $g, h \in \mathcal{H}$ and norm $\|g\| = \langle g, g\rangle^{1/2}$, $g \in \mathcal{H}$
$L^p((a, d), W(dx))$	$= (L^p((a, d), w(x)dx))$—Banach space of functions $g : (a, d) \to \mathbb{R}$ satisfying $\int_a^d \|g(x)\|^p\, W(dx) < \infty$ for $p \ge 1$ and a nonnegative measure distribution function $W : (a, d) \to \mathbb{R}$ (with density function $w(x)dx$) and norm $\|g\|_p = [\int_a^d \|g(x)\|^p W(dx)]^{1/p}$
$L^2((a, d), W(dx))$	Hilbert space of functions $g : (a, d) \to \mathbb{R}$ satisfying $\int_a^d \|g(x)\|^2\, W(dx) < \infty$ with inner product $\langle g, h\rangle = \int_a^d g(x)h(x)W(dx)$
$L^\infty((a, d), W(dx))$	Banach space of functions $g : (a, d) \to \mathbb{R}$ bounded W-almost everywhere with norm $\|g\|_\infty = \text{ess sup}\|g(x)\| = \inf\{M \ge 0 : \|g(x)\| < M,\ W - \text{a.e.}\}$
q	$= \begin{cases} +\infty & \text{for } p = 1 \\ \frac{p}{p-1} & \text{for } 1 < p < \infty \\ 1 & \text{for } p = +\infty \end{cases}$ —conjugate coefficient of p

$T_h(g)$	continuous linear functional on Banach space $(\mathcal{B}, \|\cdot\|)$, with $h \in (\mathcal{B}^*, \|\cdot\|_*)$, the conjugate Banach space, and norm $\|T_h\| = \|h\|_*$
$T_h(g)$	$= \int_a^d g(x)h(x)W(dx)$, $g \in L^p((a,d), W(dx))$, $h \in L^q((a,d), W(dx))$—continuous linear functional on Banach space $L^p((a,d), W(dx))$ with norm $\|T_h\| = \|h\|_q$
$T_h(g)$	$= \langle h, g\rangle$, $h, g \in \mathcal{H}$—continuous linear functional on Hilbert space $(\mathcal{H}, \langle\cdot,\cdot\rangle)$ with norm $\|T_h\| = \langle h, h\rangle^{1/2}$
$T_h(g)/\|g\|$	normalized continuous linear functional
$\{\phi_k(u)\}_{k=0}^{\infty}$	complete orthonormal system in a Hilbert space $\mathcal{H}$
$\phi_k(u)$	$= \frac{\sqrt{2k+1}}{k!}\frac{d^k}{du^k}\{u^k(u-1)^k\}$, for $k = 0, 1, \ldots$, the sequence of complete orthonormal Legendre polynomials in $L^2((0,1), dx)$
α_k	Fourier coefficient associated with $\phi_k(u)$
$\alpha, \beta, \ldots$	scalars, parameters of functions
$\alpha_*, \beta_*, \ldots$	optimal scalars
$g(x), h(x), \ldots$	functions, elements of function spaces
$G(x), H(x), \ldots$	antiderivatives of $g(x), h(x), \ldots$, respectively
$g \circ H(x)$	$= g(H(x))$—composition of functions H and g
$\underline{H}(x)$	greatest convex minorant of $H(x)$
$\underline{h}(x)$	(right) derivative of $\underline{H}(x)$—projection of h onto the family of nondecreasing functions in $L^2((a,d), dx)$
$\overline{H}(x)$	smallest concave majorant of $H(x)$
$\overline{h}(x)$	(right) derivative of $\overline{H}(x)$—projection of h onto the family of nonincreasing functions in $L^2((a,d), dx)$
$\tilde{x}$	$= (x_1, \ldots, x_n) \in \mathbb{R}^n$, $n \in \mathbb{N}$—(finite) sequence of numbers
$\tilde{c}$	$= (c_1, \ldots, c_n) \in \mathbb{R}^n$—(finite) sequence of coefficients
$\tilde{c}'$	$= (c_n, \ldots, c_1)$
$\tilde{c}^s$	$= (c_1 - c_n, \ldots, c_n - c_1)$
$\underline{\tilde{c}}$	$= (\underline{c}_1, \ldots, \underline{c}_n)$—projection of $\tilde{c} \in \mathbb{R}^n$ onto the family of nondecreasing sequences in $l^2(\mathbb{R}^n)$
$\overline{\tilde{c}}$	$= (\overline{c}_1, \ldots, \overline{c}_n)$—projection of $\tilde{c} \in \mathbb{R}^n$ onto the family of nonincreasing sequences in $l^2(\mathbb{R}^n)$
Π	$= \{x_1 \le \cdots \le x_N\}$—finite population of N ordered numbers
$\bar{x}$	$= \frac{1}{N}\sum_{k=1}^{N} x_k$—population mean
s_p^p	$= \frac{1}{N}\sum_{k=1}^{N} \|x_k - \bar{x}\|^p$, $1 \le p < +\infty$—population p-th central absolute moment
s_p	$= \left(s_p^p\right)^{\frac{1}{p}}$, pth root of pth absolute central moment
s_∞	$= \sup\{\|x_k - \bar{x}\| : k = 1, \ldots, N\} = \max\{x_N - \bar{x}, \bar{x} - x_1\}$
Π_0	$= \{1, \ldots, N\}$—standard uniform finite population of size N
$\mathbb{P}(A)$	probability of event A

$X, X_1, X_2, \ldots, X_n$	random variables
$F(x)$	$= \mathbb{P}(X \le x)$—distribution function of X
$F(x-)$	$= \lim_{y \nearrow x} F(y) = \mathbb{P}(X < x)$
$F^{-1}(x)$	$= \sup\{y : F(y) \le x\}, 0 \le x < 1$—upper quantile function of X
$F_{\|y}(x)$	$= \frac{F(x)-F(y)}{1-F(y)}, x \ge y$—distribution function of left-truncated X at y
$F_{y\|}(x)$	$= \frac{F(x)}{F(y)}, x \le y$—distribution function of right-truncated X at y
$f(x)$	$= \frac{dF}{dx}(x)$—density function of X
$p(x)$	$= \mathbb{P}(X = x)$—probability mass function of discrete X
$\lambda(x)$	$= \frac{f(x)}{1-F(x)}$—hazard (failure) rate of X
$\Lambda(x)$	$= -\ln(1 - F(x))$—cumulative hazard (failure) function of X
μ	$= \mathbb{E}X = \int_0^1 F^{-1}(x)dx$—expectation (mean) of X
m_p^p	$= \mathbb{E}\|X\|^p = \int_0^1 \|F^{-1}(x)\|^p dx$—$p$th absolute moment of X
m_p	$= (\mathbb{E}\|X\|^p)^{\frac{1}{p}}$
$\sigma^2 = \sigma_2^2$	$= \mathbb{V}ar\, X = \mathbb{E}\|X - \mu\|^2 = \int_0^1 \|F^{-1}(x) - \mu\|^2\, dx$—variance of X
σ_p^p	$= \mathbb{E}\|X - \mu\|^p = \int_0^1 \|F^{-1}(x) - \mu\|^p\, dx,\ p \ge 1$—$p$-th central absolute moment of X
σ_1	$= \mathbb{E}\|X - \mathbb{E}X\|$—mean absolute deviation from the mean of X
$\sigma = \sigma_2$	$\sqrt{\sigma^2} = \sqrt{\sigma_2^2} = \sqrt{\mathbb{V}ar\, X}$—standard deviation of X
σ_∞	$= \text{ess sup}\|X - \mu\| = \sup\{\|F^{-1}(x) - \mu\| : 0 \le x < 1\}$
$\mu(x)$	$= \mathbb{E}(X - x \| X > x)$—mean residual life of X
$\mathbb{C}ov(X, Y)$	$= \mathbb{E}(X - \mathbb{E}X)(Y - \mathbb{E}Y)$—covariance of X and Y
Δ	$= \mathbb{E}\|X_1 - X_2\|$, where X_1, X_2 are iid—Gini mean difference
$W(x)$	a (distinguished) distribution function which is weight distribution function in $L^p((a, d), W(dx))$ with support $(a, d) = (a_W, d_W)$
$w(x)$	density function of W which is weight density function in $L^p((a, d), w(x)dx)$
$U(x) = W_1(x)$	$= x, 0 \le x \le 1$,—distribution function of standard uniform distribution (uniform distribution on interval $(0, 1)$)
$Z(x) = W_0(x)$	$= 1 - \exp(-x), x \ge 0$,—standard exponential distribution function
$V(x) = V_0(x)$	$= \exp(x), x \le 0$, standard negative exponential distribution function
$W_\alpha(x)$	$= 1 - (1 - \alpha x)^{\frac{1}{\alpha}}, \alpha \ne 0$, α-generalized Pareto distribution
$V_\alpha(x)$	$= (1 + \alpha x)^{\frac{1}{\alpha}}, \alpha \ne 0$, α-generalized negative Pareto distribution

$\bar{X}$	$= \bar{X}_n = \frac{1}{n}\sum_{i=1}^{n} X_i$—sample mean
S_p^p	$= \frac{1}{n}\sum_{i=1}^{n} \|X_i - \bar{X}\|^p$, $p \geq 1$—sample p-th absolute central moment
S_∞	$= \max\{\|X_i - \bar{X}\| : i = 1, \ldots, n\}$
$X_{r:n}$	r-th order statistic of $X_1, \ldots, X_n$, $1 \leq r \leq n$
$L_{\bar{c}:n}$	$= \sum_{r=1}^{n} c_r X_{r:n}$—$L$-statistic of $X_1, \ldots, X_n$ with coefficient sequence $\bar{c} = (c_1, \ldots, c_n)$
$R_{i,j:n}$	$= X_{j:n} - X_{i:n}$, $1 \leq i < j \leq n$—quasi-range, difference of jth and ith order statistics
R_n	$= R_{1,n:n} = X_{n:n} - X_{1:n}$—sample range
S_j	$= R_{j,j+1:n} = X_{j+1:n} - X_{j:n}$—spacing
$M_{1,n:n}$	$= \frac{1}{2}(X_{1:n} + X_{n:n})$, $1 \leq i < j \leq n$—midrange
$M_{j,n+1-:n}$	$= \frac{1}{2}(X_{j:n} + X_{n+1-:n})$, $1 \leq i < j \leq n$—jth quasi-midrange
$T_{i,j:n}$	$= \frac{1}{j+1-i}\sum_{r=i}^{j} X_{r:n}$, $1 \leq i \leq j \leq n$—trimmed mean
$W_{i,j:n}$	$= \frac{i}{n} X_{i:n} + \frac{1}{n}\sum_{r=i+1}^{j-1} X_{r:n} + \frac{1}{n+1-j} X_{j:n}$, $1 \leq i \leq j \leq n$—Winsorized mean
$\underline{S}_{j:n}$	$= T_{1,j:n} - \bar{X}$, $1 \leq j \leq n$—lower selection differential
$\bar{S}_{j:n}$	$= T_{n+1-j,n:n} - \bar{X}$, $1 \leq j \leq n$—upper selection differential
$\mathrm{MED}_n(\tilde{X})$	$= \begin{cases} X_{\frac{n+1}{2}:n}, & n \text{ odd}, \\ \frac{1}{2}(X_{\frac{n}{2}:n} + X_{\frac{n}{2}+1:n}), & n \text{ even} \end{cases}$
$\mathrm{MAD}_n(\tilde{X})$	$= \frac{1}{n}\sum_{i=1}^{n} \|X_i - \mathrm{MED}_n(\tilde{X})\| = \frac{1}{n}\sum_{i\geq\frac{n}{2}+1} X_{i:n} - \frac{1}{n}\sum_{i\leq\frac{n}{2}} X_{i:n}$—sample median absolute deviation
$\mathrm{GMD}_n(\tilde{X})$	$= \frac{2}{n(n-1)}\sum_{1\leq i<j\leq n} \|X_i - X_j\| = \frac{2}{n(n-1)}\sum_{i=1}^{n}(2i - n - 1)X_{i:n}$—sample Gini mean difference

The counterparts of the above statistics acting on deterministic samples are written in small letters.

$B_{r,n}(x)$	$= \binom{n}{r} x^r (1-x)^{n-r}$, $0 \leq x \leq 1$, $r = 0, 1, \ldots, n$—Bernstein polynomials of degree $n \in \mathbb{N}$
$f_{r:n}(x)$	$= nB_{r-1,n-1}(x)$, $0 \leq x \leq 1$—density function of the r-th order statistic from the iid standard uniform sample
$F_{r:n}(x)$	$= \sum_{i=r}^{n} B_{i,n}(x)$, $0 \leq x \leq 1$—distribution function of the r-th order statistic from the iid standard uniform sample
$f_{\bar{c}:n}(x)$	$= \sum_{i=1}^{n} c_i\, f_{i:n}(x)$, $\bar{c} = (c_1, \ldots, c_n) \in \mathbb{R}^n$, $0 \leq x \leq 1$
$F_{\bar{c}:n}(x)$	$= \sum_{i=1}^{n} c_i\, F_{i:n}(x)$
$f_{i,j:n}(x)$	$= f_{j:n}(x) - f_{i:n}(x)$
$F_{i,j:n}(x)$	$= F_{j:n}(x) - F_{i:n}(x)$
$\bar{f}_{i,j:n}(x)$	$= \frac{1}{j+1-i}\sum_{r=i}^{j} f_{r:n}(x)$
$\bar{F}_{i,j:n}(x)$	$= \frac{1}{j+1-i}\sum_{r=i}^{j} F_{r:n}(x)$

$p_{r:n}(k)$	$= \frac{\binom{k-1}{r-1}\binom{N-k}{n-r}}{\binom{N}{n}}$, $1 \le r \le n \le N$, $r \le k \le N$—probability mass function of r-th order statistic of sample of size n drawn without replacement from standard population Π_0 of size N
$p_{i,j:n}(k)$	$= p_{j:n}^{(k)} - p_{i:n}^{(k)}$, $1 \le i < j \le n$
$\bar{p}_{i,j:n}(k)$	$= \frac{1}{j+1-i} \sum_{r=i}^{j} p_{r:n}(k)$, $1 \le i \le j \le n$
$g_{r:n}(k)$	$= n\binom{n-1}{r-1} \frac{k^{r-1}(N-k+1)^{n-r}}{N^n}$, $1 \le r \le n$, $1 \le k \le N$—probability mass function of r-th order statistic of sample of size n drawn with replacement from standard population Π_0 of size N
$g_{i,j:n}(k)$	$= g_{j:n}(k) - g_{i:n}(k)$, $1 \le i < j \le n$
$\bar{g}_{i,j:n}(k)$	$= \frac{1}{j+1-i} \sum_{r=i}^{j} g_{r:n}(k)$, $1 \le i \le j \le n$
$g_{r:n}^{\tilde{p}}(k)$	$= n\binom{n-1}{r-1}(\sum_{i=1}^{k} p_i)^{r-1} p_k (\sum_{i=k+1}^{n} p_i)^{r-r}$, $1 \le r \le n$, $1 \le k \le N$, $\tilde{p} = (p_1, \ldots, p_N) \in \mathbb{S}^N$—probability mass function of r-th order statistic of iid sample drawn from Π_0 with common probability mass function $\mathbb{P}(X_i = k) = p_k$
$\preceq_{st}$	stochastic order of distribution functions: $F \succeq_{st} W$ if $F(x) \le W(x)$ on $\mathbb{R}$
$\preceq_c$	convex transform order of distribution functions: $F \succeq_c W$ if $F^{-1} \circ W(x)$ is convex on (a, d)
$\preceq_*$	star order of distribution functions: $F \succeq_* W$ if $F^{-1} \circ W(x)$ is starshaped, i.e., $(F^{-1} \circ W(x) - a_F)/(x - a_W)$ is nondecreasing on (a_W, d_W), where $(a_G, d_G) = \{x : 0 < G(x) < 1\}$
$\mathcal{C}^{\nearrow}$	$= \{g \in L^2((0,1), dx) : g$ is nondecreasing$\}$, family of quantile functions $F^{-1}(x)$ of arbitrary distributions functions $F(x)$ with finite variance (A)
$\mathcal{C}^{\searrow}$	$= \{g \in L^2((0,1), dx) : g$ is nonincreasing$\}$
$\mathcal{C}^0$	$= \{g \in \mathcal{C}^{\nearrow} : \int_0^1 g(x)dx = 0\}$, family of centered quantile functions $F^{-1}(x)$—μ, family of quantile functions of random variables with mean $\mu = 0$
$\mathcal{C}^+$	$= \{g \in \mathcal{C}^{\nearrow} : g(0) = 0\}$—family of quantile functions of life distributions
$\mathcal{C}^s$	$= \{g \in \mathcal{C}^{\nearrow} : g(x) = -g(1 - x-)\}$—family of centered quantile functions of symmetric distributions (S)
$\mathcal{C}_{\succeq_c W}$	$= \{g \in L^2((a,d), W(dx)) : g$ is nondecreasing and convex on $(a,d)\}$, family of compositions $F^{-1} \circ W$ for $F \succeq_c W$
$\mathcal{C}_{\preceq_c W}$	$= \{g \in L^2((a,d), W(dx)) : g$ is nondecreasing and concave on $(a,d)\}$, family of compositions $F^{-1} \circ W$ for $F \preceq_c W$
$\mathcal{C}_{\succeq_* W}$	$= \{g \in L^2((a,d), W(dx)) : g$ is nondecreasing and starshaped on $(a,d)\}$, family of compositions $F^{-1} \circ W$ for $F \succeq_* W$
$\mathcal{C}_{\preceq_* W}$	$= \{g \in L^2((a,d), W(dx)) : g$ is nondecreasing and antistarshaped on $(a,d)\}$, family of compositions $F^{-1} \circ W$ for $F \preceq_* W$

$\mathcal{C}_{\succeq_c U}$	$= \{g \in L^2((0,1), dx) : g$ is nondecreasing and convex on $(0,1)\}$, family of quantile functions F^{-1} for F with decreasing density functions (DD)
$\mathcal{C}_{\preceq_c U}$	$= \{g \in L^2((0,1), dx) : g$ is nondecreasing and concave on $(0,1)\}$, family of quantile functions F^{-1} for F with increasing density functions (ID)
$\mathcal{C}_{\succeq_s U}$	$= \{g \in \mathcal{C}^s : g$ is convex on $\left(\frac{1}{2}, 1\right)\}$, family of quantile functions F^{-1} for F of symmetric unimodal distributions (SU)
$\mathcal{C}_{\succeq_c Z}$	$= \{g \in L^2((0,\infty), e^{-x}dx) : g$ is nondecreasing and convex on $(0,\infty)\}$, family of compositions $F^{-1} \circ Z$ for F with decreasing failure rates (DFR)
$\mathcal{C}_{\preceq_c Z}$	$= \{g \in L^2((0,\infty), e^{-x}dx) : g$ is nondecreasing and concave on $(0,\infty)\}$, family of compositions $F^{-1} \circ Z$ for F with increasing failure rates (IFR)
$\mathcal{C}_{\succeq_c V}$	$= \{g \in L^2((-\infty,0), e^{x}dx) : g$ is nondecreasing and convex on $(-\infty,0)\}$, family of compositions $F^{-1} \circ V$ for F with decreasing reversed failure rates (DRFR)
$\mathcal{C}_{\preceq_c V}$	$= \{g \in L^2((-\infty,0), e^{x}dx) : g$ is nondecreasing and concave on $(-\infty,0)\}$, family of compositions $F^{-1} \circ V$ for F with increasing reversed failure rates (IRFR)
$\mathcal{C}_{\succeq_* U}$	$= \{g \in L^2((0,1), dx) : g$ is nondecreasing and starshaped on $(0,1)\}$, family of quantile functions F^{-1} for F with decreasing density functions on the average (DDA)
$\mathcal{C}_{\preceq_* U}$	$= \{g \in L^2((0,1), dx) : g$ is nondecreasing and antistarshaped on $(0,1)\}$, family of quantile functions F^{-1} for F with increasing density functions on the average (IDA)
$\mathcal{C}_{\succeq_* Z}$	$= \{g \in L^2((0,\infty), e^{-x}dx) : g$ is nondecreasing and starshaped on $(0,\infty)\}$, family of compositions $F^{-1} \circ Z$ for F with decreasing failure rates on the average (DFRA)
$\mathcal{C}_{\preceq_* Z}$	$= \{g \in L^2((0,\infty), e^{-x}dx) : g$ is nondecreasing and antistarshaped on $(0,\infty)\}$, family of compositions $F^{-1} \circ Z$ for F with increasing failure rates on the average (IFRA)
$\mathcal{C}_{\bullet}^{\bullet}$	any of above defined convex cones
$P_{\bullet}^{\bullet}h$	projection of function $h \in L^2((a,d), W(dx))$ onto the cone $\mathcal{C}_{\bullet}^{\bullet} \subset L^2((a,d), W(dx))$
B	sharp upper bound

1.4 Epilogue

At the time when we planned to prepare a book detailing all the up-to-date developments concerning sharp inequalities for ordered random variables, upon going through the extensive published literature on the topic, it became evident to us that one single volume will not be sufficient to describe all the results. So, we decided to prepare a two-volume set with the first volume focusing only on results for standard order statistics. There are also numerous results available for other forms of ordered data such as record statistics, progressively Type-II censored order statistics, generalized order statistics, and so on. All these developments would form the basis for our second volume to be completed next!

Chapter 2
Analytic Inequalities and Other Tools

2.1 Introduction

Here we present analytic inequalities and other useful results which are applied in the following chapters of the book. Mainly, the inequalities evaluate sharply either finite sums of products of elements of two sequences or integrals of products of two functions. These represent linear continuous functionals on Euclidean spaces and various function spaces. We frequently consider weighted sums, when each term of the sum is multiplied by a positive weight coefficient, and weighted integrals, when the standard Lebesgue integral is replaced by the Lebesgue-Stieltjes one. Sometimes we also present classic unweighted versions of inequalities, which are commonly known, and referred to by the names of inventors. For each inequality, we provide the necessary and sufficient conditions for attainability. We do not try to present the most general result first, and deduce simpler cases from them. Therefore we first present, mostly elementary proofs of inequalities for sums, and then we prove analogous results for integrals. We do not pretend to indicate the original sources of the results presented here. Many of them can be found in classical monographs by Mitrinović (1970), Mitrinović et al. (1991), Marshall et al. (2011), and Pečarić et al. (1992).

We also gather here some other analytic tools which are useful our investigations. We start with presenting some basic notions of functional analysis. The results presented here can be found in the classic textbooks and monographs, see, e.g., Lindenstrauss and Tzafriri (1977), Lindenstrauss and Tzafriri (1979), and Rudin (1991). We also prove the Descartes rule of signs whose various modifications are applied in our analysis. We finally present basic properties of the convex transform and star orders which are used in our studies. For more detailed description of the orders, we refer to the monograph of Shaked and Shanthikumar (2007).

N. Balakrishnan, T. Rychlik, *Sharp Inequalities for Ordered Random Variables in Statistics and Reliability*, Frontiers in Probability and the Statistical Sciences,
https://doi.org/10.1007/978-3-031-61347-0_2

2.2 Function Spaces

2.2.1 *Linear Spaces*

We consider only real linear spaces. A linear (vector) space over the real field $\mathbb{R}$ is a set $\mathcal{V}$ equipped with two operations

- vector addition $+ : \mathcal{V} \times \mathcal{V} \mapsto \mathcal{V}$ denoted by $(g, h) \mapsto g + h$,
- scalar multiplication $\cdot : \mathbb{R} \times \mathcal{V} \mapsto \mathcal{V}$ denoted by $(\alpha, h) \mapsto \alpha g$,

which satisfy the following axioms. Firstly, $(\mathcal{V}, +)$ is an abelian group, i.e., the addition operator is associative $f + (g + h) = (f + g) + h$, $f, g, h \in \mathcal{V}$, and commutative $g + h = h + g$, $g, h \in \mathcal{V}$, set $\mathcal{V}$ contains a neutral (zero) element 0 satisfying $g + 0 = g$ for all $g \in \mathcal{V}$, and each $g \in \mathcal{V}$ has an inverse element $-g \in \mathcal{V}$ satisfying $g + (-g) = 0$. Moreover, the multiplication operator satisfies $\alpha(\beta g) = (\alpha\beta)g$, $\alpha, \beta \in \mathbb{R}$, $g \in \mathcal{V}$, and $1g = g$ for all $g \in \mathcal{V}$. Finally, the scalar multiplication is distributive with respect to vector and scalar addition, i.e., $\alpha(g + h) = \alpha g + \alpha h$, and $(\alpha + \beta)g = \alpha g + \beta g$, respectively.

A classic example of a linear space is the family $\mathbb{R}^n$ of n-dimensional vectors with the addition and multiplication operators defined as follows

$$\tilde{x} + \tilde{y} = (x_1, \dots, x_n) + (y_1, \dots, y_n) = (x_1 + y_1, \dots, x_n + y_n), \tag{2.1}$$

$$\alpha\tilde{x} = \alpha(x_1, \dots, x_n) = (\alpha x_1, \dots, \alpha x_n), \tag{2.2}$$

for arbitrary $\tilde{x}, \tilde{y} \in \mathbb{R}^n$ and $\alpha \in \mathbb{R}$. We also consider the linear space of real functions defined on an interval $[a, b]$, where the operations are defined in the following way

$$(g + h)(x) = g(x) + h(x), \tag{2.3}$$

$$(\alpha g)(x) = \alpha g(x) \tag{2.4}$$

for all $x \in [a, b]$, where functions f, g and constant $\alpha \in \mathbb{R}$ are arbitrary.

A subset $\mathcal{S}$ of a linear space $\mathcal{V}$ is called a convex set iff for all $g, h \in \mathcal{S}$ and all $0 \le \alpha \le 1$ vector $\alpha g + (1 - \alpha)h \in \mathcal{S}$. Set $\mathcal{C} \subset \mathcal{V}$ is called a convex cone iff $\alpha g + \beta h \in \mathcal{C}$ for all $g, h \in \mathcal{C}$ and all $\alpha, \beta \ge 0$. Finally, we call $\mathcal{L} \subset \mathcal{V}$ a linear subspace iff $\alpha g + \beta h \in \mathcal{L}$ for all $g, h \in \mathcal{L}$ and all $\alpha, \beta \in \mathbb{R}$. Obviously, every linear subspace is a convex cone, and every convex cone is a convex set. The reversed implications are not true. The simplest examples of convex set, cone and linear subspace, generated by a single non-zero element g of $\mathcal{V}$, are the line segment $\mathcal{S} = \{\alpha g = \alpha g + (1 - \alpha)0 : 0 \le \alpha \le 1\}$, half-line $\mathcal{C} = \{\alpha g : \alpha \ge 0\}$, and line $\mathcal{L} = \{\alpha g : \alpha \in \mathbb{R}\}$, respectively. We say that e is an extreme point of set $\mathcal{S} \subset \mathcal{V}$ iff the relation $e = \alpha g + (1 - \alpha)h$ for some $g, h \in \mathcal{S}$ and $0 < \alpha < 1$ implies $e = g = h$.

A real-valued function on a linear space is called a functional. We say that the functional $T : \mathcal{V} \mapsto \mathbb{R}$ is linear iff

$$T(\alpha g + \beta h) = \alpha T(g) + \beta T(h) \tag{2.5}$$

for every $\alpha, \beta \in \mathbb{R}$ and $g, h \in \mathcal{V}$.

2.2.2 Banach Spaces

A nonnegative real function $||\cdot||$ acting on a real linear space $(\mathcal{B}, +, \cdot)$ is called a norm if it satisfied three axioms

$$||g|| = 0 \qquad \text{iff} \qquad g = 0, \tag{2.6}$$

$$||\alpha g|| = |\alpha|\,||g||, \qquad \alpha \in \mathbb{R},\; g \in \mathcal{B}, \tag{2.7}$$

$$||g + h|| \leq ||g|| + ||h||, \qquad g, h \in \mathcal{B}. \tag{2.8}$$

The last condition implies

$$\max\Big\{||g|| - ||h||, ||h|| - ||g||\Big\} = \Big|\,||g|| - ||h||\,\Big| \leq ||h - g|| \,. \tag{2.9}$$

Replacing h in the above formula by elements g_n of a sequence tending to g, we conclude that the norm functional is continuous. Moreover, it is convex as well. We say that a functional $T : \mathcal{B} \mapsto \mathbb{R}$ is convex if

$$T(\alpha g + (1 - \alpha)h) \leq \alpha T(g) + (1 - \alpha)T(h) \tag{2.10}$$

for all $g, h \in \mathcal{B}$ and $0 \leq \alpha \leq 1$. The convexity of the norm follows from the relations

$$||\alpha g + (1 - \alpha)h|| \leq ||\alpha g|| + ||(1 - \alpha)h|| = \alpha||g|| + (1 - \alpha)||h||. \tag{2.11}$$

A linear space with a norm function is called a normed space. This is a metric space with the metric function $\varrho(g, h) = ||g - h||$. A sequence $(g_n)_{n\in\mathbb{N}}$ of the normed space elements satisfies the Cauchy condition if

$$\forall \varepsilon > 0 \quad \exists n_0 \in \mathbb{N} \quad \forall m, n \geq n_0 \qquad ||g_m - g_n|| \leq \varepsilon. \tag{2.12}$$

Sequence $(g_n)_{n\in\mathbb{N}}$ has a limit $g \in \mathcal{B}$ if

$$\forall \varepsilon > 0 \quad \exists n_0 \in \mathbb{N} \quad \forall n \geq n_0 \qquad ||g_n - g|| \leq \varepsilon. \tag{2.13}$$

A normed space is complete if every sequence $(g_n)_{n\in\mathbb{N}}$ of its points satisfying the Cauchy condition has a limit $g \in \mathcal{B}$. If a normed space is complete we call it a Banach space. Exemplary Banach spaces are presented below. For $p \geq 1$ and $\tilde{w} = (w_1, \ldots, w_n)$ denoting a vector in $\mathbb{R}^n$ with all positive coordinates, a function on $\mathbb{R}^n$ defined as

$$||\tilde{x}||_{p,\tilde{w}} = \left(\sum_{i=1}^{n} |x_i|^p w_i\right)^{1/p} \tag{2.14}$$

is a norm on the linear space $\mathbb{R}^n$. We call it the pth norm on $\mathbb{R}^n$ with weight vector $\tilde{w}$. We denote $\mathbb{R}^n$ equipped with this norm as $l^p(\mathbb{R}^n, \tilde{w})$. If $p = 2$ and $w_i = 1$, $i = 1, \ldots, n$, we obtain the classic n-dimensional Euclidean space with the distance $\varrho(\tilde{x}, \tilde{y}) = \sqrt{\sum_{i=1}^{n}(x_i - y_i)^2}$. For any $\tilde{w}$, as $p \to \infty$, we get

$$\lim_{p\to\infty} ||\tilde{x}||_{p,\tilde{w}} = \max_{1\leq i\leq n} |x_i|, \tag{2.15}$$

which is a norm on $\mathbb{R}^n$ as well. The respective Banach space will be denoted by $l^\infty(\mathbb{R}^n)$.

Analogously, for $p \geq 1$ and nondecreasing bounded function $W : [a, b] \mapsto \mathbb{R}$, we define the norm

$$||g||_{p,W} = \left(\int_a^b |g(x)|^p W(dx)\right)^{1/p} \tag{2.16}$$

for the Banach space $L^p([a, b], W(dx))$ of functions $g : [a, b] \mapsto \mathbb{R}$ for which the integral in the above formula is finite. For defining the Banach space $L^\infty([a, b], W(dx))$, a functional counterpart of $l_n^\infty(\mathbb{R}^n)$, we need the notion of essential supremum. Let μ_W denote a positive measure on the Borel subsets of $[a, b]$ generated by W, and satisfying $\mu_W([c, d]) = W(d) - W(c)$ for all intervals $[c, d] \subset [a, b]$. The essential supremum with respect to μ_W of a measurable function $g : [a, b] \mapsto \mathbb{R}$ is then defined as

$$\operatorname{ess\,sup}_W g = \inf\Big\{c \in \mathbb{R} : \mu_W(\{x \in [a, b] : g(x) > c\}) = 0\Big\}. \tag{2.17}$$

For instance, if $W(x) = x$, $x \in [0, 1]$, and $g : [0, 1] \mapsto \mathbb{R}$, equal to 1 at the rational numbers, and 0 elsewhere, we have $\sup g = 1$ and $\operatorname{ess\,sup}_W g = 0$. The Banach space $L^\infty([a, b], W(dx))$ has the norm

$$||g||_{\infty,W} = \operatorname{ess\,sup}_W |g|, \tag{2.18}$$

and consists of all μ_W-essentially bounded functions on $[a, b]$, i.e., ones for which the RHS of the above formula is finite.

Precisely, the elements of the L^p spaces are the families of functions whose elements are identical except for some sets of μ_W measure zero. For simplicity of presentation, we identify the families with their single representatives.

A linear functional $T : \mathcal{B} \mapsto \mathbb{R}$ on a Banach space is bounded if there exists a finite constant C such that $|T(g)| \leq C||g||$ for all $g \in \mathcal{B}$. A minimal such number is called the norm of the functional and it is denoted by

$$||T|| = \sup_{\mathcal{B} \ni g \neq 0} \frac{|T(g)|}{||g||} = \sup_{||g||=1} |T(g)|. \tag{2.19}$$

Note that $|T(g)|$ can be replaced by $T(g)$ in the above formula, because $T(-g) = -T(g)$. Every linear functional on a Banach space is continuous. Continuity implies boundedness as well.

Obviously, every linear functional on n-dimensional real vectors can be written as

$$T_{\tilde{c}}(\tilde{x}) = \sum_{i=1}^{n} c_i x_i, \qquad \tilde{x} \in \mathbb{R}^n, \tag{2.20}$$

for some $\tilde{c} = (c_1, \ldots, c_n) \in \mathbb{R}^n$. Its norm depends on the choice of the norm acting on the domain space. If the norm depends on weights $\tilde{w} = (w_1, \ldots, w_n)$, it is convenient to rewrite the above definition as

$$T_{\tilde{d}}(\tilde{x}) = \sum_{i=1}^{n} d_i x_i w_i \tag{2.21}$$

with $d_i = \frac{c_i}{w_i}$, $i = 1, \ldots, n$. Every continuous linear functional on the Banach space $L^p([a, b], W(dx))$ for some $p \geq 1$ has the form

$$T(g) = T_h(g) = \int_a^b g(x)h(x)\, W(dx), \tag{2.22}$$

where h is an element of the space $L^q([a, b], W(dx))$ with q being the conjugate parameter to p defined as $q = \frac{p}{p-1}$ for $p > 1$, and $q = +\infty$ for $p = 1$. Its norm is $||T_h|| = ||h||_{q,W}$ which follows from the Hölder inequality presented below. Representation (2.22) was proven for $p > 1$ by Riesz (1909), and extended to the case $p = 1$ by Steinhaus (1919). Linear functionals on $L^\infty([a, b], W(dx))$ of the form (2.22) with $h \in L^1([a, b], W(dx))$ are also well-defined, and they have norms $||T_h|| = ||h||_{1,w}$. However, formula (2.22) does not represent all the possible constructions of bounded linear functionals on the space $L^\infty([a, b], W(dx))$.

Let $\mathcal{S}$ be a convex compact subset of a Banach space $\mathcal{B}$. By the Krein-Milman theorem, $\mathcal{S}$ is a convex hull of its extreme points (in fact, Krein and Milman (1940) proved the claim for much more general topological spaces). Let T be a continuous convex functional on $\mathcal{B}$. Then $T(\mathcal{S})$ is a closed bounded subset of $\mathbb{R}$, and it contains extreme values of T over $\mathcal{S}$. Relations

$$T(\alpha g + (1-\alpha)h) \le \alpha T(g) + (1-\alpha)T(h) \le \max\{T(g), T(h)\} \tag{2.23}$$

for arbitrary $0 < \alpha < 1$ and all $g, h \in \mathcal{S}$ show that while searching for the global maximum of T over $\mathcal{S}$ we can eliminate all the interior points of all line segments contained in $\mathcal{S}$, and focus only on their end-points. Since every non-extreme point of $\mathcal{S}$ is an interior point of some interval $\mathcal{I} \subset \mathcal{S}$, we arrive to the conclusion that the maximum has to be attained at an extreme point of $\mathcal{S}$. Generally, such a procedure admits to overlook some maximizing points, but provides at least one solution. We shall apply the approach many times in the sequel.

2.2.3 Hilbert Spaces

For a real linear space $\mathcal{H}$, a mapping $\langle \cdot, \cdot \rangle : \mathcal{H} \times \mathcal{H} \mapsto \mathbb{R}$ is called an inner product if it is symmetric, i.e., $\langle g, h \rangle = \langle h, g \rangle$, linear in the first argument $\langle \alpha g + \beta h, f \rangle = \alpha \langle g, f \rangle + \beta \langle h, f \rangle$, and positive definite: $\langle g, g \rangle > 0$ for all $g \in \mathcal{H} \setminus \{0\}$. The conditions imply that the inner product is linear in the second argument as well, and $\langle 0, 0 \rangle = 0$. The pair $(\mathcal{H}, \langle \cdot, \cdot \rangle)$ is called a unitary space. Function $\mathcal{H} \ni g = \sqrt{\langle g, g \rangle}$ defines a norm in the space. If $\mathcal{H}$ with the norm $||g|| = \sqrt{\langle g, g \rangle}$ is complete, then it is called the Hilbert space.

The Riesz-Fréchet representation theorem asserts that every linear functional on the Hilbert space has the form

$$T(g) = T_h(g) = \langle g, h \rangle, \qquad g \in \mathcal{H}, \tag{2.24}$$

for some fixed element h of the same Hilbert space $\mathcal{H}$, and then

$$||T_h|| = ||h|| = \sqrt{\langle h, h \rangle}. \tag{2.25}$$

The theorem was proved independently by Fréchet (1907) and Riesz (1907).

In particular, the Banach spaces $l^2(\mathbb{R}^n, \tilde{w})$ and $L^2([a, b], W(dx))$ are Hilbert spaces as well. The continuous linear functionals on them have the form

$$T_{\tilde{c}}(\tilde{x}) = \sum_{i=1}^{n} c_i x_i w_i, \tag{2.26}$$

$$T_h(g) = \int_a^b g(x)h(x)W(dx) \tag{2.27}$$

for some $\tilde{c} \in \mathbb{R}^n$ and $h \in L^2([a, b], W(dx))$, respectively. The respective norms are

$$||T_{\tilde{c}}|| = ||\tilde{c}||_{2,\tilde{w}} = \sqrt{\sum_{i=1}^{n} c_i^2 w_i}, \tag{2.28}$$

$$||T_h|| = ||h||_{2,W} = \sqrt{\int_a^b h^2(x)W(dx)}. \tag{2.29}$$

2.2.4 Orthonormal Systems in Hilbert Spaces

Definition 2.1 Let $\mathcal{V}$ be an inner product space and $\phi, \xi \in \mathcal{V}$. Then, ϕ and ξ are said to be *orthogonal* if the inner product $\langle \phi, \xi \rangle = 0$.

Definition 2.2 A collection of elements $\{\phi_k\}_k$, with $\phi_k \in \mathcal{V}$, is said to be an *orthonormal system* if $\langle \phi_i, \phi_j \rangle = 0$ whenever $i \neq j$ and $\langle \phi_i, \phi_i \rangle = 1$ for all i.

With this notion of orthonormal system, the following lemma provides the best approximation for any element $\phi \in \mathcal{V}$.

Lemma 2.1 *Let $\mathcal{V}$ be an inner product space and $\{\phi_k\}_{k=0}^K$ be an orthonormal system in $\mathcal{V}$. Further, let $(\alpha_k)_{k=1}^K$ be a finite sequence of scalars. Then, for any element $\phi \in V$,*

$$\left\| \phi - \sum_{k=0}^{K} \alpha_k \phi_k \right\|^2 = \|\phi\|^2 - \sum_{k=0}^{K} |\langle \phi, \phi_k \rangle|^2 + \sum_{k=0}^{K} |\alpha_k - \langle \phi, \phi_k \rangle|^2$$

From the above lemma, we readily observe the following.

Theorem 2.1 *For any $\phi \in \mathcal{V}$, we have*

(i) $\left\| \sum_{k=0}^K \alpha_k \phi_k \right\|^2 = \sum_{k=0}^K |\alpha_k|^2$;

(ii) $\left\| \phi - \sum_{k=0}^K \langle \phi, \phi_k \rangle \phi_k \right\|^2 = \|\phi\|^2 - \sum_{k=0}^K |\langle \phi, \phi_k \rangle|^2$;

(iii) $\sum_{k=0}^K |\langle \phi, \phi_k \rangle|^2 \leq \|\phi\|^2$, *which is referred to as Bessel's inequality;*

(iv) The series $\sum_{k=0}^\infty |\langle \phi, \phi_k \rangle|^2$ *converges, for every* $\phi \in \mathcal{V}$, *and is, in fact, bounded above by* $\|\phi\|^2$ *when* $\{\phi_k\}_{k=0}^\infty$ *is an orthonormal system in* $\mathcal{V}$.

Now, we introduce the above notions in the context of a Hilbert Space.

Definition 2.3 A collection of elements $\{\phi_k\}_{k \in S}$ in a Hilbert space $\mathcal{H}$ is said to be *complete* if $\langle \phi, \phi_k \rangle = 0$ for all $k \in S$ implies that $\phi = 0$.

Then, an *orthonormal basis* of $\mathcal{H}$ is a complete orthonormal system. We then have the following result.

Theorem 2.2 *Let $\{\phi_k\}_{k=0}^{\infty}$ be an orthonormal system in a Hilbert space $\mathcal{H}$. Then, the following statements are all equivalent:*

(i) The system $\{\phi_k\}_{k=0}^{\infty}$ is complete;
(ii) For any $\phi \in H$, we have $\phi = \sum_{k=0}^{\infty}\langle\phi, \phi_k\rangle\phi_k$;
(iii) $\|\phi\|^2 = \sum_{k=0}^{\infty} |\langle\phi, \phi_k\rangle|^2$;
(iv) For any $\phi, \xi \in H$, we have $\langle\phi, \xi\rangle = \sum_{k=0}^{\infty}\langle\phi, \phi_k\rangle\overline{\langle\xi, \phi_k\rangle}$.

Remark 2.1 Let us consider the space of all Lebesgue measurable square-integrable functions on [0,1], denoted by $L^2((0,1), dx)$. An orthonormal basis in this case is the sequence of Legendre polynomials $\{\phi_k(u)\}_{k=0}^{\infty}$, where $\phi_k(u) = \frac{\sqrt{2k+1}}{k!}\frac{d^k}{du^k}\{u^k(u-1)^k\}$. In fact, these can be obtained through Gram-Schmidt orthogonalization applied on the sequence of monomials $\{1, u, u^2, \ldots\}$; see Sansone (1959), for example. As a matter of fact, this orthonormal basis of Legendre polynomials for $L^2((0,1), dx)$ is used later in Chap. 4 to develop orthogonal inverse expansions and then applied specifically to obtain bounds for moments of order statistics.

Another classic example of an orthonormal basis for $L^2((0,1), dx)$ is the trigonometric system $\{e^{i2\pi ku} : k \in \mathbb{Z}\}$. In this case, the expansion of a function $\phi \in L^2((0,1), dx)$ in this basis is the so-called *Fourier series expansion*; see Tolstov (1962), for example.

2.3 Inequalities for Monotone Sequences and Functions

2.3.1 Chebyshev Inequalities

Theorem 2.3 *Let $\tilde{c} = (c_1, \ldots, c_n)$ and $\tilde{x} = (x_1, \ldots, x_n)$ be two real sequences such that $\tilde{c}$ is nonincreasing and $\tilde{x}$ is nondecreasing. Then the following inequality holds*

$$\sum_{i=1}^{n} c_i x_i \leq \frac{1}{n}\sum_{i=1}^{n} c_i \sum_{i=1}^{n} x_i. \tag{2.30}$$

The inequality becomes equality iff either $\tilde{c}$ or $\tilde{x}$ is constant.

Relation (2.30) is called the Chebyshev inequality. Below we present a more general weighted version of the inequality.

Theorem 2.4 *Let $\tilde{c}$ and $\tilde{x}$ satisfy the assumptions of Theorem 2.3, and $\tilde{w} = (w_1, \ldots, w_n)$ be a sequence of positive weights. Then*

$$\sum_{i=1}^{n} c_i x_i w_i \leq \frac{1}{\sum_{i=1}^{n} w_i} \sum_{i=1}^{n} c_i w_i \sum_{i=1}^{n} x_i w_i \tag{2.31}$$

is true, and becomes equality iff either $\tilde{c}$ or $\tilde{x}$ is constant.

Proof Observe that

$$\sum_{i=1}^{n}\sum_{j=1}^{n} c_i(x_i - x_j) w_i w_j = \left(\sum_{i=1}^{n} w_i\right)\left(\sum_{i=1}^{n} c_i x_i w_i\right) - \left(\sum_{i=1}^{n} c_i w_i\right)\left(\sum_{i=1}^{n} x_i w_i\right) \tag{2.32}$$

and similarly,

$$\sum_{i=1}^{n}\sum_{j=1}^{n} c_j(x_j - x_i) w_i w_j = \left(\sum_{i=1}^{n} w_i\right)\left(\sum_{i=1}^{n} c_i x_i w_i\right) - \left(\sum_{i=1}^{n} c_i w_i\right)\left(\sum_{i=1}^{n} x_i w_i\right). \tag{2.33}$$

Summing (2.32) and (2.33), we obtain

$$\begin{aligned} &2\left(\sum_{i=1}^{n} w_i\right)\left(\sum_{i=1}^{n} c_i x_i w_i\right) - \left(\sum_{i=1}^{n} c_i w_i\right)\left(\sum_{i=1}^{n} x_i w_i\right) \\ &= \sum_{i=1}^{n}\sum_{j=1}^{n} (c_i x_i - c_i x_j - c_j x_i + c_i x_j) w_i w_j \\ &= \sum_{i=1}^{n}\sum_{j=1}^{n} (c_i - c_j)(x_i - x_j) w_i w_j \leq 0. \end{aligned} \tag{2.34}$$

The last sum is nonpositive, because so is each term of it. Relations (2.31) and (2.34) are equivalent. It is evident that the equality holds if either of sequences is constant. Otherwise $c_1 > c_n$, $x_i < x_n$, and $(c_1 - c_n)(x_1 - x_n)$ is strictly negative. In consequence, the Chebyshev inequality becomes strict. This completes the proof. □

Relations (2.30) and (2.31) become true if $\tilde{c}$ and $\tilde{x}$ are interchanged so that $\tilde{c}$ is nondecreasing and $\tilde{x}$ is nonincreasing. If both the sequences are either nondecreasing or nonincreasing, then inequalities (2.30) and (2.31) are reversed. Certainly, the assumptions can be slightly relaxed by admitting nonnegative weights with some $w_i = 0$. Then the respective terms are simply dropped from the sums, and no conditions on respective x_i and c_i are imposed. Now and further on, we do not consider this apparent generalization, and restrict ourselves to the case of all weights being strictly positive.

An integral counterpart of the weighted Chebyshev inequality is the following.

Theorem 2.5 *Suppose that W is a nondecreasing and nonconstant bounded function on an interval $[a, b]$, g and h are nondecreasing and nonincreasing, respectively, and g, h and gh are W-integrable on $[a, b]$. Then*

$$\int_a^b g(x)h(x)W(dx) \le \frac{1}{W(b) - W(a)} \int_a^b g(x)W(dx) \int_a^b h(x)W(dx) \tag{2.35}$$

holds, with the equality iff either of g and h becomes constant W-almost everywhere.

Inequality (2.35) *is reversed if both the functions are either nondecreasing or nonincreasing, and the equality condition is the constancy of either of them.*

Proof Like in the proof of Theorem 2.4, we write

$$\begin{aligned}&\int_a^b \int_a^b g(x)[h(x) - h(y)]W(dx)W(dy)\\ &= \int_a^b g(x)h(x)W(dx) \int_a^b W(dx) - \int_a^b g(x)W(dx) \int_a^b h(x)W(dx)\end{aligned} \tag{2.36}$$

and

$$\begin{aligned}&\int_a^b \int_a^b g(y)[h(y) - h(x)]W(dx)W(dy)\\ &= \int_a^b g(x)h(x)W(dx) \int_a^b W(dx) - \int_a^b g(x)W(dx) \int_a^b h(x)W(dx).\end{aligned} \tag{2.37}$$

Therefore

$$\begin{aligned}&2\left\{[W(b) - W(a)] \int_a^b g(x)h(x)W(dx) - \int_a^b g(x)W(dx) \int_a^b h(x)W(dx)\right\}\\ &= \int_a^b \int_a^b [g(x) - g(y)][h(x) - h(y)]W(dx)W(dy) \le 0,\end{aligned} \tag{2.38}$$

because

$$[g(x) - g(y)][h(x) - h(y)] \le 0. \tag{2.39}$$

If one of the functions is constant except for a set of W-measure zero, then (2.39) vanishes on its completion, and so does (2.38). Otherwise (2.39) is strictly negative on a set of positive W-measure, and (2.35) is strict. This proves the theorem. □

The standard integral Chebyshev inequality has the form:

$$\int_a^b g(x)h(x)dx \leq \frac{1}{b-a}\int_a^b g(x)dx \int_a^b h(x)dx. \tag{2.40}$$

2.3.2 *Abel Identity and Inequality*

The following equality

$$\sum_{i=1}^{n} c_i x_i = x_n \sum_{i=1}^{n} c_i - \sum_{m=1}^{n-1} (x_{m+1} - x_m) \sum_{i=1}^{m} c_i \tag{2.41}$$

that holds for arbitrary $\tilde{c} = (c_1, \ldots, c_n) \in \mathbb{R}^n$ and $\tilde{x} = (x_1, \ldots, x_n) \in \mathbb{R}^n$, is called the Abel identity. It is easily verified by writing consecutive transformations

$$\begin{aligned}
\sum_{i=1}^{n} c_i x_i &= \sum_{i=3}^{n} c_i x_i + x_2(c_1 + c_2) - (x_2 - x_1)c_1 \\
&= \sum_{i=4}^{n} c_i x_i + x_3(c_1 + c_2 + c_3) - (x_3 - x_2)(c_1 + c_2) - (x_2 - x_1)c_1 \\
&\vdots \\
&= x_n \sum_{i=1}^{n} c_i - \sum_{m=1}^{n-1} (x_{m+1} - x_m) \sum_{i=1}^{m} c_i
\end{aligned} \tag{2.42}$$

that consist in substitutions

$$c_{m+1}x_{m+1} + x_m \sum_{i=1}^{m} c_i = x_{m+1} \sum_{i=1}^{m+1} c_i - (x_{m+1} - x_m) \sum_{i=1}^{m} c_i \tag{2.43}$$

for $m = 1, \ldots, n-1$. The Abel identity (2.41) is directly used for proving inequalities presented in this and the following two sections under some conditions imposed on sequences $\tilde{c}$ and $\tilde{x}$.

The Abel inequality is presented in the following theorem.

Theorem 2.6 *Let $\tilde{x} = (x_1, \dots, x_n)$ be a nonincreasing sequence of nonnegative numbers. Then for arbitrary $\tilde{c} = (c_1, \dots, c_n) \in \mathbb{R}^n$, we have*

$$\min_{1\le j\le n}\left(\sum_{i=1}^{j} c_i\right) x_1 \le \sum_{i=1}^{n} c_i x_i \le \max_{1\le j\le n}\left(\sum_{i=1}^{j} c_i\right) x_1. \tag{2.44}$$

With the notation $x_{n+1} = 0$, the former inequality becomes equality iff either

$$\sum_{i=1}^{m} c_i = \min_{1\le j\le n}\sum_{i=1}^{j} c_i \quad \text{or} \quad x_m = x_{m+1}, \quad m = 1, \dots, n. \tag{2.45}$$

The latter inequality in (2.44) *becomes equality iff either*

$$\sum_{i=1}^{m} c_i = \max_{1\le j\le n}\sum_{i=1}^{j} c_i \quad \text{or} \quad x_m = x_{m+1}, \quad m = 1, \dots, n. \tag{2.46}$$

Proof Using Abel identity (2.41), and the facts that $x_1 \ge 0$ as well as all the increments $x_{m+1} - x_m$, $m = 1, \dots, n-1$, are nonpositive, we obtain

$$\begin{aligned}
\sum_{i=1}^{n} c_i x_i &= x_n \sum_{i=1}^{n} c_i + \sum_{m=1}^{n-1}(x_m - x_{m+1})\sum_{i=1}^{m} c_i \\
&\le \max_{1\le m\le n}\left(\sum_{i=1}^{m} c_i\right)\left[x_n + \sum_{m=1}^{n-1}(x_m - x_{m+1})\right] \\
&= \max_{1\le m\le n}\left(\sum_{i=1}^{m} c_i\right) x_1.
\end{aligned} \tag{2.47}$$

From the relations, we immediately conclude the necessary and sufficient conditions (2.46) for the equality. The other inequality is derived, as we replace the maximal sum by the minimal one and reverse the inequality. □

Accordingly, we obtain the following evaluations of linear combinations of nondecreasing, nonnecessarily nonpositive sequences $\tilde{x} = (x_1, \dots, x_n)$.

$$\begin{aligned}
x_n \sum_{i=1}^{n} c_i - \max_{1\le j\le n}\left(\sum_{i=1}^{j} c_i\right)(x_n - x_1) &\le \sum_{i=1}^{n} c_i x_i \\
\le x_n \sum_{i=1}^{n} c_i - \min_{1\le j\le n}\left(\sum_{i=1}^{j} c_i\right)(x_n - x_1)
\end{aligned} \tag{2.48}$$

and

$$\left[\sum_{i=1}^{n} c_i - \max_{1\le j\le n}\left(\sum_{i=1}^{j} c_i\right)\right](x_n - x_1) \le \sum_{i=1}^{n} c_i(x_i - x_1)$$
$$\le \left[\sum_{i=1}^{n} c_i - \min_{1\le j\le n}\left(\sum_{i=1}^{j} c_i\right)\right](x_n - x_1), \tag{2.49}$$

once we apply the Abel inequality to the sequence, $x_n - x_i$, $i = 1, \ldots, n$. The requirements for the attainability of the lower and upper bounds in (2.48) and (2.49) are given in (2.45) and (2.46), respectively.

2.3.3 Stochastic Ordering Inequalities

Below we impose some restrictions on the sequence of coefficients $\tilde{c} = (c_1, \ldots, c_n)$.

Theorem 2.7 *If*

$$\max_{1\le j\le n} \sum_{i=1}^{j} c_i = \sum_{i=1}^{n} c_i = 0, \tag{2.50}$$

then for arbitrary nondecreasing sequence $\tilde{x} = (x_1, \ldots, x_n)$ *yields*

$$\sum_{i=1}^{n} c_i x_i \ge 0. \tag{2.51}$$

This becomes equality iff either

$$\sum_{i=1}^{m} c_i = 0 \quad or \quad x_m = x_{m+1}, \qquad m = 1, \ldots, n. \tag{2.52}$$

Similarly, the linear combination of (2.51) is nonpositive if each partial sum $\sum_{i=1}^{j} c_i$ $j = 1, \ldots, n$, and the last one amounts to zero. The combination vanishes under conditions (2.52). Both the relations are immediate consequences of (2.48).

A slight modification is useful for comparisons of different linear combinations.

Theorem 2.8 *Let sequences* $\tilde{a} = (a_1, \ldots, a_n) \in \mathbb{R}^n$, $\tilde{b} = (b_1, \ldots, b_n) \in \mathbb{R}^n$ *satisfy*

$$\sum_{i=1}^{m} a_i \leq (\geq) \sum_{i=1}^{m} b_i, \qquad m = 1, \ldots, n-1, \tag{2.53}$$

$$\sum_{i=1}^{n} a_i = \sum_{i=1}^{n} b_n. \tag{2.54}$$

Then for arbitrary nondecreasing sequence $\tilde{x} = (x_1, \ldots, x_n)$, *we have*

$$\sum_{i=1}^{n} a_i x_i \geq (\leq) \sum_{i=1}^{n} b_i x_i, \tag{2.55}$$

which becomes equality iff either

$$\sum_{i=1}^{m} a_i = \sum_{i=1}^{m} b_i \quad \text{or} \quad x_m = x_{m+1}, \quad m = 1, \ldots, n. \tag{2.56}$$

If $\tilde{a}$ and $\tilde{b}$ are nonincreasing, and satisfy (2.53) and (2.54), we say that $\tilde{a}$ majorizes $\tilde{b}$. The definition is extended to any permutation of the vectors. There are known numerous inequalities based on the notion of majorization. The most comprehensive reference is the monograph by Marshall et al. (2011).

An integral analogue of Theorem 2.7 is the following.

Theorem 2.9 *Let* H *be a function of a bounded variation on some interval* $[a, b]$, *satisfying*

$$H(a) = \lim_{x \searrow a} H(x) = \lim_{x \nearrow b} H(x) = H(b). \tag{2.57}$$

If

$$H(x) \leq H(a) = H(b), \qquad x \in [a, b], \tag{2.58}$$

then for arbitrary nondecreasing H*-integrable function* g *on* $[a, b]$, *we have*

$$\int_a^b g(t) H(dt) \geq 0. \tag{2.59}$$

Conversely, relation (2.59) *for all nondecreasing functions* g *implies* (2.58). *Moreover, the integral of* (2.59) *vanishes for a nondecreasing* g *iff this is constant on all the intervals of the set*

$$\mathcal{H}^* = \{\max\{H(t-), H(t+)\} < H(a)\} \tag{2.60}$$

and

$$g(t) = \begin{cases} g(t-), & \text{if } H(t-) < H(t+), \\ g(t+), & \text{if } H(t+) < H(t-), \end{cases} \tag{2.61}$$

at every discontinuity point (if any) of H.

The assumptions imply that H is either continuous or has a finite or denumerable set of discontinuity points, and both the left and right limits at them. Therefore formulae (2.60) and (2.61) make sense. Moreover, (2.60) is an open subset of $[a, b]$, and, if nonempty, consists of countably many open intervals at most. The proof of the first two statements is based on Marshall and Proschan (1970, pp. 166–167). The proof of equality conditions is postponed to Sect. 2.3.5, where more general results are derived (cf. Theorems 2.15 and 2.18) by slightly more sophisticated arguments.

Proof For every nondecreasing indicator function, we have

$$\int_a^b \mathbf{1}_{[x,b]}(t)H(dt) = H(b) - H(x) \geq 0, \qquad a \leq x \leq b. \tag{2.62}$$

The same relation holds for a positive combination of nondecreasing indicators

$$\int_a^b \sum_{i=1}^n a_i \mathbf{1}_{[x_i,b]}(t)H(dt) = \sum_{i=1}^n a_i[H(b) - H(x_i)] \geq 0. \tag{2.63}$$

For every nondecreasing nonnegative function g_+ we have

$$\sum_{i=1}^n a_i \mathbf{1}_{[x_i,b]}(t) \nearrow g_+(t), \qquad \text{as } n \to \infty \tag{2.64}$$

for some $a_i \geq 0$, $a \leq x_i \leq b$, $i = 1, \ldots, n$, and all $a \leq t \leq b$. By the Lebesgue monotone convergence theorem and (2.63), inequality (2.59) holds for g_+. On the other hand, we get

$$\int_a^b \sum_{i=1}^n a_i \mathbf{1}_{[a,x_i]}(t)H(dt) = \sum_{i=1}^n a_i[H(x_i) - H(a)] \leq 0 \tag{2.65}$$

if all a_i are nonnegative. By the same arguments, (2.59) is reversed for all nonnegative nonincreasing functions g_-. Since every nondecreasing function may be decomposed as

$$g(t) = g_+(t) - g_-(t) = \max\{g(t), 0\} - \max\{-g(t), 0\}, \tag{2.66}$$

we immediately establish the first statement. Now assume that $H(x) > H(b)$ for some $a < x < b$. Then for a simple indicator function yields

$$\int_a^b \mathbf{1}_{[x,b]}(t)H(dt) = H(b) - H(x) < 0, \tag{2.67}$$

which contradicts (2.58). □

Theorem 2.9 has a well known application in probability.

Theorem 2.10 *Let real random variables X and Y have distribution functions F and G, respectively. Then the following conditions are equivalent:*

(i) $F(x) \le G(x)$ *for all real x,*
(ii) $Eg(X) \ge Eg(Y)$ *for all nondecreasing functions g for which the above expectations exist.*

The conditions define a partial ordering among random variables and distribution functions, which is called the stochastic order. This is written as $X \succcurlyeq_{st} Y$ and $F \succcurlyeq_{st} G$. The stochastic order is the most natural and popular among the partial orderings utilized in probability and its applications.

2.3.4 Some Generalizations

Another application of the Abel identity is the following result (see Rychlik, 1992).

Theorem 2.11 *Suppose that* $\tilde{c} = (c_1, \ldots, c_n)$ *satisfies*

$$\frac{1}{m}\sum_{i=1}^{m} c_i \ge \frac{1}{n}\sum_{i=1}^{n} c_i, \qquad m = 1, \ldots, n-1. \tag{2.68}$$

Then for every nondecreasing sequence $\tilde{x} = (x_1, \ldots, x_n)$, *the following inequality holds*

$$\sum_{i=1}^{n} c_i x_i \le \frac{1}{n}\sum_{i=1}^{n} c_i \sum_{i=1}^{n} x_i. \tag{2.69}$$

The equality is attained iff either

$$\frac{1}{m}\sum_{i=1}^{n} c_i = \frac{1}{n}\sum_{i=1}^{n} c_i \quad \text{or} \quad x_m = x_{m+1},\ m = 1, \ldots, n-1. \tag{2.70}$$

Proof By (2.41) again and (2.68),

$$\sum_{i=1}^{n} c_i x_i = x_n \sum_{i=1}^{n} c_i - \sum_{m=1}^{n-1} (x_{m+1} - x_m) \sum_{i=1}^{m} c_i$$
$$\leq \frac{1}{n} \sum_{i=1}^{n} c_i \left[n x_n - \sum_{i=1}^{n-1} m (x_{m+1} - x_m) \right] = \frac{1}{n} \sum_{i=1}^{n} c_i \sum_{i=1}^{n} x_i .$$

The equality holds iff

$$\frac{1}{n} \sum_{i=1}^{n} c_i (x_{m+1} - x_m) = \frac{1}{m} \sum_{i=1}^{m} c_i (x_{m+1} - x_m), \quad m = 1, \ldots, n-1, \tag{2.71}$$

which is equivalent to (2.70). □

Reversing the inequalities in (2.68) implies the same for (2.69). The conditions for equality are identical with (2.70). Furthermore, if we set $\frac{1}{n} \sum_{i=1}^{n} c_i = 0$, then we conclude the stochastic ordering inequality for sequences presented in Theorem 2.7. Also, Theorem 2.10, generalizes the Chebyshev inequality (2.30). Indeed, for nonincreasing $\tilde{c} = (c_1, \ldots, c_n)$, we have

$$c_1 \geq \frac{1}{2}(c_1 + c_2) \geq \cdots \geq \frac{1}{n} \sum_{i=1}^{n} c_i, \tag{2.72}$$

and assumptions (2.68) are satisfied. Validity of relations (2.72) follows from the identity

$$\frac{1}{m} \sum_{i=1}^{m} c_i = \frac{j}{m} \left(\frac{1}{j} \sum_{i=1}^{j} c_i \right) + \left(1 - \frac{j}{m} \right) \left(\frac{1}{m-j} \sum_{i=j+1}^{m} c_i \right), \tag{2.73}$$

which means that the mean of a set of numbers is a proper convex combination of the means of its subsets. Accordingly, including a new element in the mean, we obtain a value between the previous mean and the new element. If the sequence is nonincreasing in particular, the sequence of consecutive partial mans in nonincreasing as well, although the rate of its decrease is obviously slower than that of the original sequence. We shall frequently use the above reasoning in the sequel.

There is also an analogous version of Theorem 2.10 for integrals.

Theorem 2.12 *Let H be a function of bounded variation on an interval $[a, b]$, continuous at both endpoints, and satisfying*

$$\frac{H(x) - H(a)}{x - a} \geq \frac{H(b) - H(a)}{b - a}, \qquad x \in (a, b]. \tag{2.74}$$

Then for every nondecreasing integrable function g on $[a, b]$ we have

$$\int_a^b g(t)H(dt) \leq \frac{H(b) - H(a)}{b - a} \int_a^b g(t)dt. \tag{2.75}$$

The equality in (2.75) holds iff g is constant on each interval of the set

$$\mathcal{H}_* = \left\{ t \in [a, b] : \min\{H(t-), H(t+)\} > \frac{H(b) - H(a)}{b - a}(x - a) + H(a) \right\} \tag{2.76}$$

and

$$g(t) = \begin{cases} g(t-), & \text{if } H(t-) < H(t+), \\ g(t+), & \text{if } H(t+) < H(t-), \end{cases} \tag{2.77}$$

at every discontinuity point of H.

Theorem 2.12 is a special case of Theorem 2.15, which is proved in the next section.

2.3.5 *Moriguti Inequalities*

Although the Moriguti inequalities are not so famous as the other ones present in this chapter, they are the fundamental tools that are often used in this book. The general integral version was proved in Moriguti (1953). We first write a version for the finite sequences whose proof can be found in Rychlik (1992).

Theorem 2.13 *Given an arbitrary sequence $\tilde{c} = (c_1, \ldots, c_n) \in \mathbb{R}^n$, define $\underset{\sim}{c} = (\underset{\sim}{c}_1, \ldots, \underset{\sim}{c}_n) \in \mathbb{R}^n$ by means of the following recurrent algorithm. Set*

$$\begin{aligned} j_0 &= 0, \\ j_m &= \min \left\{ j_{m-1} < j \leq n : \frac{1}{j - j_{m-1}} \sum_{r=j_{m-1}+1}^{j} c_r \right. \\ &\qquad\qquad \left. = \min_{j_{m-1} < k \leq n} \frac{1}{k - j_{m-1}} \sum_{r=j_{m-1}+1}^{k} c_r \right\}, \end{aligned} \tag{2.78}$$

for $m = 1, \ldots, M$ with some $1 \le M \le n$ such that $j_M = n$, and

$$\underline{c}_i = \frac{1}{j_m - j_{m-1}} \sum_{r=j_{m-1}+1}^{j_m} c_r, \quad i = j_{m-1}+1, \ldots, j_m, \; m = 1, \ldots, M. \tag{2.79}$$

Then for arbitrary nondecreasing sequence $\tilde{x} = (x_1, \ldots, x_n) \in \mathbb{R}^n$, we have

$$\sum_{i=1}^{n} c_i x_i \le \sum_{i=1}^{n} \underline{c}_i x_i, \tag{2.80}$$

which is equality iff

$$x_i = x_{j_m}, \qquad i = j_{m-1}+1, \ldots, j_m, \; m = 1, \ldots, M. \tag{2.81}$$

Proof Partitioning the LHS of (2.80) into M subsums over the set of indices $\{j_{m-1}+1, \ldots, j_m\}$, $m = 1, \ldots, M$, and applying Theorem 2.11 to each subsum, we obtain

$$\begin{aligned}
\sum_{i=1}^{n} c_i x_i &= \sum_{m=1}^{M} \sum_{i=j_{m-1}+1}^{j_m} c_i x_i \\
&\le \sum_{m=1}^{M} \frac{1}{j_m - j_{m-1}} \left(\sum_{r=j_{m-1}+1}^{j_m} c_r \right) \sum_{i=j_{m-1}+1}^{j_m} x_i \\
&= \sum_{m=1}^{M} \underline{c}_{j_m} \sum_{i=j_{m-1}+1}^{j_m} x_i = \sum_{i=1}^{n} \underline{c}_i x_i.
\end{aligned} \tag{2.82}$$

Indeed, the inequality holds true, because, by (2.78),

$$\frac{1}{j - j_{m-1}} \sum_{r=j_{m-1}+1}^{j} c_r > \frac{1}{j_m - j_{m-1}} \sum_{r=j_{m-1}+1}^{j_m} c_r, \tag{2.83}$$

for all $j = j_{m-1}+1, \ldots, j_m - 1$, and $m = 1, \ldots, M$. The last two equalities in (2.82) follow from definition (2.79). Observe finally that, due to (2.83), the former relations in (2.70) never hold for the subsequences of $(c_{j_{m-1}+1}, \ldots, c_{j_m})$ so that the latter we needed for the equality. This shows that (2.81) are actually the necessary and sufficient conditions for the equality in (2.80). □

Algorithm (2.78) and (2.79) starts with determining the minimal means of the first elements of $\tilde{c}$. It assigns the mean to the smallest set of elements that attains the minimum. Then the procedure is repeated with respect to the remaining part of the sequence $\tilde{c}$ until at M steps, say, the minimal mean contains the last element of the sequence. Evidently, the partition of sequence $\tilde{c} = (c_1, \dots, c_n)$, defined by the algorithm of Theorem 2.13, depends on the relations among its elements. If all the relations (2.68) are sharp (for instance, the sequence is decreasing), we have $M = 1$ and $\underline{c}_i = \frac{1}{n}\sum_{j=1}^{n} c_j$, $i = 1, \dots, n$, and Theorem 2.13 reduces to Theorem 2.10. The constant $\tilde{x}$'s assure attainability then. In fact, sharpness of inequalities (2.68) is not needed for the conclusion. If some of them are equalities, then we still get (2.69), but then $M > 1$, and some nonconstant $\tilde{x}$ provide the equality. On the other hand, for nondecreasing sequences $\tilde{c}$, we have

$$c_i \le \frac{1}{j+1-i}\sum_{r=i}^{j} c_r, \qquad 1 \le i < j \le n. \tag{2.84}$$

Therefore $M = n$, and $\underline{\tilde{c}} = \tilde{c}$, and relation (2.80) is a trivial identity then. Note that $\underline{\tilde{c}} \neq \tilde{c}$ iff $\tilde{c}$ is not nondecreasing. In fact, we check that $\underline{\tilde{c}}$ is nondecreasing for arbitrary $\tilde{c}$. On the contrary, assume that

$$\underline{\tilde{c}}_{j_m} = \frac{1}{j_m - j_{m-1}}\sum_{r=j_{m-1}+1}^{j_m} c_r > \underline{\tilde{c}}_{j_{m+1}} = \frac{1}{j_{m+1} - j_m}\sum_{r=j_m+1}^{j_{m+1}} c_r \tag{2.85}$$

for some $1 \le m \le M-1$. Then we also have

$$\frac{1}{j_m - j_{m-1}}\sum_{r=j_{m-1}+1}^{j_m} c_r > \frac{1}{j_{m+1} - j_{m-1}}\sum_{r=j_{m-1}+1}^{j_{m+1}} c_r > \frac{1}{j_{m+1} - j_m}\sum_{r=j_m+1}^{j_{m+1}} c_r \tag{2.86}$$

(cf. (2.73)), and the former inequality in (2.86) contradicts (2.78). We further prove that $\underline{\tilde{c}}$ is the element of the family of nondecreasing sequences closest to $\tilde{c}$ in the standard Euclidean norm in $\mathbb{R}^n$.

Observe that some of relations $\underline{\tilde{c}}_{j_m} \le \underline{\tilde{c}}_{j_{m+1}}$, $m = 1, \dots, M-1$, may not be strict, and the number of different values M_0, say, in $\underline{\tilde{c}}$ may be less than M. We determine M_0, modifying algorithm (2.78) and (2.79) by defining j_m as the maximal $j > j_{m-1}$ minimizing the subsums starting from $c_{j_{m-1}+1}$. The modification defines the same $\underline{\tilde{c}}$, but expressing the conditions of attainability in (2.80) becomes more complicated. In the particular case of constant $\tilde{c}$, we have $M_0 = 1 < M = n$, and conditions (2.81) are satisfied by any nondecreasing sequence (in fact, (2.80) is the trivial identity then, valid for any other sequences as well). We see that Theorem 2.13 is of any use if $\tilde{c}$ is a nondecreasing sequence.

We finally give a geometric interpretation of the procedure described in Theorem 2.13. Let $\underline{C} : [0, n] \mapsto \mathbb{R}$ be the greatest convex function vanishing at 0, and satisfying

$$\underline{C}(i) \le \sum_{r=1}^{i} c_r, \qquad i = 1, \dots, n. \tag{2.87}$$

The graph of $\underline{C}$ is a broken straight line with breaks at $\left(j'_m, \sum_{r=1}^{j'_m} c_r\right)$, $m = 1, \dots, M_0 - 1$, defined by the modified procedure. It passes through all the points $\left(j_m, \sum_{r=1}^{j_m} c_r\right)$, $m = 1, \dots, M$, and runs below all the other points $\left(i, \sum_{r=1}^{i} c_r\right)$ for $i \notin \{j_1, \dots, j_M\}$. We define $\underline{c}_i$ as n times the slope of the line segment of the graph of $\underline{C}$ on the interval $[i-1, i]$, which is actually the slope of $\underline{C}$ on some $[j'_{m-1}, j'_m]$, $[i-1, i]$, $m \in \{1, \dots, M_0\}$. We leave to the reader verifying equivalence of the recursive and geometric definitions.

From the latter construction, we can conclude that modifications of the recursive definition are possible. Namely, instead of determining consecutive minimal means of the first elements, we can start from the right looking for the maximal means of the best elements of the sequence. Moreover, mixing the steps of the forward and backward procedures does not affect the outcome.

In Sect. 5.2 we use the following weighted version of the Moriguti inequality of Theorem 2.13. The proof is analogous to that of Theorem 2.13, and we omit it here.

Theorem 2.14 *Given an arbitrary sequences $\tilde{c} = (c_1, \dots, c_n) \in \mathbb{R}^n$ and $\tilde{w} = (w_1, \dots, w_n)$ with all positive elements we define $\tilde{\underline{c}} = (\underline{c}_1, \dots, \underline{c}_n) \in \mathbb{R}^n$ by the following recurrent algorithm. Set*

$$\begin{aligned} j_0 &= 0, \\ j_m &= \min\left\{ j_{m-1} < j \le n : \frac{\sum_{r=j_{m-1}+1}^{j} c_r w_r}{\sum_{r=j_{m-1}+1}^{j} w_r} \right. \\ &\qquad\qquad \left. = \min_{j_{m-1} < k \le n} \frac{\sum_{r=j_{m-1}+1}^{k} c_r w_r}{\sum_{r=j_{m-1}+1}^{k} w_r} \right\}, \end{aligned} \tag{2.88}$$

for $m = 1, \dots, M$ with some $1 \le M \le n$ such that $j_M = n$, and

$$\underline{c}_i = \frac{\sum_{r=j_{m-1}+1}^{j_m} c_r w_r}{\sum_{r=j_{m-1}+1}^{j} w_r}, \qquad i = j_{m-1}+1, \dots, j_m, \quad m = 1, \dots, M. \tag{2.89}$$

Then for arbitrary nondecreasing sequence $\tilde{x} = (x_1, \dots, x_n) \in \mathbb{R}^n$ we have

$$\sum_{i=1}^{n} c_i x_i w_i \le \sum_{i=1}^{n} \underline{c}_i x_i w_i, \tag{2.90}$$

which is equality iff

$$x_i = x_{j_m}, \qquad i = j_{m-1} + 1, \ldots, j_m,\; m = 1, \ldots, M. \tag{2.91}$$

The algorithm presented above is called the pool adjacent-violators algorithm. It has various applications in order restricted statistical inference, e.g., in the problems of isotonic regression (see e.g., Barlow et al., 1972; Robertson et al., 1988).

Integral versions of the inequality are based on the notion of greatest convex minorants. We first present the results of Moriguti (1953), and then describe some implications. Suppose that function H is of bounded variation on an interval $[a, b]$ and continuous at the endpoints. It follows that H has an empty, finite or countable set $\mathcal{D}$ of discontinuity points, and the left and right limits at them.

Let $\underline{H}$ be the supremum over the family of all convex functions on $[a, b]$ that are nowhere greater than H. Then $\underline{H}$ is also a member of the family, and is called the greatest convex minorant (GCM) of H. Assumptions on the original function imply that $\underline{H}$ is continuous and has finite variation. For

$$H_*(t) = \min\{H(t-), H(t+)\}, \tag{2.92}$$

the set

$$\mathcal{H}_* = \{t \in [a, b] : \underline{H}(t) < H_*(t)\} \tag{2.93}$$

is open (possibly empty), and consists of countably many disjoint open intervals at most. In each interval, function $\underline{H}$ is linear.

Let g be a nondecreasing function on $[a, b]$. This has a countable (at most) set of jump points, and satisfies

$$g(t-) \le g(t) \le g(t+). \tag{2.94}$$

We evaluate the integral of g with respect to H by the corresponding one with respect to its GCM.

Theorem 2.15 *If H has a bounded variation on $[a, b]$, and is continuous at the endpoints, and $\underline{H}$ is its greatest convex minorant, then*

$$\int_a^b g(t)H(dt) \le \int_a^b g(t)\underline{H}(dt) \tag{2.95}$$

for every nondecreasing function g for which the integrals are finite. The equality in (2.95) *holds iff g is constant in every interval contained in* (2.93)*, and at every discontinuity point $t \in \mathcal{D}$ (if any) of $H(t)$ we have*

$$g(t) = \begin{cases} g(t-), & \text{if } H(t-) > H(t+), \\ g(t+), & \text{if } H(t+) > H(t-). \end{cases} \tag{2.96}$$

Proof We first assume that g is bounded in $[a, b]$, and define its modification g_* satisfying relations (2.96) at all the points where H has jumps Applying the integration by parts formula, we write

$$\int_a^b g_*(t)H(dt) = g(b)H(b) - g(a)H(a) - \int_a^b H_*(t)g(dt) \tag{2.97}$$

and

$$\int_a^b g(t)\underline{H}(dt) = g(b)\underline{H}(b) - g(a)\underline{H}(a) - \int_a^b \underline{H}(t)g(dt). \tag{2.98}$$

Since H and $\underline{H}$ coincide at a and b, subtracting (2.97) and (2.98) gives

$$\begin{aligned}\int_a^b g_*(t)H(dt) - \int_a^b g(t)\underline{H}(dt) &= \int_a^b [H_*(t) - H(t)]g(dt) \\ &= \int_{\mathcal{H}_*} [H_*(t) - \underline{H}(t)]g(dt) \leq 0. \end{aligned} \tag{2.99}$$

because the integrand is negative and g does not decrease. Since g and g_* differ merely at the discontinuity points of H, we have

$$\int_a^b [g(t) - g_*(t)]H(dt) = \sum_{t \in \mathcal{D}} [g(t) - g_*(t)][H(t+) - H(t-)] \leq 0, \tag{2.100}$$

because so is each summand, due to (2.94) and the definition of g_*. Adding (2.99) and (2.100), we obtain (2.95). This becomes equality if both (2.99) and (2.100) vanish. For the former it is necessary and sufficient that g does not increase on any subinterval of $\mathcal{H}_*$. The latter holds iff g_* is identical with g which can be written as (2.96).

The results hold in the case of unbounded functions g as well. Then we restrict ourselves to some narrower intervals $[a_n, b_n] \subset [a, b]$ which contain all discontinuity points of h. Function g is bounded there, and we obtain the desired conclusions. We extend them to the general case as we choose $a_n \searrow a$, $b_n \nearrow b$, and pass to the limit. □

The greatest convex minorant $\underline{H}$ has nondecreasing left and right derivatives in (a, b) which are identical except for at countable many points at most, where the left derivative is strictly less than the right one. Taking any version (say, the right one for definiteness), and denoting it by $\underline{h}$, we can write (2.95) as

$$\int_a^b g(t)H(dt) \leq \int_a^b g(t)\underline{h}(t)dt. \tag{2.101}$$

If H is absolute continuous with respect to the Lebesgue measure, Theorem 2.15 takes on the form.

Theorem 2.16 *Suppose that h is an integrable function on $[a, b]$, and $H(x) = \int_a^x h(t)dt$ stands for its antiderivative. Let $\underline{h}$ denote the right derivative of the GCM $\underline{H}$ of H. Then*

$$\int_a^b g(t)h(t)dt \le \int_a^b g(t)\underline{h}(t)dt \tag{2.102}$$

for every nondecreasing function g for which the integrals are finite. The equality in (2.102) *holds iff g is constant on every interval contained in the set*

$$\mathcal{H}_* = \{t : \underline{H}(t) < H(t)\}. \tag{2.103}$$

Note that Theorem 2.16 is out of practical use if h is nondecreasing. Then H is the GCM of itself, $\underline{h} = h$, and (2.102) becomes a trivial identity.

Likewise we describe analogous lower bounds for linear functions acting on nondecreasing sequences and functions. They are based on standard sign change transformations.

Theorem 2.17 *For arbitrary $\tilde{c} = (c_1, \ldots, c_n) \in \mathbb{R}^n$, define $\bar{\tilde{c}} = (\bar{c}_1, \ldots, \bar{c}_n) \in \mathbb{R}^n$ as follows*

$$\begin{aligned} j_0 &= 0, \\ j_m &= \min\left\{ j_{m-1} < j \le n : \frac{1}{j - j_{m-1}} \sum_{r=j_{m-1}+1}^{j} c_r \right. \\ &\qquad \left. = \max_{j_{m-1}<k\le n} \frac{1}{k - j_{m-1}} \sum_{r=j_{m-1}+1}^{k} c_r \right\} \end{aligned} \tag{2.104}$$

for $m = 1, \ldots, M$ with $1 \le M \le n$ such that $j_M = n$, and

$$\bar{c}_i = \frac{1}{j_m - j_{m-1}} \sum_{r=j_{m-1}+1}^{j_m} c_r \tag{2.105}$$

for $i = j_{m-1} + 1, \ldots, j_m$, and $m = 1, \ldots, M$. Then for arbitrary nondecreasing $\tilde{x} = (x_1, \ldots, x_m) \in \mathbb{R}^n$ we have

$$\sum_{i=1}^{n} c_i x_i \ge \sum_{i=1}^{n} \bar{c}_i x_i, \tag{2.106}$$

which becomes equality for

$$x_i = x_{j_m}, \quad i = j_{m-1} + 1, \ldots, j_m, \quad m = 1, \ldots, M. \tag{2.107}$$

Procedure (2.104) and (2.105) consists in determining the slopes of the smallest concave majorant of points $\left(i, \sum_{r=1}^{i} c_r\right)$, $i = 0, \ldots, n$, and multiplying them by n.

Theorem 2.18 *Let $\bar{H}$ denote the smallest concave majorant of a function H that has a bounded variation on $[a, b]$, and is continuous at a and b. Then*

$$\int_a^b g(t)H(dt) \geq \int_a^b g(t)\bar{H}(dt) \tag{2.108}$$

for every nondecreasing H-integrable function g. Relation (2.108) is sharp and becomes equality iff g is constant on every interval of the set

$$\mathcal{H}^* = \{t \in [a, b] : \bar{H}(t) > \max\{H(t-), H(t+)\}\}, \tag{2.109}$$

and

$$g(t) = \begin{cases} g(t-), & \text{if } H(t-) < H(t+), \\ g(t+), & \text{if } H(t+) < H(t-), \end{cases} \tag{2.110}$$

at every discontinuity point of H.

Theorem 2.9 is a special case of Theorem 2.18. Condition (2.58) implies that $\bar{H}(t) = \text{const} = H(a)$, and the RHS of (2.108) vanishes. Similarly, Theorem 2.12 can be deduced from Theorem 2.15, because, by (2.74), yields

$$\underline{H}(x) = \frac{H(b) - H(a)}{b - a}(x - a) + H(a). \tag{2.111}$$

Under the assumption of absolute continuity of H with respect to the Lebesgue measure, Theorem 2.18 simplifies to the following.

Theorem 2.19 *Suppose that h is an integrable function on $[a, b]$ and $H(x) = \int_a^x h(t)dt$ is its antiderivative. Let $\overline{h}$ denote the right derivative of the smallest concave majorant $\overline{H}$ of H. Then*

$$\int_a^b g(t)h(t)dt \geq \int_a^b g(t)\overline{h}(t)dt \tag{2.112}$$

for every nondecreasing function g for which the integrals are finite. The equality holds here iff g is constant on every interval contained in the set $\{t : \overline{H}(t) > H(t)\}$.

2.4 Norm Inequalities

2.4.1 Schwarz Inequalities

The classic version of the Schwarz inequality for sequences has the form

$$\sum_{i=1}^{n} c_i x_i \le \left(\sum_{i=1}^{n} c_i^2\right)^{1/2} \left(\sum_{i=1}^{n} x_i^2\right)^{1/2}. \tag{2.113}$$

It is often called the Cauchy inequality. It holds for arbitrary n-dimensional sequences $\tilde{c} = (c_1, \ldots, c_n)$ and $\tilde{x} = (x_1, \ldots, x_n)$. Below we prove a more general weighted Cauchy inequality.

Theorem 2.20 *Let $\tilde{c} = (c_1, \ldots, c_n) \in \mathbb{R}^n$, $\tilde{x} = (x_1, \ldots, x_n) \in \mathbb{R}^n$, and $w_i > 0$ for $i = 1, \ldots, n$. Then*

$$\sum_{i=1}^{n} c_i x_i w_i \le \left(\sum_{i=1}^{n} c_i^2 w_i\right)^{1/2} \left(\sum_{i=1}^{n} x_i^2 w_i\right)^{1/2}. \tag{2.114}$$

The equality holds if either of sequences $\tilde{c}$ and $\tilde{x}$ is zero or

$$x_i = \alpha c_i, \qquad i = 1, \ldots, n, \tag{2.115}$$

for some positive α.

Proof Consider

$$\sum_{i=1}^{n} (\alpha c_i - x_i)^2 w_i = \alpha^2 \sum_{i=1}^{n} c_i^2 w_i - 2\alpha \sum_{i=1}^{n} c_i x_i w_i + \sum_{i=1}^{n} x_i^2 w_i. \tag{2.116}$$

This is nonnegative for all real α, which is equivalent to

$$4\left(\sum_{i=1}^{n} c_i x_i w_i\right)^2 - 4\sum_{i=1}^{n} c_i^2 w_i \sum_{i=1}^{n} x_i^2 w_i \ge 0. \tag{2.117}$$

This yields

$$\sum_{i=1}^{n} c_i x_i w_i \le \left|\sum_{i=1}^{n} c_i x_i w_i\right| \le \left(\sum_{i=1}^{n} c_i^2 w_i\right)^{1/2} \left(\sum_{i=1}^{n} x_i^2 w_i\right)^{1/2}, \tag{2.118}$$

and becomes a trivial identity if either $\tilde{c}$ or $\tilde{x}$ vanishes. Otherwise, the later inequality is attained if (2.116) is zero for some real α. This means that we have (2.115), and

$\alpha = 0$ is excluded here. If $\alpha > 0$, then $c_i x_i w_i \geq 0$, $i = 1, \dots, r$, the first sum in (2.118) is positive, and the former inequality becomes equality as well. For $\alpha < 0$, the first sum is strictly negative and differs from its absolute value. □

From the above proof we immediately deduce that the lower bound

$$\sum_{i=1}^{n} c_i x_i w_i \geq -\left(\sum_{i=1}^{n} c_i^2 w_i\right)^{1/2} \left(\sum_{i=1}^{n} x_i^2 w_i\right)^{1/2} \tag{2.119}$$

is sharp, and is attained if either $\tilde{c}$ or $\tilde{x}$ is zero, or (2.115) holds for some negative α. The integral counterpart of the Cauchy inequality is called the Buniakowski-Schwarz inequality.

Theorem 2.21 *Let g and h be square integrable functions on an interval* $[a, b]$ *with respect to a nondecreasing function* W *on* $[a, b]$. *Then*

$$\int_a^b g(x)h(x)W(dx) \leq \left[\int_a^b g^2(x)W(dx)\right]^{1/2} \left[\int_a^b h^2(x)W(dx)\right]^{1/2}. \tag{2.120}$$

The inequality is sharp, and becomes equality iff either $g(x) = 0$ *or* $h(x) = 0$ *or* $g(x) = \alpha h(x)$ *for some* $\alpha > 0$ *W-almost everywhere.*

Proof Proof is similar to that of Theorem 2.20. Relations

$$\begin{aligned} 0 &\leq \int_a^b [g(x) - \alpha h(x)]^2 W(dx) \\ &= \int_a^b g^2(x)W(dx) - 2\alpha \int_a^b g(x)h(x)W(dx) + \alpha^2 \int_a^b h^2(x)W(dx), \end{aligned} \tag{2.121}$$

valid for arbitrary $\alpha \in \mathbb{R}$, imply

$$\begin{aligned} \int_a^b g(x)h(x)W(dx) &\leq \left|\int_a^b g(x)h(x)W(dx)\right| \\ &\leq \left[\int_a^b g^2(x)W(dx)\right]^{1/2} \left[\int_a^b h^2(x)W(dx)\right]^{1/2} \end{aligned} \tag{2.122}$$

If one of the functions vanishes W-almost everywhere, then either of factors in the second line becomes zero, and so do the integrals in the first line. If both the factors are positive, then

$$g(x) = \alpha h(x) \quad W\text{-almost everywhere} \tag{2.123}$$

with nonzero α is the necessary and sufficient condition for equality in the latter inequality of (2.122). The former equality is also obtained if α is positive. □

The lower counterpart of (2.120) is

$$\int_a^b g(x)h(x)W(dx) \geq -\left[\int_a^b g^2(x)W(dx)\right]^{1/2}\left[\int_a^b h^2(x)W(dx)\right]^{1/2}. \tag{2.124}$$

The necessary and sufficient conditions for equality here are identical with those of Theorem 2.21, with $\alpha > 0$ replaced by a negative one.

Finally, we present the Hilbert space version of the Schwarz inequality.

Theorem 2.22 *Let $(\mathcal{H}, \langle\cdot,\cdot\rangle)$ be a real Hilbert space. Then for every $g, h \in \mathcal{H}$ the following inequality holds*

$$\langle g, h\rangle \leq \langle g, g\rangle^{1/2}\langle h, h\rangle^{1/2} = \|g\|\|h\|. \tag{2.125}$$

This is an equality if either $g = 0$ or $h = 0$ or $g = \alpha h$ for some positive α.

Proof If $g = 0$ or $h = 0$, then both the sides of (2.125) are zero so that (2.125) holds and is equality. Otherwise $\|h\|^2 = \langle h, h\rangle > 0$, and we can write

$$\begin{aligned} 0 &\leq \left\| g - \frac{\langle g, h\rangle}{\|h\|^2} h \right\|^2 \|h\|^2 \\ &= \left\langle g - \frac{\langle g, h\rangle}{\langle h, h\rangle} h, g - \frac{\langle g, h\rangle}{\langle h, h\rangle} h \right\rangle \langle h, h\rangle = \langle g, g\rangle\langle h, h\rangle - \langle g, h\rangle^2, \end{aligned} \tag{2.126}$$

which gives

$$|\langle g, h\rangle| \leq \langle g, g\rangle^{1/2}\langle h, h\rangle^{1/2}, \tag{2.127}$$

and so (2.125). Relations (2.126) and (2.127) are equalities iff

$$g = \frac{\langle g, h\rangle}{\langle h, h\rangle} h. \tag{2.128}$$

If $\alpha = \frac{\langle g,h\rangle}{\langle h,h\rangle} > 0$, then $\langle g, h\rangle = |\langle g, h\rangle|$, and (2.125) becomes equality. Otherwise, we attain the lower bound

$$\langle g, h\rangle \geq -\langle g, g\rangle^{1/2}\langle h, h\rangle^{1/2}, \tag{2.129}$$

that is also concluded from (2.127). □

The inequalities of Theorems 2.20 and 2.21 are special cases of (2.125). The sums and integrals represent the inner products in the Hilbert spaces $\ell^2(\mathbb{R}^n, \tilde{w})$ and $L^2([a, b], W(dx))$, respectively. In particular, the standard Cauchy inequality (2.113) provides a bound on the inner product in the standard Euclidean space of size n. We finally point out that the Schwarz inequality (2.125) holds in incomplete inner product spaces as well.

2.4.2 Projection Inequalities

Due to the Riesz representation theorem, every linear continuous functional acting on elements g of a Hilbert space $\mathcal{H}$ may be represented as the inner product of g with a fixed element h of the space, i.e.,

$$\exists\, h \in \mathcal{H} \quad \forall\, g \in \mathcal{H} \qquad T(g) = T_h(g) = \langle g, h \rangle. \tag{2.130}$$

The norm of the functional, due to the Schwarz inequality (2.125), is equal to

$$\|T_h\| = \sup_{g \neq 0} T_h\left(\frac{g}{\|g\|}\right) = \sup_{g \neq 0} \frac{\langle g, h \rangle}{\|g\|} = \|h\|, \tag{2.131}$$

and is attained at all $g = \alpha h$ with $\alpha > 0$. Numerous statistical problems consist in precise evaluating normalized linear functionals $\frac{T_h(g)}{\|g\|}$ over restricted subsets of Hilbert spaces. If αh for some $\alpha > 0$ belongs to the subset, then bound (2.131) is attained. Otherwise the bound is not sharp. However, if the subset under consideration is a convex cone $\mathcal{C}$ in $\mathcal{H}$, which often happens in practice, we obtain a sharp bound replacing the original h by its projection $P_{\mathcal{C}}h$ onto the cone. We define the projection of $h \in \mathcal{H}$ onto a subset $\mathcal{S} \subset \mathcal{H}$ as the $P_{\mathcal{S}} \in \mathcal{S}$ satisfying

$$\|P_{\mathcal{S}}h - h\| = \inf_{g \in \mathcal{S}} \|g - h\|. \tag{2.132}$$

For every closed convex subset $\mathcal{S}$ of a Hilbert space we can uniquely define the projection operator $P_{\mathcal{S}}$ that ascribes the projection $P_{\mathcal{S}}h$ to arbitrary $h \in \mathcal{H}$.

Theorem 2.23 *If h is an arbitrary element of a real Hilbert space $\mathcal{H}$, and $\mathcal{S}$ is a closed convex subset of $\mathcal{H}$, then there exists a unique projection $P_{\mathcal{S}}h$ of h onto $\mathcal{S}$.*

Proof If $h \in \mathcal{S}$, then for $P_{\mathcal{S}}h = h$ we have $\|P_{\mathcal{S}}h - h\| = 0$ whereas $\|g - h\| > 0$ for all other $g \in \mathcal{H} \setminus \mathcal{S}$. Otherwise there exists a sequence $g_n \in \mathcal{S}$, $n = 1, 2, \ldots$, such that

$$\lim_{n \to \infty} \|g_n - h\| = \inf_{g \in \mathcal{S}} \|g - h\| = D, \tag{2.133}$$

say, which is strictly positive, because $\mathcal{S}$ is closed. Applying the elementarily verified parallelogram law

$$\|g-h\|^2+\|g+h\|^2=2(\|g\|^2+\|h\|^2), \qquad g,h\in\mathcal{H}$$

for arbitrary two elements g_n, g_m of the sequence, we obtain

$$\left\|\frac{g_n-h}{2}+\frac{g_m-h}{2}\right\|^2+\left\|\frac{g_n-h}{2}-\frac{g_m-h}{2}\right\|^2=\frac{1}{2}\left(\|g_n-h\|^2+\|g_m-h\|^2\right). \tag{2.134}$$

If n and m are large enough, then for arbitrary $\varepsilon>0$, the RHS of (2.134) does not exceed $D^2+\varepsilon$. Since $\frac{1}{2}(g_n+g_m)\in\mathcal{S}$, the first term of the LHS of (2.134) is not less than D^2. Therefore

$$\frac{1}{4}\,||g_n-g_m||^2=\left\|\frac{g_n-g_m}{2}\right\|^2\leq\varepsilon,$$

which by completeness of $\mathcal{H}$ and closedness of $\mathcal{S}$ implies that g_n, $n=1,2,\ldots,$ has a unique limit point $P_{\mathcal{S}}h\in\mathcal{S}$, say. The limit is the desired projection, because

$$\|P_{\mathcal{S}}h-h\|\leq\|P_{\mathcal{S}}h-g_n\|+\|g_n-h\|\to D,\ \text{as } n\to\infty \tag{2.135}$$

implies $\|P_{\mathcal{S}}h-h\|=D$. □

The following two theorems characterize projections onto convex sets and cones. Ideas of proving Theorems 2.23, 2.24 and 2.25 come from Balakrishnan (1981, Section 1.4). A comprehensive study of optimal approximation problems in inner product spaces can be found in Deutsch (2001).

Theorem 2.24 *Element $P_{\mathcal{S}}h\in\mathcal{S}$ is the projection of $h\in\mathcal{H}$ onto a convex set $\mathcal{S}$ in a real Hilbert space $\mathcal{H}$ iff*

$$\forall\, g\in\mathcal{S}\quad\langle g,h-P_{\mathcal{S}}h\rangle\leq\langle P_{\mathcal{S}}h,h-P_{\mathcal{S}}h\rangle. \tag{2.136}$$

Proof Take $P_{\mathcal{S}}h$ and any other $g\in\mathcal{S}$, and for $0<\alpha<1$, define

$$\begin{aligned}D_g(\alpha)&=\|h-(1-\alpha)P_{\mathcal{S}}h-\alpha g\|^2\\&=\|h-P_{\mathcal{S}}h\|^2+2\alpha\langle h-P_{\mathcal{S}}h,P_{\mathcal{S}}h-g\rangle+\alpha^2\|P_{\mathcal{S}}h-g\|^2.\end{aligned} \tag{2.137}$$

We have

$$D_g'(\alpha)=2\langle h-P_{\mathcal{S}}h,P_{\mathcal{S}}h-g\rangle+2\alpha\|P_{\mathcal{S}}h-g\|^2, \tag{2.138}$$

$$D_g''(\alpha)=2\,\|P_{\mathcal{S}}h-g\|^2\geq 0. \tag{2.139}$$

If $P_{\mathcal{S}}h$ is the projection of h, then $D'_g(0) \geq 0$ for all $g \in \mathcal{S}$, which is identical with (2.136). On the other hand, (2.136) with (2.139) imply that $D'_g(\alpha) \geq 0$ for all $0 \leq \alpha \leq 1$, and so

$$D_g(0) = \|h - P_{\mathcal{S}}h\|^2 \leq D_g(1) = \|h - g\|^2, \quad g \in \mathcal{S}, \tag{2.140}$$

which proves that $P_{\mathcal{S}}h$ is the projection of h onto $\mathcal{S}$. □

Theorem 2.25 *Element $P_{\mathcal{C}} \in \mathcal{C}$ is the projection of $h \in \mathcal{H}$ onto a convex cone $\mathcal{C} \subset \mathcal{H}$ iff the following relations hold*

$$\forall\, g \in \mathcal{C} \quad \langle g, h\rangle \leq \langle g, P_{\mathcal{C}}h\rangle, \tag{2.141}$$

$$\langle P_{\mathcal{C}}h, h\rangle = \langle P_{\mathcal{C}}h, P_{\mathcal{C}}h\rangle. \tag{2.142}$$

Proof If $P_{\mathcal{C}}h$ is the projection, then for $g = \alpha P_{\mathcal{C}}h \in \mathcal{C}$, $\alpha \geq 0$, (2.136) holds. This gives

$$(\alpha - 1)\langle P_{\mathcal{C}}h, h - P_{\mathcal{C}}h\rangle \leq 0. \tag{2.143}$$

Relation is true for all $\alpha \geq 0$ iff (2.142) is satisfied. Substituting (2.142) in (2.136) gives (2.141). □

We show that $\|P_{\mathcal{C}}h\|$ provides the sharp bound on $\frac{T_h(g)}{\|g\|}$) over the nonzero elements g of a convex cone $\mathcal{C}$. If $P_{\mathcal{C}}h$ is nonzero, then, by (2.125) and (2.141), we have

$$\forall\, \mathbf{0} \neq g \in \mathcal{C} \quad T_h\left(\frac{g}{\|g\|}\right) = \frac{\langle g, h\rangle}{\|g\|} \leq \frac{\langle g, P_{\mathcal{C}}h\rangle}{\|g\|} \leq \|P_{\mathcal{C}}h\| > 0. \tag{2.144}$$

For $g = \alpha P_{\mathcal{C}}h$, $\alpha > 0$, in particular, (2.142) implies

$$T_h\left(\frac{\alpha P_{\mathcal{C}}h}{\|\alpha P_{\mathcal{C}}h\|}\right) = \frac{\langle P_{\mathcal{C}}h, h\rangle}{\|P_{\mathcal{C}}h\|} = \frac{\langle P_{\mathcal{C}}h, P_{\mathcal{C}}h\rangle}{\|P_{\mathcal{C}}h\|} = \|P_{\mathcal{C}}h\|. \tag{2.145}$$

If $P_{\mathcal{C}}h = 0$, we also apply (2.141) and (2.142) to derive

$$\forall\, g \in \mathcal{C} \quad T_h(g) = \langle g, h\rangle \leq \langle g, P_{\mathcal{C}}h\rangle = 0 = T_h(P_{\mathcal{C}}h). \tag{2.146}$$

Note that in the case of convex cones (and linear subspaces, in particular) $P_{\mathcal{C}}h$ and $h - P_{\mathcal{C}}h$ are orthogonal (see (2.142)). This is not generally true for the projections on other convex sets. Relation (2.142) implies

$$\begin{aligned}\|h\|^2 &= \langle h - P_{\mathcal{C}}h, h - P_{\mathcal{C}}h\rangle - 2\langle P_{\mathcal{C}}h, h - P_{\mathcal{C}}h\rangle + \langle P_{\mathcal{C}}h, P_{\mathcal{C}}h\rangle \\ &= \|h - P_{\mathcal{C}}h\|^2 - \|P_{\mathcal{C}}h\|^2 \geq \|P_{\mathcal{C}}h\|^2,\end{aligned} \tag{2.147}$$

and the inequality is strict iff $P_{\mathcal{C}}h \neq h$. It follows that in the case $h \notin \mathcal{C}$, the projection bound (2.144) for normed functionals is actually better than the general one (2.131).

In the many applications, we consider functionals over nonclosed convex cones. Then we determine projections onto the closures of the cones. The projection is a boundary point that can be approximated by sequences of elements of the original cone. This approach provides the sharp bounds on functionals which are not attained by any element of the cone.

Below we apply Theorem 2.25 for checking that the constructions of Theorems 2.13 and 2.16, based on the notion of greatest convex minorant determine projection operators onto the convex cones of nondecreasing sequences and functions, acting on arbitrary elements of the sequence and functions spaces $\ell^2(\mathbb{R}^n)$, and $L^2([a,b],dx)$ with uniform weights, respectively. The projections onto nonincreasing sequences and functions are determined by means of the smallest concave majorant (cf. Theorems 2.17 and 2.18).

Theorem 2.26 *For arbitrary $\tilde{c} = (c_1, \dots, c_n) \in \mathbb{R}^n$, define $\tilde{\underline{c}} = (\underline{c}_1, \dots, \underline{c}_n) \in \mathbb{R}^n$ by means of procedure* (2.78) *and* (2.79)*. Then $\tilde{\underline{c}}$ is the unique projection $P^{\nearrow}\tilde{c}$ of $\tilde{c}$ onto the convex cone of nondecreasing sequences in Hilbert space $\ell^2(\mathbb{R}^n)$.*

Proof Since (2.80) coincides here with (2.141), it merely suffices to check if (2.142) holds. Indeed, we have

$$
\begin{aligned}
(\tilde{\underline{c}}, \tilde{c}) &= \sum_{i=1}^{n} \underline{c}_i c_i = \sum_{m=1}^{M} \sum_{i=j_{m-1}+1}^{j_m} \underline{c}_i c_i = \sum_{m=1}^{M} \underline{c}_{j_m} \sum_{i=j_{m-1}+1}^{j_m} c_i \\
&= \sum_{m=1}^{M} (j_m - j_{m-1}) \underline{c}_{j_m} \underline{c}_{j_m} = \sum_{i=1}^{n} \bar{c}_i^2 = (\tilde{\underline{c}}, \tilde{\underline{c}}),
\end{aligned} \tag{2.148}
$$

which ends the proof. □

Theorem 2.27 *For a fixed function $h \in L^2([a,b],dx)$, define $\underline{h}$ as the right-hand side derivative of the greatest convex minorant $\underline{H}$ of the antiderivative H of h. Then $\underline{h}$ is the unique projection $P^{\nearrow}h$ of h onto the convex cone of nondecreasing functions in $L^2([a,b],dx)$.*

Proof Since (2.102) holds, it remains to check the orthogonality condition (2.142). Set $\mathcal{H}_*$ defined in (2.103), where H and $\underline{H}$ differ is either empty or consists of either finite or countable family of disjoint intervals (a_i, b_i), say. For each of them,

$$
\underline{h}(t) = \frac{H(b_i) - H(a_i)}{b_i - a_i}, \qquad a_i \le t \le b_i, \tag{2.149}
$$

and $\underline{h}(t) = h(t)$ on $[a, b] \setminus \mathcal{H}_*$. Therefore

$$\begin{aligned}
(h, \underline{h}) &= \int_a^b \underline{h}(t)h(t)dt \\
&= \sum_i \int_{a_i}^{b_i} \frac{H(b_i) - H(a_i)}{b_i - a_i} h(t)dt + \int_{[a,b]\setminus\mathcal{H}} h^2(t)dt \\
&= \sum_i \int_{a_i}^{b_i} \left[\frac{H(b_i) - H(a_i)}{b_i - a_i}\right]^2 dt + \int_{[a,b]\setminus\mathcal{H}_*} \underline{h}^2(t)dt \\
&= \int_a^b \underline{h}^2(t)dt = (\underline{h}, \underline{h}),
\end{aligned} \tag{2.150}$$

which completes the proof. □

In view of Theorems 2.26 and 2.27, the Chebyshev inequalities (2.30) and (2.40) imply that the projections of nonincreasing sequences and functions onto the families of nondecreasing ones are constants equal to the means of originals. Less restrictive conditions for constant projections are given in assumptions (2.68) and (2.74) of Theorems 2.11 and 2.12, respectively. Actually, the conditions are necessary and sufficient.

2.4.3 Orthogonal Expansion Inequalities

Let us use $\mathcal{H}$ to denote an inner product space and $\{\phi_k\}_{k=0}^{\infty}$ to denote an orthonormal system in $\mathcal{H}$. Let us use $\|f\|$ and $\langle g, h\rangle$ to denote the norm of f and the inner product of g and h, for any elements f, g and h in $\mathcal{H}$. Further, let $L^2((0, 1), dx)$ and $L^2(\mathcal{C}_k, d\tilde{u}) = L^2(\mathcal{C}_k, du_1 \ldots, du_k)$ denote the inner product space of all square integrable functions in (0,1) and in the k-dimensional hypercube $\mathcal{C}_k = \{(u_1, \ldots, u_k) : 0 \le u_1, \ldots, u_k \le 1\}$, respectively.

Then, we readily have the following two lemmas.

Lemma 2.2 *With $\alpha_k = \langle g, \phi_k\rangle$ and $\beta_k = \langle h, \phi_k\rangle$ denoting the Fourier coefficients of the elements g and h in $\mathcal{H}$ with respect to the orthonormal system $\{\phi_k\}_{k=0}^{\infty}$, respectively, we have for any $K = 0, 1, \ldots,$*

$$\left|\langle g, h\rangle - \sum_{k=0}^{K} \alpha_k\beta_k\right| \le \left\{\|g\|^2 - \sum_{k=0}^{K} \alpha_k^2\right\}^{1/2} \left\{\|h\|^2 - \sum_{k=0}^{K} \beta_k^2\right\}^{1/2}; \tag{2.151}$$

further, the equality in (2.151) *holds when and only when $g, h, \phi_0, \phi_1, \ldots, \phi_K$ are linearly dependent (see Sugiura, 1962).*

Lemma 2.3 *Suppose* $\{\phi_k\}_{k=0}^{\infty}$ *is a complete orthonormal system in the space* $L^2((0,1),dx)$. *Then,* $\{\phi_{i_1}(u_1)\phi_{i_2}(u_2)\cdots\phi_{i_k}(u_k)\}_{i_1,\ldots,i_k=0}^{\infty}$ *is a complete orthonormal system in* $L^2(\mathcal{C}_k, d\tilde{u})$*, where* $\mathcal{C}_k$ *is the* k*-dimensional hypercube defined above.*

Observe now that $L^2(\mathcal{C}_k, d\tilde{u})$ *can be decomposed into* 2^k *sub-spaces as follows:*

$$\begin{aligned} L^2_{j_1,\ldots,j_k}(\mathcal{C}_k, d\tilde{u}) = \{\eta(u_1,\ldots,u_k) \ :\ \eta \in L^2(\mathcal{C}_k, d\tilde{u}) \text{ with} \\ \eta(u_1,\ldots,u_k) &= (-1)^{j_1}\eta(1-u_1,u_2,\ldots,u_k) \\ &= (-1)^{j_2}\eta(u_1,1-u_2,u_3,\ldots,u_k) \\ &= \cdots \\ &= (-1)^{j_k}\eta(u_1,\ldots,u_{k-1},1-u_k)\}, \end{aligned} \tag{2.152}$$

where $j_m = 0, 1$ *for* $m = 1, \ldots, k$.

Then, we have the following result corresponding to the complete orthonormality of 2^k sub-systems in (2.152) (see Mathai, 1975).

Theorem 2.28 *Suppose* $\{\phi_k\}_{k=0}^{\infty}$ *is a complete orthonormal system in the space* $L^2((0,1),dx)$ *satisfying the property that* $\phi_k(u) = (-1)^k\phi_k(1-u)$ *for all* $u \in (0,1)$. *Then, for* $i_1,\ldots,i_k = 0,1,\ldots$ *and* $j_m \in \{0,1\}$ *for* $m = 1,\ldots,k$, $\{\phi_{2i_1+j_1}(u_1)\cdots\phi_{2i_k+j_k}(u_k)\} \in L^2_{j_1,\ldots,j_k}(\mathcal{C}_k, d\tilde{u})$*, and moreover each of the* 2^k *subsystems is complete and orthonormal in the corresponding sub-spaces.*

Proof Let us first define

$$\eta(u_1,\ldots,u_k) = \phi_{2i_1+j_1}(u_1)\cdots\phi_{2i_k+j_k}(u_k).$$

It is then clear that $\eta(u_1,\ldots,u_k) \in L^2(\mathcal{C}_k, d\tilde{u})$. Further, we observe that

$$\begin{aligned} &\eta(u_1,\ldots,u_{m-1},1-u_m,u_{m+1},\ldots,u_k) \\ &= \phi_{2i_1+j_1}(u_1)\cdots\phi_{2i_{m-1}+j_{m-1}}(u_{m-1})\phi_{2i_m+j_m}(1-u_m) \\ &\qquad \times\phi_{2i_{m+1}+j_{m+1}}(u_{m+1})\cdots\phi_{2i_k+j_k}(u_k) \\ &= (-1)^{j_m}\phi_{2i_1+j_1}(u_1)\cdots\phi_{2i_m+j_m}(u_m)\cdots\phi_{2i_k+j_k}(u_k) \\ &= (-1)^{j_m}\eta(u_1,\ldots,u_m,\ldots,u_k), \end{aligned}$$

for $m = 1, 2, \ldots, k$. Therefore, $\eta(u_1,\ldots,u_k) \in L^2_{j_1,\ldots,j_k}(\mathcal{C}_k, d\tilde{u})$, as required.

To prove the second claim of the theorem, let us consider an arbitrary function $\omega(u_1, \ldots, u_k) \in L^2_{j_1,\ldots,j_k}(\mathcal{C}_k, d\tilde{u})$. Then, by definition, we have

$$\|\omega\|^2 = \sum_{i_1=0}^{\infty} \cdots \sum_{i_k=0}^{\infty} \langle \omega, \phi_{i_1}(u_1) \cdots \phi_{i_k}(u_k) \rangle^2 \tag{2.153}$$

due to the completeness of the system $\{\phi_{i_1}(u_1) \cdots \phi_{i_k}(u_k)\}_{i_1,\ldots,i_k=0}^{\infty}$ in the space $L^2(\mathcal{C}_k, d\tilde{u})$, as stated in Lemma 2.3. From (2.153), let us write

$$\|\omega\|^2 = \sum_{j_1'=0}^{1} \sum_{i_1=0}^{\infty} \cdots \sum_{i_k=0}^{\infty} \langle \omega, \phi_{2i_1+j_1'}(u_1)\phi_{i_2}(u_2) \cdots \phi_{i_k}(u_k) \rangle^2.$$

Suppose now $\omega \in L^2_{0,j_2,\ldots,j_k}(\mathcal{C}_k, d\tilde{u})$. Then,

$$\begin{aligned}
&\langle \omega, \phi_{2i_1+1}(u_1)\phi_{i_2}(u_2) \cdots \phi_{i_k}(u_k) \rangle \\
&= \int_0^1 \cdots \int_0^1 \omega(u_1, \ldots, u_k)\phi_{2i_1+1}(u_1)\phi_{i_2}(u_2) \cdots \phi_{i_k}(u_k)du_1 \cdots du_k \\
&= \int_0^1 \cdots \int_0^1 \omega(1-u_1, u_2, \ldots, u_k)\phi_{2i_1+1}(1-u_1)\phi_{i_2}(u_2) \cdots \phi_{i_k}(u_k)du_1 \cdots du_k \\
&= -\int_0^1 \cdots \int_0^1 \omega(1-u_1, u_2, \ldots, u_k)\phi_{2i_1+1}(u_1)\phi_{i_2}(u_2) \cdots \phi_{i_k}(u_k)du_1 \cdots du_k \\
&= -\langle \omega, \phi_{2i_1+1}(u_1)\phi_{i_2}(u_2) \cdots \phi_{i_k}(u_k) \rangle
\end{aligned}$$

since $\omega \in L^2_{0,j_2,\ldots,j_k}(\mathcal{C}_k, d\tilde{u})$. This means

$$\langle \omega, \phi_{2i_1+1}(u_1)\phi_{i_2}(u_2) \cdots \phi_{i_k}(u_k) \rangle = 0,$$

which implies that when $\omega \in L^2_{0,j_2,\ldots,j_k}(\mathcal{C}_k, d\tilde{u})$, we have

$$\|\omega\|^2 = \sum_{i_1=0}^{\infty} \cdots \sum_{i_k=0}^{\infty} \langle \omega, \phi_{2i_1}(u_1)\phi_{i_2}(u_2) \cdots \phi_{i_k}(u_k) \rangle^2. \tag{2.154}$$

Proceeding similarly, when $\omega \in L^2_{1,j_2,\ldots,j_k}(\mathcal{C}_k, d\tilde{u})$, we can show that

$$\|\omega\|^2 = \sum_{i_1=0}^{\infty} \cdots \sum_{i_k=0}^{\infty} \langle \omega, \phi_{2i_1+1}(u_1)\phi_{i_2}(u_2) \cdots \phi_{i_k}(u_k) \rangle^2. \tag{2.155}$$

Thus, we have $\{\phi_{2i_1+j_1}(u_1)\phi_{2i_2+j_2}(u_2)\cdots\phi_{2i_k+j_k}(u_k)\}_{i_1,\ldots,i_k=0}^{\infty}$ to be a complete orthonormal system in $L^2_{j_1,\ldots,j_k}(\mathcal{C}_k, d\tilde{u})$, completing the proof of the theorem. □

Example 2.1 Let us consider the sequence of Legendre polynomials $\{\phi_k(u)\}_{k=0}^{\infty}$, with

$$\phi_k(u) = \frac{\sqrt{2k+1}}{k!}\frac{d^k}{du^k}\{u^k(u-1)^k\} \tag{2.156}$$

for $k = 0, 1, \ldots$. This sequence is known to be a complete orthonormal system in $L^2((0, 1), dx)$ (see Sansone, 1959; Szegö, 1959). Then, with $g, h \in L^2(0, 1)$ and

$$\alpha_k = \int_0^1 g(u)\phi_k(u)du \quad \text{and} \quad \beta_k = \int_0^1 h(u)\phi_k(u)du,$$

for $k = 0, 1, \ldots$, as the corresponding Fourier coefficients with respect to Legendre polynomials in (2.156), Lemma 2.2 yields a sequence of inequalities

$$\left|\int_0^1 g(u)h(u)du - \sum_{k=0}^{K}\alpha_k\beta_k\right| \leq \left\{\int_0^1 g^2(u)du - \sum_{k=0}^{K}\alpha_k^2\right\}^{1/2} \times \left\{\int_0^1 h^2(u)du - \sum_{k=0}^{K}\beta_k^2\right\}^{1/2}, \tag{2.157}$$

for $K = 0, 1, \ldots$.

Furthermore, observe from (2.156) that the sequence of Legendre polynomials satisfies the property that $\phi_k(u) = (-1)^k\phi_k(1-u)$ for all $u \in (0, 1)$. Hence, the results stated in Theorem 2.28 hold in this case; see Mathai (1975).

2.4.4 Hölder Inequalities

The basic Hölder inequality for sequences of real numbers $\tilde{c} = (c_1, \ldots, c_n)$ and $\tilde{x} = (x_1, \ldots, x_n)$ with $p, q > 1$ satisfying $\frac{1}{p} + \frac{1}{q} = 1$ has the form

$$\sum_{i=1}^{n} c_i x_i \leq \left(\sum_{i=1}^{n}|c_i|^q\right)^{1/q}\left(\sum_{i=1}^{n}|x_i|^p\right)^{1/p}. \tag{2.158}$$

For $p = q = 2$, we obtain the Schwarz inequality (2.113). We prove more general weighted series and integral version which generalize (2.114) and (2.120), respectively. The proofs are based on the following inequality.

Theorem 2.29 *If $p, q > 1$ and $\frac{1}{p} + \frac{1}{q} = 1$, then for every $x, y \geq 0$, we have*

$$xy \leq \frac{x^p}{p} + \frac{y^q}{q}. \tag{2.159}$$

This is equality iff $x^p = y^q$.

Proof For a fixed $u \geq 0$, define function

$$f_u(t) = t^{1/p}u^{1/q} - \frac{t}{p} - \frac{u}{q}, \qquad t \geq 0. \tag{2.160}$$

We aim at maximizing (2.160) with respect to $t \geq 0$.

For $u = 0$, we have

$$f_0(t) = -\frac{t}{p} \leq 0 = f_0(0). \tag{2.161}$$

If $u > 0$, then

$$f_u'(t) = \frac{1}{p}\left[\left(\frac{u}{t}\right)^{1/q} - 1\right], \tag{2.162}$$

$$f_u''(t) = -\frac{1}{pq}\frac{u^{1/q}}{t^{1/q+1}} < 0, \tag{2.163}$$

so that $f_u(t)$ is concave, increasing on $[0, u]$ and decreasing on $[u, +\infty)$, and, in consequence

$$f_u(t) \leq f_u(u) = u^{\frac{1}{p}+\frac{1}{q}} - \left(\frac{1}{p} + \frac{1}{q}\right)u = 0. \tag{2.164}$$

Therefore, for all $u, t \geq 0$, we have

$$t^{\frac{1}{p}}u^{\frac{1}{q}} \leq \frac{t}{p} + \frac{u}{q}, \tag{2.165}$$

which become equality for $t = u$. Setting $x = t^{1/p}$, $y = u^{1/q}$, we complete the proof. □

Theorem 2.30 *If $p, q > 1$ satisfy $\frac{1}{p} + \frac{1}{q} = 1$, then for arbitrary $\tilde{c} = (c_1, \ldots, c_n)$, $\tilde{x} = (x_1, \ldots, x_n)$ and $\tilde{w} = (w_1, \ldots, w_n)$ with $w_i > 0$, $i = 1, \ldots, n$, we have*

$$\sum_{i=1}^{n} c_i x_i w_i \leq \left(\sum_{i=1}^{n} |c_i|^q w_i\right)^{1/q} \left(\sum_{i=1}^{n} |x_i|^p w_i\right)^{1/p}. \tag{2.166}$$

The Hölder inequality (2.166) *becomes equality iff either all c_i or all x_i are zero or*

$$x_i = \alpha |c_i|^{q/p} \mathrm{sgn}\{c_i\}, \quad i = 1, \ldots, n, \ \alpha > 0. \tag{2.167}$$

Proof The proof is nontrivial if the RHS of (2.166) is nonzero. Let

$$\xi_i = \frac{|x_i|}{\left(\sum_{r=1}^n |x_r|^p w_r\right)^{1/p}}, \quad \eta_i = \frac{|c_i|}{\left(\sum_{r=1}^n |c_r|^q w_r\right)^{1/q}}, \tag{2.168}$$

for $i = 1, \ldots, n$. By (2.159), we have

$$\frac{|x_i|\,|c_i|}{\left(\sum_{r=1}^n |x_r|^p w_r\right)^{1/p} \left(\sum_{r=1}^n |c_r|^q w_r\right)^{1/q}} \le \frac{1}{p} \frac{|x_i|^p}{\sum_{r=1}^n |x_r|^p w_r} + \frac{1}{q} \frac{|c_i|^q}{\sum_{r=1}^n |c_r|^q w_r}$$

for $i = 1, \ldots, n$. Multiplying the LHS and RHS expressions by w_i, $i = 1, \ldots, n$, and summing them up, we obtain

$$\frac{\sum_{i=1}^n |x_i||c_i| w_i}{\left(\sum_{i=1}^n |x_i|^p w_i\right)^{1/p} \left(\sum_{i=1}^n |c_i|^q w_i\right)^{1/q}} \le \frac{1}{p} + \frac{1}{q} = 1, \tag{2.169}$$

which together with

$$\sum_{i=1}^n x_i c_i w_i \le \sum_{i=1}^n |x_i||c_i| w_i \tag{2.170}$$

yields (2.166). Relation (2.169) is the equality iff $\xi_i^p = \eta_i^q$, $i = 1, \ldots, n$, which is equivalent to

$$|x_i| = \alpha |c_i|^{q/p}, \qquad i = 1, \ldots, n \tag{2.171}$$

for some positive α. Bound (2.170) is attained iff all c_i and x_i have the same signs, which together with (2.171) gives (2.167). □

Similarly we prove

Theorem 2.31 *If $p, q > 1$ and $\frac{1}{p} + \frac{1}{q} = 1$, and W is a nondecreasing function on $[a, b]$, then*

$$\int_a^b g(x)h(x)W(dx) \le \left[\int_a^b |g(x)|^p W(dx)\right]^{1/p} \left[\int_a^b |h(x)|^q W(dx)\right]^{1/q} \tag{2.172}$$

holds for arbitrary functions g, h for which the integrals of the RHS of (2.172) *exist. Moreover,* (2.172) *is the equality iff either of functions is 0, or*

$$g(x) = \alpha |h(x)|^{q/p} \mathrm{sgn}\{h(x)\}, \qquad \alpha > 0, \tag{2.173}$$

W-almost everywhere.

Proof We confine ourselves to the case of nonzero functions. Due to (2.159),

$$\frac{|g(x)||h(x)|}{\left[\int_a^b |g(x)|^p W(dx)\right]^{1/p} \left[\int_a^b |h(x)|^q W(dx)\right]^{1/q}}$$
$$\leq \frac{1}{p} \frac{|g(x)|^p}{\int_a^b |g(x)|^p W(dx)} + \frac{1}{q} \frac{|h(x)|^q}{\int_a^b |h(x)| W(dx)}. \tag{2.174}$$

Integrating both the sides with respect to W provides 1 for the RHS and, in consequence, we conclude (2.172) for the absolute values of functions. The rest of the proof is standard, and we left it to the reader. □

The lower bounds in Theorems 2.30 and 2.31 are the negatives of the RHS of (2.166) and (2.172), respectively, and are attained in the nonzero cases with negative proportionality coefficients α in (2.167) and (2.173). The results of Theorem 2.30 can be extended to infinite sequences.

The Hölder inequalities play an important role in functional analysis. They provide evaluations of linear continuous functionals over the Banach spaces $\ell^p(\mathbb{R}^n, \tilde{w})$ (n may be finite or not), and $L^p([a, b], W(dx))$ that consist of sequences and functions satisfying

$$\sum_{i=1}^{n} |x_i|^p w_i < \infty \tag{2.175}$$

(that is only to verify in the case $n = \infty$), and

$$\int_a^b |g(x)|^p W(dx) < \infty, \tag{2.176}$$

with the norms expressed by pth roots of the LHS of (2.175) and (2.176), respectively. Then all the linear continuous functionals over the spaces are represented by the LHS of (2.166) and (2.172) for $\tilde{c} \in \mathbb{R}^n$ and $h : [a, b] \to \mathbb{R}$ being the elements of respective $\ell^q(\mathbb{R}^n, \tilde{w})$ and $L^q([a, b], W(dx))$ Banach spaces with the conjugate exponents $q = \frac{p}{p-1}$. Theorems 2.30 and 2.31 assert that the norms of functionals are identical with the qth norms of their representatives.

The theory is naturally extended to the cases $p = 1$ and $p = \infty$. In the latter one,

$$\|\tilde{x}\|_\infty = \lim_{p\to\infty}\left(\sum_{i=1}^{n}|x_i|^p w_i\right)^{1/p} = \sup_{1\le i\le n}|x_i|, \tag{2.177}$$

$$\begin{aligned}\|g\|_\infty &= \lim_{p\to\infty}\left(\int_a^b |g(x)|^p W(dx)\right)^{1/p} = \text{ess}\sup\nolimits_W |g(x)| \\ &= \inf\{c \ge 0 : \mu_W(\{x : |g(x)| > c\}) = 0\}.\end{aligned} \tag{2.178}$$

We now state without proofs the relevant versions of the Hölder inequality.

Theorem 2.32 *Inequality*

$$\sum_{i=1}^{n} c_i x_i w_i \le \sup_{1\le i\le n}|c_i| \sum_{i=1}^{n}|x_i| w_i \tag{2.179}$$

is sharp, and is attained iff

$$x_i = \begin{cases} |x_i|\,\text{sgn}\{c_i\}, & \textit{if } |c_i| = \sup_{1\le r\le n}|c_r|, \\ 0, & \textit{otherwise}. \end{cases} \tag{2.180}$$

Also,

$$\sum_{i=1}^{n} c_i x_i w_i \le \sum_{i=1}^{n}|c_i| w_i \sup_{1\le i\le n}|x_i| \tag{2.181}$$

holds, and becomes equality iff

$$x_i \begin{cases} = \alpha\,\text{sgn}\{c_i\}, & \textit{if } c_i \ne 0, \\ \in [-\alpha, \alpha] & \textit{otherwise}, \end{cases} \tag{2.182}$$

for some $\alpha \ge 0$.

Theorem 2.33 *We have*

$$\int_a^b g(x)h(x)W(dx) \le \text{ess}\sup\nolimits_W |h(x)| \int_a^b |g(x)|W(dx), \tag{2.183}$$

which becomes equality for

$$g(x) = \begin{cases} |g(x)|\,\text{sgn}\{h(x)\}, & \textit{if } |h(x)| = \text{ess}\sup_W |h(x)|, \\ 0, & \textit{otherwise} \end{cases} \tag{2.184}$$

W-almost everywhere.

Moreover,

$$\int_a^b g(x)h(x)W(dx) \leq \int_a^b |h(x)|W(dx) \text{ ess sup}_W |g(x)| \tag{2.185}$$

holds, and this is the equality iff

$$g(x) = \begin{cases} = \alpha \operatorname{sgn}\{h(x)\}, & \text{if } h(x) \neq 0, \\ \in [-\alpha, \alpha], & \text{otherwise.} \end{cases} \tag{2.186}$$

W-almost everywhere for some $\alpha \geq 0$.

Finally, it is worth pointing out that for the conjugate exponents p, q satisfying either $p < 0 < q < 1$ or $q < 0 < p < 1$, inequalities (2.166) and (2.172) are reversed whereas the respective equality conditions remain unchanged (cf. Mitrinović, 1970, pp. 51–52).

2.4.5 Minkowski Inequalities

Minkowski inequalities are used for proving the triangle inequalities in the Banach spaces of ℓ^p and L^p type.

Theorem 2.34 *Let $1 < p < \infty$. For every $\tilde{x}, \tilde{y} \in \mathbb{R}^n$ and $\tilde{w} \in \mathbb{R}^n$ with positive coordinates*

$$\left(\sum_{i=1}^n |x_i + y_i|^p w_i\right)^{1/p} \leq \left(\sum_{i=1}^n |x_i|^p w_i\right)^{1/p} + \left(\sum_{i=1}^n |y_i|^p w_i\right)^{1/p}. \tag{2.187}$$

The equality in (2.187) holds if either $\tilde{x} = \tilde{0}$ or $\tilde{y} = \tilde{0}$ or $\tilde{y} = \alpha\tilde{x}$ for some positive constant α.

Proof The equality and so the inequality follow immediately if we assume that either $\tilde{x} = \tilde{0}$ or $\tilde{y} = \tilde{0}$. Otherwise we have

$$\sum_{i=1}^n |x_i + y_i|^p w_i \leq \sum_{i=1}^n |x_i|\,|x_i + y_i|^{p-1} w_i + \sum_{i=1}^n |y_i|\,|x_i + y_i|^{p-1} w_i. \tag{2.188}$$

By the Hölder inequality (2.166)

$$\sum_{i=1}^{n}|x_i+y_i|^p w_i \le \left[\left(\sum_{i=1}^{n}|x_i|^p w_i)\right)^{1/p} + \left(\sum_{i=1}^{n}|y_i|^p w_i\right)^{1/p}\right]$$
$$\times \left(\sum_{i=1}^{n}(|x_i+y_i|^{p-1})^{\frac{p}{p-1}} w_i\right)^{\frac{p-1}{p}}. \tag{2.189}$$

Dividing (2.189) by the expression in the second line we obtain (2.187). By (2.167), the equality in (2.189) holds if $x_i = \alpha_1(|x_i + y_i|^{p-1})^{q/p} = \alpha_1(x_i + y_i)$ and $y_i = \alpha_2(x_i + y_i)$, $i = 1, \ldots, n$, for some $\alpha_1, \alpha_2 > 0$. By simple calculations we obtain a single condition $y_i = \alpha x_i$, $i = 1, \ldots, n$, for some $\alpha > 0$. The equality condition in (2.188) is $x_i y_i \ge 0$, $i = 1, \ldots, n$, which is weaker than proportionality condition of $\tilde{x}$ and $\tilde{y}$ with a positive coefficient. □

Simple extensions of the Minkowski inequality to the cases $p = 1$ and $p = +\infty$ are following. The proofs are immediate.

Theorem 2.35 *For n-dimensional vectors $\tilde{x}$, $\tilde{y}$ and positive $\tilde{w}$ we have*

$$\sum_{i=1}^{n}|x_i+y_i|w_i \le \sum_{i=1}^{n}|x_i|w_i + \sum_{i=1}^{n}|y_i|w_i, \tag{2.190}$$

and the equality holds if $x_i y_i \ge 0$, $i = 1, \ldots, n$. Moreover,

$$\max_{i=1,\ldots,n}|x_i+y_i| \le \max_{i=1,\ldots,n}|x_i| + \max_{i=1,\ldots,n}|y_i|, \tag{2.191}$$

and the equality holds if there exists $1 \le j \le n$ such that $x_j = \max_{i=1,\ldots,n}|x_i|$, $y_j = \max_{i=1,\ldots,n}|y_i|$, and $x_j y_j \ge 0$.

An integral version of the Minkowski inequality has the following form.

Theorem 2.36 *Let $W : [a, b] \mapsto \mathbb{R}$ be nondecreasing and bounded, and $1 < p < \infty$. Then for every $g, h : [a, b] \mapsto \mathbb{R}$*

$$\left(\int_a^b |g(x)+h(x)|^p W(dx)\right)^{\frac{1}{p}}$$
$$\le \left(\int_a^b |g(x)|^p W(dx)\right)^{\frac{1}{p}} + \left(\int_a^b |h(x)|^p W(dx)\right)^{\frac{1}{p}} \tag{2.192}$$

under the condition that the integrals of the RHS are finite. The equality in (2.192) holds of either $g(x) = 0$ or $h(x) = 0$ or $h(x) = \alpha g(x)$ W-almost everywhere for some positive constant α.

The theorem can be proven with use of slightly modified arguments from the proof of Theorem 2.34. The integral Minkowski inequalities for $p = 1$ and $p = \infty$ look similar to those presented in Theorem 2.35. In particular, we have

Theorem 2.37 *For bounded and nondecreasing $W : [a, b] \mapsto \mathbb{R}$ and W-integrable $g, h : [a, b] \mapsto \mathbb{R}$ we have*

$$\int_a^b |g(x) + h(x)| W(dx) \leq \int_a^b |g(x)| W(dx) + \int_a^b |h(x)| W(dx), \tag{2.193}$$

and the equality holds in (2.193) *if $g(x)h(x) \geq 0$ W-almost everywhere in $[a, b]$.*

2.4.6 Young Inequalities

If we replace symmetric power functions $x \to |x|^p$ with $p \geq 1$ by other convex, symmetric and increasing on $\mathbb{R}_+$ gauge functions measuring distances among sequences and functions, we are able to evaluate linear functions over Banach spaces more general than ℓ^p- and L^p-spaces. They are called the Orlicz spaces. We introduce some basic notions without pretending to presentation of the general theory. For more details, we refer the readers to, e.g., Rao and Ren (1991). We say that $\Phi : \mathbb{R} \to \mathbb{R}_+$ is a Young function if it is convex, symmetric about 0, and satisfies $\Phi(0) = 0$, and $\lim_{|x|\to\infty} \Phi(x) = +\infty$. Let φ stand for the left hand derivative of Φ on the positive halfline, with $\varphi(0) = 0$. This is nonnegative nondecreasing, and has countably many jumps at most. There exists a left-continuous version of the inverse of φ, defined as

$$\psi(v) = \inf\{u : \varphi(u) > v\}, \quad v \geq 0. \tag{2.194}$$

This is also nonnegative nondecreasing with $\psi(0) = 0$. Write

$$\Psi(y) = \int_0^y \psi(v) dv, \quad y \geq 0, \tag{2.195}$$

and $\Psi(-y) = \Psi(y)$ for $y < 0$. Then Ψ is also a Young function. It can be shown that the same procedure applied to Ψ gives Φ, and we say that Φ and Ψ are complementary Young functions. They satisfy the following Young inequality.

Theorem 2.38 *If $\Phi, \Psi : \mathbb{R} \to \mathbb{R}_+$ are complementary Young functions with the left continuous versions of derivatives φ and ψ, respectively, then for every $x, y \geq 0$, we have*

$$xy \leq \Phi(x) + \Psi(y), \tag{2.196}$$

with equality in (2.196) *iff*

$$\text{either} \quad y = \varphi(x) \quad \text{or} \quad x = \psi(y). \tag{2.197}$$

Another way of expressing (2.197) is

$$\varphi(x) \leq y \leq \varphi(x+). \tag{2.198}$$

If either of the functions φ and ψ is strictly increasing, then it suffices to write one condition of (2.197) for the other. If both are continuous increasing, the conditions are equivalent.

Proof For fixed $x, y \geq 0$, we write

$$xy = \iint_{\{0 \leq u \leq x,\, 0 \leq v \leq y\}} du\, dv, \tag{2.199}$$

and splitting the rectangular into two surfaces separated by the curve $v = \varphi(u)$ (or equivalently $u = \psi(v)$), we have

$$\begin{aligned} xy &= \int_0^x \left(\int_0^{\min\{y, \varphi(u)\}} dv \right) du + \int_0^y \left(\int_0^{\min\{x, \psi(v)\}} du \right) dv \\ &\leq \int_0^x \varphi(u) du + \int_0^y \psi(v) dv = \Phi(x) + \Psi(y). \end{aligned} \tag{2.200}$$

The equality holds iff $y \geq \varphi(u)$ for all $u \leq x$ and $x \geq \psi(v)$ for all $v \leq y$. In other words $y \geq \varphi(x)$ and $x \geq \psi(y)$. Since $\psi(\varphi(x)) \leq x$ and $\varphi(\psi(x)) \leq x$, we arrive at the desired equality conditions (2.197). □

The Young inequality has a nice graphical interpretation. For simplicity of explanation, we assume that φ is a strictly increasing continuous function vanishing at 0 and increasing on $\mathbb{R}_+$. Then its inverse ψ is also increasing and continuous. If $y = \varphi(x)$ and so $x = \psi(y)$, then the sum of the surface between the graph of $\varphi(u)$, $0 \leq u \leq x$ and the line segment $\{(u, 0) : 0 \leq u \leq x\}$ and the surface between the graph of $\psi(v)$, $0 \leq v \leq y$ and the line segment $\{(0, v) : 0 \leq v \leq y\}$ coincide with the rectangular $\{(u, v) : 0 \leq u \leq x, 0 \leq v \leq y\}$. Therefore the sum of the respective areas is

$$\int_0^x \varphi(u) du + \int_0^y \psi(v) dv = \Phi(x) + \Psi(y) = xy.$$

If either $y < \varphi(x)$ or $y > \varphi(x)$ (either $x > \psi(y)$ or $x < \psi(y)$, respectively), then we have the Young inequality (2.196). Theorem 2.38 extends the above arguments to possibly nondecreasing and discontinuous functions with their generalized inverses.

Theorem 2.29 is a particular case of the Young inequality with

$$\Phi(x) = \frac{|x|^p}{p}, \tag{2.201}$$

$$\Psi(y) = \frac{|y|^q}{q}, \tag{2.202}$$

$$\varphi(x) = x^{p-1}, \tag{2.203}$$

$$\psi(y) = y^{q-1}. \tag{2.204}$$

Then conditions (2.197) coincide with $x^p = y^q$. Another exemplary pair of complementary Young functions is

$$\Phi(x) = e^{|x|} - |x| - 1, \tag{2.205}$$

$$\Psi(y) = (1 + |y|)\ln(1 + |y|) - |y|, \tag{2.206}$$

with mutually inverse derivatives

$$\varphi(x) = e^x - 1, \tag{2.207}$$

$$\psi(y) = \ln(1 + y), \tag{2.208}$$

on the positive halfline. By Theorem 2.38, we have the inequality

$$xy \le e^x + (1 + y)\ln(1 + y) - x - y - 1, \quad x, y \ge 0, \tag{2.209}$$

which is attained iff $y = e^x - 1$.

Below we confine ourselves to normalized complementary Young functions that satisfy

$$\Phi(1) + \Psi(1) = 1. \tag{2.210}$$

We show that any complementary pair (Φ_0, Ψ_0) can be transformed into a normalized one. By the definition, we have

$$\text{either} \quad \varphi_0(1) \le 1 \le \psi_0(1) \quad \text{or} \quad \psi_0(1) \le 1 \le \varphi_0(1). \tag{2.211}$$

We concentrate on the former case, the reasoning for the other being similar. Then there exists $c \ge 1$ such that

$$\varphi_0(c) \le 1 \le \varphi_0(c+), \tag{2.212}$$

$$\psi_0\left(\frac{1}{c}\right) \le 1 \le \psi_0\left(\frac{1}{c}+\right). \tag{2.213}$$

In fact (2.212) and (2.213) imply that either $\varphi_0(c) = 1$ or $\psi_0\left(\frac{1}{c}\right) = 1$. Now, defining

$$\varphi(x) = \varphi_0(cx), \tag{2.214}$$

$$\psi(x) = \psi_0\left(\frac{x}{c}\right), \tag{2.215}$$

we easily verify (2.210). We also see that (2.201) and (2.202) is a normalized pair, whereas (2.205) and (2.206) is not. However, our constructions shows that

$$\Phi(x) = e^{c|x|} - c|x| - 1, \tag{2.216}$$

$$\Psi(x) = \left(\frac{|y|}{c} + 1\right)\ln\left(\frac{|y|}{c} + 1\right) - \frac{|y|}{c}, \tag{2.217}$$

with the unique $c \approx 1.23998$ satisfying

$$c\ln(1+c) = 1 \tag{2.218}$$

satisfies (2.210).

Now we provide bounds on weighted combinations $\sum_{i=1}^{n} c_i x_i w_i$ of some sequences $\tilde{c}, \tilde{x} \in \mathbb{R}^n$ with positive weights w_i, $i = 1, \ldots, n$, by means of normalized complementary Young functions. Consider functions

$$\alpha \to \sum_{i=1}^{n} \Phi\left(\frac{|x_i|}{\alpha}\right) w_i, \qquad \alpha > 0, \tag{2.219}$$

$$\beta \to \sum_{i=1}^{n} \Psi\left(\frac{|c_i|}{\beta}\right) w_i, \qquad \beta > 0, \tag{2.220}$$

for nonzero sequences $\tilde{x}$ and $\tilde{c}$. They are continuous decreasing from $+\infty$ at 0 to 0 at $+\infty$. Let $N_{\Phi,\tilde{w}}(\tilde{x}) > 0$ and $N_{\Psi,\tilde{w}}(\tilde{c}) > 0$ denote the unique solutions to the equations

$$\sum_{i=1}^{n} \Phi\left(\frac{|x_i|}{\alpha}\right) w_i = \Phi(1), \tag{2.221}$$

$$\sum_{i=1}^{n} \Psi\left(\frac{|c_i|}{\beta}\right) w_i = \Psi(1), \tag{2.222}$$

respectively. Put $N_{\Phi,\tilde{w}}(\tilde{x}) = 0$ if $\tilde{x} = \mathbf{0}$ and $N_{\Psi,\tilde{w}}(\tilde{c}) = 0$ if $\tilde{c} = \mathbf{0}$, for completeness. Then we have the following.

Theorem 2.39 *For every normalized complementary pair of Young functions* (Φ, Ψ) *with respective derivatives* φ *and* ψ, *the following inequality holds:*

$$\sum_{i=1}^{n} c_i x_i w_i \leq N_{\Psi,\tilde{w}}(\tilde{c}) N_{\Phi,\tilde{w}}(\tilde{x}). \tag{2.223}$$

If either of sequences is zero, then (2.223) *is a trivial identity. Otherwise, the equality holds iff*

$$\psi\left(\frac{|c_i|}{N_{\Psi,\tilde{w}}(\tilde{c})}\right) \leq \frac{|x_i|}{N_{\Phi,\tilde{w}}(\tilde{x})} \leq \psi\left(\frac{|c_i|}{N_{\Psi,\tilde{w}}(\tilde{c})}+\right), \tag{2.224}$$

$$\operatorname{sgn}\{x_i\} = \operatorname{sgn}\{c_i\}, \qquad i = 1, \ldots, n. \tag{2.225}$$

Proof We assume that the RHS of (2.223) is strictly positive, and apply the Young inequality (2.196) for

$$\frac{|c_i|}{N_{\Psi,\tilde{w}}(\tilde{c})} \frac{|x_i|}{N_{\Phi,\tilde{w}}(\tilde{x})} \leq \Psi\left(\frac{|c_i|}{N_{\Psi,\tilde{w}}(\tilde{c})}\right) + \Phi\left(\frac{|x_i}{N_{\Phi,\tilde{w}}(\tilde{x})}\right) \tag{2.226}$$

and all $i = 1, \ldots, n$. Multiplying each of (2.226) by the corresponding w_i, and summing the products up, we obtain

$$\begin{aligned}\frac{\sum |c_i|\,|x_i| w_i}{N_{\Psi,\tilde{w}}(\tilde{c}) N_{\Phi,\tilde{w}}(\tilde{x})} &\leq \sum_{i=1}^{n} \Psi\left(\frac{|c_i|}{N_{\Psi,\tilde{w}}(\tilde{c})}\right) w_i + \sum_{i=1}^{n} \Phi\left(\frac{|x\ |}{N_{\Phi,\tilde{w}}(\tilde{x})}\right) w_i \\ &= \Psi(1) + \Phi(1) = 1. \end{aligned} \tag{2.227}$$

This is (2.223) for $(|x_1|, \ldots, |x_n|)$ and $(|c_1|, \ldots, |c_n|)$, and becomes equality iff (2.224) holds. The bound for the original sequences follows immediately, and (2.225) is the necessary and sufficient addition that together with (2.224) ensure attainability. □

The integral version of Theorem 2.39 needs some modified notions. To avoid pathologies, we assume that Φ satisfies Δ_2-condition, which means that

$$\exists\, K > 0 \quad \forall\, x > 0 \quad \Phi(2x) \leq K\Phi(x). \tag{2.228}$$

Note that (2.201) and (2.202) satisfy (2.228), but (2.205) does not. Let $L^{\Phi}([a, b], W(dx))$ denote the family of functions $g : [a, b] \to \mathbb{R}$ such that

$$\int_a^b \Phi(g(x)) W(dx) < \infty. \tag{2.229}$$

If Φ satisfies Δ_2-condition (2.228), and $g(x) \neq 0$ W-almost everywhere, then the function

$$\alpha \to \int_a^b \Phi\left(\frac{g(x)}{\alpha}\right) W(dx), \qquad \alpha > 0, \tag{2.230}$$

is finite, positive, continuous and decreasing for all α. Furthermore, family $L^{\Phi}([a,b], W(dx))$ is closed under linear combinations. Let $N_{\Phi,W}(g)$ denote the solution to

$$\int_a^b \Phi\left(\frac{g(x)}{\alpha}\right) W(dx) = \Phi(1). \tag{2.231}$$

Set also $N_{\Phi,W}(g) = 0$ iff $g(x) = 0$ W-almost everywhere. It can be proven that family $L^{\Phi}([a,b]W(dx))$ forms a Banach space with norm $N_{\Phi,W}(g)$, which generalizes the notions of $L^p([a,b], W(dx))$ spaces. Likewise, we define space $L^{\Psi}([a,b], W(dx))$ with norm $N_{\Psi,W}(g)$ for the complement Ψ and Ψ that satisfies condition (2.228) as well.

Theorem 2.40 *For every complementary normalized pair of Young functions* (Φ, Ψ) *satisfying* Δ_2*-condition* (2.228)*, the inequality*

$$\int_a^b g(x)h(x)W(dx) \leq N_{\Psi,W}(h)N_{\Phi,W}(g) \tag{2.232}$$

holds for all $g \in L^{\Phi}([a,b], W(dx))$*, and* $h \in L^{\Psi}([a,b], W(dx))$*. For nonzero functions,* (2.232) *becomes equality, iff*

$$\psi\left(\frac{|h(x)|}{N_{\Psi,W}(h)}\right) \leq \frac{|g(x)|}{N_{\Phi,W}(g)} \leq \psi\left(\frac{|h(x)|}{N_{\Psi,W}(h)}+\right) \tag{2.233}$$

and

$$\operatorname{sgn}\{g(x)\} = \operatorname{sgn}\{h(x)\} \tag{2.234}$$

W*-almost everywhere.*

Proof Applying the Young inequality for integrals, we have

$$\begin{aligned}\int_a^b \frac{g(x)h(x)}{N_{\Phi,W}(g)N_{\Psi,W}(h)} W(dx) &\leq \int_a^b \frac{|g(x)|\,|h(x)|}{N_{\Phi,W}(g)N_{\Psi,W}(h)} W(dx)\\ \leq \int_a^b \Phi\left(\frac{g(x)}{N_{\Phi,W}(g)}\right) W(dx) &+ \int_a^b \Psi\left(\frac{h(x)}{N_{\Psi,W}(h)}\right) W(dx)\\ = \Phi(1) + \Psi(1) = 1.\end{aligned} \tag{2.235}$$

Mimicking the arguments from the proof of Theorem 2.39, we verify necessity and sufficiency of (2.233) and (2.234) for equality. □

There are very few exceptions for which bounds (2.223) and (2.231) are expressed in explicit analytic forms.

2.4.7 Inequalities Derived by Norm Maximization

Let T be a continuous linear functional on a Banach space $\mathcal{B}$ with norm $||\cdot||$, and let $\mathcal{C} \subset \mathcal{B}$ be a closed convex cone. Suppose that for every nonzero $g \in \mathcal{C}$ we have $T(g) < 0$. We aim at maximizing the norm of the functional

$$||T||_{\mathcal{C}} = \sup_{0 \neq g \in \mathcal{C}} \frac{T(g)}{||g||} = \sup_{g \in \mathcal{C}, ||g||=1} T(g)$$

restricted to the cone $\mathcal{C}$. Suppose that $g_1, g_2 \in \mathcal{C}$, $||g_1|| = ||g_2|| = 1$ and $T(g_1) = -c_1 > T(g_2) = -c_2$ for some $0 < c_1 < c_2$. Note that $T\left(\frac{g_1}{c_1}\right) = T\left(\frac{g_2}{c_2}\right) = -1$, and $\left|\left|\frac{g_1}{c_1}\right|\right| = \frac{1}{c_1} > \left|\left|\frac{g_2}{c_2}\right|\right| = \frac{1}{c_2}$. This implies that the problem of maximizing $T(g)$ over

$$\mathcal{S}_1 = \{g \in \mathcal{C} : ||g|| = 1\}$$

is equivalent to maximizing $||g||$ over

$$\mathcal{S}_2 = \{g \in \mathcal{C} : T(g) = -1\}. \tag{2.236}$$

The latter set is convex. If it is compact as well, then by the Krein-Milman theorem it is a convex hull of its extreme points.

The norm is a convex function, and it attains its maximum at some extreme points of $\mathcal{S}_2$. Indeed, every point $g \in \mathcal{S}_2$ can be approximated by a sequence $g_n = \sum_{i=1}^{n} \alpha_i e_i$ of convex combinations ($\alpha_i \geq 0$, $\sum_{i=1}^{n} \alpha_i = 1$) of some extreme points of $\mathcal{S}_2$ so that $||g_n - g|| \to 0$. Note that

$$||g_n|| \leq \sum_{i=1}^{n} \alpha_i ||e_i|| \leq \max_{1 \leq i \leq n} ||e_i||.$$

By continuity of the norm function

$$||g|| \leq \sup_{e \in \mathcal{E}} ||e||,$$

where $\mathcal{E}$ denotes the set of all extreme points of $\mathcal{S}_2$. Accordingly, the dual norm maximization problem over (2.236) can be reduced to the maximization over the set of extreme points. In many practical problems, the reduction substantially simplifies the calculations. Below we use the above norm maximization technique for establishing negative upper (and equivalently positive lower) bounds on some linear functionals based on ordered random data. In specific problems, we are able to exploit the method under weaker assumptions than compactness of (2.236).

We finally mention that the linear functional on Hilbert spaces taking negative values on some convex cones have the projections equal to zero elements of the spaces. Consequently, the projection method described above in Sect. 2.3.4 provides trivial zero bounds attainable by zero elements. The norm maximization method allows us to determine strongly negative sharp bounds over all the non-zero elements of the cones.

2.5 Steffensen Inequality

Theorem 2.41 *Assume that h is an integrable nondecreasing function on $[a, b]$, and g is nonnegative bounded with*

$$0 \le g(t) \le A, \qquad a \le t \le b, \tag{2.237}$$

for some positive A. Put

$$0 \le \lambda = \frac{1}{A} \int_a^b g(t)dt \le b - a. \tag{2.238}$$

Then

$$A \int_a^{a+\lambda} h(t)dt \le \int_a^b g(t)h(t)dt \le A \int_{b-\lambda}^b h(t)dt. \tag{2.239}$$

The former inequality in (2.239) holds when

$$g(t) \begin{cases} = A, & \text{if } h(t) < h(a+\lambda), \\ \in [0, A], & \text{if } h(t) = h(a+\lambda), \\ = 0, & \text{if } h(t) > h(a+\lambda). \end{cases} \tag{2.240}$$

The latter holds when

$$g(t) \begin{cases} = 0, & \text{if } h(t) < h(b-\lambda), \\ \in [0, A], & \text{if } h(t) = h(b-\lambda), \\ = A, & \text{if } h(t) > h(b-\lambda), \end{cases} \tag{2.241}$$

and, on both the cases for

$$g(t) = 0 \quad a \le t \le b, \quad or \quad g(t) = A, \quad a \le t \le b. \tag{2.242}$$

Proof Under (2.242), the relations are immediately checked. These is the only cases, when $\lambda = 0$ or $b - a$. Otherwise, we prove the upper Steffensen bound. We have

$$\begin{aligned}
& A \int_{b-\lambda}^{b} h(t)dt - \int_{a}^{b} g(t)h(t)dt \\
&= \int_{b-\lambda}^{b} [A - g(t)]h(t)dt - \int_{a}^{b-\lambda} g(t)h(t)dt \\
&\ge h(b-\lambda) \left\{ \int_{b-\lambda}^{b} [A - g(t)]dt \right\} - \int_{a}^{b-\lambda} g(t)h(t)dt \\
&= h(b-\lambda) \left[\lambda A - \int_{b-\lambda}^{b} g(t)dt \right] - \int_{a}^{b-\lambda} g(t)h(t)dt \\
&= h(b-\lambda) \left[\int_{a}^{b} g(t)dt - \int_{b-\lambda}^{b} g(t)dt \right] - \int_{a}^{b-\lambda} g(t)h(t)dt \\
&= \int_{a}^{b-\lambda} [h(b-\lambda) - h(t)]g(t)dt \ge 0,
\end{aligned} \tag{2.243}$$

which is the desired claim.

There are two conditions for equality in (2.243):

$$[h(t) - h(b-\lambda)][A - g(t)] = 0 \quad \text{for } b - \lambda < t < b, \tag{2.244}$$

$$[h(t) - h(b-\lambda)]g(t) = 0 \quad \text{for } a < t < b - \lambda. \tag{2.245}$$

The former says that $g(t) = A$ for all $t > b - \lambda$ such that $h(t) > h(b - \lambda)$. Due to the latter, $g(t) = 0$ for all $t < b - \lambda$ for which $h(t) < h(b - \lambda)$. This implies that g may take arbitrary values only on the set where $h(t) = h(b - \lambda)$. The conclusions are summarized in (2.241).

For getting the lower bound we simply rewrite the upper one applied to

$$\hat{g}(t) = A - g(t), \tag{2.246}$$

$$\hat{\lambda} = \frac{1}{A} \int_{a}^{b} \hat{g}(t)dt = b - a - \lambda. \tag{2.247}$$

The details are left to the reader. $\square$

Given a bounded g, the conditions on the nondecreasing h, equivalent to (2.244) and (2.245) (see also (2.241)) are

$$h(t) = h(b - \lambda) \text{ if either } g(t) > 0,\ a < t < b - \lambda, \text{ or } g(t) < A,\ b - \lambda < t < b. \tag{2.248}$$

Similarly, (2.240) could be replaced by

$$h(t) = h(a + \lambda) \text{ if either } g(t) < A,\ a < t < a + \lambda, \text{ or } g(t) > 0,\ a + \lambda < t < b. \tag{2.249}$$

together with nondecrease of h.

2.6 Inequalities for Convex Functions

2.6.1 Jensen Inequalities

The classic Jensen inequality is presented below.

Theorem 2.42 *If $\varphi : [a, b] \to \mathbb{R}$ is convex, then for arbitrary $n \in \mathbb{N}$, $a \le x_i \le b_i$, $\alpha_i \ge 0$, $i = 1, \ldots, n$, such that $\sum_{i=1}^{n} \alpha_i = 1$, we have*

$$\varphi\left(\sum_{i=1}^{n} \alpha_i x_i\right) \le \sum_{i=1}^{n} \alpha_i \varphi(x_i), \tag{2.250}$$

which is equality if φ is linear.

Proof We prove (2.250) by induction, referring to the definition of convexity. Since (2.250) is trivially true for $n = 1$, and coincides with the definition for $n = 2$, we assume (2.250) for some $n \ge 2$, and try to conclude the same for $n+1$. It suffices to consider the case $0 < \alpha_{n+1} < 1$ as follows

$$\begin{aligned}
\varphi\left(\sum_{i=1}^{n+1} \alpha_i x_i\right) &= \varphi\left(\sum_{i=1}^{n} \alpha_i \frac{\sum_{i=1}^{n} \alpha_i x_i}{\sum_{i=1}^{n} \alpha_i} + \alpha_{n+1} x_{n+1}\right) \\
&\le \sum_{i=1}^{n} \alpha_i \varphi\left(\sum_{i=1}^{n} \frac{\alpha_i}{\sum_{i=1}^{n} \alpha_i} x_i\right) + \alpha_{n+1} \varphi(x_{n+1}) \\
&\le \sum_{i=1}^{n} \alpha_i \varphi(x_i) + \alpha_{n+1} \varphi(x_{n+1}) = \sum_{i=1}^{n+1} \alpha_i \varphi(x_i).
\end{aligned} \tag{2.251}$$

The equality condition is obvious. □

In fact, equality conditions depend on function φ and both sequences $\tilde{x}$ and $\tilde{\alpha}$, and relations among them. It is natural to exclude from consideration all $\alpha_i = 0$, because respective x_i and $\varphi(x_i)$ do not affect the relations. If $x_1, \ldots, x_n$ are fixed, and all $\alpha_i > 0$, then (2.250) becomes equality iff φ is linear on $[\min_{1\le i\le n} x_i, \max_{1\le i\le n} x_i]$. If φ is strictly convex, then the equality holds iff all x_i are equal except for those whose coefficients vanish.

Taking $\alpha_i = \frac{c_i}{\sum_{r=1}^n c_r}$, $i = 1, \ldots, n$ with $\sum_{r=1}^n c_r \neq 0$, we conclude a slight generalization of (2.250)

$$\varphi\left(\frac{\sum_{i=1}^n c_i x_i}{\sum_{i=1}^n c_i}\right) \le \frac{\sum_{i=1}^n c_i \varphi(x_i)}{\sum_{i=1}^n c_i} \tag{2.252}$$

that is valid for all convex φ and $a \le x_i \le b$ with

$$0 \le c_i \le \sum_{r=1}^n c_r, \quad \text{if } \sum_{r=1}^n c_r > 0, \tag{2.253}$$

$$\sum_{r=1}^n c_r \le c_i \le 0, \quad \text{if } \sum_{r=1}^n c_r < 0. \tag{2.254}$$

An integral counterpart of (2.250) has the form

Theorem 2.43 *If $\varphi : [a, b] \to \mathbb{R}$ is convex, then for arbitrary nondecreasing $W : [a, b] \to \mathbb{R}$ satisfying $W(b) - W(a) = 1$, we have*

$$\varphi\left(\int_a^b x W(dx)\right) \le \int_a^b \varphi(x) W(dx) \tag{2.255}$$

under the condition that the integrals exist and are finite. The equality is attained if φ is linear W-almost everywhere.

A probabilistic interpretation of (2.255) is the following

$$\varphi(\mathbb{E}X) \le \mathbb{E}\varphi(X) \tag{2.256}$$

for arbitrary convex function φ and random variable X for which the expectations exist. If W is merely bounded nondecreasing on $[a, b]$, relation (2.255) is rewritten to

$$\varphi\left(\frac{\int_a^b x W(dx)}{W(b) - W(a)}\right) \le \frac{\int_a^b \varphi(x) W(dx)}{W(b) - W(a)}. \tag{2.257}$$

Proof of Theorem 2.43 Put

$$c = \int_a^b x W(dx) \in [a, b]. \tag{2.258}$$

By convexity of φ, we can write

$$\varphi(c) + \lambda_c(x - c) \leq \varphi(x) \tag{2.259}$$

for all $x \in [a, b]$ with $\varphi'(c-) \leq \lambda_c \leq \varphi'(c+)$. Integrating (2.259), and applying (2.258), we obtain

$$\begin{aligned} &\varphi(c) \int_a^b W(dx) + \lambda_c \left[\int_a^b x W(dx) - c \int_a^b W(dx) \right] \\ &= \varphi\left(\int_a^b x W(dx) \right) \leq \int_a^b \varphi(x) W(dx). \end{aligned} \tag{2.260}$$

The bound (2.260) is attained iff the equality in (2.259) holds W-almost everywhere (i.e., φ is linear W-almost everywhere). □

In fact, (2.255) holds true if the integrals exists and are possibly infinite (cf. Pečarić et al., 1992, pp. 45–47).

2.6.2 Jensen–Steffensen Inequalities

Theorem 2.42 is still true if we restrict ourselves to nondecreasing sequences $x_1 \leq \cdots \leq x_n$. In this case we can relax assumptions on coefficients $\alpha_1, \ldots, \alpha_n$ and obtain the Jensen–Steffensen inequality.

Theorem 2.44 *If $\varphi : [a, b] \to \mathbb{R}$ is convex, then for arbitrary $n \in \mathbb{N}$, $a \leq x_1 \leq \cdots \leq x_n \leq b$ and (nonnecessarily nonnegative) α_i, $i = 1, \ldots, n$, such that*

$$0 \leq \sum_{i=1}^{j} \alpha_i \leq \sum_{i=1}^{n} \alpha_i = 1, \qquad j = 1, \ldots, n-1, \tag{2.261}$$

we get

$$\varphi\left(\sum_{i=1}^{n} \alpha_i x_i \right) \leq \sum_{i=1}^{n} \alpha_i \varphi(x_i). \tag{2.262}$$

Relation (2.262) *becomes equality if φ is linear.*

Proof Consider the Abel identities

$$\sum_{i=1}^{m} \alpha_i y_i = y_m \sum_{r=1}^{m} \alpha_r - \sum_{i=1}^{m-1} (y_{i+1} - y_i) \sum_{r=1}^{i} \alpha_r, \tag{2.263}$$

for $m = 1, \ldots, n$, (cf. (2.41)) and their modifications arising by renumbering of indices

$$\sum_{i=m+1}^{n} \alpha_i y_i = y_{m+1} \sum_{r=m+1}^{n} \alpha_r - \sum_{i=m+1}^{n-1} (y_i - y_{i+1}) \sum_{r=i-1}^{n} \alpha_r, \tag{2.264}$$

for $m = 0, \ldots, n-1$. Putting $y_i = x_i$, and taking (2.263) with $m = n$, and (2.264) with $m = 0$, we obtain

$$x_1 \le \sum_{i=1}^{n} \alpha_i x_i \le x_n. \tag{2.265}$$

Set $c = \sum_{i=1}^{n} \alpha_i x_i$, and suppose that $x_m \le c \le x_{m+1}$ for some $m = 1, \ldots, n-1$. Given $x, y \in [a, b]$, for $\varphi'(x-) \le \lambda_x \le \varphi'(x+)$. and $\varphi'(y-) \le \lambda_y \le \varphi'(y+)$, we have

$$\forall\, a \le x \le y \le b \qquad \lambda_x (y - x) \le \varphi(y) - \varphi(x) \le \lambda_y (y - x). \tag{2.266}$$

Therefore

$$\forall\, a \le x \le y \le c \qquad \varphi(y) - \varphi(x) \le \lambda_y (y - x) \le \lambda_c (y - x) \tag{2.267}$$

and

$$\forall\, c \le x \le y \le b \qquad \varphi(y) - \varphi(x) \ge \lambda_x (y - x) \ge \lambda_c (y - x). \tag{2.268}$$

In particular, we have

$$\varphi(x_{i+1}) - \varphi(x_i) \le \lambda_c (x_{i+1} - x_i), \qquad i = 1, \ldots, m-1, \tag{2.269}$$

$$\varphi(c) - \varphi(x_m) \le \lambda_c (c - x_m), \tag{2.270}$$

$$\varphi(x_{m+1}) - \varphi(c) \ge \lambda_c (x_{m+1} - c), \tag{2.271}$$

$$\varphi(x_{i+1}) - \varphi(x_i) \ge \lambda_c (x_{i+1} - x_i), \qquad i = m+1, \ldots, n-1. \tag{2.272}$$

Replacing y_i by $\varphi(x_i)$ and x_i, $i = 1, \dots, n$, in (2.263) and (2.264), we obtain

$$\begin{aligned}
&\varphi\left(\sum_{i=1}^{n} \alpha_i x_i\right) - \sum_{i=1}^{n} \alpha_i \varphi(x_i)\\
&= \sum_{i=1}^{m} \alpha_i \varphi(c) + \sum_{i=m+1}^{n} \alpha_i \varphi(c) - \sum_{i=1}^{m} \alpha_i \varphi(x_i) - \sum_{i=m+1}^{n} \alpha_i \varphi(x_i)\\
&= \sum_{i=1}^{m} \alpha_i \varphi(c) + \sum_{i=m+1}^{n} \alpha_i \varphi(c) - \varphi(x_m) \sum_{r=1}^{m} \alpha_r + \sum_{i=1}^{m-1} [\varphi(x_{i+1}) - \varphi(x_i)] \sum_{r=1}^{i} \alpha_r\\
&- \varphi(x_{m+1}) \sum_{r=m+1}^{n} \alpha_r + \sum_{i=m+1}^{n-1} [\varphi(x_i) - \varphi(x_{i+1})] \sum_{r=i+1}^{n} \alpha_r\\
&+ \lambda_c \left[x_m \sum_{r=1}^{m} \alpha_r - \sum_{i=1}^{m-1} (x_{i+1} - x_i) \sum_{r=1}^{i} \alpha_r - \sum_{i=1}^{m} \alpha_i x_i \right]\\
&+ \lambda_c \left[x_{m+1} \sum_{r=m+1}^{n} \alpha_r - \sum_{i=m+1}^{n-1} (x_i - x_{i+1}) \sum_{r=i+1}^{n} \alpha_r - \sum_{i=m+1}^{n} \alpha_i x_i \right]\\
&= \sum_{i=1}^{m-1} [\varphi(x_{i+1}) - \varphi(x_i) - \lambda_c (x_{i+1} - x_i)] \sum_{r=1}^{i} \alpha_r\\
&+ [\varphi(c) - \varphi(x_m) + \lambda_c (x_m - c)] \sum_{r=1}^{m} \alpha_r\\
&+ [\varphi(c) - \varphi(x_{m+1}) + \lambda_c (x_{m+1} - c)] \sum_{r=m+1}^{n} \alpha_r\\
&+ \sum_{i=m+1}^{n-1} [\varphi(x_i) - \varphi(x_{i+1}) - \lambda_c (x_i - x_{i+1})] \sum_{r=i+1}^{n} \alpha_r. \qquad (2.273)
\end{aligned}$$

This is nonpositive and proves (2.262), because, by (2.269) to (2.272), so is each expression in brackets whereas all partial sums of coefficients $\sum_{r=1}^{i} \alpha_i$, $\sum_{r=i+1}^{n} \alpha_i$ are nonnegative.

Relation (2.262) becomes equality iff either of factors in each product of the last expression in (2.273) is zero. Choosing α_i so that $0 < \sum_{r=1}^{i} \alpha_r < 1$, we force each expression in brackets to vanish for arbitrary choice of $x_1, \dots, x_n$. This implies that φ is linear. □

There is also a version of the Jensen–Steffensen inequality for unnormalized sequences of coefficients

$$\varphi\left(\frac{\sum_{i=1}^{n} c_i x_i}{\sum_{i=1}^{n} c_i}\right) \leq \frac{\sum_{i=1}^{n} c_i \varphi(x_i)}{\sum_{i=1}^{n} c_i} \tag{2.274}$$

when either

$$0 \leq \sum_{i=1}^{j} c_i \leq \sum_{i=1}^{n} c_i > 0, \qquad j = 1, \ldots, n-1, \tag{2.275}$$

or

$$0 > \sum_{i=1}^{n} c_i \leq \sum_{i=1}^{j} c_i \leq 0, \qquad j = 1, \ldots, n-1. \tag{2.276}$$

For integrals, we have

Theorem 2.45 *If W has a bounded variation on $[a, b]$, and satisfies*

$$\forall\, x \in [a, b] \quad W(a) \leq W(x) \leq W(b) = W(a) + 1, \tag{2.277}$$

then for arbitrary convex function φ on $[a, b]$, the inequality holds

$$\varphi\left(\int_a^b x W(dx)\right) \leq \int_a^b \varphi(x) W(dx) \tag{2.278}$$

which becomes equality if φ is linear W-almost everywhere.

Proof Without loss of generality we assume $W(a) = 0$. Set

$$c = \int_a^b x W(dx) \in [a, b], \tag{2.279}$$

and take $\lambda_c \in [\varphi'(c-), \varphi'(c+)]$. Then function

$$\Lambda_c(x) = \lambda_c x - \varphi(x) \tag{2.280}$$

is nondecreasing for $x \in [a, c]$ and nonincreasing for $x \in [b, c]$. We also have

$$\int_a^c x W(dx) = cW(c) - \int_a^c W(x)dx, \tag{2.281}$$

$$\int_c^b x W(dx) = c[1 - W(c)] + \int_c^b [1 - W(x)]dx, \tag{2.282}$$

$$\int_a^c \varphi(x)W(dx) = \varphi(c)W(c) - \int_a^c W(x)\varphi(dx), \tag{2.283}$$

$$\int_c^b \varphi(x)W(dx) = \varphi(c)[1 - W(c)] + \int_c^b [1 - W(x)]\varphi(dx). \tag{2.284}$$

Therefore, by (2.279) and (2.281) to (2.284), we obtain

$$\begin{aligned}
&\int_a^b \varphi(x)W(dx) = \int_a^c \varphi(x)W(dx) + \int_c^b \varphi(x)W(dx)\\
&= \varphi(c)W(c) - \int_a^c W(x)\varphi(dx) + \varphi(c)[1 - W(c)] + \int_c^b [1 - W(x)]\varphi(dx)\\
&+ \lambda_c\left[\int_a^c xW(dx) - cW(c) + \int_a^c W(x)dx\right]\\
&+ \lambda_c\left[\int_c^b xW(dx) - c[1 - W(c)] - \int_c^b [1 - W(x)]dx\right]\\
&= \lambda_c \int_a^c W(x)dx - \int_a^c W(x)\varphi(dx) + \int_c^b [1 - W(x)]\varphi(dx)\\
&- \lambda_c \int_c^b [1 - W(x)]dx + \varphi(c) - \lambda_c\left[\int_a^b xW(dx) - c\right]\\
&= \int_a^c W(x)\Lambda_c(dx) - \int_c^b [1 - W(x)]\Lambda_c(dx) + \varphi(c)\\
&\geq \varphi(c) = \varphi\left(\int_a^b xW(dx)\right).
\end{aligned} \tag{2.285}$$

The inequality holds here, because both W and $1 - W$ are nonnegative, and Λ_c is nondecreasing up to c and then nonincreasing so that we have

$$\int_c^b [1 - W(x)]\Lambda_c(dx) \leq 0 \leq \int_a^c W(x)\Lambda_c(dx). \tag{2.286}$$

The equality in (2.285) holds iff both the integrals in (2.286) vanish. Function Λ_c is nonconstant except for a (possibly degenerate) interval containing (2.279), where φ is linear with slope λ_c. Therefore the equalities hold in (2.286) iff $W(x) = 0$ on the left of the interval and $W(x) = 1$ on the right. This establishes the attainability conditions. □

Obviously, when

$$0 \neq W(b) - W(a) \neq 1, \tag{2.287}$$

we have

$$\varphi\left(\frac{\int_a^b x W(dx)}{W(b)-W(a)}\right) \le \frac{\int_a^b \varphi(x) W(dx)}{W(b)-W(a)} \tag{2.288}$$

under either of conditions

$$\forall\, x \in [a,b] \quad W(a) \le W(x) \le W(b), \tag{2.289}$$

$$\forall\, x \in [a,b] \quad W(a) \ge W(x) \ge W(b). \tag{2.290}$$

2.6.3 Bernoulli Inequality

We here use the Bernoulli inequality.

Theorem 2.46 *For every $r > 1$ we have*

$$(1+x)^r \ge 1 + rx, \qquad x > -1. \tag{2.291}$$

For $0 < r < 1$ the inequality is reversed. In both the cases the equality is attained for $x = 0$.

For $r = 1$ (2.291) becomes a trivial equality.

Proof Consider the continuously differentiable function

$$f(x) = (1+x)^r, \qquad x > -1. \tag{2.292}$$

For $r > 1$ it is strictly convex. Accordingly,

$$f(x) - f(0) = (1+x)^r - 1 \ge f'(0)\,x = r\,x, \tag{2.293}$$

and the inequality is sharp for all $x \ne 0$.

For $0 < r < 1$ the function is strictly concave, and inequality (2.293) is reversed and strict for $x \ne 0$. □

2.7 Descartes Rule of Signs

Below we present a stronger version of the Descartes rule of signs proved in Komornik (2006) in a short and elegant way.

Theorem 2.47 *Let $\tilde{a} = (a_1, \ldots, a_n) \in \mathbb{R}^n$, $n \ge 1$, be a vector of reals with non-zero elements, and $b_1 < \ldots < b_n$. Then the number $z(p)$ of positive zeros (even* *counted with multiplicities) of the function*

$$p(x) = \sum_{i=1}^{n} a_i x^{b_i}, \qquad x > 0, \tag{2.294}$$

does not exceed the number of sign changes $v(p)$ in the vector of coefficients $\tilde{a}$.

Proof It is performed by induction with respect to $v(p)$. If $v(p) = 0$, i.e., all the components of $\tilde{a}$ have the same sign, then (2.294) has the same sign for all $x > 0$, and $z(p) = 0$.

Suppose that $v(p) > 0$, and $a_j a_{j+1} < 0$ for some $1 \le j \le n-1$. Without loss of generality we can assume that $b_j < 0 < b_{j+1}$, because the transformation $x^{-b} p(x)$ for some $b_j < b < b_{j+1}$ does not change the number and order of the roots of the original function. Observe that for

$$p'(x) = \sum_{i=1}^{n} a_i b_i x^{b_i - 1}, \qquad x > 0, \tag{2.295}$$

we have

$$v(p') = v(p) - 1, \tag{2.296}$$

because

$$\operatorname{sgn}(a_i b_i) = \begin{cases} -\operatorname{sgn}(a_i), & i \le j, \\ \operatorname{sgn}(a_i), & i > j. \end{cases} \tag{2.297}$$

If p has k different positive roots $x_1 < \ldots < x_k$ with respective multiplicities $m_1, \ldots, m_k$, then they are the roots of p' with multiplicities $m_1 - 1, \ldots m_k - 1$, respectively. Besides $p(x_i+)$ and $p(x_{i+1}-)$, $i = 1, \ldots, k-1$, have different signs, which in view of the Rolle theorem implies that each interval (x_i, x_{i+1}), $i = 1, \ldots, k-1$ contains at least one root of the derivative (2.295). Therefore

$$z(p') \ge \sum_{i=1}^{k} (m_i - 1) + (k-1) = \sum_{i=1}^{k} m_i - 1 = z(p) - 1. \tag{2.298}$$

By the induction hypothesis we also have

$$v(p') \le z(p'). \tag{2.299}$$

Combining (2.296), (2.298) and (2.299) we obtain

$$z(p) \leq z(p') + 1 \leq v(p') + 1 \leq v(p), \tag{2.300}$$

which ends the proof. □

Descartes in 1637 (see Descartes, 1954) stated without proof that the above theorem holds under the restrictions that all the exponents $b_1, \ldots, b_n$ are positive integers. This was rigorously proved by Segner (1728). Since this paper was not available for years, and it was found in 1977, till then the paper of Segner (1756) was treated as the origin source of the formal proof. The same result can be deduced from more general considerations (see, e.g., Karlin & Studden, 1966, Chapter I, Corollary 4.4).

Corollary 2.1 *Under the assumptions and notation of Theorem* 2.47 *the number of sign changes of the function* (2.294) *on the positive half-axis is not greater than the number of sign changes in the sequence of coefficients* $\tilde{a} = (a_1, \ldots a_n)$. *The first and the ultimate signs of* (2.294) *coincide with the signs of* a_1 *and* a_n, *respectively.*

The above property is called the variation diminishing property of linear combinations of monomials. The last statement can be immediately concluded. This remains formally true if we admit that some (but not all) elements of $\tilde{a}$ are equal to 0, because these zeros do not affect the number of signs changes of the vector and the function.

2.8 Transform Orders

2.8.1 Convex Transform Order

The convex transform order is a partial ordering of one-dimensional distribution functions which was introduced by van Zwet (1964). We say that F precedes G in the order (denoted as $F \preceq_c G$) iff $G^{-1} \circ F$ is a convex function on the support of F, or equivalently, iff $F^{-1} \circ G$ is concave on the support of G. It can be easily proved that the relation is reflexive, antisymmetric, and transitive. Obviously, there are many pairs F and G which are not related in the convex transform order. Moreover, $F \preceq_c G$ and $G \preceq_c F$ if both $F^{-1} \circ G$ and $G^{-1} \circ F$ are linear, which means that $G(x) = F(ax + b)$ for some $a > 0$ and $b \in \mathbb{R}$. Accordingly, we can say that the convex transform order is location and scale invariant. Conventionally, the notion is used for random variables as well. We say that $X \preceq_c Y$ if the corresponding distribution functions F and G satisfy $F \preceq_c G$.

Various nonparametric families of distributions, widely used in applications, can be defined as being related with a fixed distribution function W in the convex transform order. For instance, let W be an element of the family of absolutely continuous generalized Pareto distribution functions

$$W_\alpha(x) = \begin{cases} 1-(1-\alpha x)^{1/\alpha}, & 0 < x < \infty,\ \alpha < 0, \\ 1-\exp(-x), & 0 < x < \infty,\ \alpha = 0, \\ 1-(1-\alpha x)^{1/\alpha}, & 0 < x < \frac{1}{\alpha},\ \alpha > 0. \end{cases} \tag{2.301}$$

When α is negative, then W_α is a Lomax distribution (translated Pareto one so that the left-end point of the support is equal to 0) with the shape parameter $-\frac{1}{\alpha}$. For $\alpha > 0$ we have a linear transformation of a power distribution with the exponent $\frac{1}{\alpha}$. In particular, $W_1 = U$ is the standard uniform distribution function. Obviously, $W_0 = Z$ is the standard exponential distribution function. The quantile functions of (2.301) have the forms

$$W_\alpha^{-1}(x) = \begin{cases} \frac{1}{\alpha}[1-(1-x)^\alpha], & \alpha \neq 0, \\ -\ln(1-x), & \alpha = 0. \end{cases} \tag{2.302}$$

If F has a density function f, then $F \preceq_c W_\alpha$ ($F \succeq_c W_\alpha$) implies that

$$\frac{d}{dx} W_\alpha^{-1} \circ F(x) = [1-F(x)]^{\alpha-1} f(x) \tag{2.303}$$

is nondecreasing (nonincreasing, respectively) on the support of F. The RHS of (2.303) is called α-generalized failure rate of the distribution function F (and any random variable with distribution function F).

In particular, the assumption $F \preceq_c W_0$ ($F \succeq_c W_0$) combined with absolute continuity of F imply that F has nondecreasing (nonincreasing, respectively) failure rate. By a popular convention, we say that F is an element of increasing (decreasing, respectively) failure rate family, and write shortly $F \in$ IFR (DFR, respectively).

The other particular case is that F has a density function and $F \preceq_c W_1$ ($F \succeq_c W_1$, respectively). It means that the corresponding density function is nondecreasing (nonincreasing, respectively). Following the convention, we say that F is a member of the increasing (decreasing, respectively) density family, and write $F \in$ ID (DD, respectively). Generally, we say that F has an increasing α-generalized failure rate and write $F \in$ IGFR(α), if the RHS of (2.303) is nondecreasing. Similarly we define the DGFR(α) families.

Proposition 2.1

(a) If $\alpha_1 < \alpha_2$, then $W_{\alpha_1} \succ_c W_{\alpha_2}$.
(b) In consequence, the families $IGFR(\alpha)$, $\alpha \in \mathbb{R}$, are strictly decreasing with respect to α, and $DGFR(\alpha)$ are strictly increasing.

Proof

(a) Take first nonzero α_1 and α_2. Then

$$W_{\alpha_1}^{-1} \circ W_{\alpha_2}(x) = \frac{1}{\alpha 1}\left[1 - (1 - \alpha_2 x)^{\frac{\alpha_1}{\alpha_2}}\right]$$

has the second derivative

$$\frac{d^2}{dx^2} W_{\alpha_1}^{-1} \circ W_{\alpha_2}(x) = (\alpha_2 - \alpha_1)(1 - \alpha_2 x)^{\frac{\alpha_1}{\alpha_2} - 2}.$$

Since the second factor is positive on the support of W_{α_2}, we have $W_{\alpha_1} \succ_c W_{\alpha_2}$ if $\alpha_2 > \alpha_1$.

Put now $\alpha_1 = 0 \neq \alpha_2$. Then

$$W_0^{-1} \circ W_{\alpha_2}(x) = -\frac{1}{\alpha_2} \ln(1 - \alpha_2 x)$$

and

$$\frac{d^2}{dx^2} W_0^{-1} \circ W_{\alpha_2}(x) = \alpha_2(1 - \alpha_2 x)^{-2}.$$

If $\alpha_2 > 0$, then the second derivative is positive, and $W_0 \succ_c W_{\alpha_2}$. If $\alpha_2 < 0$, we obtain the opposite conclusion.

(b) If F is IGFR(α_2), i.e., $F \preceq_c W_{\alpha_2}$, then by statement (a) and transitivity of the convex transform order $F \preceq_c W_{\alpha_1}$ as well, and so it has increasing α_1-generalized failure rate. Note that IGFR(α_1) contains W_{α_1}, which does not belong to IGFR(α_2), because $W_{\alpha_1}^{-1} \circ W_{\alpha_2}$ is strictly convex, and its inverse cannot be convex as well. This implies IGFR(α_2) $\subsetneq$ IGFR(α_1).

The second claim is proved analogously.

□

Other natural shape restricted nonparametric families of distributions are ones obtained by the convex transform comparison with the negative generalized Pareto distribution functions. These are the distribution functions of the negatives of generalized Pareto random variables. They have the forms

$$V_\alpha(x) = \begin{cases} (1 + \alpha x)^{1/\alpha}, & -\infty < x < 0,\ \alpha < 0, \\ \exp(x), & -\infty < x < 0,\ \alpha = 0, \\ (1 + \alpha x)^{1/\alpha}, & -\frac{1}{\alpha} < x < 0,\ \alpha > 0, \end{cases} \tag{2.304}$$

and the respective quantile functions are

$$V_\alpha^{-1}(x) = \begin{cases} \frac{1}{\alpha}(x^\alpha - 1), & \alpha \neq 0, \\ \ln x, & \alpha = 0. \end{cases} \tag{2.305}$$

If F has a density function f, then

$$\frac{d}{dx} V_\alpha^{-1} \circ F(x) = F^{\alpha-1}(x) f(x). \tag{2.306}$$

For $\alpha = 0$, this is just the reversed failure rate function. Therefore relation $F \preceq_c V_0 = V$ ($F \succeq_c V_0 = V$) for the absolutely continuous distribution function F coincides with the fact that F has a nondecreasing (nonincreasing, respectively) reversed failure rate. We call the RHS of (2.306) the α-generalized reverse failure rate. The families of distribution functions with monotone (2.306) shall be called the increasing and decreasing α-generalized reverse failure rate families, and denoted by IGRFR(α) and DGRFR(α), respectively. Observe that the following equalities hold

$$\text{IGRFR}(1) = \text{IGFR}(1) = \text{ID},$$
$$\text{DGRFR}(1) = \text{DGFR}(1) = \text{DD}.$$

They become obvious in view of the equality $V_1(x) = W_1(x + 2)$, and translation invariance of the convex transform order.

Mimicking the arguments of the proof of Proposition 2.1 we can determine its analog for distributions with monotone α-generalized reversed failure rates.

Proposition 2.2

(a) If $\alpha_1 < \alpha_2$, then $V_{\alpha_1} \prec_c V_{\alpha_2}$.
(b) Moreover, the families $IGRFR(\alpha)$, $\alpha \in \mathbb{R}$, are strictly increasing with respect to α, and $DGFR(\alpha)$ are strictly decreasing.

For instance, while searching for the bounds on the expectations of order statistics from iid samples with decreasing failure rate, and for the distribution functions $F \in$ DFR which attain the bounds, we maximize some integral transformations of compositions $F^{-1} \circ Z$ belonging to some Banach spaces $L^p((0, \infty), e^{-x}dx)$. By definition, these functions are nondecreasing and convex, but they do not constitute a closed subset of the respective space. On the other hand, the family of nondecreasing and convex functions is a closed and convex cone in the space, and the functional maximum is certainly attained there. However, each element of the family can be approximated by differentiable almost everywhere compositions $F_n^{-1} \circ Z$, $n \to \infty$, where F_n have density functions and nonincreasing (or even strictly decreasing) failure rates. It follows that even if the bound is attained by a discontinuous $F \succeq_c Z$, it is attained in the limit by sequences for parent distributions with decreasing failure rates. The same remark applies to the problems of evaluating the expectations of order statistics and other functions of ordered random variables from the families with other monotone α-generalized (possibly reversed) failure rates

Propositions 2.1 and 2.2 provide some intuition concerning the sharp bounds on expectations of functions of ordered random variables (single order and record statistics in particular) under the shape restrictions on the families of parent distributions. For instance, consider the families with decreasing α-generalized failure rates which increase when parameter α increases. If $\alpha_1 < \alpha_2$, we can expect greater (or at least not less) upper bound in the DGFR(α_2) case than for the DGFR(α_1) family. Similarly the lower bounds should become smaller for greater α. In particular, the upper (lower) bounds for the decreasing density family should be greater (smaller, respectively) than the ones for the decreasing failure rate family. The opposite tendency should be observed for the increasing α-generalized failure rate families for varying parameter α. For instance, the upper bounds in the ID families should be less than those for the IFR case. These hypotheses are confirmed by direct calculations presented in the further chapters.

2.8.2 Star Order

The star ordering for one-dimensional distribution functions is mostly studied under the assumption that the distributions are supported on the positive half-axis (see Barlow & Proschan, 1966). Then we say that F precedes G in the star order (and write $F \preceq_* G$) if the composition $\frac{G^{-1} \circ F(x)}{x}$ is nondecreasing for $x > 0$. This property on the function $G^{-1} \circ F$ is called starshapedness. The notion was introduced and studied in Bruckner and Ostrow (1962). Equivalently, $F \preceq_* G$ if $F^{-1} \circ G$ is antistarshaped, i.e., $\frac{F^{-1} \circ G(x)}{x}$, $x > 0$, is nonincreasing. By convention, the star relation for distribution functions is naturally inherited by the random variables which have the respective distribution functions.

Various properties of the star ordering of positive random variables are presented in Shaked and Shanthikumar (2007, Section 4.B). In particular, $F \preceq_* G \preceq_* F$ holds if $G^{-1} \circ F(x) = cx$ for some positive c which means that the order is scale invariant. It is also easy to see that every convex (concave) function on $\mathbb{R}_+$ that vanishes at 0 is starshaped (antistarshaped, respectively). It follows that in the families of positive random variables and their distributions, the convex order implies the star one.

In our investigations we use a more general so-called support dependent version of the star ordering (see, e.g., Loh, 1984; Rychlik, 2002; Szymkowiak, 2020). This is defined for the distribution functions (and random variables) with the supports bounded on the left. For two distribution functions F and G with finite left-end support points ℓ_F and ℓ_G, respectively, we say that F precedes G in the (support dependent) star order if

$$x \mapsto \frac{G^{-1} \circ F(x) - \ell_G}{x - \ell_F}$$

is nondecreasing on the support of F, or equivalently,

$$x \mapsto \frac{F^{-1} \circ G(x) - \ell_F}{x - \ell_G}$$

is nonincreasing on the support of G. The above definition makes the order location and scale invariant. Also, the support dependent star order is implied by the convex transform order.

The most popular families of distributions determined by star ordering with a fixed distribution function (the exponential one in this case) are the families of distributions with increasing and decreasing failure rates on the average (IFRA and DFRA, for short). We have $F \preceq_* W_0 = Z$ ($F \succeq_* Z$) if the function

$$x \mapsto \frac{-\ln[1 - F(x)]}{x - \ell_F}, \qquad x \in \operatorname{supp} F, \tag{2.307}$$

is nondecreasing (nonincreasing, respectively). Note that for an absolutely continuous F with density f, the numerator

$$-\ln[1 - F(x)] = \int_{\ell_F}^{x} \frac{f(t)}{1 - F(t)}\, dt$$

is the cumulative failure rate of F at x, whereas (2.307) stands for the average of the failure rate over $[\ell_F, x]$. This justifies the names IFRA and DFRA for the classes of distributions preceding and succeeding the exponential distribution in the star order.

Other two natural nonparametric families of distributions are determined by the star order relation with the uniform distribution function $W_1 = U$. We have $F \preceq_* U$ ($F \succeq_* U$) iff

$$\frac{F(x)}{x - \ell_F} = \frac{1}{x - \ell_F} \int_{\ell_F}^{x} f(t)\, dt, \qquad x \in \operatorname{supp} F,$$

(the latter formula holds in the absolutely continuous case) is nondecreasing (nonincreasing, respectively). We call the distributions preceding and succeeding U in the star order the increasing and decreasing density on the average distributions, respectively, denoting them by IDA and DDA, respectively.

Similarly, we can determine the monotone α-generalized failure rate on the average distribution functions as the ones being related in the star ordering with the generalized Pareto distribution functions W_α, $\alpha \in \mathbb{R}$ (see (2.301)), and denote them by IGFRA(α) and DGFRA(α). Using the arguments of the proof of Proposition 2.1, we can conclude that the families IGFRA(α) and DGFRA(α) are decreasing and increasing, respectively, with respect to α.

Notice that we can analogously define the families with α-generalized reversed failure rates on the average. However, the notion of reversed failure rate on the average does not have a useful interpretation in the lifetime analysis, and it has not appeared in the literature by now. Therefore we shall not develop this idea here.

Chapter 3
Deterministic Bounds

In Sect. 3.1 we define order statistics $X_{i:n}$, $i = 1, \ldots, n$, (n stands here for the sample size) and their linear combinations $\sum_{i=1}^{n} c_i X_{i:n}$, called shortly L-statistics. We describe particular properties of L-statistics useful in statistical inference, and some popular examples of L-statistics. Generally, this chapter is devoted to evaluations of L-statistics with fixed deterministic values $x_{1:n}, \ldots\ x_{n:n}$ of order statistics. In Sect. 3.2 we present simple sharp upper bounds on the values of L-statistics with nondecreasing sequences $c_1, \ldots, c_n$ of coefficients, represented in sample standard deviation units. The best reference here is David (1988). Respective nonnegative upper bounds for arbitrary L-statistics are presented in Sect. 3.3. The results are generalized in Sect. 3.4 to more general scale units constructed on the basis of sample central absolute moments of various orders. These results were described in Rychlik (1993). Section 3.5 contains sharp upper negative bounds for negative L-statistics, and analogous lower positive bounds for positive L-statistics. The main contribution to the topic is given in Goroncy and Rychlik (2006). Section 3.6 contains generalizations of the results of Sect. 3.4. They consist in replacing scale units based on sample moments by more general ones.

3.1 Order Statistics and L-Statistics, Basic Properties and Examples

Let $X_1, \ldots, X_n$ be an arbitrary finite sequence of real random variables. In the statistical literature, it is called a sample of size n. Let $X_{1:n} \leq X_{2:n} \leq \cdots X_{n:n}$ denote the reordering of the original sample in the nondecreasing sequence. The jth smallest variable $X_{j:n}$, $1 \leq j \leq n$, is called the jth order statistic of the sample. Accordingly, the sample minimum and maximum are denoted by $X_{1:n}$ and $X_{n:n}$, respectively. If $X_1, \ldots, X_n$ are exchangeable random variables, that

N. Balakrishnan, T. Rychlik, *Sharp Inequalities for Ordered Random Variables in Statistics and Reliability*, Frontiers in Probability and the Statistical Sciences
https://doi.org/10.1007/978-3-031-61347-0_3

means the distributions of all permutations $X_{\pi(1)}, \ldots, X_{\pi(n)}$ of the original sample are identical (for instance, in the case of independent and identically distributed samples) then the sequence of order statistic $(X_{1:n}, \ldots, X_{n:n})$ is a sufficient statistic. This means in practice that in statistical inference it suffices to construct procedures based on values of ordered samples without bothering about the order they appear in the original sample. It strongly depends on the problem and statistical model which function of order statistics should be taken into account. In some problems it is possible to determine explicitly optimal statistical procedures which are certainly based on order statistics. Otherwise we try to restrict the class of functions of order statistics and choose a solution in the restricted class. The family of linear combinations of order statistics

$$L_{\tilde{c}:n} = L_{\tilde{c}:n}(X_1, \ldots, X_n) = \sum_{i=1}^{n} c_i X_{i:n}, \quad \tilde{c} = (c_1, \ldots, c_n) \in \mathbb{R}^n, \tag{3.1}$$

called shortly L-statistics, is the most popular restricted class of functions. Observe that the problem of choosing optimal L-statistic reduces to determining a solution in a finite dimensional space of coefficients $\tilde{c} \in \mathbb{R}^n$. Specific L-statistics proved optimal estimates of parameters in various parametric models. Other appeared efficient in problems of robust and nonparametric statistics. Below we present examples of L-statistics that are useful in nonparametric estimation of location and dispersion of statistical populations.

We first confine ourselves to the location estimates. A desired property of the estimates is equivariance with respect to translations. In the case of L-statistics, the equivariance condition

$$\begin{aligned} L_{\tilde{c}:n}(X_1 + c, \ldots, X_n + c) &= \sum_{i=1}^{n} c_i X_{i:n} + c \sum_{i=1}^{n} c_i \\ &= L_{\tilde{c}}(X_1, \ldots, X_n) + c = \sum_{i=1}^{n} c_i X_{i:n} + c \end{aligned} \tag{3.2}$$

for arbitrary values $X_1, \ldots, X_n$ of the sample and all real constants c is satisfied iff

$$\sum_{i=1}^{n} c_i = 1. \tag{3.3}$$

Another important requirement is monotonicity

$$X_i \le Y_i, \; i = 1, \ldots, n, \quad \Rightarrow \quad L_{\tilde{c}:n}(X_1, \ldots, X_n) \le L_{\tilde{c}:n}(Y_1, \ldots, Y_n). \tag{3.4}$$

Note that the predecessor implies

$$X_{i:n} \leq Y_{i:n}, \quad i = 1, \ldots, n. \tag{3.5}$$

The successor coincides with

$$\sum_{i=1}^{n} c_i (Y_{i:n} - X_{i:n}) \geq 0. \tag{3.6}$$

If

$$c_i \geq 0, \quad i = 1, \ldots, n, \tag{3.7}$$

then (3.6) holds. We show that this is also necessary. Indeed, if $c_k < 0$ for some $k = 1, \ldots, n$, then taking $X_{i:n} = Y_{i:n}$ for all $1 \leq i \leq n$ except for $i = k$, we conclude the contradiction of (3.6). Therefore, the convex combinations of order statistics are the only translation equivariant and monotone L-estimates of location. The most popular examples are the single order statistics $X_{j:n}$, $1 \leq j \leq n$, (called also the sample pth quantiles if $j = \lfloor np \rfloor + 1$ for $p \in (0, 1)$), trimmed means

$$T_{j,k:n} = \frac{1}{k + 1 - j} \sum_{i=j}^{k} X_{i:n}, \quad 1 \leq j \leq k \leq n, \tag{3.8}$$

arising by averaging the order statistics that remained after rejecting $j - 1$ smallest and $n - k$ largest observations, and the sample mean

$$T_{1,n:n} = \bar{X}_n = \bar{X} = \frac{1}{n} \sum_{i=1}^{n} X_{i:n} \tag{3.9}$$

in particular, and the Winsorized means

$$W_{j,k:n} = \frac{j}{n} X_{j:n} + \frac{1}{n} \sum_{i=j+1}^{k-1} X_{i:n} + \frac{n + 1 - k}{n} X_{k:n}, \ 1 \leq j < k \leq n, \tag{3.10}$$

arising from the sample mean by replacing $j - 1$ smallest and $n - k$ largest observations by the respective nearest neighbors $X_{j:n}$ and $X_{k:n}$.

If the random variables are symmetrically distributed about a symmetry center which is the population median (and mean assumed exists), a reasonable property of location estimates is antisymmetry. The property

$$L_{\tilde{c}:n}(-X_1, \dots, -X_n) = -\sum_{i=1}^{n} c_i X_{n+1-i:n} = -\sum_{i=1}^{n} c_{n+1-i} X_{i:n}$$
$$= -L_{\tilde{c}:n}(X_1, \dots, X_n) = -\sum_{i=1}^{n} c_i X_{i:n} \tag{3.11}$$

holds for all sample values iff the sequence of coefficients is symmetric, i.e.,

$$c_{n+1-i} = c_i, \quad i = 1, \dots, n. \tag{3.12}$$

To show that (3.12) is necessary for (3.11), it suffices, in the case $c_{n+1-k} \neq c_k$ for some pair $(k, n+1-k)$ to verify falsity of (3.11) by taking the sample values satisfying $X_{n+1-i:n} = -X_{i:n}$ for all the pairs $(i, n+1-i)$ except for $(k, n+1-k)$. The property (3.11) holds, for instance, for the sample median

$$\text{MED}_n(\tilde{X}) = \begin{cases} X_{\frac{n+1}{2}:n}, & \text{if } n \text{ is odd}, \\ \frac{1}{2}(X_{\frac{n}{2}:n} + X_{\frac{n}{2}+1;n}) & \text{if } n \text{ is even}, \end{cases} \tag{3.13}$$

sample midrange

$$M_{1,n:m} = \frac{1}{2}(X_{1:n} + X_{n:n}), \tag{3.14}$$

jth quasi midranges

$$M_{j,n+1-j:n} = \frac{1}{2}(X_{j:n} + X_{n+1-j:n}), \quad 2 \le j \le \frac{n}{2}, \tag{3.15}$$

as well as the symmetric trimmed and Winsorized means (3.8) and (3.10) with $1 \le j = n+1-k \le \frac{n}{2}$. Other exemplary antisymmetric location L-estimates, based on few order statistics are the Tukey trimean

$$\text{TUK}_n(\tilde{X}) = \frac{1}{2}\text{MED}_n(\tilde{X}) + \frac{1}{2}M_{j,n+1-j:n}, \qquad j = \lfloor n/4 \rfloor, \tag{3.16}$$

and Gastwirth mean

$$\text{GAS}_n(\tilde{X}) = \frac{2}{5}\,\text{MED}_n(\tilde{X}) + \frac{3}{5}M_{j,n+1-j:n}, \qquad j = \lfloor n/3 \rfloor, \tag{3.17}$$

that proved efficient in robust statistical inference.

Now we establish conditions on the L-estimates of scale. First we notice that all the L-statistics are scale equivariant, that means

$$L_{\tilde{c}:n}(cX_1, \ldots, cX_n) = cL_{\tilde{c}:n}(X_1, \ldots, X_n) \tag{3.18}$$

holds for all $X_1, \ldots, X_n$ and positive c. A natural requirement for the scale estimates is nonnegativity

$$L_{\tilde{c}:n}(X_1, \ldots, X_n) \geq 0 \tag{3.19}$$

for all possible values of the sample. This holds iff

$$\sum_{i=1}^{j} c_i \leq \sum_{i=1}^{n} c_i = 0, \quad j = 1, \ldots, n-1. \tag{3.20}$$

The sufficiency of (3.20) follows from Theorem 2.7. The arguments for necessity are presented below. If $\sum_{i=1}^{n} c_i > 0$ and $X_{1:n} = \cdots = X_{n:n} < 0$, then $L_{\tilde{c}:n}(X_1, \ldots, X_n) < 0$. Similarly $\sum_{i=1}^{n} c_i < 0$ with $X_{1:n} = \cdots = X_{n:n} > 0$ implies the same and so condition $\sum_{i=1}^{n} c_i = 0$ is necessary for (3.19). If $\sum_{i=1}^{k} c_i > \sum_{i=1}^{n} c_i = 0$ for some $1 \leq k \leq n-1$ and $X_{1:n} = X_{k:n} < X_{k+1:n} = X_{n:n}$, then

$$\begin{aligned} L_{\tilde{c}:n}(X_1, \ldots, X_n) &= \sum_{i=1}^{k} c_i X_{k:n} + \sum_{i=k+1}^{n} c_i X_{k+1:n} \\ &= -(X_{k+1:n} - X_{k:n}) \sum_{i=1}^{k} c_i < 0, \end{aligned} \tag{3.21}$$

which provides the contradiction. Observe that nonnegativity of L-statistics implies translation invariance

$$\begin{aligned} L_{\tilde{c}:n}(X_1 + c, \ldots, X_n + c) &= L_{\tilde{c}:n}(X_1, \ldots, X_n) + c \sum_{i=1}^{n} c_i \\ &= L_{\tilde{c}:n}(X_1, \ldots, X_n) \end{aligned} \tag{3.22}$$

that is equivalent to the last condition in (3.20). The simplest nonnegative L-statistics are the differences

$$R_{j,k:n} = X_{k:n} - X_{j:n}, \quad 1 \leq j \leq n, \tag{3.23}$$

and the sample range

$$R_n = R_{1,n:n} = X_{n:n} - X_{1:n}, \tag{3.24}$$

and spacings

$$S_{j:n} = R_{j,j+1:n} = X_{j+1:n} - X_{j:n}, \quad 1 \le j \le n-1, \tag{3.25}$$

in particular. Other examples are the sample upper selection differentials

$$\bar{S}_{k:n} = T_{n+1-k,n:n} - T_{1,n:n} = \frac{1}{k} \sum_{i=n+1-k}^{n} X_{i:n} - \bar{X}_n, \quad 1 \le k < n, \tag{3.26}$$

(their lower counterparts

$$\underline{S}_{j:n} = T_{1,j:n} - T_{1,n:n} = \frac{1}{j} \sum_{i=1}^{j} X_{i:n} - \bar{X}_n, \quad 1 \le j < n, \tag{3.27}$$

are nonpositive) and the differences of sample differentials

$$\bar{\underline{S}}_{j,k:n} = \bar{S}_{k:n} - \underline{S}_{j:n} = \frac{1}{k} \sum_{i=n+1-k}^{n} X_{i:n} - \frac{1}{j} \sum_{i=1}^{j} X_{i:n}, \quad 1 \le j, k \le n. \tag{3.28}$$

The notions have important interpretations in sampling problems from finite populations. If the population size is n, then (3.26) and (3.27) stand for the extreme possible derivations of samples (with sizes k and j) from the population mean. They are frequently used, for instance, in the genetics studies. Mallows and Richter (1969) called (3.28) the generalized ranges.

In analysis of symmetrically distributed populations, it is useful to confine ourselves to symmetric estimates of scale. For L-statistics, condition

$$\begin{aligned} L_{\bar{c}:n}(-X_1, \ldots, -X_n) &= -\sum_{i=1}^{n} c_i X_{n+1-i:n} = -\sum_{i=1}^{n} c_{n+1-i} X_{i:n} \\ &= L(X_1, \ldots, X_n) = \sum_{i=1}^{n} c_i X_{i:n} \end{aligned} \tag{3.29}$$

is satisfied iff the coefficients are antisymmetric

$$c_{n+1-i} = -c_i, \quad i = 1, \ldots, n. \tag{3.30}$$

The most natural nonnegative and symmetric L-estimates of scale have the coefficients satisfying

$$c_{n+1-i} = -c_i \ge 0, \quad i \le \frac{n+1}{2}. \tag{3.31}$$

These are, for example, the sample range (3.24) and jth quasi ranges

$$R_{j,n+1-j:n:n} = X_{n+1-j:n} - X_{j:n} \quad j \leq n/2, \tag{3.32}$$

the median absolute deviation

$$\mathrm{MAD}_n(\tilde{x}) = \frac{1}{n}\sum_{i=1}^{n} |X_{i:n} - \mathrm{MED}_n(\tilde{X})| = \frac{1}{n}\sum_{i \geq \frac{n}{2}+1} X_{i:n} - \frac{1}{n}\sum_{i \leq \frac{n}{2}} X_{i:n}, \tag{3.33}$$

and the Gini mean differences

$$\mathrm{GMD}_n(\tilde{X}) = \frac{1}{n(n-1)}\sum_{i,j=1}^{n} |X_i - X_j| = \frac{2}{n(n-1)}\sum_{i=1}^{n}(2i-n-1)X_{i:n}. \tag{3.34}$$

Observe that nonnegativity and symmetry properties are preserved if we multiply L-statistics by positive constants. Multiplying the L-estimates of scale presented above by properly chosen constants allows us to determine the estimates which are consistent in specific statistical models.

Below we present optimal deterministic bounds on values of L-statistics

$$L_{\bar{c}:n}(x_1, \ldots, x_n) = \sum_{i=1}^{n} c_i x_{i:n}, \tag{3.35}$$

where $x_{1:n} \leq x_{2:n} \leq \cdots \leq x_{n:n}$ arise by arranging values $x_1, \ldots, x_n$ of random variables $X_1, \ldots, X_n$ is nondecreasing order. The bounds do not depend on the joint distribution of $X_1, \ldots, X_n$ which can be nonidentically distributed and/or arbitrarily dependent. In fact, it would suffice to write analogous results for $x_1, \ldots, x_n$ themselves under condition that they do not decrease. We decided to express them in terms of $x_{i:n}$, $i = 1, \ldots, n$, because of their natural interpretation in statistics the more that they are expressed by means of other statistical notions as, for instance, the sample mean

$$\bar{x} = \frac{1}{n}\sum_{i=1}^{n} x_i \tag{3.36}$$

and sample variance

$$s_2^2 = s^2 = \frac{1}{n}\sum_{i=1}^{n}(x_i - \bar{x})^2. \tag{3.37}$$

On the other hand, we use small letters in order to point out that the results are purely deterministic.

In Sect. 3.2 we derive the most popular bounds based on the Schwarz inequality. We distinguish the cases for which the Schwarz bounds are sharp. The best reference is here David (1988). Most of the results presented in the next two subsections can be found in Rychlik (1993, 1998). In Sect. 3.3 we provide sharp bounds for the other cases based on combination of the Moriguti and Schwarz inequalities. Section 3.4 is devoted to more general bounds expressed in scale units generated by absolute sample moments of various orders. They are derived by means of the Moriguti and Hölder inequalities. Further generalizations are presented in Sect. 3.6. This contains bounds for other statistics in terms of generalized sample moments defined by means of general convex gauge functions and some extensions, due to Beesack (1973, 1976). There are also the Orlicz space norm bounds for L-statistics based the Young inequality and abstract bounds based on general symmetric norms, due to Rychlik (1992).

For the most of results it is possible to present more general weighted versions. We restricted ourselves to the uniform unweighted representations by several reasons. Firstly, the formulae with weights become by far more complicated. Also, the uniform versions are more natural in analysing deterministic samples. Moreover, introducing weights imposes replacing standard natural notions like sample moments by sophisticated and meaningless weighted counterparts based on ordered samples.

3.2 Schwarz Bounds

We start with the classic problem of how deviant an observation x_i (precisely, the sample maximum $x_{n:n}$) can be deviant from the sample mean $\bar{x}$ in the standard deviation units s (cf. (3.36) and (3.37)).

Theorem 3.1 *For arbitrary* $\tilde{x} = (x_1, \ldots, x_n)$ *we have*

$$x_{n:n} - \bar{x} \le \sqrt{n-1}s, \tag{3.38}$$

and the equality holds if

$$x_{i:n} = \begin{cases} \bar{x} - s\frac{1}{\sqrt{n-1}}, & i = 1, \ldots, n-1, \\ \bar{x} + s\sqrt{n-1}\,, & i = n. \end{cases} \tag{3.39}$$

Proof The result is trivial if $x_i = \bar{x}$, $i = 1, \ldots, n$, and $s = 0$. If $s > 0$, then with the notation $\delta_{ij} = 1$ if $i = j$ and 0 otherwise, we have

$$x_{n:n} - \bar{x} = \sum_{i=1}^{n} \delta_{in}(x_{i:n} - \bar{x}) = \sum_{i=1}^{n} (\delta_{in} - \frac{1}{n})(x_{i:n} - \bar{x})$$

$$\leq \left[\sum_{i=1}^{n} \left(\delta_{in} - \frac{1}{n} \right)^2 \sum_{i=1}^{n} (x_{i:n} - \bar{x})^2 \right]^{1/2}$$

$$= \left[\frac{n-1}{n^2} + \frac{(n-1)^2}{n^2} \right]^{1/2} \left[\sum_{i=1}^{n} (x_i - \bar{x}) \right]^{1/2} = \sqrt{n-1}\, s, \quad (3.40)$$

by the Schwarz inequality, where the equality holds if

$$x_{i:n} - \bar{x} = \alpha \left(\delta_{in} - \frac{1}{n} \right) = \begin{cases} -\alpha \frac{1}{n}, & i = 1, \ldots, n-1, \\ \alpha \frac{n-1}{n}, & i = n, \end{cases} \quad (3.41)$$

for a positive α. The only factor α for which

$$\frac{1}{n} \alpha^2 \sum_{i=1}^{n} (\delta_{in} - 1)^2 = s^2 \quad (3.42)$$

for a fixed s^2 is

$$\alpha = \frac{n}{\sqrt{n-1}} s. \quad (3.43)$$

Combining (3.41) with (3.43), we obtain (3.39). □

The result seems to be derived first by Scott (1936). After rediscovering it by Sansone (1959), a number of alternative proofs and generalizations appeared in literature. They were reviewed by Arnold and Balakrishnan (1989, Section 3.2) and Olkin (1992).

Mimicking the proof of Theorem 3.1 for general L-statistics, we obtain

$$\begin{aligned} \sum_{i=1}^{n} c_i (x_{i:n} - \bar{x}) &= \sum_{i=1}^{n} (c_i - c)(x_{i:n} - \bar{x}) \\ &\leq \left[\sum_{i=1}^{n} (c_i - c)^2 \sum_{i=1}^{n} (x_{i:n} - \bar{x})^2 \right]^{1/2} \\ &= \left[n \sum_{i=1}^{n} (c_i - c)^2 \right]^{1/2} \left[\frac{1}{n} \sum_{i=1}^{n} (x_i - \bar{x})^2 \right]^{1/2} \\ &= B(c) s, \end{aligned} \quad (3.44)$$

say, for arbitrary real c. The bound is minimized for

$$c = \bar{c} = \frac{1}{n}\sum_{i=1}^{n} c_i. \tag{3.45}$$

The smallest bound

$$B(\bar{c}) = \left[n\sum_{i=1}^{n} c_i^2 - (n\bar{c})^2\right]^{1/2} \tag{3.46}$$

is attained iff

$$x_{i:n} = \bar{x} + \alpha(c_i - \bar{c}), \quad i = 1, \ldots, n, \tag{3.47}$$

for

$$\alpha = \frac{ns}{B(\bar{c})} = \left(\frac{n}{\sum_{i=1}^{n} c_i - n\bar{c}^2}\right)^{1/2} s. \tag{3.48}$$

For the single order statistic in particular, we have

$$x_{i:n} - \bar{x} \le \sqrt{n-1}\, s \tag{3.49}$$

(cf. (3.38)), and the equality conditions are

$$x_{i:n} = \bar{x} + s\sqrt{n-1}, \tag{3.50}$$

$$x_{j:n} = \bar{x} - s\frac{1}{\sqrt{n-1}}, \quad j \ne i. \tag{3.51}$$

Note that this is impossible for $i \ne n$, because $x_{i:n}$ are greater than $x_{n:n}$ then. Likewise, bounds (3.39) cannot be attained by (3.47) with (3.48) unless the sequence of coefficients is nondecreasing. Therefore (3.44) with $c = \bar{c}$ is sharp only for the L-statistics with nondecreasing coefficients.

Theorem 3.2 *Suppose that $c_1 \le \cdots \le c_n$. Then*

$$\sum_{i=1}^{n} c_i(x_{i:n} - \bar{x}) \le \left(n\sum_{i=1}^{n} c_i^2 - n^2\bar{c}^2\right)^{1/2} s \tag{3.52}$$

holds for all $\tilde{x} \in \mathbb{R}^n$ and, under condition $c_1 < c_n$ it becomes equality for

$$x_{i:n} = \bar{x} + s\frac{c_i - \bar{c}}{(\frac{1}{n}\sum_{i=1}^{n} c_i^2 - \bar{c}^2)^{1/2}}, \quad i = 1, \ldots, n. \tag{3.53}$$

If $c_1 = c_n$, then both sides of (3.52) vanish for arbitrary $\bar{x} \in \mathbb{R}^n$. In particular, the sharp bounds for the sample ranges are

$$r_n = x_{n:n} - x_{1:n} \le \sqrt{2n}\, s \tag{3.54}$$

and these are attained iff

$$x_{i:n} = \begin{cases} \bar{x} - s\sqrt{\frac{n}{2}}, & i = 1, \\ \bar{x}, & i = 2, \dots, n-1, \\ \bar{x} + s\sqrt{\frac{n}{2}}, & i = n. \end{cases} \tag{3.55}$$

The bounds were determined by Nair (1948) and Thomson (1955). Note that here and, more generally, for translation invariant L-statistics with $\sum_{i=1}^{n} c_i = 0$, the sample mean $\bar{x}$ does not need to be subtracted in the LHS of (3.52). Below we list several further implications of Theorem 3.2. For the upper selection differentials (3.26) with nondecreasing coefficients

$$c_i = \begin{cases} -\frac{1}{n}, & i = 1, \dots, n-k, \\ \frac{1}{k} - \frac{1}{n}, & i = n+1-k, \dots, n, \end{cases} \tag{3.56}$$

we have

$$\bar{s}_{k:n} = \frac{1}{k} \sum_{i=n-1-k}^{n} x_{i:n} - \bar{x} \le \sqrt{\frac{n-k}{k}} s \tag{3.57}$$

which becomes equality for

$$x_{i:n} = \begin{cases} \bar{x} - s\sqrt{\frac{k}{n-k}}, & i = 1, \dots, n-k, \\ \bar{x} + s\sqrt{\frac{n-k}{k}}, & i = n+1-k, \dots, n, \end{cases} \tag{3.58}$$

(see Mallows & Richter, 1969; Nagaraja, 1981). In the particular case $k = 1$ we clearly obtain the Scott-Samuelson bound (3.38) for the sample maximum. Similarly, considering the differences of upper and lower differentials we generalize the range bound (3.54) as follows

$$\bar{\bar{s}}_{j,k:n} = \frac{1}{k} \sum_{i=n+1-k}^{n} x_{i:n} - \frac{1}{j} \sum_{i=1}^{j} x_{i:n} \le \sqrt{\frac{(j+k)n}{jk}}\, s. \tag{3.59}$$

The equality is derived here when

$$x_{i:n} = \begin{cases} \bar{x} - s\sqrt{\frac{kn}{j(j+k)}}, & i = 1, \dots, j, \\ \bar{x}, & i = j+1, \dots, n-k, \\ \bar{x} + s\sqrt{\frac{jn}{k(j+k)}}, & i = n+1-k, \dots, n, \end{cases} \tag{3.60}$$

(see Mallows & Richter, 1969; Nagaraja, 1981). In the case of median absolute deviation with coefficients

$$c_i = \begin{cases} -\frac{1}{n}\,, & i = 1, \dots, \left\lfloor \frac{n}{2} \right\rfloor, \\ 0\,, & i = \frac{n+1}{2} \text{ (if } n \text{ is odd)}, \\ +\frac{1}{n}\,, & i = \left\lfloor \frac{n}{2} \right\rfloor + 1, \dots, n, \end{cases} \tag{3.61}$$

bound (3.52) takes on the form

$$\mathrm{MAD}_n(\tilde{x}) = \frac{1}{n}\sum_{i=1}^{n} |x_i - \mathrm{MED}_n(\tilde{x})| \le \sqrt{\frac{2}{n}\left\lfloor \frac{n}{2} \right\rfloor} s = \begin{cases} s, & \text{if } n \text{ is even}, \\ \sqrt{\frac{n-1}{n}} s, & \text{if } n \text{ is odd}, \end{cases} \tag{3.62}$$

with the equality conditions

$$x_{i:n} = \begin{cases} \bar{x} - s\sqrt{\frac{n}{2\lfloor n/2 \rfloor}}, & i = 1, \dots, \lfloor n/2 \rfloor, \\ \bar{x}, & i = \frac{n+1}{2} \text{ (if } n \text{ is odd)}, \\ \bar{x} + s\sqrt{\frac{n}{2\lfloor n/2 \rfloor}}, & i = \lfloor n/2 \rfloor + 1, \dots, n. \end{cases} \tag{3.63}$$

If n is even, then (3.63) simplifies to

$$x_{i:n} = \begin{cases} \bar{x} - s, & i = 1, \dots, \frac{n}{2}, \\ \bar{x} + s, & i = \frac{n}{2} + 1, \dots, n, \end{cases} \tag{3.64}$$

whereas for n odd we have

$$x_{i:n} = \begin{cases} \bar{x} - s\sqrt{\frac{n}{n-1}}, & i = 1, \dots, \frac{n-1}{2}, \\ \bar{x}, & i = \frac{n+1}{2}, \\ \bar{x} + s\sqrt{\frac{n}{n-1}}, & i = \frac{n+3}{2}, \dots, n. \end{cases} \tag{3.65}$$

Observe that (3.62) and (3.63) are easily obtained from (3.59) and (3.60), respectively, because $\mathrm{MAD}_n(\tilde{x}) = \frac{\lfloor n/2 \rfloor}{n} \underline{s}_{j,k:n}$ with $j = k = \left\lfloor \frac{n}{2} \right\rfloor$.

Finally, we present the deterministic Schwarz bounds for the Gini mean differences

$$\mathrm{MAD}_n(\tilde{x}) = \frac{1}{n(n-1)} \sum_{\substack{i,j=1 \\ i \neq j}}^{n} |x_i - x_j| \leq \sqrt{\frac{4(n+1)}{3(n-1)}}\, s. \tag{3.66}$$

This becomes equality iff

$$x_{i:n} = \bar{x} + s\sqrt{\frac{3}{n^2-1}}(2i-n-1), \quad i = 1, \ldots, n, \tag{3.67}$$

form an arithmetic sequence. Equation (3.66) is derived by use of Theorem 3.2 for the L-statistic (3.34) representing Gini mean differences with linearly increasing coefficients $c_i = \frac{2(2i-n-1)}{n(n-1)}$, $i = 1, \ldots, n$, combined with the identity

$$\sum_{i=1}^{n}(2i-n-1)^2 = \frac{(n-1)n(n+1)}{3}. \tag{3.68}$$

We recommend the reader verification of other particular results presented above as well.

The Schwarz inequality can also be applied for evaluating L-statistics of nonnegative samples in terms of the square root of the raw sample second moment

$$m_2 = \left(\frac{1}{n}\sum_{i=1}^{n} x_i^2\right)^{1/2}. \tag{3.69}$$

The conditions for sharpness are here less stringent then in Theorem 3.2. Monotonicity of sequence $\tilde{c}$ is not necessary. It merely suffices that the positive elements are nondecreasing and succeed all the nonpositive ones. In other words, we only need that $\tilde{c}_+ = ((c_1)_+, \ldots, (c_n)_+)$, $(c_i)_+ = \max\{c_i, 0\}$, $i = 1, \ldots, n$, is nondecreasing. This holds, for instance, for the coefficients of differences of order statistics (3.23) with $k = n$ which are not nondecreasing themselves.

Theorem 3.3 *If $\tilde{c} \in \mathbb{R}^n$ and $\tilde{x} \in \mathbb{R}^n_+$, then*

$$\sum_{i=1}^{n} c_i x_{i:n} \leq \left[n \sum_{i=1}^{n} (c_i)_+^2\right]^{1/2} m_2. \tag{3.70}$$

If $\tilde{c}_+ \neq \tilde{0}$ is nondecreasing then (3.70) is sharp and becomes equality for

$$x_{i:n} = m_2 \frac{(n^{1/2} c_i)_+}{[\sum_{i=1}^{n} (c_i)_+]^{1/2}}, \quad i = 1, \ldots, n. \tag{3.71}$$

Proof For (3.70) it suffices to write

$$\sum_{i=1}^{n} c_i x_{i:n} \leq \sum_{i=1}^{n} (c_i)_+ x_{i:n} \leq \left[\sum_{i=1}^{n} (c_i)_+^2 \sum_{i=1}^{n} x_{i:n}^2 \right]^{1/2}. \tag{3.72}$$

The equality condition in the Schwarz inequality is that $x_{i:n}$ are proportional to $(c_i)_+$, $i = 1, \ldots, n$, which is possible iff $\tilde{c}_+$ is nondecreasing. The proportionality coefficient is determined by the moment condition. Also, note that the first inequality in (3.72) becomes equality iff $x_{i:n}$ vanish when respective c_i are negative. This is guaranteed by (3.71) as well. □

In consequence, we easily obtain the following optimal inequalities for specific L-statistics of nonnegative samples

$$x_{n:n} \leq \sqrt{n}\, m_2, \tag{3.73}$$

$$x_{n:n} - x_{j:n} \leq \sqrt{n}\, m_2, \quad j = 1, \ldots, n-1, \tag{3.74}$$

attained by

$$x_{i:n} = \begin{cases} 0, & i = 1, \ldots, n-1, \\ \sqrt{n}\, m_2, & i = n, \end{cases} \tag{3.75}$$

and

$$\bar{s}_{k:n} \leq \frac{n-k}{\sqrt{nk}}\, m_2, \quad k = 1, \ldots, n, \tag{3.76}$$

$$\bar{\underline{s}}_{j,k:n} \leq \frac{n-k}{\sqrt{nk}}\, m_2, \quad 1 \leq j < k \leq n, \tag{3.77}$$

attained by

$$x_{i:n} = \begin{cases} 0, & i = 1, \ldots, n-k, \\ \sqrt{\frac{n}{k}}\, m_2, & i = n+1-k, \ldots, n. \end{cases} \tag{3.78}$$

For the median absolute deviation we have

$$\mathrm{MAD}_n(\tilde{x}) \leq \begin{cases} \frac{1}{\sqrt{2}}\, m_2, & \text{if } n \text{ is even}, \\ \sqrt{\frac{n-1}{2n}}\, m_2, & \text{if } n \text{ is odd}, \end{cases} \tag{3.79}$$

which is an equality for

$$x_{i:n} = \begin{cases} 0, & i = 1, \dots, \frac{n}{2}, \\ \sqrt{2}\, m_2, & i = \frac{n}{2} + 1, \dots, n, \end{cases} \tag{3.80}$$

and

$$x_{i:n} = \begin{cases} 0, & i = 1, \dots, \frac{n+1}{2}, \\ \sqrt{\frac{2n}{n-1}}, & i = \frac{n+3}{2}, \dots, n, \end{cases} \tag{3.81}$$

for the samples of even and odd sizes, respectively. Analogous bounds and equality conditions for the Gini mean differences are as follows:

$$\mathrm{GMD}_n(\tilde{x}) \le \sqrt{\frac{2(n+1)}{3(n-1)}}\, m_2 \tag{3.82}$$

and

$$x_{i:n} = \begin{cases} 0, & \text{if } i \le \frac{n+1}{2}, \\ \sqrt{\frac{6}{n^2-1}}\, m_2 & \text{if } i > \frac{n+1}{2}. \end{cases} \tag{3.83}$$

It is hard to compare bounds (3.52) and (3.70) in general. Up to factor $\sqrt{n}$, they coincide with the Euclidean norms of vectors $\tilde{c} - \bar{c}\tilde{1}$ and $\tilde{c}_+$. We only know that

$$\|\tilde{c} - \bar{c}\tilde{1}\| \le \|\tilde{c}\|, \tag{3.84}$$

because $\tilde{c} - \bar{c}\tilde{1}$ is the projection of $\tilde{c}$ onto the linear subspace orthogonal to $\tilde{1}$, and

$$\|\tilde{c}_+\| \le \|\tilde{c}\|, \tag{3.85}$$

because $\tilde{c}_+$ is the projection of $\tilde{c}$ onto the convex cone of nonnegative vectors. Both inequalities are strict unless the respective projections are identical with $\tilde{c}$. If $\tilde{c} = \tilde{c}_+ \neq \tilde{0}$ (that means the respective L-statistic has the monotonicity property (3.4), then

$$\|\tilde{c} - \bar{c}\tilde{1}\| = (\|\tilde{c}_+\|^2 - \bar{c}^2)^{1/2} < \|\tilde{c}_+\| \tag{3.86}$$

(cf., for instance, (3.38) and (3.73)). We also have $s_2 < m_2$. On the other hand,

$$\sum_{i=1}^{n} c_i(x_{i:n} - \bar{x}) < \sum_{i=1}^{n} c_i x_{i:n} \tag{3.87}$$

for $\tilde{0} \neq \tilde{x} \in \mathbb{R}^n_+$. If $\tilde{c} = \tilde{c} - \bar{c}\tilde{1} \neq \tilde{0}$ (that means the L-statistic satisfies the translation invariance principle (3.22)), then

$$\|c_+\| = (\|\tilde{c}\|^2 - \|\tilde{c}_-\|^2)^{1/2} < \|\tilde{c}\| = \|\tilde{c} - \bar{c}\tilde{1}\| \tag{3.88}$$

(cf. the pairs (3.54) with (3.74) in the case $j = n$, (3.57) with (3.76), and (3.59) with (3.77) and (3.62) with (3.79), and (3.66) with (3.82)). Observe that in the case $\bar{c} = 0$, the LHS of (3.70) does not change under the translations and the RHS does through the values of m_2. It is smallest if $x_{1:n} = 0$ or, equivalently if m_2 is replaced by

$$m_2(x_{1:n}) = \left(\frac{1}{n}\sum_{i=1}^{n}(x_i - x_{1:n})^2\right)^{1/2}. \tag{3.89}$$

This justifies replacing (3.70) by

$$\sum_{i=1}^{n} c_i(x_{i:n} - x_{1:n}) \leq \left(n\sum_{i=2}^{n}(c_i)_+^2\right)^{1/2} m_2(x_{1:n}) \tag{3.90}$$

which holds for arbitrary $\tilde{x} \in \mathbb{R}^n$ and is attained for

$$x_{i:n} = x_{1:n} + m_2(x_{1:n})\,\frac{n^{1/2}(c_i)_+}{\left[\sum_{i=2}^{n}(c_i)_+^2\right]^{1/2}} \qquad i = 2, \ldots, n, \tag{3.91}$$

provided that

$$(c_2)_+ \leq \cdots \leq (c_n)_+. \tag{3.92}$$

The crucial difference consists actually in changing the locations and scale parameters. The sample mean and standard deviation are substituted by the sample minimum and (3.89), correspondingly. Formally, it is possible to determine analogues of (3.90) and (3.91) with location and scale parameters based on other statistics.

Notice that (3.92) is less restrictive than the sharpness condition of Theorem 3.2. Below we extend the results of this subsection to arbitrary L-statistics.

3.3 Moriguti-Schwarz Bounds

In order to establish sharp deterministic bounds for arbitrary L-statistics of centered samples

$$L_{\tilde{c}:n}(x_1-\bar{x},\ldots,x_n-\bar{x}) = \sum_{i=1}^{n} c_i(x_{i:n}-\bar{x}), \qquad \tilde{c} = (c_1,\ldots,c_n) \in \mathbb{R}^n, \quad (3.93)$$

in the sample standard deviation units

$$s = \left[\frac{1}{n}\sum_{i=1}^{n}(x_i - \bar{x})^2\right]^{1/2}, \qquad (3.94)$$

we combine the Morguti and Schwarz inequalities.

3.3.1 Upper Bounds

Theorem 3.4 *For a fixed $\tilde{c} = (c_1,\ldots,c_n) \in \mathbb{R}^n$, define the projection $\underline{\tilde{c}} = P^{\nearrow}\tilde{c} = (\underline{c}_1,\ldots,\underline{c}_n)$ of $\tilde{c}$ onto the convex cone of nondecreasing sequences in the standard Euclidean Hilbert space $\ell^2(\mathbb{R}^n)$ by means of algorithm (2.78) and (2.79). Then*

$$\sum_{i=1}^{n} c_i(x_{i:n} - \bar{x}) \le \left[n\sum_{i=1}^{n}\underline{c}_i^2 - n^2\bar{c}^2\right]^{1/2} s \qquad (3.95)$$

for all $\tilde{x}$, and the equality holds iff

$$x_{i:n} = \bar{x} + s\frac{\underline{c}_i - \bar{c}}{(\frac{1}{n}\sum_{i=1}^{n}\underline{c}_i^2 - \bar{c}^2)^{1/2}}, \qquad i = 1,\ldots,n, \qquad (3.96)$$

provided that $\underline{c}_1 < \underline{c}_n$.

Relation $\underline{c}_1 < \underline{c}_n$ assures that the RHS bound of (3.95) is strictly positive and (3.96) are well-defined. It coincides with the condition

$$\min_{1\le j\le n-1}\frac{1}{j}\sum_{i=1}^{j}c_i < \frac{1}{n}\sum_{i=1}^{n}c_i = \bar{c} \qquad (3.97)$$

expressed in terms of the original coefficients $c_1,\ldots,c_n$.

Proof Using the Moriguti inequality (2.80) first, and the Schwarz inequality (2.113) then, we obtain

$$\begin{aligned}\sum_{i=1}^{n} c_i(x_{i:n}-\bar{x}) &= \sum_{i=1}^{n}(c_i-\bar{c})(x_{i:n}-\bar{x}) \le \sum_{i=1}^{n}(\underline{c}_i-\bar{c})(x_{i:n}-\bar{x}) \\ &\le \left[n\sum_{i=1}^{n}(\underline{c}_i-\bar{c})^2\right]^{1/2}\left[\frac{1}{n}\sum_{i=1}^{n}(x_{i:n}-\bar{x})^2\right]^{1/2} \\ &= \left(n\sum_{i=1}^{n}\underline{c}_i^2-n^2\bar{c}^2\right)^{1/2} s. \end{aligned} \tag{3.98}$$

By (2.115), the necessary and sufficient condition for equality in the latter inequality is

$$x_{i:n}-\bar{x}=\alpha(\underline{c}_i-\bar{c}), \qquad \alpha>0,\ i=1,\ldots,n, \tag{3.99}$$

which in combination with (3.94) implies that

$$\alpha=\frac{s}{(\frac{1}{n}\sum_{i=1}^{n}\underline{c}_i^2-\bar{c}^2)^{1/2}}, \tag{3.100}$$

and (3.96) holds in consequence. Note finally that, due to (2.81), relations (3.99) are sufficient for attaining the equality in the first inequality of (3.98). □

In particular, for the single order statistics $x_{j:n}$, $j=1,\ldots,n$, with coefficients

$$c_i=\begin{cases}1, & i=j,\\ 0, & i\ne j,\end{cases} \tag{3.101}$$

and $\bar{c}=\frac{1}{n}$, procedure (2.78) and (2.79) provides

$$\bar{c}_i=\begin{cases}0, & i=1,\ldots,j-1,\\ \frac{1}{n+1-j}, & i=j,\ldots,n.\end{cases} \tag{3.102}$$

Therefore, by Theorem 3.4, we have

Corollary 3.1 *For $j=1,\ldots,n$ and all $\tilde{x}\in\mathbb{R}^n$, inequality*

$$x_{j:n}-\bar{x}\le\sqrt{\frac{j-1}{n+1-j}}\,s \tag{3.103}$$

holds and is sharp for $j = 2, \ldots, n$ with the equality conditions

$$x_{i:n} = \begin{cases} \bar{x} - s\sqrt{\frac{n+1-j}{j-1}}, & i = 1, \ldots, j-1, \\ \bar{x} + s\sqrt{\frac{j-1}{n+1-j}}, & i = j, \ldots, n. \end{cases} \tag{3.104}$$

Bound (3.103) for the case $j = n$ is the classic Scott-Samuelson bound (3.38) of Theorem 3.1. For the other cases, it was determined by Boyd (1971) and Hawkins (1971). For $i = 1$, (3.103) is a trivial zero bound which is attained when all x_i are identical and $s = 0$ as well.

Observe that for the trimmed means (3.8) and Winsorized means (3.10), the projections of the coefficient vectors onto nondecreasing sequences coincide with (3.102), and consequently the sharp bounds

$$t_{j,k:n} - \bar{x} = \frac{1}{k+1-j}\sum_{i=j}^{k} x_{i:n} - \bar{x} \le \sqrt{\frac{j-1}{n+1-j}}\, s, \tag{3.105}$$

$$w_{j,k:n} - \bar{x} = \frac{j}{n}x_{j:n} + \frac{1}{n}\sum_{i=j+1}^{k-1} x_{i:n} + \frac{n+1-k}{n}x_{k:n} - \bar{x} \le \sqrt{\frac{j-1}{n+1-j}}\, s, \tag{3.106}$$

$1 \le j < k \le n$, are identical with those for the jth order statistics (3.103) and are attained under the same conditions (3.104). On the other hand, inequalities (3.103), (3.105) and (3.106) can be obtained without referring to the projection methods. For instance, we get

$$x_{j:n} - \bar{x} \le \frac{1}{n+1-j}\sum_{i=j}^{n} x_{i:n} - \bar{x} \le \sqrt{\frac{j-1}{n+1-j}}\, s \tag{3.107}$$

by the direct application of the Schwarz inequality to the upper selection differential (see (3.57)). Condition (3.58) for an equality in the second inequality of (3.107) implies the equality in the first one. Similarly, we obtain (3.105) and (3.106) by majorizing the trimmed and Winsorized means

$$\frac{1}{k+1-j}\sum_{i=j}^{n} x_{i:n} \le \frac{1}{n+1-j}\sum_{i=j}^{n} x_{i:n}, \tag{3.108}$$

$$\begin{aligned} \frac{j}{n}x_{j:n} + \frac{1}{n}\sum_{i=j+1}^{k-1} x_{i:n} + \frac{n+1-k}{n}x_{k:n} &\le \frac{j}{n}x_{j:n} + \frac{1}{n}\sum_{i=j+1}^{n} x_{i:n} \\ &\le \frac{1}{n+1-j}\sum_{i=j}^{n} x_{i:n} \end{aligned} \tag{3.109}$$

and using (3.57). Equalities in (3.108) and (3.109) are attained iff $x_{j:n} = \ldots = x_{n:n}$ which is guaranteed by (3.58).

Observe that bound (3.103) and attainability conditions (3.104) with $j = \lfloor \frac{n+1}{2} \rfloor$ hold for the sample median (3.13) for both the even and odd sample size cases. It is left to the reader to check that the results extend to other central quasi-midranges (3.15) under the restriction that $\frac{n+1}{3} \leq j \leq \frac{n}{2}$. Otherwise we have

$$m_{j,n+1-j:n} - \bar{x} = \frac{1}{2}x_{j:n} + \frac{1}{2}x_{n+1-j:n} - \bar{x} \leq Bs, \tag{3.110}$$

with

$$B = \left[\frac{n(n+1-j)}{4j(n+1-2j)} - 1\right]^{1/2}, \tag{3.111}$$

and the bound is attained by

$$x_{i:n} = \begin{cases} \bar{x} - \frac{s}{B}, & i = 1, \ldots, j-1, \\ \bar{x} + \frac{s}{B}\left(\frac{n}{2(n+1-j)} - 1\right), & i = j, \ldots, n-j, \\ \bar{x} + \frac{s}{B}\left(\frac{n}{2j} - 1\right), & i = n+1-j, \ldots, n. \end{cases} \tag{3.112}$$

For the standard midrange (3.14), formulae (3.110) and (3.112) simplify to

$$m_{1,n:n} - \bar{x} = \frac{1}{2}(x_{1:n} + x_{n:n}) - \bar{x} \leq \frac{n-2}{2\sqrt{n-1}}\, s \tag{3.113}$$

and (3.39), respectively.

Applying Theorem 3.4 to the differences of order statistics yields

Corollary 3.2 (Fahmy and Proschan (1981)) *For arbitrary $\bar{x} \in \mathbb{R}^n$ and $1 \leq j < k \leq n$, we have*

$$x_{k:n} - x_{j:n} \leq \sqrt{\frac{n(n+1-k+j)}{j(n+1-k)}}\, s, \tag{3.114}$$

which becomes equality iff

$$x_{i:n} = \begin{cases} \bar{x} - s\sqrt{\frac{(n+1-k)n}{j(n+1-k+j)}}, & i = 1, \ldots, j, \\ \bar{x}, & i = j+1, \ldots, k-1, \\ \bar{x} + s\sqrt{\frac{jn}{(n+1-k)(n+1-k+j)}}, & i = k, \ldots, n. \end{cases} \tag{3.115}$$

In the special cases of spacings and quasi-ranges we have

$$s_{j,n} = x_{j+1:n} - x_{j:n} \le \frac{n}{\sqrt{j(n-j)}}\, s, \tag{3.116}$$

$$r_{j,n+1-j:n} = x_{n+1-j:n} - x_{j:n} \le \sqrt{\frac{2n}{j}}\, s, \tag{3.117}$$

respectively, and (3.54) for the standard range in particular. Relation (3.114) can also be derived by writing

$$x_{k:n} - x_{j:n} \le \frac{1}{n+1-k} \sum_{i=k}^{n} x_{i:n} - \frac{1}{j} \sum_{i=1}^{j} x_{i:n} = \bar{\underline{s}}_{j,k:n} \tag{3.118}$$

and applying the Schwarz inequality to the generalized range (3.28) (comp. (3.107)–(3.109)). This is in fact an implicit reference to the projection method and can be used for arbitrary L-statistics. It suffices to replace properly chosen subsums of the linear combination by majorizing products of averaged subsums of respective subsequences. The original method of Fahmy and Proschan (1981) consisted in minimizing the sample variance s^2 with respect to $x_{1:n} \le \cdots \le x_{n:n}$.

3.3.2 Lower Bounds

We now discuss the optimal lower deterministic Moriguti-Schwarz bounds on L-statistics. To this end, the projection method of Theorem 3.4 is applicable as well. Indeed, by

$$\inf_{\tilde{x}\in\mathbb{R}^n} L_{\tilde{c}}\left(\frac{x_1-\bar{x}}{s}, \ldots, \frac{x_n-\bar{x}}{s}\right) = \inf_{\tilde{x}\in\mathbb{R}^n} \sum_{i=1}^{n} c_i \frac{x_{i:n}-\bar{x}}{s}$$
$$= -\sup_{\tilde{x}\in\mathbb{R}^n} \sum_{i=1}^{n} (-c_i) \frac{x_{i:n}-\bar{x}}{s} = -\sup_{\tilde{x}\in\mathbb{R}^n} L_{-\tilde{c}}\left(\frac{x_1-\bar{x}}{s}, \ldots, \frac{x_n-\bar{x}}{s}\right), \tag{3.119}$$

we conclude that the lower bound is just the negative of the upper one for the L-statistic with the coefficients being the negatives of the original ones. The method consists in the use of projection procedure (2.78) and (2.79) for $-\tilde{c} = (-c_1, \ldots, -c_n)$ and calculating respective bounds (3.94) and conditions (3.96) for the equality. Clearly, this is the negative of the projection $\bar{\bar{c}} = P^{\searrow}\tilde{c}$ of the original $\tilde{c}$ onto the convex cone $C^{\searrow}$ of nonincreasing sequences in $\mathbb{R}^n$, described in Theorem 2.17.

Theorem 3.5 *For arbitrary* $\tilde{c} \in \mathbb{R}^n$ *define* $\bar{\tilde{c}} = P^{\searrow}\tilde{c} \in \mathbb{R}^n$ *by use of (2.104) and (2.105). Then for every* $\tilde{x} \in \mathbb{R}^n$ *yields*

$$\sum_{i=1}^{n} c_i(x_{i:n} - \bar{x}) \geq -\left(n\sum_{i=1}^{n} \bar{c}_i^2 - n^2\bar{c}^2\right)^{1/2} s. \tag{3.120}$$

Moreover, if

$$\max_{1\leq j\leq n-1} \frac{1}{j}\sum_{i=1}^{j} c_i > \frac{1}{n}\sum_{i=1}^{n} c_i = \bar{c}, \tag{3.121}$$

then inequality (3.120) is sharp and becomes equality iff

$$x_{i:n} = \bar{x} - s\frac{\bar{c}_i - \bar{c}}{(\frac{1}{n}\sum_{i=1}^{n} \bar{c}_i^2 - \bar{c}^2)^{1/2}}, \quad i = 1, \ldots, n. \tag{3.122}$$

The proof is similar to that of Theorem 3.4. If (3.121) holds, then $\bar{c}_1 > \bar{c}_n$, bound (3.120) is strictly negative and points (3.122) are well defined.

Some of lower bounds may be concluded from the results presented earlier without calculating the projections $\bar{\tilde{c}} = P^{\searrow}\tilde{c}$ by use of the following arguments. Define a one-to-one transformation of $\mathbb{R}^n$ onto itself by

$$y_i = 2\bar{x} - x_i, \quad i = 1, \ldots, n. \tag{3.123}$$

Then $y_{i:n} = 2\bar{x} - x_{n+1-i:n}$, $i = 1, \ldots, n$, $\bar{y} = \bar{x}$, and

$$s^2(\tilde{y}) = \frac{1}{n}\sum_{i=1}^{n}(y_i - \bar{y})^2 = \frac{1}{n}\sum_{i=1}^{n}(x_i - \bar{x})^2 = s^2(\tilde{x}) = s^2 \tag{3.124}$$

with

$$\begin{aligned}\inf_{\tilde{x}\in\mathbb{R}^n} \sum_{i=1}^{n} c_i\frac{x_{i:n} - \bar{x}}{s} &= -\sup_{\tilde{x}\in\mathbb{R}^n} \sum_{i=1}^{n} c_i\frac{\bar{x} - x_{i:n}}{s} \\ &= -\sup_{\tilde{y}\in\mathbb{R}^n} \sum_{i=1}^{n} c_i\frac{y_{n+1-i:n} - \bar{y}}{s} = -\sup_{\tilde{y}\in\mathbb{R}^n} \sum_{i=1}^{n} c_{n+1-i}\frac{y_{i:n} - \bar{y}}{s}.\end{aligned} \tag{3.125}$$

The conclusion is also obvious by the following argument. If we take the reverse $\tilde{c}' = (c_n, \ldots, c_1)$ of a given sequence $\tilde{c} = (c_1, \ldots, c_n)$, the reverse $(P^{\nearrow}\tilde{c}')'$ of the projection $P^{\nearrow}\tilde{c}'$ of $\tilde{c}'$ onto the convex cone $C^{\nearrow}$ of nondecreasing sequences determines the projection $P^{\searrow}\tilde{c}$ of the original $\tilde{c}$ onto the convex cone $C^{\searrow}$ of nonincreasing vectors. Therefore, due to (3.103) and (3.104) combined with (3.125),

$$\inf_{\tilde{x}\in\mathbb{R}^n} \frac{x_{j:n} - \bar{x}}{s} = - \sup_{\tilde{y}\in\mathbb{R}^n} \frac{y_{n+1-j:n} - \bar{y}}{s} = -\sqrt{\frac{n-j}{j}}, \tag{3.126}$$

$j = 1, \ldots, n-1$, holds and it is attained by

$$x_{i:n} = 2\bar{x} - y_{n+1-i:n} = \begin{cases} \bar{x} - s\sqrt{\frac{n-j}{n}}, & i = 1, \ldots, j, \\ \bar{x} + s\sqrt{\frac{j}{n-j}}, & i = j+1, \ldots, n. \end{cases} \tag{3.127}$$

Similarly, the lower bounds for the trimmed means

$$t_{j,k:n} - \bar{x} = \frac{1}{k+1-j} \sum_{i=j}^{k} (x_{i:n} - \bar{x}) \geq -\sqrt{\frac{n-k}{k}}\, s, \tag{3.128}$$

$1 \leq j < k \leq n$, are optimal and attained by (3.127) with j replaced by k. The same holds for the respective Winsorized means (3.10). Moreover, we deduce from (3.125) that for L-statistics with symmetric coefficients $c_i = c_{n+1-i}$, $i = 1, \ldots, n$, the lower and upper Moriguti-Schwarz bounds are symmetric about zero. In particular, these are the cases of the sample medians (both for the odd and even sample sizes), quasi midranges and symmetric trimmed and Winsorized means.

Now we concentrate on the cases for which

$$\bar{c} = \max_{1\leq j\leq n} \frac{1}{j} \sum_{i=1}^{j} c_i \tag{3.129}$$

and hence $\bar{c}_1 = \bar{c}_n = \bar{c}$ hold true. This happens for all L-statistics with nondecreasing coefficients, but also for other ones including the differences of order statistics. Then $P^{\searrow}\tilde{c} = \bar{c}\tilde{1}$ and the Moriguti-Schwarz inequality provides the trivial zero lower bounds, which hold true for the sample maximum, range and generalized ranges, quasi ranges, spacings, upper selection differentials, median absolute deviation and Gini mean differences in particular. The zero bound is attained if $s = 0$, but otherwise is not necessarily optimal.

Once we have sharp lower and upper bounds on L-estimates of scale is standard deviation units, we are able to deduce sharp evaluations of the standard deviation in terms of various scale L-estimates by means of reciprocals of the original bounds. Firstly, by (3.57) and (3.293), we have

$$\sqrt{\frac{k}{n-k}}\, \bar{s}_{k:n} \leq s \leq \max\left\{1, \frac{k}{n-k}\right\} \sqrt{n-1}\bar{s}_{k:n} \tag{3.130}$$

and for $k = 1$

$$\frac{1}{\sqrt{n-1}}(x_{n:n} - \bar{x}) \le s \le \sqrt{n-1}(x_{n:n} - \bar{x}) \tag{3.131}$$

in particular. Furthermore, due to (3.59) and (3.307), we obtain

$$\sqrt{\frac{jk}{n(j+k)}}\,\bar{\underline{s}}_{j,k:n} \le s \le \frac{1}{n}\max\left\{\sqrt{n-1}\max\{j,k\}, \sqrt{\left\lfloor\frac{n}{2}\right\rfloor\left(n-\left\lfloor\frac{n}{2}\right\rfloor\right)}\right\}\bar{\underline{s}}_{j,k:n} \tag{3.132}$$

and particularly for $j = k = 1$

$$\frac{1}{\sqrt{2n}}r_n \le s \le \frac{1}{n}\sqrt{\left\lfloor\frac{n}{2}\right\rfloor\left(n-\left\lfloor\frac{n}{2}\right\rfloor\right)}\,r_n. \tag{3.133}$$

These inequalities were especially useful before the computer era, because r_n and $\bar{\underline{s}}_{j,k:n}$ for small j and k require by far less calculations than s. It is impossible, however, to evaluate the standard deviation by the differences of trimmed means (or order statistics) which do not include both the extremes $x_{1:n}$ and $x_{n:n}$. The reason is that s depends on them whereas the L-estimate does not. Hence, by increasing the absolute value of the extreme, we can arbitrarily increase the sample variance keeping the L-estimate value fixed.

3.3.3 Nonnegative Samples

Now we proceed to the nonnegative samples. Since

$$\sum_{i=1}^{n} c_i x_{i:n} \le \sum_{i=1}^{n} \underline{c}_i x_{i:n} \le \sum_{i=1}^{n} (\underline{c}_i)_+ x_{i:n}, \tag{3.134}$$

we are in a position to formulate

Theorem 3.6 *For all $\tilde{c} \in \mathbb{R}^n$ and $\tilde{x} \in \mathbb{R}^n_+$*

$$\sum_{i=1}^{n} c_i x_{i:n} \le \left[n\sum_{i=1}^{n}(\underline{c}_i)_+^2\right]^{1/2} m_2 \tag{3.135}$$

which becomes an equality iff

$$x_{i:n} = m_2 \frac{n^{1/2}(c_i)_+}{[\sum_{i=1}^n (c_i)_+^2]^{1/2}}, \quad i = 1, \ldots, n, \tag{3.136}$$

provided that $\underline{\tilde{c}}_+ \neq \tilde{0}$.

In view of Theorems 3.3 and 3.4 the proof is simple.

The remarks at the end of the previous section on the mutual relations between bounds in s_2 and m_2 units are here valid as well, because the projection operator $\tilde{c} \to \underline{\tilde{c}}$ preserves nonnegativity and orthogonality to constants of original sequences. Below we show several implications of Theorem 3.6. Firstly, the bounds

$$x_{j:n} - \bar{x} \leq \sqrt{\frac{n}{n+1-j}}\, m_2 \tag{3.137}$$

are attained for all $j \leq n$ including $j = 1$ by

$$x_{i:n} = \begin{cases} 0, & \text{if } i = 1, \ldots, j-1, \\ \sqrt{\frac{n}{n+1-j}}\, m_2, & \text{if } i = j, \ldots, n. \end{cases} \tag{3.138}$$

Substituting k for j in (3.137) and (3.138), we obtain analogous results for the differences $r_{j,k:n} = x_{k:n} - x_{j:n}$, $1 \leq j < k \leq n$. For the midranges, the inequalities

$$m_{j,n+1-j:n} = \frac{1}{2}(x_{j:n} + x_{n+1-j:n}) \leq \begin{cases} \frac{1}{2}\sqrt{\frac{n(n+1-j)}{j(n+1-2j)}}\, m_2, & \text{if } j < \frac{n+1}{3}, \\ \sqrt{\frac{n}{n+1-j}}\, m_2, & \text{if } \frac{n+1}{3} \leq j \leq \frac{n}{2}, \end{cases}$$

become equalities if

$$x_{i:n} = \begin{cases} 0, & i = 1, \ldots, j-1, \\ \sqrt{\frac{nj}{(n+1-2j)(n+1-j)}}\, m_2, & i = j, \ldots, n-j, \\ \sqrt{\frac{n(n+1-2j)}{j(n+1-j)}}\, m_2, & i = n+1-j, \ldots, n, \end{cases} \tag{3.139}$$

and (3.138) hold, respectively.

In order to get an analogue of (3.90), we need to make a more significant modification. Given $\tilde{c} = (c_1, \ldots, c_n) \in \mathbb{R}^n$ define first $\tilde{c}' = (c_2, \ldots, c_n) \in \mathbb{R}^{n-1}$, then the projection $P^{\nearrow}\tilde{c}' = \underline{\tilde{c}}' = (\underline{c}_2', \ldots, \underline{c}_n')$ of $\tilde{c}'$ onto nondecreasing vectors in $\mathbb{R}^{n-1}$ and finally the positive part of $\underline{\tilde{c}}_+' = ((\underline{c}_2')_+, \ldots, (\underline{c}_n')_+)$, of the projection. Then

$$\sum_{i=1}^n c_i(x_{i:n} - x_{1:n}) \leq \left[n \sum_{i=2}^n (\underline{c}_i')_+^2\right]^{1/2} m_2(x_{1:n}) \tag{3.140}$$

and, when $\underline{\tilde{c}}_+ \neq \tilde{0}$, the equality conditions are

$$x_{i:n} = x_{1:n} + m_2(x_{1:n})n^{1/2} \frac{(\underline{c}'_i)_+}{[\sum_{i=2}^n (\underline{c}'_i)_+^2]^{1/2}} \quad i = 2, \ldots, n. \tag{3.141}$$

For completeness, we write down respective lower bounds

$$\sum_{i=1}^n c_i x_{i:n} \geq - \left[n \sum_{i=1}^n (\underline{-c_i})_+^2 \right]^{1/2} m_2, \quad x \in \mathbb{R}_+^n \tag{3.142}$$

$$\sum_{i=1}^n c_i (x_{i:n} - x_{1:n}) \geq - \left[n \sum_{i=2}^n (\underline{-c_i}')_+^2 \right]^{1/2} m_2(x_{1:n}). \tag{3.143}$$

Here $\underline{-c}_i$ and $\underline{-c}'_i$ stand for the coordinates of projections of the vectors $-\tilde{c} = (-c_1, \ldots, -c_n)$ and $-\tilde{c}' = (-c_2, \ldots, -c_n)$ onto the classes of nondecreasing sequences. The lower bounds are rarely used in practice, because they vanish for many important examples. Indeed, for nondecreasing L-estimates of location (see (3.4) and (3.7)) with $-\tilde{c} \in \mathbb{R}_-^n$ yields $\underline{-\tilde{c}} \in \mathbb{R}_-^n$ and $\underline{-\tilde{c}}_+ = \tilde{0}$. For nonnegative L-estimates of scale, we have

$$0 = \sum_{i=1}^n (-c_i) \leq \sum_{i=1}^j (-c_i) \quad j = 1, \ldots, n-1, \tag{3.144}$$

(cf., (3.20)), and in consequence $\underline{-\tilde{c}}_+ = \underline{-\tilde{c}} = \tilde{0}$.

3.4 Moriguti-Hölder Bounds

Here we extend the deterministic bounds on L-statistics of the previous section by replacing the sample standard deviations (3.94) by the scale units generated by other sample absolute central moments. We consider

$$s_p = \left(\frac{1}{n} \sum_{i=1}^n |x_i - \bar{x}|^p \right)^{1/p}, \quad 1 \leq p < +\infty, \tag{3.145}$$

and

$$s_\infty = \lim_{p \to +\infty} \left(\frac{1}{n} \sum_{i=1}^n |x_i - \bar{x}|^p \right)^{1/p} = \max\{|x_i - \bar{x}| : i = 1, \ldots, n\}$$
$$= \max\{x_{n:n} - \bar{x}, \bar{x} - x_{1:n}\}. \tag{3.146}$$

The bounds are based on the Hölder inequality. Actually, a direct application of the Hölder inequality provides the optimal bounds for the L-statistics with nondecreasing coefficients (cf. Theorem 3.2 for special case $p = 2$). Otherwise the inequality should be combined with the Moriguti inequality (cf. Theorem 3.4) which is the crucial tool in numerous evaluations of nondecreasing sequences presented here. We first describe the bounds expressed in terms of (3.145) with $p > 1$.

3.4.1 General Samples

Theorem 3.7 *For given $\tilde{c} \in \mathbb{R}^n$, $1 < p < \infty$, and $q = p/(p-1)$, define $\underline{\tilde{c}} = P^{\nearrow}\tilde{c}$ by (2.78) and (2.79), and let $\underline{c}_1 \le \bar{\underline{c}}_q \le \underline{c}_n$ be the unique solution to*

$$\sum_{i=1}^{n} |\underline{c}_i - c|^{q-1} \mathrm{sgn}\{\underline{c}_i - c\} = 0. \tag{3.147}$$

Then for arbitrary $\tilde{x} \in \mathbb{R}^n$ inequality

$$\sum_{i=1}^{n} c_i(x_{i:n} - \bar{x}) \le n^{1/p} \left(\sum_{i=1}^{n} |\underline{c}_i - \bar{\underline{c}}_q|^q \right)^{1/q} s_p \tag{3.148}$$

holds. Moreover, if

$$\min_{1 \le j \le n-1} \frac{1}{j} \sum_{i=1}^{j} c_i < \frac{1}{n} \sum_{i=1}^{n} c_i, \tag{3.149}$$

then (3.148) is positive and sharp, and becomes equality iff

$$x_{i:n} = \bar{x} + s_p n^{1/p} \frac{|\underline{c}_i - \bar{\underline{c}}_q|^{q/p} \mathrm{sgn}\{\underline{c}_i - \bar{\underline{c}}_q\}}{(\sum_{i=1}^{n} |\underline{c}_i - \bar{\underline{c}}_q|^q)^{1/p}}, \quad i = 1, \ldots, n. \tag{3.150}$$

Proof For arbitrary $c \in \mathbb{R}$, by the Moriguti and Hölder inequalities yields

$$\begin{aligned} \sum_{i=1}^{n} c_i(x_{i:n} - \bar{x}) &= \sum_{i=1}^{n} (c_i - c)(x_{i:n} - \bar{x}) \le \sum_{i=1}^{n} (\underline{c}_i - c)(x_{i:n} - \bar{x}) \\ &\le n^{1/p} \left(\sum_{i=1}^{n} |\underline{c}_i - c|^q \right)^{1/q} \left(\frac{1}{n} \sum_{i=1}^{n} |x_{i:n} - \bar{x}|^p \right)^{1/p}. \end{aligned} \tag{3.151}$$

Function

$$c \mapsto \sum_{i=1}^{n} |\underline{c}_i - c|^q \tag{3.152}$$

is strictly convex and attains its unique minimum at the point satisfying (3.147) which is the unique zero of its derivative. Therefore the bound with $c = \bar{\underline{c}}_q$ is the best possible among bounds (3.151). This is zero iff $\underline{c}_1 = \underline{c}_n$ and strictly positive otherwise that means if (3.149) holds. In the latter case, the second inequality in (3.151) becomes equality iff $x_{i:n} - x$ are proportional to $|\underline{c}_i - \bar{\underline{c}}_q|^{q/p}\mathrm{sgn}\{\underline{c}_i - \bar{\underline{c}}_q\}$, $i = 1, \ldots, n$ (see (2.167)). Note that (3.150) sum up to $n\bar{x}$ by (3.147) and the proportionality factor

$$\alpha = s_p \left(\frac{1}{n} \sum_{i=1}^{n} |\underline{c}_i - \bar{\underline{c}}_q|^q \right)^{-1/p} \tag{3.153}$$

is chosen so that the pth sample absolute central moment condition is satisfied. Finally, we observe that $\underline{c}_i = \underline{c}_j$ implies $x_{i:n} = x_{j:n}$ in (3.150) which is the sufficient condition for the equality in the first Moriguti inequality of (3.151). This ends the proof. □

In the special case of single order statistics we have

Corollary 3.3 *For given $1 \le j \le n$ and $1 < p < +\infty$ and arbitrary $\tilde{x} \in \mathbb{R}^n$ we have*

$$x_{j:n} - \bar{x} \le \left\{ \frac{n(j-1)^{p-1}}{(n+1-j)[(n+1-j)^{p-1} + (j-1)^{p-1}]} \right\}^{1/p} s_p \tag{3.154}$$

and this becomes equality for $2 \le j \le n$ iff

$$x_{i:n} = \begin{cases} \bar{x} - s_p \left\{ \frac{n(n+1-j)^{p-1}}{(j-1)[(n+1-j)^{p-1} + (j-1)^{p-1}]} \right\}^{1/p} & i = 1, \ldots, j-1, \\ \bar{x} + s_p \left\{ \frac{n(j-1)^{p-1}}{(n+1-j)[(n+1-j)^{p-1} + (j-1)^{p-1}]} \right\}^{1/p} & i = j, \ldots, n. \end{cases} \tag{3.155}$$

Proof Recall that in the case of jth order statistic

$$\underline{c}_i = \begin{cases} 0, & i = 1, \ldots, j-1, \\ \frac{1}{n+1-j}, & i = j, \ldots, n, \end{cases} \tag{3.156}$$

which are identical with $\underline{c}_i = \bar{c} = \frac{1}{n}$ in the case $j = 1$. This implies the zero bound by Theorem 3.7 then. For $2 \le j \le n$ Eq. (3.147) with (3.156) has the unique solution

$$\underline{\bar{c}}_q = \frac{(n+1-j)^{p-2}}{(j-1)^{p-1} + (n+1-j)^{p-1}} \tag{3.157}$$

(which also amounts to $\bar{c} = \frac{1}{n}$ in the case $p = 2$.) Accordingly, the bound of (3.148) takes on the form

$$n^{1/p} \left(\sum_{i=1}^{n} |\underline{c}_i - \underline{\bar{c}}_q|^q \right)^{1/q}$$

$$= n^{1/p} \left[(j-1)\underline{\bar{c}}_q^q + (n+1-j) \left(\frac{1}{n+1-j} - \underline{\bar{c}}_q \right)^q \right]^{1/q}$$

which after some calculations simplifies to the RHS of (3.154). Combining (3.150) with (3.156) and (3.157), we obtain

$$x_{i:n} = \begin{cases} \bar{x} - s_p \frac{n^{1/p} \underline{\bar{c}}_q^{q/p}}{[(j-1)\underline{\bar{c}}_q^q + (n+1-j)(\frac{1}{n+1-j} - \underline{\bar{c}}_q)^q]^{1/p}}, & i = 1, \ldots, j-1, \\ \bar{x} + s_p \frac{n^{1/p} (\frac{1}{n+1-j} - \underline{\bar{c}}_q)^{q/p}}{[(j-1)\underline{\bar{c}}_q^q + (n+1-j)(\frac{1}{n+1-j} - \underline{\bar{c}}_q)^q]^{1/p}}, & i = j, \ldots, n. \end{cases} \tag{3.158}$$

Using a standard algebra, we transform (3.158) into (3.155). This completes the proof. □

The results of Corollary 3.3 were established by Beesack (1973) in a different way.

Bounds (3.154) are also the best ones for the trimmed and Winsorized means with $x_{j:n}$ being the smallest value included in the means, because (3.156) are the projection coefficients then as well. For L-statistics with $\underline{c}_1 = \underline{c}_n$ bounds (3.148) are zero. If $\underline{c}_i$, $i = 1, \ldots, n$, take on two values, then the constants $\underline{\bar{c}}_q$ defined implicitly in (3.147) can be determined analytically. Except for the Winsorized and trimmed means (and single order statistics in particular) this also holds for the central quasi midranges (3.15) with $j \ge \frac{n+1}{3}$, Gastwirth means (3.17) and spacings (3.25). Otherwise equations (3.147) have to be solved numerically. However, explicit Hölder bounds for arbitrary L-statistics can be established in the extreme cases $p = 1$ and $p = +\infty$. Now we examine the first one.

Theorem 3.8 *For a given $\bar{c} \in \mathbb{R}^n$, under the notation of (2.78) and (2.79), we have*

$$\sum_{i=1}^{n} c_i (x_{i:n} - \bar{x}) \le \frac{n}{2} (\underline{c}_n - \underline{c}_1) s_1 \tag{3.159}$$

for arbitrary $\bar{x} \in \mathbb{R}^n$*. If (3.149) holds, then the bound is sharp and becomes equality for*

$$x_{i:n} = \begin{cases} \bar{x} - s_1 \frac{n}{2j_1}, & i = 1, \ldots, j_1, \\ \bar{x}, & i = j_1 + 1, \ldots, j_{M-1}, \\ \bar{x} + s_1 \frac{n}{2(n - j_{M-1})}, & i = j_{M-1} + 1, \ldots, n. \end{cases} \tag{3.160}$$

Proof We have

$$\sum_{i=1}^{n} c_i(x_{i:n} - \bar{x}) \le \sum_{i=1}^{n} \underline{c}_i(x_{i:n} - \bar{x}) = \sum_{i=1}^{n} \left(\underline{c}_i - \frac{\underline{c}_1 + \underline{c}_n}{2} \right)(x_{i:n} - \bar{x})$$

$$\le \max_{1 \le i \le n} \left| \underline{c}_i - \frac{\underline{c}_1 + \underline{c}_n}{2} \right| \sum_{i=1}^{n} |x_{i:n} - \bar{x}| = n \frac{\underline{c}_n - \underline{c}_1}{2} s_1. \tag{3.161}$$

Under (3.149) $\underline{c}_1 < \underline{c}_n$ and $M \ge 2$ and (3.160) are well defined. It is easy to check that they obey both the moment conditions and provide the equality in the second inequality of (3.161). They also satisfy condition (2.81) that is sufficient for the equality in the former inequality. □

Conditions (3.160) are necessary and sufficient when $\underline{c}_{j_1} < \underline{c}_{j_2}$ and $\underline{c}_{j_{M-1}} < \underline{c}_{j_M}$. If $\underline{c}_{j_1} = \underline{c}_{j_k}$ for some $k > 1$ and/or $\underline{c}_{j_\ell} = \underline{c}_{j_M}$ for some $\ell < M$, one can also take $x_{j_k:n}$ and/or $x_{j_\ell:n}$ different from $\bar{x}$ provided that monotonicity, moment and (2.81) conditions are preserved.

In the special cases of single order statistics (together with respective trimmed and Winsorized means) and differences of order statistics, the following statements hold.

Corollary 3.4 *For* $1 < j \le n$ *and* $\bar{x} \in \mathbb{R}^n$ *inequality*

$$x_{j:n} - \bar{x} \le \frac{n}{2(n+1-j)} s_1 \tag{3.162}$$

is tight and becomes equality for

$$x_{i:n} = \begin{cases} \bar{x} - s_1 \frac{n}{2(j-1)}, & i = 1, \ldots, j-1, \\ \bar{x} + s_1 \frac{n}{2(n+1-j)}, & i = j, \ldots, n. \end{cases} \tag{3.163}$$

Corollary 3.5 *For* $1 \le j < k \le n$ *and* $\bar{x} \in \mathbb{R}^n$ *we have*

$$x_{k:n} - x_{j:n} \le \frac{n(n+1-k+j)}{2j(n+1-k)} s_1 \tag{3.164}$$

which is attained by

$$x_{i:n} = \begin{cases} \bar{x} - s_1 \frac{n}{2j}, & i = 1, \ldots, j, \\ \bar{x}, & i = j+1, \ldots, k-1, \\ \bar{x} + s_1 \frac{n}{2(n-1-k)}, & i = k, \ldots, n. \end{cases} \tag{3.165}$$

In particular, we get

$$x_{n:n} - \bar{x} \leq \frac{n}{2} s_1, \tag{3.166}$$

$$x_{n:n} - x_{1:n} \leq n s_1, \tag{3.167}$$

$$x_{n+1-j:n} - x_{j:n} \leq \frac{n}{j} s_1, \tag{3.168}$$

$$x_{j+1:n} - x_{j:n} \leq \frac{n^2}{2j(n-j)} s_1. \tag{3.169}$$

One can also easily verify the following tight deterministic bounds for the sample median

$$\mathrm{MED}_n(\tilde{x}) - \bar{x} \leq \begin{cases} \frac{n}{n+1} s_1, \text{ if } n \text{ is odd}, \\ \frac{n}{n+2} s_1, \text{ if } n \text{ is even}, \end{cases} \tag{3.170}$$

quasi midranges

$$m_{j,n+1-j:n} - \bar{x} \leq \begin{cases} \frac{n}{4j} s_1, & 1 \leq j \leq \frac{n+1}{3}, \\ \frac{n}{2(n+1-j)} s_1, & \frac{n+1}{3} \leq j \leq \frac{n}{2}, \end{cases} \tag{3.171}$$

upper selection differentials

$$\bar{s}_{k:n} \leq \frac{n}{2k} s_1, \tag{3.172}$$

generalized ranges

$$\bar{\underline{s}}_{j,k:n} \leq \frac{n(j+k)}{2jk} s_1, \tag{3.173}$$

median absolute deviation

$$\mathrm{MAD}_n(\tilde{x}) \leq s_1, \tag{3.174}$$

and Gini mean differences

$$\mathrm{GMD}_n(\tilde{x}) \leq 2 s_1. \tag{3.175}$$

We recommend the reader to establish the samples which obtain the bounds (3.170) to (3.175). Now we proceed to the case $p = +\infty$.

Theorem 3.9 *For a given $\tilde{c} \in \mathbb{R}^n$ define $\underline{\tilde{c}} = P^{\nearrow}\tilde{c}$ by means of (2.78) and (2.79) and set*

$$\mathrm{MED}_n(\underline{\tilde{c}}) = \begin{cases} \underline{c}_{(n+1)/2}, & \text{if } n \text{ is odd}, \\ \frac{1}{2}(\underline{c}_{n/2} + \underline{c}_{n/2+1}), & \text{if } n \text{ is even}. \end{cases} \tag{3.176}$$

Then for arbitrary $\tilde{x} \in \mathbb{R}^n$ inequality

$$\sum_{i=1}^{n} c_i(x_{i:n} - \bar{x}) \le \sum_{i=1}^{n} |\underline{c}_i - \mathrm{MED}_n(\underline{\tilde{c}})| s_\infty \tag{3.177}$$

holds and is sharp under (3.149).

In this case, if $\mathrm{MED}_n(\underline{\tilde{c}}) = \underline{c}_{j_k}$ for some $1 \le k \le M$ (which always happens for odd sized samples), then the equality holds in (3.177) for

$$x_{i:n} = \begin{cases} \bar{x} - s_\infty, & i = 1, \ldots, j_{k-1}, \\ \bar{x} + \frac{j_k + j_{k-1} - n}{j_k - j_{k-1}} s_\infty, & i = j_{k-1} + 1, \ldots, j_k, \\ \bar{x} + s_\infty, & i = j_k + 1, \ldots, n. \end{cases} \tag{3.178}$$

Otherwise (3.177) becomes equality for

$$x_{i:n} = \begin{cases} \bar{x} - s_\infty, & i = 1, \ldots, n/2, \\ \bar{x} + s_\infty, & i = \frac{n}{2} + 1, \ldots, n. \end{cases} \tag{3.179}$$

Proof Using standard arguments, we obtain

$$\sum_{i=1}^{n} c_i(x_{i:n} - \bar{x}) \le \sum_{i=1}^{n} (\underline{c}_i - c)(x_{i:n} - \bar{x}) \le \sum_{i=1}^{n} |\underline{c}_i - c| s_\infty \tag{3.180}$$

for any real c. The bound is a piecewise linear convex function $B(c)$, say, with knots at c_{j_k}, $k = 1, \ldots, M$, which has the derivative

$$B'(c) = \#\{i : \underline{c}_i < c\} - \#\{i : \underline{c}_i > c\} \tag{3.181}$$

except for at the knots. Therefore this is minimized by (3.176) (if $\underline{c}_{n/2} < \underline{c}_{n/2+1}$ for odd sample sizes then the minimum is attained at all points of $[\underline{c}_{n/2}, \underline{c}_{n/2+1}]$). For the latter equality in (3.180) with $c = \mathrm{MED}_n(\underline{\tilde{c}})$, we need

$$x_{i:n} - \bar{x} = s_\infty \operatorname{sgn}\{\underline{c}_i - \mathrm{MED}_n(\underline{\tilde{c}})\} \qquad \text{iff } c_i \ne \mathrm{MED}_n(\underline{\tilde{c}}). \tag{3.182}$$

The former inequality becomes equality iff the $x_{i:n}$ are equal when the respective $\underline{c}_i$ are. If $\underline{c}_{n/2} < \underline{c}_{n/2+1}$, then (3.179) is sufficient (and necessary). Otherwise we also need $x_{j_{k-1}+1:n} = x_{j_k:n}$. The first moment condition combined with (3.182) imply that they amount to the value of the middle line of (3.178).

By definition, $2j_{k-1} \leq n \leq 2j_k$ which gives

$$-1 \leq \frac{j_k + j_{k-1} - n}{j_k - j_{k-1}} \leq 1 \tag{3.183}$$

and

$$|x_{i:n} - \bar{x}| \leq s_\infty, \quad i = j_{k-1}+1, \ldots, j_k. \tag{3.184}$$

This completes the proof. □

If $\mathrm{MED}_n(\underline{\tilde{c}}) = \underline{c}_{j_k}$ for some $1 \leq k \leq M$, and such k is unique, then (3.178) is also necessary for the equality in (3.177). Otherwise some different constructions are also possible. In the special case of single order statistics we conclude the following.

Corollary 3.6 *For arbitrary $\tilde{x} \in \mathbb{R}^n$ and $1 \leq j \leq n$ we have*

$$x_{j:n} - \bar{x} \leq \begin{cases} \frac{j-1}{n+1-j}\, s_\infty, & \text{if } j \leq \frac{n+1}{2}, \\ s_\infty, & \text{if } j \geq \frac{n+1}{2}. \end{cases} \tag{3.185}$$

If $2 \leq j \leq \frac{n+1}{2}$, then the equality holds for

$$x_{i:n} = \begin{cases} \bar{x} - s_\infty, & i = 1, \ldots, j-1, \\ \bar{x} + s_\infty \frac{j-1}{n+1-j}, & i = j, \ldots, n. \end{cases} \tag{3.186}$$

If $\frac{n+1}{2} \leq j \leq n$, then the equality holds for

$$x_{i:n} = \begin{cases} \bar{x} - s_\infty \frac{n-1-j}{j-1}, & i = 1, \ldots, j-1, \\ \bar{x} + s_\infty, & i = j, \ldots, n. \end{cases} \tag{3.187}$$

These are also the best bounds for properly chosen trimmed and Winsorized means. Setting $j = \lfloor \frac{n+1}{2} \rfloor$, we also obtain the optimal evaluations of the sample median. Then (3.185) transforms into

$$\mathrm{MED}_n(\tilde{x}) - \bar{x} \leq \begin{cases} \frac{n-1}{n+1}\, s_\infty, & \text{if } n \text{ is odd} \\ \frac{n-2}{n+2}\, s_\infty, & \text{if } n \text{ is even.} \end{cases} \tag{3.188}$$

Further we present specifications of Theorem 3.9 for differences of adjacent and nonadjacent order statistics.

Corollary 3.7 *For arbitrary $\tilde{x} \in \mathbb{R}^n$ and $1 \le j \le n-1$*

$$x_{j+1:n} - x_{j:n} \le \begin{cases} \frac{n}{n-j}\, s_\infty, & \text{if } j \le \frac{n}{2}, \\ \frac{n}{j}\, s_\infty, & \text{if } j \ge \frac{n}{2}. \end{cases} \tag{3.189}$$

The equalities are attained if

$$x_{i:n} = \begin{cases} \bar{x} - s_\infty, & i = 1, \dots, j, \\ \bar{x} + \frac{j}{n-j}\, s_\infty, & i = j+1, \dots, n, \end{cases} \tag{3.190}$$

and

$$x_{i:n} = \begin{cases} \bar{x} - \frac{n-j}{j}\, s_\infty, & i = 1, \dots, j, \\ \bar{x} + s_\infty, & i = j+1, \dots, n, \end{cases} \tag{3.191}$$

respectively.

In the case $j = \frac{n}{2}$, the bounds attain the maximal value $2s_\infty$ and formulae (3.190) and (3.191) coincide then.

Corollary 3.8 *If $\tilde{x} \in \mathbb{R}^n$ and $1 \le j < k-1 \le n-1$, then*

$$x_{k:n} - x_{j:n} \le \begin{cases} \frac{n}{n+1-k}\, s_\infty, & \text{if } k-1 \le \frac{n}{2}, \\ 2\, s_\infty, & \text{if } j+1 \le \frac{n}{2} \le k-2, \\ \frac{n}{j}\, s_\infty, & \text{if } \frac{n}{2} \le j. \end{cases} \tag{3.192}$$

The equalities in the consecutive inequalities are attained iff

$$x_{i:n} = \begin{cases} \bar{x} - s_\infty, & i = 1, \dots, k-1, \\ \bar{x} + s_\infty \frac{k-1}{n+1-k}, & i = k, \dots, n, \end{cases} \tag{3.193}$$

$$x_{i:n} = \begin{cases} \bar{x} - s_\infty, & i = 1, \dots, j, \\ \bar{x} + s_\infty \frac{k+j-n-1}{k-j-1}, & i = j+1, \dots, k-1, \\ \bar{x} + s_\infty, & i = k, \dots, n, \end{cases} \tag{3.194}$$

and

$$x_{i:n} = \begin{cases} \bar{x} - s_\infty \frac{n-j}{j}, & i = 1, \dots, j, \\ \bar{x} + s_\infty, & i = j+1, \dots, n, \end{cases} \tag{3.195}$$

respectively.

In particular, for the range and quasi ranges the middle bound of (3.192) is valid, and the respective extreme samples (3.194) are symmetric about $\bar{x}$. We also calculate the bounds in s_∞ units for the median absolute deviation

$$\mathrm{MAD}_n(\tilde{x}) \le \begin{cases} s_\infty, & \text{if } n \text{ is even}, \\ \frac{n-1}{n}\, s_\infty, & \text{if } n \text{ is odd}, \end{cases} \tag{3.196}$$

and the Gini mean differences

$$\mathrm{GMD}_n(\tilde{x}) \le \begin{cases} \frac{n}{n-1}\, s_\infty, & \text{if } n \text{ is even}, \\ \frac{n-1}{n}\, s_\infty, & \text{if } n \text{ is odd}. \end{cases} \tag{3.197}$$

Both these bounds are obtained by

$$x_{i:n} = \begin{cases} \bar{x} - s_\infty, & i = 1, \dots, \frac{n}{2}, \\ \bar{x} + s_\infty, & i = \frac{n}{2} + 1, \dots, n, \end{cases} \tag{3.198}$$

for the even samples and

$$x_{i:n} = \begin{cases} \bar{x} - s_\infty, & i = 1, \dots, \frac{n-1}{2}, \\ \bar{x}, & i = \frac{n+1}{2}, \\ \bar{x} + s_\infty, & i = \frac{n+3}{2}, \dots, n, \end{cases} \tag{3.199}$$

for the odd ones.

Combining arguments for Theorems 3.5 and 3.7 we are in a position to describe the sharp lower bounds for L-statistics in s_p units. Assume that $\tilde{c} \in \mathbb{R}^n$ is fixed and $P^{\searrow}\tilde{c} = \bar{\tilde{c}} = (\bar{c}_1, \dots, \bar{c}_n)$ is its projection onto the family of nonincreasing sequences in $\mathbb{R}^n$ constructed by means of (2.104) and (2.105).

Theorem 3.10 *Let $1 < p < \infty$ and $q = p/(p-1)$. Define $\bar{\bar{c}}_q \in [\bar{c}_n, \bar{c}_1]$ as the unique solution to*

$$\sum_{i=1}^{n} |\bar{c}_i - c|^{q-1} \mathrm{sgn}\{\bar{c}_i - c\} = 0. \tag{3.200}$$

Then for every $\tilde{x} \in \mathbb{R}^n$ we obtain the inequality

$$\sum_{i=1}^{n} c_i (x_{i:n} - \bar{x}) \ge -n^{1/p} \left(\sum_{i=1}^{n} |\bar{c}_i - \bar{\bar{c}}_q|^q \right)^{1/q} s_p \tag{3.201}$$

which under condition

$$\max_{1 \le j \le n-1} \frac{1}{j} \sum_{i=1}^{j} c_i > \frac{1}{n} \sum_{i=1}^{n} c_i = \bar{c} \tag{3.202}$$

is sharp and then becomes equality iff

$$x_{i:n} = \bar{x} - s_p n^{1/p} \frac{|\bar{c}_i - \bar{\bar{c}}_q|^{q/p} \mathrm{sgn}\{\bar{c}_i - \bar{\bar{c}}_q\}}{(\sum_{i=1}^{n} |\bar{c}_i - \bar{\bar{c}}_q|^q)^{1/p}}, \quad i = 1, \ldots, n. \tag{3.203}$$

We also state without proofs analogous results for $p = 1$ and $p = +\infty$.

Theorem 3.11 *For every $\tilde{x} \in \mathbb{R}^n$*

$$\sum_{i=1}^{n} c_i (x_{i:n} - \bar{x}) \geq -\frac{n}{2}(\bar{c}_1 - \bar{c}_n) s_1. \tag{3.204}$$

Under (3.202) inequality (3.204) is tight and becomes equality when conditions (3.160) are satisfied.

Notice that here integers $1 \leq j_1 \leq \cdots \leq j_M = n$, determined by (2.104) are different from those of (3.160) which were based on (2.78). The same reservation concerns the second statement of the next theorem.

Theorem 3.12 *The inequality*

$$\sum_{i=1}^{n} c_i (x_{i:n} - \bar{x}) \geq -\sum_{i=1}^{n} |\bar{c}_i - \mathrm{MED}_n(\bar{c})| s_\infty \tag{3.205}$$

holds for all $\tilde{x} \in \mathbb{R}^n$ and is tight when (3.202) is true. Then the equality conditions are (3.178) if $\mathrm{MED}_n(\bar{c}) = \bar{c}_{j_k}$ for some $1 \leq k \leq M$ and (3.179) otherwise.

In particular, for $1 \leq j \leq n-1$, we have

$$x_{j:n} - \bar{x} \geq -\left\{ \frac{n(n-j)^{p-1}}{j[j^{p-1} + (n-j)^{p-1}]} \right\}^{1/p} s_p \tag{3.206}$$

for $1 \leq p < +\infty$ and

$$x_{j:n} - \bar{x} \geq \begin{cases} -s_\infty, & \text{if } j \leq \frac{n}{2}, \\ -\frac{n-j}{j} s_\infty, & \text{if } j \geq \frac{n}{2}, \end{cases} \tag{3.207}$$

for $p = +\infty$. The equality conditions for the equalities in (3.206) and in consecutive inequalities in (3.207) are

$$x_{i:n} = \begin{cases} \bar{x} - s_p \left\{ \frac{n(n-j)^{p-1}}{j[j^{p-1}+(n-j)^{p-1}]} \right\}^{1/p}, & i = 1, \ldots, j, \\ \bar{x} + s_p \left\{ \frac{nj^{p-1}}{(n-j)[j^{p-1}+(n-j)^{p-1}]} \right\}^{1/p}, & i = j+1, \ldots, n, \end{cases} \tag{3.208}$$

$$x_{i:n} = \begin{cases} \bar{x} - s_\infty, & i = 1, \ldots, j, \\ \bar{x} + \frac{j}{n-j} s_\infty, & i = j+1, \ldots, n, \end{cases} \tag{3.209}$$

and

$$x_{i:n} = \begin{cases} \bar{x} - \frac{n-j}{j} s_\infty, & i = 1, \ldots, j, \\ \bar{x} + s_\infty, & i = j+1, \ldots, n, \end{cases} \tag{3.210}$$

respectively. Bounds (3.206) were established by Beesack (1973). It is easy to check that the lower bounds (3.206), and (3.207) are identical with the negatives of the corresponding upper bounds for the $(n+1-j)$th order statistics. They also hold true for the trimmed means $t_{k,j:n}$ and Winsorized means $w_{k,j:n}$ with $x_{j:n}$ being the largest order statistic included in the means. Clearly, the lower and upper bounds for symmetric trimmed and Winsorized means are symmetric about zero and so does occur for all midranges and other L-statistics with symmetric coefficients.

3.4.2 *Nonnegative Samples*

Now we confine ourselves to nonnegative samples. Using first the projection inequalities in the cases of cones of nondecreasing and nonnegative sequences, and then the Hölder inequality, we conclude

$$\begin{aligned} \sum_{i=1}^n c_i x_{i:n} &\le \sum_{i=1}^n \underline{c}_i x_{i:n} \le \sum_{i=1}^n (\underline{c}_i)_+ x_{i:n} \\ &\le n^{1/p} \left[\sum_{i=1}^n (\underline{c}_i)_+^q \right]^{1/q} m_p, \quad 1 < p < -\infty, \end{aligned} \tag{3.211}$$

where

$$m_p = \left(\frac{1}{n} \sum_{i=1}^n x_i^p \right)^{1/p} \tag{3.212}$$

denotes the pth root of the sample raw moment of degree p. Bound (3.211) is strictly positive and sharp iff $\underline{c}_n > 0$, or equivalently

$$\max_{1 \le j \le n} \frac{1}{j} \sum_{i=n+1-j}^n c_i > 0. \tag{3.213}$$

It is easy to verify that under (3.213) the following conditions for the equality in the last Hölder inequality of (3.211)

$$x_{i:n} = m_p n^{1/p} \frac{(\underline{c}_i)_+^{q/p}}{[\sum_{i=1}^n (\underline{c}_i)_+^q]^{1/q}}, \qquad i = 1, \ldots, n, \tag{3.214}$$

provide the equalities in the preceding ones. Equation (3.213) assures positivity and sharpness of the Moriguti-Hölder bounds in the remaining cases $p = 1$ and $p = +\infty$.

For $p = 1$ with $m_1 = \bar{x}$, inequality

$$\sum_{i=1}^n c_i x_{i:n} \le n(\underline{c}_n)_+ \bar{x} \tag{3.215}$$

is attained for

$$x_{i:n} = \begin{cases} 0, & i = 1, \ldots, j_{M-1}, \\ \frac{n}{n-j_{M-1}}\,\bar{x}, & i = j_{M-1}+1, \ldots, n. \end{cases} \tag{3.216}$$

For $p = +\infty$ with $m_\infty = x_{n:n}$, we get

$$\sum_{i=1}^n c_i x_{i:n} \le \sum_{i=1}^n (\underline{c}_i)_+ x_{n:n} \tag{3.217}$$

with the equality condition

$$x_{i:n} = \begin{cases} 0, & i = 1, \ldots, j_{k_*-1}, \\ x_{n:n}, & i = j_{k_*-1}+1, \ldots, n \end{cases} \tag{3.218}$$

where

$$k_* = \min\{1 \le k \le M : \underline{c}_{j_k} > 0\}. \tag{3.219}$$

In the special case of single order statistics, the results are simplified to

$$x_{j:n} \le \begin{cases} \left(\frac{n}{n+1-j}\right)^{1/p} m_p, & \text{if } 1 \le p < \infty, \\ x_{n:n}, & \text{if } p = +\infty. \end{cases} \tag{3.220}$$

Relations (3.220) become equalities iff $x_{i:n} = 0$ for $i = 1, \ldots, j-1$ and $x_{i:n}$ amount to the respective bounds for $i = j, \ldots, n$. Note that here the bounds on the sample minima are strictly positive although they are trivial and obtained by trivial identical samples $x_1 = \cdots = x_n$. As in all the previous cases, the results extend to the trimmed means $t_{j,k:n}$ and Winsorized means $w_{j,k:n}$. Analogous results for the differences $r_{j,k:n} = x_{k:n} - x_{j:n}$, $1 \leq j < k \leq n$ coincide with those for the single kth order statistic. The bounds on order statistics and differences in the mean units (case $p = 1$) were established by Groeneveld (1982).

A slight modification of (3.211) leads us to evaluations of L-statistics in the scale units generated by pth moments centered about the sample minimum

$$m_p(x_{1:n}) = \begin{cases} \left[\frac{1}{n}\sum_{i=1}^{n}(x_i - x_{1:n})^p\right]^{1/p}, & 1 \leq p < +\infty, \\ x_{n:n} - x_{1:n}, & p = +\infty. \end{cases} \tag{3.221}$$

Let $\underline{\tilde{c}}' = (\underline{c}_2', \ldots, \underline{c}_n') \in \mathbb{R}^{n-1}$ denote the ℓ^2-projection of $\tilde{c}' = (c_2, \ldots, c_n)$ onto $C^{\nearrow} \subset \mathbb{R}^{n-1}$. Then the inequalities

$$\sum_{i=1}^{n} c_i(x_{i:n} - x_{1:n}) \leq n(\underline{c}_n')_+ m_1(x_{1:n}), \tag{3.222}$$

$$\sum_{i=1}^{n} c_i(x_{i:n} - x_{1:n}) \leq n^{1/p}\left[\sum_{i=2}^{n}(\underline{c}_i')_+^q\right]^{1/q} m_p(x_{1:n}), \; 1 < p < \infty, \tag{3.223}$$

and

$$\sum_{i=1}^{n} c_i(x_{i:n} - x_{1:n}) \leq \sum_{i=2}^{n}(\underline{c}_i')_+ r_n, \tag{3.224}$$

true for arbitrary $\tilde{c}, \tilde{x} \in \mathbb{R}^n$, are positive and sharp when

$$\max_{1\leq j\leq n-1} \frac{1}{j} \sum_{i=n+1-j}^{n} c_i > 0. \tag{3.225}$$

The respective attainability conditions are straightforward modifications of (3.216), (3.214) and (3.218), respectively. Firstly, $x_{i:n}$, $i = 1, \ldots, n$ and m_p are replaced by $x_{i:n} - x_{1:n}$, $i = 2, \ldots, n$ and $m_p(x_{1:n})$. Secondly, the bounds are based on the nondecreasing projections of $\tilde{c}' = (c_2, \ldots, c_n)$ instead of the original $\tilde{c} = (c_1, \ldots, c_n)$. The most natural and interesting are the bounds (3.224) expressed in the range units. In many special cases of interest, including almost all the examples presented in Sect. 3.1, we have

$$((\underline{c}_1)_+, (\underline{c}_2)_+, \ldots, (\underline{c}_n)_+) = (0, (\underline{c}_2')_+, \ldots, (\underline{c}_n')_+). \tag{3.226}$$

Then the optimal constants in (3.222) to (3.224) coincide with those of (3.215), (3.211) and (3.217), respectively. An exemplary exception is the sample midrange with the bounds

$$m_{1,n:n} = \frac{1}{2}(x_{1:n} + x_{n:n}) \leq \begin{cases} \frac{n}{2}\,\bar{x}, & p = 1, \\ \left(\frac{n}{n-1}\right)^{1/p} \frac{[(n-1)^{q-1}+1]^{1/q}}{2}\, m_p, & 1 < p < \infty, \\ x_{n:n}, & p = +\infty, \end{cases}$$

for $\tilde{x} \in \mathbb{R}^n_+$ and

$$m_{1,n:n} - x_{1:n} = \frac{1}{2}(x_{1:n} - x_{n:n}) \leq \begin{cases} \frac{n}{2}^{1/p}\, m_p(x_{1:n}), & 1 \leq p < \infty, \\ \frac{1}{2}(x_{n:n} - x_{1:n}), & p = +\infty, \end{cases}$$

for arbitrary $\tilde{x} \in \mathbb{R}^n$.

Groeneveld (1982) estimated the extreme deviations of order statistics from the sample mean in the range units.

Theorem 3.13 *For every $\tilde{x} \in \mathbb{R}^n$ the following inequalities are sharp*

$$-\frac{n-1}{n}\, r_n \leq x_{1:n} - \bar{x} \leq -\frac{1}{n}\, r_n, \tag{3.227}$$

$$-\frac{n-j}{n}\, r_n \leq x_{j:n} - \bar{x} \leq \frac{j-1}{n}\, r_n, \quad 2 \leq j \leq n-1, \tag{3.228}$$

$$\frac{1}{n}\, r_n \leq x_{n:n} - \bar{x} \leq \frac{n-1}{n}\, r_n. \tag{3.229}$$

Moreover, let

$$x_{i:n} = \begin{cases} \bar{x} - \frac{k}{n}\, r_n, & i = 1, \ldots, k, \\ \bar{x} + \frac{n-k}{n}\, r_n, & i = k+1, \ldots, n, \end{cases} \tag{3.230}$$

for some $k = 1, \ldots, n-1$. Then the RHS bounds in (3.227) to (3.229) are attained by (3.230) with $k = 1, n+1-j$, and 1, respectively. The respective LHS bounds are attained by (3.230) for $k = n-1$, $n-j$ and $n-1$.

Proof We only prove the upper bounds. The lower ones can be derived in much the same way. To avoid trivialities we assume that $r_n > 0$ and define

$$y_i = \frac{x_i - x_{1:n}}{r_n}, \quad i = 1, \ldots, n. \tag{3.231}$$

The problem is to maximize

$$y_{j:n} - \bar{y} = \frac{n}{n-1}\, y_{j:n} - \frac{1}{n} \sum_{i \neq j} y_{i:n} \tag{3.232}$$

under the constraints

$$0 = y_{1:n} \le y_{2:n} \le \cdots \le y_{n-1:n} \le y_{n:n} = 1. \tag{3.233}$$

For $j = 1$, we have

$$y_{1:n} - \bar{y} = -\frac{1}{n}\left(\sum_{i=2}^{n-1} y_{i:n} + 1\right) \le -\frac{1}{n} \tag{3.234}$$

which is the equality for

$$0 = y_{1:n} = \cdots = y_{n-1:n} < y_{n:n} = 1. \tag{3.235}$$

For $j \ge 1$, we write

$$\begin{aligned} y_{j:n} - \bar{y} &= \frac{j-1}{n} y_{j:n} - \frac{1}{n}\left[\sum_{i=1}^{j-1} y_{i:n} + \sum_{i=j+1}^{n-1} (y_{i:n} - y_{j:n}) + (1 - y_{j:n})\right] \\ &\le \frac{j-1}{n} y_{j:n} \le \frac{j-1}{n} \end{aligned} \tag{3.236}$$

which becomes the equality iff

$$0 = y_{1:n} = \cdots y_{j-1:n} < y_{j:n} = \cdots = y_{n:n} = 1. \tag{3.237}$$

Coming back to the original variables gives the desired claims. □

Mărgăritescu and Nicolae (1990) extended the Groeneveld results replacing the standard ranges $r_n = \bar{\underline{s}}_{1,1:n}$ by the generalized ones

$$\bar{\underline{s}}_{j,k:n} = \frac{1}{k} \sum_{i=n+1-k}^{n} x_{i:n} - \frac{1}{j} \sum_{i=1}^{j} x_{i:n}, \quad 1 \le j, k \le n, \; j + k < 2n. \tag{3.238}$$

Theorem 3.14 *For arbitrary $\tilde{x} \in \mathbb{R}^n$ the following inequalities are sharp*

$$-\frac{j(n-1)}{n} \bar{\underline{s}}_{j,k:n} \le x_{1:n} - \bar{x} \le -\frac{k}{n} \bar{\underline{s}}_{j,k:n}, \tag{3.239}$$

$$\frac{j(n-i)}{ni} \bar{\underline{s}}_{j,k:n} \le x_{i:n} - \bar{x} \le \frac{j}{n} \bar{\underline{s}}_{j,k:n}, \quad 2 \le i \le j, \tag{3.240}$$

$$-\frac{n-i}{n} \bar{\underline{s}}_{j,k:n} \le x_{i:n} - \bar{x} \le \frac{i-1}{n} \bar{\underline{s}}_{j,k:n}, \quad j + 1 \le i \le n - k, \tag{3.241}$$

$$-\frac{k}{n}\,\underline{\bar{s}}_{j,k:n} \le x_{i:n} - \bar{x} \le \frac{k(i-1)}{n(n+1-i)}\,\underline{\bar{s}}_{j,k:n}, \quad n+1-k \le i \le n-1, \tag{3.242}$$

$$\frac{j}{n}\,\underline{\bar{s}}_{j,k:n} \le x_{n:n} - \bar{x} \le \frac{k(n-1)}{n}\,\underline{\bar{s}}_{j,k:n}. \tag{3.243}$$

3.5 Norm Maximization Bounds

3.5.1 Negative Upper Bounds and Positive Lower Bounds

In the previous sections we evaluated centered L-statistics $\sum_{i=1}^{n} c_i(x_{i:n} - \bar{x})$ in various dispersion units s_p. The optimal upper bounds were positive under the assumption

$$\min_{1\le j\le n-1} \frac{1}{j}\sum_{i=1}^{j} c_i < \bar{c} = \frac{1}{n}\sum_{i=1}^{n} c_i$$

on the vector of coefficients $\tilde{c}$. If the assumption is violated, i.e.,

$$\min_{1\le j\le n-1} \frac{1}{j}\sum_{i=1}^{j} c_i \ge \bar{c} = \frac{1}{n}\sum_{i=1}^{n} c_i \tag{3.244}$$

holds then the projection of $\tilde{c}$ onto the family of nondecreasing vectors is the constant vector $\underline{\tilde{c}} = (\bar{c}, \dots, \bar{c})$ (see (2.78), (2.79)). Due to (2.80)

$$\sum_{i=1}^{n} c_i(x_{i:n} - \bar{x}) \le \sum_{i=1}^{n} \underline{c}_i(x_{i:n} - \bar{x}) = \sum_{i=1}^{n} (\underline{c}_i - \bar{c})(x_{i:n} - \bar{x}) = 0 = 0\, s_p \tag{3.245}$$

for arbitrary $\tilde{x} \in \mathbb{R}^n$ and $1 \le p \le \infty$. Note that for the constant $\tilde{x}$ we obtain the equality here.

We try to answer whether the trivial zero bound can be improved if we exclude constant $\tilde{x}$'s. The problem was solved in Goroncy and Rychlik (2006), and two cases should be treated separately.

Theorem 3.15 *Assume that for some* $1 \le k \le n-1$ *we have*

$$\frac{1}{k}\sum_{i=1}^{k} c_i = \min_{1\le j\le n-1} \frac{1}{j}\sum_{i=1}^{j} c_i = \bar{c} = \frac{1}{n}\sum_{i=1}^{n} c_i. \tag{3.246}$$

Then for every nonconstant $\tilde{x} \in \mathbb{R}^n$ and $1 \leq p \leq \infty$ the following bound

$$\sum_{i=1}^{n} c_i \frac{x_{i:n} - \bar{x}}{s_p} \leq 0 \tag{3.247}$$

is optimal, and it is attained when

$$x_{i:n} = \begin{cases} \bar{x} - \frac{n-k}{k} a, & i = 1, \ldots, k, \\ \bar{x} + a, & i = k+1, \ldots, n, \end{cases} \tag{3.248}$$

for some arbitrary $a > 0$.

Proof Under the assumption (3.246) $\tilde{c}$ is constant, and we have (3.245) which can be rewritten as (3.247), because $s_p > 0$. Due to (2.78) k belongs to $\{j_1, \ldots, j_M\}$, and by (2.81) the equality in (3.245) holds when $x_{1:n} = \ldots = x_{k:n}$ and $x_{k+1:n} = \ldots = x_{n:n}$. If $x_{k:n} < x_{k+1:n}$, then $s_p > 0$ for all $1 \leq p \leq \infty$. In particular (3.248) has the mean $\bar{x}$. □

Observe that if several integers $1 \leq k \leq n-1$ satisfy (3.246) then the elements of $\tilde{x}$ attaining the bound in (3.247) may take on more different values. In particular for constant $\tilde{c}$ algorithm (2.78) determines $j_k = k$, $k = 1, \ldots, M = n$, and Theorem 2.13 asserts the equality in (3.245) for all nondecreasing $\tilde{x}$ (in fact it is true for unordered vectors as well). Theorem 3.15 has a counterpart for the lower zero bounds.

Theorem 3.16 *If for some $1 \leq k \leq n-1$*

$$\frac{1}{k} \sum_{i=1}^{k} c_i = \max_{1 \leq j \leq n-1} \frac{1}{j} \sum_{i=1}^{j} c_i = \bar{c} = \frac{1}{n} \sum_{i=1}^{n} c_i. \tag{3.249}$$

Then for every nonconstant $\tilde{x} \in \mathbb{R}^n$ and $1 \leq p \leq \infty$ the following bound

$$\sum_{i=1}^{n} c_i \frac{x_{i:n} - \bar{x}}{s_p} \geq 0 \tag{3.250}$$

for all nonconstant $\tilde{x} \in \mathbb{R}^n$ and $1 \leq p \leq \infty$, and the equality holds in (3.250) for (3.248) with k defined in (3.249).

Corollary 3.9 *Let $1 \leq j < k \leq n$ and $k - j \leq n - 2$. Then for arbitrary nonconstant $\tilde{x} \in \mathbb{R}^n$ and $1 \leq p \leq \infty$*

$$\frac{x_{k:n} - x_{j:n}}{s_p} \geq 0, \tag{3.251}$$

and the equality holds when

$$x_{1:n} \le \ldots \le x_{j:n} = \ldots = x_{k:n} \le \ldots \le x_{n:n} > x_{1:n}.$$

Now we determine strictly negative sharp upper bounds on the centered L-statistics.

Theorem 3.17 *Suppose that* $\tilde{c} \in \mathbb{R}^n$ *satisfies*

$$\min_{1 \le j \le n-1} \frac{1}{j} \sum_{i=1}^{j} c_i > \bar{c} = \frac{1}{n} \sum_{i=1}^{n} c_i \tag{3.252}$$

and $1 \le p < \infty$ *is fixed. Determine*

$$U_p^{(j)}(\tilde{c}) = \left[\frac{n^{p+1}}{j(n-j)^p + j^p(n-j)} \right]^{1/p} \sum_{i=1}^{j} (c_i - \bar{c}), \qquad j = 1, \ldots, n-1, \tag{3.253}$$

and define two-valued vectors $\tilde{x}^{(j)} \in \mathbb{R}^n$, $j = 1, \ldots, n-1$, *satisfying*

$$x_{i:n}^{(j)} = \begin{cases} \bar{x} - s_p(n-j) \left[\frac{n}{j(n-j)^p + j^p(n-j)} \right]^{1/p}, & i = 1, \ldots, j, \\ \bar{x} + s_p j \left[\frac{n}{j(n-j)^p + j^p(n-j)} \right]^{1/p}, & i = j+1, \ldots, n, \end{cases} \tag{3.254}$$

for some fixed $\bar{x} \in \mathbb{R}$ *and* $s_p > 0$. *Suppose that* $1 \le j_* \le n-1$ *minimizes* (3.253). *Then*

$$B_p(\tilde{c}) = \sup_{\bar{x}\mathbf{1} \ne \tilde{x} \in \mathbb{R}^n} \sum_{i=1}^{n} c_i \frac{x_{i:n} - \bar{x}}{s_p} = \sum_{i=1}^{n} c_i \frac{x_{i:n}^{(j_*)} - \bar{x}}{s_p} = -U_p^{(j_*)}(\tilde{c}). \tag{3.255}$$

Proof Note first that under assumption (3.252) we have $\underline{\tilde{c}} = (\bar{c}, \ldots, \bar{c})$, and (3.245) holds. However, in this case algorithm (2.78)–(2.79) terminates in one step. It follows that the equality in (3.245) is attained iff $x_{1:n} = \ldots = x_{n:n}$. For all nonconstant $\tilde{x}$'s we have $\sum_{i=1}^{n} c_i \frac{x_{i:n} - \bar{x}}{s_p} < 0$.

We further check that all nonconstant $\tilde{x}^{(j)}$ defined in (3.254) have the sample mean $\bar{x}$ and pth absolute central moment s_p^p. They satisfy

$$\sum_{i=1}^n c_i \frac{x_{i:n}^{(j)} - \bar{x}}{s_p} = \left[-(n-j)\sum_{i=1}^j c_i + j \sum_{i=j+1}^n c_i \right] \left[\frac{n}{j(n-j)^p + j^p(n-j)} \right]^{1/p}$$
$$= -U_p^{(j)}(\tilde{c}) \le -U_p^{(j_*)}(\tilde{c}) < 0.$$

We aim now at showing that

$$\sum_{i=1}^n c_i \frac{x_{i:n} - \bar{x}}{s_p} = \sum_{i=1}^n (c_i - \bar{c}) \frac{x_{i:n} - \bar{x}}{s_p} \le -U_p^{(j_*)}(\tilde{c}) \tag{3.256}$$

for every nonconstant $\tilde{x} \in \mathbb{R}^n$. For simplicity of notation we introduce $d_i = c_i - \bar{c}$, $i = 1, \dots, n$, which satisfy

$$\sum_{i=1}^j d_i > \sum_{i=1}^n d_i = 0, \qquad j = 1 \dots, n-1. \tag{3.257}$$

We also note that it suffices to focus on standardized vectors $\tilde{x}$ such that $\bar{x} = 0$ and $s_p = 1$. Summing up, for fixed $\tilde{c} \in \mathbb{R}^n$ satisfying (3.252) we minimize

$$T_{\tilde{c}}(\tilde{x}) = \sum_{i=1}^n d_i x_{i:n}, \qquad \tilde{x} \in \mathcal{X}_0, \tag{3.258}$$

over the closed and bounded set

$$\mathcal{X}_0 = \left\{ \tilde{x} \in R^n : \sum_{i=1}^n x_i = 0, \sum_{i=1}^n |x_i|^p = n \right\}. \tag{3.259}$$

Obviously continuous function (3.258) restricted to (3.259) attains its finite and strictly negative extremes there.

For every $\tilde{x} \in \mathcal{X}_0$ satisfying $T_{\tilde{c}}(\tilde{x}) = -a$ for some positive a, define a nondecreasing vector $\tilde{y} \in R^n$ by $y_i = y_i(\tilde{x}) = x_{i:n}/a$, $i = 1, \dots, n$. Then

$$\sum_{i=1}^n y_i = 0, \tag{3.260}$$

$$\sum_{i=1}^n d_i y_i = -1, \tag{3.261}$$

and

$$\sum_{i=1}^{n} |y_i|^p = \frac{n}{a^p} \leq \frac{n}{|B_p(\tilde{c})|^p}.$$

Our original problem will be solved once we determine

$$\max\left\{\sum_{i=1}^{n} |y_i|^p : \tilde{y} \in \mathcal{Y}\right\} \tag{3.262}$$

(which amounts to $n/|B_p(\tilde{c})|^p$), where

$$\mathcal{Y} = \left\{\tilde{y} \in R^n : y_1 \leq \ldots \leq y_n, \sum_{i=1}^{n} y_i = 0, \sum_{i=1}^{n} d_i y_i = -1\right\}$$

for fixed $d_1, \ldots, d_n$ satisfying (3.257). For this purpose, we change the variables

$$\begin{aligned} z_0 &= y_1 \in \mathbb{R}, \\ z_i &= y_{i+1} - y_i \geq 0, \qquad i = 1, \ldots, n-1. \end{aligned}$$

The inverse transformation is

$$y_i = \sum_{r=0}^{i-1} z_r, \qquad i = 1, \ldots, n. \tag{3.263}$$

Combining (3.263) with (3.257), (3.260), and (3.261), yields

$$\sum_{i=1}^{n} y_i = \sum_{i=0}^{n-1} (n-i) z_i = 0, \tag{3.264}$$

$$\sum_{i=1}^{n} d_i y_i = \sum_{i=0}^{n-1} \Big(\sum_{r=i+1}^{n} d_r \Big) z_i = -\sum_{i=1}^{n} D_i z_i = -1,$$

where

$$D_i = -\sum_{r=i+1}^{n} d_r = \sum_{r=1}^{i} d_r = \sum_{r=1}^{i} (c_r - \bar{c}) > 0, \qquad i = 1, \ldots, n-1. \tag{3.265}$$

By (3.263) and (3.264) we get

$$
\begin{bmatrix} y_1 \\ \cdot \\ \cdot \\ \cdot \\ y_n \end{bmatrix}
=
\begin{bmatrix}
\frac{1-n}{n} & \frac{2-n}{n} & \frac{3-n}{n} & \cdots\cdots & \frac{j-n}{n} & \frac{j-n+1}{n} & \cdots & -\frac{2}{n} & -\frac{1}{n} \\
\frac{1}{n} & \frac{2-n}{n} & \frac{3-n}{n} & \cdots\cdots & \frac{j-n}{n} & \frac{j-n+1}{n} & \cdots & -\frac{2}{n} & -\frac{1}{n} \\
\frac{1}{n} & \frac{2}{n} & \frac{3-n}{n} & \cdots\cdots & \frac{j-n}{n} & \frac{j-n+1}{n} & \cdots & -\frac{2}{n} & -\frac{1}{n} \\
 & & \cdots\cdots & \cdots\cdots & \cdots\cdots & \cdots\cdots & & & \\
 & \cdots\cdots & \cdots\cdots & & \frac{j-n}{n} & & \cdots\cdots & & \\
\frac{1}{n} & \frac{2}{n} & \frac{3}{n} & \cdots \; \frac{j-1}{n} & \frac{j}{n} & \frac{j-n+1}{n} & \cdots & -\frac{2}{n} & -\frac{1}{n} \\
 & & \cdots\cdots & \cdots\cdots & \cdots\cdots & \cdots\cdots & & & \\
\frac{1}{n} & \frac{2}{n} & \frac{3}{n} & \cdots\cdots & \frac{j}{n} & \frac{j-n+1}{n} & \cdots & \frac{n-2}{n} & \frac{n-1}{n}
\end{bmatrix}
\cdot
\begin{bmatrix} z_1 \\ \cdot \\ \cdot \\ \cdot \\ z_{n-1} \end{bmatrix}
$$

We denote the above $n \times (n-1)$ matrix by A, and its columns by $A_1, \ldots, A_{n-1}$, respectively. We restate problem (3.262) as

$$
\max \left\{ \|A\tilde{z}\|_p^p = \sum_{i=1}^{n} |(Az)_i|^p : \tilde{z} \in \mathcal{Z} \right\}
$$

where

$$
\mathcal{Z} = \left\{ \tilde{z} \in R^{n-1} : z_i \geq 0, \; i = 1, \ldots, n-1, \; \sum_{i=1}^{n} D_i z_i = 1 \right\}, \tag{3.266}
$$

and $D_i, i = 1, \ldots, n-1$, are fixed positive numbers defined in (3.265).

Obviously the composition of the linear transformation with the convex norm function $\tilde{z} \mapsto \|A\tilde{z}\|_p$ is a convex function as well, and (3.266) is a closed convex and bounded body in $\mathbb{R}^{n-1}$. A convex function attains its maximum over a closed bounded and convex set of finite-dimensional space $\mathbb{R}^{n-1}$ at some extreme point of the set (see, e.g., Rockafellar, 1970, p. 344, Corollary 32.3.1). Here the function is nonnegative so that raising it to power p does not affect the maximizers. We also easily see that the extreme points of (3.266) amount to $\frac{1}{D_i}\tilde{e}_i$, $i = 1, \dots, n-1$, where $\tilde{e}_i$ denotes the ith standard versor in $\mathbb{R}^{n-1}$. Therefore

$$\frac{n}{|U_p(\tilde{c})|^p} = \max_{\tilde{z}\in\mathcal{Z}} \|A\tilde{z}\|_p^p = \max_{1\le j\le n-1} \|\frac{1}{D_j}A_j\|_p^p$$
$$= \max_{1\le j\le n-1}\left[\frac{j(n-j)^p + (n-j)j^p}{(nD_j)^p}\right].$$

This, together with (3.253) and (3.265), establish (3.256) and complete the proof.

□

Theorem 3.17 asserts that for determining the negative upper bounds on centered L-statistics it suffices to calculate n expressions (3.253), and compare them. The respective formulae have much simpler forms in the most interesting cases $p = 1$ and 2:

$$U_1^{(j)}(\tilde{c}) = \frac{n^2}{2j(n-j)}\sum_{i=1}^{j}(c_i - \bar{c}), \qquad j = 1, \dots, n-1, \tag{3.267}$$

$$U_2^{(j)}(\tilde{c}) = \frac{n}{\sqrt{j(n-j)}}\sum_{i=1}^{j}(c_i - \bar{c}), \qquad j = 1, \dots, n-1. \tag{3.268}$$

Negative bounds of Theorem 3.17 are not unique if maximal values of (3.253) are attained by several $j \in \{1, \dots, n-1\}$. It is possible to construct vectors $\tilde{c}$ with arbitrarily chosen positive values of (3.253). Given (3.253) we first determine respective $D_j = \sum_{i=1}^{j}(c_i - \bar{c})$, $j = 1, \dots, n-1$, then calculate the increments $c_i - \bar{c}$, $i = 1, \dots, n$, and possibly add an arbitrary constant $\bar{c}$ at the end. However, the bounds cannot be achieved by points different from (3.254). This were possible for convex combinations of extreme maximizers under condition that

$$\left\|\frac{\alpha}{D_i}A_i + \frac{1-\alpha}{D_j}A_j\right\|_p = \frac{\alpha}{D_i}\|A_i\|_p + \frac{1-\alpha}{D_j}\|A_j\|_p \tag{3.269}$$

for some $i \neq j$ and $0 < \alpha < 1$. If $p > 1$, the equality conditions in the Minkowski inequality (cf. Theorem 2.34) force A_i and A_j to be proportional. This is impossible, because each A_j has the first j values equal to $\frac{j-n}{n} < 0$, and the other $n - j$ ones are $\frac{j}{n} > 0$. In the case $p = 1$, (3.269) holds if the signs of consecutive elements of A_i coincide with their counterparts for A_j, which is also impossible (see Theorem 2.35).

We obtain analogous bounds for $p = +\infty$ slightly modifying arguments of the proof. Verification of details is left to the reader.

Theorem 3.18 *Assume (3.252) and put*

$$U_\infty^{(j)}(\tilde{c}) = \frac{n}{\max\{j, n-j\}} \sum_{i=1}^{j} (c_i - \bar{c}), \qquad j = 1, \ldots, n-1, \tag{3.270}$$

and choose nonconstant $\tilde{x}^{(j)} \in \mathbb{R}^n$, $j = 1, \ldots, n-1$, satisfying

$$x_{i:n}^{(j)} = \begin{cases} \bar{x} - s_\infty, & i = 1, \ldots, j, \\ \bar{x} + s_\infty \frac{j}{n-j}, & i = j+1, \ldots, n, \end{cases} \qquad 1 \le j \le \frac{n}{2}, \tag{3.271}$$

$$x_{i:n}^{(j)} = \begin{cases} \bar{x} - s_\infty \frac{n-j}{j}, & i = 1, \ldots, j, \\ \bar{x} + s_\infty, & i = j+1, \ldots, n, \end{cases} \qquad \frac{n}{2} \le j \le n-1. \tag{3.272}$$

with arbitrarily chosen mean $\bar{x} \in \mathbb{R}$ and maximal mean deviation $s_\infty > 0$. If $1 \le j_* \le n-1$ minimizes (3.270), then

$$B_\infty(\tilde{c}) = \sup_{\bar{x}\tilde{1} \neq \tilde{x} \in \mathbb{R}^n} \sum_{i=1}^{n} c_i \frac{x_{i:n} - \bar{x}}{s_\infty} = \sum_{i=1}^{n} c_i \frac{x_{i:n}^{(j_*)} - \bar{x}}{s_\infty} = -U_\infty^{(j_*)}(\tilde{c}). \tag{3.273}$$

For describing all possible conditions of attainability of the bounds of Theorem 3.18, we should check the requirements for (3.269) with $p = \infty$. The supremum norm of the sum of two vectors amounts to the sum of the norms iff both the vectors attain their maximal absolute value at a common coordinate, and the signs of the respective values coincide (see Theorem 2.35). In particular, this property holds for the pairs A_i, A_j if either $i, j \le \frac{n}{2}$ or $i, j \ge \frac{n}{2}$. This means that if (3.270) is minimized at some $1 \le j_1^* < \ldots < j_\kappa^* \le n-1$, then bound (3.273) is attained by convex combinations of (3.271) for some $1 \le j_i^* \le \frac{n}{2}$ and by convex combinations of (3.272) with $\frac{n}{2} \le j_i^* \le n-1$.

Corollary 3.10 *For arbitrary $\tilde{x} \in \mathbb{R}^n$ and $1 \le p < +\infty$ the inequality*

$$\frac{x_{1:n} - \bar{x}}{s_p} \le -\left[\frac{n}{n-1+(n-1)^p}\right]^{1/p} \tag{3.274}$$

is valid and becomes equality for

$$x_{i:n} = \begin{cases} \bar{x} - s_p \left[\frac{n}{n-1+(n-1)^p}\right]^{1/p} & i = 1, \ldots, n-1, \\ \bar{x} + s_p \left[\frac{n(n-1)^p}{n-1+(n-1)^p}\right]^{1/p} & i = n. \end{cases} \tag{3.275}$$

Moreover,

$$\frac{x_{1:n}-\bar{x}}{s_\infty} \le -\frac{1}{n-1}, \tag{3.276}$$

and the equality holds for (3.272) with $j = n-1$.

Proof If $\tilde{c} = \tilde{e}_1 = (1, 0, \ldots, 0)$ and $1 \le p < \infty$, then

$$U_p^{(j)}(\tilde{e}_1) = \left[\frac{n(n-j)^p}{j(n-j)^p + j^p(n-j)}\right]^{1/p} = f_p^{-1/p}\left(\frac{j}{n}\right), \tag{3.277}$$

where

$$f_p(x) = x + x^p(1-x)^{1-p}, \qquad 0 < x < 1.$$

They have derivatives

$$f_p'(x) = 1 + (p-x)x^{p-1}(1-x)^{-p} > 1, \qquad p > 1,$$

and $f_1'(x) = 2$, which means that each $f_p(x)$ is increasing, and its negative power restricted to arguments $\frac{i}{n}$, $i = 1, \ldots, n-1$, attains its minimum (3.274) at $\frac{n-1}{n}$. Plugging $j = n-1$ into (3.254) we obtain (3.275). For $p = +\infty$

$$U_\infty^{(j)}(\tilde{e}_1) = \min\left\{\frac{n-j}{j}, 1\right\}.$$

is minimized at $j = n-1$ as well, and the minimal value is $\frac{1}{n-1}$. □

Especially, for $p = 1$ and $p = 2$ we have

$$\frac{x_{1:n}-\bar{x}}{s_1} \le -\frac{n}{2(n-1)}, \tag{3.278}$$

$$\frac{x_{1:n}-\bar{x}}{s_2} \le -\frac{1}{\sqrt{n-1}}. \tag{3.279}$$

Inequality (3.274) was proved by Beesack (1973) in a different way. Earlier, (3.279) was shown independently by Boyd (1971) and Hawkins (1971) by arguments different from ours as well.

Much more practically useful are positive lower bounds for positive (centered) L-statistics. For instance, L-estimates of scale are positive. We deduce these bounds from Theorems 3.17 and 3.18 by the following standard arguments. If a vector $\tilde{x} = (x_1, \ldots, x_n)$ has mean $\bar{x}$ and sample pth absolute central moment $s_p^p = s_p^p(\tilde{x}) = \frac{1}{n}\sum_{i=1}^n |x_i - \bar{x}|^p > 0$, then $\tilde{y}$ defined as $y_i = 2\bar{x} - x_i$, $i = 1, \ldots, n$, satisfies $\bar{y} = \bar{x}$,

$s_p^p(\tilde{y}) = s_p^p(\tilde{x})$ and $y_{i:n} = 2\bar{x} - x_{n+1-i:n}$, $i = 1, \ldots, n$. Therefore

$$\sum_{i=1}^{n} c_i \frac{y_{i:n} - \bar{y}}{s_p(\tilde{y})} = \sum_{i=1}^{n} c_i \frac{\bar{x} - x_{n+1-i:n}}{s_p(\tilde{x})} = -\sum_{i=1}^{n} c_{n+1-i} \frac{x_{i:n} - \bar{x}}{s_p(\tilde{x})}$$

and

$$\inf_{\bar{y}\mathbf{1} \neq \tilde{y} \in \mathbb{R}^n} \sum_{i=1}^{n} c_i \frac{y_{i:n} - \bar{y}}{s_p(\tilde{y})} = - \sup_{\bar{x}\mathbf{1} \neq \tilde{x} \in \mathbb{R}^n} \sum_{i=1}^{n} c_{n+1-i} \frac{x_{i:n} - \bar{x}}{s_p(\tilde{x})}.$$

If the RHS supremum is attained by some $\tilde{x}^* = (x_1^*, \ldots, x_n^*)$ then the infimum of the LHS is attained by $\tilde{y}^* = (2\bar{x} * -x_1^*, \ldots, 2\bar{x}^* - x_n^*)$.

Theorem 3.19 *Assume that*

$$\max_{1 \leq j \leq n-1} \frac{1}{j} \sum_{i=1}^{j} c_i < \bar{c} = \frac{1}{n} \sum_{i=1}^{n} c_i \tag{3.280}$$

and $1 \leq p < \infty$. *Determine*

$$L_p^{(j)}(\tilde{c}) = \left[\frac{n^{p+1}}{j(n-j)^p + j^p(n-j)} \right]^{1/p} \sum_{i=1}^{n-j} (\bar{c} - c_i), \qquad j = 1, \ldots, n-1, \tag{3.281}$$

and define $\tilde{x}^{(j)} \in \mathbb{R}^n$, $j = 1, \ldots, n-1$, *as follows*

$$x_{i:n}^{(j)} = \begin{cases} \bar{x} - s_p j \left[\frac{n}{j(n-j)^p + j^p(n-j)} \right]^{1/p}, & i = 1, \ldots, n-j, \\ \bar{x} + s_p(n-j) \left[\frac{n}{j(n-j)^p + j^p(n-j)} \right]^{1/p}, & i = n+1-j, \ldots, n, \end{cases} \tag{3.282}$$

for some fixed $\bar{x} \in \mathbb{R}$ *and* $s_p > 0$. *If* $1 \leq j_* \leq n-1$ *minimizes* (3.281), *then*

$$b_p(\tilde{c}) = \inf_{\bar{x}\mathbf{1} \neq \tilde{x} \in \mathbb{R}^n} \sum_{i=1}^{n} c_i \frac{x_{i:n} - \bar{x}}{s_p} = \sum_{i=1}^{n} c_i \frac{x_{i:n}^{(j_*)} - \bar{x}}{s_p} = L_p^{(j_*)}(\tilde{c}). \tag{3.283}$$

Theorem 3.20 *Assume that (3.280) holds and set*

$$L_\infty^{(j)}(\tilde{c}) = \frac{n}{\max\{j, n-j\}} \sum_{i=1}^{n-j} (\bar{c} - c_i), \qquad j = 1, \ldots, n-1, \tag{3.284}$$

and choose nonconstant $\tilde{x}^{(j)} \in \mathbb{R}^n$, $j = 1, \ldots, n-1$, *satisfying*

$$x_{i:n}^{(j)} = \begin{cases} \bar{x} - s_\infty \frac{j}{n-j}, & i = 1, \ldots, n-j, \\ \bar{x} + s_\infty, & i = n-j+1, \ldots, n, \end{cases} \qquad 1 \le j \le \frac{n}{2}, \tag{3.285}$$

$$x_{i:n}^{(j)} = \begin{cases} \bar{x} - s_\infty, & i = 1, \ldots, n-j, \\ \bar{x} + s_\infty \frac{n-j}{j}, & i = n-j+1, \ldots, n, \end{cases} \qquad \frac{n}{2} \le j \le n-1. \tag{3.286}$$

with arbitrarily chosen mean $\bar{x} \in \mathbb{R}$ *and maximal mean deviation* $s_\infty > 0$. *If* $1 \le j_* \le n-1$ *minimizes* (3.284), *then*

$$b_\infty(\tilde{c}) = \inf_{\bar{x}\tilde{\mathbf{1}} \ne \tilde{x} \in \mathbb{R}^n} \sum_{i=1}^n c_i \frac{x_{i:n} - \bar{x}}{s_\infty} = \sum_{i=1}^n c_i \frac{x_{i:n}^{(j_*)} - \bar{x}}{s_\infty} = L_\infty^{(j_*)}(\tilde{c}). \tag{3.287}$$

Corollary 3.11 *For the sample maximum we have*

$$x_{n:n} - \bar{x} \ge \left[\frac{n}{n-1+(n-1)^p} \right]^{1/p} s_p, \qquad 1 \le p < +\infty, \tag{3.288}$$

which is attained by

$$x_{i:n} = \begin{cases} \bar{x} - s_p \left[\frac{n(n-1)^p}{n-1+(n-1)^p} \right]^{1/p}, & i = 1, \\ \bar{x} + s_p \left[\frac{n}{n-1+(n-1)^p} \right]^{1/p}, & i = 2, \ldots, n. \end{cases} \tag{3.289}$$

Moreover,

$$x_{n:n} - \bar{x} \ge \frac{s_\infty}{n-1} \tag{3.290}$$

with the equality attained by

$$x_{i:n} = \begin{cases} \bar{x} - s_\infty, & i = 1, \\ \bar{x} + \frac{1}{n-1} s_\infty, & i = 2, \ldots, n. \end{cases}$$

It is easier to prove the corollary referring to the arguments preceding Theorem 3.19 and the statements of Corollary 3.10 rather than directly to Theorems 3.19 and 3.20. Inequality (3.288) was established by Beesack (1973). Its special case for $p = 2$ has the form

$$x_{n:n} - \bar{x} \ge \frac{s}{\sqrt{n-1}}, \tag{3.291}$$

and it is attained by

$$x_{i:n} = \begin{cases} \bar{x} - \sqrt{n-1}\, s, \ i = 1, \\ \bar{x} + \frac{s}{\sqrt{n-1}}, \quad i = 2, \ldots, n. \end{cases} \tag{3.292}$$

It was generalized by Mallows and Richter (1969) to the case of upper selection differentials

$$\bar{s}_{k:n} \geq \begin{cases} \frac{s}{\sqrt{n-1}}, & \text{if } k \leq n/2, \\ \frac{n-k}{k} \frac{s}{\sqrt{n-1}}, & \text{if } k \geq \frac{n}{2}\,. \end{cases} \tag{3.293}$$

Bounds (3.293) become equalities when (3.292) and

$$x_{i:n} = \begin{cases} \bar{x} - \frac{s}{\sqrt{n-1}}, & i = 1, \ldots, n-1, \\ x + s\sqrt{n-1}, \ i = n, \end{cases} \tag{3.294}$$

hold, respectively.

According to Corollary 3.9 all the differences $x_{k:n} - x_{j:n}$, $j < k$, except for the sample ranges, have trivial lower bounds. The bounds for ranges are positive, and have quite complicated forms.

Corollary 3.12 *For* $1 \leq p < \infty$ *define*

$$L_p^{(r_n)}(j) = \left[\frac{n^{p+1}}{j(n-j)^p + j^p(n-j)} \right]^{1/p}. \tag{3.295}$$

(a) If $1 \leq p \leq 3$ *then*

$$\frac{r_n}{s_p} = \frac{x_{n:n} - x_{1:n}}{s_p} \geq \begin{cases} 2, & \text{if } n \text{ is even}, \\ \left\{ \frac{(2n)^{p+1}}{(n^2-1)[(n+1)^{p-1}+(n-1)^{p-1}]} \right\}^{1/p}, & \text{if } r \text{ is odd}. \end{cases} \tag{3.296}$$

If n is even (odd, respectively) then the equality holds in (3.296) for (3.282) with $j = \frac{n}{2}$ *(either* $j = \frac{n-1}{2}$ *or* $j = \frac{n+1}{2}$*, respectively).*

(b) For $3 < p < \infty$ *equation*

$$(1-x)^p - px(1-x)^{p-1} + px^{p-1}(1-x) - x^p = 0. \tag{3.297}$$

has a unique solution $x_p \in \left(0, \frac{1}{2}\right)$. *Then*

$$\frac{r_n}{s_p} = \frac{x_{n:n} - x_{1:n}}{s_p} \geq \begin{cases} \left[\frac{n^{p+1}}{(n-1)^p+(n-1)}\right]^{1/p}, & \text{if } x_p \leq \frac{1}{n}, \\ \min\{L_p^{(r_n)}(\lfloor nx_p \rfloor), L_p^{(r_n)}(\lceil nx_p \rceil)\}, & \text{otherwise}. \end{cases} \tag{3.298}$$

In the former case, the equality is attained by (3.282) with $j = 1$ *and* $j=n-1$. *In the latter one we have the following. When* $L_p^{(r_n)}(\lfloor nx_p \rfloor) \leq L_p^{(r_n)}(\lceil nx_p \rceil)$, *then the equality is attained by (3.282) with* $j = \lfloor nx_p \rfloor$ *and* $j = n - \lfloor nx_p \rfloor$. *If* $L_p^{(r_n)}(\lfloor nx_p \rfloor) \geq L_p^{(r_n)}(\lceil nx_p \rceil)$, *then the equality holds for (3.282) with* $j = \lceil nx_p \rceil$ *and* $j = n - \lceil nx_p \rceil$.

(c) If $p = +\infty$, *then*

$$\frac{r_n}{s_\infty} = \frac{x_{n:n} - x_{1:n}}{s_\infty} \geq \frac{n}{n-1}, \tag{3.299}$$

and the equality holds for (3.285) with $j = 1$ *and (3.286) with* $j = n - 1$.

Proof Take $1 \leq p < \infty$. The coefficient vector of the range $\tilde{c} = \tilde{e}_n - \tilde{e}_1$ has mean $\bar{c} = 0$ and $\sum_{i=1}^{j}(\bar{c} - c_i) = 1$, $i = 1, \ldots, n-1$. Therefore respective expressions (3.281) are equal to (3.295) which can be represented as $L_p^{(r_n)}(j) = f_p^{-1/p}\left(\frac{j}{n}\right)$, $j = 1, \ldots, n-1$, for

$$f_p(x) = x(1-x)^p + x^p(1-x), \qquad 0 < x < 1. \tag{3.300}$$

Therefore we analyze variability of the functions and then find their maximal values among the arguments $x = \frac{j}{n}$, $j = 1, \ldots, n-1$. Function (3.300) is continuous, positive, vanishing at the ends, and symmetric about $\frac{1}{2}$. Accordingly, it has a local extremum at $\frac{1}{2}$, and possibly several pairs of local extremes located symmetrically about $\frac{1}{2}$. Moreover, (3.300) has derivatives

$$f_p'(x) = (1-x)^p - px(1-x)^{p-1} + px^{p-1}(1-x) - x^p, \tag{3.301}$$

$$f_p''(x) = -2p[(1-x)^{p-1} + x^{p-1}] + p(p-1)[x(1-x)^{p-2} + x^{p-2}(1-x)]$$

which satisfy $f_p'\left(\frac{1}{2}\right) = 0$, and $f_p''\left(\frac{1}{2}\right) = \frac{p(p-3)}{2^{p-2}}$ in particular. It follows that $\frac{1}{2}$ is a local maximum point of (3.300) for $p < 3$, and a local minimum for $p > 3$. For $p = 3$ we have $f_3'(x) = (1-2x)^3$ which implies that f_3 has a maximum at $\frac{1}{2}$. Dividing (3.301) by $(1-x)^p$, and changing the variable $y = \frac{x}{1-x}$ we obtain function

$$g_p(y) = 1 - py + py^{p-1} - y^p, \qquad 0 < y < \infty, \tag{3.302}$$

which has the same sequences of intervals of arguments with given signs of the values as (3.301). Due to

$$g_p''(y) = p(p-1)y^{p-3}(p-2-y)$$

(3.302) is strictly convex in $(0, p-2)$, and strictly concave in $(p-2, \infty)$, and it has at most two zeros in each interval. On the other hand, the number of zeros should be identical with the number of local extremes of (3.300) which is odd. This leads us to the following conclusions. If $1 \leq p \leq 3$, then $x = \frac{1}{2}$ is the global maximum of f_p, because it is increasing and decreasing near 0 and 1, respectively, and cannot have extra extremes in $\left(0, \frac{1}{2}\right)$ and $\left(\frac{1}{2}, 1\right)$. If $p > 3$ and $x = \frac{1}{2}$ is a local minimum, there must exist two symmetrically located about $\frac{1}{2}$ local and global maxima. The smaller one $0 < x_p < \frac{1}{2}$ is the unique solution to $f_p(x) = 0$ in the interval (see (3.297)). The rest of the reasoning is standard. If $1 \leq p \leq 3$ and n is even then $\max_{j=1,\ldots,n-1} f_p\left(\frac{j}{n}\right) = f_p\left(\frac{n/2}{n}\right)$ which determines bound $L_p^{(r_n)}\left(\frac{n}{2}\right) = 2$. If n is odd then $\max_{j=1,\ldots,n-1} f_p\left(\frac{j}{n}\right) = f_p\left(\frac{n-1}{2n}\right) = f_p\left(\frac{n-1}{2n}\right)$ which gives the latter expression in (3.296).

If $p > 3$ and $x_p \leq \frac{1}{n}$, then $\max_{j=1,\ldots,n-1} f_p\left(\frac{j}{n}\right) = f_p\left(\frac{1}{n}\right) = f_p\left(\frac{n-1}{n}\right)$. If $x_p > \frac{1}{n}$, the restricted maximum is attained at either of pairs $\frac{\lfloor nx_p \rfloor}{n}$ with $1 - \frac{\lfloor nx_p \rfloor}{n}$ or $\frac{\lceil nx_p \rceil}{n}$ with $1 - \frac{\lceil nx_p \rceil}{n}$. This allows us to conclude the statement of Corollary 3.12(b).

For $p = \infty$ (3.284) takes on the form

$$L_\infty^{(r_n)}(j) = \frac{n}{\max\{j, n-j\}}, \qquad j = 1, \ldots, n-1.$$

Its minimal value is

$$L_\infty^{(r_n)}(1) = L_\infty^{(r_n)}(n-1) = \frac{n}{n-1}.$$

This immediately gives the final claim of the corollary. □

In particular we have

$$r_n = x_{n:n} - x_{1:n} \geq \begin{cases} 2s, & \text{if } n \text{ is even}, \\ \frac{2n}{\sqrt{n^2-1}}\, s, & \text{if } n \text{ is odd}. \end{cases} \tag{3.303}$$

for every nonconstant $\tilde{x}$ which become equality if

$$x_{i:n} = \begin{cases} \bar{x} - s, \ i = 1, \ldots, \frac{n}{2}, \\ \bar{x} + s, \ i = \frac{n}{2} + 1, \ldots, n, \end{cases} \tag{3.304}$$

if n is even and in either of the below cases

$$x_{i:n} = \begin{cases} \bar{x} - \sqrt{\frac{n+1}{n-1}}\, s, & i = 1, \ldots, \frac{n-1}{2}, \\ \bar{x} + \sqrt{\frac{n-1}{n+1}}\, s, & i = \frac{n+1}{2}, \ldots, n, \end{cases} \tag{3.305}$$

$$x_{i:n} = \begin{cases} \bar{x} - \sqrt{\frac{n-1}{n+1}}\, s, & i = 1, \ldots, \frac{n+1}{2}, \\ \bar{x} + \sqrt{\frac{n+1}{n-1}}\, s, & i = \frac{n+3}{2}, \ldots, n, \end{cases} \tag{3.306}$$

when n is odd. This was first proved by Thomson (1955).

Analogous evaluations for the generalized ranges were derived by Mallows and Richter (1969)

$$\underline{\bar{s}}_{j,k:n} \geq \begin{cases} \frac{n}{k}\frac{s}{\sqrt{n-1}}, & \text{if } k = \max\left\{k, j, \frac{\lfloor n/2 \rfloor}{\sqrt{n-1}}\right\}, \\ \frac{n}{j}\frac{s}{\sqrt{n-1}}, & \text{if } j = \max\left\{k, j, \frac{\lfloor n/2 \rfloor}{\sqrt{n-1}}\right\}, \\ \frac{ns}{\sqrt{\lfloor n/2 \rfloor (n - \lfloor n/2 \rfloor)}}, & \text{otherwise.} \end{cases} \tag{3.307}$$

In the first two cases, the equalities are achieved if (3.294) and (3.292) hold, respectively. In the last one, the standard range is included. The bound is identical with (3.303), and (3.304) to (3.306) provide the equality.

3.5.2 Lower Bounds for Scale Estimates

Corollary 3.13 *For the Gini mean differences*

$$\mathrm{GMD}_n(\tilde{x}) = \frac{1}{n(n-1)} \sum_{i,j=1}^{n} |x_i - x_j| = \frac{2}{n(n-1)} \sum_{i=1}^{n} (2i - n - 1) x_{i:n}$$

of nonconstant $\tilde{x} \in \mathbb{R}^n$ *we have*

$$\frac{\mathrm{GMD}_n(\tilde{x})}{s_p} \geq \begin{cases} \frac{n}{n-1}, & p = 1, \\ 2\left[\frac{n}{(n-1)^p + n - 1}\right]^{1/p}, & 1 < p < \infty, \\ \frac{2}{n-1}, & p = +\infty. \end{cases} \tag{3.308}$$

In the first case the equality is attained by (3.282) with $p = 1$ for all $j = 1, \ldots, n-1$. For $1 < p < +\infty$ the bound in (3.308) is attained by (3.282) with $j = 1$ and $j = n-1$. Finally, for $p = +\infty$ we have the equality for (3.285) with $j = 1$ and (3.286) with $j = n-1$.

Proof For the coefficient vector $\tilde{c}$ of the L-statistic representation of the Gini mean differences we have $\bar{c} = 0$, and $\sum_{i=1}^{j}(\bar{c} - c_i) = \frac{2j(n-j)}{n(n-1)}$, $j = 1, \ldots, n-1$. By (3.281) and (3.284), we need to minimize

$$L_p^{(\mathrm{GMD})}(j) = \frac{2n^{1/p} j(n-j)}{(n-1)[j(n-j)^p + j^p(n-j)]^{1/p}}, \qquad 1 \le p < \infty, \tag{3.309}$$

and

$$L_\infty^{(\mathrm{GMD})}(j) = \frac{2j(n-j)}{(n-1)\max\{j, n-j\}} = \frac{2}{n-1}\min\{j, n-j\} \tag{3.310}$$

for $j = 1, \ldots, n-1$. Once we determine j minimizing, the rest of the proof becomes a routine work: we determine the bounds plugging the minimizing arguments into (3.309) and (3.310), and recall respective attainability conditions of Theorems 3.19 and 3.20. We omit these steps.

Not first that (3.309) for $p = 1$ is constant, and amounts to $\frac{n}{n-1}$. For other p we can represent (3.309) as $L_p^{(\mathrm{GMD})}(j) = \frac{2n}{n-1} f_p^{-1/p}\left(\frac{j}{n}\right)$, $j = 1, \ldots, n-1$, for symmetric function

$$f_p(x) = x^{1-p} + (1-x)^{1-p}, \qquad 0 < x < 1,$$

with the derivative

$$f_p'(x) = \frac{p-1}{x^p(1-x)^p}[x^p - (1-x)^p].$$

This implies that f_p is decreasing on $\left(0, \frac{1}{2}\right)$ and increasing on $\left(\frac{1}{2}, 1\right)$, and restricted to the points $\frac{j}{n}$, $j = 1, \ldots, n-1$, it is minimized at both $\frac{1}{n}$ and $\frac{n-1}{n}$. This allows us to determine the middle inequality in (3.308) and respective attainability conditions. Obviously, (3.310) is also minimized at $j = 1$ and $n-1$, which provides the last claim. □

Formula (3.308) for $p = 2$ has the form

$$\frac{\mathrm{GMD}_n(\tilde{x})}{s_2} \ge \frac{2}{\sqrt{n-1}}.$$

Condition that $j = 1$ and $j = n-1$ minimize $L_p^{(j)}(\tilde{c})$ practically means that the bound is attained by a sample with $n-1$ identical elements, and one different from them. If all $j = 1, \ldots, n-1$ are the minimizing arguments, then the bound is attained by any two-valued samples with arbitrary numbers of elements in each group.

Corollary 3.14 *For the median absolute deviation (see (3.33)) the optimal lower bounds are*

$$\frac{\mathrm{MAD}_n(\tilde{x})}{s_p} \geq \begin{cases} \left[\frac{n}{(n-1)^p+n-1}\right]^{1/p}, & 1 \leq p < \infty, \\ \frac{1}{n-1}, & p = \infty, \end{cases} \tag{3.311}$$

The equality is attained in the above inequality by (3.282) with $j = 1$ and $j = n-1$ for $1 \leq p < \infty$, and by (3.285) with $j = 1$ and (3.286) with $j = n-1$ for $p = +\infty$.

Proof For the median absolute deviation we have $\bar{c} = 0$ and

$$\sum_{i=1}^{j}(\bar{c} - c_i) = \begin{cases} \frac{j}{n}, & j \leq \frac{n}{2}, \\ \frac{n-1}{2n}, & j = \frac{n+1}{2}, \\ 1 - \frac{j}{n}, & j \geq \frac{n}{2} + 1, \end{cases}$$

(the middle term appears only for odd n). We first focus on the cases $p < \infty$. Then

$$L_p^{(\mathrm{MAD})}(j) = \frac{n^{1/p}\min\{j, n-j\}}{[j(n-j)^p + j^p(n-!j)]^{1/p}} = \begin{cases} \frac{1}{\left[\frac{j}{n}\left(\frac{n-j}{j}\right)^p + 1 - \frac{j}{n}\right]^{1/p}}, & j \leq \frac{n}{2}, \\ \frac{1}{\left[\left(1-\frac{j}{n}\right))\left(\frac{j}{n-j}\right)^p + \frac{j}{n}\right]^{1/p}}, & j \geq \frac{n}{2} + 1, \end{cases} \tag{3.312}$$

and

$$L_p^{(\mathrm{MAD})}\left(\frac{n+1}{2}\right) = \left\{\frac{2n(n-1)^{p-1}}{(n+1)[(n-1)^{p-1} + (n+1)^{p-1}]}\right\}^{1/p} \tag{3.313}$$

for the sample median in the odd sized samples. We minimize (3.312) and compare its minimal value with (3.313). Note that (3.312) can be written as $L_p^{(\mathrm{MAD})}(j) = f_p^{-1/p}\left(\frac{j}{n}\right)$ with

$$f_p(x) = x^{1-p}(1-x)^p + 1 - x, \qquad 0 < x < \frac{1}{2}. \tag{3.314}$$

For $p = 1$ we have $f_1(x) = 2 - 2x$ which is obviously decreasing. For $p > 1$

$$f_p'(x) = (1-p)\left(\frac{1-x}{x}\right)^p - p\left(\frac{1-x}{x}\right)^{p-1} - 1.$$

Changing the variable $y = \frac{x}{1-x}$ we transform the above into

$$g_p(y) = (1-p)y^{-p} - py^{1-p} - 1 = -(py + p - 1)y^{-p} - 1 < -1. \tag{3.315}$$

Moreover,

$$g'_p(y) = p(p-1)\frac{1+y}{y^{p+1}} > 0.$$

Consequently, (3.315) is negative increasing in (0, 1), and (3.314) is decreasing convex in $\left(0, \frac{1}{2}\right)$. It follows that the symmetric about $\frac{n}{2}$ sequence (3.312) is first increasing, and then decreasing. It attains minimal value at $j = 1$ and $j = n - 1$, and this amounts to

$$L_p^{(\mathrm{MAD})}(1) = L_p^{(\mathrm{MAD})}(n-1) = \left[\frac{n}{(n-1)^p + n - 1}\right]^{1/p}.$$

This is equal to (3.313) for $n = 3$ (note that then $\frac{n+1}{2} = n - 1$) and strictly less for all odd $n > 3$. This gives the final conclusion for $1 \le p < +\infty$.

For $p = \infty$ we have

$$L_\infty^{(\mathrm{MAD})}(j) = \begin{cases} \frac{j}{n-j}, & j \le \frac{n}{2}, \\ \frac{n-1}{n+1}, & j = \frac{n+1}{2}, \\ \frac{n-j}{j}, & j \ge \frac{n}{2} + 1. \end{cases}$$

We easily note that the minimum of the above expression is attained at $j = 1$ and $j = n - 1$, and the minimal value is $\frac{1}{n-1}$. □

3.6 Generalizations

Beesack (1973) generalized the Moriguti-Hölder bounds on single order statistics, presented in Corollaries 3.3, 3.10 and 3.4, by replacing the standard central absolute moments s_p^p by generalized ones

$$s_\Phi(\tilde{x}) = \frac{1}{n}\sum_{i=1}^{n} \Phi(|x_i - \bar{x}|) \tag{3.316}$$

for some functions $\Phi : \mathbb{R}_+ \to \mathbb{R}_+$ that vanish at 0 and increase on $\mathbb{R}_+$.

Theorem 3.21 *Let $\Phi : \mathbb{R}_+ \to \mathbb{R}_+$ be strictly increasing and convex and satisfy $\Phi(0) = 0$. Under notation (3.316), for arbitrary nonconstant $\tilde{x} \in \mathbb{R}^n$ define $\alpha_1 = \alpha_1(s_\Phi(\tilde{x})) > 0$ as the unique positive solution to*

$$(n-1)\Phi(\alpha) + \Phi((n-1)\alpha) = s_\Phi(\tilde{x}), \tag{3.317}$$

and $\alpha_j = \alpha_j(s_\Phi(\tilde{x})) > 0$, $j = 2, \ldots, n$, *as the unique positive solutions to*

$$(n+1-j)\Phi(\alpha) + (j-1)\Phi\left(\frac{n+1-j}{j-1}\alpha\right) = ns_\Phi(\tilde{x}). \quad (3.318)$$

Then the following inequalities are satisfied

$$x_{1:n} - \bar{x} \le -\alpha_1(s_\Phi(\tilde{x})), \quad (3.319)$$

$$x_{j:n} - \bar{x} \le \alpha_j(s_\Phi(\tilde{x})), \quad j = 2, \ldots, n, \quad (3.320)$$

and become equalities for

$$x_{i:n} = \begin{cases} \bar{x} - \alpha_1, & i = 1, \ldots, n-1, \\ \bar{x} + (n-1)\alpha_1, & i = n, \end{cases} \quad (3.321)$$

and

$$x_{i:n} = \begin{cases} \bar{x} - \frac{n+1-j}{j-1}\alpha_j, & i = 1, \ldots, j-1, \\ \bar{x} + \alpha_j, & i = j, \ldots, n, \end{cases} \quad j = 2, \ldots, n, \quad (3.322)$$

respectively.

Furthermore, the lower bound for the jth order statistic amounts to the negative of the upper one for $x_{n+1-j:n}$. Setting $\Phi(x) = x^p$, $1 \le p < \infty$, we obtain the results of Corollaries 3.3, 3.4 and 3.10. Beesack (1976) established analogous results for the ordered distances of observations from the sample mean. Given $\tilde{x} = (x_1, \ldots, x_n) \in \mathbb{R}^n$ define $\tilde{y} = (y_1, \ldots, y_n) = (|x_1 - \bar{x}|, \ldots, |x_n - \bar{x}|)$ and arrange it in the ascending order $y_{1:n} \le y_{2:n} \le y_{n:n}$. Note that $s_\Phi(\tilde{y}) = s_\Phi(\tilde{x})$.

Theorem 3.22 *Under the notation of Theorem* 3.21 *for nonconstant* $\tilde{x} \in \mathbb{R}^n$ *define* $\beta_j = \beta_j(s_\Phi(\tilde{x}))$, $j = 1, \ldots, n$, *as follows:*

(i) if $n + 1 - j$ *is even, then* β_j *is the unique positive solution to*

$$(n+1-j)\Phi(\beta) = ns_\Phi(\tilde{x}), \quad (3.323)$$

(ii) if $n + 1 - j$ *is odd and* $j > 1$, *then* β_j *is the unique positive solution to*

$$(j-1)\Phi\left(\frac{\beta}{j-1}\right) + (n-j+1)\Phi(\beta) = ns_\Phi(\tilde{x}), \quad (3.324)$$

(iii) if $j = 1$ *and* n *is odd, then*

$$\beta_1 = \max\left\{\beta_{1,k} : \frac{n+1}{2} \le k \le n-1\right\}, \quad (3.325)$$

where $\beta_{1,k}$ is the unique positive root of the equation

$$k\Phi(\beta) + (n-k)\Phi\left(\frac{kx}{n-k}\right) = ns_{\Phi}(\tilde{x}). \tag{3.326}$$

Then the following inequalities are optimal

$$y_{j:n} \leq \beta_j, \qquad j = 1, \ldots, n, \tag{3.327}$$

and they are attained by

$$x_{i:n} = \begin{cases} \bar{x} - \beta_j, & i = 1, \ldots, \frac{n-j+1}{2}, \\ \bar{x}, & i = \frac{n-j+3}{2}, \ldots, \frac{n-1+j}{2}, \\ \bar{x} + \beta_j, & i = \frac{n+1=j}{2}, \ldots, n, \end{cases} \tag{3.328}$$

in case (i),

$$x_{i:n} = \begin{cases} \bar{x} - \beta_j, & i = 1, \ldots, \frac{n-j}{2}, \\ \bar{x} - \frac{\beta_j}{j-1}, & i = \frac{n-j}{2} + 1, \ldots, \frac{n+j}{2} - 1, \\ \bar{x} + \beta_j, & i = \frac{n+j}{2}, \ldots, n, \end{cases} \tag{3.329}$$

or $x'_{i:n} = 2\bar{x} - x_{n+1-i:n}$, $i = 1, \ldots, n$, in case (ii), and

$$x_{i:n} = \begin{cases} \bar{x} - \frac{k_*}{n-k_*}\beta_{1k_*}, & i = 1, \ldots, k_*, \\ \bar{x} + \beta_{1k_*}, & i = k_* + 1, \ldots, n, \end{cases} \tag{3.330}$$

or $x'_{i:n} = 2\bar{x} - x_{n+1-i:n}$, $i = 1, \ldots, n$, for $\frac{n+1}{2} \leq k_* \leq n-1$ satisfying $\beta_{1k_*} = \beta_1$ in case (iii).

In the special cases $\Phi(x) = x^p$, $p \geq 1$, we have

$$y_{j:n} \leq \begin{cases} \left(\frac{n}{n+1-j}\right)^{1/p} s_p, & n+1-j \text{ even}, \\ \left(\frac{n}{n+1-j+(j-1)^p}\right)^{1/p} s_p, & n+1-j \text{ odd}, \ j > 1, \\ (n-1)\left(\frac{2n}{(n-1)(n+1)^p + (n+1)(n-1)^p}\right)^{1/p} s_p, & j = 1 \text{ and } n \text{ odd}. \end{cases} \tag{3.331}$$

Inequalities (3.331) for $p = 2$ are due to Scott (1936). Beesack (1973, 1976) determined analogous results for concave gauge functions Φ. We only present the sharp bounds in the special cases $\Phi(x) = x^p, 0 < p < 1$,

$$-\left(\frac{n}{2}\right)^{1/p} s_p \leq x_{1:n} - \bar{x} \leq \left(\frac{n}{\max_{1\leq k\leq n-1}\{n-k+k^{1-p}(n-k)^p\}}\right)^{1/p} s_p,$$

$$-\left(\frac{n}{j+j^p}\right)^{1/p} s_p \leq x_{j:n} - \bar{x} \leq \left(\frac{n}{n+1-j+(n+1-j)^p}\right)^{1/p} s_p,$$
$$j = 2, \ldots, n-1, \tag{3.332}$$

$$\left(\frac{n}{\max_{1\leq k\leq n-1}\{k+k^p(n-k)^{1-p}\}}\right)^{1/p} s_p \leq x_{n:n} - \bar{x} \leq \left(\frac{n}{2}\right)^{1/p} s_p \tag{3.333}$$

and

$$y_{j:n} \leq \begin{cases} \left(\frac{n}{n+1-j}\right)^{1/p} s_p, & \text{if } n+1-j \text{ is even,} \\ \left(\frac{n}{n+2^p-j}\right)^{1/p} s_p, & \text{if } n+1-j \text{ is odd,} \end{cases} \quad j = 1, \ldots, n-1, \tag{3.334}$$

with

$$y_{n:n} \leq \left(\frac{n}{2}\right)^{1/p} s_p. \tag{3.335}$$

Now we describe sharp bounds for L-statistics based on the Moriguti and Young inequalities. Let $\Phi : \mathbb{R} \mapsto \mathbb{R}_+$ be a fixed symmetric convex function such that $\Phi(0) = 0$. For simplicity of presentation we also assume that its derivative on the positive halfaxis $\varphi : \mathbb{R}_+ \to \mathbb{R}_+$ is continuous and strictly increasing. Set

$$\psi(x) = \varphi^{-1}(x), \qquad x > 0, \tag{3.336}$$

and

$$\Psi(x) = \int_0^x \psi(t)dt, \quad x \geq 0, \tag{3.337}$$

with

$$\Psi(x) = \Psi(-x), \quad x < 0. \tag{3.338}$$

Let $\|\tilde{x}\|_\Phi$ for $\tilde{0} \neq \tilde{x} \in \mathbb{R}^n$ be defined as the unique positive root of the equation

$$\sum_{i=1}^{n} \Phi\left(\frac{x_i}{\alpha}\right) = \Phi(1) \tag{3.339}$$

and set $\|\tilde{0}\|_\Phi = 0$. Function $\tilde{x} \to \|\tilde{x}\|_\Psi$, $\tilde{x} \in \mathbb{R}^n$, is defined in the same way. Both $\|\cdot\|_\Phi$ and $\|\cdot\|_\Psi$ are norms on $\mathbb{R}^n$ symmetric under rearrangements of coordinates.

Theorem 3.23 *For a given $\tilde{c} \in \mathbb{R}^n$ let $\underline{c}_\Psi \tilde{1}$ be the metric projection of $\underline{\tilde{c}} = P^{\nearrow}\tilde{c}$ onto the linear subspace of constant vectors in $\mathbb{R}^n$ with respect to norm $\|\cdot\|_\Psi$. Then for arbitrary $\tilde{x} \in \mathbb{R}^n$ the following inequality holds*

$$\sum_{i=1}^{n} c_i(x_{i:n} - \bar{x}) \le \|\underline{\tilde{c}} - \underline{c}_\Psi \tilde{1}\|_\Psi \|\tilde{x} - \bar{x}\tilde{1}\|_\Phi. \tag{3.340}$$

Moreover, if

$$\min_{1 \le j \le n-1} \frac{1}{j} \sum_{i=1}^{j} c_i < \bar{c}, \tag{3.341}$$

then (3.340) is sharp and positive and becomes equality for

$$x_{i:n} = \bar{x} + \frac{\|\tilde{x} - \bar{x}\tilde{1}\|_\Phi}{\|\tilde{\psi}(\underline{\tilde{c}} - \underline{c}_\Psi \tilde{1})\|_\Phi} \operatorname{sgn}(\underline{c}_i - \underline{c}_\Psi)\psi\left(\frac{|\underline{c}_i - \underline{c}_\Psi|}{\|\underline{\tilde{c}} - \underline{c}_\Psi \tilde{1}\|_\Psi}\right), \quad i = 1, \ldots, n, \tag{3.342}$$

where

$$\tilde{\psi}(\tilde{c}) = \left(\psi\left(\frac{|c_1|}{\|\tilde{c}\|_\Psi}\right), \ldots, \psi\left(\frac{|c_n|}{\|\tilde{c}\|_\Psi}\right)\right), \quad \tilde{c} = (c_1, \ldots, c_n) \in \mathbb{R}^n. \tag{3.343}$$

Proof Using the Moriguti and Young inequalities (2.80) and (2.223), respectively, we write

$$\begin{aligned}\sum_{i=1}^{n} c_i(x_{i:n} - \bar{x}) &= \sum_{i=1}^{n}(c_i - \underline{c}_\Psi)(x_{i:n} - \bar{x}) \le \sum_{i=1}^{n}(\underline{c}_i - \underline{c}_\Psi)(x_{i:n} - \bar{x}) \\ &\le \|\underline{\tilde{c}} - \underline{c}_\Psi \tilde{1}\|_\Psi \|\tilde{x} - \bar{x}\tilde{1}\|_\Phi.\end{aligned} \tag{3.344}$$

Under (3.341), $\|\tilde{\underline{c}} - \bar{\underline{c}}_\Psi \tilde{1}\|_\Psi > 0$ and the Young inequality in (3.344) becomes equality iff

$$x_{i:n} = \bar{x} + \alpha \mathrm{sgn}(\underline{c}_i - \bar{\underline{c}}_\Psi)\psi\left(\frac{|\underline{c}_i - \bar{\underline{c}}_\Psi|}{\|\tilde{\underline{c}} - \bar{\underline{c}}_\Psi \tilde{1}\|_\Psi}\right), \quad i = 1, \ldots, n, \ \alpha > 0. \tag{3.345}$$

Sequence (3.345) is nondecreasing. Since $\bar{\underline{c}}_\Psi$ is chosen so to minimize $\|\tilde{\underline{c}} - c\tilde{1}\|_\Psi$, $c \in \mathbb{R}$, which is defined implicitly by

$$\sum_{i=1}^{n} \Psi\left(\frac{|\underline{c}_i - c|}{\|\tilde{\underline{c}} - c\tilde{1}\|_\Psi}\right) = \Psi(1), \tag{3.346}$$

the following is satisfied

$$\sum_{i=1}^{n} \psi\left(\frac{|c_i - \bar{\underline{c}}_\Psi\|}{\|\tilde{\underline{c}} - \bar{\underline{c}}_\Psi \tilde{1}\|_\Psi}\right) \mathrm{sgn}(\underline{c}_i - \bar{\underline{c}}_\Psi) = 0 \tag{3.347}$$

and, in consequence, (3.345) has the desired mean. Moreover, it satisfies the norm condition iff

$$\alpha = \frac{\|\tilde{x} - \bar{x}\tilde{1}\|_\Phi}{\|\tilde{\psi}(\tilde{\underline{c}} - \bar{\underline{c}}_\Psi \tilde{1}\|_\Phi}\,. \tag{3.348}$$

Finally, observe that $\underline{c}_i = \underline{c}_j$ implies $x_{i:n} = x_{j:n}$ which is necessary for replacing the former inequality in (3.344) by the equality. □

In the special cases of single order statistics $x_{j:n}$, $2 \le j \le n$, we have

Theorem 3.24 *For a given pair of complementary Young functions Φ and Ψ with continuous increasing derivatives φ and ψ, respectively, and $2 \le j \le n$, define $0 \le c = c_\Psi(j, n) \le \frac{1}{n+1-j}$ and $\alpha = \alpha_\Psi(j, n) > 0$ as the unique pair of solutions to*

$$\begin{cases} (j-1)\Psi\left(\frac{c}{\alpha}\right) + (n+1-j)\Psi\left(\frac{1-(n+1-j)c}{(n+1-j)\alpha}\right) = \Psi(1), \\ (j-1)\psi\left(\frac{c}{\alpha}\right) = (n+1-j)\psi\left(\frac{1-(n+1-j)c}{(n+1-j)\alpha}\right). \end{cases} \tag{3.349}$$

Then for arbitrary non-zero $\tilde{x} \in \mathbb{R}^n$ we have

$$x_{j:n} - \bar{x} \le \alpha_\Psi(j, n)\beta_\Phi(\tilde{x}), \tag{3.350}$$

where $\beta = \beta_\Phi(\tilde{x}) > 0$ *is the unique solution to*

$$\sum_{i=1}^{n} \Phi\left(\frac{|x_i - \bar{x}|}{\beta}\right) = \Phi(1). \tag{3.351}$$

Equation (3.350) becomes equality iff

$$x_{i:n} = \begin{cases} \bar{x} - \beta_\Phi(\tilde{x})\gamma_{\Phi\Psi}(j,n)\psi\left(\frac{c_\Psi(j,n)}{\alpha_\Psi(j,n)}\right), & i = 1, \ldots, j-1, \\ \bar{x} + \beta_\Phi(\tilde{x})\gamma_{\Phi\Psi}(j,n)\psi\left(\frac{1-(n+1-j)c_\Psi(j,n)}{(n+1-j)\alpha_\Psi(j,n)}\right), & i = j, \ldots, n. \end{cases} \tag{3.352}$$

where $\gamma = \gamma_{\Phi\Psi}(j,n)$ *is the unique positive solution to*

$$(j-1)\Phi\left(\gamma\psi\left(\frac{c}{\alpha}\right)\right) + (n+1-j)\Phi\left(\gamma\psi\left(\frac{1-(n+1-j)c}{(n+1-j)\alpha}\right)\right) = \Phi(1) \tag{3.353}$$

with c *and* α *determined from (3.349).*

Proof Recall first that the projection of the vector of coefficients $\tilde{c} = \tilde{e}_j = (0, \ldots, 0, 1, 0, \ldots, 0) \in \mathbb{R}^n$ with 1 at the jth position onto nondecreasing vectors has the coordinates

$$\underline{c}_i = \begin{cases} 0, & i = 1, \ldots, j-1, \\ \frac{1}{n+1-j}, & i = j, \ldots, n. \end{cases} \tag{3.354}$$

Therefore the first equation in (3.349) defines

$$\alpha(c) = \alpha_\Psi(j,n,c) = \|\tilde{\underline{c}} - c\tilde{1}\|_\Psi, \quad c \in \mathbb{R}. \tag{3.355}$$

Now the problem is to minimize (3.355). It is clear that the optimal $c = c_\Psi(j,n)$ does not go beyond the range of $\tilde{\underline{c}}$. Consider $\alpha(c)$ as the implicit function defined by the former equation in (3.349). The function has a unique solution. Since the latter equation in (3.349) provides the criterion for vanishing of $\alpha'(c)$, the optimal $c = c_\Psi(j,n)$ with the minimal norm $\alpha(c_\Psi(j,n)) = \alpha_\Psi(j,n)$ are uniquely determined by (3.349). The norm is the best constant in inequality (3.350) whereas $\beta_\Phi(\tilde{x})$, that is defined in (3.351), describes the dispersion of $\tilde{x}$ from its mean. Observe finally that γ determined from (3.353) is the reciprocal of $\|\tilde{\psi}(\tilde{\underline{c}} - \bar{\underline{c}}_\Psi\tilde{1})\|_\Phi$ with $\tilde{\underline{c}}$ and $\bar{\underline{c}}_\Psi = c_\Psi(j,n)$ defined in (3.354) and (3.349), respectively. This allows us to rewrite the equality conditions of the general form (3.342) in the specific one (3.352). This completes the proof. □

If $\Phi(x) = |x|^p$, $p \geq 1$, then (3.320) and (3.350) provide identical bounds presented explicitly in (3.154). Set now, for example

$$\Phi(x) = (1 + |x|)\ln(1 + |x|) - |x| \tag{3.356}$$

(cf. (2.205) to (2.208)). Then the Beesack bound (3.320) is the solution $\alpha = \alpha(\beta, \tilde{x}, n)$ with $\beta = \frac{j-1}{n}$ and $\tilde{0} \neq \tilde{x} \in \mathbb{R}^n$ to

$$\begin{aligned}&(1-\beta)(1+\alpha)\ln(1+\alpha) + (\alpha - \alpha\beta + \beta)\ln\frac{\alpha - \alpha\beta + \beta}{\beta} - 2\alpha(1-\beta)\\ &= \frac{1}{n}\sum_{i=1}^{n}(1 + |x_i - \bar{x}|)\ln(1 + |x_i - \bar{x}|) - \frac{1}{n}\sum_{i=1}^{n}|x_i - \bar{x}|.\end{aligned} \tag{3.357}$$

In order to derive the Moriguti-Young bound (3.350), we take $\alpha = \alpha_\Psi(\beta, n)$, $\beta = \frac{j-1}{n}$, that together with $\gamma = \frac{c_\Psi(j,n)}{\alpha_\Psi(j,n)}$ solve the system of two equations

$$\begin{cases} \beta\exp(2\gamma) + (1-2\beta)\exp(\gamma) - (1-\beta)\exp\left(\frac{1}{n(1-\beta)\alpha}\right) = 0, \\ 2\beta\exp(\gamma) + (1-2\beta)(1+2\gamma) - \frac{1/\alpha + e - 2}{n} = 0, \end{cases} \tag{3.358}$$

and multiply it by the solution $\beta = \beta_\Phi(\tilde{x})$, $\tilde{x} \neq 0$, to the equation

$$\sum_{i=1}^{n}(\beta + |x_i - \bar{x}|)\ln\frac{\beta + |x_i - \bar{x}|}{\beta} - \sum_{i=1}^{n}|x_{i:n} - \bar{x}| = (2\ln 2 - 1)\beta. \tag{3.359}$$

Clearly, it is easier to solve the single equation (3.357) than (3.358) and (3.359) and so is for the gauge functions different from (3.356). Also, one should solve yet another equation to determine the equality conditions in the Moriguti-Young case. However, the Beesack bound (3.320) depends on $\tilde{x}$ through its generalized moment (3.316) in the implicit way described in (3.318). The Moriguti-Young bound (3.350) depends on the dispersion of $\tilde{x}$, measured in $\beta_\Phi(\tilde{x}) = \|\tilde{x} - \bar{x}\tilde{1}\|_\Phi$ units (see (3.351)) up to a universal multiplicative constant $\alpha_\Psi(j, n)$ (see (3.349)) that does not depend on $\tilde{x}$. Another advantage of the Moriguti-Young bounds is that they are applicable for arbitrary combinations of order statistics.

In an obvious way we determine the lower Moriguti-Young bounds

$$\sum_{i=1}^{n} c_i(x_{i:n} - \bar{x}) \geq -\|\bar{\tilde{c}} - \bar{\bar{c}}_\Psi\tilde{1}\|_\Psi \|\tilde{x} - \bar{x}\tilde{1}\|_\Phi \tag{3.360}$$

which are negative and tight iff

$$\max_{1\leq j\leq n-1}\frac{1}{j}\sum_{i=1}^{j} c_i > \bar{c} \tag{3.361}$$

and attained by

$$x_{i:n} = \bar{x} + \frac{\|\tilde{x} - \bar{x}\tilde{1}\|_\Phi}{\|\tilde{\psi}(\bar{\bar{c}} - \bar{\bar{c}}_\Psi \tilde{1})\|_\Phi} \operatorname{sgn}(\bar{\bar{c}}_\Psi - \bar{c}_i)\psi\left(\frac{|\bar{c}_i - \bar{\bar{c}}_\Psi|}{\|\bar{\bar{c}} - \bar{\bar{c}}_\Psi \tilde{1}\|_\Psi}\right), \; i = 1, \ldots, n, \tag{3.362}$$

where $\bar{\bar{c}} = P^{\searrow}\bar{c}$ is the ℓ^2-projection of $\bar{c}$ onto the class of nonincreasing sequences and $\bar{\bar{c}}_\Psi$ minimizes $\|\bar{\bar{c}} - c\tilde{1}\|_\Psi$. Clearly, the lower and upper bounds for the jth and $(n + 1 - j)$th order statistics, respectively, are equal up to the sign.

For the sake of completeness, we write down the Moriguti-Young inequalities for arbitrary L-statistics, expressed in $\|\tilde{x}\|_\Phi$ and $\|\tilde{x} - x_{1:n}\tilde{1}\|_\Phi$ units. The former hold for nonnegative samples and take on the form

$$\sum_{i=1}^{n} c_i x_{i:n} \le \|\underline{\tilde{c}}_+\|_\Psi \|\tilde{x}\|_\Phi. \tag{3.363}$$

They are attained for

$$x_{i:n} = \frac{\|\tilde{x}\|_\Phi}{\|\tilde{\psi}(\underline{\tilde{c}}_+)\|_\Phi} \psi\left(\frac{(\underline{c}_i)_+}{\|\underline{\tilde{c}}_+\|_\Psi}\right), \quad i = 1, \ldots, n, \tag{3.364}$$

under condition that $\underline{\tilde{c}}_+ \neq \tilde{0}$. Furthermore, with $\tilde{c}' = (c_2, \ldots, c_n)$ we have

$$\sum_{i=1}^{n} c_i (x_{i:n} - x_{1:n}) \le \|\underline{\tilde{c}}'_+\|_\Psi \|\tilde{x} - x_{1:n}\tilde{1}\|_\Phi, \tag{3.365}$$

which are attained by

$$x_{i:n} = x_{1:n} + \frac{\|\tilde{x} - x_{1:n}\tilde{1}\|_\Phi}{\|\tilde{\psi}(\underline{\tilde{c}}'_+)\|_\Phi} \psi\left(\frac{(\underline{c}'_i)_+}{\|\underline{\tilde{c}}'_+\|_\Psi}\right), \quad i = 2, \ldots, n, \tag{3.366}$$

provided that $\underline{\tilde{c}}'_+ \neq \tilde{0}$. For the single order statistics, $x_{j:n}$, $2 \le j \le n$, the optimal constants in (3.363) and (3.365) are identical and coincide with the solutions $\alpha = \alpha_\Psi(j, n)$ to equations

$$(n + 1 - j)\Psi\left(\frac{1}{(n + 1 - j)\alpha}\right) = \Psi(1). \tag{3.367}$$

We conclude the subsection with an abstract generalization of the Moriguti-Young bounds. Here we replace the scale units expressed in terms of Orlicz norms $\|\tilde{x}\|_\Phi$, symmetric with respect to all rearrangements of coordinates of $\tilde{x}$, by general symmetric norms on $\mathbb{R}^n$. We recall some basic notions of the theory of Banach spaces. Every continuous linear functional acting on the elements of $\mathbb{R}^n$ endowed with a norm $\|\cdot\|$ has the form

$$\tilde{x} = (x_1, \ldots, x_n) \to \sum_{i=1}^{n} c_i x_i, \quad \tilde{x} \in \mathbb{R}^n, \tag{3.368}$$

for some $\tilde{c} = (c_1, \ldots, c_n) \in \mathbb{R}^n$ representing the functional. The family of functionals, represented by $\tilde{c} \in \mathbb{R}^n$, may be also endowed with a norm $\|\cdot\|_*$ defined by

$$\|\tilde{c}\|_* = \sup_{\tilde{x} \neq \tilde{0}} \frac{|\sum_{i=1}^{n} c_i x_i|}{\|x\|} = \sup_{\|x\|=1} \left| \sum_{i=1}^{n} c_i x_i \right| \tag{3.369}$$

which is called conjugate to $\|\cdot\|$. Consequently, the following inequality is sharp

$$\left| \sum_{i=1}^{n} c_i x_i \right| \leq \|\tilde{c}\|_* \|\tilde{x}\|, \quad \tilde{x} \in \mathbb{R}^n. \tag{3.370}$$

Completeness of $(\mathbb{R}^n, \|\cdot\|)$ implies completeness of $(\mathbb{R}^n, \|\cdot\|_*)$. General evaluations of L-statistics are based on the following theorem presented in Rychlik (1992).

Theorem 3.25 *Let $\underline{\tilde{c}} = P^{\nearrow}\tilde{c}$ denote the ℓ^2-projection of a given $\tilde{c} \in \mathbb{R}^n$ onto the convex cone of nondecreasing sequences in $\mathbb{R}^n$. For every nondecreasing $\tilde{x} \in \mathbb{R}^n$ we have*

$$\sum_{i=1}^{n} c_i x_i \leq \|\underline{\tilde{c}}\|_* \|\tilde{x}\|. \tag{3.371}$$

Moreover, if $\|\cdot\|$ is symmetric with respect to the rearrangements of coordinates and $\underline{\tilde{c}} \neq \tilde{0}$, then inequality (3.371) is sharp.

Proof Equation (3.371) immediately follows from (2.80) and (3.370). Now we assume that $\|\cdot\|$ is symmetric and $\|\underline{\tilde{c}}\|_*$ is positive. For a given $0 < \varepsilon < \|\underline{\tilde{c}}\|_*$, let $\tilde{x} = \tilde{x}(\varepsilon)$ be an element of the unit ball with respect to $\|\cdot\|$ that satisfies

$$0 < \sum_{i=1}^{n} \underline{c}_i x_i \geq \|\underline{\tilde{c}}\|_* - \varepsilon. \tag{3.372}$$

Our proof consists in constructing a nondecreasing x^* with $\|x^*\| \leq 1$ that satisfies

$$\sum_{i=1}^{n} c_i x_i^* = \sum_{i=1}^{n} \underline{c}_i x_i. \tag{3.373}$$

Let $0 = j_0 < j_1 < \cdots < j_M = n$ be the partition of $\{0, \ldots, n\}$ defined in (2.78). Put

$$x'_i = \frac{1}{j_k - j_{k-1}} \sum_{r=j_{k-1}+1}^{j_k} x_r, \quad i = j_{k-1}+1, \ldots, j_k, \ k = 1, \ldots, M. \tag{3.374}$$

Then, by (2.79) and (3.374)

$$\sum_{i=1}^{n} \underline{c}_i x'_i = \sum_{k=1}^{M} \underline{c}_{j_k} \sum_{i=j_{k-1}+1}^{j_k} x'_i = \sum_{k=1}^{M} \underline{c}_{j_k} \sum_{i=j_{k-1}+1}^{j_k} x_i = \sum_{i=1}^{n} \underline{c}_i x_i. \tag{3.375}$$

Now we apply (2.78) and (2.79) for defining the projection $\tilde{x}^* = P^{\nearrow}\tilde{x}'$ of $\tilde{x}'$ onto the convex cone of nondecreasing sequences. We have

$$x^*_i = \frac{1}{\ell_m - \ell_{m-1}} \sum_{r=\ell_{m-1}+1}^{\ell_m} x'_r, \quad i = \ell_{m-1}+1, \ldots, \ell_m, \ m = 1, \ldots, N, \tag{3.376}$$

for some $0 = \ell_0 < \ell_1 < \cdots < \ell_N = n$. A crucial fact is here that

$$\{\ell_m : m = 0, \ldots, N\} = \{j_{k_m} : m = 0, \ldots, N\} \subset \{j_k : k = 0, \ldots, M\}. \tag{3.377}$$

One consequence is

$$\sum_{i=1}^{n} \underline{c}_i x^*_i = \sum_{i=1}^{n} \underline{c}_i x'_i \tag{3.378}$$

(cf. (3.375)) and the other, due to (2.81), is

$$\sum_{i=1}^{n} c^*_i x^*_i = \sum_{i=1}^{n} \underline{c}_i x^*_i. \tag{3.379}$$

Combining (3.375), (3.377) and (3.379), we obtain (3.373).

It remains to verify that $\|\tilde{x}^*\| \leq \|\tilde{x}\|$. Observe that $\tilde{x}^*$ arised from $\tilde{x}$ by twofold application of the operation averaging some subsets of coordinates. The operation that averages coordinates $y_{i_1}, \ldots, y_{i_k}$ of a vector $\tilde{y} \in \mathbb{R}^n$ consists in taking a uniform mixture of all the modifications of $\tilde{y}$ such that $y_{i_1}, \ldots, y_{i_k}$ are replaced by a permutation of them, and the other coordinates remain unchanged. All the modifications have the same norm, because this is symmetric with respect to rearrangements. By convexity, the norm of a mixture does not exceed the norm of the original vector. This ends the proof. □

As the consequences of Theorem 3.25 we obtain the following sharp evaluations of L-statistics

$$\sum_{i=1}^{n} c_i(x_{i:n} - \bar{x}) \leq \|\tilde{\underline{c}} - \bar{\underline{c}}_* \tilde{1}\|_* \|\tilde{x} - \bar{x}\tilde{1}\|, \tag{3.380}$$

$$\sum_{i=1}^{n} c_i x_{i:n} \leq \|\tilde{\underline{c}}_+\|_* \|\tilde{x}\|, \qquad \tilde{x} \in \mathbb{R}_+^n, \tag{3.381}$$

$$\sum_{i=1}^{n} c_i(x_{i:n} - x_{1:n}) \leq \|\tilde{\underline{c}}'_+\|_* \|\tilde{x} - x_{1:n}\tilde{1}\|, \tag{3.382}$$

where $\bar{\underline{c}}_*$ minimizes $\|\tilde{\underline{c}} - c\tilde{1}\|_*$, and the remaining notation is standard.

Chapter 4
General, Symmetric and Life IID Samples

In this chapter we consider order statistics based on iid samples. In Sect. 4.1 we gathered basic formulae representing the distribution and density functions of order statistics and their expectations. There is also presented the variation diminishing property of the Bernstein polynomials which is very useful for analyzing variability of distribution and density functions of order statistics and their combinations. Section 4.2 contains the most classic bounds for the expectations of L-statistics with monotone vectors of coefficients. In particular, the fundamental inequality of Gumbel (1954) and Hartley and David (1954) for the sample maximum, and that of Plackett (1947) for the sample range are presented there. Section 4.3 contains analogous results for arbitrary L-statistics. They are based on the seminal paper by Moriguti (1953) who determined sharp bounds on the expectations of single order statistics and their differences. Moriguti (1953) provided efficient tools for determining sharp bounds on the expectations of arbitrary L-statistics in various scale units. In particular, the sharp bounds on properly centered L-statistics measured in the scale units based on central absolute population moments are studied in Sect. 4.4. Nonpositive upper bounds for negative L-statistics and nonnegative lower ones for positive L-statistics are presented in Sect. 4.5.

More refined bounds for L-statistics coming from symmetric and life populations are established in Sects. 4.6 and 4.7, respectively. In the latter case, the scale units are constructed on the basis of raw population moments. Higher moments of order statistics are evaluated as well. The most comprehensive studies of these topics were presented in Rychlik (2021, 2022), respectively. Bounds on the expectations of L-statistics gauged in the Gini mean difference units, determined in Kozyra and Rychlik (2017a) are described in Sect. 4.8. Section 4.9 contains bounds on the expectations of L-statistics expressed in terms of the mean residual life function. Inequalities for moments of order statistics based on the orthogonal expansions are considered in Sect. 4.10. The last section is devoted to optimal bounds on variances of order and L-statistics. A special case of order statistics based on symmetrically distributed iid samples is treated in particular. The main results on the topic were

N. Balakrishnan, T. Rychlik, *Sharp Inequalities for Ordered Random Variables in Statistics and Reliability*, Frontiers in Probability and the Statistical Sciences,
https://doi.org/10.1007/978-3-031-61347-0_4

presented in Papadatos (1995, 1997), Jasiński and Rychlik (2013) and Kozyra and Rychlik (2017b).

4.1 Basic Formulae

We consider order statistics $X_{1:n} \leq \ldots \leq X_{n:n}$ based on independent identically distributed samples (iid, for short) $X_1, \ldots, X_n$ with a parent distribution function denoted by F. We start with presenting some well known formulae. The event $\{X_{j:n} \leq x\}$ is identical with the fact that at least j among $X_1, \ldots, X_n$ are not greater than x. The latter can be represented as the sum of disjoint events that precisely i of $X_1, \ldots, X_n$ are not greater than x, and the remaining $n-i$ exceed x, where i takes on the values from $\{j, \ldots, n\}$. Therefore

$$\begin{aligned} &\mathbb{P}(X_{j:n} \leq x) \\ &= \sum_{i=j}^{n} \sum_{1 \leq k_1 < \ldots < k_i \leq n} \mathbb{P}(X_k \leq x,\ k \in \{k_1, \ldots, k_i\},\ X_k > x,\ k \notin \{k_1, \ldots, k_i\}) \\ &= \sum_{i=j}^{n} \sum_{1 \leq k_1 < \ldots < k_i \leq n} F^i(x)[1-F(x)]^{n-i} = \sum_{i=j}^{n} \binom{n}{i} F^i(x)[1-F(x)]^{n-i}. \end{aligned} \tag{4.1}$$

In particular, if $X_1, \ldots, X_n$ are standard uniform, then

$$\mathbb{P}(X_{j:n} \leq x) = F_{j:n}(x) = \sum_{i=j}^{n} \binom{n}{i} x^i (1-x)^{n-i} = \sum_{i=j}^{n} B_{i,n}(x), \qquad 0 < x < 1, \tag{4.2}$$

where

$$B_{i,n}(x) = \binom{n}{i} x^i (1-x)^{n-i}, \qquad 0 < x < 1, \tag{4.3}$$

denotes the ith Bernstein polynomial of degree n, $i = 0, \ldots, n$. Note that (4.1) is the composition of the distribution function of the jth order statistics from the iid standard uniform sample with the parent distribution function F

$$\mathbb{P}(X_{j:n} \leq x) = F_{j:n} \circ F(x) = \sum_{i=j}^{n} B_{i,n} \circ F(x) \tag{4.4}$$

If F has a density function f, then each $X_{j:n}$, $j = 1, \dots, n$, has the density function

$$\frac{d}{dx} F_{j:n} \circ F(x) = f_{j:n} \circ F(x)\, f(x) = nB_{j-1,n-1} \circ F(x)\, f(x), \tag{4.5}$$

where

$$f_{j:n}(x) = n\binom{n-1}{j-1}x^{j-1}(1-x)^{n-j}, \qquad j = 1, \dots, n, \tag{4.6}$$

are the density functions of consecutive order statistics in the iid standard uniform sample.

It is well known that if $U_1, \dots, U_n$ are iid standard uniform, then $F^{-1}(U_1)$, ..., $F^{-1}(U_n)$ are iid with the marginal distribution function F. Since the quantile transformations preserve the ordering, it follows that $F^{-1}(U_{1:n}) \le \dots \le F^{-1}(U_{n:n})$ represent the order statistics based on iid samples with the baseline distribution function F. The above fact implies the following formula

$$\mathbb{E}X_{j:n} = \mathbb{E}F^{-1}(U_{j:n}) = \int_0^1 F^{-1}(x) f_{j:n}(x)dx, \qquad j = 1, \dots, n, \tag{4.7}$$

which we frequently use in the sequel. In particular we also have

$$\mathbb{E}X_1 = \mathbb{E}F^{-1}(U_1) = \int_0^1 F^{-1}(x)dx. \tag{4.8}$$

If X_1 has an expectation $\mu \in \mathbb{R}$ and a finite second moment, we can also write

$$\sigma^2 = \mathbb{V}ar\, X_1 = \mathbb{E}[F^{-1}(U_1) - \mu]^2 = \int_0^1 [F^{-1}(x) - \mu]^2 dx \tag{4.9}$$

More generally, we express central absolute moments of various positive orders p as follows

$$\sigma_p^p = \mathbb{E}|X_1 - \mu|^p = \int_0^1 |F^{-1}(x) - \mu|^p dx. \tag{4.10}$$

We further use scale units based on absolute central moments defined as follows

$$\sigma_p = (\sigma_p^p)^{1/p} = \left(\mathbb{E}|X_1 - \mu|^p\right)^{1/p}, \qquad p > 0, \tag{4.11}$$

under the assumption that the pth moment of X_1 exists, and

$$\sigma_\infty = \lim_{p \to +\infty} \sigma_p = \text{ess sup}|X_1 - \mu| = \max\{F^{-1}(1) - \mu, \mu - F^{-1}(0)\}, \tag{4.12}$$

where $F^{-1}(0)$ and $F^{-1}(1)$ denote the left- and right-end points of the support of X_1 under the condition that the support is bounded.

We also consider linear combinations $L_{\tilde{c}:n} = \sum_{i=1}^{n} c_i X_{i:n}$ of order statistics based on iid samples with arbitrary vectors of coefficients $\tilde{c} = (c_1, \dots, c_n) \in \mathbb{R}^n$. We call them shortly L-statistics. Given $\tilde{c}$, we define the following linear combinations of distribution and density functions

$$F_{\tilde{c}:n}(x) = \sum_{i=1}^{n} c_i F_{i:n}(x), \tag{4.13}$$

$$f_{\tilde{c}:n}(x) = \sum_{i=1}^{n} c_i f_{i:n}(x), \tag{4.14}$$

We obviously have

$$\mathbb{E} \sum_{i=1}^{n} c_i X_{i:n} = \sum_{i=1}^{n} c_i \int_0^1 F^{-1}(x) f_{i:n}(x) dx = \int_0^1 F^{-1}(x) f_{\tilde{c}:n}(x) dx. \tag{4.15}$$

As we see the expectation depends on the parent distribution function as well as the linear combination (4.14). The latter is obviously a combination of Bernstein polynomials of degree $n-1$. Note that (4.13) is a linear combination of Bernstein polynomials of degree n. A useful tool for analyzing variability of (4.13) and (4.14) applied frequently in our further investigations is so called variation diminishing property (VDP, for short) of Bernstein polynomials. It is presented in the following lemma.

Lemma 4.1 *The number of sign changes of an arbitrary linear combination*

$$B(x) = \sum_{i=0}^{n} a_i B_{i,n}(x), \qquad 0 < x < 1, \tag{4.16}$$

of Bernstein polynomials of a fixed degree n does not exceed the number of sign changes of the vector $\tilde{a} = (a_0, \dots, a_n)$ *of combination coefficients. Moreover, the signs of the combination (4.16) in a right neighborhood of* 0 *and a left neighborhood of* 1 *coincide with the signs of the first and last non-zero element of the vector* $\tilde{a}$.

Proof It can be immediately concluded from the variation diminishing property of the combinations of power functions acting on the positive half-axis (see Theorem 2.47 and the following Corollary 2.1). Recall that the number of sign changes of

$$p(x) = \sum_{i=1}^{n} a_i x^{b_i}, \qquad x > 0, \tag{4.17}$$

where $a_i \in \mathbb{R}$, $i = 1, \ldots, n$, $b_1 < \ldots < b_n$, is not greater than the number of signs changes in the sequence $a_1, \ldots, a_n$, and the signs of (4.17) in the neighborhoods of 0 and $+\infty$ are identical with the signs of the first and last nonzero elements of $a_1, \ldots, a_n$, respectively. Indeed, we can write

$$B(x) = \sum_{i=0}^{n} a_i \binom{n}{i} x^i (1-x)^{n-i} = (1-x)^n \sum_{i=0}^{n} a_i \binom{n}{i} \left(\frac{x}{1-x}\right)^i. \tag{4.18}$$

Using the increasing transformation changing the variable $(0, 1) \ni x \mapsto y = \frac{x}{1-x} \in \mathbb{R}_+$, we obtain

$$P(y) = \frac{1}{(1+y)^n} \sum_{i=0}^{n} a_i \binom{n}{i} y^i, \qquad y > 0, \tag{4.19}$$

and preserve the number and order of intervals of positivity and negativity. The VDP for the power functions applies to the sum, and the positive factor preceding it does not affect the number and order of the sign changes. □

In Sect. 4.2 we describe sharp negative lower and upper positive evaluations of the expectations of standardized L-statistics $\mathbb{E} \sum_{i=1}^{n} c_i \frac{X_{i:n}-\mu}{\sigma}$ with nonincreasing and nondecreasing, respectively vectors of coefficients. They are obtained with use of the Schwarz inequality. In particular, we present the inequalities for the sample range and extremes due to Plackett (1947), Gumbel (1954) and Hartley and David (1954). These are the oldest results in the subject. In Sect. 4.3 we determine the positive upper and negative lower mean-variance bounds on the expectations of general L-statistics. They are based on the combination of the Schwarz inequality and the convex minorant method proposed in Moriguti (1953), where the particular bounds for the single order statistics and their differences were presented. The generalization to the bounds expressed in various scale units based on the absolute central moments are described in Sect. 4.4 (see also Rychlik, 1998). Analogous nonnegative lower and nonpositive upper bounds are contained in Sect. 4.5. They were presented in Goroncy (2009). The above results are sharpened in Sects. 4.6 and 4.7 under the extra restrictions that the original random variables are symmetrically distributed and supported in the positive half-axis, respectively. They are based on the papers by Rychlik (2021, 2022), respectively. Section 4.8 contains the results of Kozyra and Rychlik (2017a) who established bounds on the centered L-statistics expressed in the Gini mean difference units $\mathbb{E}|X_1 - X_2| = \mathbb{E}(X_{2:2} - X_{1:2})$. In Sect. 4.9 we find evaluations of expected L-statistics represented in terms of the mean residual life of the parent distribution. Section 4.10 is devoted to inequalities based on orthogonal expansions. The bounds on variances of order statistics and spacings presented in Sect. 4.11 are based on the papers by Moriguti (1951), Papadatos (1995, 1997), Jasiński and Rychlik (2013), and Kozyra and Rychlik (2017b).

4.2 Positive Upper and Negative Lower Schwarz Bounds

4.2.1 Sample Maximum and Range

We first present sharp upper bounds on the expectations of sample maxima based on iid samples. The result was independently derived by Gumbel (1954) and Hartley and David (1954).

Theorem 4.1 *Let $X_1, \ldots, X_n$ be iid with arbitrarily fixed mean $\mu \in \mathbb{R}$ and variance $\sigma^2 > 0$. Then*

$$\mathbb{E}\frac{X_{n:n} - \mu}{\sigma} \le \frac{n-1}{\sqrt{2n-1}}, \tag{4.20}$$

and the equality holds if the parent distribution function has the form

$$F(x) = \left[\frac{1}{n}\left(\frac{n-1}{\sqrt{2n-1}}\,\frac{x-\mu}{\sigma} + 1\right)\right]^{\frac{1}{n-1}}, \quad -\frac{\sqrt{2n-1}}{n-1} < \frac{x-\mu}{\sigma} < \sqrt{2n-1}. \tag{4.21}$$

Proof Using (4.7), we can write

$$\mathbb{E}\frac{X_{n:n} - \mu}{\sigma} = \int_0^1 \frac{F^{-1}(x) - \mu}{\sigma} f_{n:n}(x)\,dx = \int_0^1 \frac{F^{-1}(x) - \mu}{\sigma} n x^{n-1}\,dx. \tag{4.22}$$

By (4.8), the Schwarz inequality (2.120) and (4.9) we obtain

$$\begin{aligned}
\mathbb{E}\frac{X_{n:n} - \mu}{\sigma} &= \int_0^1 \frac{F^{-1}(x) - \mu}{\sigma}[n x^{n-1} - 1]\,dx \\
&\le \left[\int_0^1 \frac{[F^{-1}(x) - \mu]^2}{\sigma^2}\,dx\right]^{\frac{1}{2}} \left[\int_0^1 [n x^{n-1} - 1]^2 dx\right]^{\frac{1}{2}} \\
&= \left[\int_0^1 n^2 x^{2n-2}\,dx - 1\right]^{\frac{1}{2}} = \frac{n-1}{\sqrt{2n-1}}.
\end{aligned} \tag{4.23}$$

The equality in the Schwarz inequality is attained if

$$\frac{F^{-1}(x) - \mu}{\sigma} = \alpha[f_{n:n}(x) - 1], \qquad 0 < x < 1, \tag{4.24}$$

for some positive α. Note that the LHS and RHS are nondecreasing, and have integrals over [0, 1] equal to 0. The LHS has the L^2-norm equal to 1. The RHS has the same norm if

$$\alpha = \frac{1}{||f_{n:n} - 1||_2} = \frac{\sqrt{2n-1}}{n-1}. \tag{4.25}$$

Combining (4.24) and (4.25) we obtain a unique representation of the quantile function of the unique marginal distribution with mean μ and variance σ^2 which attains the equality in (4.20). Simple calculations allow us to determine the explicit form (4.21) of the respective distribution function. □

Observe that (4.21) is the distribution function of a linear transformation of the power distribution function with the shape parameter $\frac{1}{n-1}$.

Even older are the analogous bounds for the sample ranges determined by Plackett (1947).

Theorem 4.2 *Under the assumptions of Theorem* 4.1 *we have*

$$\mathbb{E}\frac{X_{n:n} - X_{1:n}}{\sigma} \leq n\sqrt{\frac{2}{2n-1}\left(1 - \frac{1}{\binom{2n-2}{n-1}}\right)}. \tag{4.26}$$

The inequality becomes the equality if the parent distribution function is determined from the equality

$$\frac{F^{-1}(x) - \mu}{\sigma} = \sqrt{\frac{2n-1}{2}\frac{\binom{2n-2}{n-1}}{\binom{2n-2}{n-1} - 1}}\,[x^{n-1} - (1-x)^{n-1}]. \tag{4.27}$$

Relations (4.27) for $n = 2$ and 3 describe the uniform distribution on the interval $[\mu - \sqrt{3}\sigma, \mu + \sqrt{3}\sigma]$. For the other $n \geq 4$ formula (4.27) does not determine the distribution functions which have explicit representations.

Proof The sample range is an example of L-statistic with the vector of coefficients $\tilde{c} = (-1, 0 \ldots, 0, 1)$. Then (4.14) takes on the form

$$f_{\tilde{c}:n}(x) = f_{n:n}(x) - f_{1:n}(x) = n[x^{n-1} - (1-x)^{n-1}]. \tag{4.28}$$

Consequently,

$$\begin{aligned}
\mathbb{E}\frac{X_{n:n} - X_{1:n}}{\sigma} &= n\int_0^1 \frac{F^{-1}(x) - \mu}{\sigma}[x^{n-1} - (1-x)^{n-1}]dx \\
&\leq n\left[\int_0^1 \frac{[F^{-1}(x) - \mu]^2}{\sigma^2}dx\right]^{\frac{1}{2}}\left[\int_0^1 [x^{n-1} - (1-x)^{n-1}]^2 dx\right]^{\frac{1}{2}}
\end{aligned}$$

$$= n\left[\int_0^1 [x^{2n-2} + (1-x)^{2n-2} - 2x^{n-1}(1-x)^{n-1}]dx\right]^{\frac{1}{2}}$$

$$= n\left[\frac{2}{2n-1} - 2\frac{[(n-1)!]^2}{(2n-1)!}\right]^{\frac{1}{2}} = n\sqrt{\frac{2}{2n-1}\left(1 - \frac{1}{\binom{2n-2}{n-1}}\right)}. \tag{4.29}$$

The Schwarz inequality is the equality, and the moment conditions are satisfied if

$$\frac{F^{-1}(x) - \mu}{\sigma} = \frac{f_{n:n}(x) - f_{1:n}(x)}{||f_{n:n} - f_{1:n}||_2} = \sqrt{\frac{2n-1}{2}\frac{\binom{2n-2}{n-1}}{\binom{2n-2}{n-1} - 1}}\,[x^{n-1} - (1-x)^{n-1}]. \tag{4.30}$$

This ends the proof. □

4.2.2 Imputation-Type Bounds

Observe that the bounds presented in Theorems 4.1 and 4.2 are valid for any arbitrary distribution with fixed mean $\mu \in \mathbb{R}$ and variance $\sigma^2 > 0$. They do not assume any other knowledge about the underlying distribution. However, for many common distributions, moments of order statistics (such as means, variances and covariances) have been tabulated rather extensively; see, for example, Harter and Balakrishnan (1996) wherein tables of these quantities are presented for numerous distributions for small sample sizes (often up to 20) except for normal and exponential distributions for which they are available up to large sample sizes.

Thence, the natural question that arises is whether for a particular distribution of interest, these tabulated quantities of moments of order statistics (from small or large sample sizes, whichever available) could be utilized to develop improved imputation-type bounds. The answer to this question is affirmative, as carried out by Balakrishnan (1990).

To fix the ideas, let $X_1, \ldots, X_n$ be iid with arbitrarily fixed mean $\mu \in \mathbb{R}$ and variance $\sigma^2 > 0$. Without loss of any generality, let us take $\mu = 0$ and $\sigma = 1$. Further, suppose $\mathbb{E}X_{m:m}$ and $\mathbb{E}X_{p:p}$ are available exactly for two sample sizes m and p (for example, from two sample sizes from the tables of Harter and Balakrishnan (1996)). Then, the imputation-type bound derived by Balakrishnan (1990) is as presented in the following theorem.

Theorem 4.3 *Under the assumptions of Theorem 4.1 with mean $\mu = 0$ and standard deviation $\sigma = 1$, without loss of any generality, we have*

$$\mathbb{E}X_{n:n} \leq (n-1)\left\{\frac{1}{2n-1} - \tau_{n,m,p}\right\}^{1/2}, \tag{4.31}$$

where

$$\tau_{n,m,p} \tag{4.32}$$
$$= \frac{\left\{\frac{m-1}{m+n-1}\mathbb{E}X_{p:p} - \frac{p-1}{p+n-1}\mathbb{E}X_{m:n}\right\}^2}{\frac{(m-1)^2}{2m-1}(\mathbb{E}X_{p:p})^2 + \frac{(p-1)^2}{2p-1}(\mathbb{E}X_{m:m})^2 - \frac{2(m-1)(p-1)}{m+p-1}\mathbb{E}X_{m:m}\,\mathbb{E}X_{p:p}},$$

with the exact values of $\mathbb{E}X_{m:m}$ and $\mathbb{E}X_{p:p}$ being available (for some m and p) to be used for imputational purpose. Furthermore, $\tau_{n,m,p}$ is strictly in the interval $\left(0, \frac{1}{2n-1}\right)$.

Proof For an arbitrary distribution with $\mathbb{E}X = \mu = 0$ and $\mathbb{E}X^2 = \sigma^2 = 1$, let us first write

$$\mathbb{E}X_{n:n} = \int_0^1 F^{-1}(x)[nx^{n-1} - C\{mx^{m-1} - F^{-1}(x)\mathbb{E}X_{m:m}\}$$
$$-D\{px^{p-1} - F^{-1}(x)\mathbb{E}X_{p:p}\} - \lambda]dx, \tag{4.33}$$

where C, D and λ are Lagrangian multipliers to be suitably determined. Now, by applying the Cauchy-Schwarz inequality in (2.120) to (4.33), we obtain

$$\mathbb{E}X_{n:n}$$
$$\leq \left[\int_0^1 \{F^{-1}(x)\}^2 dx\right]^{1/2} \left[\int_0^1 \left\{nx^{n-1} - C(mx^{m-1} - F^{-1}(x)\mathbb{E}X_{m:m})\right.\right.$$
$$\left.\left.-D(px^{p-1} - F^{-1}(x)\mathbb{E}X_{p:p}) - \lambda\right\}^2 dx\right]^{1/2}$$
$$= \left[\frac{n^2}{2n-1} + C^2\left\{\frac{m^2}{2m-1} - (\mathbb{E}X_{m:m})^2\right\}\right.$$
$$+D^2\left\{\frac{p^2}{2p-1} - (\mathbb{E}X_{p:p})^2\right\}$$
$$+\lambda^2 - 2C\left\{\frac{mn}{m+n-1} - \mathbb{E}X_{n:n}\,\mathbb{E}X_{m:m}\right\}$$

$$
\begin{aligned}
&-2D\left\{\frac{pn}{p+n-1}-\mathbb{E}X_{n:n}\,\mathbb{E}X_{p:p}\right\}\\
&+2CD\left\{\frac{mp}{m+p-1}-\mathbb{E}X_{m:m}\,\mathbb{E}X_{p:p}\right\}\\
&\left.-2\lambda+2\lambda C+2\lambda D\right]^{1/2}. \qquad (4.34)
\end{aligned}
$$

Realizing that the quantity $\mathbb{E}X_{n:n}$ is present on the RHS of (4.34) and that it is what we are seeking a bound for, we set

$$
C\mathbb{E}X_{m:m}+D\mathbb{E}X_{p:p}=0 \quad \text{or} \quad D=-C\frac{\mathbb{E}X_{m:m}}{\mathbb{E}X_{p:p}}
$$

in which case the bound on the RHS of (4.34) reduces to

$$
\begin{aligned}
&\left[\frac{n^2}{2n-1}+C^2\left\{\frac{m^2}{2m-1}-(\mathbb{E}X_{m:m})^2\right\}\right.\\
&\quad+C^2\frac{(\mathbb{E}X_{m:m})^2}{(\mathbb{E}X_{p:p})^2}\left\{\frac{p^2}{2p-1}-(\mathbb{E}X_{p:p})^2\right\}\\
&\quad+\lambda^2-2C\frac{mn}{m+n-1}+2C\frac{pn}{p+n-1}\frac{\mathbb{E}X_{m:m}}{\mathbb{E}X_{p:p}}\\
&\quad-2C^2\left\{\frac{mp}{m+p-1}-\mathbb{E}X_{m:m}\,\mathbb{E}X_{p:p}\right\}\frac{\mathbb{E}X_{m:m}}{\mathbb{E}X_{p:p}}\\
&\quad\left.-2\lambda+2\lambda C-2\lambda C\frac{\mathbb{E}X_{m:m}}{\mathbb{E}X_{p:p}}\right]^{1/2}. \qquad (4.35)
\end{aligned}
$$

Upon minimizing (4.35) with respect to λ, we obtain

$$
2\lambda-2+2C-2C\frac{\mathbb{E}X_{m:m}}{\mathbb{E}X_{p:p}}=0 \quad \text{or} \quad \lambda=1-C+C\frac{\mathbb{E}X_{m:m}}{\mathbb{E}X_{p:p}}
$$

which simply yields

$$
\lambda^2-2\lambda+2\lambda C-2\lambda C\frac{\mathbb{E}X_{m:m}}{\mathbb{E}X_{p:p}}=-\left(1-C+C\frac{\mathbb{E}X_{m:m}}{\mathbb{E}X_{p:p}}\right)^2.
$$

Substituting this in (4.35) and simplifying the resulting expression, we obtain the upper bound for $\mathbb{E}X_{n:n}$ to be

$$\left[\frac{(n-1)^2}{2n-1}+C^2\left\{\frac{(m-1)^2}{2m-1}-(\mathbb{E}X_{m:m})^2\right\}\right.$$
$$+C^2\left\{\frac{(p-1)^2}{2p-1}-(\mathbb{E}X_{p:p})^2\right\}\frac{(\mathbb{E}X_{m:m})^2}{(\mathbb{E}X_{p:p})^2}$$
$$-2C\frac{(m-1)(n-1)}{m+n-1}+2C\frac{(p-1)(n-1)}{p+n-1}\frac{\mathbb{E}X_{m:m}}{\mathbb{E}X_{p:p}}$$
$$\left.-2C^2\left\{\frac{(m-1)(p-1)}{m+p-1}-\mathbb{E}X_{m:m}\,\mathbb{E}X_{p:p}\right\}\frac{\mathbb{E}X_{m:m}}{\mathbb{E}X_{p:p}}\right]^{1/2}. \tag{4.36}$$

By minimizing (4.36) with respect to C, we find the optimal choice of C to be

$$C=\frac{\frac{(m-1)(n-1)}{m+n-1}-\frac{(p-1)(n-1)}{p+n-1}\frac{\mathbb{E}X_{m:m}}{\mathbb{E}X_{p:p}}}{\frac{(m-1)^2}{2m-1}+\frac{(p-1)^2}{2p-1}\frac{(\mathbb{E}X_{m:m})^2}{(\mathbb{E}X_{p:p})^2}-2\frac{(m-1)(p-1)}{m+p-1}\frac{\mathbb{E}X_{m:m}}{\mathbb{E}X_{p:p}}}. \tag{4.37}$$

Upon substituting the expression of C obtained in (4.37) into (4.36) and simplifying the resulting expression, we readily derive the bound in (4.31).

Moreover, the fact that $\tau_{n,m,p}$ in (4.32) is strictly in the interval $\left(0,\frac{1}{2n-1}\right)$ follows immediately from (4.34). □

A simple comparison of the bounds in Theorems 4.1 and 4.5 readily reveal that the upper bound in the latter is an improvement over the bound in the former for any choice of m and p due to the fact that $\tau_{n,m,p}$ in (4.32) is positive.

In Table 4.1, the values of the upper bound in (4.20) (which is an universal upper bound) and the one in (4.31) for the case of standard normal distribution (with the choice of $m=10$ and $p=20$ and the corresponding values of $\mathbb{E}X_{m:m}=1.53875$ and $\mathbb{E}X_{p:p}=1.86748$ taken from Harter and Balakrishnan (1996)) are presented for $n=30(10)100$, along with the corresponding exact values of $\mathbb{E}X_{n:n}$ for comparative purposes. The latter are taken from the extensive tables of expected values of standard normal order statistics presented by Harter and Balakrishnan (1996).

The results in the table readily reveal that the imputation-type bound in Theorem 4.5 provides a significant improvement over the bound in Theorem 4.1 for all values of n considered. Yet, a comparison of it with the exact values shows that further improvement may be possible!

As can be seen from Table 4.1, even though the bound in (4.31) is a significant improvement over the Hartley-David-Gumbel bound in (4.20), it loses its closeness to the exact value of $\mathbb{E}X_{n:n}$ as n increases. One way to improve this is to impute with more expected values of order statistics taken from small sample sizes, rather than just two with $\mathbb{E}X_{m:m}$ and $\mathbb{E}X_{p:p}$. This has been carried out recently by Balakrishnan

Table 4.1 Values of the upper bounds in (4.20) and (4.31) for $\mathbb{E}X_{n:n}$ for the case of standard normal distribution, along with exact values

n	Bound in (4.20)	Bound in (4.31)	Exact value
30	3.77548	2.50616	2.04276
40	4.38784	2.65241	2.16078
50	4.92469	2.87579	2.24907
60	5.40852	3.16532	2.31928
70	5.85250	3.49676	2.37736
80	6.26511	3.85007	2.42677
90	6.65217	4.21199	2.46970
100	7.01792	4.57442	2.50759

and Samanta (2023) who developed the bound by making use of a set of k expected values of order statistics from small sample sizes, say $\mathbb{E}X_{m_i:m_i}$ for $i = 1, \ldots, k$.

To describe this result, let us use the following notation:

$$
\begin{aligned}
S_{m_i,m_j} &= \frac{(m_i-1)(m_j-1)}{m_i+m_j-1} - \mathbb{E}X_{m_i:m_i}\,\mathbb{E}X_{m_j:m_j}, \quad 1 \le i, j \le k,\\
T_{n,m_i} &= \frac{(n-1)(m_i-1)}{n+m_i-1}, \quad 1 \le i \le k,\\
U_{m_i,m_j} &= S_{m_i,m_j} + \frac{\mathbb{E}X_{m_i:m_i}\,\mathbb{E}X_{m_j:m_j}}{(\mathbb{E}X_{m_k:m_k})^2} S_{m_k,m_k}\\
&\quad - \frac{\mathbb{E}X_{m_i:m_i}}{\mathbb{E}X_{m_k:m_k}} S_{m_j,m_k} - \frac{\mathbb{E}X_{m_j:m_j}}{\mathbb{E}X_{m_k:m_k}} S_{m_i,m_k}, \quad 1 \le i, j \le k-1,\\
V_{m_i} &= T_{n,m_i} \frac{\mathbb{E}X_{m_i:m_i}}{\mathbb{E}X_{m_k:m_k}} T_{n,m_k}, \quad 1 \le i \le k-1.
\end{aligned}
$$

Then, Balakrishnan and Samanta (2023) established the following result providing an upper bound for $\mathbb{E}X_{n:n}$.

Theorem 4.4 *Under the assumptions of Theorem* 4.1 *and with* $\mu = 0$ *and* $\sigma = 1$, *without loss of generality, we have*

$$
\mu_{n:n} \le \left\{ \frac{(n-1)^2}{2n-1} - \boldsymbol{V}^T \boldsymbol{U}^{-1} \boldsymbol{V} \right\}^{1/2}, \tag{4.38}
$$

where $\boldsymbol{V} = (V_{m_1}, \ldots, V_{m_{k-1}})^T$ *and* $\boldsymbol{U} = ((U_{m_i,m_j}))_{i,j=1}^{k-1}$, *with the exact values of* $\mathbb{E}X_{m_1:m_1}, \ldots, \mathbb{E}X_{m_k:m_k}$ *being available (for some* $m_1, \ldots, m_k$*) to be used for imputational purpose. Furthermore,* $\boldsymbol{V}^T \boldsymbol{U}^{-1} \boldsymbol{V}$ *is strictly in the interval* $(0, \frac{(n-1)^2}{2n-1})$.

The proof of Theorem 4.4 proceeds along the same lines as of Theorem 4.3 and is therefore not presented here. If, in particular, we take $k = 2$ in (4.38) with $m_1 = m$ and $m_2 = p$, we simply deduce the bound in (4.31). By using some moments of order statistics known in some small sample sizes, one can similarly

Table 4.2 Values of the upper bounds in (4.20), (4.31) and (4.38) for $\mathbb{E}X_{n:n}$ for the case of standard normal distribution, along with exact values

n	Bound in (4.20)	Bound in (4.31)	Bound in (4.38)	Exact value
50	4.92469	2.87579	2.3641	2.24907
60	5.40852	3.16532	2.4389	2.31928
70	5.85250	3.49676	2.5027	2.37736
80	6.26511	3.85007	2.5608	2.42677
90	6.65217	4.21199	2.6178	2.46970
100	7.01792	4.57442	2.6778	2.50759

derive imputation-type bounds for other quantities as well such as expected value of range, expected value of quasi-range and also more generally for expected value of a linear combination of order statistics.

In Table 4.2, the values of the Hartley-David-Gumbel bound in (4.20), the bound in (4.31) with the choice of $m = 10$ and $p = 20$, and the bound in (4.38) for the case of standard normal distribution for $k = 7$ with the choice of $m_1 = 5, m_2 = 10, m_3 = 15, m_4 = 20, m_5 = 25, m_6 = 30$ and $m_7 = 35$ and the corresponding values of $\mathbb{E}X_{5:5} = 1.16296, \mathbb{E}X_{10:10} = 1.53875, \mathbb{E}X_{15:15} = 1.73591, \mathbb{E}X_{20:20} = 1.86748, \mathbb{E}X_{25:25} = 1.96531, \mathbb{E}X_{30:30} = 2.04276$ and $\mathbb{E}X_{35:35} = 2.10661$ taken from the tables of Harter and Balakrishnan (1996), are all presented for $n = 50(10)100$, along with the corresponding exact values of $\mathbb{E}X_{n:n}$ for comparative purposes. The results in the table show that the bound in Theorem 4.4 is a significant improvement over the bound in Theorem 4.3 for all values of n considered, which is to be expected as the former uses more values of $\mathbb{E}X_{m_i:m_i}$ for imputational purpose!

4.2.3 Other L-statistics

The theorems of Subsect. 4.2.1 are generalized as follows.

Theorem 4.5 *Let $X_1, \ldots, X_n$ be iid with mean $\mu \in \mathbb{R}$ and variance $\sigma^2 > 0$. Assume that $\tilde{c} = (c_1, \ldots, c_n) \in \mathbb{R}^n$ is chosen so that the function*

$$f_{\tilde{c}:n}(x) = \sum_{i=1}^{n} c_i f_{i:n}(x), \qquad 0 < x < 1, \tag{4.39}$$

is nondecreasing. Then the bound

$$\mathbb{E}\sum_{i=1}^{n} c_i \frac{X_{i:n} - \mu}{\sigma} \leq B, \tag{4.40}$$

where

$$B^2 = \frac{n^2}{2n-1}\sum_{i=1}^{n}\sum_{j=1}^{n} c_i c_j \frac{\binom{n-1}{i-1}\binom{n-1}{j-1}}{\binom{2n-2}{i+j-2}} - \left(\sum_{i=1}^{n} c_i\right)^2 \tag{4.41}$$

is sharp, and it is attained if

$$\frac{F^{-1}(x)-\mu}{\sigma} = \frac{f_{\tilde{c}:n}(x) - \sum_{i=1}^{n} c_i}{B}. \tag{4.42}$$

Proof We have

$$\begin{aligned}\mathbb{E}\sum_{i=1}^{n} c_i \frac{X_{i:n}-\mu}{\sigma} &= \int_0^1 \frac{F^{-1}(x)-\mu}{\sigma}\left[f_{\tilde{c}:n}(x) - \sum_{i=1}^{n} c_i\right] dx \\ &\le \left[\int_0^1 \frac{[F^{-1}(x)-\mu]^2}{\sigma^2}\, dx\right]^{\frac{1}{2}} \left[\int_0^1 \left(f_{\tilde{c}:n}(x) - \sum_{i=1}^{n} c_i\right)^2 dx\right]^{\frac{1}{2}}.\end{aligned} \tag{4.43}$$

The first integral in the second line amounts to 1, and the other one, denoted by B^2, is evaluated as follows

$$\begin{aligned}B^2 &= \int_0^1 f^2_{\tilde{c}:n}(x)\, dx - 2\sum_{i=1}^{n} c_i \int_0^1 f_{\tilde{c}:n}(x)\, dx + \left(\sum_{i=1}^{n} c_i\right)^2 \\ &= \int_0^1 \left[\sum_{i=1}^{n} c_i f_{i:n}(x)\right]\left[\sum_{j=1}^{n} c_j f_{j:n}(x)\right] dx - \left(\sum_{i=1}^{n} c_i\right)^2 \\ &= n^2 \sum_{i=1}^{n}\sum_{j=1}^{n} c_i c_j \binom{n-1}{i-1}\binom{n-1}{j-1} \int_0^1 x^{i+j-2}(1-x)^{2n-i-j} dx - \left(\sum_{i=1}^{n} c_i\right)^2 \\ &= n^2 \sum_{i=1}^{n}\sum_{j=1}^{n} c_i c_j \binom{n-1}{i-1}\binom{n-1}{j-1} \frac{(j+i-2)!(2n-i-j)!}{(2n-1)!} - \left(\sum_{i=1}^{n} c_i\right)^2,\end{aligned} \tag{4.44}$$

which actually coincides with (4.41).

We obtain the equality in (4.43) if both the factors in the first integral are mutually proportional

$$\frac{F^{-1}(x)-\mu}{\sigma}=\alpha\left[f_{\tilde{c}:n}(x)-\sum_{i=1}^{n}c_i\right],\qquad 0<x<1, \tag{4.45}$$

and have the same signs. Observe that the LHS and RHS of (4.45) are nondecreasing (which is necessary for the linearly transformed quantile function), and their integrals over [0, 1] vanish. Moreover, the RHS has the norm identical with the norm of the LHS, being equal to 1, if

$$\alpha=\frac{1}{\left\| f_{\tilde{c}:n}-\sum_{i=1}^{n}c_i\right\|_2}=\frac{1}{B}. \tag{4.46}$$

This provides the attainability condition (4.42). □

There is not known a method of characterizing all the vectors $\tilde{c}=(c_1,\ldots,c_n)$ such that (4.39) is nondecreasing. We present a simple sufficient condition: if $(c_1,\ldots,c_n)$ is nondecreasing, then so is (4.39). Indeed,

$$\begin{aligned} f'_{\tilde{c}:n}(x) &= n\sum_{i=1}^{n}c_i B'_{i-1,n-1}(x)\\ &= n(n-1)\sum_{i=1}^{n}c_i[B_{i-2,n-2}(x)-B_{i-1,n-2}(x)]\\ &= n(n-1)\sum_{i=0}^{n-2}(c_{i+2}-c_{i+1})B_{i,n-2}(x). \end{aligned} \tag{4.47}$$

It suffices now to refer to the variation diminishing property of the Bernstein polynomials (see Lemma 4.1) in order to get the desired claim. The above observation leads to the following conclusion.

Corollary 4.1 *Let $X_1,\ldots,X_n$ satisfy the assumptions of Theorem 4.5, and $\tilde{c}=(c_1,\ldots,c_n)$ be nondecreasing. Then (4.40) with (4.41) hold, and the former one becomes the equality under the condition (4.42).*

Note that the sample maximum and range are the examples of L-statistics with nondecreasing vectors of coefficients. Another example of L-statistic satisfying the assumptions of Corollary 4.1 is the sample median absolute deviation (see (3.33)). Then we have

$$f_{\tilde{c}:n}(x)-\sum_{i=1}^{n}c_i=f_{\tilde{c}:n}(x)=\sum_{i=\lceil\frac{n}{2}\rceil+1}^{n}B_{i-1,n-1}(x)-\sum_{i=1}^{\lfloor\frac{n}{2}\rfloor}B_{i-1,n-1}(x), \tag{4.48}$$

and

$$\int_0^1 f_{\tilde{c}:n}^2(x)\,dx = \int_0^1 \left[\sum_{i=\lceil \frac{n}{2} \rceil+1}^{n} B_{i-1,n-1}(x) \right]^2 dx + \int_0^1 \left[\sum_{i=1}^{\lfloor \frac{n}{2} \rfloor} B_{i-1,n-1}(x) \right]^2 dx$$

$$- 2\int_0^1 \sum_{i=1}^{\lfloor \frac{n}{2} \rfloor} \sum_{j=\lceil \frac{n}{2} \rceil+1}^{n} B_{i-1,n-1}(x) B_{j-1,n-1}(x) dx. \tag{4.49}$$

Since the first two integrals are identical, we finally obtain

$$\mathbb{E}\frac{\mathrm{MAD}_n(\tilde{X})}{\sigma} \leq B, \tag{4.50}$$

with

$$B^2 = \frac{2}{2n-1} \left[\sum_{i=1}^{\lfloor \frac{n}{2} \rfloor} \sum_{j=1}^{\lfloor \frac{n}{2} \rfloor} \frac{\binom{n-1}{i-1}\binom{n-1}{j-1}}{\binom{2n-2}{i+j-2}} - \sum_{i=1}^{\lfloor \frac{n}{2} \rfloor} \sum_{j=\lceil \frac{n}{2} \rceil+1}^{n} \frac{\binom{n-1}{i-1}\binom{n-1}{j-1}}{\binom{2n-2}{i+j-2}} \right] \tag{4.51}$$

with the equality condition

$$\frac{F^{-1}(x) - \mu}{\sigma} = \frac{1}{B} \left[\sum_{i=\lceil \frac{n}{2} \rceil+1}^{n} B_{i-1,n-1}(x) - \sum_{i=1}^{\lfloor \frac{n}{2} \rfloor} B_{i-1,n-1}(x) \right]. \tag{4.52}$$

For the Gini mean differences we have simpler evaluations because

$$\mathbb{E}\,\mathrm{GMD}_n(\tilde{X}) = \frac{1}{n(n-1)} \mathbb{E}|X_1 - X_2| = \mathbb{E}(X_{2:2} - X_{1:2}). \tag{4.53}$$

Therefore, by Theorem 4.2 we have the bound

$$\mathbb{E}\frac{\mathrm{GMD}_n(\tilde{X})}{\sigma} \leq \frac{2}{\sqrt{3}}, \tag{4.54}$$

independent of the sample size, which is attained by uniform distributions.

The lower bounds based on the Schwarz inequality are presented in the following analog of Theorem 4.5.

Theorem 4.6 *Suppose that $X_1, \ldots, X_n$ are iid with a finite second moment and $\mu \in \mathbb{R}$, and $\sigma^2 > 0$ denoting the respective mean and variance. If $\tilde{c} = (c_1, \ldots, c_n)$ is such that (4.39) is nonincreasing, then*

$$\mathbb{E} \sum_{i=1}^{n} c_i \frac{X_{i:n} - \mu}{\sigma} \geq -B, \tag{4.55}$$

where B^2 is determined by the formula (4.41). The equality in (4.55) holds if

$$\frac{F^{-1}(x) - \mu}{\sigma} = -\frac{f_{\tilde{c}:n}(x) - \sum_{i=1}^{n} c_i}{B}. \tag{4.56}$$

Proof The following holds

$$\begin{aligned}
\mathbb{E} \sum_{i=1}^{n} c_i \frac{X_{i:n} - \mu}{\sigma} &= \int_0^1 \frac{F^{-1}(x) - \mu}{\sigma} \left[f_{\tilde{c}:n}(x) - \sum_{i=1}^{n} c_i \right] dx \\
&\geq - \left[\int_0^1 \left[f_{\tilde{c}:n}(x) - \sum_{i=1}^{n} c_i \right]^2 dx \right]^{\frac{1}{2}} \\
&= - \left[\int_0^1 f_{\tilde{c}:n}^2(x)\, dx - \left(\sum_{i=1}^{n} c_i \right)^2 dx \right]^{\frac{1}{2}}
\end{aligned} \tag{4.57}$$

with the equality condition

$$\frac{F^{-1}(x) - \mu}{\sigma} = -\alpha \left[f_{\tilde{c}:n}(x) - \sum_{i=1}^{n} c_i \right] \tag{4.58}$$

for some $\alpha > 0$. Notice that the RHS is nondecreasing by the assumption, and has the integral equal to 0. The variance condition for the LHS implies that $\alpha = \frac{1}{B}$. □

By (4.47) and the VDP of the Bernstein polynomials, we assumptions of Theorem 4.6. are satisfied if $(c_1, \ldots, c_n)$ is nonincreasing. In particular, we conclude that

$$\mathbb{E} \frac{X_{1:n} - \mu}{\sigma} \geq -\frac{n-1}{\sqrt{2n-1}}. \tag{4.59}$$

with the equality attained by the parent distribution function

$$F(x) = 1 - \left[\frac{1}{n}\left(1 - \frac{n-1}{\sqrt{2n-1}}\frac{x-\mu}{\sigma}\right)\right]^{\frac{1}{n-1}},$$

$$-\sqrt{2n-1} < \frac{x-\mu}{\sigma} < \frac{\sqrt{2n-1}}{n-1}. \tag{4.60}$$

We finally notice that the following inequalities (cf. (4.41))

$$-\sqrt{\frac{n^2}{2n-1}\sum_{i=1}^{n}\sum_{j=1}^{n} c_i c_j \frac{\binom{n-1}{i-1}\binom{n-1}{j-1}}{\binom{2n-2}{i+j-2}} - \left(\sum_{i=1}^{n} c_i\right)^2}$$

$$\leq \mathbb{E}\sum_{i=1}^{n} c_i \frac{X_{i:n}-\mu}{\sigma} \leq \sqrt{\frac{n^2}{2n-1}\sum_{i=1}^{n}\sum_{j=1}^{n} c_i c_j \frac{\binom{n-1}{i-1}\binom{n-1}{j-1}}{\binom{2n-2}{i+j-2}} - \left(\sum_{i=1}^{n} c_i\right)^2} \tag{4.61}$$

are valid for arbitrary $\tilde{c} = (c_1, \ldots, c_n) \in \mathbb{R}^n$, but if it does not satisfy the assumptions of Theorems 4.5 and 4.6 then the respective bound is not optimal.

4.3 Positive Upper and Negative Lower Schwarz-Moriguti Bounds

Here we present sharp positive upper and negative lower bounds on the expectations of linear combinations of order statistics under much weaker conditions than monotonicity of the coefficient vectors. We first consider the upper ones.

4.3.1 L-statistics

Theorem 4.7 *Assume that $X_1, \ldots, X_n$ are iid with mean μ and finite positive variance σ^2. For $\tilde{c} = (c_1, \ldots, c_n) \in \mathbb{R}^n$, let $\underline{f}_{\tilde{c}:n}$ denote the derivative of the greatest convex minorant $\underline{F}_{\tilde{c}:n}$ of*

$$F_{\tilde{c}:n}(x) = \sum_{i=1}^{n} c_i F_{i:n}(x), \qquad 0 < x < 1. \tag{4.62}$$

Then

$$\mathbb{E}\sum_{i=1}^{n} c_i \frac{X_{i:n}-\mu}{\sigma} \le \left\| \underline{f}_{\bar{c}:n} - \sum_{i=1}^{n} c_i \right\|_2 = \left[\int_0^1 \underline{f}^2_{\bar{c}:n}(x)\,dx - \left(\sum_{i=1}^{n} c_i \right)^2 \right]^{\frac{1}{2}}. \tag{4.63}$$

If

$$F_{\bar{c}:n}(x) < \sum_{i=1}^{n} c_i\, x \qquad \textit{for some } 0 < x < 1, \tag{4.64}$$

then (4.63) becomes the equality under the condition

$$\frac{F^{-1}(x)-\mu}{\sigma} = \frac{\underline{f}_{\bar{c}:n}(x) - \sum_{i=1}^{n} c_i}{\left\| \underline{f}_{\bar{c}:n} - \sum_{i=1}^{n} c_i \right\|_2}. \tag{4.65}$$

Note that $F_{\bar{c}:n}$ is a polynomial of degree n at most, and it has a continuous derivative. Its greatest convex minorant $\underline{F}_{\bar{c}:n}$ is tangent to $F_{\bar{c}:n}$ at all the border points of the set $\{x : \underline{F}_{\bar{c}:n}(x) = F_{\bar{c}:n}(x)\}$. except possibly 0 and 1. Therefore $\underline{f}_{\bar{c}:n}$ is a well-defined continuous function.

Proof We start from the classic representation

$$\mathbb{E}\sum_{i=1}^{n} c_i \frac{X_{i:n}-\mu}{\sigma} = \int_0^1 \frac{F^{-1}(x)-\mu}{\sigma} f_{\bar{c}:n}(x)\,dx, \tag{4.66}$$

and, due to nondecrease of the first factor, we apply the Moriguti inequality (see Theorem 2.27)

$$\begin{aligned}\mathbb{E}\sum_{i=1}^{n} c_i \frac{X_{i:n}-\mu}{\sigma} &= \int_0^1 \frac{F^{-1}(x)-\mu}{\sigma} f_{\bar{c}:n}(x)\,dx \le \int_0^1 \frac{F^{-1}(x)-\mu}{\sigma} \underline{f}_{\bar{c}:n}(x)\,dx \\ &= \int_0^1 \frac{F^{-1}(x)-\mu}{\sigma} \left[\underline{f}_{\bar{c}:n}(x) - \sum_{i=1}^{n} c_i \right] dx \end{aligned} \tag{4.67}$$

We can subtract the sum for the latter term (as well as any other constant) because

$$\int_0^1 \frac{F^{-1}(x)-\mu}{\sigma}\,dx = 0.$$

By the Schwarz inequality

$$\int_0^1 \frac{F^{-1}(x)-\mu}{\sigma}\left[\underline{f}_{\tilde{c}:n}(x)-\sum_{i=1}^n c_i\right]dx$$

$$\leq \left[\int_0^1 \frac{[F^{-1}(x)-\mu]^2}{\sigma^2}\,dx\right]^{\frac{1}{2}}\left[\int_0^1\left[\underline{f}_{\tilde{c}:n}(x)-\sum_{i=1}^n c_i\right]^2 dx\right]^{\frac{1}{2}}$$

$$=\left[\int_0^1\left[\underline{f}_{\tilde{c}:n}(x)-\sum_{i=1}^n c_i\right]^2 dx\right]^{\frac{1}{2}}. \tag{4.68}$$

Since

$$\int_0^1 \underline{f}_{\tilde{c}:n}(x)\,dx = \underline{F}_{\tilde{c}:n}(1) = F_{\tilde{c}:n}(1) = \sum_{i=1}^n c_i, \tag{4.69}$$

we have

$$\left[\int_0^1\left[\underline{f}_{\tilde{c}:n}(x)-\sum_{i=1}^n c_i\right]^2 dx\right]^{\frac{1}{2}} = \left[\int_0^1 \underline{f}^2_{\tilde{c}:n}(x)\,dx - \left(\sum_{i=1}^n c_i\right)^2\right]^{\frac{1}{2}}, \tag{4.70}$$

which is the final form of our bound.

Assume now that (4.64) holds. This implies that $\underline{F}_{\tilde{c}:n}(x) < \sum_{i=1}^n c_i\, x$ for all $0 < x < 1$, and $\underline{f}_{\tilde{c}:n} \not\equiv \sum_{i=1}^n c_i$. This means that $\left\|\underline{f}_{\tilde{c}:n} - \sum_{i=1}^n c_i\right\|_2 > 0$. The equality in the Schwarz inequality (4.68) holds if both the factors in the first line are proportional with a positive proportionality factor α:

$$\frac{F^{-1}(x)-\mu}{\sigma} = \alpha\left[\underline{f}_{\tilde{c}:n}(x)-\sum_{i=1}^n c_i\right]. \tag{4.71}$$

Note that the RHS is nondecreasing and has the integral over [0, 1] equal to zero, which well defines the quantile function of a distribution function with the mean μ. Moreover, the distribution has the variance equal to σ^2 if both the expressions have L^2-norms equal to 1, which forces that $\alpha = \frac{1}{\left\|\underline{f}_{\tilde{c}:n}-\sum_{i=1}^n c_i\right\|_2}$. This provides the formula (4.65). We finally observe that relation (4.65) implies the equality in the Moriguti inequality (4.67) as well. Indeed, by Theorem 2.27 the equality holds there if $\frac{F^{-1}(x)-\mu}{\sigma}$ is constant on every interval contained in $\{x : \underline{F}_{\tilde{c}:n}(x) < F_{\tilde{c}:n}(x)\}$. Note that $\underline{F}_{\tilde{c}:n}(x)$ is linear and $\underline{f}_{\tilde{c}:n}(x)$ is constant on every such interval. By (4.65)

$\frac{F^{-1}(x)-\mu}{\sigma}$ inherits all the constancy intervals of $\underline{f}_{c:n}(x)$. This implies that (4.65) assures the equality in (4.63) under the assumption (4.64). □

If

$$\sum_{i=1}^{n} c_i F_{i:n}(x) \geq \sum_{i=1}^{n} c_i\, x, \qquad 0 < x < 1, \tag{4.72}$$

then $\underline{F}_{\bar{c}:n}(x) = \sum_{i=1}^{n} c_i\, x$, $\underline{f}_{\bar{c}:n}(x) = \sum_{i=1}^{n} c_i$, and $\left\| \underline{f}_{\bar{c}:n} - \sum_{i=1}^{n} c_i \right\|_2 = 0$. The inequality (4.63) is still valid. However, the equality condition (4.71) in the Schwarz inequality cannot be fulfilled, because the LHS and RHS of (4.71) have different norms, equal to 1 and 0, respectively. The upper bounds under the assumption (4.72) are studied in Sect. 4.5.

Since

$$x = \frac{1}{n} \sum_{i=1}^{n} F_{i:n}(x) = \frac{1}{n} \sum_{i=1}^{n} \sum_{j=i}^{n} B_{i,n}(x), \qquad 0 < x < 1, \tag{4.73}$$

relation (4.64) is equivalent to

$$\sum_{i=1}^{n} \left(\sum_{j=1}^{i} (c_i - \bar{c}) \right) B_{i,n}(x) < 0 \quad \text{for some } 0 < x < 1, \tag{4.74}$$

which, by the VDP of Bernstein polynomials implies that

$$\sum_{j=1}^{i} (c_i - \bar{c}) < 0 \quad \text{for some } 1 \leq i \leq n-1. \tag{4.75}$$

The reversed condition

$$\sum_{j=1}^{i} (c_i - \bar{c}) \geq 0, \qquad i = 1, \ldots, n-1. \tag{4.76}$$

implies (4.72), and

$$\mathbb{E} \sum_{i=1}^{n} c_i (X_{i:n} - \mu) \leq 0. \tag{4.77}$$

This is obvious because (4.76) forces

$$\sum_{i=1}^{n} c_i(x_{i:n} - \bar{x}) \leq 0. \tag{4.78}$$

for arbitrary deterministic $x_{1:n} \leq \ldots \leq x_{n:n}$ (cf. (3.244) and Theorems 3.15 and 3.17).

It is not possible to express the assumptions (4.64) analytically in terms of some conditions on the elements of $(c_1, \ldots, c_n)$ only. Obviously, the condition

$$\sum_{j=1}^{i} (c_i - \bar{c}) \leq 0, \qquad i = 1, \ldots, n-1. \tag{4.79}$$

with the sharp inequality for at least one i gives

$$\sum_{i=1}^{n} c_i(x_{i:n} - \bar{x}) > 0, \tag{4.80}$$

and a positive bound in (4.63) in consequence, but it is satisfied by a relatively small proportion of vectors $\tilde{c}$.

4.3.2 Single Order Statistics

We present more precise specifications of Theorem 4.7 in the cases of single order statistics and their differences. We focus on the nonextreme order statistics $X_{j:n}$, $j = 2, \ldots, n-1$. For $j = n$ the bounds were presented in Theorem 4.1. For $j = 1$ the assumption (4.64) is not satisfied, because $1 - (1-x)^n > x$ for $0 < x < 1$ and $n \geq 2$. In this case Theorem 4.7 provides the trivial bound $\mathbb{E}\frac{X_{1:n}-\mu}{\sigma} \leq 0$, but it does not guarantee its optimality.

Theorem 4.8 *Let $X_1, \ldots, X_n$ be iid with mean $\mu \in \mathbb{R}$ and variance $0 < \sigma^2 < \infty$. For $j = 2, \ldots, n-1$ the equation*

$$f_{j:n}(x) = \frac{1 - F_{j:n}(x)}{1 - x}, \qquad 0 < x < 1, \tag{4.81}$$

has a unique solution $\alpha_* = \alpha_*(j, n)$ located in $\left(0, \frac{j-1}{n-1}\right)$. Moreover,

$$\mathbb{E}\frac{X_{j:n} - \mu}{\sigma} \leq B = B(j, n), \tag{4.82}$$

where

$$B^2(j,n) = n\,\frac{\binom{2j-2}{j-1}\binom{2n-2j}{n-j}}{\binom{2n-1}{n}}\,F_{2j-1:2n-1}(\alpha_*) + (1-\alpha_*)f_{j:n}^2(\alpha_*) - 1. \tag{4.83}$$

The equality in (4.82) is attained by

$$F(x) = \begin{cases} 0, & \frac{x-\mu}{\sigma} \le -\frac{1}{B}, \\ f_{j:n}^{-1}\left(1 + B\,\frac{x-\mu}{\sigma}\right), & -\frac{1}{B} \le \frac{x-\mu}{\sigma} < \frac{f_{j:n}(\alpha_*)-1}{B}, \\ 1, & \frac{x-\mu}{\sigma} \ge \frac{f_{j:n}(\alpha_*)-1}{B}. \end{cases} \tag{4.84}$$

Function $f_{j:n}$ is increasing on $\left(0, \frac{j-1}{n-1}\right)$ and decreasing on $\left(\frac{j-1}{n-1}, 1\right)$. The inverse function $f_{j:n}^{-1}$ appearing in (4.84) is actually the inverse of the increasing part of $f_{j:n}$. A more detailed justification is presented in the proof.

Proof Relation $F'_{j:n}(x) = f_{j:n}(x) > 0$, $0 < x < 1$, implies that function $F_{j:n}$ is strictly increasing. It obviously satisfies $F_{j:n}(0) = 0$ and $F_{j:n}(1) = 1$. Since

$$f'_{j:n}(x) = n(n-1)[B_{j-2,n-2}(x) - B_{j-1,n-2}(x)], \tag{4.85}$$

we immediately conclude from the VDP of Bernstein polynomials (see Lemma 4.1), that $F_{j:n}$ is first convex and then concave. Simple calculations show that the tangency point is equal to $\frac{j-1}{n-1}$ which is the maximum point of $f_{j:n}$.

The greatest convex minorant of any increasing, first convex and the concave function on [0, 1] may have two possible shapes: either it is linear on [0, 1], or it is first convex and identical with the original function, and then linear. The first case is excluded, because the slope of the line connecting the end-points (0, 0) and (1, 1) of the graph of $F_{j:n}$, equal to 1, is greater than the its derivative at 0, equal to $f_{j:n}(0) = 0$. This implies that the line runs above $F_{j:n}$ on some neighborhood of 0, and cannot be its minorant.

In the latter case, the transition point of the minorant from strict convexity to linearity is determined by Eq. (4.31). This means that this is the point where the line $\frac{1-F_{j:n}(\alpha_*)}{1-\alpha_*}(x-1)+1$, connecting the points $(\alpha_*, F_{j:n}(\alpha_*))$ and $(1, F_{j:n}(1)) = (1, 1)$ is tangent to the graph of $F_{j:n}$. Equation (4.81) means that the slope of the line is equal to the derivative of $F_{j:n}$ at α_*. Note the specified line runs below $F_{j:n}$ on $(\alpha_*, 1)$, and α_* belongs to the convexity interval $\left(0, \frac{j-1}{n-1}\right)$ of $F_{j:n}$. The line, extended by $F_{j:n}$ on $(0, \alpha_*)$ creates a convex function greater than any other convex function minorizing $F_{j:n}$.

Summing up, we concluded that the greatest convex minorant of $F_{j:n}$, and its derivative have the forms

$$\underline{F}_{j:n}(x) = \begin{cases} F_{j:n}(x), & 0 < x \leq \alpha_*, \\ \frac{1-F_{j:n}(\alpha_*)}{1-\alpha_*}(x-1)+1 = f_{j:n}(\alpha_*)(x-1)+1, & \alpha_* \leq x < 1, \end{cases} \tag{4.86}$$

and

$$\underline{f}_{j:n}(x) = \begin{cases} f_{j:n}(x), & 0 < x \leq \alpha_*, \\ f_{j:n}(\alpha_*), & \alpha_* \leq x < 1, \end{cases} \tag{4.87}$$

respectively. We have

$$\begin{aligned} \int_0^{\alpha_*} f_{j:n}^2(x)\,dx &= n\frac{\binom{2j-2}{j-1}\binom{2n-2j}{n-j}}{\binom{2n-1}{n}} \int_0^{\alpha_*} f_{2j-1:2n-1}(x)\,dx \\ &= n\frac{\binom{2j-2}{j-1}\binom{2n-2j}{n-j}}{\binom{2n-1}{n}} F_{2j-1:2n-1}(\alpha_*), \end{aligned} \tag{4.88}$$

and

$$\begin{aligned} ||\underline{f}_{j:n} - 1||_2^2 &= \int_0^1 \underline{f}_{j:n}^2(x)\,dx - 1 \\ &= n\frac{\binom{2j-2}{j-1}\binom{2n-2j}{n-j}}{\binom{2n-1}{n}} F_{2j-1:2n-1}(\alpha_*) + (1-\alpha_*) f_{j:n}^2(\alpha_*) - 1. \end{aligned} \tag{4.89}$$

By Theorem 4.7, the square root of (4.89) is the upper bound for the standardized expectation $\mathbb{E}\frac{X_{j:n}-\mu}{\sigma}$ of the jth order statistic (note that here $\sum_{i=1}^n c_i = 1$). The attainability condition is written as

$$\frac{F^{-1}(x)-\mu}{\sigma} = \frac{\underline{f}_{j:n}(x)-1}{B} = \frac{1}{B}\begin{cases} f_{j:n}(x)-1, & 0 < x \leq \alpha_*, \\ f_{j:n}(\alpha_*)-1, & \alpha_* \leq x < 1. \end{cases} \tag{4.90}$$

Observe that $f_{j:n}$ is strictly increasing on $(0, \alpha_*) \subset \left(0, \frac{j-1}{n-1}\right)$. Denoting by $f_{j:n}^{-1}$ the inverse of the increasing part of $f_{j:n}$, we can present the explicit form of the marginal distribution function attaining the bound (4.82) with use of the formula (4.84). □

The parent distribution function (4.90) has an atom of size $1-\alpha_*$, and a continuous term left to it. However, one cannot describe it in an analytic way, because the inverse of function $f_{j:n}(x) = n\binom{n-1}{j-1}x^{j-1}(1-x)^{n-j}$ cannot be represented analytically with a few exceptions: $j = 1$ and $j = n$ for any n (which are not considered in Theorem 4.8), and the pairs $(j, n) = (2, 3)$, $(2, 4)$ and $(3, 4)$. In the first case we need to find the inverse of a quadratic function which is a relatively simple task. For $n = 4$, we need to solve a family of cubic equations $4\binom{3}{j-1}x^{j-1}(1-x)^{4-j} = y$, $j = 2, 3$, with varying y, which gives very complicated analytic forms of solutions.

We also observe that in general the bounds (see (4.83)) cannot be represented explicitly, because Eq. (4.81) has to be solved numerically. Balakrishnan (1993) noticed that (4.81) has an analytic solution for small j. Indeed, it can be reduced to the polynomial equation of degree $j - 1$

$$n\binom{n-1}{j-1}x^{j-1} = \sum_{i=0}^{j-1}\binom{n}{i}x^i(1-x)^{j-i-1}. \tag{4.91}$$

In particular, for $j = 2$ and 3 we have linear and quadratic equations, respectively, with the following solutions

$$\alpha_*(2, n) = \frac{1}{(n-1)^2}, \tag{4.92}$$

$$\alpha_*(3, n) = \frac{n - 2 + \sqrt{(n-2)n(2n-3)}}{(n-1)^2(n-2)}. \tag{4.93}$$

These together in Theorem 4.8 imply the following.

Corollary 4.2 *Under the assumptions of Theorem* 4.8 *the following bounds*

$$\mathbb{E}\frac{X_{2:n} - \mu}{\sigma} \le B = B(2, n), \tag{4.94}$$

where

$$B^2(2, n) = \frac{n^2(n-1)}{(2n-3)(2n-1)} - \frac{1}{(2n-3)(2n-1)}\frac{[n(n-2)]^{2n-1}}{(n-1)^{4n-5}} - 1, \tag{4.95}$$

are sharp and they are attained by the distribution functions F *satisfying*

$$\frac{F^{-1}(x) - \mu}{\sigma} = \frac{1}{B}\begin{cases} [n(n-1)x(1-x)^{n-2} - 1], & 0 < x \le \frac{1}{(n-1)^2}, \\ \left[\frac{n}{n-1}\left(1 - \frac{1}{(n-1)^2}\right)^{n-2} - 1\right], & \frac{1}{(n-1)^2} \le x < 1. \end{cases} \tag{4.96}$$

Corollary 4.3 *Under the assumptions of Theorem* 4.8 *the following bounds*

$$\mathbb{E}\frac{X_{3:n} - \mu}{\sigma} \le B = B(3, n), \tag{4.97}$$

hold, where

$$
\begin{aligned}
B^2(3,n) = {} & \frac{15n^2(n-1)(n-2)}{(2n-5)(2n-3)(2n-1)} - n(n-1)(n-2)[1-\alpha_*(3,n)]^{n-2} \\
& \times \left[\frac{n-3}{n-2} + \frac{n-1}{n}[1-\alpha_*(3,n)]^2 - 2\frac{n-2}{n-1}[1-\alpha_*(3,n)]\right] \\
& + \frac{n^2(n-1)^2(n-2)^2}{4}\left[2\frac{n-3}{2n-5}[1-\alpha_*(3,n)]^{2n-5}\right. \\
& - 2\frac{2n-5}{n-2}[1-\alpha_*(3,n)]^{2n-4} + 12\frac{n-2}{2n-3}[1-\alpha_*(3,n)]^{2n-3} \\
& \left. - 2\frac{2n-3}{n-1}[1-\alpha_*(3,n)]^{2n-2} + 2\frac{n-1}{2n-1}[1-\alpha_*(3,n)]^{2n-1}\right] - 1,
\end{aligned}
\tag{4.98}
$$

and $\alpha_*(3, n)$ is defined in (4.93). They are attained by the distribution functions F satisfying

$$
\begin{aligned}
\frac{F^{-1}(x)-\mu}{\sigma} = {} & \frac{1}{B}\left[\frac{1}{2}n(n-1)(n-2)(\min\{x,\alpha_*(3,n)\})^2\right. \\
& \left. \times (1-\min\{x,\alpha_*(3,n)\})^{n-3} - 1\right].
\end{aligned}
\tag{4.99}
$$

We observe that the parent distributions (4.96) and (4.99) attaining the bounds in the case $j = 2$ and 3 have small contributions (4.92) and (4.93), respectively, of continuous parts, with the masses decreasing to 0 as n increases.

4.3.3 Differences of Order Statistics

Now we evaluate the expectations of differences of order statistics $X_{j:n} - X_{i:n}$, $1 \le i < j \le n$. Note that for $i > j$ the condition (4.76) holds, and Theorem 4.7 provides trivial zero bounds.

Theorem 4.9 *Let $\alpha_* = \alpha_*(i, j, n)$ be equal to 0 for $i = 1$, and solve the equation*

$$
\frac{F_{j:n}(x) - F_{i:n}(x)}{x} = f_{j:n}(x) - f_{i:n}(x), \qquad 0 < x < 1, \tag{4.100}
$$

otherwise. Moreover, let $\beta_* = \beta_*(i, j, n)$ be equal to 1 for $j = n$, and solve the equation

$$\frac{F_{i:n}(x) - F_{j:n}(x)}{1-x} = f_{j:n}(x) - f_{i:n}(x), \qquad 0 < x < 1. \tag{4.101}$$

for $1 \le i < j \le n-1$. Then

$$\mathbb{E}\frac{X_{j:n} - X_{i:n}}{\sigma} \le B = B(i, j, n), \tag{4.102}$$

with

$$\begin{aligned} B^2 = {} & \alpha_*[f_{j:n}(\alpha_*) - f_{i:n}(\alpha_*)]^2 + (1-\beta_*)[f_{j:n}(\beta_*) - f_{i:n}(\beta_*)]^2 \\ & + n\frac{\binom{2i-2}{i-1}\binom{2n-2i}{n-i}}{\binom{2n-1}{n}}[F_{2i-1:2n-1}(\beta_*) - F_{2i-1:2n-1}(\alpha_*)] \\ & + n\frac{\binom{2j-2}{j-1}\binom{2n-2j}{n-j}}{\binom{2n-1}{n}}[F_{2j-1:2n-1}(\beta_*) - F_{2j-1:2n-1}(\alpha_*)] \\ & - 2n\frac{\binom{i+j-2}{i-1}\binom{2n-i-j}{n-i}}{\binom{2n-1}{n}}[F_{i+j-1:2n-1}(\beta_*) - F_{i+j-1:2n-1}(\alpha_*)]. \end{aligned} \tag{4.103}$$

The equality in (4.102) holds if

$$\frac{F^{-1}(x) - \mu}{\sigma} = \frac{1}{B}\begin{cases} [f_{j:n}(\alpha_*) - f_{i:n}(\alpha_*)], & 0 < x \le \alpha_*, \\ [f_{j:n}(x) - f_{i:n}(x)], & \alpha_* \le x \le \beta_*, \\ [f_{j:n}(\beta_*) - f_{i:n}(\beta_*)], & \beta_* \le x < 1. \end{cases} \tag{4.104}$$

For the sample range with $i = 1$ and $j = n$ the above result simplifies to Theorem 4.2. The distribution functions (4.104) attaining the bounds in (4.102) have an atom of size $\alpha_*(i, j, n)$ on the left end-point of the support (except for $i = 1$) and/or another atom on size $1 - \beta_*(i, j, n)$ (except for $j = n$) on the right, and a density function in between.

Proof The basic problem consists in determining the greatest convex minorants of the functions

$$F_{\tilde{c}:n}(x) = F_{j:n}(x) - F_{i:n}(x) = -\sum_{k=i}^{j-1} B_{k,n}(x), \qquad 1 \le i < j \le n, \tag{4.105}$$

with $\tilde{c}$ denoting here the vector with -1 at the ith position, and 1 at the jth position, and zeros elsewhere. They are obviously negative on $(0, 1)$ and vanish at the ends of the interval. We first concentrate on the cases $2 \le i < j \le n-1$. Since

$$F'_{j:n}(x) - F'_{i:n}(x) = n[-B_{i-1,n-1}(x) + B_{j-1,n-1}(x)], \tag{4.106}$$

they are first decreasing, and then increasing. Moreover,

$$\begin{aligned} F''_{j:n}(x) - F''_{:in}(x) = n(n-1)[-B_{i-2,n-2}(x) + B_{i-1,n-2}(x) \\ + B_{j-2,n-2}(x) + B_{j-1,n-2}(x)], \end{aligned} \tag{4.107}$$

which together with the VDP imply that (4.105) is either first concave, then convex, and ultimately concave, or concave on the whole interval. However, every smooth function is convex about its minimum point which means that the latter possibility is excluded.

Summing up, for every $2 \le i < j \le n-1$ the functions are negative on $(0, 1)$, equal to 0 at 0 and 1, first concave decreasing, then convex decreasing, convex increasing, and finally concave increasing. We immediately conclude that their greatest convex minorants are linear on the end parts of the interval, and identical with the original functions on single intervals contained in the convexity domains of (4.105) and containing the minimum points. The right end of the first linearity interval, and the left end of the latter one are the points such that the lines tangent to (4.105) there run through $(0, 0)$ and $(1, 0)$, respectively. These are just the points determined by Eqs. (4.100) and (4.101), respectively. It is easy to notice that $\alpha_*(i, j, n)$ and $\beta_*(i, j, n)$ are the unique solutions to Eqs. (4.100) and (4.101), respectively, in $(0, 1)$ for given i, j, and n. It follows that the greatest convex minorants have the forms

$$\underline{F}_{\tilde{c}:n}(x) = \begin{cases} [f_{j:n}(\alpha_*) - f_{i:n}(\alpha_*)]x, & 0 < x \le \alpha_*, \\ F_{j:n}(x) - F_{i:n}(x), & \alpha_* \le x \le \beta_*, \\ \left[f_{j:n}(\beta_*) - f_{i:n}(\beta_*)\right](x-1), & \beta_* \le x < 1. \end{cases} \tag{4.108}$$

and

$$\underline{f}_{\tilde{c}:n}(x) = \begin{cases} [f_{j:n}(\alpha_*) - f_{i:n}(\alpha_*)], & 0 < x \le \alpha_*, \\ f_{j:n}(x) - f_{i:n}(x), & \alpha_* \le x \le \beta_*, \\ \left[f_{j:n}(\beta_*) - f_{i:n}(\beta_*)\right], & \beta_* \le x < 1. \end{cases} \tag{4.109}$$

are the respective derivatives. Some elementary calculations show that the squared norms of (4.109) are equal (4.103). Since the sum of coefficient in $\sum_{k=1}^{n} c_k X_{k:n} = X_{j:n} - X_{i:n}$ amounts to 0, by Theorem 4.7 (see (4.63) and (4.65)) we conclude the bounds $B = B(i, j, n)$ determined from (4.103), and the attainability conditions described in (4.104).

If $i = 1 < j \le n-1$, we can rewrite (4.105), (4.106) and (4.107) with the only modification that $B_{i-2,n-2}$ is dropped in the last formula. This implies that (4.105) is first convex decreasing, the convex increasing, and eventually concave increasing. The convex minorant is linear on a left neighborhood of 1, but it does not have a linear part on the left. The single transition point $\beta_*(1, j, n)$ is determined by Eq. (4.101). Putting $\alpha_*(1, j, n) = 0$, we can formally apply the functions (4.108), (4.109) and (4.104) removing the first lines in the respective formulae. Also, we can simplify (4.103) by skipping the first term, and $F_{k,2n-1}(\alpha_*) = 0$, $k = 2i-1, 2j-1, i+j-1$, in the last three summands.

Analogous simplifications can be done for $1 \le i < j = n$. Then (4.105) does not have a concave increasing part in the neighborhood of 1. Therefore we can apply (4.103) and (4.104) with $\beta_*(i, n, n) = 1$. Obviously, for $i = 1$ and $j = n$ the function (4.105) is convex on (0, 1), and so it is identical with its greatest convex minorant there. Writing (4.103) and (4.104) with $\alpha_*(1, n, n) = 0$ and $\beta_*(1, n, n) = 1$ we obtain (4.26) and (4.27), respectively, of Theorem 4.2. □

One can reduce the degree $n-1$ of the polynomial equations (4.100) and (4.101) by writing

$$-\sum_{k=i}^{j-1}\binom{n}{k}x^{k-i}(1-x)^{j-k} = n\left[\binom{n-1}{j-1}x^{j-i} - \binom{n-1}{i-1}(1-x)^{j-i}\right], \tag{4.110}$$

$$\sum_{k=i}^{j-1}\binom{n}{k}x^{k+1-i}(1-x)^{j-k-1} = n\left[\binom{n-1}{j-1}x^{j-i} - \binom{n-1}{i-1}(1-x)^{j-i}\right], \tag{4.111}$$

respectively. These are the polynomial equations of degree $j-i$. The simplification was used by Ludwig (1973) who evaluated numerically the expectations of differences of order statistics of small and moderate iid samples. In the case of spacings with $j = i+1$, Eqs. (4.110) and (4.111) are linear and have simple solutions

$$\alpha_*(i, i+1, n) = \frac{i-1}{n-1}, \tag{4.112}$$

$$\beta_*(i, i+1, n) = \frac{i}{n-1}, \qquad i = 1, \ldots, n-1. \tag{4.113}$$

This allows us to present explicit bounds for the expectations of spacings.

Corollary 4.4 *For the iid random variables $X_1, \ldots, X_n$ with the mean μ and variance σ^2 we have*

$$\mathbb{E}\frac{X_{i+1:n} - X_{i:n}}{\sigma} \le B = B(i, i+1, n), \tag{4.114}$$

with

$$
\begin{aligned}
B^2 = {} & \frac{i-1}{n-1}\left[f_{i+1:n}\left(\frac{i-1}{n-1}\right) - f_{i:n}\left(\frac{i-1}{n-1}\right)\right]^2 \\
& + n\,\frac{\binom{2i-2}{i-1}\binom{2n-2i}{n-i}}{\binom{2n-1}{n}}\left[F_{2i-1,2n-1}\left(\frac{i}{n-1}\right) - F_{2i-1:2n-1}\left(\frac{i-1}{n-1}\right)\right] \\
& + n\,\frac{\binom{2i}{i}\binom{2n-2i-2}{n-i-1}}{\binom{2n-1}{n}}\left[F_{2i+1,2n-1}\left(\frac{i}{n-1}\right) - F_{2i+1:2n-1}\left(\frac{i-1}{n-1}\right)\right] \\
& - 2n\,\frac{\binom{2i-1}{i-1}\binom{2n-2i-1}{n-i}}{\binom{2n-1}{n}}\left[F_{2i,2n-1}\left(\frac{i}{n-1}\right) - F_{2i:2n-1}\left(\frac{i-1}{n-1}\right)\right]. \\
& + \frac{n-i}{n-1}\left[f_{i+1:n}\left(\frac{i}{n-1}\right) - f_{i:n}\left(\frac{i}{n-1}\right)\right]^2. \qquad (4.115)
\end{aligned}
$$

The equality in (4.114) holds if

$$
\frac{F^{-1}(x)-\mu}{\sigma} = \frac{1}{B}\begin{cases}
\left[f_{i+1:n}\left(\frac{i-1}{n-1}\right) - f_{i:n}\left(\frac{i-1}{n-1}\right)\right], & 0 < x \le \frac{i-1}{n-1}, \\
\left[f_{i+1:n}(x) - f_{i:n}(x)\right], & \frac{i-1}{n-1} \le x \le \frac{i}{n-1}, \\
\left[f_{i+1:n}\left(\frac{i}{n-1}\right) - f_{i:n}\left(\frac{i}{n-1}\right)\right], & \frac{i}{n-1} \le x < 1.
\end{cases} \qquad (4.116)
$$

Obviously, (4.115) and (4.116) slightly simplify in the cases $i = 1$ and $n-1$. Moreover, one can determine analogous explicit formulas in the case $j = i+2$, but they look complicated, and we do not present them here.

4.3.4 Negative Lower Bounds

Now we present lower Moriguti-Schwarz bounds on the expectations of L-statistics form iid samples.

Theorem 4.10 *Assume that $X_1, \ldots, X_n$ are iid with mean $\mu \in \mathbb{R}$ and variance $\sigma^2 \in (0, \infty)$. For $\tilde{c} = (c_1, \ldots, c_n)$, let $\overline{f}_{\tilde{c}:n}$ denote the derivative of the smallest concave majorant $\overline{F}_{\tilde{c}:n}$ of the function $F_{\tilde{c}:n}(x) = \sum_{i=1}^n c_i F_{i:n}(x)$, $0 < x < 1$. Then*

$$
\mathbb{E}\sum_{i=1}^n c_i\,\frac{X_{i:n}-\mu}{\sigma} \ge -\left\|\overline{f}_{c:n} - \sum_{i=1}^n c_i\right\|_2 = -\left[\int_0^1 \overline{f}^{\,2}_{\tilde{c}:n}(x)dx - \left(\sum_{i=1}^n c_i\right)^2\right]^{\frac{1}{2}}. \qquad (4.117)
$$

If

$$F_{\bar{c}:n}(x) > \sum_{i=1}^{n} c_i\, x \qquad for\ some\ 0 < x < 1, \tag{4.118}$$

then the bound (4.117) is strictly negative and sharp. It is attained by the distribution function F satisfying

$$\frac{F^{-1}(x) - \mu}{\sigma} = \frac{\sum_{i=1}^{n} c_i - \overline{f}_{\bar{c}:n}(x)}{\|\overline{f}_{c:n} - \sum_{i=1}^{n} c_i\|_2}. \tag{4.119}$$

Proof By the reversed Moriguti and Schwarz inequalities (see Theorem 2.19 and (2.124))

$$\begin{aligned}
\mathbb{E}\sum_{i=1}^{n} c_i \frac{X_{i:n} - \mu}{\sigma} &= \int_0^1 \frac{F^{-1}(x) - \mu}{\sigma}\left[f_{\bar{c}:n}(x) - \sum_{i=1}^{n} c_i\right] dx \\
&\geq \int_0^1 \frac{F^{-1}(x) - \mu}{\sigma}\left[\overline{f}_{\bar{c}:n}(x) - \sum_{i=1}^{n} c_i\right] dx \\
&\geq -\left[\int_0^1 \frac{[F^{-1}(x) - \mu]^2}{\sigma^2} dx\right]^{\frac{1}{2}} \left[\int_0^1 \left[\overline{f}_{\bar{c}:n}(x) - \sum_{i=1}^{n} c_i\right]^2 dx\right]^{\frac{1}{2}} \\
&= -\left[\int_0^1 \overline{f}^2_{\bar{c}:n}(x) dx - \left(\sum_{i=1}^{n} c_i\right)^2\right]^{\frac{1}{2}}
\end{aligned} \tag{4.120}$$

which proves (4.117).

If (4.118) is satisfied then $\overline{F}_{\bar{c}:n}(x) > \sum_{i=1}^{n} c_i\, x$, $0 < x < 1$, and the norm $\|\overline{f}_{c:n} - \sum_{i=1}^{n} c_i\|_2$ is positive. The equality in the latter inequality of (4.120) holds, and the monotonicity and moment conditions are satisfied if

$$\frac{F^{-1}(x) - \mu}{\sigma} = -\frac{\overline{f}_{\bar{c}:n}(x) - \sum_{i=1}^{n} c_i}{\|\overline{f}_{c:n} - \sum_{i=1}^{n} c_i\|_2}. \tag{4.121}$$

This also assures the equality in the first inequality of (4.120), because the LHS is constant over all the intervals where so is the RHS, and they contain (in fact coincide with) all the intervals where $\overline{F}_{\bar{c}:n}(x) > F_{\bar{c}:n}(x)$. This ends the proof. □

The condition (4.118) can be represented as

$$\sum_{i=1}^{n}\left(\sum_{j=1}^{i}(c_i - \bar{c})\right) B_{i,n}(x) > 0 \qquad \text{for some } 0 < x < 1, \tag{4.122}$$

which is possible if

$$\sum_{j=1}^{i}(c_i - \bar{c}) > 0 \qquad \text{for some } 1 \le i \le n-1. \tag{4.123}$$

The condition reversed to (4.123) assures $\sum_{i=1}^{n} c_i(x_{i:n} - \bar{x}) \ge 0$ for all ordered $x_{1:n} \le \ldots \le x_{n:n}$ which guarantees that the lower bound on (4.117) is necessarily nonnegative.

If $X_1, \ldots, X_n$ are iid, and have mean μ and variance σ^2, then $Y_i = 2\mu - x_i$, $i = 1, \ldots, n$, have the same mean and variance, and

$$\mathbb{E}\sum_{i=1}^{n} c_i \frac{X_{i:n} - \mu}{\sigma} = -\mathbb{E}\sum_{i=1}^{n} c_i \frac{Y_{n+1-i:n} - \mu}{\sigma} = -\mathbb{E}\sum_{i=1}^{n} c_{n+1-i} \frac{Y_{i:n} - \mu}{\sigma}, \tag{4.124}$$

because $Y_{i:n} = 2\mu - X_{n+1-i:n}$, $i = 1, \ldots, n$. One can expect that the lower bounds on the expectation of the LHS is the negative of the upper bound on the expectation of the RHS where the original vector of coefficients $\tilde{c} = (c_1, \ldots, c_n)$ is replaced by the reversed vector $\tilde{c}' = (c_n, \ldots, c_1)$. The conjecture is confirmed by the following corollary where the lower bounds and their attainability conditions are expressed in terms of $\underline{f}_{\tilde{c}':n}$.

Corollary 4.5 *Let $X_1, \ldots, X_n$ be iid with mean μ and variance σ^2. Then for every $\tilde{c} = (c_1, \ldots, c_n) \in \mathbb{R}^n$ we have*

$$\mathbb{E}\sum_{i=1}^{n} c_i \frac{X_{i:n} - \mu}{\sigma} \ge -\left\| \underline{f}_{\tilde{c}':n} - \sum_{i=1}^{n} c_i \right\|_2 = -\left[\int_0^1 \underline{f}^2_{\tilde{c}':n}(x)dx - \left(\sum_{i=1}^{n} c_i \right)^2 \right]^{\frac{1}{2}}, \tag{4.125}$$

where $\tilde{c}' = (c_n, \ldots, c_1)$, and $\underline{f}_{\tilde{c}':n}$ is the derivative of the greatest convex minorant of $F_{\tilde{c}':n} = \sum_{i=1}^{n} c_{n+1-i} F_{i:n}$. Moreover, under condition (4.118) we obtain the equality in (4.125) if

$$\frac{F^{-1}(x) - \mu}{\sigma} = \frac{\sum_{i=1}^{n} c_i - \underline{f}_{\tilde{c}':n}(1-x)}{\left\| \underline{f}_{\tilde{c}':n} - \sum_{i=1}^{n} c_i \right\|_2}. \tag{4.126}$$

Proof Consider $Y_i = 2\mu - X_i$, $i = 1, \ldots, n$, and denote their parent distribution function by G. Then

$$\mathbb{E}\sum_{i=1}^{n} c_i \frac{X_{i:n} - \mu}{\sigma} = -\mathbb{E}\sum_{i=1}^{n} c_i \frac{Y_{n+1-i:n} - \mu}{\sigma} = -\mathbb{E}\sum_{i=1}^{n} c_{n+1-i} \frac{Y_{i:n} - \mu}{\sigma}. \tag{4.127}$$

By Theorem 4.7 (see (4.63)) we have

$$\mathbb{E}\sum_{i=1}^{n} c_{n+1-i} \frac{Y_{i:n} - \mu}{\sigma} \leq \left\| \underline{f}_{\bar{c}':n} - \sum_{i=1}^{n} c_i \right\|_2, \tag{4.128}$$

which together with (4.127) gives (4.125).

If we show that $\overline{f}_{\bar{c}:n}(x) = \underline{f}_{\bar{c}':n}(1-x)$, we prove that the bounds (4.117) and (4.128) are identical. Since

$$F_{i:n}(x) = \sum_{k=i}^{n} B_{k,n}(x) = \sum_{k=0}^{n-i} B_{k,n}(1-x) = 1 - F_{n+1-i:n}(1-x), \tag{4.129}$$

we have

$$F_{\bar{c}:n}(x) = \sum_{i=1}^{n} c_i[1 - F_{n+1-i:n}(1-x)] = \sum_{i=1}^{n} c_i - F_{\bar{c}':n}(1-x) \tag{4.130}$$

We the notation $\overleftarrow{H}(x) = H(1-x)$, we check that

$$\begin{aligned} \overline{F}_{\bar{c}:n} &= \inf\left\{ g - \text{concave} : g \geq \sum_{c=1}^{n} c_i - \overleftarrow{F}_{\bar{c}':n} \right\} \\ &= \sum_{i=1}^{n} c_i - \sup\left\{ g - \text{convex} : g \leq \overleftarrow{F}_{\bar{c}':n} \right\}. \end{aligned} \tag{4.131}$$

Obviously, if $\underline{H}$ is the greatest convex minorant of H, then $\overleftarrow{\underline{H}}$ is the greatest convex minorant of $\overleftarrow{H}$. Therefore

$$\overline{F}_{\bar{c}:n}(x) = \sum_{i=1}^{n} c_i - \underline{F}_{\bar{c}':n}(1-x), \tag{4.132}$$

and, in consequence,

$$\overline{f}_{\bar{c}:n}(x) = \underline{f}_{\bar{c}':n}(1-x) \tag{4.133}$$

This shows that the bounds of Theorem 4.10 and Corollary 4.5 coincide.

The condition (4.118) implies

$$\sum_{i=1}^{n} c_i F_{i:n}(x) = \sum_{i=1}^{n} c_i[1 - F_{n+1-i:n}(1-x)] = \sum_{i=1}^{n} c_i - F_{\bar{c}':n}(1-x) > \sum_{i=1}^{n} c_i\, x \tag{4.134}$$

for some $0 < x < 1$, and consequently

$$F_{\bar{c}':n}(y) < \sum_{i=1}^{n} c_i\, y, \qquad \text{for some } 0 < y = 1 - x < 1. \tag{4.135}$$

Owing to the second statement of Theorem 4.7 the bound in (4.128) is positive, and it is attained if

$$\frac{G^{-1}(x) - \mu}{\sigma} = \frac{\underline{f}_{\bar{c}':n}(x) - \sum_{i=1}^{n} c_i}{\left\| \underline{f}_{\bar{c}':n} - \sum_{i=1}^{n} c_i \right\|_2}. \tag{4.136}$$

By the assumption, the marginal distribution functions F and G of X_i and Y_i, $i = 1, \ldots, n$, respectively satisfy $F(x) = 1 - G(2\mu - x-)$ and $F^{-1}(x) - \mu = -[G^{-1}(1-x) - \mu]$, which implies

$$\frac{F^{-1}(x) - \mu}{\sigma} = \frac{\sum_{i=1}^{n} c_i - \underline{f}_{\bar{c}':n}(1-x)}{\left\| \underline{f}_{\bar{c}':n} - \sum_{i=1}^{n} c_i \right\|_2}. \tag{4.137}$$

Due to (4.133) the equality conditions (4.119) and (4.126) for attainability of the bounds (4.117) and (4.125) are equivalent. □

In particular, Corollary 4.5 asserts that the lower bound on the expectation of standardized jth smallest order statistic $\mathbb{E}\frac{X_{j:n}-\mu}{\sigma}$ amounts to the negative of the upper bound on expectation of standardized jth greatest order statistic $\mathbb{E}\frac{X_{n+1-j:n}-\mu}{\sigma}$.

Corollary 4.6 *Let $X_1, \ldots, X_n$ be iid with mean μ and variance σ^2. For $j = 2, \ldots, n-1$, with $\alpha_*(j, n)$ and $B(j, n)$ defined in (4.81) and (4.83), respectively, we have*

$$\mathbb{E}\frac{X_{j:n} - \mu}{\sigma} \geq -B(n+1-j, n), \tag{4.138}$$

with the equality for the baseline distribution function

$$F(x) = \begin{cases} 0, & \frac{x-\mu}{\sigma} < \frac{1-f_{j:n}(1-\alpha_*(n+1-j,n))}{B(n+1-j,n)}, \\ f_{j:n}^{-1}\left(1 - B(n+1-j,n)\frac{x-\mu}{\sigma}\right), & \frac{1-f_{j:n}(1-\alpha_*(n+1-j,n))}{B(n+1-j,n)} \\ & \le \frac{x-\mu}{\sigma} \le \frac{1}{B(n+1-j,n)}, \\ 1, & \frac{x-\mu}{\sigma} \ge \frac{1}{B(n+1-j,n)}, \end{cases} \tag{4.139}$$

where $f_{j:n}^{-1}$ denotes the inverse of the decreasing part of $f_{j:n}$.

Distribution function (4.139) has a pole of height $1 - \alpha_*(n+1-j,n)$ at the left end of its support. Theorem 4.10 gives the trivial lower bound for the sample maximum. For the minimum, we have the inequality (4.59) with the attainability condition (4.60). Corollaries 4.2 and 4.3 combined with Corollary 4.5 provide analogous results for $j = n-1$ and $j = n-2$.

Corollary 4.7 *Under the assumptions of Theorem 4.10 we have*

$$\mathbb{E}\frac{X_{n-1:n} - \mu}{\sigma} \ge -B(2,n), \tag{4.140}$$

where $B^2(2,n)$ is defined in (4.95). The inequality (4.140) becomes the equality if

$$\frac{F^{-1}(x) - \mu}{\sigma} = \frac{1}{B(2,n)} \begin{cases} \left[1 - \frac{n}{(n-1)^{2n-3}}\left(1 - \frac{1}{(n-1)^2}\right)\right], & 0 < x < 1 - \frac{1}{(n-1)^2}, \\ \left[1 - n(n-1)x^{n-2}(1-x)\right], & 1 - \frac{1}{(n-1)^2} \le x < 1. \end{cases} \tag{4.141}$$

Corollary 4.8 *Under the assumptions of Theorem 4.10 with $\alpha_*(3,n)$, $B^2(3,n)$ defined in (4.93) and (4.98), respectively, the following inequality*

$$\mathbb{E}\frac{X_{n-2:n} - \mu}{\sigma} \ge -B(3,n) \tag{4.142}$$

holds. This becomes the equality for the distribution functions F satisfying

$$\frac{F^{-1}(x) - \mu}{\sigma} = \frac{1}{B(3,n)}\left[1 - \frac{1}{2}n(n-1)(n-2)(1 - \max\{x, \alpha_*(3,n)\}^2 \right.$$
$$\left. \times (\max\{x, \alpha_*(3,n)\})^{n-3}\right]. \tag{4.143}$$

4.4 Positive Upper and Negative Lower Hölder-Moriguti Bounds

4.4.1 Positive Upper Bounds in σ_p-Units, $1 < p < \infty$

Here we generalize the results of the previous section presenting bounds on the expectations of centered L-statistics $\mathbb{E}\sum_{i=1}^{n} c_i(X_{i:n}-\mu)$ expressed in various scale unites $\sigma_p = (\mathbb{E}|X_1-\mu|^p)^{1/p}$, $p \geq 1$, based on absolute central moments. We first consider $1 < p < \infty$. The case $p = 1$ needs a slightly different approach, and is treated separately later on.

Theorem 4.11 *Suppose that $X_1, \dots, X_n$ are iid and have a finite moment of order $p > 1$. For a given $\tilde{c} = (c_1, \dots, c_n) \in \mathbb{R}^n$, let $\underline{f}_{\tilde{c}:n}$ denote the derivative of the greatest convex minorant $\underline{F}_{\tilde{c}:n}$ of $F_{\tilde{c}:n} = \sum_{i=1}^{n} c_i F_{i:n}$, and $0 < \gamma_* = \gamma_*(p, \tilde{c}) < 1$ be any solution to the equation*

$$\int_0^{\gamma} [\underline{f}_{\tilde{c}:n}(\gamma) - \underline{f}_{\tilde{c}:n}(x)]^{q-1} dx = \int_{\gamma}^{1} [\underline{f}_{\tilde{c}:n}(x) - \underline{f}_{\tilde{c}:n}(\gamma)]^{q-1} dx, \tag{4.144}$$

where $q = \frac{p}{p-1}$. Then

$$\begin{aligned}\mathbb{E}\sum_{i=1}^{n} c_i \frac{X_{i:n}-\mu}{\sigma_p} &\leq ||\underline{f}_{\tilde{c}:n} - \underline{f}_{\tilde{c}:n}(\gamma_*)||_q \\ &= \left(\int_0^1 |\underline{f}_{\tilde{c}:n}(x) - \underline{f}_{\tilde{c}:n}(\gamma_*)|^q dx \right)^{\frac{1}{q}}.\end{aligned} \tag{4.145}$$

If, moreover,

$$F_{\tilde{c}:n}(x) < \sum_{i=1}^{n} c_i\, x \qquad \text{for some } 0 < x < 1 \tag{4.146}$$

(cf. (4.64)), then the bound (4.145) is positive and sharp, and it is attained by the parent distribution function F satisfying

$$\frac{F^{-1}(x)-\mu}{\sigma_p} = \frac{|\underline{f}_{\tilde{c}:n}(x) - \underline{f}_{\tilde{c}:n}(\gamma_*)|^{q/p} \operatorname{sgn}\{\underline{f}_{\tilde{c}:n}(x) - \underline{f}_{\tilde{c}:n}(\gamma_*)\}}{||\underline{f}_{\tilde{c}:n} - \underline{f}_{\tilde{c}:n}(\gamma_*)||_q^{q/p}}. \tag{4.147}$$

Proof We start in the standard way, applying the Moriguti inequality, and than the Hölder inequality (see Theorem 2.31) which is more general than the Schwarz one

$$
\begin{aligned}
\mathbb{E}\sum_{i=1}^{n} c_i \frac{X_{i:n}-\mu}{\sigma_p} &\le \int_0^1 \frac{F^{-1}(x)-\mu}{\sigma_p}\,[\underline{f}_{\tilde{c}:n}(x)-c]\,dx \\
&\le \left[\int_0^1 \frac{|F^{-1}(x)-\mu|^p}{\sigma_p^p}\,dx\right]^{\frac{1}{p}} \left[\int_0^1 |\underline{f}_{\tilde{c}:n}(x)-c|^q\,dx\right]^{\frac{1}{q}} \\
&= ||\underline{f}_{\tilde{c}:n}-c||_q,
\end{aligned}
\tag{4.148}
$$

where c is an arbitrarily chosen real number. We determine the sharpest bound of this type if we choose c minimizing the norm. Note that $\underline{f}_{\tilde{c}:n}$ is a continuous and differentiable, bounded and nondecreasing function. If we choose c being beyond the range of $\underline{f}_{\tilde{c}:n}(x)$, $0 < x < 1$, we can decrease the norm moving the constant towards the range. Accordingly, we can restrict ourselves to the constants of the form $c = \underline{f}_{\tilde{c}:n}(\gamma)$ for some $0 \le \gamma \le 1$, and minimize the function

$$
\begin{aligned}
D(\gamma) &= \int_0^1 |\underline{f}_{\tilde{c}:n}(x)-\underline{f}_{\tilde{c}:n}(\gamma)|^q dx \\
&= \int_0^\gamma [\underline{f}_{\tilde{c}:n}(\gamma)-\underline{f}_{\tilde{c}:n}(x)]^q dx + \int_\gamma^1 [\underline{f}_{\tilde{c}:n}(x)-\underline{f}_{\tilde{c}:n}(\gamma)]^q dx.
\end{aligned}
\tag{4.149}
$$

Differentiating the function, we obtain

$$
\begin{aligned}
D'(\gamma) = q\underline{f}'_{\tilde{c}:n}(\gamma)\Bigg[&\int_0^\gamma [\underline{f}_{\tilde{c}:n}(\gamma)-\underline{f}_{\tilde{c}:n}(x)]^{q-1}dx \\
&-\int_\gamma^1 [\underline{f}_{\tilde{c}:n}(x)-\underline{f}_{\tilde{c}:n}(\gamma)]^{q-1}dx\Bigg].
\end{aligned}
\tag{4.150}
$$

The derivative $\underline{f}'_{\tilde{c}:n}(\gamma)$ is nonnegative, possibly vanishing on some intervals. We analyze more thoroughly the expression in the brackets. The first integral is a continuous nondecreasing function of γ that is equal to 0 at 0, and $\int_0^1 [\underline{f}_{\tilde{c}:n}(1)-\underline{f}_{\tilde{c}:n}(x)]^{q-1}dx > 0$ at 1. The other one continuously nondecreases from $\int_0^1 [\underline{f}_{\tilde{c}:n}(x)-\underline{f}_{\tilde{c}:n}(0)]^{q-1}dx > 0$ at 0 to 0 at 1. It follows that the difference of the integrals is first negative, and ultimately positive. It vanishes at some point $\gamma_* = \gamma_*(p,\tilde{c}) \in (0,1)$. The point solving Eq. (4.144) may be unique, but the equation may have many solutions that would form a nondegenerate closed interval.

In particular, this happens if

$$\underline{F}_{\tilde{c}:n}(x) \geq \sum_{i=1}^{n} c_i\, x, \qquad 0 < x < 1. \tag{4.151}$$

Then $\underline{F}_{\tilde{c}:n}(x) = \sum_{i=1}^{n} c_i\, x$, and $\underline{f}_{\tilde{c}:n}(x) = \sum_{i=1}^{n} c_i$, and obviously

$$||\underline{f}_{\tilde{c}:n} - \underline{f}_{\tilde{c}:n}(\gamma)||_q = \left\| \underline{f}_{\tilde{c}:n}(x) - \sum_{i=1}^{n} c_i \right\|_q = 0 \tag{4.152}$$

for every $0 < \gamma < 1$, which implies that the bound (4.145) is simply zero then.

Otherwise, i.e., when (4.146) holds, we show that the bound (4.145) is positive and the best possible. Note first that $\underline{f}_{\tilde{c}:n}$ is not constant then, and $||\underline{f}_{\tilde{c}:n} - c||_q > 0$ for every $c \in \mathbb{R}$. The equality in the Hölder inequality of (4.148) with $c = \underline{f}_{\tilde{c}:n}(\gamma_*)$ holds if

$$\frac{F^{-1}(x) - \mu}{\sigma_p} = \alpha |\underline{f}_{\tilde{c}:n}(x) - \underline{f}_{\tilde{c}:n}(\gamma_*)|^{q/p} \text{sgn}\{\underline{f}_{\tilde{c}:n}(x) - \underline{f}_{\tilde{c}:n}(\gamma_*)\} \tag{4.153}$$

for some $\alpha > 0$ (see (2.164)). Observe that the RHS is nondecreasing. Since $\frac{q}{p} = q-1$ and (4.144) holds, the integral of the RHS over [0, 1] vanishes. This guarantees that the LHS defines the quantile function of a distribution with the mean equal to μ. The respective pth absolute central moment amounts to σ_p^p if the LHS has the pth norm equal to 1, which is satisfied for $\alpha = \frac{1}{||\underline{f}_{\tilde{c}:n} - \underline{f}_{\tilde{c}:n}(\gamma_*)||_q^{q/p}}$. Note finally that the condition (4.147) provides the equality in the first inequality of (4.148), because the formula (4.147) asserts that the respective quantile function is constant on every interval contained in the set $\{x \in (0, 1) : \underline{F}_{\tilde{c}:n}(x) < F_{\tilde{c}:n}(x)\}$. This completes the proof of attainability of the bound (4.145) by (4.147) under the assumption (4.146). □

It is impossible to provide analytic expressions for $\gamma_*(p, \tilde{c})$, and formulae (4.145) and (4.147) in consequence, except for the case $p = q = 2$ for which $\gamma_*(2, \tilde{c}) = \sum_{i=1}^{n} c_i$. The respective simplification of Theorem 4.11 is described in Theorem 4.7. In view of Theorems 4.8 and 4.9, we present the Moriguti-Hölder bounds more precisely in the cases of order statistics and their differences.

Corollary 4.9 *For $j = 2, \ldots, n$ with $\alpha_* = \alpha_*(j, n)$ defined in (4.81) for $j = 2, \ldots, n-1$, and $\alpha_* = \alpha_*(n, n) = 1$, $\underline{f}_{j:n}(x) = f_{j:n}(\min\{x, \alpha_*\})$ and $0 < \gamma_* = \gamma_*(p, j, n) < \alpha_*(j, n)$ being the unique solution to the equation*

$$\int_0^{\gamma} [f_{j:n}(\gamma) - f_{j:n}(x)]^{q-1} dx = \int_{\gamma}^{\alpha_*} [f_{j:n}(x) - f_{j:n}(\gamma)]^{q-1} dx$$
$$+ (1 - \alpha_*)[f_{j:n}(\alpha_*) - f_{j:n}(\gamma)]^{q-1}, \tag{4.154}$$

we have

$$\mathbb{E}\frac{X_{j:n}-\mu}{\sigma_p} \le ||\underline{f}_{j:n} - f_{j:n}(\gamma_*)||_q, \tag{4.155}$$

where

$$\begin{aligned} ||\underline{f}_{j:n} - \underline{f}_{j:n}(\gamma_*)||_q^q &= \int_0^{\alpha_*} |f_{j:n}(x) - f_{j:n}(\gamma_*)|^q dx \\ &= (1-\alpha_*)[f_{j:n}(\alpha_*) - f_{j:n}(\gamma_*)]^q. \end{aligned} \tag{4.156}$$

The bound is attained by the distribution function satisfying

$$\frac{F^{-1}(x)-\mu}{\sigma_p} = \begin{cases} \frac{|f_{j:n}(x)-f_{j:n}(\gamma_*)|^{q/p}\text{sgn}\{f_{j:n}(x)-f_{j:n}(\gamma_*)\}}{||\underline{f}_{j:n}-f_{j:n}(\gamma_*)||_q^{q/p}}, & 0 < x \le \alpha_*, \\ \frac{[f_{j:n}(\alpha_*)-f_{j:n}(\gamma_*)]^{q/p}}{||\underline{f}_{j:n}-f_{j:n}(\gamma_*)||_q^{q/p}}, & \alpha_* \le x < 1. \end{cases} \tag{4.157}$$

The statement of Corollary 4.9 in the special case $j = n$ was presented in Arnold (1985).

Corollary 4.10 *For $n \ge 2$, $1 < p < \infty$, $q = \frac{p}{p-1}$ and $\gamma_* = \gamma_*(n, p)$ satisfying the equation*

$$\int_0^{\gamma} [\gamma^{n-1} - x^{n-1}]^{q-1} dx = \int_{\gamma}^1 [x^{n-1} - \gamma^{n-1}]^{q-1} dx, \qquad 0 < \gamma < 1, \tag{4.158}$$

we have

$$\mathbb{E}\frac{X_{n:n}-\mu}{\sigma_p} \le B = B(n,p) = n\left(\int_0^1 |x^{n-1} - \gamma_*^{n-1}|^q \, dx\right)^{\frac{1}{q}} \tag{4.159}$$

with the equality condition

$$\frac{F^{-1}(x)-\mu}{\sigma_p} = \left(\frac{n|x^{n-1} - \gamma_*^{n-1}|}{B}\right)^{\frac{q}{p}} \text{sgn}\{x - \gamma_*\}, \qquad 0 < x < 1. \tag{4.160}$$

For $2 \le j \le n-1$ the function $\underline{f}_{j:n}$ is constant only on $[\alpha_*, 1)$, where it is maximal, and the optimal constant $c_* = \underline{f}_{j:n}(\gamma_*)$ belongs to the interior of the range of $\underline{f}_{j:n}$. Therefore γ_* is determined uniquely, and $\underline{f}_{j:n}(\gamma_*) = f_{j:n}(\gamma_*)$. Note that the distribution function (4.157) has a single atom with the mass $1-\alpha_*$ located at the right end of its support. It is impossible to establish analytic formulae for the bounds and their attainability conditions for general p and $q = \frac{p}{p-1}$ even in the seemingly simplest case of sample maximum for which $\underline{f}_{n:n}(x) = f_{n:n}(x) = nx^{n-1}$ and $\alpha_*(n,n) = 1$, because $\gamma_*(p,n,n)$ has to be determined numerically. The exception is $\gamma_*(2,n,n) = 1$ (and, more generally $\gamma_*(2,j,n) = 1$) which enables us to present Theorem 4.1 (and Theorem 4.8, respectively) in a simpler form. For the sample minima with $j = 1$, Theorem 4.11 provides the zero bounds.

Corollary 4.11 *For $1 \le i < j \le n$ denote $f_{i,j:n} = f_{j:n} - f_{i:n}$ for brevity, and $\underline{f}_{i,j,n}(x) = f_{i,j:n}(\max\{\alpha_*, \min\{x, \beta_*\}\})$ with $\alpha_* = \alpha_*(i,j,n)$ and $\beta_* = \beta_*(i,j,n)$ defined in Theorem 4.9. Let $\alpha_*(i,j,n) < \gamma_* = \gamma_*(p,i,j,n) < \beta_*(i,j,n)$ be the unique solution to*

$$\alpha_*[f_{i,j:n}(\gamma_*) - f_{i,j:n}(\alpha_*)]^{q-1} + \int_{\alpha_*}^{\gamma_*} [f_{i,j:n}(\gamma_*) - f_{i,j:n}(x)]^{q-1} dx$$
$$= \int_{\gamma_*}^{\beta_*} [f_{i,j:n}(x) - f_{i,j:n}(\gamma_*)]^{q-1} dx + (1-\beta_*)[f_{i,j:n}(\beta_*) - f_{i,j:n}(\gamma_*)]^{q-1}. \tag{4.161}$$

Then

$$\mathbb{E}\frac{X_{j:n} - X_{i:n}}{\sigma_p} \le ||\underline{f}_{i,j:n} - f_{i,j:n}(\gamma_*)||_q \tag{4.162}$$

with

$$||\underline{f}_{i,j:n} - f_{i,j:n}(\gamma_*)||_q^q = \alpha_*[f_{i,j:n}(\gamma_*) - f_{i,j:n}(\alpha_*)]^q + \int_{\alpha_*}^{\gamma_*} |f_{i,j:n}(x) - f_{i,j:n}(\gamma_*)|^q dx + (1-\beta_*)[f_{i,j:n}(\beta_*) - f_{i,j:n}(\gamma_*)]^q. \tag{4.163}$$

The equality holds in (4.162) if

$$\frac{F^{-1}(x) - \mu}{\sigma_p} = \begin{cases} -\frac{[f_{i,j:n}(\gamma_*) - f_{i,j:n}(\alpha_*)]^{q/p}}{||\underline{f}_{i,j:n} - f_{i,j:n}(\gamma_*)||_q^{q/p}}, & 0 < x \le \alpha_*, \\ \frac{|f_{i,j:n}(x) - f_{i,j:n}(\gamma_*)|^{q/p} \operatorname{sgn}\{f_{i,j:n}(x) - f_{i,j:n}(\gamma_*)\}}{||\underline{f}_{i,j:n} - f_{i,j:n}(\gamma_*)||_q^{q/p}}, & \alpha_* \le x \le \beta_*, \\ \frac{[f_{i,j:n}(\beta_*) - f_{i,j:n}(\gamma_*)]^{q/p}}{||\underline{f}_{i,j:n} - f_{i,j:n}(\gamma_*)||_q^{q/p}}, & \beta_* \le x < 1. \end{cases} \tag{4.164}$$

Distribution functions (4.164) have atoms of sizes α_* and $1-\beta_*$ at the ends of their interval supports (with the exception of the cases $i = 1$ and $j = n$, where the left and right atom, respectively, disappear). Since $\underline{f}_{i,j:n}(\alpha_*) < \underline{f}_{i,j:n}(\gamma_*) < \underline{f}_{i,j:n}(\beta_*)$, the parameter $\gamma_*(p,i,j,n)$ is determined uniquely, and $\underline{f}_{i,j:n}(\gamma_*) = f_{i,j:n}(\gamma_*)$.

4.4.2 Positive Upper Bounds in σ_1-Units

Now we focus on the bounds expressed in the mean absolute deviation scale units $\sigma_1 = \mathbb{E}|X_1 - \mu|$.

Theorem 4.12 *Let $X_1, \ldots, X_n$ be iid with the mean μ and $0 < \sigma_1 = \mathbb{E}|X_1 - \mu| < \infty$. Then for arbitrary $\tilde{c} = (c_1, \ldots, c_n) \in \mathbb{R}^n$, with $\underline{f}_{\tilde{c}:n}$ denoting the derivative of the greatest convex minorant of $F_{\tilde{c}:n}$, the following inequality holds*

$$\mathbb{E}\sum_{i=1}^{n} c_i \frac{X_{i:n} - \mu}{\sigma_1} \leq \frac{1}{2}[\underline{f}_{\tilde{c}:n}(1) - \underline{f}_{\tilde{c}:n}(0)] \tag{4.165}$$

If

$$F_{\tilde{c}:n}(x) < \sum_{i=1}^{n} c_i\, x \qquad for\ some\ 0 < x < 1, \tag{4.166}$$

(cf. (4.64) and (4.146)), then the bound (4.165) is positive and optimal.

If, moreover,

$$F_{\tilde{c}:n}(x) \leq nc_1 x \qquad for\ some\ 0 < x < 1, \tag{4.167}$$

and

$$\sum_{i=1}^{n} c_i - F_{\tilde{c}:n}(x) \geq nc_n(1 - x) \qquad for\ some\ 0 < x < 1, \tag{4.168}$$

then

$$\alpha_* = \alpha_*(1, \tilde{c}) = \sup\{0 \leq x \leq 1 : \underline{f}_{\tilde{c}:n}(x) = \underline{f}_{\tilde{c}:n}(0)\}, \tag{4.169}$$

$$\beta_* = \beta_*(1, \tilde{c}) = \inf\{0 \leq x \leq 1 : \underline{f}_{\tilde{c}:n}(x) = \underline{f}_{\tilde{c}:n}(1)\}, \tag{4.170}$$

satisfy $0 < \alpha_* < \beta_* < 1$, and the equality in (4.165) is attained by the three-point marginal distribution

$$\mathbb{P}\left(X_1 = \mu - \frac{\sigma_1}{2\alpha_*}\right) = \alpha_*,$$
$$\mathbb{P}(X_1 = \mu) = \beta_* - \alpha_*,$$
$$\mathbb{P}\left(X_1 = \mu + \frac{\sigma_1}{2(1 - \beta_*)}\right) = 1 - \beta_*. \tag{4.171}$$

However, if

$$F_{\tilde{c}:n}(x) > nc_1 x, \qquad 0 < x < 1, \tag{4.172}$$

and/or

$$\sum_{i=1}^{n} c_i - F_{\tilde{c}:n}(x) < nc_n(1-x) \qquad 0 < x < 1, \tag{4.173}$$

then the bound in (4.165) is attained in the limit with α_* by the distributions of the form (4.171) with $0 < \alpha_k \to 0$ and /or β_* by $1 > \beta_k \to 1$.

Proof For arbitrary $\tilde{c} \in \mathbb{R}^n$ and $c \in \mathbb{R}$, we have

$$\begin{aligned}
\mathbb{E}\sum_{i=1}^{n} c_i \frac{X_{i:n}-\mu}{\sigma_1} &\le \int_0^1 \frac{F^{-1}(x)-\mu}{\sigma_1}[\underline{f}_{\tilde{c}:n}(x) - c]\,dx \\
&\le \sup_{0<x<1} |\underline{f}_{\tilde{c}:n}(x) - c| \int_0^1 \frac{|F^{-1}(x)-\mu|}{\sigma_1}\,dx \\
&= \max\{|\underline{f}_{\tilde{c}:n}(0) - c|, |\underline{f}_{\tilde{c}:n}(1) - c|\}.
\end{aligned} \tag{4.174}$$

The maximum is minimized over $c \in \mathbb{R}$ at $c_* = \frac{1}{2}[\underline{f}_{\tilde{c}:n}(1) + \underline{f}_{\tilde{c}:n}(0)]$, and the respective minimum amounts to $\frac{1}{2}[\underline{f}_{\tilde{c}:n}(1) - \underline{f}_{\tilde{c}:n}(0)]$. This implies (4.165).

Under the assumption (4.166), $\underline{f}_{\tilde{c}:n}(x)$, $0 < x < 1$, is nonconstant, and the RHS of (4.165) is positive. We check now that it is optimal then.

Assume now that (4.167) and (4.168) hold. Since $nc_1 = f_{\tilde{c}:n}(0)$, the former implies that $\underline{F}_{\tilde{c}:n}$ is linear, and $\underline{f}_{\tilde{c}:n}$ is constant on $[0, \alpha_*]$. Similarly, (4.168) implies that $\underline{f}_{\tilde{c}:n}$ is constant on $[\beta_*, 1]$. Since $\underline{f}_{\tilde{c}:n}$ is nonconstant, $\alpha_* \le \beta_*$. In fact, the inequality is sharp, because $\underline{F}_{\tilde{c}:n}$ has the derivatives $\underline{f}_{\tilde{c}:n}(\alpha_*) = \underline{f}_{\tilde{c}:n}(0)$, and $\underline{f}_{\tilde{c}:n}(\beta_*) = \underline{f}_{\tilde{c}:n}(1)$ at α_* and β_*, respectively, and these are different by nonconstancy of $\underline{f}_{\tilde{c}:n}$ on the whole $[0, 1]$.

Note that the probability distribution (4.171) has the mean μ and mean absolute deviation σ_1, and the quantile function satisfying

$$\frac{F^{-1}(x)-\mu}{\sigma_1} = \begin{cases} -\frac{1}{2\alpha_*}, & 0 < x < \alpha_*, \\ 0, & \alpha_* \le x < \beta_*, \\ \frac{1}{2(1-\beta_*)}, & \beta_* \le x < 1. \end{cases} \tag{4.175}$$

This implies that (4.175) assures the equality in the latter inequality of (4.174). Actually, the first inequality becomes the equality as well then. Observe that (4.175) is constant on the first and last intervals where $\underline{F}_{\tilde{c}:n}$ is linear. Most likely, these are the first and last intervals where $\underline{F}_{\tilde{c}:n} < F_{\tilde{c}:n}$. It is rather unlikely, but still possible that the intervals $(0, \alpha_*)$ and $(\beta_*, 1)$ contain separate points where $\underline{F}_{\tilde{c}:n} = F_{\tilde{c}:n}$.

However, this only means that $(0, \alpha_*)$ and/or $(\beta_*, 1)$ contain several intervals where $\underline{F}_{\tilde{c}:n} < F_{\tilde{c}:n}$ separated only by single points. The remaining intervals where $\underline{F}_{\tilde{c}:n}$ and $F_{\tilde{c}:n}$ differ, if any, are contained in the interval (α_*, β_*), where (4.175) is constant as well. This means that the conditions of Theorem 2.16 assuring the equality in the first inequality of (4.174) are satisfied.

Consider now the situations when (4.168) and (4.172) hold so that $\alpha_* = 0$ and $0 < \beta_* < 1$. It follows that $\underline{F}_{\tilde{c}:n} = F_{\tilde{c}:n}$ and $\underline{f}_{\tilde{c}:n} = f_{\tilde{c}:n}$ on some right neighborhood of 0. Let $\alpha_k > 0$ belong to this neighborhood. This implies that all the intervals where $\underline{F}_{\tilde{c}:n} < F_{\tilde{c}:n}$ are contained in (α_k, β_*) and $(\beta_*, 1)$, where the standardized quantile function (4.175) with α_* replaced by α_k is constant. Therefore the former inequality in (4.174) becomes the equality then. Letting $\alpha_k \to 0$, we get in equality in the latter inequality.

Similar arguments we apply in the cases $0 < \alpha_* < \beta_* = 1$, and $\alpha_* = 0$ with $\beta_* = 1$. They allow us to determine asymptotic attainability of the bounds in (4.165) when either of conditions (4.172) and (4.173) holds. □

Particular forms of Theorem 4.12 for single order statistics and their differences are presented below.

Corollary 4.12 *Under the assumptions of Theorem* 4.12, *for* $j = 2, \ldots, n-1$ *with* $\alpha_* = \alpha_*(j, n)$ *defined in (4.81) of Theorem* 4.8 *the inequality yields*

$$\mathbb{E}\frac{X_{j:n} - \mu}{\sigma_1} \leq \frac{1}{2} f_{j:n}(\alpha_*), \tag{4.176}$$

which is the equality attained in the limit by the sequence of parent distributions

$$\begin{aligned} \mathbb{P}\left(X_1 = \mu - \frac{\sigma_1}{2\alpha_k}\right) &= \alpha_k, \\ \mathbb{P}(X_1 = \mu) &= \alpha_* - \alpha_k, \\ \mathbb{P}\left(X_1 = \mu + \frac{\sigma_1}{2(1-\alpha_*)}\right) &= 1 - \alpha_*, \end{aligned} \tag{4.177}$$

for $\alpha_k \to 0$ *as* $k \to \infty$.

For $j = n$ *we have*

$$\mathbb{E}\frac{X_{n:n} - \mu}{\sigma_1} \leq \frac{n}{2}, \tag{4.178}$$

with the equality attained in the limit by the family of distributions of the form (4.177) with α_* *replaced by* β_k, *when* $\alpha_k \to 0$, *and* $\beta_k \to 1$ *as* $k \to \infty$.

The formulae (4.92) and (4.93) enable us to present explicit counterparts of (4.176) for the second and third order statistics

$$\mathbb{E}\frac{X_{2:n}-\mu}{\sigma_1} \le \frac{n}{2(n-1)}\left[1-\frac{1}{(n-1)^2}\right]^{n-2}, \tag{4.179}$$

$$\mathbb{E}\frac{X_{3:n}-\mu}{\sigma_1} \le \frac{n(n-1)(n-2)}{4}\left[\frac{n-2+\sqrt{(n-2)n(2n-3)}}{(n-1)^2(n-2)}\right]^2 \times\left[1-\frac{n-2+\sqrt{(n-2)n(2n-3)}}{(n-1)^2(n-2)}\right]^{n-3}. \tag{4.180}$$

Corollary 4.13 *Under the assumptions of Theorem* 4.12, *for* $1 \le i < j \le n$ *with the notation* $\alpha_* = \alpha_*(i, j, n)$ *and* $\beta_* = \beta_*(i, j, n)$ *of Theorem* 4.9 *we have*

$$\mathbb{E}\frac{X_{j:n}-X_{i:n}}{\sigma_1} \le \frac{1}{2}[f_{i,j:n}(\beta_*) - f_{i,j:n}(\alpha_*)] \tag{4.181}$$

with the equality conditions for $2 \le i < j \le n-1$

$$\mathbb{P}\left(X_1 = \mu - \frac{\sigma_1}{2\alpha_*}\right) = \alpha_*,$$
$$\mathbb{P}(X_1 = \mu) = \beta_* - \alpha_*,$$
$$\mathbb{P}\left(X_1 = \mu + \frac{\sigma_1}{2(1-\beta_*)}\right) = 1-\beta_*. \tag{4.182}$$

If $i = 1$, *then* $f_{i,j:n}(\alpha_*)$ *in (4.181) is replaced by* 0, *and* α_* *in (4.182) is replaced by* $\alpha_k \to 0$. *If* $j = n$, *then* $f_{i,j:n}(\beta_*)$ *in (4.181) is replaced by* n, *and* β_* *in (4.182) is replaced by* $\beta_k \to 1$.

The proof is similar to that of Corollary 4.12, and it is omitted here. In particular, for the sample range we have

$$\mathbb{E}\frac{X_{n:n}-X_{1:n}}{\sigma_1} \le n. \tag{4.183}$$

This bound is attained in the limit.

4.4.3 Positive Upper Bounds in σ_∞-Units

Now we determine the greatest possible expectations of centered L-statistics based on iid samples from bounded populations, measuring their values in the following units

$$\sigma_\infty = \operatorname{ess\,sup}|X_1 - \mu| = \max\{\mu - F^{-1}(0), F^{-1}(1) - \mu\}. \tag{4.184}$$

Theorem 4.13 *Let $X_1, \ldots, X_n$ be iid with mean μ and $0 < \sigma_\infty < \infty$. For every $\tilde{c} = (c_1, \ldots, c_n) \in \mathbb{R}^n$, and $\underline{F}_{\tilde{c}:n}$ denoting the greatest convex minorant of $F_{\tilde{c}:n}$, we have*

$$\mathbb{E}\sum_{i=1}^n c_i \frac{X_{i:n} - \mu}{\sigma_\infty} \le \sum_{i=1}^n c_i - 2\underline{F}_{\tilde{c}:n}\left(\frac{1}{2}\right). \tag{4.185}$$

Under the condition (4.64), the equality in the above inequality is attained by the marginal distribution

$$\begin{aligned} \mathbb{P}(X_1 = \mu - \sigma_\infty) &= \alpha_*, \\ \mathbb{P}\left(X_1 = \mu + \sigma_\infty \frac{\alpha_* + \beta_* - 1}{\beta_* - \alpha_*}\right) &= \beta_* - \alpha_*, \\ \mathbb{P}(X_1 = \mu + \sigma_\infty) &= 1 - \beta_*, \end{aligned} \tag{4.186}$$

where

$$\alpha_* = \alpha_*(\infty, \tilde{c}) = \min\left\{0 \le x \le \frac{1}{2} : \underline{f}_{\tilde{c}:n}(x) = \underline{f}_{\tilde{c}:n}\left(\frac{1}{2}\right)\right\}, \tag{4.187}$$

$$\beta_* = \beta_*(\infty, \tilde{c}) = \max\left\{\frac{1}{2} \le x \le 1 : \underline{f}_{\tilde{c}:n}(x) = \underline{f}_{\tilde{c}:n}\left(\frac{1}{2}\right)\right\}, \tag{4.188}$$

and $\underline{f}_{\tilde{c}:n}$ is the derivative of the greatest convex minorant $\underline{F}_{\tilde{c}:n}$.

Notice that the parent distribution is supported on three point only if $0 < \alpha_* < \beta_* < 1$. If $\alpha_* = \frac{1}{2}$ and $\beta_* = \frac{1}{2}$, then (4.186) describes a symmetric two-point distribution. We obtain an asymmetric two-point distribution if either $\alpha_* = 0$ and $\frac{1}{2} < \beta_* < 1$ or $0 < \alpha_* < \frac{1}{2}$, and $\beta_* = 1$. If $\alpha_* = 0$ and $\beta_* = 1$, we obtain $\sigma_\infty = 0$, which contradicts our assumptions.

Proof For every $c \in \mathbb{R}$ we have

$$\begin{aligned} \mathbb{E}\sum_{i=1}^n c_i \frac{X_{i:n} - \mu}{\sigma_\infty} &\le \int_0^1 \frac{F^{-1}(x) - \mu}{\sigma_\infty}[\underline{f}_{\tilde{c}:n}(x) - c]\,dx \\ &\le \sup_{0<x<1} \frac{|F^{-1}(x) - \mu|}{\sigma_\infty} \int_0^1 |\underline{f}_{\tilde{c}:n}(x) - c|\,dx. \end{aligned} \tag{4.189}$$

The first factor amounts to 1, and we minimize the latter one with respect to c. We immediately notice that it suffices to focus on the values belonging to the range of

$\underline{f}_{\tilde{c}:n}$ only. Using the notation $c = \underline{f}_{\tilde{c}:n}(\gamma)$, $0 \le \gamma \le 1$, we calculate

$$\begin{aligned} D(\gamma) &= \int_0^\gamma [\underline{f}_{\tilde{c}:n}(\gamma) - \underline{f}_{\tilde{c}:n}(x)]dx + \int_\gamma^1 [\underline{f}_{\tilde{c}:n}(x) - \underline{f}_{\tilde{c}:n}(\gamma)]dx \\ &= (2\gamma - 1)\underline{f}_{\tilde{c}:n}(\gamma) + \underline{F}_{\tilde{c}:n}(1) - 2\underline{F}_{\tilde{c}:n}(\gamma), \end{aligned} \tag{4.190}$$

and the respective derivative

$$D'(\gamma) = (2\gamma - 1)\underline{f}'_{\tilde{c}:n}(\gamma). \tag{4.191}$$

Since $\underline{f}'_{\tilde{c}:n} \ge 0$, we conclude that (4.190) is nonincreasing on $\left(0, \frac{1}{2}\right)$ and nondecreasing on $\left(\frac{1}{2}, 1\right)$ (and possibly constant on some intervals). We rewrite now the relations (4.189) with the optimal $c_* = \underline{f}_{\tilde{c}:n}\left(\frac{1}{2}\right)$

$$\begin{aligned} \mathbb{E}\sum_{i=1}^n c_i \frac{X_{i:n} - \mu}{\sigma_\infty} &\le \int_0^1 \frac{F^{-1}(x) - \mu}{\sigma_\infty}\left[\underline{f}_{\tilde{c}:n}(x) - \underline{f}_{\tilde{c}:n}\left(\frac{1}{2}\right)\right]dx \\ &\le \sup_{0<x<1} \frac{|F^{-1}(x) - \mu|}{\sigma_\infty} \int_0^1 \left|\underline{f}_{\tilde{c}:n}(x) - \underline{f}_{\tilde{c}:n}\left(\frac{1}{2}\right)\right| dx \\ &= \underline{F}_{\tilde{c}:n}(1) - 2\underline{F}_{\tilde{c}:n}\left(\frac{1}{2}\right) = \sum_{i=1}^n c_i - 2\underline{F}_{\tilde{c}:n}\left(\frac{1}{2}\right). \end{aligned} \tag{4.192}$$

If $F_{\tilde{c}:n}(x) \ge \sum_{i=1}^n c_i\, x$, $0 < x < 1$, i.e., (4.64) does not hold, then $\underline{F}_{\tilde{c}:n}(x) = \sum_{i=1}^n c_i\, x$, $0 < x < 1$, and the bound in (4.192) amounts to 0.

Suppose, however, that the condition (4.64) is satisfied. Then $\underline{F}_{\tilde{c}:n}(x) < \sum_{i=1}^n c_i\, x$, $0 < x < 1$, and the bound is positive. We show that it is attainable. Function $\frac{F^{-1}(x)-\mu}{\sigma_\infty}$, $0 < x < 1$, has all values contained in $[-1, 1]$. The latter inequality in (4.192) becomes the equality if

$$\begin{aligned} \frac{F^{-1}(x) - \mu}{\sigma_\infty} & \begin{cases} = -1, & \underline{f}_{\tilde{c}:n}(x) < \underline{f}_{\tilde{c}:n}\left(\frac{1}{2}\right), \\ \in [-1, 1], & \underline{f}_{\tilde{c}:n}(x) = \underline{f}_{\tilde{c}:n}\left(\frac{1}{2}\right), \\ = +1, & \underline{f}_{\tilde{c}:n}(x) > \underline{f}_{\tilde{c}:n}\left(\frac{1}{2}\right), \end{cases} \\ & \begin{cases} = -1, & 0 < x < \alpha_*, \\ \in [-1, 1], & \alpha_* \le x \le \beta_*, \\ = +1, & \beta_* < x < 1, \end{cases} \end{aligned} \tag{4.193}$$

Moreover, a necessary condition for the equality in the first inequality of (4.192) is that $\frac{F^{-1}(x)-\mu}{\sigma_\infty}$ = const on the set $\left\{x : \underline{f}_{\tilde{c}:n}(x) = \underline{f}_{\tilde{c}:n}\left(\frac{1}{2}\right)\right\} = [\alpha_*, \beta_*]$ if it is a nondegenerate interval. The first moment condition implies that $\frac{F^{-1}(x)-\mu}{\sigma_\infty} = \frac{\alpha_*+\beta_*-1}{\beta_*-\alpha_*}$ then. Since $0 \le \alpha_* \le \frac{1}{2} \le \beta_* \le 1$, this value actually belongs to $[-1, 1]$. This means that (4.193) satisfies both the range and nondecrease conditions. Observe also that all the intervals where $\underline{F}_{\tilde{c}:n} < F_{\tilde{c}:n}$ and $\underline{f}_{\tilde{c}:n}$ is a constant different from $\underline{f}_{\tilde{c}:n}\left(\frac{1}{2}\right)$ (if any) belong to the intervals $[0, \alpha_*)$ and $(\beta_*, 1]$, where (4.193) is constant. This asserts the equality in the first inequality of (4.192), and completes the proof. □

Now we specify the general results for single order statistics and their differences.

Corollary 4.14 *Let the assumptions of Theorem* 4.13 *hold. For the sample maximum, we have*

$$\mathbb{E}\frac{X_{n:n}-\mu}{\sigma_\infty} \le 1 - \frac{1}{2^{n-1}}, \tag{4.194}$$

and the equality holds for the marginal distribution

$$\mathbb{P}(X_1 = \mu - \sigma_\infty) = \mathbb{P}(X_1 = \mu + \sigma_\infty) = \frac{1}{2}. \tag{4.195}$$

If

$$2\left[1 - F_{j:n}\left(\frac{1}{2}\right)\right] \ge f_{j:n}\left(\frac{1}{2}\right) \tag{4.196}$$

for some $j = 2, \ldots, n-1$, *then*

$$\mathbb{E}\frac{X_{j:n}-\mu}{\sigma_\infty} \le 1 - 2F_{j:n}\left(\frac{1}{2}\right), \tag{4.197}$$

and the equality holds for the marginal distribution (4.195).

If $2 \le j \le n-1$ *and*

$$2\left[1 - F_{j:n}\left(\frac{1}{2}\right)\right] \le f_{j:n}\left(\frac{1}{2}\right) \tag{4.198}$$

then

$$\mathbb{E}\frac{X_{j:n}-\mu}{\sigma_\infty} \le f_{j:n}(\alpha_*) - 1, \tag{4.199}$$

with the equality condition

$$\mathbb{P}(X_1 = \mu - \sigma_\infty) = \alpha_* = 1 - \mathbb{P}\left(X_1 = \mu + \sigma_\infty \frac{\alpha_*}{1-\alpha_*}\right), \tag{4.200}$$

where $\alpha_* = \alpha_*(j, n)$ is defined in Theorem 4.8 as the solution to Eq. (4.81).

Proof We use the results of Theorem 4.13 for the specific vector $\tilde{c}$ corresponding to the single order statistics. The case $j = n$ provides the simplest solution. Then $\underline{F}_{n:n}(x) = F_{n:n}(x) = x^n$, which implies (4.194). Moreover $\underline{f}_{n:n}(x) = f_{n:n}(x) = nx^{n-1}$, and we have $\alpha_*(\infty, j, n) = \beta_*(\infty, j, n) = \frac{1}{2}$ (comp. (4.187) and (4.188))) Plugging them into (4.186) we obtain (4.195).

Consider now $2 \le j \le n-1$. Since $\underline{f}_{j:n}(x) = f_{j:n}(\min\{x, \alpha_*\})$, the crucial point of our reasoning is determining the ordering relation between α_* and $\frac{1}{2}$. We observe that the difference $\frac{1-F_{j:n}(x)}{1-x} - f_{j:n}(x)$, $0 < x < 1$, vanishing at α_*, is first positive, and then negative, owing to the following relations

$$\begin{aligned} 1 - F_{j:n}(x) - f_{j:n}(x)(1-x) &= \sum_{i=0}^{j-1} B_{i,n}(x) - nB_{j-1,n-1}(x)(1-x) \\ &= \sum_{i=0}^{j-2} B_{i,n}(x) - (n-j)B_{j-1,n}(x), \end{aligned} \tag{4.201}$$

and the variation diminishing property. Accordingly, if (4.196) then $\alpha_*(j,n) \ge \frac{1}{2}$. In this case $\underline{f}_{j:n}\left(\frac{1}{2}\right) = f_{j:n}\left(\frac{1}{2}\right)$ and $\underline{F}_{j:n}\left(\frac{1}{2}\right) = F_{j:n}\left(\frac{1}{2}\right)$. This gives (4.197). Moreover, $\alpha_*(\infty, j, n) = \beta_*(\infty, j, n) = \frac{1}{2}$ then (see (4.187) and (4.188))), and, by (4.186), the equality condition is represented by (4.195).

If (4.198) holds, then $\alpha_*(j,n) \le \frac{1}{2}$, which means that $\underline{f}_{j:n}\left(\frac{1}{2}\right) = f_{j:n}(\alpha_*)$ and $\underline{F}_{j:n}\left(\frac{1}{2}\right) = 1 - \frac{1}{2} f_{j:n}(\alpha_*)$ which gives (4.199). In this case $\alpha_*(\infty, j, n) = \alpha_*(j, n)$ and $\beta_*(\infty, j, n) = 1$, and we use (4.186) for determining the parent distribution (4.200). □

Both the bounds in (4.197) and (4.199) are positive. The first one holds under the condition (4.196) which implies $\frac{1}{2} < \alpha_*$, $\underline{f}_{j:n}\left(\frac{1}{2}\right) = f_{j:n}\left(\frac{1}{2}\right)$, and $\underline{F}_{j:n}\left(\frac{1}{2}\right) = F_{j:n}\left(\frac{1}{2}\right)$. Since $F_{j:n}(x) < x$, $j = 2, \ldots, n-1$, on some neighborhoods of 0 (comp. (4.64)), $\underline{F}_{j:n}(x) < x$, $0 < x < 1$, and so $\underline{F}_{j:n}\left(\frac{1}{2}\right) = F_{j:n}\left(\frac{1}{2}\right) < \frac{1}{2}$. In the latter case (4.198) we use the fact that $f_{j:n}(\alpha_*)$ is the essentially maximal value of the density function $\underline{f}_{j:n}$ supported on [0, 1]. However, every probability density function concentrated on [0, 1], except for the uniform one, has the essential supremum greater than 1. If we have the equality in (4.198), then $\alpha_* = \frac{1}{2}$, and the bounds (4.197) and (4.199) and their attainability conditions coincide.

Observe that $\alpha_*(j,n) < \frac{j-1}{n-1}$. This means that if $\frac{j-1}{n-1} \le \frac{1}{2}$, then (4.199) holds. Also, note that (4.194) is a particular case of (4.197) with $j = n$. These observations allow us to guess that the former bound (4.197) is valid for the order statistics with large indices, and the latter one (4.199) holds for small j's.

For determining the bounds for the differences of order statistics $X_{j:n} - X_{i:n}$, $1 \le i < j \le n$, we need the values $\alpha_* = \alpha_*(i,j,n)$ and $\beta_* = \beta_*(i,j,n)$ defined in Theorem 4.9 (see (4.100) and (4.101)). We also use the notation $f_{i,j:n} = f_{j:n} - f_{i:n}$ and $F_{i,j:n} = F_{j:n} - F_{i:n}$ for brevity. We get the following bounds

$$\mathbb{E}\frac{X_{j:n} - X_{i:n}}{\sigma_\infty} = \begin{cases} -f_{i,j:n}(\alpha_*), & \text{if } \alpha_* \ge \frac{1}{2}, \\ -2F_{i,j:n}\left(\frac{1}{2}\right), & \text{if } \alpha_* \le \frac{1}{2} \le \beta_*, \\ f_{i,j:n}(\beta_*), & \text{if } \beta_* \le \frac{1}{2}, \end{cases} \tag{4.202}$$

with the respective attainability conditions

$$\mathbb{P}\left(X_1 = \mu - \sigma_\infty \frac{1-\alpha_*}{\alpha_*}\right) = \alpha_* = 1 - \mathbb{P}(X_1 = \mu + \sigma_\infty), \quad \alpha_* \ge \frac{1}{2}, \tag{4.203}$$

$$\mathbb{P}(X_1 = \mu - \sigma_\infty) = \frac{1}{2} = \mathbb{P}(X_1 = \mu + \sigma_\infty), \quad \alpha_* \le \frac{1}{2} \le \beta_*, \tag{4.204}$$

$$\mathbb{P}(X_1 = \mu - \sigma_\infty) = \beta_* = 1 - \mathbb{P}\left(X_1 = \mu + \sigma_\infty \frac{\beta_*}{1-\beta_*}\right), \quad \beta_* \le \frac{1}{2}. \tag{4.205}$$

Obviously, $\alpha_*(1,j,n) = 0 < \frac{1}{2}$, and $\beta_*(1,j,n) = 1 > \frac{1}{2}$. For $1 < i < j \le n$ we have $\alpha_*(i,j,n) \ge (\le, \text{ resp.}) \frac{1}{2}$ iff

$$2F_{i,j:n}\left(\frac{1}{2}\right) \ge (\le, \text{ resp.}) f_{i,j:n}\left(\frac{1}{2}\right). \tag{4.206}$$

For $1 \le i < j < n$ the relation $\beta_*(i,j,n) \le (\ge, \text{ resp.}) \frac{1}{2}$ holds iff

$$-2F_{i,j:n}\left(\frac{1}{2}\right) \le (\ge, \text{ resp.}) f_{i,j:n}\left(\frac{1}{2}\right). \tag{4.207}$$

We have a simple analytic bound

$$\mathbb{E}\frac{X_{n:n} - X_{1:n}}{\sigma_\infty} \le 2 - \frac{1}{2^{n-2}} \tag{4.208}$$

for the sample range, and this is attained by the symmetric two-point distribution (4.204). Verification of the above formulae is left to the interested readers.

4.4.4 Negative Lower Bounds

Now we present without proofs analogous lower bounds on the expectations of L-statistics. The proofs of the following three theorems are similar to the proofs of Theorems 4.11, 4.12 and 4.13, respectively.

Theorem 4.14 *Let $X_1, \ldots, X_n$ be iid and have a finite moment of order $p > 1$, and let for a given $\tilde{c} = (c_1, \ldots, c_n) \in \mathbb{R}^n$, function $\overline{f}_{\tilde{c}:n}$ denote the derivative of the smallest concave majorant $\overline{F}_{\tilde{c}:n}$ of $F_{\tilde{c}:n} = \sum_{i=1}^n c_i F_{i:n}$. Define $0 < \gamma_{**} = \gamma_{**}(p, \tilde{c}) < 1$ as any solution to the equation*

$$\int_0^\gamma [\overline{f}_{\tilde{c}:n}(x) - \overline{f}_{\tilde{c}:n}(\gamma)]^{q-1} dx = \int_\gamma^1 [\overline{f}_{\tilde{c}:n}(\gamma) - \overline{f}_{\tilde{c}:n}(x)]^{q-1} dx. \tag{4.209}$$

Then

$$\begin{aligned} \mathbb{E} \sum_{i=1}^n c_i \frac{X_{i:n} - \mu}{\sigma_p} &\geq -||\overline{f}_{\tilde{c}:n} - \overline{f}_{\tilde{c}:n}(\gamma_{**})||_q \\ &= -\left(\int_0^1 |\overline{f}_{\tilde{c}:n}(x) - \overline{f}_{\tilde{c}:n}(\gamma_{**})|^q dx \right)^{\frac{1}{q}}. \end{aligned} \tag{4.210}$$

If, moreover,

$$F_{\tilde{c}:n}(x) > \sum_{i=1}^n c_i\, x \qquad \text{for some } 0 < x < 1 \tag{4.211}$$

(comp. (4.118)), then the bound (4.210) is strictly negative and sharp, and it is attained by the parent distribution function F satisfying

$$\frac{F^{-1}(x) - \mu}{\sigma_p} = \frac{|\overline{f}_{\tilde{c}:n}(x) - \overline{f}_{\tilde{c}:n}(\gamma_{**})|^{q/p} \operatorname{sgn}\{\overline{f}_{\tilde{c}:n}(\gamma_{**}) - \overline{f}_{\tilde{c}:n}(x)\}}{||\overline{f}_{\tilde{c}:n} - \overline{f}_{\tilde{c}:n}(\gamma_{**})||_q^{q/p}}. \tag{4.212}$$

Theorem 4.15 *Let $X_1, \ldots, X_n$ be iid with the finite mean μ and positive mean absolute deviation $\sigma_1 = \mathbb{E}|X_1 - \mu|$. Then for arbitrary $\tilde{c} = (c_1, \ldots, c_n) \in \mathbb{R}^n$, with $\overline{f}_{\tilde{c}:n}$ denoting the derivative of the smallest concave majorant of $F_{\tilde{c}:n}$, the following holds*

$$\mathbb{E} \sum_{i=1}^n c_i \frac{X_{i:n} - \mu}{\sigma_1} \geq -\frac{1}{2}[\overline{f}_{\tilde{c}:n}(0) - \overline{f}_{\tilde{c}:n}(1)] \tag{4.213}$$

If

$$F_{\tilde{c}:n}(x) > \sum_{i=1}^{n} c_i\, x \qquad for\ some\ 0 < x < 1, \tag{4.214}$$

(cf. (4.118) and (4.211)), then the bound (4.213) is negative and sharp.

If, moreover,

$$F_{\tilde{c}:n}(x) \geq nc_1 x \qquad for\ some\ 0 < x < 1, \tag{4.215}$$

and

$$\sum_{i=1}^{n} c_i - F_{\tilde{c}:n}(x) \leq nc_n(1-x) \qquad for\ some\ 0 < x < 1, \tag{4.216}$$

then

$$\alpha_{**} = \alpha_{**}(1, \tilde{c}) = \sup\{0 \leq x \leq 1 : \overline{f}_{\tilde{c}:n}(x) = \overline{f}_{\tilde{c}:n}(0)\}, \tag{4.217}$$

$$\beta_{**} = \beta_{**}(1, \tilde{c}) = \inf\{0 \leq x \leq 1 : \overline{f}_{\tilde{c}:n}(x) = \overline{f}_{\tilde{c}:n}(1)\}, \tag{4.218}$$

satisfy $0 < \alpha_* < \beta_* < 1$, *and the equality in (4.213) is attained by the three-point parent distribution*

$$\begin{aligned} \mathbb{P}\left(X_1 = \mu - \frac{\sigma_1}{2\alpha_{**}}\right) &= \alpha_{**}, \\ \mathbb{P}\,(X_1 = \mu) &= \beta_{**} - \alpha_{**}, \\ \mathbb{P}\left(X_1 = \mu + \frac{\sigma_1}{2(1-\beta_{**})}\right) &= 1 - \beta_{**}. \end{aligned} \tag{4.219}$$

However, if

$$F_{\tilde{c}:n}(x) < nc_1 x, \qquad 0 < x < 1, \tag{4.220}$$

and/or

$$\sum_{i=1}^{n} c_i - F_{\tilde{c}:n}(x) > nc_n(1-x) \qquad 0 < x < 1, \tag{4.221}$$

then the bound in (4.213) is attained in the limit by the distributions of the form (4.219) with α_{**} *and/or* β_{**} *replaced by* $0 < \alpha_k \to 0$ *and/or* $1 > \beta_k \to 1$.

Theorem 4.16 *Assume that* $X_1, \ldots, X_n$ *are iid with mean* μ *and* $0 < \sigma_\infty < \infty$. *For every* $\tilde{c} = (c_1, \ldots, c_n) \in \mathbb{R}^n$, *and* $\overline{F}_{\tilde{c}:n}$ *denoting the smallest concave majorant* *of $F_{\tilde{c}:n}$, we have*

$$\mathbb{E}\sum_{i=1}^{n} c_i \frac{X_{i:n}-\mu}{\sigma_\infty} \geq \sum_{i=1}^{n} c_i - 2\overline{F}_{\tilde{c}:n}\left(\frac{1}{2}\right). \tag{4.222}$$

Under the condition (4.118), the RHS of (4.222) is negative, and the equality is attained there by the marginal distribution

$$\begin{aligned}\mathbb{P}(X_1 = \mu - \sigma_\infty) &= \alpha_{**},\\ \mathbb{P}\left(X_1 = \mu + \sigma_\infty \frac{\alpha_{**}+\beta_{**}-1}{\beta_{**}-\alpha_{**}}\right) &= \beta_{**}-\alpha_{**},\\ \mathbb{P}(X_1 = \mu + \sigma_\infty) &= 1-\beta_{**},\end{aligned} \tag{4.223}$$

where

$$\alpha_{**} = \alpha_{**}(\infty, \tilde{c}) = \min\left\{0 \leq x \leq \frac{1}{2} : \overline{f}_{\tilde{c}:n}(x) = \overline{f}_{\tilde{c}:n}\left(\frac{1}{2}\right)\right\}, \tag{4.224}$$

$$\beta_{**} = \beta_{**}(\infty, \tilde{c}) = \max\left\{\frac{1}{2} \leq x \leq 1 : \overline{f}_{\tilde{c}:n}(x) = \overline{f}_{\tilde{c}:n}\left(\frac{1}{2}\right)\right\}, \tag{4.225}$$

and $\overline{f}_{\tilde{c}:n}$ is the derivative of the smallest concave majorant $\overline{F}_{\tilde{c}:n}$.

The lower bounds of Theorems 4.14, 4.15 and 4.16 are just the negatives of the upper bounds of Theorems 4.11, 4.12 and 4.13, respectively, calculated for the L-statistics $\sum_{i=1}^{n} c_{n+1-i}X_{i:n}$ with the reversed vector of coefficients $\tilde{c}' = (c_n, \ldots, c_1)$. This follows from the relations

$$\overline{F}_{\tilde{c}:n}(x) = \sum_{i=1}^{n} c_i - \underline{F}_{\tilde{c}':n}(1-x), \tag{4.226}$$

$$\overline{f}_{\tilde{c}:n}(x) = \underline{f}_{\tilde{c}':n}(1-x), \tag{4.227}$$

(comp. (4.132) and (4.133)). These further imply that for every $1 < p < \infty$ the parameter $\gamma_{**}(p, \tilde{c})$ of Theorem 4.14 coincides with $1-\gamma_*(p, \tilde{c}')$ of Theorem 4.11, and

$$\alpha_{**}(p, \tilde{c}) = 1 - \beta_*(p, \tilde{c}'), \tag{4.228}$$

$$\beta_{**}(p, \tilde{c}) = 1 - \alpha_*(p, \tilde{c}'), \qquad p = 1, \infty, \tag{4.229}$$

(see (4.169), (4.170), (4.187), (4.188), and (4.217), (4.218), (4.224), (4.225)). In particular, the lower bounds for the jth bottom order statistic $X_{j:n}$ are identical with the negatives of respective upper bounds for the jth top order statistic $X_{n+1-j:n}$ of Corollaries 4.9, 4.12 and 4.14. Relations $\gamma_{**}(p, \tilde{c}) = 1 - \gamma_*(p, \tilde{c}')$, (4.228), and (4.229) enable us to determine the attainability conditions for the lower bounds as well.

4.5 Nonpositive Upper and Nonnegative Lower Bounds

4.5.1 Nonpositive Upper Bounds

In the previous sections we presented sharp positive upper bounds on the expectations of centered L-statistics, expressed in various scale units under the condition that $\sum_{i=1}^n c_i F_{i:n}(x) < \sum_{i=1}^n c_i x$ for some $0 < x < 1$. Now we describe nonpositive upper bounds when the condition is not satisfied, i.e., when the function

$$L_{\tilde{c}:n}(x) = \sum_{i=1}^{n} c_i [F_{i:n}(x) - x], \qquad 0 < x < 1, \tag{4.230}$$

is nonnegative. Notice that it vanishes at 0 and 1. We first consider the case

$$\exists\, 0 < \alpha_* = \alpha_*(\tilde{c}) < 1 \qquad L_{\tilde{c}:n}(\alpha_*) = 0 \le L_{\tilde{c}:n}(x), \qquad 0 < x < 1. \tag{4.231}$$

Theorem 4.17 *For every iid $X_1, \ldots, X_n$ with a finite pth moment for some $1 \le p < \infty$, and $\tilde{c} \in \mathbb{R}^n$ satisfying the condition (4.231) the bound*

$$\mathbb{E} \sum_{i=1}^{n} c_i \frac{X_{i:n} - \mu}{\sigma_p} \le 0 \tag{4.232}$$

is sharp, and it is attained by the two-point marginal distribution

$$\mathbb{P}\left(X_1 = \mu - \sigma_p \frac{1 - \alpha_*}{[\alpha_*^p (1 - \alpha_*) + \alpha_* (1 - \alpha_*)^p]^{1/p}} \right) = \alpha_*,$$
$$\mathbb{P}\left(X_1 = \mu + \sigma_p \frac{\alpha_*}{[\alpha_*^p (1 - \alpha_*) + \alpha_* (1 - \alpha_*)^p]^{1/p}} \right) = 1 - \alpha_*. \tag{4.233}$$

Proof Observe that under the assumption (4.231) $\underline{F}_{\tilde{c}:n}(x) = \sum_{i=1}^n c_i x$, and $\underline{f}_{\tilde{c}:n}(x) = \sum_{i=1}^n c_i$, $0 < x < 1$. Therefore

$$\mathbb{E} \sum_{i=1}^{n} c_i \frac{X_{i:n} - \mu}{\sigma_p} \le \int_0^1 \frac{F^{-1}(x) - \mu}{\sigma_p} \left[\underline{f}_{\tilde{c}:n}(x) - \sum_{i=1}^{n} c_i \right] = 0. \tag{4.234}$$

Also, condition (4.231) implies that

$$\underline{F}_{\tilde{c}:n}(\alpha_*) = F_{\tilde{c}:n}(\alpha_*) = \sum_{i=1}^{n} c_i\, \alpha_*. \tag{4.235}$$

It follows that every interval where $\underline{F}_{\tilde{c}:n} < F_{\tilde{c}:n}$ is contained either in $(0, \alpha_*)$ or in $(\alpha_*, 1)$. It follows (see Theorem 2.16) that the equality in (4.232) is attained if $\frac{F^{-1}(x)-\mu}{\sigma_p}$ takes on two different values on $(0, \alpha_*)$ and $[\alpha_*, 1)$, respectively. The corresponding two-point distribution is nondegenerate, and $0 < \sigma_p < \infty$ for all $p \geq 1$. The moment conditions imply that

$$\frac{F^{-1}(x) - \mu}{\sigma_p} = \begin{cases} -\frac{1-\alpha_*}{[\alpha_*^p(1-\alpha_*)+\alpha_*(1-\alpha*)^p]^{1/p}}, & 0 < x < \alpha_*, \\ \frac{\alpha_*}{[\alpha_*^p(1-\alpha_*)+\alpha_*(1-\alpha_*)^p]^{1/p}}, & \alpha_* \leq x < 1, \end{cases} \tag{4.236}$$

which determines the probability distribution (4.233). □

In a similar way we can prove that following.

Theorem 4.18 *If $X_1, \ldots, X_n$ are iid, and have a bounded support, and $\tilde{c}$ satisfies (4.231), then the bound*

$$\mathbb{E} \sum_{i=1}^{n} c_i \frac{X_{i:n} - \mu}{\sigma_\infty} \leq 0 \tag{4.237}$$

holds, and it is attained by the parent distribution

$$\mathbb{P}(X_1 = \mu - \sigma_\infty) = \alpha_* = 1 - \mathbb{P}\left(X_1 = \mu + \sigma_\infty \frac{\alpha_*}{1 - \alpha_*}\right), \tag{4.238}$$

if $0 < \alpha_* \leq \frac{1}{2}$, and

$$\mathbb{P}\left(X_1 = \mu - \sigma_\infty \frac{1 - \alpha_*}{\alpha_*}\right) = \alpha_* = 1 - \mathbb{P}(X_1 = \mu + \sigma_\infty), \tag{4.239}$$

if $\frac{1}{2} \leq \alpha_* < 1$.

If (4.230) is nonnegative, and amounts to 0 at $k \geq 2$ points, then the zero bounds of Theorems 4.17 and 4.18 are attained by various discrete distributions with at most $k + 1$ support points.

Assume now that

$$L_{\tilde{c}:n}(x) = \sum_{i=1}^{n} c_i [F_{i:n}(x) - x] > 0, \qquad 0 < x < 1. \tag{4.240}$$

It follows that $\underline{F}_{\tilde{c}:n}(x) = \sum_{i=1}^{n} c_i x < F_{\tilde{c}:n}(x)$, $0 < x < 1$, and the zero bounds (4.232) and (4.237) cannot be attained by any nondegenerate parent distributions. However, for $1 < p \leq \infty$, these bounds are attained in the limit by sequences of two-point distributions tending to a degenerate one.

Theorem 4.19 *Let $X_1, \ldots, X_n$ be iid, and (4.240) hold. Then for every $1 < p < \infty$*

$$\mathbb{E} \sum_{i=1}^{n} c_i \frac{X_{i:n} - \mu}{\sigma_p} \leq 0, \tag{4.241}$$

and the bound is attained in the limit by the sequences of two point distributions of the form (4.233) with α_* replaced by α such that either $\alpha \to 0$ or $\alpha \to 1$.

Proof For every $p > 1$ consider the family of two-point distributions with the mean μ and pth absolute central moment σ_p^p. Their quantile functions satisfy

$$\frac{F_\alpha^{-1}(x) - \mu}{\sigma_p} = \begin{cases} -\frac{1-\alpha}{[\alpha^p(1-\alpha)+\alpha(1-\alpha)^p]^{1/p}}, & 0 < x < \alpha, \\ \frac{\alpha}{[\alpha^p(1-\alpha)+\alpha(1-\alpha)^p]^{1/p}}, & \alpha \leq x < 1, \end{cases} \qquad 0 < \alpha < 1. \tag{4.242}$$

Then

$$\begin{aligned} \mathbb{E}_\alpha \sum_{i=1}^{n} c_i \frac{X_{i:n} - \mu}{\sigma_p} &= \int_0^1 \frac{F_\alpha^{-1}(x) - \mu}{\sigma_p} \left[f_{\tilde{c}:n}(x) - \sum_{i=1}^{n} c_i \right] dx \\ &= \frac{-(1-\alpha) \int_0^\alpha \left[f_{\tilde{c}:n}(x) - \sum_{i=1}^n c_i \right] dx + \alpha \int_\alpha^1 \left[f_{\tilde{c}:n}(x) - \sum_{i=1}^n c_i \right] dx}{[\alpha^p(1-\alpha) + \alpha(1-\alpha)^p]^{1/p}} \\ &= -\frac{L_{\tilde{c}:n}(\alpha)}{[\alpha^p(1-\alpha) + \alpha(1-\alpha^p]^{1/p}} = -\frac{L_{\tilde{c}:n}(\alpha)}{M_p(\alpha)}, \end{aligned} \tag{4.243}$$

say (see (4.230)). We have

$$\lim_{\alpha \to 0} \frac{L_{\tilde{c}:n}(\alpha)}{\alpha} = \lim_{\alpha \to 0} \sum_{i=1}^{n} c_i \left(\frac{F_{i:n}(\alpha)}{\alpha} - 1 \right) = nc_1 - \sum_{i=1}^{n} c_i, \tag{4.244}$$

$$\begin{aligned} \lim_{\alpha \to 1} \frac{L_{\tilde{c}:n}(\alpha)}{1-\alpha} &= \lim_{\alpha \to 1} \sum_{i=1}^{n} c_i \frac{1 - \alpha - F_{n+1-i:n}(1-\alpha)}{1-\alpha} \\ &= \lim_{\alpha \to 1} \sum_{i=1}^{n} c_{n+1-i} \left[1 - \frac{F_{i:n}(\alpha)}{\alpha} \right] = \sum_{i=1}^{n} c_i - nc_n, \end{aligned} \tag{4.245}$$

and

$$\lim_{\alpha\to 0}\frac{\alpha}{M_p(\alpha)}=\lim_{\alpha\to 0}\alpha^{1-1/p}=0, \tag{4.246}$$

$$\lim_{\alpha\to 1}\frac{1-\alpha}{M_p(\alpha)}=\lim_{\alpha\to 1}(1-\alpha)^{1-1/p}=0. \tag{4.247}$$

Therefore

$$\lim_{\alpha\to 0}\frac{L_{\tilde{c}:n}(\alpha)}{M_p(\alpha)}=\lim_{\alpha\to 0}\frac{L_{\tilde{c}:n}(\alpha)}{\alpha}\frac{\alpha}{M_p(\alpha)}=0, \tag{4.248}$$

$$\lim_{\alpha\to 1}\frac{L_{\tilde{c}:n}(\alpha)}{M_p(\alpha)}=\lim_{\alpha\to 1}\frac{L_{\tilde{c}:n}(\alpha)}{1-\alpha}\frac{1-\alpha}{M_p(\alpha)}=0, \tag{4.249}$$

which in view of (4.243) provides the desired statement. □

Theorem 4.20 *Let $X_1,\ldots,X_n$ be iid, have a bounded support, mean μ, and satisfy* ess sup $|X_1-\mu|=\sigma_\infty$. *For every $\tilde{c}\in\mathbb{R}^n$ satisfying (4.240) we have*

$$\mathbb{E}\sum_{i=1}^{n}c_i\frac{X_{i:n}-\mu}{\sigma_\infty}\le 0, \tag{4.250}$$

The zero bound is attained in the limit by the distributions of the form (4.238) and (4.239) with α_* replaced by $\alpha\to 0$, and $\alpha\to 1$, respectively.

Proof The inequality follows from the fact that the greatest convex minorant $\underline{F}_{\tilde{c}:n}$ is linear and has the derivative $\underline{f}_{\tilde{c}:n}=\sum_{i=1}^{n}c_i$ (cf. (4.234)). Consider (4.238) and (4.239) with α_* replaced by arbitrary $0<\alpha\le\frac{1}{2}$ and $\frac{1}{2}\le\alpha<1$, respectively. They satisfy the moment and support conditions. The respective standardized quantile functions have the forms

$$\frac{F_\alpha^{-1}(x)-\mu}{\sigma_\infty}=\begin{cases}-1, & 0<x<\alpha,\\ \frac{\alpha}{1-\alpha}, & \alpha\le x<1,\end{cases}\qquad 0<\alpha\le\frac{1}{2}, \tag{4.251}$$

and

$$\frac{F_\alpha^{-1}(x)-\mu}{\sigma_\infty}=\begin{cases}-\frac{1-\alpha}{\alpha}, & 0<x<\alpha,\\ 1, & \alpha\le x<1,\end{cases}\qquad \alpha\le\frac{1}{2}<1. \tag{4.252}$$

In the first case we have

$$\begin{aligned}
\mathbb{E}_\alpha \sum_{i=1}^n c_i \frac{X_{i:n}-\mu}{\sigma_\infty} &= \int_0^1 \frac{F_\alpha^{-1}(x)-\mu}{\sigma_\infty}\left[f_{\tilde{c}:n}(x) - \sum_{i=1}^n c_i\right] dx \\
&= -\int_0^\alpha \left[f_{\tilde{c}:n}(x) - \sum_{i=1}^n c_i\right] dx + \frac{\alpha}{1-\alpha}\int_\alpha^1 \left[f_{\tilde{c}:n}(x) - \sum_{i=1}^n c_i\right] dx \\
&= -L_{\tilde{c}:n}(\alpha) - \frac{\alpha}{1-\alpha} L_{\tilde{c}:n}(\alpha) = -\frac{L_{\tilde{c}:n}(\alpha)}{1-\alpha}, \qquad (4.253)
\end{aligned}$$

which tends to 0 as $\alpha \to 0$. Similarly, for (4.252) yields

$$\mathbb{E}_\alpha \sum_{i=1}^n c_i \frac{X_{i:n}-\mu}{\sigma_\infty} = -\frac{L_{\tilde{c}:n}(\alpha)}{\alpha}, \qquad \frac{1}{2} \le \alpha < 1, \tag{4.254}$$

and letting $\alpha \to 1$ we obtain 0 in the limit. □

If $p = 1$, then the limits in (4.246) and (4.247) amount to 1, which implies that those of (4.248) and (4.249) do not necessarily vanish. This allows us to hope that the bounds expressed in $\sigma_1 = \mathbb{E}|X_1 - \mu|$ units may be different from 0. The suspicion is confirmed in the following theorem.

Theorem 4.21 *Assume that $X_1, \ldots, X_n$ are iid, have a finite mean μ and positive mean absolute deviation from the mean $\sigma_1 = \mathbb{E}|X_1 - \mu|$. Let $\tilde{c} \in \mathbb{R}^n$ satisfy (4.240). Then*

$$\mathbb{E} \sum_{i=1}^n c_i \frac{X_{i:n}-\mu}{\sigma_1} \le -\min_{0\le\alpha\le1} U_{\tilde{c}:n}(\alpha), \tag{4.255}$$

where

$$U_{\tilde{c}:n}(\alpha) = \begin{cases} \frac{L_{\tilde{c}:n}(\alpha)}{2\alpha(1-\alpha)}, & 0 < \alpha < 1, \\ \lim_{\alpha\to0} \frac{L_{\tilde{c}:n}(\alpha)}{2\alpha} = \frac{1}{2}\left(nc_1 - \sum_{i=1}^n c_i\right), & \alpha = 0, \\ \lim_{\alpha\to1} \frac{L_{\tilde{c}:n}(\alpha)}{2(1-\alpha)} = \frac{1}{2}\left(\sum_{i=1}^n c_i - nc_n\right), & \alpha = 1. \end{cases} \tag{4.256}$$

If $\alpha_* = \alpha_*(\tilde{c})$ minimizing (4.255) belongs to $(0, 1)$, then the bound (4.255) is attained by the two-point marginal distribution

$$\mathbb{P}\left(X_1 = \mu - \frac{\sigma_1}{2\alpha_*}\right) = \alpha_* = 1 - \mathbb{P}\left(X_1 = \mu + \frac{\sigma_1}{2(1-\alpha_*)}\right). \tag{4.257}$$

If the minimum is attained at $\alpha_* = 0$ ($\alpha_* = 1$, respectively), then the equality in (4.255) is attained in the limit once we replace α_* in (4.257) by $\alpha \to 0$ ($\alpha \to 1$, respectively).

Proof Under the assumption (4.240), we have

$$T_h(g) = \int_0^1 g(x)h(x)dx = \int_0^1 \frac{F^{-1}(x) - \mu}{\sigma_1} \left[f_{\tilde{c}:n}(x) - \sum_{i=1}^{n} c_i \right] dx < 0 \tag{4.258}$$

for fixed $h(x) = f_{\tilde{c}:n}(x) - \sum_{i=1}^{n} c_i$ and every $g(x) = \frac{F^{-1}(x)-\mu}{\sigma_1}$ which is necessarily nondecreasing, and satisfies

$$T_1(g) = \int_0^1 g(x)\,dx = \int_0^1 \frac{F^{-1}(x) - \mu}{\sigma_1}\,dx = 0, \tag{4.259}$$

$$||g||_1 = \int_0^1 |g(x)|dx = \int_0^1 \frac{|F^{-1}(x) - \mu|}{\sigma_1}\,dx = 1. \tag{4.260}$$

We apply here the norm maximization method presented in Subsect. 2.4.7. Under the simple transformation $\tilde{g} = \frac{g}{-T_h(g)}$, we obtain a convex set $\mathcal{S}_2$ of nondecreasing functions such that

$$T_1(\tilde{g}) = -\frac{T_1(g)}{T_h(g)} = 0, \tag{4.261}$$

$$T_h(\tilde{g}) = -\frac{T_h(g)}{T_h(g)} = -1, \tag{4.262}$$

with the respective norms

$$||\tilde{g}||_1 = \frac{||g||_1}{|T_h(g)|} = -\frac{1}{T_h(g)}. \tag{4.263}$$

We see that our original problem of maximizing (4.258) is equivalent to maximizing the norm over the functions $\tilde{g}$ from the convex set $\mathcal{S}_2$.

Every element of $\mathcal{S}_2$ can be approximated with arbitrary accuracy by stepwise nondecreasing elements of $\mathcal{S}_2$. Indeed, it suffices to apply a standard piecewise constant interpolations g_k, $k = 1, 2, \ldots$, with an increasing number of knots, and the uniformly decreasing distances among them, and take their linear transformations $\tilde{g}_k$ assuring that $T_1(\tilde{g}_k) = 0$ and $T_h(\tilde{g}_k) = -1$. Every stepwise element of $\mathcal{S}_2$ can be represented as a convex mixture of two-valued elements of $\mathcal{S}_2$. Since the norm functional is convex, the norm of a mixture cannot exceed the maximal norm of its elements. It follows that maximizing the norm over $\mathcal{S}_2$ we can focus only on two-valued members of the set. Equivalently, we can just maximize the original functional (4.258) over the family of two-valued standardized quantile functions

$$g_\alpha(x) = \frac{F_\alpha^{-1}(x) - \mu}{\sigma_1} = \begin{cases} -\frac{1}{2\alpha}, & 0 < x < \alpha, \\ \frac{1}{2(1-\alpha)}, & \alpha \le x < 1. \end{cases} \qquad 0 < \alpha < 1. \tag{4.264}$$

Then we have

$$
\begin{aligned}
T_h(g_\alpha) &= \int_0^1 \frac{F_\alpha^{-1}(x)-\mu}{\sigma_1}\left[f_{\tilde{c}:n}(x)-\sum_{i=1}^n c_i\right]dx \\
&= -\frac{1}{2\alpha}\int_0^\alpha \left[f_{\tilde{c}:n}(x)-\sum_{i=1}^n c_i\right]dx + \frac{1}{2(1-\alpha)}\int_\alpha^1 \left[f_{\tilde{c}:n}(x)-\sum_{i=1}^n c_i\right]dx \\
&= -\frac{L_{\tilde{c}:n}(\alpha)}{2\alpha(1-\alpha)} = -U_{\tilde{c}:n}(\alpha), \qquad 0<\alpha<1. \qquad (4.265)
\end{aligned}
$$

The function is continuous over the interval $(0, 1)$, and has finite limits at 0 and 1, equal to $\frac{1}{2}\left(nc_1 - \sum_{i=1}^n c_i\right)$ and $\frac{1}{2}\left(\sum_{i=1}^n c_i - nc_n\right)$, respectively. Minimizing (4.256) over $[0, 1]$, we obtain the upper bound (4.255). The attainability condition follows from (4.264). □

Corollary 4.15 *For $X_1, \ldots, X_n$ satisfying the assumptions of Theorem 4.21 we have*

$$\mathbb{E}\frac{X_{1:n}-\mu}{\sigma_1} \le -\frac{1}{2}. \qquad (4.266)$$

For $n = 2$ the bound is attained by (4.257) with arbitrary $0 < \alpha_* < 1$. For $n \ge 3$, this is attained by the distribution of the form (4.257) with α_* replaced by $\alpha \to 1$.

Proof In this case

$$L_{\tilde{c}:n}(\alpha) = L_{1:n}(\alpha) = F_{1:n}(\alpha) - \alpha = 1 - \alpha - (1-\alpha)^n, \qquad (4.267)$$

and

$$U_{\tilde{c}:n}(\alpha) = U_{1:n}(\alpha) = \frac{L_{1:n}(\alpha)}{2\alpha(1-\alpha)} = \frac{1}{2}\sum_{i=0}^{n-2}(1-\alpha)^i. \qquad (4.268)$$

For $n = 2$ the function is constant, equal to $\frac{1}{2}$. For $n \ge 3$ it is decreasing, and its minimal value is $U_{1:n}(1) = \frac{1}{2}$. The reference to Theorem 4.21 provides the conclusions. □

4.5.2 Nonnegative Lower Bounds

Applying the standard transformation $Y_i = 2\mu - X_i$, $i = 1, \ldots, n$, which preserves the iid property, and the values of moments μ and σ_p, $1 \le p \le \infty$, and provides the identity $\sum_{i=1}^n \frac{X_{i:n}-\mu}{\sigma_p} = -\sum_{i=1}^n c_{n+1-i}\frac{Y_{i:n}-\mu}{\sigma_p}$, we obtain analogous lower bounds.

We assume that vector $\tilde{c} = (c_1, \dots, c_n)$ satisfies

$$L_{\tilde{c}:n}(x) = \sum_{i=1}^{n} c_i [F_{i:n}(x) - x] < 0, \qquad 0 < x < 1. \tag{4.269}$$

Observe that this is equivalent to

$$L_{\tilde{c}':n}(x) = \sum_{i=1}^{n} c_{n+1-i} [F_{i:n}(x) - x] > 0, \qquad 0 < x < 1, \tag{4.270}$$

(cf. (4.240)).

Corollary 4.16 *For $X_1, \dots, X_n$ being iid with the expectation $\mu \in \mathbb{R}$ and $0 < \sigma_p < \infty$ for some $1 < p \leq \infty$, and $\tilde{c}$ satisfying (4.269) we have*

$$\mathbb{E} \sum_{i=1}^{n} c_i \frac{X_{i:n} - \mu}{\sigma_p} \geq 0. \tag{4.271}$$

The attainability conditions are described in Theorem 4.19 for $1 < p < \infty$, and in Theorem 4.20 for $p = \infty$.

Corollary 4.17 *Under the assumptions of Corollary 4.16 with $p = 1$, we have*

$$\mathbb{E} \sum_{i=1}^{n} c_i \frac{X_{i:n} - \mu}{\sigma_1} \geq \min_{0 \leq \alpha \leq 1} U_{\tilde{c}':n}(\alpha) \tag{4.272}$$

for $\tilde{c}' = (c_n, \dots, c_1)$ and $U_{\tilde{c}:n}$ defined in (4.256). The attainability conditions are concluded from Theorem 4.21.

Combining Corollaries 4.15 and 4.17 we obtain the following.

Corollary 4.18 *Let $X_1, \dots, X_n$ be nondegenerate iid random variables with finite expectation μ and first absolute central moment σ_1. Then*

$$\mathbb{E} \frac{X_{n:n} - \mu}{\sigma_1} \geq \frac{1}{2}. \tag{4.273}$$

For $n = 2$, the equality is attained by the two-point distributions

$$\mathbb{P}\left(X_1 = \mu - \frac{\sigma_1}{2\alpha}\right) = \alpha = 1 - \mathbb{P}\left(X_1 = \mu + \frac{\sigma_1}{2(1-\alpha)}\right) \tag{4.274}$$

for arbitrary $0 < \alpha < 1$. For $n \geq 3$ the equality in (4.273) holds in the limit for the parent distributions (4.274) with $\alpha \to 1$.

Corollary 4.19 *Under the assumptions of Corollary* 4.18 *the following yields*

$$\mathbb{E}\frac{X_{n:n} - X_{1:n}}{\sigma_1} \geq 2 - \frac{1}{2^{n-2}}. \tag{4.275}$$

If $n = 2$ *and* 3*, then the bound is attained by (4.274) with any* $0 < \alpha < 1$*. For* $n \geq 4$*, the equality in (4.275) holds for*

$$\mathbb{P}(X_1 = \mu - \sigma_1) = \mathbb{P}(X_1 = \mu + \sigma_1) = \frac{1}{2}. \tag{4.276}$$

Proof Our purpose is to minimize the function (4.256) which in the particular case of sample range has the form

$$U_{1,n:n}(\alpha) = \frac{1 - (1-\alpha)^n - \alpha^n}{2\alpha(1-\alpha)}. \tag{4.277}$$

We easily check that for $n = 2$ and 3 the function is constant, and equal to 1 and $\frac{3}{2}$, respectively. Therefore in these cases the equality in (4.275) is attained by arbitrary distribution (4.274).

For $n \geq 4$ we use the following representation

$$\begin{aligned} U_{1,n:n}(\alpha) &= \frac{\sum_{i=1}^{n-1} B_{i,n}(\alpha)}{2\alpha(1-\alpha)} = \frac{1}{2}\sum_{i=1}^{n-1}\binom{n}{i}\alpha^{i-1}(1-\alpha)^{n-i-1} \\ &= \frac{n(n-1)}{2}\sum_{i=0}^{n-2}\frac{B_{i,n-2}(\alpha)}{(i+1)(n-i-1)}. \end{aligned} \tag{4.278}$$

It has the derivative

$$\begin{aligned} U'_{1,n:n}(\alpha) = & \frac{n(n-1)(n-2)}{2} \\ & \times \sum_{i=0}^{n-3}\left[\frac{1}{(i+2)(n-i-2)} - \frac{1}{(i+1)(n-i-1)}\right] B_{i,n-3}(\alpha). \end{aligned} \tag{4.279}$$

Since the function $(0, n) \ni x \mapsto \frac{1}{x(n-x)}$ is first decreasing and then increasing, the sequence of coefficients $a_{i,n} = \frac{1}{(i+2)(n-i-2)} - \frac{1}{(i+1)(n-i-1)}$, $i = 0, \ldots, n-3$, is first negative, and then positive. The variation diminishing property of the Bernstein polynomials implies that (4.279) is first negative and then positive, and so (4.278) is first decreasing and then increasing. Since (4.278) is symmetric about $\frac{1}{2}$, its minimum point is necessarily located at $\frac{1}{2}$. We have $U_{1,n:n}\left(\frac{1}{2}\right) = 2 - \frac{1}{2^{n-2}}$, and the equality in (4.275) is attained by (4.274) with $\alpha = \frac{1}{2}$, which is just (4.276). □

Since for the Gini mean differences we have $\mathbb{E}\,\mathrm{GMD}_n(\tilde{X}) = \mathbb{E}(X_{2:2} - X_{1:2})$, we immediately deduce that

$$\mathbb{E}\,\mathrm{GMD}_n(\tilde{X}) \geq 1, \tag{4.280}$$

and the equality is attained by any two-point distribution (4.274) with arbitrary parameter $0 < \alpha < 1$.

Mimicking the arguments of the proofs of Theorems 4.17 and 4.18 we can also show that the bounds

$$\mathbb{E} \sum_{i=1}^{n} c_i \frac{X_{i:n} - \mu}{\sigma_p} \geq 0, \qquad 1 \leq p \leq \infty, \tag{4.281}$$

are optimal and attained by some two-point parent distributions for all $\tilde{c}$ such that

$$\exists\, 0 < \alpha_{**} < 1 \quad L_{\tilde{c}:n}(\alpha_{**}) = 0 \geq L_{\tilde{c}:n}(x), \qquad 0 < x < 1, \tag{4.282}$$

(comp. (4.231)).

4.6 Symmetric Distributions

4.6.1 Positive Bounds

Symmetric distribution functions with the mean μ satisfy

$$F^{-1}(x) - \mu = \mu - F^{-1}(1 - x-), \qquad 0 < x < 1. \tag{4.283}$$

It follows that $F^{-1}(x) - \mu$ is nonpositive for $0 < x < \frac{1}{2}$ and nonnegative for $\frac{1}{2} \leq x < 1$. Moreover, for every $\tilde{c} \in \mathbb{R}^n$ we have

$$\begin{aligned}
&\int_0^{\frac{1}{2}} [F^{-1}(x) - \mu] \sum_{i=1}^{n} c_i f_{i:n}(x)\, dx \\
&= \int_{\frac{1}{2}}^{1} [F^{-1}(1 - x) - \mu] \sum_{i=1}^{n} c_i f_{i:n}(1 - x)\, dx \\
&= \int_{\frac{1}{2}}^{1} [F^{-1}(1 - x-) - \mu] \sum_{i=1}^{n} c_i f_{n+1-i:n}(x)\, dx \\
&= -\int_{\frac{1}{2}}^{1} [F^{-1}(x) - \mu] \sum_{i=1}^{n} c_{n+1-i} f_{i:n}(x)\, dx, \qquad (4.284)
\end{aligned}$$

and, in consequence,

$$\mathbb{E}\sum_{i=1}^{n} c_i(X_{i:n}-\mu) = \int_0^1 [F^{-1}(x)-\mu]\sum_{i=1}^{n} c_i f_{i:n}(x)\,dx$$
$$= \int_{\frac{1}{2}}^1 [F^{-1}(x)-\mu]\sum_{i=1}^{n}(c_i - c_{n+1-i}) f_{i:n}(x)\,dx$$
$$= \int_{\frac{1}{2}}^1 [F^{-1}(x)-\mu] f_{\tilde{c}^s:n}(x)\,dx \tag{4.285}$$

where $\tilde{c}^s = (c_1 - c_n, c_2 - c_{n-1}, \ldots, c_n - c_1)$. Notice that

$$f_{\tilde{c}^s:n}(x) = \sum_{i=1}^{n}(c_i - c_{n+1-i}) f_{i:n}(x) = \sum_{i=1}^{n} c_i[f_{i:n}(x) - f_{i:n}(1-x)] \tag{4.286}$$

is antisymmetric about $\frac{1}{2}$, where it vanishes. It is the derivative of the function

$$F_{\tilde{c}^s:n}(x) = \sum_{i=1}^{n}(c_i - c_{n+1-i}) F_{i:n}(x) = \sum_{i=1}^{n} c_i[F_{i:n}(x) + F_{i:n}(1-x) - 1] \tag{4.287}$$

which is symmetric about $\frac{1}{2}$, and vanishes at 0 and 1. These observations allow us to prove the following theorem.

Theorem 4.22 *Suppose that $X_1, \ldots, X_n$ are iid symmetrically distributed about μ, and they have a finite moment of order $1 < p < \infty$. Then*

$$\mathbb{E}\sum_{i=1}^{n} c_i \frac{X_{i:n}-\mu}{\sigma_p} \le \frac{1}{2^{1/p}}\left[\int_{\frac{1}{2}}^1 [(\underline{f}_{\tilde{c}^s:n})_+(x)]^q dx\right]^{\frac{1}{q}}, \tag{4.288}$$

where $(\underline{f}_{\tilde{c}^s:n})_+(x) = \max\{\underline{f}_{\tilde{c}^s:n}(x), 0\}$ is the positive part of the derivative $\underline{f}_{\tilde{c}^s:n}(x)$ of the greatest convex minorant $\underline{F}_{\tilde{c}^s:n}(x)$ of (4.287) restricted to the interval $\left(\frac{1}{2}, 1\right)$. Moreover, if

$$F_{\tilde{c}^s:n}(x) < 0 \qquad for\ some\ \frac{1}{2} < x < 1, \tag{4.289}$$

then the bound (4.288) is positive and sharp. It is attained then by the distribution function F determined from the relation

$$\frac{F^{-1}(x)-\mu}{\sigma_p} = \begin{cases} -\frac{[(\underline{f}_{\tilde{c}^s:n})_+(1-x)]^{q/p}}{\left[2\int_{\frac{1}{2}}^{1}(\underline{f}_{\tilde{c}^s:n})_+^q(x)dx\right]^{1/p}}, & 0<x<\frac{1}{2}, \\ \frac{[(\underline{f}_{\tilde{c}^s:n})_+(x)]^{q/p}}{\left[2\int_{\frac{1}{2}}^{1}(\underline{f}_{\tilde{c}^s:n})_+^q(x)dx\right]^{1/p}}, & \frac{1}{2}\le x<1. \end{cases} \tag{4.290}$$

Proof The bound (4.288) follows from the relations

$$\begin{aligned} \mathbb{E}\sum_{i=1}^{n} c_i \frac{X_{i:n}-\mu}{\sigma_p} &= \int_{\frac{1}{2}}^{1} \frac{F^{-1}(x)-\mu}{\sigma_p} f_{\tilde{c}^s:n}(x)\,dx \\ &\le \int_{\frac{1}{2}}^{1} \frac{F^{-1}(x)-\mu}{\sigma_p} \underline{f}_{\tilde{c}^s:n}(x)\,dx \\ &\le \int_{\frac{1}{2}}^{1} \frac{F^{-1}(x)-\mu}{\sigma_p} (\underline{f}_{\tilde{c}^s:n})_+(x)\,dx \\ &\le \left[\int_{\frac{1}{2}}^{1} \frac{|F^{-1}(x)-\mu|^p}{\sigma_p^p}\,dx\right]^{\frac{1}{p}} \left[\int_{\frac{1}{2}}^{1} [(\underline{f}_{\tilde{c}^s:n})_+(x)]^q\,dx\right]^{\frac{1}{q}} \\ &= \frac{1}{2^{\frac{1}{p}}}\left[\int_{\frac{1}{2}}^{1} [(\underline{f}_{\tilde{c}^s:n})_+(x)]^q\,dx\right]^{\frac{1}{q}}. \end{aligned} \tag{4.291}$$

The first two inequalities follow from the facts that $\frac{F^{-1}(x)-\mu}{\sigma_p}$ is nondecreasing and nonnegative on $\left[\frac{1}{2},1\right)$.

Under the assumption (4.289) $\underline{f}_{\tilde{c}^s:n}$ is positive on some neighborhood of 1. It follows that the RHS of (4.288) is strictly positive. We also show that the corresponding bound is attainable. The equality in the last inequality of (4.291) is attained under the assumption

$$\frac{F^{-1}(x)-\mu}{\sigma_p} = \alpha[(\underline{f}_{\tilde{c}^s:n})_+(x)]^{\frac{q}{p}}, \qquad \frac{1}{2}\le x<1, \tag{4.292}$$

for some positive α. By the assumption, the LHS has the pth norm equal to $\frac{1}{2^{1/p}}$. The RHS has the same norm for

$$\alpha = \left[2\int_{\frac{1}{2}}^{1} [(\underline{f}_{\tilde{c}^s:n})_+(x)]^q\,dx\right]^{-\frac{1}{p}}. \tag{4.293}$$

Note that under the conditions (4.292) with (4.293) the second inequality in (4.291) becomes the equality, because $\frac{F^{-1}(x)-\mu}{\sigma_p}$ vanishes on the (possibly empty) interval where $\underline{f}_{\tilde{c}^s:n}(x)$ is negative. Moreover, the conditions imply the equality in the first inequality, because the constancy intervals of (4.292) contain all the intervals where $\underline{F}_{\tilde{c}^s:n}(x) < F_{\tilde{c}:n}(x)$. By symmetry assumption (4.283) and continuity of (4.292) we define the lacking part of the symmetric quantile function attaining the bound in (4.291)

$$\frac{F^{-1}(x)-\mu}{\sigma_p} = \frac{\mu - F^{-1}(1-x)}{\sigma_p} = -\frac{[(\underline{f}_{\tilde{c}^s:n})_+(1-x)]^{q/p}}{\left[2\int_{\frac{1}{2}}^{1}(\underline{f}_{\tilde{c}^s:n})_+^q(x)dx\right]^{1/p}}, \qquad 0 < x < \frac{1}{2}. \tag{4.294}$$

This shows that the parent distribution described in (4.290) actually provides the equality in (4.288). □

In the most classic subcase $p = 2$ the formulae slightly simplify.

Corollary 4.20 *Let $X_1, \ldots, X_n$ be iid symmetrically distributed about μ, and have a finite variance $\sigma_2^2 > 0$. Then*

$$\mathbb{E}\sum_{i=1}^{n} c_i \frac{X_{i:n}-\mu}{\sigma_2} \le \left[\frac{1}{2}\int_{\frac{1}{2}}^{1}[(\underline{f}_{\tilde{c}^s:n})_+(x)]^2 dx\right]^{\frac{1}{2}}. \tag{4.295}$$

Under the condition (4.289) the above bound is positive and it is attained then by the distribution function satisfying

$$\frac{F^{-1}(x)-\mu}{\sigma_2} = \begin{cases} -\dfrac{(\underline{f}_{\tilde{c}^s:n})_+(1-x)}{\left[2\int_{\frac{1}{2}}^{1}[(\underline{f}_{\tilde{c}^s:n})_+(x)]^2dx\right]^{1/2}}, & 0 < x < \frac{1}{2}, \\ \dfrac{(\underline{f}_{\tilde{c}^s:n})_+(x)}{\left[2\int_{\frac{1}{2}}^{1}[(\underline{f}_{\tilde{c}^s:n})_+(x)]^2dx\right]^{1/2}}, & \frac{1}{2} \le x < 1. \end{cases} \tag{4.296}$$

Finally, we consider the extreme cases $p = 1$ and $p = \infty$.

Theorem 4.23 *If $X_1, \ldots, X_n$ are iid symmetrically distributed about μ, and have a finite first absolute moment σ_1, and $\tilde{c} \in \mathbb{R}^n$, then*

$$\mathbb{E}\sum_{i=1}^{n} c_i \frac{X_{i:n}-\mu}{\sigma_1} \le \frac{1}{2}(\underline{f}_{\tilde{c}:n})_+(1). \tag{4.297}$$

Under condition (4.289) the RHS is positive. Define then

$$\alpha_*^s = \alpha_*^s(1, \tilde{c}) = \min\left\{\frac{1}{2} \le \alpha \le 1 : \ \underline{f}_{\tilde{c}:n}(\alpha) = \underline{f}_{\tilde{c}:n}(1)\right\}. \tag{4.298}$$

If $\alpha_*^s < 1$ then the equality in (4.297) holds if

$$\mathbb{P}\left(X_1 = \mu - \frac{\sigma_1}{2(1-\alpha_*^s)}\right) = \mathbb{P}\left(X_1 = \mu + \frac{\sigma_1}{2(1-\alpha_*^s)}\right) = 1 - \alpha_*^s,$$
$$\mathbb{P}(X_1 = \mu) = 2\alpha_*^s - 1. \tag{4.299}$$

If $\alpha_*^s = 1$, then the bound (4.297) is attained in the limit by the marginal distributions (4.299) with α_*^s replaced by $\alpha \to 1$.

Proof The inequality proof is following

$$\begin{aligned}
\mathbb{E}\sum_{i=1}^n c_i \frac{X_{i:n}-\mu}{\sigma_1} &\le \int_{\frac{1}{2}}^1 \frac{F^{-1}(x)-\mu}{\sigma_1} \underline{f}_{\bar{c}^s:n}(x)\,dx \\
&\le \int_{\frac{1}{2}}^1 \frac{F^{-1}(x)-\mu}{\sigma_1} (\underline{f}_{\bar{c}^s:n})_+(x)\,dx \\
&\le (\underline{f}_{\bar{c}^s:n})_+(1) \int_{\frac{1}{2}}^1 \frac{F^{-1}(x)-\mu}{\sigma_1} \\
&= \frac{1}{2}(\underline{f}_{\bar{c}^s:n})_+(1).
\end{aligned} \tag{4.300}$$

If (4.289) holds then $(\underline{f}_{\bar{c}^s:n})_+(1) = \underline{f}_{\bar{c}^s:n}(1) > 0$. If moreover $\alpha_*^s < 1$, then the last inequality in (4.300) is attained if

$$\frac{F^{-1}(x)-\mu}{\sigma_1} = \begin{cases} 0, & \frac{1}{2} \le x < \alpha_*^s, \\ c > 0, & \alpha_*^s \le x < 1. \end{cases} \tag{4.301}$$

Note that all the intervals where $\underline{F}_{\bar{c}^s:n}(x) < F_{\bar{c}^s:n}(x)$ and $\underline{f}_{\bar{c}^s:n}(x)$ is constant are contained in the intervals $\left(\frac{1}{2}, \alpha_*^s\right)$ and $(\alpha_*^s, 1)$. This assures the equality in the first inequality of (4.300). Moreover, $\{x : \underline{f}_{\bar{c}^s:n}(x) < 0\}$ is contained in $\left(\frac{1}{2}, \alpha_*^s\right)$ which guarantees the equality in the middle inequality. By the symmetry assumption (4.301) is extended on $\left(0, \frac{1}{2}\right)$ by

$$\frac{F^{-1}(x)-\mu}{\sigma_1} = \begin{cases} -c, & 0 < x < 1-\alpha_*^s, \\ 0, & 1-\alpha_*^s \le x < \frac{1}{2}. \end{cases} \tag{4.302}$$

The first absolute central moment condition forces $c = \frac{1}{2(1-\alpha_*^s)}$. Then the conditions (4.301) and (4.302) determine (4.299).

If $\alpha_*^s = 1$, then $(\underline{f}_{\bar{c}^s:n})_+(x) = \underline{f}_{\bar{c}^s:n}(x)$ is strictly increasing on some neighborhood of 1. For (4.299) with α_k replacing α_*^s sufficiently close to 1, all the intervals where $\underline{f}_{\tilde{c}^s:n}$ is constant and $\underline{f}_{\tilde{c}^s:n} < 0$ are contained in $\left(\frac{1}{2}, \alpha_k\right)$, which gives the equalities in the first two inequalities of (4.300), respectively. By continuity of $(\underline{f}_{\tilde{c}^s:n})_+$, the last inequality becomes the equality as well. □

Observe that in the case $\alpha_*^s = 1$, then the RHS of (4.297) takes on a simple form

$$\frac{1}{2} f_{\tilde{c}^s:n}(1) = \frac{n}{2}(c_n - c_1). \tag{4.303}$$

Theorem 4.24 *Suppose that $X_1, \ldots, X_n$ are independent, have a common distribution symmetric about μ, and a nondegenerate bounded support. Let*

$$\alpha_{**}^s = \alpha_{**}^s(\infty, \tilde{c}) = \max\left\{\frac{1}{2} \le \alpha \le 1 : \underline{f}_{\tilde{c}^s;n}(\alpha) \le 0\right\} \tag{4.304}$$

for some fixed $\tilde{c} \in \mathbb{R}^n$. Then

$$\mathbb{E} \sum_{i=1}^n c_i \frac{X_{i:n} - \mu}{\sigma_\infty} \le -F_{\tilde{c}^s:n}(\alpha_{**}^s). \tag{4.305}$$

Under the assumption (4.289) the bound is positive and it is attained by the parent distribution

$$\begin{aligned} \mathbb{P}(X_1 = \mu - \sigma_\infty) = \mathbb{P}(X_1 = \mu + \sigma_\infty) &= 1 - \alpha_{**}^s, \\ \mathbb{P}(X_1 = \mu) &= 2\alpha_{**}^s - 1. \end{aligned} \tag{4.306}$$

Proof As usual, we start with proving the inequality

$$\begin{aligned} \mathbb{E} \sum_{i=1}^n c_i \frac{X_{i:n} - \mu}{\sigma_\infty} &\le \int_{\frac{1}{2}}^1 \frac{F^{-1}(x) - \mu}{\sigma_\infty} \underline{f}_{\tilde{c}^s:n}(x)\, dx \\ &\le \int_{\frac{1}{2}}^1 \frac{F^{-1}(x) - \mu}{\sigma_\infty} (\underline{f}_{\tilde{c}^s:n})_+(x)\, dx \\ &\le \int_{\frac{1}{2}}^1 (\underline{f}_{\tilde{c}^s:n})_+(x)\, dx \\ &= -\underline{F}_{\tilde{c}^s:n}(\alpha_{**}^s) = -F_{\tilde{c}^s:n}(\alpha_{**}^s). \end{aligned} \tag{4.307}$$

Note that $\underline{f}_{\tilde{c}^s:n}(x) \ne f_{\tilde{c}^s:n}(x)$ only inside the intervals where $\underline{F}_{\tilde{c}^s:n}(x) < F_{\tilde{c}^s:n}(x)$, and $\int_{\alpha_1}^{\alpha_2} \underline{f}_{\tilde{c}^s:n}(x)dx = \int_{\alpha_1}^{\alpha_2} f_{\tilde{c}^s:n}(x)dx$ if $\underline{f}_{\tilde{c}^s:n}(\alpha_i) = f_{\tilde{c}^s:n}(\alpha_i)$, $i = 1, 2$. Therefore $\underline{F}_{\tilde{c}^s:n}(\alpha_{**}^s) = F_{\tilde{c}^s:n}(\alpha_{**}^s)$.

The ultimate and penultimate inequalities are attained under the conditions

$$\frac{F^{-1}(x)-\mu}{\sigma_\infty}=1 \quad \text{if} \quad \underline{f}_{\tilde{c}^s:n}(x)>0, \tag{4.308}$$

$$\frac{F^{-1}(x)-\mu}{\sigma_\infty}=0 \quad \text{if} \quad \underline{f}_{\tilde{c}^s:n}(x)<0, \tag{4.309}$$

respectively. These intervals together with $\{x : \underline{f}_{\tilde{c}^s:n}(x) = 0\}$ (if nondegenerate) contain all the intervals where $\underline{F}_{\tilde{c}^s:n}(x) < F_{\tilde{c}^s:n}(x)$. This means that under the conditions (4.308) and (4.309) the bound (4.307) is attained.

Assume now that (4.289) holds, and the condition (4.308) is satisfied on a nondegenerate interval (otherwise the bound is zero, and it is attained by a degenerate parent distribution which contradicts our assumptions). Then the bound is positive, and

$$\frac{F^{-1}(x)-\mu}{\sigma_\infty}=\begin{cases}0, & \frac{1}{2}\le x<\alpha^s_{**},\\ 1, & \alpha^s_{**}\le x<1,\end{cases} \tag{4.310}$$

satisfies (4.308) and (4.309), which means the bound (4.304) is attained by (4.306). □

Observe that if $\underline{f}_{\tilde{c}:n}(x) > 0$, $\frac{1}{2} < x < 1$, then the optimal marginal distribution (4.306) is supported on two points. We cannot exclude the possibility that $\underline{f}_{\tilde{c}:n}(x) = 0$ on some nondegenerate interval $[\beta^s_{**}, \alpha^s_{**}]$, say. Then the bound is attained by the five-point symmetric distributions

$$\begin{aligned}\mathbb{P}(X_1=\mu)&=2\beta^s_{**}-1,\\ \mathbb{P}(X_1=\mu\mp c)&=\alpha^s_{**}-\beta^s_{**},\\ \mathbb{P}(X_1=\mu\mp\sigma_\infty)&=1-\alpha^s_{**},\end{aligned} \tag{4.311}$$

for any $0 < c < 1$ as well. Obviously, the point μ is reduced if $\beta^s_{**} = \frac{1}{2}$.

Now we focus on the special case of single order statistics. For $X_{j:n}$, $1 \le j \le n$, we have

$$\begin{aligned}f^s_{j:n}(x)&=f_{j:n}(x)-f_{n+1-j:n}(x)\\ &=n\binom{n-1}{j-1}[x^{j-1}(1-x)^{n-j}-x^{n-j}(1-x)^{j-1}].\end{aligned} \tag{4.312}$$

This is negative, equal to 0, and positive on $\left(\frac{1}{2}, 1\right)$ for $j \le \frac{n}{2}$, $j = \frac{n+1}{2}$ and $j \ge \frac{n}{2}+1$, respectively. It always vanishes at $\frac{1}{2}$. It follows that the respective antiderivative

$$F^s_{j:n}(x)=F_{j:n}(x)-F_{n+1-j:n}(x), \qquad \frac{1}{2}\le x\le 1, \tag{4.313}$$

is nonnegative for $j \le \frac{n+1}{2}$, and we can obtain sharp positive upper bounds of Theorems 4.22, 4.23 and 4.24 only for $j \ge \frac{n}{2}+1$.

If $j = n$, then

$$f^s_{n:n}(x) = n[x^{n-1} - (1-x)^{n-1}], \qquad \frac{1}{2} \le x \le 1, \tag{4.314}$$

is positive and increasing, which implies

$$(\underline{f}^s_{n:n})_+(x) = f^s_{n:n}(x) = n[x^{n-1} - (1-x)^{n-1}]. \tag{4.315}$$

For $\frac{n}{2}+1 \le j \le n-1$, (4.312) is negative on $\left(0, \frac{1}{2}\right)$, positive on $\left(\frac{1}{2}, 1\right)$, and equal to 0 at 0, $\frac{1}{2}$, and 1. It has at least one local maximum in $\left(\frac{1}{2}, 1\right)$, and its derivative changes the sign from plus to minus there. Also, it has at least one minimum in $\left(0, \frac{1}{2}\right)$, and its derivative is negative and positive left and right to it, respectively. We have

$$\begin{aligned}(f^s_{j:n})'(x) &= n(n-1) \\ &\times[-B_{n-j-1,n-2}(x) + B_{n-j,n-2}(x) + B_{j-2,n-2}(x) - B_{j-1,n-2}(x)].\end{aligned} \tag{4.316}$$

By the VDP theorem $F^s_{j:n}(x)$ is first concave, then convex, and ultimately concave, because it has at least one convexity interval. Since it is symmetric about $\frac{1}{2}$, it is first convex and then concave on $\left(\frac{1}{2}, 1\right)$. Its greatest convex minorant $\underline{F}^s_{j:n}$ is first identical with $F^s_{j:n}$ and then linear. Linearity of $\underline{F}^s_{j:n}$ on the whole interval is excluded, because $F^s_{j:n}\left(\frac{1}{2}\right) < F^s_{j:n}(1)$ and $f^s_{j:n}\left(\frac{1}{2}\right) = 0$. The shape changing point $\alpha^s_* = \alpha^s_*(j,n)$ is determined by the equation

$$\begin{aligned}f^s_{j:n}(\alpha) &= f_{j:n}(\alpha) - f_{n+1-j:n}(\alpha) \\ &= \frac{F^s_{j:n}(1) - F^s_{j:n}(\alpha)}{1-\alpha} = \frac{F_{n+1-j:n}(\alpha) - F_{j:n}(\alpha)}{1-\alpha}.\end{aligned} \tag{4.317}$$

Consequently,

$$\begin{aligned}(f^s_{j:n})_+(x) &= \underline{f}^s_{j:n}(x) = f^s_{j:n}(\min\{x, \alpha_*\}) \\ &= f_{j:n}(\min\{x, \alpha_*\}) - f_{n+1-j:n}(\min\{x, \alpha_*\}).\end{aligned} \tag{4.318}$$

This leads us to the following conclusions.

Corollary 4.21 *Assume that $X_1, \ldots, X_n$ are iid, and symmetrically distributed about μ, and $0 < \sigma_p < \infty$ for some $1 \le p \le \infty$. Then for $p = 1$*

$$\mathbb{E}\frac{X_{n:n} - \mu}{\sigma_1} \le \frac{n}{2}, \tag{4.319}$$

and the equality is attained in the limit by the marginal distributions

$$\mathbb{P}\left(X_1 = \mu \mp \frac{\sigma_1}{2(1-\alpha)}\right) = 1 - \alpha,$$
$$\mathbb{P}(X_1 = \mu) = 2\alpha - 1, \tag{4.320}$$

for $\alpha \to 1$.

For $1 < p < \infty$

$$\mathbb{E}\frac{X_{n:n} - \mu}{\sigma_p} \le \frac{n}{2^{1/p}} \left[\int_{\frac{1}{2}}^{1} [x^{n-1} - (1-x)^{n-1}]^q dx\right]^{1/q}, \tag{4.321}$$

and the equality condition is

$$\frac{F^{-1}(x) - \mu}{\sigma_p} = \begin{cases} -\dfrac{[(1-x)^{n-1} - x^{n-1}]^{q/p}}{\left[2\int_{\frac{1}{2}}^{1} [x^{n-1} - (1-x)^{n-1}]^q dx\right]^{1/p}}, & 0 < x < \frac{1}{2}, \\ \dfrac{[x^{n-1} - (1-x)^{n-1}]^{q/p}}{\left[2\int_{\frac{1}{2}}^{1} [x^{n-1} - (1-x)^{n-1}]^q dx\right]^{1/p}}, & \frac{1}{2} \le x < 1. \end{cases} \tag{4.322}$$

For $p = \infty$

$$\mathbb{E}\frac{X_{n:n} - \mu}{\sigma_\infty} \le 1 - \frac{1}{2^{n-1}}, \tag{4.323}$$

with the equality condition

$$\mathbb{P}(X_1 = \mu \mp \sigma_\infty) = \frac{1}{2}. \tag{4.324}$$

The statement of the corollary for $1 < p < \infty$ can be found in Arnold (1985). In the case $p = 2$ inequality (4.321) simplifies to

$$\mathbb{E}\frac{X_{n:n} - \mu}{\sigma_2} \le n\sqrt{\frac{1}{4n-2}\left(1 - \frac{[(n-1)!]^2}{(2n-2)!}\right)}. \tag{4.325}$$

This inequality was established by Moriguti (1951).

Corollary 4.22 *Under the assumptions of Corollary* 4.21 *for* $\frac{n}{2}+1 \le j \le n-1$ *the following inequalities are sharp.*

For $p = 1$ *we have*

$$\mathbb{E}\frac{X_{j:n}-\mu}{\sigma_1} \le \frac{1}{2}\, f^s_{j:n}(\alpha^s_*), \tag{4.326}$$

which becomes the equality for

$$\mathbb{P}\left(X_1 = \mu \mp \frac{\sigma_1}{2(1-\alpha^s_*)}\right) = 1-\alpha^s_*,$$
$$\mathbb{P}(X_1=\mu) = 2\alpha^s_* - 1, \tag{4.327}$$

where α^s_* *is defined in Eq. (4.317).*

For $1 < p < \infty$

$$\mathbb{E}\frac{X_{j:n}-\mu}{\sigma_p} \le \frac{B_q}{2^{1/p}}, \tag{4.328}$$

where

$$B_q^q = \int_{\frac{1}{2}}^{\alpha^s_*} [f_{j:n}(x) - f_{n+1-j:n}(x)]^q dx + (1-\alpha^s_*)[f_{j:n}(\alpha^s_*) - f_{n+1-j:n}(\alpha^s_*)]^q. \tag{4.329}$$

The parent distribution assuring the equality in (4.328) is defined by the relations

$$\frac{F^{-1}(x)-\mu}{\sigma_p} = \frac{1}{(2B_q^q)^{1/p}} \times \begin{cases} [f_{n+1-j:n}(\alpha^s_*) - f_{j:n}(\alpha^s_*)]^{q/p}, & 0 < x < 1-\alpha^s_*, \\ [f_{n+1-j:n}(x) - f_{j:n}(x)]^{q/p}, & 1-\alpha^s_* \le x < \frac{1}{2}, \\ [f_{j:n}(x) - f_{n+1-j:n}(x)]^{q/p}, & \frac{1}{2} \le x < \alpha^s_*, \\ [f_{j:n}(\alpha^s_*) - f_{n+1-j:n}(\alpha^s_*)]^{q/p}, & \alpha^s_* \le x < 1. \end{cases} \tag{4.330}$$

For $p = \infty$

$$\mathbb{E}\frac{X_{j:n}-\mu}{\sigma_\infty} \le \sum_{i=n+1-j}^{j-1} B_{i,n}\left(\frac{1}{2}\right), \tag{4.331}$$

and the equality holds for

$$\mathbb{P}(X_1 = \mu \mp \sigma_\infty) = \frac{1}{2}. \tag{4.332}$$

4.6.2 Nonpositive Bounds

Now we consider the linear combinations of order statistics with coefficient vectors $\tilde{c}$ which do not satisfy (4.289), i.e., for which

$$F_{\tilde{c}^s:n}(x) = \sum_{i=1}^{n}(c_i - c_{n+1-i})F_{i:n}(x) \geq 0, \qquad \frac{1}{2} \leq x < 1. \tag{4.333}$$

Assume first that (4.333) holds together with

$$F_{\tilde{c}^s:n}(\alpha_*) = 0, \qquad \text{for some } \frac{1}{2} \leq \alpha_* = \alpha_*(\tilde{c}) < 1. \tag{4.334}$$

Since $F_{\tilde{c}^s:n}(1) = 0$ as well, we obtain $\underline{F}_{\tilde{c}^s:n}(x) = \underline{f}_{\tilde{c}^s:n}(x) = 0$, $\alpha_* \leq x \leq 1$, and this is the maximal value of the derivative of the greatest convex minorant. We have

$$\begin{aligned}
\mathbb{E}\sum_{i=1}^{n}\frac{X_{i:n}-\mu}{\sigma_p} &= \int_{\frac{1}{2}}^{1}\frac{F^{-1}(x)-\mu}{\sigma_p}f_{\tilde{c}:n}(x)dx \\
&\leq \int_{\frac{1}{2}}^{1}\frac{F^{-1}(x)-\mu}{\sigma_p}\underline{f}_{\tilde{c}:n}(x)dx \\
&\leq \int_{\frac{1}{2}}^{1}\frac{F^{-1}(x)-\mu}{\sigma_p}(\underline{f}_{\tilde{c}:n})_+(x)dx = 0
\end{aligned} \tag{4.335}$$

for arbitrary $1 \leq p \leq \infty$ for which σ_p is finite and positive.

If $F_{\tilde{c}^s:n}\left(\frac{1}{2}\right) = 0$, then we take $(\underline{f}_{\tilde{c}:n})_+(x) = \underline{f}_{\tilde{c}:n}(x) = 0$, $\frac{1}{2} \leq x < 1$, and the last inequality is the equality. The first one is the equality as well if $\frac{F^{-1}(x)-\mu}{\sigma_p}$ is constant on $\left[\frac{1}{2}, 1\right)$. By assumption, it is nonnegative. If we assume that it is strictly positive, we define a two-point symmetric distribution supported on $\mu - c$ and $\mu + c$, which satisfies $0 < \sigma_p < \infty$ for every $1 \leq p \leq \infty$.

Otherwise we take the minimal $\alpha_* > \frac{1}{2}$ which satisfies $F_{\tilde{c}^s:n}(\alpha_*) = 0$. Then $\underline{f}_{\tilde{c}^s:n}(x)$ is negative and equal to 0 on the intervals $\left[\frac{1}{2}, \alpha_*\right)$ and $[\alpha_*, 1)$, respectively. The latter inequality in (4.335) becomes the equality if $\frac{F^{-1}(x)-\mu}{\sigma_p} = 0$ for $\frac{1}{2} \leq x < \alpha_*$. So does the previous one if $\frac{F^{-1}(x)-\mu}{\sigma_p} = \text{const}$ for $\alpha_* \leq x < 1$. If the constant is strictly positive, taking into account the symmetry condition (4.283) we determine the three-point symmetric distribution

$$\begin{aligned}
\mathbb{P}(X_1 = \mu \mp c) &= 1 - \alpha_*, \\
\mathbb{P}(X_1 = \mu) &= 2\alpha_* - 1,
\end{aligned} \tag{4.336}$$

which has a bounded support (and all the moments finite in consequence), and attains the zero bound in (4.335).

The construction of the distribution functions attaining the zero bounds presented above are unique up to the scale transformations if there is a single $\frac{1}{2} \leq \alpha_* < 1$ satisfying (4.334). If there are more such points then the extreme distributions may have more support points. A specific family of coefficients consists of ones that satisfy $\tilde{c}' = \tilde{c}$, i.e., $c_i = c_{n+1-i}$, $i = 1, \ldots, n$. The corresponding L-statistics are linear combinations of quasi-midranges $X_{i:n} + X_{n+1-i:n}$. Then $f_{\tilde{c}^s:n}(x) = 0$, $\frac{1}{2} \leq x < 1$ (see (4.286)), and the equality

$$\mathbb{E} \sum_{i=1}^{n} c_i \frac{X_{i:n} - \mu}{\sigma_p} = \int_{\frac{1}{2}}^{1} \frac{F^{-1}(x) - \mu}{\sigma_p} f_{\tilde{c}^s:n}(x) dx = 0 \tag{4.337}$$

is attained by arbitrary symmetric parent distribution with mean μ and $0 < \sigma_p < \infty$. In particular, for all symmetric populations the expectation of the sample median (equal to $X_{\frac{n+1}{2}:n}$ and $\frac{1}{2}(X_{\frac{n}{2}:n} + X_{\frac{n}{2}+1:n})$, when the sample size is odd and even, respectively) is identical with the population mean.

It remains the case of L-statistics with the coefficients $c_1, \ldots, c_n$ satisfying the condition

$$F_{\tilde{c}^s:n}(x) = \sum_{i=1}^{n} (c_i - c_{n+1-i}) F_{i:n}(x) > 0, \qquad \frac{1}{2} \leq x < 1. \tag{4.338}$$

Theorem 4.25 *Suppose that $X_1, \ldots, X_n$ are independent identically symmetrically distributed about $\mu \in \mathbb{R}$ nondegenerate random variables with a finite moment of order $1 < p < \infty$. If the vector $\tilde{c} = (c_1, \ldots, c_n)$ satisfies the condition (4.338), then the bound*

$$\mathbb{E} \sum_{i=1}^{n} c_i \frac{X_{i:n} - \mu}{\sigma_p} \leq 0 \tag{4.339}$$

is sharp. It is attained in the limit by the family of three-point distributions

$$\mathbb{P}\left(X_1 = \mu \mp \frac{\sigma_p}{[2(1-\alpha)]^{1/p}}\right) = 1 - \alpha,$$
$$\mathbb{P}(X_1 = \mu) = 2\alpha - 1, \tag{4.340}$$

as $\alpha \to 1$.

Proof The upper bound for the LHS of (4.339) under the assumption (4.338) cannot be greater than 0 due to Theorem 4.22. Our aim is to show that the zero bound cannot be improved. We consider the parametric family of one-dimensional probability distributions with the quantile functions satisfying

$$F_{\alpha}^{-1}(x) - \mu = \begin{cases} -1, & 0 < x < 1 - \alpha, \\ 0, & 1 - \alpha \le x < \alpha, \\ 1, & \alpha \le x < 1. \end{cases} \qquad \frac{1}{2} < \alpha < 1. \tag{4.341}$$

The respective pth absolute central moments and their pth roots are equal to $\sigma_p^p = \mathbb{E}_\alpha |X_1 - \mu|^p = 2(1-\alpha)$ and $\sigma_p = [2(1-\alpha)]^{1/p}$, respectively. Furthermore

$$\begin{aligned} \mathbb{E}_\alpha \sum_{i=1}^{n} c_i \frac{X_{i:n} - \mu}{\sigma_p} &= \int_{\frac{1}{2}}^{1} \frac{F_\alpha^{-1}(x) - \mu}{\sigma_p} f_{\tilde{c}^s:n}(x)dx \\ &= \frac{\int_\alpha^1 f_{\tilde{c}^s:n}(x)dx}{[2(1-\alpha)]^{1/p}} = -\frac{F_{\tilde{c}^s:n}(\alpha)}{[2(1-\alpha)]^{1/p}}. \end{aligned} \tag{4.342}$$

Note that this is negative by (4.338). We represent the numerator as follows

$$\begin{aligned} F_{\tilde{c}^s:n}(\alpha) &= \sum_{i=1}^{n} (c_i - c_{n+1-i}) \sum_{j=i}^{n} B_{j,n}(\alpha) \\ &= \sum_{j=1}^{n} \left[\sum_{i=1}^{j} (c_i - c_{n+1-i}) \right] B_{j,n}(\alpha) \\ &= \sum_{j=1}^{n-1} \left[\sum_{i=1}^{j} (c_i - c_{n+1-i}) \right] \binom{n}{j} \alpha^j (1-\alpha)^{n-j}. \end{aligned} \tag{4.343}$$

The last summand can be dropped because $\sum_{i=1}^{n} (c_i - c_{n+1-i}) = 0$. Therefore

$$\begin{aligned} \mathbb{E}_\alpha \sum_{i=1}^{n} c_i \frac{X_{i:n} - \mu}{\sigma_p} &= -\frac{F_{\tilde{c}^s:n}(\alpha)}{[2(1-\alpha)]^{1/p}} \\ = -\frac{1}{2^{1/p}} \sum_{j=1}^{n-1} &\left[\sum_{i=1}^{j} (c_i - c_{n+1-i}) \right] \binom{n}{j} \alpha^j (1-\alpha)^{n-j-1/p}. \end{aligned} \tag{4.344}$$

tends to 0 as $\alpha \to 1$, because $j < n - \frac{1}{p}$ for all $j = 1, \ldots, n-1$. Since our bound problem is location-scale invariant, we simply normalize the distributions satisfying (4.341) transforming them into (4.340) which satisfy moment conditions. □

Corollary 4.23 *Under the assumptions of Theorem 4.25 with an extra condition that $X_1, \ldots, X_n$ have a bounded support, the following yields*

$$\mathbb{E} \sum_{i=1}^{n} c_i \frac{X_{i:n} - \mu}{\sigma_\infty} \leq 0 \tag{4.345}$$

with the attainability condition

$$\begin{aligned} \mathbb{P}\left(X_1 = \mu \mp \sigma_p\right) &= 1 - \alpha, \\ \mathbb{P}(X_1 = \mu) &= 2\alpha - 1, \end{aligned} \tag{4.346}$$

as $\alpha \to 1$.

Proof It is easily deduced from the above one. For the distribution functions defined in (4.341) we have $\sigma_\infty = 1$. In consequence

$$\mathbb{E}_\alpha \sum_{i=1}^{n} c_i \frac{X_{i:n} - \mu}{\sigma_p} = -F_{\tilde{c}^s:n}(\alpha). \tag{4.347}$$

(cf. (4.342)). Looking at (4.343) we immediately conclude that (4.347) tends to 0 when α approaches 1. □

Theorem 4.26 *Under the assumptions of Theorem 4.25 with $p = 1$ we have*

$$\mathbb{E} \sum_{i=1}^{n} c_i \frac{X_{i:n} - \mu}{\sigma_1} \leq - \min_{\frac{1}{2} \leq \alpha \leq 1} U_{\tilde{c}^s:n}(\alpha), \tag{4.348}$$

where

$$U_{\tilde{c}^s:n}(\alpha) = \frac{n}{2} \sum_{j=1}^{n-1} \frac{1}{n-j} \left[\sum_{i=1}^{j} (c_i - c_{n+1-i}) \right] B_{j,n-1}(\alpha). \tag{4.349}$$

If the minimum is attained at $\frac{1}{2} < \alpha_* < 1$ ($\alpha_* = \frac{1}{2}$, respectively) then the bound is attained by the three-point (two-point, respectively) symmetric marginal distribution

$$\begin{aligned} \mathbb{P}\left(X_1 = \mu \mp \frac{\sigma_1}{2(1-\alpha_*)}\right) &= 1 - \alpha_*, \\ \mathbb{P}(X_1 = \mu) &= 2\alpha_* - 1. \end{aligned} \tag{4.350}$$

If 1 is the minimum point of (4.349), then the bound (4.348) is attained in the limit by (4.350) with α_* replaced by $\alpha \to 1$.

Proof For $\tilde{c}$ satisfying (4.338) we have

$$\mathbb{E}\sum_{i=1}^{n} c_i \frac{X_{i:n} - \mu}{\sigma_1} < 0 \tag{4.351}$$

for every symmetric parent distribution with mean μ. Therefore we apply the norm maximization method described in Subsect. 2.4.7. The arguments of the first part of the proof are identical with these of the proof of Theorem 4.21. The only difference is that we consider the family of functions $g(x) = \frac{F^{-1}(x)-\mu}{\sigma_1}$ which have the domain $\left[\frac{1}{2}, 1\right)$, are nonnegative and nondecreasing, and their L^1-norms are equal to $\frac{1}{2}$. It can be shown that every element of the family is the limit of convex combinations of two-valued functions with 0 on the left, and a positive value on the right. The norm maximization method assures that the supremum of the standardized expectation of the L-statistic $\mathbb{E}\sum_{i=1}^{n} c_i \frac{X_{i:n}-\mu}{\sigma_1}$ coincides with the supremum over the parent distributions whose quantile function has two values on $\left[\frac{1}{2}, 1\right)$.

It follows that we can focus our attention on the symmetric distributions with quantile function satisfying (4.341). Then $\sigma_1 = 2(1-\alpha)$ and

$$\mathbb{E}_\alpha \sum_{i=1}^{n} c_i \frac{X_{i:n} - \mu}{\sigma_1} = -\frac{F_{\tilde{c}^s:n}(\alpha)}{2(1-\alpha)}$$

$$= -\frac{n}{2}\sum_{j=1}^{n-1} \frac{1}{n-j}\left[\sum_{i=1}^{j}(c_i - c_{n+1-i})\right] B_{j,n-1}(\alpha). \tag{4.352}$$

This assures that the formulae of RHS of (4.348) with (4.349) represent the upper bound for the LHS of (4.348). The attainability condition is easily established then. □

Corollary 4.24 *For $X_1, \ldots, X_n$ satisfying the assumptions of Theorem 4.26 the following bound*

$$\mathbb{E}\frac{X_{1:n} - \mu}{\sigma_1} \leq -1 + \frac{1}{2^{n-1}} \tag{4.353}$$

is valid, and it is attained by the two-point symmetric distribution

$$\mathbb{P}(X_1 = \mu - \sigma_1) = \mathbb{P}(X_1 = \mu + \sigma_1) = \frac{1}{2}. \tag{4.354}$$

Proof We apply the conclusion of Theorem 4.26 for $\tilde{c} = (1, 0, \ldots, 0)$. Then

$$\sum_{i=1}^{j}(c_i - c_{n+1-i}) = 1, \qquad i = 1, \ldots, n-1, \tag{4.355}$$

and we need to maximize the function

$$- U_{1:n}^{s}(\alpha) = -\frac{n}{2}\sum_{j=1}^{n-1}\frac{1}{n-j}B_{j,n-1}(\alpha), \qquad \frac{1}{2} \le \alpha \le 1. \tag{4.356}$$

Elementary calculations show that

$$- (U_{1:n}^{s})'(\alpha) = -\frac{n}{2}B_{0,n-2}(\alpha) - \frac{n(n-1)}{2}\sum_{j=1}^{n-2}\frac{1}{(n-j)(n-j-1)}B_{j,n-2}(\alpha), \tag{4.357}$$

which is negative on the whole standard unit interval [0, 1]. It follows that (4.356) is maximized at $\frac{1}{2}$. For calculating the respective minimal value it is more convenient to use the representation of the first line of (4.352) which in the case of sample minimum takes on the form

$$-\frac{F_{1:n}^{s}(\alpha)}{2(1-\alpha)} = -\frac{F_{1:n}(\alpha) - F_{n:n}(\alpha)}{2(1-\alpha)} = -\frac{1-(1-\alpha)^n - \alpha^n}{2(1-\alpha)}, \tag{4.358}$$

and whose value at $\frac{1}{2}$ amounts to the RHS of (4.353). The attainability condition immediately follows from Theorem 4.26. □

The above result was proved in Rychlik (2009c) with use of a different method.

Corollary 4.25 *Assume that $X_1, \ldots, X_n$ are independent and identically symmetrically distributed about its mean μ. Suppose that $\bar{c} = (c_1, \ldots, c_n)$ satisfies (4.338), and moreover $c_1 = c_n$. Then*

$$\mathbb{E}\sum_{i=1}^{n} c_i \frac{X_{i:n} - \mu}{\sigma_1} \le 0, \tag{4.359}$$

and this bound is attained in the limit by the parent distributions

$$\mathbb{P}\left(X_1 = \mu \mp \frac{\sigma_1}{2(1-\alpha)}\right) = 1 - \alpha,$$
$$\mathbb{P}(X_1 = \mu) = 2\alpha - 1, \tag{4.360}$$

as $\alpha \to 1$.

Proof Under the assumption

$$\sum_{i=1}^{n-1}(c_i - c_{n+1-i}) = c_1 - c_n = 0, \tag{4.361}$$

and (4.352) can be represented as

$$U_{\tilde{c}^s:n}(\alpha) = \frac{n}{2} \sum_{j=1}^{n-2} \frac{1}{n-j} \left[\sum_{i=1}^{j} (c_i - c_{n+1-i}) \right] B_{j,n-1}(\alpha). \quad (4.362)$$

Notice that the function tends to 0 as $\alpha \to 1$, because so do all $B_{j,n-1}(\alpha)$, $j = 1, \ldots, n-2$. □

Corollary 4.26 *For iid random variables $X_1, \ldots, X_n$ symmetrically distributed about its mean μ, for every $2 \le j \le \frac{n}{2}$ the following inequality is sharp*

$$\mathbb{E}\frac{X_{j:n} - \mu}{\sigma_1} \le 0. \quad (4.363)$$

The bound is attained in the limit by the marginal distribution (4.360) as $\alpha \to 1$.

We finally notice that the transformations $Y_i = 2\mu - X_i$, $i = 1, \ldots, n$, neither change the marginal symmetric distribution nor affect the independence. We also have $Y_{i:n} - \mu = -(X_{n+1-i:n} - \mu)$, $i = 1, \ldots, n$. It follows that the lower bound on $\mathbb{E}\sum_{i=1}^{n} c_i \frac{X_{i:n}-\mu}{\sigma_p}$ is just the negative of the upper bound on $\mathbb{E}\sum_{i=1}^{n} c_{n+1-i} \frac{X_{i:n}-\mu}{\sigma_p}$, and their attainability conditions are identical. Therefore we do not present the lower bounds explicitly.

4.6.3 Imputation-Type Bounds

Along the lines of Subsect. 4.2.2, one can proceed to develop improved imputation-type bounds for moments of maximal order statistics from a symmetric distribution by making use of moments of maxima from small sample sizes.

To begin with, let us assume, without loss of generality, that the underlying distribution is symmetric about $\mu = 0$. Then, with the choice of the functions

$$g(x) = F^{-1}(x) \quad \text{and} \quad h(x) = \frac{n}{2}\left\{x^{n-1} - (1-x)^{n-1}\right\}, \ 0 < x < 1,$$

and by proceeding along the lines of Theorem 4.1 with the use of Cauchy-Schwarz inequality, the Hartley-David-Gumbel upper bound for the symmetric distribution case can be derived as

$$\mathbb{E}X_{n:n} \le \frac{n}{\sqrt{2}}\left\{\frac{1}{2n-1} - B(n,n)\right\}^{1/2}, \quad (4.364)$$

where $B(a,b)$ denotes the complete beta function.

Then, one can proceed exactly as done earlier in Subsect. 4.2.2 to derive improved upper bounds for $\mathbb{E}X_{n:n}$ in the case of symmetric distributions by imputation based on the knowledge of some moments of order statistics from small sample sizes.

To fix the ideas, let us assume the population distribution to be continuous, symmetric about 0, with finite mean $\mu = 0$ and variance $\sigma^2 = 1$, without loss of any generality. Let m and p be two sample sizes for which $\mathbb{E}X_{m:m}$ and $\mathbb{E}X_{p:p}$ are both available exactly. Then, by proceeding as in Theorem 4.5, Balakrishnan (1990) derived the following imputation-type bound for $\mathbb{E}X_{n:n}$.

Theorem 4.27 *For a continuous symmetric distribution with mean $\mu = 0$ and standard deviation $\sigma = 1$, without loss of generality, we have*

$$\mathbb{E}X_{n:n} \le \frac{n}{\sqrt{2}}\left\{\frac{1}{2n-1} - B(n,n) - \tau^*_{n,m,p}\right\}^{1/2}, \tag{4.365}$$

where

$$\tau^*_{n,m,p} = \frac{\{m\mathbb{E}X_{p:p}T_{n,m} - p\mathbb{E}X_{m:m}T_{n,p}\}^2}{m^2(\mathbb{E}X_{p:p})^2T_{m,m} + p^2(\mathbb{E}X_{m:m})^2T_{p,p} - 2mp\mathbb{E}X_{m:m}\,\mathbb{E}X_{p:p}T_{m,p}}, \tag{4.366}$$

with

$$T_{i,j} = \frac{1}{i+j-1} - B(i,j),$$

$B(\cdot,\cdot)$ being the complete beta function, and $\mathbb{E}X_{m:m}$ and $\mathbb{E}X_{p:p}$ being exactly available (for some m and p) to be used for imputational purpose. Furthermore, $\tau^*_{n,m,p}$ is strictly in the interval $\left(0, \frac{1}{2n-1} - B(n,n)\right)$.

Proof Upon writing

$$\begin{aligned}\mathbb{E}X_{n:n} = \int_0^1 F^{-1}(x)\Bigg[&\frac{1}{2}\left\{nx^{n-1} - n(1-x)^{n-1}\right\}\\ &-\frac{C}{2}\left\{mx^{m-1} - m(1-x)^{m-1} - 2F^{-1}(x)\mathbb{E}X_{m:m}\right\}\\ &-\frac{D}{2}\left\{px^{p-1} - p(1-x)^{p-1} - 2F^{-1}(x)\mathbb{E}X_{p:p}\right\} - \lambda\Bigg]dx,\end{aligned} \tag{4.367}$$

applying the Cauchy-Schwarz inequality in (4.367) and proceeding exactly the same way as in Theorem 4.5, the proof can be completed. □

Table 4.3 Values of the upper bounds in (4.364) and (4.365) for $\mathbb{E}X_{n:n}$ for the case of standard normal distribution, along with exact values

n	Bound in (4.364)	Bound in (4.365)	Exact value
30	2.76172	2.07153	2.04276
40	3.18223	2.20673	2.16078
50	3.55335	2.37658	2.24907
60	3.88922	2.57937	2.31928
70	4.19832	2.80437	2.37736
80	4.48618	3.04170	2.42677
90	4.75665	3.28429	2.46970
100	5.01255	3.52755	2.50759

The fact that $\tau^*_{n,m,p}$ in (4.366) is positive shows readily that the bound in Theorem 4.27 is an improvement of the Hartley-David-Gumbel bound in (4.364) for any choice of m and p.

In Table 4.3, the values of the upper bound in (4.364) (which is an universal upper bound) and the one in (4.365) for the case of standard normal distribution (with the choice of $m = 10$ and $p = 20$ and the corresponding values of $\mathbb{E}X_{m:m} = 1.53875$ and $\mathbb{E}X_{p:p} = 1.86748$ taken from Harter and Balakrishnan (1996)) are presented for $n = 30(10)100$, along with the corresponding exact values of $\mathbb{E}X_{n:n}$ for comparative purposes.

The results in Table 4.3 readily reveal that the imputation-type bound in Theorem 4.27 provides a significant improvement over the bound in (4.364) for all values of n considered. Yet, as compared to the exact values, we observe that the improved bound in (4.365) needs to be improved further. This has been done by Balakrishnan and Samanta (2023) by developing an upper bound for $\mathbb{E}X_{n:n}$ by making use of a set of k expected values of order statistics from small sample sizes, say, $\mathbb{E}X_{m_i:m_i}$ for $i = 1, \ldots, k$.

In this regard, let us use the following notation:

$$
\begin{aligned}
S^*_{m_i:m_j} &= \frac{2m_i m_j}{m_i + m_j - 1} - 2m_i m_j B(m_i, m_j) \\
&\quad -4\mathbb{E}X_{m_i:m_i}\mathbb{E}X_{m_j:m_j}, \quad 1 \le i, j \le k, \\
T^*_{n,m_i} &= \frac{2nm_i}{n + m_i - 1} - 2nm_i B(m_i, n), \ 1 \le i \le k, \\
U^*_{m_i,m_j} &= S^*_{m_i,m_j} + \frac{\mathbb{E}X_{m_i:m_i}\mathbb{E}X_{m_j:m_j}}{(\mathbb{E}X_{m_k:m_k})^2} S^*_{m_k,m_k} \\
&\quad - \frac{\mathbb{E}X_{m_i:m_i}}{\mathbb{E}X_{m_k:m_k}} S^*_{m_j,m_k} - \frac{\mathbb{E}X_{m_j:m_j}}{\mathbb{E}X_{m_k:m_k}} S^*_{m_i,m_k}, \ 1 \le i, j \le k-1, \\
V^*_{m_i} &= T^*_{n,m_i} - \frac{\mathbb{E}X_{m_i:m_i}}{\mathbb{E}X_{m_k:m_k}} T^*_{n,m_k}, \quad 1 \le i \le k-1.
\end{aligned}
$$

Then, Balakrishnan and Samanta (2023) established the following result providing an upper bound for $\mathbb{E}X_{n:n}$.

Theorem 4.28 *Under the assumptions of Theorem 4.27, for any symmetric (about 0) continuous distribution, we have*

$$\mu_{n:n} \leq \frac{1}{2}\left[2n^2\left\{\frac{1}{2n-1} - B(n,n)\right\} - \boldsymbol{V}^{*^T}\boldsymbol{U}^{*^{-1}}\boldsymbol{V}^*\right]^{1/2}, \tag{4.368}$$

where $B(\cdot,\cdot)$ *is the complete beta function,* $\boldsymbol{V}^* = (V^*_{m_1},\ldots,V^*_{m_{k-1}})$ *and* $\boldsymbol{U}^* = ((U_{m_i,m_j}))_{i,j=1}^{k-1}$ *with the exact values of* $\mathbb{E}X_{m_1:m_1},\ldots,\mathbb{E}X_{m_k:m_k}$ *being available (for some* $m_1,\ldots,m_k$*) to be used for imputational purpose. Furthermore,* $\boldsymbol{V}^{*^T}\boldsymbol{U}^{*^{-1}}\boldsymbol{V}^*$ *is strictly in the interval* $\left(0, 2n^2\{\frac{1}{2n-1} - B(n,n)\}\right)$.

Proof The result can be proved along the lines of Theorem 4.27 and, in particular, if we take $k = 2$ with $m_1 = m$ and $m_2 = p$, the result in Theorem 4.27 will be deduced. □

In Table 4.4, the values of the Hartley-David-Gumbel bound in (4.364), the bound in (4.365) with the choice of $m = 10$ and $p = 20$, the bound in (4.368) for the case of standard normal distribution for $k = 7$ with the choice of $m_1 = 5$, $m_2 = 10$, $m_3 = 15$, $m_4 = 20$, $m_5 = 25$, $m_6 = 30$ and $m_7 = 35$ and the corresponding values of $\mathbb{E}X_{5:5} = 1.16296$, $\mathbb{E}X_{10:10} = 1.53875$, $\mathbb{E}X_{15:15} = 1.73591$, $\mathbb{E}X_{20:20} = 1.86748$, $\mathbb{E}X_{25:25} = 1.96531$, $\mathbb{E}X_{30:30} = 2.04276$ and $\mathbb{E}X_{35:35} = 2.10661$ taken from the tables of Harter and Balakrishnan (1996), are all presented for $n = 50(10)100$, along with the corresponding exact values of $\mathbb{E}X_{n:n}$ for comparative purposes. The results in the table reveal that the bound in Theorem 4.28 is a significant improvement over the bound in Theorem 4.27 for all values of n considered, which is to be expected as the former uses more values of $\mathbb{E}X_{m_i:m_i}$ for imputational purpose!

By using some moments of order statistics known in some small sample sizes, one can similarly derive imputation-type bounds for other quantities as well such mean range, mean quasi-range and also more generally for expected value of a linear combination of order statistics!

Table 4.4 Values of the upper bounds in (4.364), (4.365) and (4.368) for $\mathbb{E}X_{n:n}$ for the case of standard normal distribution, along with exact values

n	Bound in (4.364)	Bound in (4.365)	Bound in (4.368)	Exact value
50	3.55335	2.37658	2.2574	2.24097
60	3.88922	2.57937	2.3269	2.31928
70	4.19832	2.80437	2.3904	2.37736
80	4.48618	3.04170	2.4555	2.42677
90	4.75665	3.28429	2.5273	2.46970
100	5.01255	3.52755	2.6090	2.50759

4.7 Life Distributions

4.7.1 Bounds on Expectations

Here we evaluate the expectations of L-statistics $\sum_{i=1}^n c_i X_{i:n}$, $\tilde{c} = (c_1, \ldots, c_n) \in \mathbb{R}^n$, based on iid nondegenerate nonnegative random variables $X_1, \ldots, X_n$. The bounds are expressed in the scale units $m_p = (m_p^p)^{1/p}$, $p \geq 1$, generated by the raw moments $m_p^p = \mathbb{E}X_1^p$ under the assumption that they are finite. Note that $m_1 = \mu = \mathbb{E}X_1 > 0$. We also consider the units $m_\infty = \sup\{x : F(x) < 1\}$, i.e., the right end-points of the support of the parent distribution function F, when we restrict ourselves to bounded random variables.

The crucial series of inequalities is the following

$$\begin{aligned}\mathbb{E}\sum_{i=1}^n c_i \frac{X_{i:n}}{m_p} &= \int_0^1 \frac{F^{-1}(x)}{m_p} f_{\tilde{c}:n}(x)\,dx \\ &\leq \int_0^1 \frac{F^{-1}(x)}{m_p} \underline{f}_{\tilde{c}:n}(x)\,dx \\ &\leq \int_0^1 \frac{F^{-1}(x)}{m_p} (\underline{f}_{\tilde{c}:n})_+(x)\,dx \\ &\leq \|(\underline{f}_{\tilde{c}:n})_+\|_q, \qquad 1 \leq p \leq \infty, \end{aligned} \tag{4.369}$$

where $q = \infty$, $\frac{p}{p-1}$ and 1 for $p = 1$, $1 < p < \infty$, and $p = \infty$, respectively. As above $f_{\tilde{c}:n}(x) = \sum_{i=1}^n c_i f_{i:n}(x)$, $\underline{f}_{\tilde{c}:n}(x)$ is the derivative of the greatest convex minorant of $F_{\tilde{c}:n}(x) = \sum_{i=1}^n c_i F_{i:n}(x)$, and $(\underline{f}_{\tilde{c}:n})_+(x)$ is the nonnegative part of $\underline{f}_{\tilde{c}:n}(x)$. Below we specify the bounds (4.369) for various $\tilde{c} \in \mathbb{R}^n$ and $1 \leq p \leq \infty$, and the conditions of their attainability.

Theorem 4.29 *Let $X_1, \ldots, X_n$ be nondegenerate nonnegative iid random variables with $m_p < \infty$ for some fixed $1 \leq p \leq \infty$. Assume that $\tilde{c} \in \mathbb{R}^n$ satisfies*

$$\frac{\sum_{i=1}^n c_i - F_{\tilde{c}:n}(\alpha)}{1-\alpha} > \left(\sum_{i=1}^n c_i\right)_+ \qquad for\ some\ 0 < \alpha < 1. \tag{4.370}$$

Then for $p = 1$ we have

$$\mathbb{E}\sum_{i=1}^n c_i \frac{X_{i:n}}{\mu} \leq \underline{f}_{\tilde{c}:n}(1). \tag{4.371}$$

If

$$\alpha_* = \alpha_*(1, \tilde{c}) = \min\{\alpha : \underline{f}_{\tilde{c}:n}(\alpha) = \underline{f}_{\tilde{c}:n}(1)\} < 1, \tag{4.372}$$

then the equality in (4.371) holds for the parent distribution

$$\mathbb{P}(X_1 = 0) = \alpha_* = 1 - \mathbb{P}\left(X_1 = \frac{\mu}{1-\alpha_*}\right). \tag{4.373}$$

If $\alpha_* = 1$, then the bound (4.371) is attained in the limit by the distributions (4.373) with α_* replaced by $\alpha \to 1$.

For $1 < p < \infty$ we obtain

$$\mathbb{E}\sum_{i=1}^{n} c_i \frac{X_{i:n}}{m_p} \le \left[\int_0^1 (\underline{f}_{\tilde{c}:n})_+^q(x)\,dx\right]^{1/q} \tag{4.374}$$

with the equality condition

$$\frac{F^{-1}(x)}{m_p} = \frac{(\underline{f}_{\tilde{c}:n})_+^{q/p}(x)}{\left[\int_0^1 (\underline{f}_{\tilde{c}:n})_+^q(x)\,dx\right]^{1/p}} \tag{4.375}$$

For $p = \infty$ (4.369) has the form

$$\mathbb{E}\sum_{i=1}^{n} c_i \frac{X_{i:n}}{m_\infty} \le \sum_{i=1}^{n} c_i - \underline{F}_{\tilde{c}:n}(\alpha_{**}), \tag{4.376}$$

where

$$\alpha_{**} = \alpha_{**}(\infty, \tilde{c}) = \max\{\alpha : \underline{f}_{\tilde{c}:n}(\alpha) \le 0\}. \tag{4.377}$$

If $\alpha_{**} > 0$, then the equality in (4.376) is attained by the marginal distribution

$$\mathbb{P}(X_1 = 0) = \alpha_{**} = 1 - \mathbb{P}(X_1 = m_\infty). \tag{4.378}$$

If $\alpha_{**} = 0$, then the equation in (4.376) is attained in the limit by the distributions (4.378) with α_{**} replaced by $\alpha \to 0$.

Proof Assume first that (4.370) holds together with $\sum_{i=1}^{n} c_i > 0$ which means than $F_{\tilde{c}:n}(\alpha) < \sum_{i=1}^{n} c_i\,\alpha$ for some $0 < \alpha < 1$. Since $\underline{F}_{\tilde{c}:n}(0) = F_{\tilde{c}:n}(0) = 0$ and $\underline{F}_{\tilde{c}:n}(1) = F_{\tilde{c}:n}(1) = \sum_{i=1}^{n} c_i$ and $\underline{F}_{\tilde{c}:n}(\alpha) \le F_{\tilde{c}:n}(\alpha) < \sum_{i=1}^{n} c_i\,\alpha$, we have $\underline{F}_{\tilde{c}:n}(x) < \sum_{i=1}^{n} c_i\,x$, $0 < x < 1$, and $\underline{f}_{\tilde{c}:n}(x) \ne \sum_{i=1}^{n} c_i$ on $(0, 1)$. Since $\underline{f}_{\tilde{c}:n}(x)$ is nondecreasing, and $\int_0^1 \underline{f}_{\tilde{c}:n}(x)\,dx = \sum_{i=1}^{n} c_i$, we conclude that $\underline{f}_{\tilde{c}:n}(x)$ is less and greater than $\sum_{i=1}^{n} c_i$ on some neighborhoods of 0 and 1, respectively. Condition

$\underline{f}_{\tilde{c}:n}(x) > 0$ on some neighborhood of 1 (it can be even the whole interval $(0, 1)$) implies that $(\underline{f}_{\tilde{c}:n})_+(x)$ is not equal to zero everywhere on $(0, 1)$.

If (4.370) and $\sum_{i=1}^n c_i \le 0$ hold, then (4.370) can be rewritten as

$$F_{\tilde{c}:n}(\alpha) < \sum_{i=1}^n c_i = \underline{F}_{\tilde{c}:n}(1) \le \underline{F}_{\tilde{c}:n}(0) = 0 \qquad \text{for some } 0 < \alpha < 1. \tag{4.379}$$

It again follows that $\underline{F}_{\tilde{c}:n}(x)$ is decreasing and increasing on some neighborhoods of 0 and 1, and $\underline{f}_{\tilde{c}:n}(x)$ is negative and positive there. We conclude that under the assumption (4.370) $(\underline{f}_{\tilde{c}:n})_+(x)$ is a nonzero and nonconstant function on $(0, 1)$. This implies positive nontrivial bounds in (4.369).

For $p = 1$ we have

$$\begin{aligned}
\mathbb{E} \sum_{i=1}^n \frac{X_{i:n}}{\mu} &\le \int_0^1 \frac{F^{-1}(x)}{\mu} (\underline{f}_{\tilde{c}:n})_+(x)\, dx \\
&\le \sup_{0<x<1} (\underline{f}_{\tilde{c}:n})_+(x) \int_0^1 \frac{F^{-1}(x)}{\mu}\, dx \\
&= \underline{f}_{\tilde{c}:n}(1).
\end{aligned} \tag{4.380}$$

If α_* defined in (4.372) is less than 1, then (4.373) provides the equality in the second line of (4.380). If not, the equality is attained in the limit by (4.373) with α_* replaced by $\alpha \to 1$, because $(\underline{f}_{\tilde{c}:n})_+(x)$ is continuous. Note that in both the cases all the intervals where $\underline{F}_{\tilde{c}:n}(x) < F_{\tilde{c}:n}(x)$ and $\underline{f}_{\tilde{c}:n}(x)$ is constant as well as the interval where $\underline{f}_{\tilde{c}:n}(x) < 0$ are contained in the intervals $(0, \alpha_*)$ and $(\alpha_*, 1)$ in the first case, and $(0, \alpha)$ for sufficiently large α in the latter one. This implies that the first two inequalities in (4.369) become the equalities (at least in the limit) for the two-point distributions described above.

For $1 < p < \infty$, the Hölder inequality (see Theorem 2.31) implies that in view of (4.369) we obtain the inequality (4.374). The condition (4.375) implies the equality in the last inequality of (4.369). The RHS of (4.375) is nondecreasing and nonnegative which implies that the respective LHS defines a distribution function of a life random variable. The special choice of the denominator in the RHS of (4.375) implies that both the sides of (4.375) have the pth norms equal to 1, which means that the respective marginal distribution has the pth raw moment equal to m_p^p. Recalling the arguments from the case $p = 1$ we notice that (4.375) guarantees the equality in the first two inequalities of (4.369).

For $p = \infty$, we have

$$\mathbb{E}\sum_{i=1}^{n}\frac{X_{i:n}}{m_\infty} \le \int_0^1 \frac{F^{-1}(x)}{m_\infty}(\underline{f}_{\tilde{c}:n})_+(x)\,dx$$
$$\le \sup_{0<x<1}\frac{F^{-1}(x)}{m_\infty}\int_0^1 (\underline{f}_{\tilde{c}:n})_+(x)\,dx$$
$$= \int_{\alpha_{**}}^1 \underline{f}_{\tilde{c}:n}(x)\,dx, \tag{4.381}$$

because the supremum in the penultimate line is equal to 1, by definition. Moreover,

$$\int_{\alpha_{**}}^1 \underline{f}_{\tilde{c}:n}(x)\,dx = \int_{\alpha_{**}}^1 f_{\tilde{c}:n}(x)\,dx = \sum_{i=1}^{n} c_i - F_{\tilde{c}:n}(\alpha_{**}). \tag{4.382}$$

To check the first equality in (4.382) (the latter is obvious) we observe that for every interval (α_1, α_2) such that $\underline{f}_{\tilde{c}:n}(\alpha_i) = f_{\tilde{c}:n}(\alpha_i)$, $\underline{F}_{\tilde{c}:n}(\alpha_i) = F_{\tilde{c}:n}(\alpha_i)$, $i = 1, 2$, and $\underline{F}_{\tilde{c}:n}(x) < F_{\tilde{c}:n}(x)$ for $\alpha_1 < x < \alpha_2$, we have

$$\int_{\alpha_1}^{\alpha_2} \underline{f}_{\tilde{c}:n}(x)\,dx = \int_{\alpha_1}^{\alpha_2} \frac{F_{\tilde{c}:n}(\alpha_2) - F_{\tilde{c}:n}(\alpha_1)}{\alpha_2 - \alpha_1}\,dx$$
$$= F_{\tilde{c}:n}(\alpha_2) - F_{\tilde{c}:n}(\alpha_1) = \int_{\alpha_1}^{\alpha_2} f_{\tilde{c}:n}(x)\,dx. \tag{4.383}$$

It follows that

$$\int_{\alpha}^1 \underline{f}_{\tilde{c}:n}(x)\,dx = \int_{\alpha}^1 f_{\tilde{c}:n}(x)\,dx \tag{4.384}$$

for all $0 < \alpha < 1$ such that $\underline{f}_{\tilde{c}:n}(\alpha) = f_{\tilde{c}:n}(\alpha)$. It remains to notice that by definition α_{**} is either the unique solution to $\underline{f}_{\tilde{c}:n}(x) = 0$ or the maximal point of the interval where $\underline{f}_{\tilde{c}:n}(x)$ vanishes. In both the cases $\underline{f}_{\tilde{c}:n}(\alpha_{**}) = f_{\tilde{c}:n}(\alpha_{**})$.

If $\alpha_* = \alpha_*(\infty, \tilde{c})$ defined in (4.377) is positive, then the equality appears in the latter inequality of (4.381) if

$$\frac{F^{-1}(x)}{m_\infty} = \begin{cases} 0, & 0 < x < \alpha_*, \\ 1, & \alpha_* \le x < 1. \end{cases} \tag{4.385}$$

Note that this defines the distribution function of a nonnegative random variable with the essential supremum equal to m_∞. If $\alpha_* = 0$, then the equality is attained in the limit by the two-point distributions supported on 0 and m_∞ with the probability of 0 equal to $\alpha \to \alpha_* = 0$. We also easily check that the above construction satisfies the equality conditions in the first two inequalities of (4.369), and in consequence, the attainability of the bound (4.376). □

The cases different from (4.370) are studied below in Theorems 4.30, 4.31 and 4.32.

Theorem 4.30 *Let $X_1, \ldots, X_n$ satisfy the assumptions of Theorem 4.29. Suppose further that*

$$\frac{\sum_{i=1}^n c_i - F_{\tilde{c}:n}(\alpha)}{1-\alpha} \le \left(\sum_{i=1}^n c_i\right)_+, \qquad 0 < \alpha < 1, \tag{4.386}$$

and the equality is attained at some $0 < \alpha_* = \alpha_*(\tilde{c}) < 1$. Then for every $1 \le p \le \infty$

$$\mathbb{E}\sum_{i=1}^n c_i \frac{X_{i:n}}{m_p} \le \left(\sum_{i=1}^n c_i\right)_+, \tag{4.387}$$

and the equality is attained by the parent distribution

$$\mathbb{P}(X_1 = 0) = \alpha_* = 1 - \mathbb{P}\left(X_1 = \frac{m_p}{(1-\alpha_*)^{1/p}}\right), \tag{4.388}$$

under the convention that $\frac{1}{\infty} = 0$.

Proof If $0 \le \sum_{i=1}^n c_i$, then (4.386) is equivalent to

$$F_{\tilde{c}:n}(\alpha) \ge \sum_{i=1}^n c_i \,\alpha, \qquad 0 < \alpha < 1, \tag{4.389}$$

with the equality at 0, α_* and 1. It follows that $\underline{F}_{\tilde{c}:n}(x) = \sum_{i=1}^n c_i\, x$,

$$\underline{f}_{\tilde{c}:n}(x) = (\underline{f}_{\tilde{c}:n})_+(x) = \sum_{i=1}^n c_i, \tag{4.390}$$

and $\|(\underline{f}_{\tilde{c}:n})_+\|_q = \sum_{i=1}^n c_i$ for all $1 \le p \le \infty$, which proves the bound. The last two inequalities in (4.369) become actually the equalities then. For the distribution (4.388) function $\frac{F^{-1}(x)}{m_p}$ is constant on the intervals $(0, \alpha_*)$ and $(\alpha_*, 1)$ which contain all the intervals where $\underline{F}_{\tilde{c}:n}(x) = \sum_{i=1}^n c_i\, x < F_{\tilde{c}:n}(x)$. This assures the equality in the first inequality of (4.369) as well.

If $0 > \sum_{i=1}^n c_i$, then (4.386) has the form

$$F_{\tilde{c}:n}(\alpha) \geq \sum_{i=1}^{n} c_i, \qquad 0 < \alpha < 1, \tag{4.391}$$

with the equality at α_* and 1. Therefore $\underline{F}_{\tilde{c}:n}(x) = \sum_{i=1}^n c_i$ on $[\alpha_*, 1]$. It follows that $\underline{f}_{\tilde{c}:n}(x) = 0$, $\alpha_* \leq x < 1$, and $\underline{f}_{\tilde{c}:n}(x) \leq 0$ elsewhere. Therefore $(\underline{f}_{\tilde{c}:n})_+(x) = 0$, and we have the zero bound in (4.369). We easily check that the parent distribution (4.388) assures that all the inequalities in (4.369) become the equalities. □

Theorem 4.31 *Let $X_1, \ldots, X_n$ follow the assumptions of Theorem 4.29. Moreover, let $\tilde{c} \in \mathbb{R}^n$ satisfy the relations*

$$\frac{\sum_{i=1}^n c_i - F_{\tilde{c}:n}(\alpha)}{1-\alpha} < \left(\sum_{i=1}^{n} c_i\right)_+ = \sum_{i=1}^{n} c_i, \qquad 0 < \alpha < 1. \tag{4.392}$$

Then for all $1 \leq p \leq \infty$

$$\mathbb{E} \sum_{i=1}^{n} c_i \frac{X_{i:n}}{m_p} \leq \sum_{i=1}^{n} c_i. \tag{4.393}$$

The bound is attained in the limit by the marginal distributions

$$\mathbb{P}_\alpha(X_1 = 0) = \alpha = 1 - \mathbb{P}_\alpha\left(X_1 = \frac{m_p}{(1-\alpha)^{1/p}}\right) \tag{4.394}$$

as $\alpha \to 0$, with $\frac{1}{\infty} = 0$.

Proof The assumption (4.392) means that

$$F_{\tilde{c}:n}(x) > \sum_{i=1}^{n} c_i\, x, \qquad 0 < x < 1. \tag{4.395}$$

Note that the equality holds at $x = 0$ and $x = 1$ in the above inequality. Therefore $\underline{F}_{\tilde{c}:n}(x) = \sum_{i=1}^n c_i\, x$ and

$$\underline{f}_{\tilde{c}:n}(x) = (\underline{f}_{\tilde{c}:n})_+(x) = \sum_{i=1}^{n} c_i \geq 0. \tag{4.396}$$

Observe that for every $1 \leq p \leq \infty$ and respective $\infty \geq q \geq 1$ we have $\| \sum_{i=1}^n c_i \|_q = \sum_{i=1}^n c_i$. In view of (4.369) this is our bound for every $1 \leq p \leq \infty$.

For the family of distributions (4.394) we have

$$\frac{F_\alpha^{-1}(x)}{m_p} = \begin{cases} 0, & 0 < x < \alpha, \\ \frac{1}{(1-\alpha)^{1/p}}, & \alpha \le x < 1, \end{cases} \tag{4.397}$$

and in consequence

$$\begin{aligned} \mathbb{E}_\alpha \sum_{i=1}^n c_i \frac{X_{i:n}}{m_p} &= \int_0^1 \frac{F_\alpha^{-1}(x)}{m_p} f_{\tilde{c}:n}(x)\,dx \\ &= \frac{1}{(1-\alpha)^{1/p}} \int_\alpha^1 f_{\tilde{c}:n}(x)\,dx \\ &= \sum_{i=1}^n c_i \frac{1 - F_{i:n}(\alpha)}{(1-\alpha)^{1/p}}. \end{aligned} \tag{4.398}$$

If $\alpha \to 0$, this tends to $\sum_{i=1}^n c_i$ which is our bound. □

Theorem 4.32 *Let $X_1, \ldots, X_n$ satisfy the assumptions of Theorem 4.29. Also, assume that the following holds*

$$\frac{\sum_{i=1}^n c_i - F_{\tilde{c}:n}(\alpha)}{1-\alpha} < 0 > \sum_{i=1}^n c_i, \qquad 0 < \alpha < 1. \tag{4.399}$$

(i) Then for every $1 < p \le \infty$

$$\mathbb{E} \sum_{i=1}^n c_i \frac{X_{i:n}}{m_p} \le 0, \tag{4.400}$$

and the equality is attained in the limit by the parent distributions (4.394) as $\alpha \to 1$.

(ii) For $p = 1$ we have

$$\mathbb{E} \sum_{i=1}^n c_i \frac{X_{i:n}}{m_p} \le \max_{1 \le \alpha \le 1} U_{\tilde{c}:n}(\alpha), \tag{4.401}$$

where

$$U_{\tilde{c}:n}(\alpha) = n \sum_{i=0}^{n-1} \left(\frac{1}{n-i} \sum_{j=i}^n c_j \right) B_{i,n-1}(\alpha). \tag{4.402}$$

If the function is minimized at some $0 < \alpha_* < 1$, then the bound (4.401) is attained by the two-point distribution (4.394) with $p = 1$ and $\alpha = \alpha_*$. If it is maximized at $\alpha = 0$ ($\alpha = 1$), then the bound is attained in the limit by (4.394) with $p = 1$ and $\alpha \to 0$ ($\alpha \to 1$, respectively).

Proof The assumption (4.399) implies that

$$F_{\bar{c}:n}(\alpha) > \sum_{i=1}^{n} c_i = F_{\bar{c}:n}(1) < 0, \qquad 0 < \alpha < 1, \tag{4.403}$$

Therefore $\underline{F}_{\bar{c}:n}(x)$ is decreasing, $\underline{f}_{\bar{c}:n}(x) < 0$, and $(\underline{f}_{\bar{c}:n})_+(x) = 0$ for all $0 < x < 1$. In consequence, we obtain the zero bound in (4.369) for all $1 \le p \le \infty$.

(i) When $1 < p \le \infty$, we show that the zero bound of (4.369) is optimal. For the family of two-point distributions (4.394) we have

$$\begin{aligned}
\mathbb{E}_\alpha \sum_{i=1}^{n} c_i \frac{X_{i:n}}{m_p} &= \frac{1}{(1-\alpha)^{1/p}} \int_\alpha^1 f_{\bar{c}:n}(x)\,dx = \frac{\sum_{i=1}^{n} c_i - F_{\bar{c}:n}(\alpha)}{(1-\alpha)^{1/p}} \\
&= \sum_{i=1}^{n} c_i \frac{1 - F_{i:n}(\alpha)}{(1-\alpha)^{1/p}} = \sum_{i=1}^{n} c_i \frac{\sum_{j=0}^{i-1} B_{j,n}(\alpha)}{(1-\alpha)^{1/p}} \\
&= \sum_{i=1}^{n} c_i \left(\sum_{j=0}^{i-1} \binom{n}{j} \alpha^j (1-\alpha)^{n-j-1/p} \right).
\end{aligned} \tag{4.404}$$

If $p > 1$ then $n - j - 1/p > 0$, $j = 0, \ldots, i-1$, $i = 1, \ldots, n$, and in consequence all the summands tend to 0 as $\alpha \to 1$.

(ii) This is not necessarily true for $p = 1$ for which we need some more elaborate arguments. Since $\underline{f}_{\bar{c}:n}(x)$ is strictly negative and $\frac{F^{-1}(x)}{\mu}$ is nonnegative and not equal to 0 almost everywhere on (0, 1), we get

$$\mathbb{E} \sum_{i=1}^{n} c_i \frac{X_{i:n}}{\mu} = \int_0^1 \frac{F^{-1}(x)}{\mu} f_{\bar{c}:n}(x)\,dx \le \int_0^1 \frac{F^{-1}(x)}{\mu} \underline{f}_{\bar{c}:n}(x)\,dx < 0 \tag{4.405}$$

for any nondegenerate F supported in $[0, \infty)$ with a finite mean. Therefore we can use here the norm maximization method (see Subsect. 2.4.7). We do not repeat all the steps of the reasoning, and recall only the conclusion. Since an arbitrary $\frac{F^{-1}(x)}{\mu}$ is nonnegative and nondecreasing, and its L^1-norm is equal to 1, and it can be represented (at least in the limit) as a convex combination of two-valued functions (where the smaller one is 0), the upper bound for the functional

$$\frac{F^{-1}(x)}{\mu} \mapsto \int_0^1 \frac{F^{-1}(x)}{\mu} f_{\tilde{c}:n}(x)\,dx \tag{4.406}$$

coincides with the upper bound restricted to the two-valued functions $\frac{F^{-1}(x)}{\mu}$. These represent the marginal distribution functions (4.394) with $p = 1$. For them, the functional (4.406) is represented by (4.404) with $p = 1$, and can be alternatively written as (4.402). Its maximum provides the desired bound.

□

The assumption (4.399) implies that

$$F_{\tilde{c}:n}(\alpha) = \sum_{i=1}^{n} \left(\sum_{j=1}^{i} c_i \right) B_{i:n}(\alpha) < 0, \qquad 0 < \alpha < 1, \tag{4.407}$$

(see also (4.403)). This is obviously satisfied if all the partial sums $\sum_{j=1}^{i} c_j$, $j = 1, \ldots, n$, are non-positive, and at least one of them is strictly negative. A necessary condition is that the first nonzero element and the last non-zero partial sum of the elements of $\tilde{c}$ are negative which assure that $F_{\tilde{c}:n}(\alpha)$ is actually negative on some neighborhoods of 0 and 1.

Corollary 4.27 *Under the assumptions of Theorem* 4.32 *with* $c_n = 0$

$$\mathbb{E} \sum_{i=1}^{n} c_i \frac{X_{i:n}}{\mu} \le 0, \tag{4.408}$$

with the equality attained in the limit by the distributions (4.394) with $p = 1$ *if* $\alpha \to 1$.

Proof It suffices to observe that (4.404) with $p = 1$ and $c_n = 0$ tends to 0 as α approaches 1. □

We apply the above results for establishing the corresponding bounds for the single order statistics. For $j = 1$ we have $f_{1:n}(x) = n(1-x)^n$, and $\underline{f}_{1:n}(x) = 1$, $0 < x < 1$. By Theorem 4.31 we get

$$\mathbb{E} \frac{X_{1:n}}{m_p} \le 1, \qquad 1 \le p \le \infty, \tag{4.409}$$

which is attained in the limit by (4.394) with $\alpha \to 0$. For $j = 2, \ldots, n-1$ the derivatives of the greatest convex minorants of $F_{j:n}(x)$ have the forms

$$\underline{f}_{j:n}(x) = f_{j:n}(\min\{x, \alpha_*\}) > 0, \tag{4.410}$$

where $\alpha_* = \alpha_*(j, n)$ are determined by the equations

$$f_{j:n}(\alpha) = \frac{1 - F_{j:n}(\alpha)}{1 - \alpha} \tag{4.411}$$

(cf. (4.87)). By Theorem 4.29 we obtain

$$\mathbb{E}\frac{X_{j:n}}{m_p} \le \begin{cases} f_{j:n}(\alpha_*), & p = 1, \\ B_q = B_q(j, n), & 1 < p < \infty, \\ 1, & p = \infty, \end{cases} \tag{4.412}$$

where

$$B_q^q(j, n) = \int_0^{\alpha_*} f_{j:n}^q(x)\, dx + (1 - \alpha_*) f_{j:n}^q(\alpha_*) \tag{4.413}$$

with $q = \frac{p}{p-1}$. The equality in the first inequality is satisfied by (4.373) (see also (4.394) with $p = 1$ and $\alpha = \alpha_*$), and that in the second line by F satisfying

$$\frac{F^{-1}(x)}{m_p} = \left[\frac{\underline{f}_{j:n}(x)}{B_q}\right]^{p/q}. \tag{4.414}$$

The last bound is attained in the limit by (4.394) with $p = \infty$ and $\alpha \to 0$. For the sample maximum with $f_{n:n}(x) = \underline{f}_{n:n}(x) = (\underline{f}_{n:n})_+(x) = nx^{n-1}$ we have the explicit formulae

$$\mathbb{E}\frac{X_{n:n}}{m_p} \le \begin{cases} n, & p = 1, \\ \frac{n}{[(n-1)q+1]^{1/q}}, & 1 < p < \infty, \\ 1, & p = \infty. \end{cases} \tag{4.415}$$

The first and last bounds are attained by (4.394) when $p = 1$ with $\alpha \to 1$, and $p = \infty$ with $\alpha \to 0$, respectively. The middle one is attained by

$$F(x) = \frac{1}{[(n-1)q + 1]^{\frac{1}{(n-1)q}}} \left(\frac{x}{m_p}\right)^{\frac{p}{(n-1)q}}, \qquad 0 < x < [(n-1)q + 1]^{1/p} m_p, \tag{4.416}$$

which is a scale transformed power distribution function with the shape parameter $\frac{p}{(n-1)q}$.

We can also provide analogous lower bounds calculating the upper ones for $\mathbb{E}\frac{-X_{j:n}}{m_p}$, $j = 1, \ldots, n$, $1 \le p \le \infty$. The condition (4.399) of Theorem 4.32 has the particular form

$$\frac{-1+F_{j:n}(\alpha)}{1-\alpha} < 0 > -1, \qquad 0<\alpha<1, \tag{4.417}$$

and it is obviously satisfied for every $j = 1,\ldots,n$. Therefore for every $j = 1,\ldots,n$ and every $1 < p \le \infty$ we have

$$\mathbb{E}\frac{X_{j:n}}{m_p} \ge 0, \tag{4.418}$$

and the equality is attained in the limit by the distributions supported on 0 and $\frac{m_p}{(1-\alpha)^{1/p}}$ as the probability mass of α of 0 tends to 1. By Corollary 4.27 the result can be extended to $p = 1$ for $j = 1,\ldots,n-1$ with the same attainability conditions. By Theorem 4.32(ii)

$$\mathbb{E}\frac{X_{n:n}}{\mu} \ge -\sup_{0<\alpha<1}\frac{-1+\alpha^n}{1-\alpha} = -\max_{0\le\alpha\le 1}\left(-\sum_{i=0}^{n-1}\alpha^i\right) = 1, \tag{4.419}$$

which is attained in the limit by (4.394) with $p = 1$ and $\alpha \to 0$.

In the similar way we can determine the sharp lower bounds on $\mathbb{E}\sum_{i=1}^n c_i \frac{X_{i:n}}{m_p}$ for arbitrary $\tilde{c} = (c_1,\ldots,c_n)$ and $1 \le p \le \infty$. It suffices to apply the above theorems for the L-statistics with the coefficient vectors $-\tilde{c} = (-c_1,\ldots,-c_n)$.

4.7.2 Bounds on Other Moments

Now we consider bounds on the moments $\mathbb{E}X_{j:n}^p$ of order statistics based on iid life random variables expressed in the units of $\mu^p = (\mathbb{E}X_1)^p$ for positive p. Due to Sen (1959) (see also Papadatos, 2021), we restrict ourselves to $0 < p \le n+1-j$, because there exist parent life distributions with finite expectations such that $\mathbb{E}X_{j:n}^p = \infty$ for $p > n+1-j$.

We first consider $0 < p < 1$. In the sample minimum case, by the Jensen inequality we have

$$\mathbb{E}X_{1:n}^p = \mathbb{E}[X_{1:n}^n]^{\frac{p}{n}} \le \left[\mathbb{E}X_{1:n}^n\right]^{\frac{p}{n}} \le [\mathbb{E}X_1\ldots X_n]^{\frac{p}{n}} = \mu^p, \tag{4.420}$$

and the equality holds for the degenerated parent distribution concentrated at μ. This can be extended to $p = 1$ as well. For other $2 \le j \le n$, the bounds are concluded from the Moriguti inequality and the Hölder inequality with parameters $\frac{1}{p}, \frac{1}{1-p} > 1$.

Theorem 4.33 *Suppose that* $X_1, \ldots, X_n$, $n \geq 2$, *are nonnegative iid random variables with a finite positive mean* μ. *Then for* $j = 2, \ldots, n$ *and* $0 < p < 1$ *we have*

$$\frac{\mathbb{E}X_{j:n}^p}{\mu^p} \leq B = B(j, n, p), \tag{4.421}$$

where

$$B^{\frac{1}{1-p}} = \int_0^{\alpha_*} f_{j:n}^{\frac{1}{1-p}}(x)\, dx + (1 - \alpha_*) f_{j:n}^{\frac{1}{1-p}}(\alpha_*), \tag{4.422}$$

and $0 < \alpha_* = \alpha_*(j, n) < \frac{j-1}{n-1}$ *is the unique solution to (4.81) for* $2 \leq j \leq n-1$ *and* $\alpha_*(n, n) = 1$. *The equality in (4.421) is attained if*

$$F^{-1}(x) = \mu \frac{f_{j:n}(\min\{x, \alpha_*\})}{B^{\frac{1}{1-p}}}. \tag{4.423}$$

Proof Since $[F^{-1}(x)]^p$ is nondecreasing

$$\begin{aligned} \mathbb{E}X_{j:n}^p &= \int_0^1 [F^{-1}(x)]^p f_{j:n}(x)\, dx \leq \int_0^1 [F^{-1}(x)]^p \underline{f}_{j:n}(x)\, dx \\ &\leq \left[\int_0^1 [F^{-1}(x)]^{p \cdot \frac{1}{p}}\, dx\right]^p \left[\int_0^1 \underline{f}_{j:n}^{\frac{1}{1-p}}(x)\, dx\right]^{1-p} = \mu^p B. \end{aligned} \tag{4.424}$$

We easily see that (4.423) satisfies the equality in the former Moriguti inequality and the latter Hölder one, and moreover, it satisfies the first moment condition as well. □

For establishing the bounds for $1 \leq p \leq n+1-j$ we need the following two Lemmas (see Corollary 4 and Lemma 2 in Papadatos (2021)).

Lemma 4.2 *Let* F *be a distribution function of a nonnegative random variable with a finite and positive mean. Then for all* $p > 1$ *yields*

$$p \int_0^\infty x^{p-1} [1 - F(x)]^p dx \leq \left[\int_0^\infty [1 - F(x)]\, dx\right]^p. \tag{4.425}$$

The equality holds if F *is either one or two-point distribution with at most one positive support point.*

For $p = 1$ (4.425) is a trivial equality attained by arbitrary distribution function with a finite mean.

Proof Note that the RHS of (4.425) is the pth power of a nonnegative random variable X, say, with the distribution function F, and it can be written as

$$[\mathbb{E}\,X]^p = \left[\int_0^1 F^{-1}(x)\,dx\right]^p. \tag{4.426}$$

The LHS is the pth moment of a random variable Y, say, with the distribution function $F_p(x) = 1 - [1 - F(x)]^p$, and then

$$\begin{aligned}\mathbb{E}\,Y^p &= p\int_0^\infty x^{p-1}[1 - F(x)]^p dx = \int_0^1 [F_p^{-1}(x)]^p dx \\ &= \int_0^1 [F^{-1}(1 - (1 - x)^{\frac{1}{p}})]^p dx = p\int_0^1 (1 - x)^{p-1}[F^{-1}(x)]^p dx.\end{aligned} \tag{4.427}$$

Since under our assumptions F^{-1} is a nonnegative nondecreasing function with a finite integral, then (4.425) is equivalent to

$$p\int_0^1 (1 - x)^{p-1}[g(x)]^p dx \le \left[\int_0^1 g(x)\,dx\right]^p \tag{4.428}$$

that should be valid for arbitrary nonnegative and nondecreasing g with a finite Lebesgue integral over $(0, 1)$. Due to the statement of the lemma, the inequality becomes the equality if either g is constant or it takes on only two values, one of them being 0.

We immediately check that (4.428) holds and becomes the equality for constant functions. Then we consider multi-valued stepwise functions

$$g(x) = \sum_{i=1}^k \left(\sum_{j=1}^i \delta_j\right) \mathbf{1}_{(s_{i-1}, s_i]}(x) \tag{4.429}$$

for $\delta_1 \ge 0 < \delta_2, \ldots, \delta_k$, and $s_0 = 0 < s_1 < \ldots < s_k = 1$. We have

$$\int_0^1 g(x)\,dx = \sum_{i=1}^k \left(\sum_{j=1}^i \delta_j\right)(s_i - s_{i-1}) = \sum_{j=1}^k (1 - s_{j-1})\delta_j \tag{4.430}$$

and

$$p\int_0^1 (1-x)^p g^p(x)\,dx = \sum_{i=1}^{k}\left(\sum_{j=1}^{i}\delta_j\right)^p [(1-s_i)^p - (1-s_{i-1})^p]$$
$$= \delta_1^p + \sum_{j=1}^{k-1}\left[\left(\sum_{i=1}^{j+1}\delta_i\right)^p - \left(\sum_{i=1}^{j}\delta_i\right)^p\right](1-s_j)^p. \tag{4.431}$$

We check now that

$$\left(\sum_{j=1}^{k}(1-s_{j-1})\delta_j\right)^p - \delta_1^p - \sum_{j=1}^{k-1}\left[\left(\sum_{i=1}^{j+1}\delta_i\right)^p - \left(\sum_{i=1}^{j}\delta_i\right)^p\right](1-s_j)^p \geq 0, \tag{4.432}$$

applying the induction method with respect to k.

For $k = 2$ we fix δ_1 and s_1, and treat the LHS of (4.432) as a function $f_1(\delta_2)$. We easily note that $f_1(0) = 0$ and

$$f_1'(\delta_2) = p(1-s_1)\{[\delta_1 + (1-s_1)\delta_2]^{p-1} - [(1-s_1)\delta_1 + (1-s_1)\delta_2]^{p-1}\} \tag{4.433}$$

is nonnegative, because $p > 1$ and $\delta_1 + (1-s_1)\delta_2 \geq (1-s_1)\delta_1 + (1-s_1)\delta_2$. Note that $f_1'(\delta_2) = 0$ and we have the equality in (4.432) in consequence iff $\delta_1 = 0 < \delta_2$.

We assume now that (4.432) holds, and we check that

$$\left(\sum_{j=1}^{k+1}(1-s_{j-1})\delta_j\right)^p - \delta_1^p - \sum_{j=1}^{k}\left[\left(\sum_{i=1}^{j+1}\delta_i\right)^p - \left(\sum_{i=1}^{j}\delta_i\right)^p\right](1-s_j)^p \geq 0 \tag{4.434}$$

as well. We fix almost all the arguments of the LHS and treat it as a function f_k of single argument δ_{k+1}. We have $f_k(0) = f_{k-1}(\delta_k) \geq 0$, and

$$f_k'(\delta_{k+1}) = p(1-s_k)\left[\left(\sum_{j=1}^{k+1}(1-s_{j-1})\delta_j\right)^{p-1} - \left(\sum_{j=1}^{k+1}(1-s_k)\delta_j\right)^{p-1}\right] > 0, \tag{4.435}$$

since $p > 1$ and $\sum_{j=1}^{k+1}(1-s_{j-1})\delta_j > \sum_{j=1}^{k+1}(1-s_k)\delta_j$, which proves that the inequality (4.432) holds for $k \geq 3$, and it is strict then.

The proof for arbitrary nondecreasing function g is standard. We take a nondecreasing approximation g_m, $m = 1, 2, \ldots$, of g so that $g_m(x) \to g(x)$, and so $p(1-x)^{p-1} g_m(x) \to p(1-x)^{p-1} g(x)$ for all $0 < x < 1$. Applying the Lebesgue monotone convergence theorem we obtain

$$p \int_0^1 (1-x)^{p-1} g^p(x)\, dx = \lim_{m \to \infty} p \int_0^1 p(1-x)^{p-1} g_m^p(x)\, dx$$

$$\leq \lim_{m \to \infty} \left(\int_0^1 g_m(x)\, dx \right)^p = \left(\int_0^1 g(x)\, dx \right)^p . \tag{4.436}$$

This ends the proof. □

Lemma 4.3 *Let* $2 \leq j \leq n$. *If* $1 \leq p < n+1-j$, *then*

$$B_p(x) = B_p(x; j, n) = \frac{1 - F_{j:n}(x)}{(1-x)^p}, \qquad 0 \leq x < 1, \tag{4.437}$$

has a unique maximum point $0 < \alpha_p = \alpha_p(j, n) < \frac{j-1}{n-p}$ *being the solution to the equation*

$$p[1 - F_{j:n}(x)] = (1-x) f_{j:n}(x), \qquad 0 < x < 1. \tag{4.438}$$

If $p = n+1-j$, *then*

$$\sup_{0 \leq x < 1} B_{n+1-j}(x) = \lim_{x \to 1} B_{n+1-j}(x) = \binom{n}{j-1}. \tag{4.439}$$

Proof In fact, for $p = 1$ the solution was presented earlier (see Theorem 4.8). We consider now the cases $1 < p < n+1-j$. We have $B_p(0) = 1$, $\lim_{x \to 1} B_p(x) = 0$ and

$$B_p'(x) = \frac{p[1 - F_{j:n}(x)] - (1-x) f_{j:n}(x)}{(1-x)^{p+1}} = \frac{g_p(x)}{(1-x)^{p+1}}, \tag{4.440}$$

say. Since

$$g_p'(x) = [(n-p)x - j + 1] \frac{f_{j:n}(x)}{x}, \tag{4.441}$$

g_p is first decreasing on $\left(0, \frac{j-1}{n-p}\right)$ and increasing on $\left(\frac{j-1}{n-p}, 1\right)$. Moreover, $g_p(0) = p > 0$ and $g_p(1) = 0$ which allows us to deduce that g_p and so B_p' are first positive on some $(0, \alpha_p) \subset \left(0, \frac{j-1}{n-p}\right)$, they have the unique zero at α_p, and are negative in $(\alpha_p, 1)$. So, α_p maximizes B_p.

If $p = n+1-j$, then $B_{n+1-j}(0) = 1$, $\lim_{x\to 1} B_{n+1-j}(x) = \binom{n}{j-1}$, and $B'_{n+1-j}(x) = \frac{g_{n+1-j}(x)}{(1-x)^{p+1}}$ (cf. (4.440)). Since $g_{n+1-j}(0) = n+1-j > 0$, $g_{n+1-j}(1) = 0$ and

$$g'_{n+1-j}(x) = (j-1)(x-1)\,\frac{f_{j:n}(x)}{x} < 0, \qquad 0 < x < 1.$$

it follows that $g_{n+1-j}(x)$, and consequently $B'_{n+1-j}(x)$, $0 \le x < 1$ are positive. Therefore $B_{n+1-j}(x)$, $0 \le x < 1$, is increasing, and attains its supremum at 1. □

Theorem 4.34 *Suppose that $X_1, \ldots, X_n$, $n \ge 3$, satisfy the assumptions of Theorem 4.33. Then for every $2 \le j \le n-1$ and $1 \le p < n+1-j$ we have*

$$\frac{\mathbb{E}X^p_{j:n}}{\mu^p} \le B_p(\alpha_p), \tag{4.442}$$

where the RHS function and its maximal argument are defined in Lemma 4.3. The equality in (4.442) holds iff

$$\mathbb{P}(X_1 = 0) = \alpha_p = 1 - \mathbb{P}\left(X_1 = \frac{\mu}{1-\alpha_p}\right). \tag{4.443}$$

Proof With F denoting the marginal distribution function of X_1, we get

$$\begin{aligned}
\mathbb{E}X^p_{j:n} &= p\int_0^\infty x^{p-1}[1 - F_{j:n}\circ F(x)]dx \\
&\le \sup_{0<x<1} \frac{1-F_{j:n}(x)}{(1-x)^p}\, p\int_0^\infty x^{p-1}[1-F(x)]^p dx \\
&\le B_p(\alpha_p)\left[\int_0^\infty [1-F(x)]dx\right]^p = B_p(\alpha_p)\mu^p,
\end{aligned} \tag{4.444}$$

where the latter inequality follows from Lemma 4.2. For $p > 1$ we obtain the equality in the first inequality if the only values of F in $\mathbb{R}_+$ are α_p and 1. This means that F is a two-valued distribution function with probability of 0, being the smaller support point, equal to α_p. For $p = 1$ this is the equality under no conditions. This two-point distribution provides the equality in the latter inequality as well. The only distribution with mean μ satisfying the above conditions is presented in (4.443). □

Theorem 4.35 *Under the assumptions of Theorem 4.33, for $2 \le j \le n$ we have*

$$\frac{\mathbb{E}X^{n+1-j}_{j:n}}{\mu^{n+1-j}} \le \binom{n}{j-1}, \tag{4.445}$$

with the equality attained in the limit by the two-point parent distributions

$$\mathbb{P}(X_1 = 0) = \alpha = 1 - \mathbb{P}\left(X_1 = \frac{\mu}{1-\alpha}\right) \tag{4.446}$$

as $\alpha \to 1$.

Proof We can rewrite the relations of (4.444) replacing p and $B_p(\alpha_p)$ by $n + 1 - j$ and $\sup_{0<x<1} B_{n+1-j}(x) = \binom{n}{j-1}$, respectively. For the marginal distributions (4.446) we obtain

$$\frac{\mathbb{E}_\alpha X_{j:n}^{n+1-j}}{\mu^{n+1-j}} = \frac{1 - F_{j:n}(\alpha)}{(1-\alpha)^{n+1-j}} \to \binom{n}{j-1} \tag{4.447}$$

as α tends to 1. □

We see that paradoxically the greatest ratios $\frac{\mathbb{E}X_{j:n}^{n+1-j}}{\mu^{n+1-j}} = \binom{n}{j-1}$ are attained by central order statistics. Applying the methods of Theorems 4.34 and 4.35 with the functions

$$B_p(x; 1, n) = (1-x)^{n-p}, \qquad 0 \le x < 1,\ 1 \le p \le n, \tag{4.448}$$

we obtain

$$\frac{\mathbb{E}X_{1:n}^p}{\mu^p} \le 1, \qquad 1 \le p \le n. \tag{4.449}$$

For $1 \le p < n$, the equality is attained by the degenerate distribution (comp. (4.420)). If $p = n$, we also get the equality for the two-point distributions (4.446) with arbitrary $0 < \alpha < 1$.

4.8 Upper Bounds in Gini Mean Difference Units

4.8.1 General L-Statistics

In this subsection we present sharp lower and upper bounds on the expectations of centered L-statistics $\mathbb{E}\sum_{i=1}^n c_i(X_{i:n} - \mu)$, with arbitrary $c_1, \ldots, c_n \in \mathbb{R}$, expressed in terms of the Gini mean difference scale units

$$\Delta = \Delta(F) = \mathbb{E}|X_1 - X_2|, \tag{4.450}$$

where X_1, X_2 are iid random variables with a parent distribution function F and finite mean μ. This is positive if F is nondegenerate. The Gini mean difference is a popular and useful measure of dispersion. Note that for defining it only finiteness of the first population moment is necessary.

Given $\tilde{c} = (c_1, \ldots, c_n) \in \mathbb{R}^n$ with the arithmetic mean $\bar{c} = \frac{1}{n}\sum_{i=1}^{n} c_i$, we define vector $\tilde{a} = \tilde{a}(\tilde{c}) = (a_0, \ldots, a_{n-2}) \in \mathbb{R}^{n-1}$ as follows

$$a_i = a_i(\tilde{c}) = \frac{n(n-1)}{2(i+1)(n-i-1)} \sum_{k=1}^{i+1} (\bar{c} - c_k), \qquad i = 0, \ldots, n-2. \tag{4.451}$$

Furthermore, we put

$$P_{\tilde{c}}(u) = \sum_{i=0}^{n-2} a_i B_{i,n-2}(u), \qquad 0 \le u \le 1, \tag{4.452}$$

where $B_{i,n}$, $i = 0, \ldots, n$, denote the Bernstein polynomials of degree n. Obviously, $P_{\tilde{c}}$ is a polynomial of degree $n-2$ at most.

The main result of this subsection is following.

Theorem 4.36 *Let $X_1, \ldots, X_n$ be nondegenerate iid random variables with the mean μ and Gini mean difference Δ. Then*

$$\min_{0 \le u \le 1} P_{\tilde{c}}(u) \le \mathbb{E} \sum_{i=1}^{n} c_i \frac{X_{i:n} - \mu}{\Delta} \le \max_{0 \le u \le 1} P_{\tilde{c}}(u). \tag{4.453}$$

If $0 < u_1 < \ldots < u_r < 1$ ($0 < v_1 < \ldots < v_\ell < 1$) are all the inner points of the unit interval $[0, 1]$, being the arguments of the maximum (minimum, respectively), then the upper (lower) bound is attained iff the parent distribution function has the form

$$F(x) = \begin{cases} 0, & x < x_0, \\ u_1, & x_0 \le x < x_1, \\ \vdots & \\ u_r, & x_{r-1} \le x < x_r, \\ 1, & x \ge x_r, \end{cases} \tag{4.454}$$

for arbitrary $x_0 \le \ldots \le x_r > x_0$ (and

$$F(x) = \begin{cases} 0, & x < x_0, \\ v_1, & x_0 \le x < x_1, \\ \vdots & \\ v_\ell, & x_{\ell-1} \le x < x_\ell, \\ 1, & x \ge x_\ell, \end{cases} \tag{4.455}$$

for arbitrary $x_0 \le \ldots \le x_\ell > x_0$, respectively).

If the maximum (minimum) amounts to $P_{\tilde{c}}(0)$*, then the upper (lower) bound is attained in the limit by any two-point distributions such that the probabilities of the smaller point tend to* 0*. Similarly, if the maximum (minimum) amounts to* $P_{\tilde{c}}(1)$*, the upper (lower) bound is attained in the limit by any two-point distributions such that the probabilities of the smaller point tend to* 1.

According to (4.454), any distribution function attaining the upper bound is discrete, and the set of its values that differ from 0 and 1 is a nonempty subset of $\{u_1, \dots, u_r\}$. The number of jumps is greater by 1 than the cardinality of the subset. A polynomial of degree $n-2$ may have $\lfloor \frac{n}{2} \rfloor - 1$ local maxima at most. It is theoretically possible, but practically very unlikely that all the maxima belong to $(0, 1)$ and provide identical values of the polynomial. For majority of L-statistics, especially these commonly used in statistical analysis, respective functions $P_{\tilde{c}}$ have either one or (quite rarely) two maxima in $(0, 1)$. It also happens that the maximum is attained at either of the border points of the unit interval. Similar remarks concern the minima of $P_{\tilde{c}}$ for various $\tilde{c}$.

Proof We first remove μ from the representation of the expectation of centered L-statistics

$$\begin{aligned}\mathbb{E}\sum_{i=1}^{n} c_i(X_{i:n} - \mu) &= \mathbb{E}\sum_{i=1}^{n} c_i\left(X_{i:n} - \frac{1}{n}\sum_{k=1}^{n} X_{k:n}\right)\\ &= \mathbb{E}\left(\sum_{i=1}^{n} c_i X_{i:n} - \frac{1}{n}\sum_{i=1}^{n} c_i \sum_{k=1}^{n} X_{k:n}\right)\\ &= \mathbb{E}\sum_{i=1}^{n}(c_i - \bar{c})X_{i:n}. \qquad (4.456)\end{aligned}$$

Since the new coefficients $c_i - \bar{c}$, $i = 1, \dots, n$, sum up to 0, we can represent the modified L-statistic $\sum_{i=1}^{n}(c_i - \bar{c})X_{i:n}$ as a linear combination of spacings

$$\sum_{i=1}^{n}(c_i - \bar{c})X_{i:n} = \sum_{i=1}^{n-1} b_i(X_{i+1:n} - X_{i:n}), \qquad (4.457)$$

where

$$b_i = b_i(\tilde{c}) = \sum_{k=1}^{i}(\bar{c} - c_k), \qquad i = 1, \dots, n-1. \qquad (4.458)$$

Now we use integral representations of the expected spacings

$$\mathbb{E}(X_{i+1:n} - X_{i:n}) = \int_{-\infty}^{\infty} B_{i,n} \circ F(x)\, dx, \qquad i = 1, \dots, n-1, \qquad (4.459)$$

first presented in Pearson (1902). This is derived with use of the integration by parts rule. It is also useful in representing the Gini mean difference

$$\Delta = \mathbb{E}|X_1 - X_2| = \mathbb{E}(X_{2:2} - X_{1:2}) = \int_{-\infty}^{\infty} B_{1,2} \circ F(x)\, dx. \tag{4.460}$$

Accordingly, we can write

$$\begin{aligned}
\mathbb{E}\sum_{i=1}^{n} c_i(X_{i:n} - \mu) &= \mathbb{E}\sum_{i=1}^{n-1}\left[\sum_{k=1}^{i}(\bar{c} - c_k)\right](X_{i+1:n} - X_{i:n}) \\
&= \int_{-\infty}^{\infty}\sum_{i=1}^{n-1}\left[\sum_{k=1}^{i}(\bar{c} - c_k)\right] B_{i,n} \circ F(x)\, dx \\
&= \int_{-\infty}^{\infty}\sum_{i=1}^{n-1}\left[\sum_{k=1}^{i}(\bar{c} - c_k)\right]\frac{n(n-1)}{2i(n-i)} B_{i-1,n-2} \circ F(x)\, B_{1,2} \circ F(x)\, dx \\
&= \int_{-\infty}^{\infty}\sum_{i=0}^{n-2}\frac{n(n-1)}{2(i+1)(n-i-1)}\left[\sum_{k=1}^{i+1}(\bar{c} - c_k)\right] B_{i,n-2} \circ F(x)\, B_{1,2} \circ F(x)\, dx \\
&= \int_{-\infty}^{\infty} P_{\tilde{c}} \circ F(x)\, B_{1,2} \circ F(x)\, dx.
\end{aligned} \tag{4.461}$$

In consequence, we obtain

$$\begin{aligned}
\mathbb{E}\sum_{i=1}^{n} c_i(X_{i:n} - \mu) &\le \sup_{-\infty < x < \infty} P_{\tilde{c}} \circ F(x) \int_{-\infty}^{\infty} B_{1,2} \circ F(x)\, dx \\
&\le \max_{0 \le u = F(x) \le 1} P_{\tilde{c}}(u)\, \Delta,
\end{aligned} \tag{4.462}$$

as desired.

We get the equality in the latter inequality if we do not exclude any $0 \le u \le 1$ from the possible values of the parent distribution function F. We also have the equality in the former one iff for almost all $x \in \mathbb{R}$ we have either $P_{\tilde{c}} \circ F(x) = \max_{0 \le u \le 1} P_{\tilde{c}}(u)$ or $B_{1,2} \circ F(x) = 0$. The latter is equivalent to either $F(x) = 0$ or $F(x) = 1$. The only possibility for attaining the bound is that the set of values of $F(x)$ is contained in $\{u_1, \ldots, u_r\} \cup \{0, 1\}$. By assumption, $\{F(x) : x \in \mathbb{R}\} \cap \{u_1, \ldots, u_r\} \neq \emptyset$.

Suppose now that $P_{\tilde{c}}(0) > P_{\tilde{c}}(u)$, $0 < u < 1$, and consider the family of parent distribution functions

$$F_u(x) = \begin{cases} 0, \; x < x_0, \\ u, \; x_0 \le x < x_1, \\ 1, \; x \ge x_1, \end{cases} \qquad 0 < u < 1, \tag{4.463}$$

for some arbitrary $x_0 < x_1$. Then

$$\mathbb{E}_u \sum_{i=1}^{n} c_i (X_{i:n} - \mu) = P_{\tilde{c}}(u)\, \Delta_u. \tag{4.464}$$

Letting $u \to 0$, by continuity of $P_{\tilde{c}}$ we obtain

$$\lim_{u \to 0} \mathbb{E}_u \sum_{i=1}^{n} c_i \frac{X_{i:n} - \mu}{\Delta_u} = P_{\tilde{c}}(0) = \max_{0 \le u \le 1} P_{\tilde{c}}(u). \tag{4.465}$$

Similarly, in the case $P_{\tilde{c}}(1) > P_{\tilde{c}}(u)$, $0 < u < 1$, the following yields

$$\lim_{u \to 1} \mathbb{E}_u \sum_{i=1}^{n} c_i \frac{X_{i:n} - \mu}{\Delta_u} = P_{\tilde{c}}(1) = \max_{0 \le u \le 1} P_{\tilde{c}}(u). \tag{4.466}$$

The proofs for the lower bound as well as for its attainability are analogous. □

Now we determine more precisely the bounds for particular L-statistics which are often used in statistical inference. We first note that for $n = 2$ we trivially obtain

$$\mathbb{E}[c_1(X_{1:2} - \mu) + c_2(X_{2:2} - \mu)] = \frac{c_2 - c_1}{2} \mathbb{E}(X_{2:2} - X_{1:2}) = \frac{c_2 - c_1}{2} \Delta \tag{4.467}$$

for arbitrary parent distribution. Therefore we further concentrate of L-statistics based on at least $n = 3$ observations. Since the polynomials $P_{\tilde{c}}$ are smooth functions, for establishing its extremes it is sufficient to analyze the sign changes of the respective derivatives

$$P'_{\tilde{c}}(u) = \sum_{i=0}^{n-3} \alpha_i B_{i,n-3}(u), \tag{4.468}$$

where

$$\alpha_i = \alpha_i(\tilde{c}) = (n-2)[a_{i+1}(\tilde{c}) - a_i(\tilde{c})] = \frac{n(n-1)(n-2)}{2}$$
$$\times \left[\frac{\sum_{k=1}^{i+2}(\bar{c}-c_k)}{(i+2)(n-i-2)} - \frac{\sum_{k=1}^{i+1}(\bar{c}-c_k)}{(i+1)(n-i-1)} \right], \qquad i = 0 \dots, n-3, \tag{4.469}$$

For this purpose we use the variation diminishing property of the Bernstein polynomials described in Lemma 4.1. The following observations shall be also helpful in our considerations.

We have $P_{\tilde{c}}(u) = -P_{\tilde{d}}(1-u)$, $0 \le u \le 1$, for some $\tilde{a}(\tilde{c}) = (a_0(\tilde{c}), \dots, a_{n-2}(\tilde{c}))$ and $\tilde{a}(\tilde{d}) = (a_0(\tilde{d}), \dots, a_{n-2}(\tilde{d}))$ iff $a_i(\tilde{c}) = a_{n-2-i}(\tilde{d})$, $i = 0, \dots, n-2$, i.e., when $\sum_{k=i+2}^{n} c_k = \sum_{k=1}^{n-i-1} d_i$, $i = 0, \dots, n-2$. This simply means that $\tilde{d} = \tilde{c}' = (c_n, \dots, c_1)$. The condition implies that the upper (lower) evaluation for $\sum_{i=1}^{n} c_i X_{i:n}$ is identical with the negative of the lower (upper respectively) evaluation for $\sum_{i=1}^{n} c_{n+1-i} X_{i:n}$. In consequence, for every L-statistics with $c_i = c_{n+1-i}$, $i = 1, \dots, n$, i.e., for every linear combination of quasi-midranges the lower bound is the negative of the upper one.

Also, we get $P_{\tilde{c}}(u) = P_{\tilde{c}}(1-u)$, $0 \le u \le 1$, iff $a_i(\tilde{c}) = a_{n-2-i}(\tilde{c})$, $i = 0, \dots, n-2$, which is consecutively equivalent to $\frac{1}{n-2-2i} \sum_{k=i+2}^{n-1-i} c_k = \bar{c}$, $i = 0, \dots, \lfloor \frac{n-3}{2} \rfloor$, and $\frac{1}{2}(c_i + c_{n+1-i}) = \bar{c}$ for $i = 1, \dots, n$. This is satisfied by linear combinations of quasi-ranges and the sample mean $\sum_{i=1}^{\lfloor n/2 \rfloor} d_i (X_{n+1-i:n} - X_{i:n}) + \frac{d}{n} \sum_{i=1}^{n} X_{i:n}$ with arbitrary real coefficients $d_1, \dots d_{\lfloor n/2 \rfloor}$ and d. Under the condition, the set of maximum (minimum) points of $P_{\tilde{c}}(u)$, $0 \le u \le 1$, is symmetric about $\frac{1}{2}$. In particular, $u = \frac{1}{2}$ is a local extremum of $P_{\tilde{c}}(u)$, and a candidate for the global one.

We finally notice that the vector transformations $\tilde{a} : \mathbb{R}^n \mapsto \mathbb{R}^{n-1}$ and $\tilde{\alpha} : \mathbb{R}^n \mapsto \mathbb{R}^{n-2}$ described by formulae (4.451) and (4.469), respectively, are linear. In consequence, functions (4.452) and (4.468) are linear operators acting on vectors of coefficients $\tilde{c} \in \mathbb{R}^n$.

4.8.2 Single Order Statistics

Now we proceed to single order statistics $X_{j:n}$, $j = 1, \dots, n$, which are obviously examples of L-statistics with the coefficient vector $\tilde{c} = \tilde{e}_j$ being the jth versor with 1 at the position j and zeros elsewhere. Simple calculations show that the corresponding vector $\tilde{a}(\tilde{e}_j)$ has the coordinates

$$a_i(\tilde{e}_j) = a_i(j) = \begin{cases} \frac{n-1}{2(n-i-1)}, & \text{if } i \leq j-2, \\ -\frac{n-1}{2(i+1)}, & \text{if } i \geq j-1. \end{cases} \tag{4.470}$$

It follows that the respective polynomials

$$P_{\mathbf{e}_j}(u) = P_j(u) = \sum_{i=0}^{n-2} a_i(j) B_{i,n-2}(u)$$

have particular forms

$$P_1(u) = -\sum_{i=0}^{n-2} \frac{n-1}{2(i+1)} B_{i,n-2}(u), \tag{4.471}$$

$$P_2(u) = \frac{1}{2} B_{0,n-2}(u) - \sum_{i=1}^{n-2} \frac{n-1}{2(i+1)} B_{i,n-2}(u), \tag{4.472}$$

$$P_j(u) = \sum_{i=0}^{j-2} \frac{n-1}{2(n-i-1)} B_{i,n-2}(u) - \sum_{i=j-1}^{n-2} \frac{n-1}{2(i+1)} B_{i,n-2}(u), \tag{4.473}$$

for $3 \leq j \leq n-2$, and

$$P_{n-1}(u) = \sum_{i=0}^{n-3} \frac{n-1}{2(n-i-1)} B_{i,n-2}(u) - \frac{1}{2} B_{n-2,n-2}(u), \tag{4.474}$$

$$P_n(u) = \sum_{i=0}^{n-2} \frac{n-1}{2(n-i-1)} B_{i,n-2}(u). \tag{4.475}$$

The corresponding derivatives are

$$P_1'(u) = \sum_{i=0}^{n-3} \frac{(n-1)(n-2)}{2(i+1)(i+2)} B_{i,n-3}(u), \tag{4.476}$$

$$P_2'(u) = -\frac{(n+1)(n-2)}{4} B_{0,n-3}(u)$$
$$+ \sum_{i=1}^{n-3} \frac{(n-1)(n-2)}{2(i+1)(i+2)} B_{i,n-3}(u), \tag{4.477}$$

$$P'_j(u) = \sum_{i=0}^{j-3} \frac{(n-1)(n-2)}{2(n-i-2)(n-i-1)} B_{i,n-3}(u) - \frac{(n^2-1)(n-2)}{2j(n+1-j)} B_{j-2,n-3}(u)$$

$$+ \sum_{i=j-1}^{n-3} \frac{(n-1)(n-2)}{2(i+1)(i+2)} B_{i,n-3}(u), \qquad 3 \le j \le n-2, \tag{4.478}$$

$$P'_{n-1}(u) = \sum_{i=0}^{n-4} \frac{(n-1)(n-2)}{2(n-i-2)(n-i-1)} B_{i,n-3}(u)$$

$$- \frac{(n+1)(n-2)}{4} B_{n-3,n-3}(u), \tag{4.479}$$

$$P'_n(u) = \sum_{i=0}^{n-3} \frac{(n-1)(n-2)}{2(n-i-2)(n-i-1)} B_{i,n-3}(u). \tag{4.480}$$

Proposition 4.1

(i) For the extreme order statistics, we have

$$-\frac{n-1}{2} \le \mathbb{E}\frac{X_{1:n}-\mu}{\Delta} \le -\frac{1}{2}, \tag{4.481}$$

$$\frac{1}{2} \le \mathbb{E}\frac{X_{n:n}-\mu}{\Delta} \le \frac{n-1}{2}. \tag{4.482}$$

The lower and upper bounds in the above relations are attained in the limit by the two-point distributions such that the probabilities of the smaller points tend to 0 *and* 1*, respectively.*

(ii) The derivatives (4.477) and (4.479) have unique zeros $v_1(2)$ *and* $u_1(n-1) = 1 - v_1(2)$*, respectively. Moreover,*

$$P_2(v_1(2)) \le \mathbb{E}\frac{X_{2:n}-\mu}{\Delta} \le \frac{1}{2}, \tag{4.483}$$

$$-\frac{1}{2} \le \mathbb{E}\frac{X_{n-1:n}-\mu}{\Delta} \le P_2(u_1(n-1)). \tag{4.484}$$

The lower bound in (4.483) and the upper bound in (4.484) are attained by the two-point distributions such that the probability of the smaller points are $v_1(2)$ *and* $u_1(n-1)$*, respectively. The upper bound in (4.483) and the lower one in (4.484) are attained in the limit by the two-point distributions such that the probabilities of the smaller point tend to* 0 *and* 1*, respectively.*

(iii) For $3 \le j \le n-2$, the derivative (4.478) has exactly two zeros $u_1(j) < v_1(j)$ in $(0, 1)$. Moreover,

$$P_j(v_1(j)) \le \mathbb{E}\frac{X_{j:n}-\mu}{\Delta} \le P_j(u_1(j)), \tag{4.485}$$

and the lower and upper bounds are attained by the two-point distributions such that the probabilities of the smaller points amount to $v_1(j)$ and $u_1(j)$, respectively.

Since $P_j(u) = P_{n+1-j}(1-u)$, we have $u_1(j) = v_1(n+1-j)$ and $P_j(v_1(j)) = -P_j(u_1(n+1-j))$.

Proof

(i) Owing to Theorem 4.36, it suffices to determine the arguments and corresponding values of particular functions (4.471)–(4.475) which provide their global extremes over the interval [0, 1]. Due to VDP, functions (4.476) and (4.480) are positive in (0, 1). Hence (4.471) and (4.475) are increasing in [0, 1]. Their extreme values are

$$P_1(0) = a_0(1) = -\frac{n-1}{2},$$

$$P_1(1) = a_{n-2}(1) = -\frac{1}{2},$$

$$P_n(0) = a_0(n) = \frac{1}{2},$$

$$P_n(1) = a_{n-2}(n) = \frac{n-1}{2}.$$

Observe that according to the remark following the proposition, the bounds for the sample maxima could be also deduced from ones for the sample minima and *vice versa*.

(ii) By (4.477) and VDP, function (4.472) is first decreasing and then increasing. Since $P_2(0) = a_0(2) = \frac{1}{2}$, and $P_2(1) = a_{n-2}(2) = -\frac{1}{2}$, we immediately conclude that the global maximum and minimum are $P_2(0) = \frac{1}{2}$ and $P_2(v_1(2)) < -\frac{1}{2}$, respectively, where $v_1(2)$ is the unique zero of (4.477) in (0, 1). The bounds for the case $j = n-1$ are derived in much the same way.

(iii) For $3 \le j \le n-2$, the formula (4.478) together with the VDP of Bernstein polynomials assert that (4.473) is either consecutively increasing, decreasing and eventually increasing in (0, 1), or simply increasing in the whole interval. Relations $P_j(0) = a_0(j) = \frac{1}{2}$ and $P_j(1) = a_{n-2}(j) = -\frac{1}{2}$ contradict the latter possibility. Therefore (4.478) has two zeros in (0, 1). The first one $u_1(j)$ provides the global maximum of (4.473), and the other gives the minimum. This ends the proof.

□

4.8.3 Differences of Order Statistics

In the sequel we consider differences of order statistics $X_{j:n} - X_{i:n}$ for $1 \le i < j \le n$. The vector of coefficients is $\tilde{c}(i,j) = \tilde{e}_j - \tilde{e}_i$, and it has the mean $\bar{c} = 0$. We specify the elements of (4.451) as follows

$$a_k(i,j) = a_k(\tilde{c}(i,j)) = \begin{cases} 0, & k = 0,\dots,i-2, \\ \frac{n(n-1)}{2(k+1)(n-k-1)}, & k = i-1,\dots,j-2, \\ 0, & k = j-1,\dots,n-2. \end{cases} \tag{4.486}$$

Note that the first and last zeros vanish if $i = 1$ and $j = n$, respectively. Furthermore the coefficients (4.452) in the Bernstein polynomial representation of the respective derivatives satisfy

$$\alpha_k(i,j) = \alpha_i(\tilde{c}(i,j)) = \begin{cases} 0, & k = 0,\dots,i-3, \\ (n-2)a_{k+1}(i,j), & k = i-2, \\ (n-2)[a_{k+1}(i,j) - a_k(i,j)], & k = i-1,\dots,j-3, \\ -(n-2)a_k(i,j), & k = j-2, \\ 0, & k = j-1,\dots,n-3. \end{cases} \tag{4.487}$$

The first and last zeros disappear when $i \le 2$ and $j \ge n-1$, respectively. The second and fourth rows vanish for $i = 1$ and $j = n$, respectively. The third one does not appear for the spacings, i.e., when $j = i+1$. We also check that under the condition $i - 1 \le k \le j - 3$ yields $a_{k+1}(i,j) < (=,>) a_k(i,j)$ iff $k < (=,>) \frac{n-3}{2}$, respectively. This implies that all the elements of the third row are non-positive and non-negative iff $j \le \frac{n+3}{2}$ and $i \ge \frac{n-1}{2}$, respectively. They are first negative and ultimately positive when $i < \frac{n-1}{2}$ and $j > \frac{n+3}{2}$. Presenting below the bounds on the order statistics differences, we consider two cases separately. In Proposition 4.2, we analyze the possibilities that the difference contains at least one of the sample extremes. The remaining ones $2 \le i < j \le n-1$ are treated in Proposition 4.3.

Proposition 4.2

(i) *For the sample range with $n \ge 4$, we have*

$$2 - \frac{1}{2^{n-2}} \le \mathbb{E}\frac{X_{n:n} - X_{1:n}}{\Delta} \le \frac{n}{2}. \tag{4.488}$$

The lower bound is attained by the symmetric two-point distributions. The upper one is attained in the limit by the two-point distributions such that the probability of one point decreases to 0.

(ii) If either $i = 1 < j < n$ or $1 < i < j = n$, then

$$0 \le \mathbb{E}\frac{X_{j:n} - X_{i:n}}{\Delta} \le \frac{n}{2}. \tag{4.489}$$

In the former (latter) case the lower and upper bounds are attained in the limit by the two-point distributions such that the probabilities of the smaller points tend to 1 *and* 0*, respectively (*0 *and* 1*, respectively).*

The relations (4.488) hold for $n = 2$ and 3 as well. What is more, then the lower and upper bounds are identical and equal to 1 and $\frac{3}{2}$, respectively. The are obviously attained by arbitrary parent distributions. For $n = 2$, the result immediately follows from the definition (see, e.g., (4.460)). For $n = 3$, the following yields

$$\begin{aligned}
\Delta &= \frac{1}{3}\left(\mathbb{E}|X_1 - X_2| + \mathbb{E}|X_1 - X_3| + \mathbb{E}|X_2 - X_3|\right) \\
&= \frac{1}{3}\,\mathbb{E}\left(X_{2:3} - X_{1:3} + X_{3:3} - X_{1:3} + X_{3:3} - X_{2:3}\right) \\
&= \frac{2}{3}\,\mathbb{E}\left(X_{3:3} - X_{1:3}\right).
\end{aligned} \tag{4.490}$$

Proof

(i) We use the notation $P_{i,j} = P_{\tilde{e}_j - \tilde{e}_i}$. By the comments following (4.487), the function $P_{1,n}$ is first decreasing and then increasing. Since $P_n(u) = P_1(1-u)$, this is symmetric about $\frac{1}{2}$. Therefore

$$\max_{0\le u\le 1} P_{1,n}(u) = P_{1,n}(0) = P_{1,n}(1) = \frac{n}{2}, \tag{4.491}$$

$$\min_{0\le u\le 1} P_{1,n}(u) = P_{1,n}\left(\frac{1}{2}\right) = 2 - \frac{1}{2^{n-2}}, \tag{4.492}$$

because

$$P_{1,n}(u) = \frac{\sum_{i=1}^{n-1} B_{i,n}(u)}{B_{1,2}(u)} = \frac{1-(1-u)^n - u^n}{2u(1-u)}. \tag{4.493}$$

(ii) Suppose that $i = 1$ and $j < n$. If $j \le \frac{n+3}{2}$, then function $P_{1,j}$ is decreasing by the VDP of its derivative. We have $P_{1,j}(0) = \frac{n}{2}$ which is the upper bound, and $P_{1,j}(1) = 0$ which is the lower one. If $j > \frac{n+3}{2}$, then $P_{1,j}$ is decreasing in some neighborhoods of 0 and 1, but it is also possible that it is increasing in some inner subinterval (v_0, u_0), say, of (0, 1). Then the right end-point of the subinterval is a possible candidate for the global maximum. Suppose that

$P_{1,j}(u_0) > P_{1,j}(0) = \frac{n}{2}$. Therefore for some two-point distribution we have

$$P_{1,j}(u_0) = \mathbb{E}\frac{X_{j:n} - X_{1:n}}{\Delta} \leq \mathbb{E}\frac{X_{n:n} - X_{1:n}}{\Delta} < \frac{n}{2}, \tag{4.494}$$

which contradicts part (i) of the proposition. Moreover, point v_0 cannot provide the global minimum less than $P_{1,j}(1) = 0$ either, because

$$\mathbb{E}\frac{X_{j:n} - X_{1:n}}{\Delta} = P_{1,j}(v_0) < P_{1,j}(1) = 0 \tag{4.495}$$

is impossible for the expectation of non-negative random function $X_{j:n} - X_{1:n}$. The proof for $1 < i < j = n$ is similar.

□

Proposition 4.3 *Assume that* $2 \leq i < j \leq n-1$.

(i) If either $i \geq \frac{n-1}{2}$ *or* $j \leq \frac{n+3}{2}$, *then the polynomial*

$$P'_{i,j}(u) = \sum_{k=0}^{n-3} \alpha_k(i,j) B_{k,n-3}(u), \qquad 0 \leq u \leq 1, \tag{4.496}$$

(see (4.487)) has a unique zero $u_1 = u_1(i,j)$, *say, in* $(0,1)$, *and*

$$0 \leq \mathbb{E}\frac{X_{j:n} - X_{i:n}}{\Delta} \leq P_{i,j}(u_1). \tag{4.497}$$

(ii) If $i < \frac{n-1}{2}$ *and* $j > \frac{n+3}{2}$, *the polynomial has either one zero* $u_1 \in (0,1)$ *or three zeros* $u_1 < v_1 < u_2$, *where* u_1 *and* u_1, u_2 *are local maxima of* $P_{i,j}$ *in the former and latter cases, respectively. In these cases we have*

$$0 \leq \mathbb{E}\frac{X_{j:n} - X_{i:n}}{\Delta} \leq P_{i,j}(u_1), \tag{4.498}$$

$$0 \leq \mathbb{E}\frac{X_{j:n} - X_{i:n}}{\Delta} \leq \max\{P_{i,j}(u_1), P_{i,j}(u_2)\}, \tag{4.499}$$

respectively.

The lower bounds in the inequalities presented in the statements (i) and (ii) are attained in the limit by the two-point distributions with the contributions of one point decreasing to 0. If the maximum of $P_{i,j}(u)$, $0 \leq u \leq 1$, *is attained at a single point* u_i, *then the upper bound is attained by the two-point distribution with the probability mass of the smaller point equal to* u_i. *If the global maximum is attained simultaneously at* u_1 *and* u_2, *then the bound is attained by the parent distribution functions of the form*

$$F(x) = \begin{cases} 0, & x < x_0, \\ u_1, & x_0 \leq x < x_1, \\ u_2, & x_1 \leq x < x_2, \\ 1, & x \geq x_2, \end{cases} \tag{4.500}$$

for arbitrary $x_0 \leq x_1 \leq x_2 > x_0$.

Existence of two global maxima is possible, e.g. for the quasi-ranges with $j = n + 1 - i$. Then $P_{i,j}(u)$ is symmetric about $\frac{1}{2}$. Moreover, for i located relatively close to 1 and far from $\frac{n}{2}$, the polynomial is bimodal. Clearly, the lower and upper bounds for $X_{i:n} - X_{j:n}$, $i < j$, are the negatives of the upper and lower bounds for $X_{j:n} - X_{i:n}$.

Proof Under the assumptions of point (i), the derivative $P'_{i,j}(u)$ changes its sign once, from $+$ to $-$. Therefore $P_{i,j}(u)$ is unimodal with the global maximum at the zero point of its derivative. The lower bound coincides with

$$P_{i,j}(0) = P_{i,j}(1) = a_0(i, j) = a_{n-2}(i, j) = 0. \tag{4.501}$$

In the case (ii), Lemma 4.1 asserts that possible sequences of signs of $P'_{i,j}$ are either $+ - +-$ or $+-$. In the latter one, we can repeat the arguments of part (i). Otherwise, there are two local maxima in (0, 1), and both are appropriate candidates for the global one. Notice that the local minimum point v_1 located between them cannot beat the interval endpoints in the contest for the global minimum, because it is impossible that for any parent distribution. Indeed,

$$\mathbb{E}\frac{X_{j:n} - X_{i:n}}{\Delta} = P_{i,j}(v_1) > 0 = P_{i,j}(0) = P_{i,j}(1). \tag{4.502}$$

This completes the proof. □

The upper bounds for the spacings (with $j = i + 1$) and second spacings ($j = i + 2$) can be calculated analytically, because then equations $P'_{i,j}(u) = 0$ can be simplified to linear and quadratic equations, respectively. Note that the solutions to the cases $i = 1$ and $j = n$ were precisely described in Proposition 4.2. Otherwise we write

$$\mathbb{E}\frac{X_{i+1:n} - X_{i:n}}{\Delta} \leq P_{i,i+1}\left(\frac{i-1}{n-2}\right) = \frac{1}{2}\binom{n}{i}\frac{(i-1)^{i-1}(n-i-1)^{n-i-1}}{(n-2)^{n-2}}, \tag{4.503}$$

$$\mathbb{E}\frac{X_{i+2:n} - X_{i:n}}{\Delta} \leq P_{i,i+2}(u_1(i, i+2)), \tag{4.504}$$

where

$$u_1(i, i+2) = \begin{cases} \frac{1}{2}, & \text{if } i = \frac{n-1}{2}, \\ \frac{n-3+2i(i-1)-\sqrt{[(n-3+2i(i-1)]^2-4(i^2-1)(n-2)(2i-n-1)}}{2(n-2)(2i-n-1)}, & \text{otherwise.} \end{cases} \tag{4.505}$$

The conclusions of Proposition 4.3(ii) can be specified more precisely for the quasi-ranges, i.e., the differences of ith largest and smallest order statistics, $2 \le i \le \frac{n}{2}$. Then function $P_{i,n+1-i}$ is symmetric about $\frac{1}{2}$, and $P'_{i,n+1-i}$ is antisymmetric. Hence the latter vanishes at $\frac{1}{2}$. If

$$P''_{i,n+1-i}\left(\frac{1}{2}\right) = \frac{n!}{(n-4)!2^{n-4}} \sum_{k=i}^{n-i} \binom{n}{k} [(2k-n)^2 - n + 2] < 0, \tag{4.506}$$

then $P_{i,n+1-i}$ has a local maximum at $\frac{1}{2}$. Since it is positive on $(0, 1)$, and vanishes at 0 and 1, and it has either one or three local extremes in $(0, 1)$. The latter possibility is excluded, and $P_{i,n+1-i}\left(\frac{1}{2}\right)$ is the global maximum. Summing up, the inequality in (4.506) implies

$$\mathbb{E}\frac{X_{n+1-i:n} - X_{i:n}}{\Delta} \le P_{i,n+1-i}\left(\frac{1}{2}\right) = \frac{1}{2^{n-1}} \sum_{k=i}^{n-i} \binom{n}{k}, \tag{4.507}$$

which can be also written as $\left(2 - \frac{1}{2^{n-2}}\right) \sum_{k=0}^{i-1} \binom{n}{k}$. The equality is attained by the symmetric two-point distributions.

If the inequality in (4.506) is reversed, then $P_{i,n+1-i}$ has a local minimum at $\frac{1}{2}$ and two global maxima attained at two arguments $0 < u_1 < \frac{1}{2} < u_2 = 1 - u_1 < 1$ located symmetrically about $\frac{1}{2}$. Accordingly,

$$\mathbb{E}\frac{X_{n+1-i:n} - X_{i:n}}{\Delta} \le P_{i,n+1-i}(u_1) = P_{i,n+1-i}(u_2), \tag{4.508}$$

where u_1 is the unique solution to

$$\frac{P'_{i,n+1-i}(u)}{n(n-1)(n-2)} = \frac{B_{i-2,n-3}(u) - B_{n-i-1,n-3}(u)}{i(n-i)} + \sum_{k=i-2}^{n-i-2} \frac{(2k+3-n)B_{k,n-3}(u)}{(k+1)(k+2)(n-k-1)(n-k-2)} = 0 \tag{4.509}$$

in $\left(0, \frac{1}{2}\right)$. This bound is attained by the families of two-point distributions where the probability of one point is u_1, and symmetric three-point distributions when the probabilities of extreme points are equal to u_1.

The elements of the sum in (4.506) are negative for k close to $\frac{n}{2}$, and positive for the remote ones. Therefore relations (4.507) and (4.508) are valid for large and small values of i, respectively. Informal arguments provide the first and second central spacings, and the sample range. Note that for $\frac{n-\sqrt{n-2}}{2} \leq i \leq \frac{n}{2}$ we have $(2k-n)^2 < n-2$ for all $i \leq k \leq n-i$, and (4.507) follows. However, the assumption is very restrictive, and (4.507) holds true for much greater range of $2 \leq i \leq \frac{n}{2}$.

4.8.4 L-Estimates of Location and Scale

We finally evaluate two popular L-estimates, one of the location parameter, and the other of the scale. The first one is the symmetric trimmed mean

$$T_{j,n+1-j:n} = \sum_{i=1}^{n} t_i(j) X_{i:n} = \frac{1}{n+2-2j} \sum_{i=j}^{n+1-j} X_{i:n}, \qquad 2 \leq j \leq \frac{n}{2}. \tag{4.510}$$

Its coefficient vector, denoted by $\tilde{t}(j)$ has the coordinates

$$t_i(j) = \begin{cases} \frac{1}{n+2-2j}, & i = j, \dots, n+1-j, \\ 0, & \text{otherwise.} \end{cases} \tag{4.511}$$

In this special case the polynomials (4.452) and (4.468)

$$P_{\tilde{t}(j)}(u) = \sum_{i=0}^{n-2} a_i(\tilde{t}(j)) B_{i,n-2}(u), \tag{4.512}$$

$$P'_{\tilde{t}(j)}(u) = \sum_{i=0}^{n-3} \alpha_i(\tilde{t}(j)) B_{i,n-3}(u), \tag{4.513}$$

have the coefficients

$$a_i(\tilde{t}(j)) = \begin{cases} \frac{n-1}{2(n-i-1)}, & i = 0, \dots, j-2, \\ \frac{(n-1)(j-1)}{2(n+2-2j)} \frac{n-2i-2}{(i+1)(n-i-1)}, & i = j-1, \dots, n-j-1, \\ -\frac{n-1}{2(i+1)}, & i = n-j, \dots, n-2, \end{cases} \tag{4.514}$$

and

$$\alpha_i(\tilde{t}(j)) = (n-2)[a_{i+1}(\tilde{t}(j)) - a_i(\tilde{t}(j))]$$
$$= \begin{cases} \frac{(n-1)(n-2)}{2(n-i-1)(n-i-2)}, & i = 0, \ldots, j-3, \\ \frac{(n-1)(n-2)}{2}\left[\frac{(j-1)(n-2j)}{(n+2-2j)j(n-j)} - \frac{1}{n-j+1}\right], & i = j-2, n-j-1, \\ -\frac{(n-1)(n-2)(j-1)}{4(n+2-2j)}\frac{(2i-n+3)^2-n^2-1}{(i+1)(i+2)(n-i-1)(n-i-2)}, & i = j-1, \ldots, n-j-2, \\ \frac{(n-1)(n-2)}{2(i+1)(i+2)}, & i = n-j, \ldots, n-3, \end{cases} \tag{4.515}$$

respectively (cf. (4.451) and (4.469)).

Proposition 4.4 *If $j = 2$, then*

$$-\frac{1}{2} \le \frac{\mathbb{E}T_{2,n-1:n} - \mu}{\Delta} \le \frac{1}{2}. \tag{4.516}$$

The lower and the upper bounds are attained in the limit by the two-point distributions such that the probability of the smaller point tends to 1 and 0, respectively.

If $j \ge 3$, then (4.513) has exactly two zeros $0 < u_1 = u_1(r) < \frac{1}{2} < v_1 = v_1(r) = 1 - u_1 < 1$, mutually symmetric about $\frac{1}{2}$, and then

$$-P_{\tilde{t}(j)}(u_1) = P_{\tilde{t}(j)}(v_1) \le \frac{\mathbb{E}T_{j,n+1-j:n} - \mu}{\Delta} \le P_{\tilde{t}(j)}(u_1). \tag{4.517}$$

The lower and the upper inequalities become equalities for the two-point distributions such that the probabilities of the greater and smaller points, respectively, amount to u_1.

It follows from the proof that for $j \ge 3$ the upper bounds are greater than $\frac{1}{2}$ and the lower ones are less than $-\frac{1}{2}$.

Proof Due to Theorem 4.36, it suffices to find the global extremes of (4.512) over the interval [0, 1]. To this end, we consider its derivative (4.513), analyzing the signs of the elements in (4.515). For $j = 2$, the first and last rows do not appear. The second one

$$\alpha_0(\tilde{t}(2)) = \frac{(n-1)(n-2)}{2}\left[\frac{n-4}{2(n-2)^2} - \frac{1}{n-1}\right] = -\frac{n^2-3n+4}{4(n-2)} \tag{4.518}$$

is negative, because so is the discriminant of the quadratic numerator in the last fraction. The elements of the third row are negative for $n \ge 5$, and do not appear for $n = 4$. Therefore $\tilde{\alpha}(\tilde{t}(2)) \in \mathbb{R}_-^{n-2}$, (4.513) is negative, and (4.512) is decreasing from $P_{\tilde{t}(j)}(0) = a_0(\tilde{t}(j)) = \frac{1}{2}$ to $P_{\tilde{t}(j)}(1) = a_{n-2}(\tilde{t}(j)) = -\frac{1}{2}$. This assures the validity of the first statement of the proposition.

If $j \geq 3$, then the elements of the first and last rows in (4.515) are positive. For $3 \leq j \leq \frac{n-1}{2}$, those of the third one are negative. Without analysis of the second row we conclude that the elements of (4.515) are first positive, then negative and eventually positive. For $3 \leq j = \frac{n}{2}$, the third line disappears, but the second one is evidently negative. So we can extend the previous conclusion to $3 \leq j \leq \frac{n}{2}$. The VDP implies that (4.512) is either increasing on the whole (0, 1) or first increasing, then decreasing and finally increasing there. The first possibility is excluded by the relations

$$P_{\tilde{t}(j)}(0) = a_0(\tilde{t}(j)) = \frac{1}{2}, \tag{4.519}$$

$$P_{\tilde{t}(j)}(1) = a_{n-2}(\tilde{t}(j)) = -\frac{1}{2}. \tag{4.520}$$

So it increases from $\frac{1}{2}$ at 0 to the global maximum $P_{\tilde{t}(j)}(u_1) > \frac{1}{2}$, then decreases to the global minimum $P_{\tilde{t}(j)}(v_1) < -\frac{1}{2}$, and ultimately increases to $-\frac{1}{2}$ at 1. Symmetry of coefficients $t_i(j) = t_{n+1-i}(j)$, $i = 1, \ldots, n$, implies $P_{\tilde{t}(j)}(u) = -P_{\tilde{t}(j)}(1-u)$, $0 \leq u \leq 1$, and, in consequence, $u_1 = 1 - v_1$ with $P_{\tilde{t}(j)}(u_1) = -P_{\tilde{t}(j)}(v_1)$. This ends the proof. □

Kozyra and Rychlik (2017a) provided analogous evaluations for the symmetric Winsorized means $W_{j,n+1-j:n}$ and for the differences $T_{j,n+1-j:n} - W_{j,n+1-j:n}$.

Now we proceed to mean absolute deviation from the median

$$\mathrm{MAD}_n(\tilde{X}) = \frac{1}{n}\sum_{i=1}^{n} |X_i - \mathrm{MED}_n(\tilde{X})| \tag{4.521}$$

which is a classic L-estimate of the scale. Since the sample median $\mathrm{MED}_n(\tilde{X})$ is defined by different formulae

$$\mathrm{MED}_n(\tilde{X}) = \begin{cases} X_{\frac{n+1}{2}:n}, & \text{if } n \text{ is odd}, \\ \frac{1}{2}\left(X_{\frac{n}{2}:n} + X_{\frac{n}{2}+1:n}\right), & \text{if } n \text{ is even}, \end{cases} \tag{4.522}$$

for odd and even-sized samples, the same concerns the mean absolute deviation from the median

$$\mathrm{MAD}_n(\tilde{X}) = \begin{cases} -\frac{1}{n}\sum_{i=1}^{\frac{n-1}{2}} X_{i:n} + \frac{1}{n}\sum_{i=\frac{n+3}{2}}^{n} X_{i:n}, & \text{if } n \text{ is odd}, \\ -\frac{1}{n}\sum_{i=1}^{\frac{n}{2}} X_{i:n} + \frac{1}{n}\sum_{i=\frac{n}{2}+1}^{n} X_{i:n}, & \text{if } n \text{ is even}. \end{cases} \tag{4.523}$$

We have the following.

Proposition 4.5 *For $n \geq 3$*

$$\frac{1}{2} \leq \frac{\mathbb{E}\mathrm{MAD}_n(\tilde{X})}{\Delta} \leq 1 - B_{\lfloor \frac{n}{2} \rfloor, n-1}\left(\frac{1}{2}\right). \tag{4.524}$$

The lower bound is attained in the limit for the two-point distributions such that the probability mass of one point decreases to zero. The upper one becomes equality for the symmetric two-point distributions.

The upper bound for each odd sample size $n = 2m + 1$ coincides with that for the proceeding even size $n = 2m$. Using the Stirling approximation of the factorial sequence we note that the upper bounds tend to 1 at the rate $\sqrt{\frac{2}{\pi n}}$.

Proof If n is odd, we immediately calculate the specific coefficient (4.451) and (4.469)

$$a_i = \begin{cases} \frac{n-1}{2(n-i-1)}, & i = 0, \ldots, \frac{n-3}{2}, \\ \frac{n-1}{2(i+1)}, & i = \frac{n-1}{2}, \ldots, n-2, \end{cases} \tag{4.525}$$

$$\alpha_i = \begin{cases} \frac{(n-1)(n-2)}{2(n-i-1)(n-i-2)}, & i = 0, \ldots, \frac{n-5}{2}, \\ 0, & i = \frac{n-3}{2}, \\ -\frac{(n-1)(n-2)}{2(i+1)(i+2)}, & i = \frac{n-1}{2}, \ldots, n-3. \end{cases} \tag{4.526}$$

of the polynomials (4.452) and (4.468), respectively. The latter together with the VDP described in Lemma 4.1 imply that in this case $P_{\tilde{c}}(u)$ is first increasing and then decreasing in $(0, 1)$. Since $a_i = a_{n-2-i}$, $i = 0, \ldots, n-2$, it is symmetric as well, and so

$$\min_{0 \leq u \leq 1} P_{\tilde{c}}(u) = P_{\tilde{c}}(0) = a_0 = P_{\tilde{c}}(1) = a_{n-2} = \frac{1}{2}, \tag{4.527}$$

$$\begin{aligned} \max_{0 \leq u \leq 1} P_{\tilde{c}}(u) &= P_{\tilde{c}}\left(\frac{1}{2}\right) \\ &= \sum_{i=0}^{\frac{n-3}{2}} \frac{n-1}{2(n-i-1)} B_{i,n-2}\left(\frac{1}{2}\right) + \sum_{i=\frac{n-1}{2}}^{n-2} \frac{n-1}{2(i+1)} B_{i,n-2}\left(\frac{1}{2}\right) \\ &= \sum_{i=0}^{\frac{n-3}{2}} B_{i,n-1}\left(\frac{1}{2}\right) + \sum_{i=\frac{n+1}{2}}^{n-1} B_{i,n-1}\left(\frac{1}{2}\right) \\ &= 1 - B_{\frac{n-1}{2},n-1}\left(\frac{1}{2}\right). \end{aligned} \tag{4.528}$$

For even n, the formulae (4.525) and (4.526) are slightly modified

$$a_i = \begin{cases} \frac{n-1}{2(n-i-1)}, & i = 0, \ldots, \frac{n}{2} - 1, \\ \frac{n-1}{2(i+1)}, & i = \frac{n}{2}, \ldots, n-2, \end{cases} \tag{4.529}$$

$$\alpha_i = \begin{cases} \frac{(n-1)(n-2)}{2(n-i-1)(n-i-2)}, & i = 0, \ldots, \frac{n}{2} - 2, \\ -\frac{(n-1)(n-2)}{2(i+1)(i+2)}, & i = \frac{n}{2} - 1, \ldots, n-3, \end{cases} \tag{4.530}$$

but the arguments remain the same, and lead to the conclusions

$$\min_{0 \le u \le 1} P_{\tilde{c}}(u) = P_{\tilde{c}}(0) = a_0 = P_{\tilde{c}}(1) = a_{n-2} = \frac{1}{2}, \tag{4.531}$$

$$\begin{aligned} \max_{0 \le u \le 1} P_{\tilde{c}}(u) &= P_{\tilde{c}}\left(\frac{1}{2}\right) \\ &= \sum_{i=0}^{\frac{n}{2}-1} \frac{n-1}{2(n-i-1)} B_{i,n-2}\left(\frac{1}{2}\right) + \sum_{i=\frac{n}{2}}^{n-2} \frac{n-1}{2(i+1)} B_{i,n-2}\left(\frac{1}{2}\right) \\ &= \sum_{i=0}^{\frac{n}{2}-1} B_{i,n-1}\left(\frac{1}{2}\right) + \sum_{i=\frac{n}{2}+1}^{n-1} B_{i,n-1}\left(\frac{1}{2}\right) \\ &= 1 - B_{\frac{n}{2},n-1}\left(\frac{1}{2}\right). \end{aligned} \tag{4.532}$$

This ends the proof. □

The sample Gini mean difference $\mathrm{GMD}_n(\tilde{X}))$ is an unbiased estimator of Δ, and so $\mathbb{E}\frac{\mathrm{GMD}_n(\tilde{X})}{\Delta} = 1$ for all parent F. Relations (4.524) show that the expectations of $\mathrm{MAD}_n(\tilde{X})$ is always smaller than the expectations of $\mathrm{GMD}_n(\tilde{X})$.

4.9 Mean Residual Life Inequalities

We present some sharp evaluations for expectations of arbitrary L-statistics of iid populations described in terms of differences between the population mean and conditional expectation that the parent random variable is greater than its quantile of a particular order. It can also be expressed by means of the mean residual life. The evaluations are based on the Moriguti inequality combined with the Steffensen inequality. This approach was proposed by Gajek and Okolewski (1997) and further developed in Gajek and Okolewski (2000a, 2000b, 2001) and Danielak and Rychlik (2003a). The general results for arbitrary L-statistics were presented in Balakrishnan and Rychlik (2006).

4.9.1 Upper Bounds

Let $X_1, X_2, \ldots, X_n$ be iid random variables with a common cdf F. No restrictions on the parent distribution are made except for finiteness of the mean μ. Let X denote another random variable with the same distribution function. Further, let $\tilde{c} = (c_1, \ldots, c_n) \in \mathbb{R}^n$ be a fixed vector. We know that

$$E\sum_{i=1}^{n} c_i(X_{i:n} - \mu) = \int_0^1 [F^{-1}(x) - \mu] f_{\tilde{c}:n}(x)dx$$
$$\leq \int_0^1 [F^{-1}(x) - \mu] \underline{f}_{\tilde{c}:n}(x)dx \tag{4.533}$$

where $\underline{f}_{\tilde{c}:n} = P^{\nearrow} f_{\tilde{c}:n}$ is the projection of $f_{\tilde{c}:n} = \sum_{i=1}^{n} c_i f_{i:n}$ onto the family of nondecreasing functions in $L^2([0, 1), dx)$. If $\underline{f}_{\tilde{c}:n}(x) = \underline{f}_{\tilde{c}:n}(0)$ on some interval $[0, \alpha]$, then α_* is defined as the greatest $0 < \alpha \leq 1$ such that

$$f_{\tilde{c}:n}(\alpha) = \frac{F_{\tilde{c}:n}(\alpha)}{\alpha} = \inf_{0<x\leq 1} \frac{F_{\tilde{c}:n}(x)}{x}. \tag{4.534}$$

If $\underline{f}_{\tilde{c}:n}(x) > \underline{f}_{\tilde{c}:n}(0)$ for all $x > 0$, we take $\alpha_* = 0$, because

$$f_{\tilde{c}:n}(0) = nc_1 = \inf_{0<x\leq 1} \frac{F_{\tilde{c}:n}(x)}{x}. \tag{4.535}$$

Similarly, we define β_* as the smallest $\beta < 1$ satisfying

$$f_{\tilde{c}:n}(\beta) = \frac{\sum_{i=1}^{n} c_i - F_{\tilde{c}:n}(\beta)}{1-\beta} = \sup_{0\leq x<1} \frac{\sum_{i=1}^{n} c_i - F_{\tilde{c}:n}(x)}{1-x} \tag{4.536}$$

and we put $\beta_* = 1$ if

$$f_{\tilde{c}:n}(1) = nc_n = \sup_{0\leq x<1} \frac{\sum_{i=1}^{n} c_i - F_{\tilde{c}:n}(x)}{1-x}. \tag{4.537}$$

To avoid trivial results, we assume here that $\underline{f}_{\tilde{c}:n}$ is not a constant function which implies that $0 \leq \alpha_* < \beta_* \leq 1$, and

$$\underline{f}_{\tilde{c}:n}(0) = f_{\tilde{c}:n}(\alpha_*) < \int_0^1 \underline{f}_{\tilde{c}:n}(x)dx = \int_0^1 f_{c:n}(x)dx$$
$$= \sum_{i=1}^{n} c_i < f_{\tilde{c}:n}(\beta_*) = \underline{f}_{\tilde{c}:n}(1). \tag{4.538}$$

Set

$$0 < \gamma_* = \frac{f_{\tilde{c}:n}(\beta_*) - \sum_{i=1}^n c_i}{f_{\tilde{c}:n}(\beta_*) - f_{\tilde{c}:n}(\alpha_*)} < 1. \tag{4.539}$$

Since $f_{\tilde{c}:n}(\alpha_*) < f_{\tilde{c}:n}(\beta_*)$, we have

$$\begin{aligned}
\beta_* f_{\tilde{c}:n}(\alpha_*) + (1-\beta_*) f_{\tilde{c}:n}(\beta_*) &= \int_0^{\beta_*} f_{\tilde{c}:n}(\alpha_*)dx + \int_{\beta_*}^1 f_{\tilde{c}:n}(\beta_*)dx \\
< \int_0^1 \underline{f}_{\tilde{c}:n}(x)dx &= \sum_{i=1}^n c_i = \gamma_* f_{\tilde{c}:n}(\alpha_*) + (1-\gamma_*) f_{\tilde{c}:n}(\beta_*) \\
< \int_0^{\alpha_*} f_{\tilde{c}:n}(\alpha_*)dx &+ \int_{\alpha_*}^1 f_{\tilde{c}:n}(\beta_*)dx = \alpha_* f_{\tilde{c}:n}(\alpha_*) + (1-\alpha_*) f_{\tilde{c}:n}(\beta_*),
\end{aligned} \tag{4.540}$$

which implies $\alpha_* < \gamma_* < \beta_*$.

Theorem 4.37 *With the above notation, we have*

$$\frac{\mathbb{E}\sum_{i=1}^n c_i (X_{i:n} - \mu)}{\mathbb{E}(X - \mu | X > F^{-1}(\gamma_*))} \le (1-\gamma_*)[f_{\tilde{c}:n}(\beta_*) - f_{\tilde{c}:n}(\alpha_*)]. \tag{4.541}$$

The above inequality becomes the equality if for some $a < b < c$

$$\begin{aligned}
\mathbb{P}(X = a) &= \alpha_*, \\
\mathbb{P}(X = b) &= \beta_* - \alpha_*, \\
\mathbb{P}(X = c) &= 1 - \beta_*.
\end{aligned} \tag{4.542}$$

Inequality (4.541) evaluates the expectations of L-statistics based on centered variables $X_i - \mu$, $i = 1, \ldots, n$, by means of the expectation of centered single observation under the condition that X exceeds a fixed quantile of the parent distribution. We can also express if by means of the mean residual life at the quantile of order γ_* as follows

$$\begin{aligned}
\mathbb{E}(X - \mu | X > F^{-1}(\gamma_*)) &= \mathbb{E}(X - F^{-1}(\gamma_*) + F^{-1}(\gamma_*) - \mu | X > F^{-1}(\gamma_*)) \\
&= \mu(F^{-1}(\gamma_*)) + F^{-1}(\gamma_*) - \mu,
\end{aligned} \tag{4.543}$$

where $m(t) = \mathbb{E}(X - t | X > t)$ denotes the mean residual life of X at time t. The order of the quantile depends only on the coefficient vector $\tilde{c} = (c_1, \ldots, c_n)$ of the L-statistic and it occurs also in the first factor of the constant in the RHS of (4.541). Another factor is the range of the projection $P^{\nearrow} f_{\tilde{c}:n}$ of polynomial $f_{\tilde{c}:n}$ that defines the functional of the expectation of L-statistic. Loosely speaking, $f_{\tilde{c}:n}(\beta_*) - f_{\tilde{c}:n}(\alpha_*)$ depends mainly on the range of coefficients $c_1, \ldots, c_n$, and $1 - \gamma_*$ is connected with their mutual relations. This means in particular that the RHS of (4.541) is independent of the baseline distribution function F. By (4.533), inequality (4.541) remains still true if

$$f_{\tilde{c}:n}(\alpha_*) = f_{\tilde{c}:n}(\beta_*) = \sum_{i=1}^{n} c_i, \tag{4.544}$$

although γ_* is meaningless then.

Actually, Eqs. (4.542) define an at most three-point distribution. If $\alpha_* = 0$ and $\beta_* = 1$ (which always happens if $\tilde{c}$ is nondecreasing and $P^{\nearrow} f_{\tilde{c}:n} = f_{\tilde{c}:n}$), then bound (4.541) is only attained in the limit by parent distribution functions tending to deterministic ones. If either $\alpha_* = 0$ or $\beta_* = 1$ and $\beta_* - \alpha_* < 1$ then (4.542) defines two-point distributions. Three atoms are only possible if $0 < \alpha_* < \beta_* < 1$. In fact, the equality would be even preserved if we stick together some atoms so that two atoms with probabilities either α_* and $1 - \alpha_*$ or β_* and $1 - \beta_*$ or a single atom with probability one are also possible. If there are other α's and/or β's that attain the extremes in (4.534) and (4.536), respectively, then other atoms may also occur. The necessary condition for an equality in (4.541) is that F takes only the values that provide the extremes in (4.534) and (4.536).

Proof of Theorem 4.37 Using (4.533), we have

$$\mathbb{E}\sum_{i=1}^{n} c_i(X_{i:n} - \mu) \leq \int_0^1 [F^{-1}(x) - \mu][\underline{f}_{\tilde{c}:n}(x) - f_{\tilde{c}:n}(\alpha_*)]\,dx, \tag{4.545}$$

where $F^{-1}(x) - \mu$ is nondecreasing and

$$0 \leq \underline{f}_{\tilde{c}:n}(x) - f_{\tilde{c}:n}(\alpha_*) \leq f_{\tilde{c}:n}(\beta_*) - f_{\tilde{c}:n}(\alpha_*) = A, \tag{4.546}$$

say, which is the maximal value of the difference. Set

$$\lambda = \frac{1}{A}\int_0^1 [\underline{f}_{\tilde{c}:n}(x) - f_{\tilde{c}:n}(\alpha_*)]dx = \frac{\sum_{i=1}^{n} c_i - f_{\tilde{c}:n}(\alpha_*)}{f_{\tilde{c}:n}(\beta_*) - f_{\tilde{c}:n}(\alpha_*)} = 1 - \gamma_*. \tag{4.547}$$

Then, by the Steffensen inequality (2.239) with nondecreasing nonnegative and bounded $g(x) = \underline{f}_{\tilde{c}:n}(x) - f_{\tilde{c}:n}(\alpha_*)$ and nondecreasing $h(x) = F^{-1}(x) - \mu$, we obtain

$$\begin{aligned}
&\int_0^1 [F^{-1}(x) - \mu][\underline{f}_{\tilde{c}:n}(x) - f_{\tilde{c}:n}(\alpha_*)]dx \\
&\leq [f_{\tilde{c}:n}(\beta_*) - f_{\tilde{c}:n}(\alpha_*)]\int_{\gamma_*}^1 [F^{-1}(x) - \mu]dx \\
&= (1 - \gamma_*)[f_{\tilde{c}:n}(\beta_*) - f_{\tilde{c}:n}(\alpha_*)]\mathbb{E}(X - \mu | X > F^{-1}(\gamma_*)).
\end{aligned} \tag{4.548}$$

By (2.248), the equality in (4.548) holds iff $F^{-1}(x) - \mu = F^{-1}(\gamma_*) - \mu$ for all $x < \gamma_*$ such that $\underline{f}_{\tilde{c}:n}(x) - f_{\tilde{c}:n}(\alpha_*) > 0$ and $x > \gamma_*$ such that $\underline{f}_{\tilde{c}:n}(x) - f_{\tilde{c}:n}(\alpha_*) < f_{\tilde{c}:n}(\beta_*) - f_{\tilde{c}:n}(\alpha_*)$. In other words

$$F^{-1}(x) - \mu = \text{const.}, \qquad \alpha_* < x < \beta_*. \tag{4.549}$$

If we also assume that this is constant on $[0, \alpha_*)$ and $(\beta_*, 1)$, we obtain an equality in (4.545). This ends the proof. □

Inequality (4.541) is especially meaningful for the scale L-estimates with antisymmetric coefficients $c_{n+1-i} = -c_i$, $i = 1, \ldots, n$. Then both $f_{\tilde{c}:n}(x)$ and $\underline{f}_{\tilde{c}:n}(x)$ are antisymmetric about $\frac{1}{2}$, $\beta_* = 1 - \alpha_*$, $f_{\tilde{c}:n}(\alpha_*) = -f_{\tilde{c}:n}(\beta_*)$ and $\gamma_* = \frac{1}{2}$. Accordingly, then

$$\frac{\mathbb{E}\sum_{i=1}^{n} c_i X_{i:n}}{\mathbb{E}\left(X - \mu \,\middle|\, X > F^{-1}\left(\frac{1}{2}\right)\right)} \le f_{\tilde{c}:n}(\beta_*). \tag{4.550}$$

In particular, we get

$$\mathbb{E}(X_{n:n} - X_{1:n}) \le n\,\mathbb{E}\left(X - \mu \,\middle|\, X > F^{-1}\left(\frac{1}{2}\right)\right) \tag{4.551}$$

for the sample range,

$$\mathbb{E}\frac{1}{k}\left(\sum_{i=n+1-k}^{n} X_{i:n} - \sum_{i=1}^{k} X_{i:n}\right) \le \frac{n}{k}\,\mathbb{E}\left(X - \mu \,\middle|\, X > F^{-1}\left(\frac{1}{2}\right)\right) \tag{4.552}$$

for the symmetric generalized range,

$$\begin{aligned} \mathbb{E}\mathrm{MAD}_n(\tilde{X}) &= \mathbb{E}\frac{1}{n}\sum_{i=1}^{n} |X_i - \mathrm{MED}_n(\tilde{X})| \\ &\le \mathbb{E}\left(X - \mu \,\middle|\, X > F^{-1}\left(\frac{1}{2}\right)\right) \end{aligned} \tag{4.553}$$

for the median absolute deviation,

$$\mathbb{E}\mathrm{GMD}_n(\tilde{X}) = \mathbb{E}\frac{1}{n(n-1)}\sum_{i,j=1}^{n} |X_i - X_j| \le 2\,\mathbb{E}\left(X - \mu \,\middle|\, X > F^{-1}\left(\frac{1}{2}\right)\right), \tag{4.554}$$

for the Gini mean differences, and

$$\begin{aligned}&\mathbb{E}(X_{n+1-j:n} - X_{j:n})\\&\leq [f_{j:n}(\alpha_*) - f_{j:n}(1-\alpha_*)]\mathbb{E}\left(X - \mu \,\middle|\, X > F^{-1}\left(\frac{1}{2}\right)\right),\end{aligned} \qquad (4.555)$$

$2 \leq j \leq \frac{n}{2}$, for the jth quasi-ranges, where $\alpha_* \in \left(0, \frac{1}{2}\right)$ is the unique solution to

$$\alpha[f_{j:n}(\alpha) - f_{j:n}(1-\alpha)] = F_{j:n}(1-\alpha) - F_{j:n}(\alpha). \qquad (4.556)$$

Bounds (4.551) to (4.554) have analytic representations valid for various sample sizes n, because the L-statistics have nondecreasing coefficients, and then $P^{\nearrow} f_{\bar{c}:n} = f_{\bar{c}:n}$, $\alpha_* = 0$, $\beta_* = 1$, $f_{\bar{c}:n}(0) = nc_1$ and $f_{\bar{c}:n}(1) = nc_n$. Similarly, we easily obtain

$$\mathbb{E}X_{n:n} - \mu \leq \mathbb{E}\left(X - \mu \,\middle|\, X > F^{-1}\left(1 - \frac{1}{n}\right)\right) \qquad (4.557)$$

for the sample maximum and

$$\mathbb{E}\left(\frac{1}{k}\sum_{i=n+1-k}^{n} X_{i:n} - \bar{X}\right) \leq \mathbb{E}\left(X - \mu \,\middle|\, X > F^{-1}\left(1 - \frac{1}{k}\right)\right) \qquad (4.558)$$

for the upper selection differentials. However, these nice bounds can be attained in the limit when the parent distributions tend to deterministic ones. On the other hand, there exist three-point symmetric distributions that provide equalities in (4.555).

For the single non-extreme order statistics $X_{j:n}$, $2 \leq j \leq n-1$, we have $\alpha_* = 0$ and $\beta_* \in (0, \frac{j-1}{n-1})$ which solves the equation

$$(1-\beta) f_{j:n}(\beta) = 1 - F_{j:n}(\beta). \qquad (4.559)$$

Then $\gamma_* = 1 - \frac{1}{f_{j:n}(\beta_*)}$ and, by (4.541),

$$\mathbb{E}X_{j:n} - \mu \leq \mathbb{E}\left(X - \mu | X > F^{-1}\left(1 - \frac{1}{f_{j:n}(\beta_*)}\right)\right). \qquad (4.560)$$

(cf. (4.557)). The equality in (4.560) is attained for two-point distributions which assign probability β_* to the smaller value. The above result in an alternative representation was originally derived in Gajek and Okolewski (2000a) who also presented a table of numerical values of $f_{j:n}(\beta_*)$ for $1 \leq j \leq n = 20$.

A generalization to trimmed means $T_{j,k:n}$, due to Danielak and Rychlik (2003a) consists in replacing $f_{j:n}$ by $f_{j,k:n} = \frac{1}{k+1-j}\sum_{i=j}^{k} f_{i:n}$ in (4.559) to (4.560). For arbitrary differences of order statistics, we have

$$\mathbb{E}(X_{k:n} - X_{j:n}) \leq -f_{j,k:n}(\alpha_*)\mathbb{E}\left(X - \mu | X > F^{-1}\left(\frac{f_{j,k:n}(\beta_*)}{f_{j,k:n}(\beta_*) - f_{j,k:n}(\alpha_*)}\right)\right) \tag{4.561}$$

where $\alpha_* \in (0, 1)$ solves the equation

$$\alpha f_{j,k:n}(\alpha) = \alpha[f_{k:n}(\alpha) - f_{j:n}(\alpha)] = F_{j,k:n}(\alpha) = F_{k:n}(\alpha) - F_{j:n}(\alpha) \tag{4.562}$$

if $j \geq 2$ and $\alpha_* = 0$ when $j = 1$, and $\beta_* \in (\alpha_*, 1)$ is uniquely determined by

$$\begin{aligned}(1-\beta)f_{j,k:n}(\beta) &= (1-\beta)[f_{k:n}(\beta) - f_{j:n}(\beta)] \\ &= -F_{j,k:n}(\beta) = F_{j:n}(\beta) - F_{k:n}(\beta)\end{aligned} \tag{4.563}$$

for $k \leq n-1$ and $\beta_* = 1$ otherwise. The equality in (4.561) is attained when (4.542) holds with α_* and β_* determined from (4.562) and (4.563), respectively.

4.9.2 Lower Bounds

We now establish lower bounds based on the Moriguti and Steffensen inequalities. Let $\overline{f}_{\tilde{c}:n} = P^{\searrow} f_{\tilde{c}:n}$ be the nonincreasing function closest to $f_{\tilde{c}:n}$ in the L^2-norm. This is determined by means of the smallest concave majorant method. Then

$$\overline{f}_{\tilde{c}:n}(0) = \sup_{0<x\leq 1} \frac{F_{\tilde{c}:n}(x)}{x} \tag{4.564}$$

and

$$\overline{f}_{\tilde{c}:n}(1) = \inf_{0\leq x<1} \frac{1 - F_{\tilde{c}:n}(x)}{1-x}. \tag{4.565}$$

Assume that $\overline{f}_{\tilde{c}:n}(0) > \overline{f}_{\tilde{c}:n}(1)$ in order to exclude trivial results. Let α'_* and β'_* denote the largest and smallest arguments for which $\overline{f}_{\tilde{c}:n}(x)$ attains the maximal and minimal values, respectively. We have

$$0 \leq \alpha'_* < \beta'_* \leq 1, \tag{4.566}$$

$$\overline{f}_{\tilde{c}:n}(0) = \overline{f}_{\tilde{c}:n}(\alpha'_*) = f_{\tilde{c}:n}(\alpha'_*), \tag{4.567}$$

$$\overline{f}_{\tilde{c}:n}(1) = \overline{f}_{\tilde{c}:n}(\beta'_*) = f_{\tilde{c}:n}(\beta'_*). \tag{4.568}$$

Moreover,

$$f_{\tilde{c}:n}(\alpha'_*) > \sum_{i=1}^{n} c_i > f_{\tilde{c}:n}(\beta'_*) \tag{4.569}$$

and $\alpha'_* < \gamma'_* < \beta'_*$ for

$$\gamma'_* = \frac{\sum_{i=1}^{n} c_i - f_{\tilde{c}:n}(\beta'_*)}{f_{\tilde{c}:n}(\alpha'_*) - f_{\tilde{c}:n}(\beta'_*)} \tag{4.570}$$

(cf. (4.537) to (4.540)).

Theorem 4.38 *With the above notation,*

$$\frac{\mathbb{E}\sum_{i=1}^{n} c_i(X_{i:n} - \mu)}{\mathbb{E}(\mu - X|X \le F^{-1}(\gamma'_*))} \ge -\gamma'_*[f_{\tilde{c}:n}(\alpha'_*) - f_{\tilde{c}:n}(\beta'_*)], \tag{4.571}$$

and the equality holds if

$$\begin{aligned} P(X = a) &= \alpha'_*, \\ P(X = b) &= \beta'_* - \alpha'_*, \\ P(X = c) &= 1 - \beta'_* \end{aligned} \tag{4.572}$$

for some $a < b < c$.

Note that $\nu(t) = \mathbb{E}(t - X|X \le t)$ is called the mean inactivity time of X at t, and $\mathbb{E}(\mu - X|X \le F^{-1}(\gamma'_*)) = \nu(F^{-1}(\gamma'_*)) + \mu - F^{-1}(\gamma'_*)$.

Proof Using the lower versions of the Moriguti and Steffensen inequalities, we obtain

$$\begin{aligned} \mathbb{E}\sum_{i=1}^{n}(X_{i:n} - \mu) &= \int_0^1 [F^{-1}(x) - \mu] f_{\tilde{c}:n}(x)dx \ge \int_0^1 [F^{-1}(x) - \mu]\overline{f}_{\tilde{c}:n}(x)dx \\ &= \int_0^1 [F^{-1}(x) - \mu][\overline{f}_{\tilde{c}:n}(x) - f_{\tilde{c}:n}(\beta'_*)]dx \\ &\ge [f_{\tilde{c}:n}(\alpha'_*) - f_{\tilde{c}:n}(\beta'_*)] \int_0^{\gamma'_*} [F^{-1}(x) - \mu]dx \\ &= \gamma'_*[f_{\tilde{c}:n}(\alpha'_*) - f_{\tilde{c}:n}(\beta'_*)]\mathbb{E}(X - \mu|X \le F^{-1}(\gamma'_*)) \\ &= -\gamma'_*[f_{\tilde{c}:n}(\alpha'_*) - f_{\tilde{c}:n}(\beta'_*)]\mathbb{E}(\mu - X|X \le F^{-1}(\gamma'_*)). \end{aligned} \tag{4.573}$$

Note that the latter conditional expectation is positive so that we can write (4.571). Due to (2.249), the equality in the second inequality is attained if $F^{-1}(x) - \mu = F^{-1}(\gamma_*') - \mu$ for all $x < \gamma_*'$ such that $\bar{f}_{\tilde{c}:n}(x) < f_{\tilde{c}:n}(\alpha_*')$ and all $x > \gamma_*'$ such that $\bar{f}_{\tilde{c}:n}(x) > f_{\tilde{c}:n}(\beta_*')$. This simply means that $F^{-1}(x)$ is constant on the whole interval (α_*', β_*'). It suffices that $F^{-1}(x)$ is constant on $(0, \alpha_*')$ and $(\beta_*', 1)$ to obtain an equality in the first inequality. This ends the proof. □

We finally point out a relation between the upper and lower bounds (4.541) and (4.571), respectively. We have

$$f_{\tilde{c}:n}(x) = f_{\tilde{c}':n}(1-x) \tag{4.574}$$

and

$$F_{\tilde{c}:n}(x) = \sum_{i=1}^{n} c_i - F_{\tilde{c}':n}(1-x) \tag{4.575}$$

for $\tilde{c}' = (c_n, c_{n-1}, \ldots, c_1)$. Therefore the smallest concave majorant of $F_{\tilde{c}:n}(x)$ is identical up to a constant with the negative of the greatest convex minorant of $F_{\tilde{c}':n}(1-x)$, and in consequence we have

$$\overline{f}_{\tilde{c}:n}(x) = \underline{f}_{\tilde{c}':n}(1-x). \tag{4.576}$$

Therefore,

$$\alpha_*' = \alpha_*'(\tilde{c}) = 1 - \beta_*(\tilde{c}'), \tag{4.577}$$

$$\beta_*' = \beta_*'(\tilde{c}) = 1 - \alpha_*(\tilde{c}'), \tag{4.578}$$

$$\gamma_*' = \gamma_*'(\tilde{c}) = 1 - \gamma_*(\tilde{c}'), \tag{4.579}$$

and

$$\gamma_*'(\tilde{c})[f_{\tilde{c}:n}(\alpha_*'(\tilde{c})) - f_{\tilde{c}:n}(\beta_*'(\tilde{c}))] = [1 - \gamma_*'(\tilde{c}')][f_{\tilde{c}':n}(\beta_*(\tilde{c}')) - f_{\tilde{c}':n}(\alpha_*(\tilde{c}'))]. \tag{4.580}$$

This means that the universal constant in the lower bound (4.571) for an L-statistic $\sum_{i=1}^{n} c_i X_{i:n}$ is just the negative of the best constant in the upper bound (4.541) for $\sum_{i=1}^{n} c_{n+1-i:n} X_{i:n}$. In particular,

$$\mathbb{E}X_{j:n} - \mu \geq -\mathbb{E}\left(\mu - X | X \leq F^{-1}\left(\frac{1}{f_{n+1-j:n}(\beta_*)}\right)\right) \tag{4.581}$$

with β_* defined by (4.559) and, more generally,

$$\mathbb{E}T_{j,k:n} - \mu \geq -\mathbb{E}\left(\mu - X|X \leq F^{-1}\left(\frac{1}{f_{n+1-k,n+1-j:n}(\beta_*)}\right)\right) \quad (4.582)$$

for $\beta_* \in (0, 1)$ solving the equation

$$(1 - \beta_*) f_{n+1-k,n+1-j:n}(\beta) = 1 - F_{n+1-k,n+1-j:n}(\beta) \quad (4.583)$$

and $f_{j,k:n} = \frac{1}{k+1-j} \sum_{i=j}^{k} f_{i:n}$. For symmetric trimmed means yields

$$-\mathbb{E}\left(\mu - X|X \leq F^{-1}\left(\frac{1}{\bar{f}_{j,n+1-j:n}(\beta_*)}\right)\right) \leq \mathbb{E}\frac{1}{n-2j} \sum_{i=j+1}^{r-j} X_{i:n}$$

$$\leq \mathbb{E}\left(X - \mu|X > F^{-1}\left(1 - \frac{1}{f_{j,n+1-j:n}(\beta_*)}\right)\right). \quad (4.584)$$

4.10 Inequalities Based on Orthogonal Expansions

4.10.1 Inequalities for Moments

Suppose the order statistics $X_{r:n}$ $(1 \leq r \leq n)$ are from an absolutely continuous distribution with mean μ and finite variance σ^2. Then, it is convenient to work with standardized variable $Z = \frac{X-\mu}{\sigma}$ and the corresponding standardized order statistics $Z_{r:n}$ $(1 \leq r \leq n)$ as any inequality established for them can be readily transformed to corresponding ones for $X_{r:n}$ through the affine relationship $X_{r:n} = \mu + \sigma Z_{r:n}$ $(1 \leq r \leq n)$. Let $F^{-1}(u)$ denote the quantile function of Z. Then, it is evident that

$$\int_0^1 F^{-1}(u)du = 0, \qquad \int_0^1 [F^{-1}(u)]^2 du = 1 \quad (4.585)$$

and, for $r = 1, \ldots, n$,

$$\mathbb{E}Z_{r:n} = \int_0^1 \{F^{-1}(u)\} b_{r,n-r+1}(u)du, \quad (4.586)$$

where $b_{p,q}(u)$ is the beta density function given by

$$b_{p,q}(u) = \frac{1}{B(p,q)} u^{p-1}(1-u)^{q-1}, \quad 0 < u < 1, \ p, q > 0, \quad (4.587)$$

with $B(p,q) = \frac{\Gamma(p)\Gamma(q)}{\Gamma(p+q)}$ $(p, q > 0)$ denoting the complete beta function.

Then, with $\{\phi_k\}_{k=0}^{\infty}$ being a complete orthonormal system in $L^2((0, 1), dx)$, let

$$\alpha_k = \int_0^1 F^{-1}(u)\phi_k(u)du \quad \text{and} \quad \beta_k = \int_0^1 b_{r,n-r+1}(u)\phi_k(u)du \qquad (4.588)$$

for $k = 0, 1, \ldots,$ be the Fourier coefficients of $F^{-1}(u)$ and $b_{r,n-r+1}(u)$, respectively.

Though the results to be developed here, based on the orthogonal expansion inequalities presented earlier in Subsect. 2.4.3, can be stated in terms of a general complete orthonormal system in $L^2(0, 1)$, it seems natural to state them all in terms of the complete orthonormal system of Legendre polynomials (see Remark 2.1 in this regard) due to the beta form of $b_{r,n-r+1}(u)$ in (4.586). Yet, all the results presented here will hold true for *any* complete orthonormal systems in $L^2((0, 1), dx)$; see Sugiura (1962).

Theorem 4.39 *For* $r = 1, \ldots, n$ *and* $K = 1, 2, \ldots,$

$$\left|\mathbb{E}Z_{r:n} - \sum_{k=1}^{K} \alpha_k\beta_k\right|$$
$$\le \left\{1 - \sum_{k=1}^{K} \alpha_k^2\right\}^{1/2} \left\{\frac{B(2r-1, 2n-2r+1)}{(B(r, n-r+1))^2} - 1 - \sum_{k=1}^{K} \beta_k^2\right\}^{1/2}, \qquad (4.589)$$

where the equality holds iff $F^{-1}(u)$, $b_{r,n-r+1}(u)$, $\phi_1(u), \ldots, \phi_K(u)$ *are linearly dependent.*

Proof In Lemma 2.2, upon choosing the functions $g, h \in L^2((0, 1), dx)$ to be

$$g(u) = F^{-1}(u) \quad \text{and} \quad h(u) = b_{r,n-r+1}(u), \qquad \text{for } u \in (0, 1),$$

we observe the following:

$$\langle g, h\rangle = \int_0^1 g(u)h(u)du = \mathbb{E}Z_{r:n},$$
$$\|g\|^2 = \int_0^1 g^2(u)du = \mathbb{E}Z^2 = 1$$

using (4.586) and (4.585), respectively, and

$$\|h\|^2 = \int_0^1 h^2(u)du = \frac{B(2r-1, 2n-2r+1)}{(B(r, n-r+1))^2}.$$

Note now that

$$\alpha_0 = \int_0^1 F^{-1}(u)\phi_0(u)du = \int_0^1 F^{-1}(u)du = 0$$

due to (4.585), and

$$\beta_0 = \int_0^1 b_{r,n-r+1}(u)\phi_0(u)du = \int_0^1 b_{r,n-r+1}(u)du = 1,$$

and so (4.589) readily follows from the bound in (2.151). □

Proceeding similarly, and assuming that the fourth moment $\mathbb{E}Z^4$ of the standardized variable Z is finite, we can similarly obtain the following inequality for $\mathbb{E}(Z_{r:n}^2)$.

Theorem 4.40 *For $r = 1, \ldots, n$ and $K = 0, 1, 2, \ldots$,*

$$\left|\mathbb{E}Z_{r:n}^2 - \sum_{k=0}^{K} \alpha_k^{(2)} \beta_k\right| \\ \leq \left\{\mathbb{E}Z^4 - \sum_{k=0}^{K} \alpha_k^{(2)^2}\right\}^{1/2} \left\{\frac{B(2r-1, 2n-2r+1)}{(B(r, n-r+1))^2} - \sum_{k=0}^{K} \beta_k^2\right\}^{1/2}, \tag{4.590}$$

where the Fourier coefficients $\alpha_k^{(2)} = \int_0^1 \{F^{-1}(u)\}^2 \phi_k(u)du$ and β_k is as given in (4.588), for $k = 0, 1, \ldots,$ and the equality holds iff $\{F^{-1}(u)\}^2$, $b_{r,n-r+1}(u)$, $\phi_0(u), \phi_1(u), \ldots, \phi_K(u)$ are linearly dependent.

Proof The result follows readily from Lemma 2.2 upon choosing the functions $g, h \in L^2(0, 1)$ to be

$$g(u) = \{F^{-1}(u)\}^2 \quad \text{and} \quad h(u) = b_{r,n-r+1}(u), \qquad \text{for } u \in (0, 1).$$

□

4.10.2 *Inequalities for Moments for Symmetric Distributions*

The inequalities presented in Theorems 4.39 and 4.40 hold true for order statistics from any continuous distribution provided the second and fourth moments exist, respectively. If, further, the distribution is assumed to be symmetric, then these inequalities can be improved as presented below (see Sugiura, 1962).

Theorem 4.41 *For any symmetric continuous distribution having finite variance* σ^2, *with the standardized variable* $Z = \frac{X-\mu}{\sigma}$ *having its quantile function satisfying* $F^{-1}(u) = -F^{-1}(1-u)$, $u \in (0, 1)$, *we have for* $r = 1, \ldots, n$ *and* $K = 0, 1, \ldots$,

$$\left|\mathbb{E}Z_{r:n} - \sum_{k=0}^{K} \alpha_{2k+1}\beta_{2k+1}\right|$$
$$\leq \left\{1 - \sum_{k=0}^{K} \alpha_{2k+1}^2\right\}^{\frac{1}{2}} \left\{\frac{B(2r-1, 2n-2r+1) - B(n, n)}{2(B(r, n-r+1))^2} - \sum_{k=0}^{K} \beta_{2k+1}^2\right\}^{\frac{1}{2}}, \tag{4.591}$$

where the Fourier coefficients α_k *and* β_k *are as in* (4.588) *and the equality holds iff* $F^{-1}(u)$, $b_{r,n-r+1}(u) - b_{n-r+1,r}(u)$, $\phi_1(u), \phi_3(u), \ldots, \phi_{2K+1}(u)$ *are linearly dependent.*

Proof In Lemma 2.2, upon choosing the functions $g, h \in L^2((0, 1), dx)$ to be

$$g(u) = F^{-1}(u) \text{ and } h(u) = \frac{1}{2}\left(b_{r,n-r+1}(u) - b_{n-r+1,r}(u)\right), \text{ for } u \in (0, 1),$$

we readily observe the following:

$$\langle g, h\rangle = \int_0^1 g(u)h(u)du = \mathbb{E}Z_{r:n}$$

due to the fact that $\mathbb{E}Z_{r:n} = -\mathbb{E}Z_{n-r+1:n}$ because of the symmetry of the distribution of Z about 0,

$$\|g\|^2 = \int_0^1 g^2(u)du = 1$$

using (4.585), and

$$\|h\|^2 = \int_0^1 h^2(u)du = \frac{1}{2}\left\{\frac{B(2r-1, 2n-2r+1) - B(n, n)}{(B(r, n-r+1))^2}\right\}.$$

The required result now follows readily from Lemma 2.2 and the fact that the sequence $\{\phi_{2k+1}\}_{k=0}^{\infty}$ forms a complete orthonormal system in the space of all odd functions in $L^2(0, 1)$ (see Theorem 2.28). Alternatively, it can also be seen directly from Lemma 2.2 by observing that

$$\begin{aligned}\alpha_{2k} &= \int_0^1 F^{-1}(u)\phi_{2k}(u)du \\ &= -\int_0^1 F^{-1}(1-u)\phi_{2k}(1-u)du = -\alpha_{2k}\end{aligned}$$

and

$$\beta_{2k} = \int_0^1 \frac{1}{2}\{b_{r,n-r+1}(u) - b_{n-r+1,r}(u)\}\phi_{2k}(u)du = -\beta_{2k},$$

which simply imply that $\alpha_{2k} = \beta_{2k} = 0$ for $k = 0, 1, \ldots$. □

In a similar manner, we can present the following result as an analogue of Theorem 4.40 for the case of a symmetric continuous distribution.

Theorem 4.42 *For any symmetric and continuous distribution having finite fourth moment, with the standardized variable $Z = \frac{X-u}{\sigma}$ having its quantile function satisfying $F^{-1}(u) = -F^{-1}(1-u)$, $u \in (0, 1)$, we have for $r = 1, \ldots, n$ and $K = 0, 1, \ldots,$*

$$\begin{aligned}\left|\mathbb{E}Z_{r:n}^2 - \sum_{k=0}^{K} \alpha_{2k}^{(2)}\beta_{2k}\right| &\leq \left\{\mathbb{E}Z^4 - \sum_{k=0}^{K}\alpha_{2k}^{(2)^2}\right\}^{1/2} \\ &\quad \times \left\{\frac{B(2r-1, 2n-2r+1) + B(n, n)}{2(B(r, n-r+1))^2} - \sum_{k=0}^{K}\beta_{2k}^2\right\}^{1/2},\end{aligned} \tag{4.592}$$

where the Fourier coefficients $\alpha_{2k}^{(2)}$ and β_{2k} are as given in Theorem 4.40, and the equality holds iff the functions $[F^{-1}(u)]^2$, $b_{r,n-r+1}(u) + b_{n-r+1,r}(u)$, and $\phi_0(u), \phi_2(u), \ldots, \phi_{2K}(u)$ are linearly dependent.

Proof The required result follows readily from Lemma 2.2 upon choosing the functions $g, h \in L^2((0, 1), dx)$ to be

$$g(u) = \{F^{-1}(u)\}^2 \text{ and } h(u) = \frac{1}{2}\left(b_{r,n-r+1}(u) + b_{n-r+1,r}(u)\right), \text{ for } u \in (0, 1),$$

and then using the fact that the sequence $\{\phi_{2k}\}_{k=0}^{\infty}$ forms a complete orthonormal system in the space of all even functions in $L^2(0, 1)$ (see Theorem 2.28). □

4.10.3 Inequalities for Moments when Mean Does Not Exist

All the above presented inequalities are for the case when the population distribution has finite second and fourth moments. However, for some distributions like Cauchy, the mean does not exist and so the bounds in Theorems 4.39–4.42 are not applicable. In this case, treating μ as location parameter (not as mean) and σ as scale parameter (not as standard deviation), and then observing in the Cauchy case that, due to the existence of moments of order less than 1, $\mathbb{E}Z_{r:n}^k$ exists if and only if $k < \min\{r, n-r+1\}$; see Sen (1959), Barnett (1966), and Stoops and Barr (1971). This means that $\mathbb{E}Z_{r:n}$ exists only for $r = 2, \ldots, n-1$, but the orthogonal inverse expansions through Lemma 2.2 can be provided only for $r = 3, \ldots, n-2$ since the upper bound requires the existence of the second moment of order statistics.

Based on the above comments, Joshi (1969) developed modified orthogonal inverse expansion inequalities as described below.

Theorem 4.43 *Let $\mathbb{E}Z_{2p+1:2p+2q+1}^2$ be finite for some non-negative integral values of p and q. Then, for $r = 1, \ldots, n$ and $K = 0, 1, \ldots,$*

$$\left|C_1\mathbb{E}Z_{r+p:n+p+q} - \sum_{k=0}^{K}\alpha_k^*\beta_k\right| \le \left\{C_2\mathbb{E}Z_{2p+1:2p+2q+1}^2 - \sum_{k=0}^{K}{\alpha_k^*}^2\right\}^{1/2} \times \left\{\frac{B(2r-1, 2n-2r+1)}{(B(r, n-r+1))^2} - \sum_{k=0}^{K}\beta_k^2\right\}^{1/2}, \tag{4.593}$$

where

$$C_1 = \frac{B(r+p, n-r+1+q)}{B(r, n-r+1)}, \qquad C_2 = B(2p+1, 2q+1),$$

the Fourier coefficients $\alpha_k^* = \int_0^1 F^{-1}(u)u^p(1-u)^q\phi_k(u)du$ and β_k is as in (4.588), and the equality holds iff $F^{-1}(u)u^p(1-u)^q$, $b_{r,n-r+1}(u)$, and $\phi_0(u), \phi_1(u), \ldots, \phi_K(u)$ are linearly dependent.

Proof The theorem follows readily from Lemma 2.2 upon choosing the functions $g, h \in L^2(0, 1)$ to be

$$g(u) = F^{-1}(u)u^p(1-u)^q \quad \text{and} \quad h(u) = b_{r,n-r+1}(u), \qquad \text{for } u \in (0, 1).$$

□

Proceeding similarly and assuming that $\mathbb{E}Z^4_{2p+1:2p+2q+1}$ is finite, we can similarly obtain the following inequality concerning $\mathbb{E}Z^2_{r+p:n+p+q}$.

Theorem 4.44 *Let $\mathbb{E}Z^4_{2p+1:2p+2q+1}$ be finite for some non-negative integral values of p and q. Then, for $r = 1, \ldots, n$ and $K = 0, 1, \ldots,$*

$$\left| C_1 \mathbb{E}Z^2_{r+p:n+p+q} - \sum_{k=0}^{K} \alpha_k^{*(2)} \beta_k \right|$$
$$\leq \left\{ C_2 \mathbb{E}Z^4_{2p+1:2p+2q+1} - \sum_{k=0}^{K} \alpha_k^{*(2)^2} \right\}^{1/2}$$
$$\times \left\{ \frac{B(2r-1, 2n-2r+1)}{(B(r, n-r+1))^2} - \sum_{k=0}^{K} \beta_k^2 \right\}^{1/2}, \tag{4.594}$$

where C_1 and C_2 are as given in Theorem 4.43, the Fourier coefficients $\alpha_k^{*(2)} = \int_0^1 \{F^{-1}(u)\}^2 u^p (1-u)^q \phi_k(u) du$ and β_k is as in (4.588), and the equality holds iff $[F^{-1}(u)]^2 u^p (1-u)^q$, $b_{r,n-r+1}(u)$, $\phi_0(u), \phi_1(u), \ldots, \phi_K(u)$ are linearly dependent.

Proof The theorem follows readily from Lemma 2.2 upon choosing the functions $g, h \in L^2((0, 1), dx)$ to be

$$g(u) = \{F^{-1}(u)\}^2 u^p (1-u)^q \qquad \text{and} \qquad h(u) = b_{r,n-r+1}(u), \text{ for } u \in (0, 1).$$

□

The inequalities presented in Theorems 4.43 and 4.44 hold true for order statistics from any continuous distribution provided the assumed moments exist. If, in addition, the distribution is assumed to be symmetric, then these inequalities can be improved as presented below (see Joshi, 1969).

Theorem 4.45 *For any symmetric continuous distribution with finite $\mathbb{E}Z^2_{2m+1:4m+1}$ for non-negative integral values of m, with the standardized variable $Z = \frac{X-\mu}{\sigma}$ having its quantile function satisfying $F^{-1}(u) = -F^{-1}(1-u)$, $u \in (0, 1)$, we have for $r = 1, \ldots, n$ and $K = 0, 1, \ldots,$*

$$\left| C_3 \mathbb{E}Z_{r+m:n+2m} - \sum_{k=0}^{K} \alpha'_{2k+1} \beta_{2k+1} \right|$$
$$\leq \{C_4 \mathbb{E}Z^2_{2m+1:4m+1} - \sum_{k=0}^{K} \alpha'^2_{2k+1}\}^{1/2}$$

$$\times \left\{ \frac{B(2r-1, 2n-2r+1) - B(n,n)}{2(B(r, n-r+1))^2} - \sum_{k=0}^{K} \beta_{2k+1}^2 \right\}^{1/2}, \tag{4.595}$$

where

$$C_3 = \frac{B(r+m, n-r+1+m)}{B(r, n-r+1)}, \qquad C_4 = B(2m+1, 2m+1),$$

the Fourier coefficients $\alpha'_k = \int_0^1 F^{-1}(u)u^m(1-u)^m\phi_k(u)du$ *and* β_k *is as in* (4.588), *and the equality holds iff* $F^{-1}(u)u^m(1-u)^m$, $b_{r,n-r+1}(u) - b_{n-r+1,r}(u)$, $\phi_1(u), \phi_3(u), \ldots, \phi_{2K+1}(u)$ *are linearly dependent.*

Proof The theorem follows readily from Lemma 2.2 upon choosing the functions $g, h \in L^2((0,1), dx)$ to be

$$g(u) = F^{-1}(u)u^m(1-u)^m$$

and

$$h(u) = \frac{1}{2}\{b_{r,n-r+1}(u) - b_{n-r+1,r}(u)\}, \text{ for } u \in (0,1).$$

□

Theorem 4.46 *For any symmetric continuous distribution with finite* $\mathbb{E}Z^4_{2m+1:4m+1}$ *for non-negative integral values of m, with the standardized variable* $Z = \frac{X-\mu}{\sigma}$ *having its quantile function satisfying* $F^{-1}(u) = -F^{-1}(1-u)$, $u \in (0,1)$, *we have for* $r = 1, \ldots, n$ *and* $K = 0, 1, \ldots$,

$$\left| C_3\mathbb{E}Z^2_{r+m:n+2m} - \sum_{k=0}^{K} \alpha'^{(2)}_{2k}\beta_{2k} \right| \leq \left\{ C_4\mathbb{E}Z^4_{2m+1:4m+1} - \sum_{k=0}^{K} \alpha'^{(2)^2}_{2k} \right\}^{1/2} \times \left\{ \frac{B(2r-1, 2n-2r+1) + B(n,n)}{(B(r, n-r+1))^2} - \sum_{k=0}^{K} \beta_{2k}^2 \right\}^{1/2}, \tag{4.596}$$

where C_3 and C_4 are as given in Theorem 4.45, the Fourier coefficients $\alpha_k'^{(2)} = \int_0^1 \{F^{-1}(u)\}^2 u^m (1-u)^m \phi_k(u) du$ and β_k is as in (4.588), and the equality holds iff $[F^{-1}(u)]^2 u^m (1-u)^m$, $b_{r,n-r+1}(u) + b_{n-r+1,r}(u)$, and $\phi_0(u), \phi_2(u), \ldots, \phi_{2K}(u)$ are linearly dependent.

Proof The theorem follows readily from Lemma 2.2 upon choosing the functions $g, h \in L^2(0,1)$ to be

$$g(u) = \{F^{-1}(u)\}^2 u^m (1-u)^m$$

and

$$h(u) = \frac{1}{2}\{b_{r,n-r+1}(u) + b_{n-r+1,r}(u)\},$$

for $u \in (0,1)$, and then observing the following facts:

$$\langle g, h \rangle = \int_0^1 g(u)h(u)du = C_3 \mathbb{E} Z_{r+m:n+2m}^2,$$

$$\|g\|^2 = \int_0^1 g^2(u)du = C_4 \mathbb{E} Z_{2m+1:4m+1}^4,$$

$$\|h\|^2 = \int_0^1 h^2(u)du = \frac{1}{2}\left\{\frac{B(2r-1, 2n-2r+1) + B(n,r)}{(B(r, n-1+1))^2}\right\},$$

$$\begin{aligned}\alpha_{2k+1}'^{(2)} &= \int_0^1 [F^{-1}(u)]^2 u^m (1-m)^m \phi_{2k+1}(u)du \\ &= -\int_0^1 [F^{-1}(u)]^2 u^m (1-m)^m \phi_{2k+1}(u)du = -\alpha_{2k+1}'^{(2)}.\end{aligned}$$

and

$$\begin{aligned}\beta_{2k+1} &= \int_0^1 \frac{1}{2}\{b_{r,n-r+1}(u) + b_{n-r+1,r}(u)\}\phi_{2k+1}(u)du \\ &= -\beta_{2k+1},\end{aligned}$$

which imply that $\alpha_{2k+1} = \beta_{2k+1} = 0$ for $k = 0, 1, \ldots$. □

4.10.4 Computation of Fourier Coefficients

All the bounds for the moments of order statistics obtained from the inequalities presented in all the above theorems require convenient computation of the Fourier coefficients involved in the inequalities. In this regard, the explicit expression of Legendre polynomials (see Remark 2.1) facilitate simple and explicit expressions for the Fourier coefficients as described below.

First, the k-th order Legendre polynomial can be readily expressed as

$$\begin{aligned}\phi_k(u) &= \frac{\sqrt{2k+1}}{k!}\frac{d^k}{du^k}\{u^k(u-1)^k\}\\ &= \sqrt{2k+1}\sum_{\ell=0}^{k}(-1)^{k-\ell}\binom{k}{\ell}\binom{k+\ell}{\ell}u^\ell \end{aligned}\tag{4.597}$$

for $k = 0, 1, 2, \ldots$. With this expression, we then find the Fourier coefficients α_k and β_k in Theorems 4.39 and 4.41 to be

$$\begin{aligned}\alpha_k &= \int_0^1 F^{-1}(u)\phi_k(u)du\\ &= \sqrt{2k+1}\sum_{\ell=0}^{k}(-1)^{k-\ell}\binom{k}{\ell}\binom{k+\ell}{\ell}\int_0^1 F^{-1}(u)u^\ell du\\ &= \sqrt{2k+1}\sum_{\ell=0}^{k}(-1)^{k-\ell}\binom{k}{\ell}\binom{k+\ell}{\ell}\frac{\mathbb{E}Z_{\ell+1:\ell+1}}{\ell+1}\end{aligned}\tag{4.598}$$

and

$$\begin{aligned}\beta_k &= \int_0^1 b_{r,n-r+1}(u)\phi_k(u)du\\ &= \sqrt{2k+1}\sum_{\ell=0}^{k}(-1)^{k-\ell}\binom{k}{\ell}\binom{k+\ell}{\ell}\frac{1}{B(r,n-r+1)}\\ &\qquad\times\int_0^1 u^{r+\ell-1}(1-u)^{n-r}du\\ &= \sqrt{2k+1}\sum_{\ell=0}^{k}(-1)^{k-\ell}\binom{k}{\ell}\binom{k+\ell}{\ell}\frac{B(r+\ell,n-r+1)}{B(r,n-r+1)}\\ &= \sqrt{2k+1}\sum_{\ell=0}^{k}(-1)^{k-\ell}\binom{k}{\ell}\binom{k+\ell}{\ell}\frac{r^{(\ell)}}{(n+1)^{(\ell)}}\end{aligned}\tag{4.599}$$

for $k = 0, 1, 2, \ldots$, where $r^{(\ell)}$ denotes the upward factorial $r^{(\ell)} = r(r+1)\cdots(r+\ell-1)$ for $\ell = 1, 2, \ldots$, and $r^{(\ell)} = 1$ for $\ell = 0$.

Similarly, the coefficients $\alpha_k^{(2)}$ in Theorems 4.40 and 4.42 are simply given by

$$\begin{aligned}\alpha_k^{(2)} &= \int_0^1 [F^{-1}(u)]^2 \phi_k(u)du \\ &= \sqrt{2k+1}\sum_{\ell=0}^{k}(-1)^{k-\ell}\binom{k}{\ell}\binom{k+\ell}{\ell}\int_0^1 \{F^{-1}(u)\}^2 u^\ell du \\ &= \sqrt{2k+1}\sum_{\ell=0}^{k}(-1)^{k-\ell}\binom{k}{\ell}\binom{k+\ell}{\ell}\frac{\mathbb{E}Z_{\ell+1:\ell+1}^2}{\ell+1}; \qquad (4.600)\end{aligned}$$

the coefficients α_k^* in Theorem 4.43 are given by

$$\begin{aligned}\alpha_k^* &= \int_0^1 F^{-1}(u)u^p(1-u)^q \phi_k(u)du \\ &= \sqrt{2k+1}\sum_{\ell=0}^{k}(-1)^{k-\ell}\binom{k}{\ell}\binom{k+\ell}{\ell}\int_0^1 F^{-1}(u)u^{p+\ell}(1-u)^q du \\ &= \sqrt{2k+1}\sum_{\ell=0}^{k}(-1)^{k-\ell}\binom{k}{\ell}\binom{k+\ell}{\ell}\frac{\mathbb{E}Z_{p+\ell+1:p+q+\ell+1}}{B(p+\ell+1,q+1)} \qquad (4.601)\end{aligned}$$

and the coefficients $\alpha_k^{*(2)}$ in Theorem 4.44 are given by

$$\begin{aligned}\alpha_k^{*(2)} &= \int_0^1 \{F^{-1}(u)\}^2 u^p(1-u)^q \phi_k(u)du \\ &= \sqrt{2k+1}\sum_{\ell=0}^{k}(-1)^{k-\ell}\binom{k}{\ell}\binom{k+\ell}{\ell}\frac{\mathbb{E}Z_{p+\ell+1:p+q+\ell+1}^2}{B(p+\ell+1,q+1)}. \qquad (4.602)\end{aligned}$$

Finally, the coefficients α_k' in Theorem 4.45 are given by

$$\begin{aligned}\alpha_k' &= \int_0^1 F^{-1}(u)u^m(1-u)^m \phi_k(u)du \\ &= \sqrt{2k+1}\sum_{\ell=0}^{k}(-1)^{k-\ell}\binom{k}{\ell}\binom{k+\ell}{\ell}\frac{\mathbb{E}Z_{m+\ell+1:2m+\ell+1}}{B(m+\ell+1,m+1)} \qquad (4.603)\end{aligned}$$

and the coefficients $\alpha_k'^{(2)}$ in Theorem 4.46 are given by

$$\alpha_k'^{(2)} = \sqrt{2k+1}\sum_{\ell=0}^{k}(-1)^{k-\ell}\binom{k}{\ell}\binom{k+\ell}{\ell}\frac{\mathbb{E}Z^2_{m+\ell+1:2m+\ell+1}}{B(m+\ell+1, m+1)}. \quad (4.604)$$

From the expressions of all the Fourier coefficients presented above, the α-coefficients require the moments of order statistics from smaller sample sizes. For example, the coefficients α_k in Theorem 4.39 make use of $\mathbb{E}Z_{1:1} = 0, \mathbb{E}Z_{2:2}, \ldots, \mathbb{E}Z_{K+1:K+1}$ to develop a K-th order bound for $\mathbb{E}Z_{r:n}$ which, therefore, would be of interest only for $n \geq K+2$. Similarly, the coefficients $\alpha_k^{(2)}$ in Theorem 4.40 make use of $\mathbb{E}Z^2_{1:1} = 1, \mathbb{E}Z^2_{2:2}, \ldots, \mathbb{E}Z^2_{K+1:K+1}$ to develop a K-th order bound for $\mathbb{E}Z^2_{r:n}$ which, therefore, would be of interest only for $n \geq K+2$.

4.10.5 Numerical Evaluation of Bounds

The orthogonal expansion bounds presented in Theorems 4.39–4.42 are quite accurate even for small K, with the accuracy increasing over increasing values of K. For the purpose of illustration, we present here some results for the case of standard normal distribution. In this situation, due to the symmetry of the standard normal distribution, we can consider the improved bounds presented in Theorems 4.41 and 4.42. For this, first of all, we need the Fourier coefficients α_{2k+1} and $\alpha_{2k}^{(2)}$, for $k = 0, 1, \ldots$, which have been computed, using the expressions in (4.598) and (4.600), respectively, as presented in Table 4.5.

By using the Fourier coefficients in Table 4.5, the bounds in Theorems 4.41 and 4.42 can be readily computed for $\mathbb{E}Z_{r:n}$ and $\mathbb{E}Z^2_{r:n}$, respectively, for the case of standard normal distribution. For example, when $n = 20$, the bounds from Theorem 4.41 for $\mathbb{E}Z_{r:n}$ are presented in Table 4.6 for $r = 11(1)20$ and $K = 0(1)2$, along with the corresponding exact values taken from Harter and Balakrishnan (1996).

From Table 4.6, it is evident that even a three-term approximation (i.e., with $K = 2$) is quite accurate for $\mathbb{E}Z_{r:n}$. It is also evident that the margin of error decreases

Table 4.5 Values of the Fourier coefficients α_{2k+1} and $\alpha_{2k}^{(2)}$ for some selected values of k for the standard normal distribution

k	α_{2k+1}	$\alpha_{2k}^{(2)}$
0	0.97721	1.00000
1	0.18301	1.23281
2	0.08170	0.52112
3	0.04773	0.30451
4	0.03188	0.20559

Table 4.6 Bounds for $\mathbb{E}Z_{r:n}$ from Theorem 4.41 for $n = 20$, $r = 11(1)20$ and $K = 0(1)2$ for the case of standard normal distribution along with the corresponding exact values

r	Bounds with			Exact value
	$K = 0$	$K = 1$	$K = 2$	
11	0.08 ∓ 0.05	0.054 ∓ 0.021	0.066 ∓ 0.008	0.06200
12	0.24 ∓ 0.14	0.163 ∓ 0.053	0.197 ∓ 0.020	0.18696
13	0.40 ∓ 0.19	0.281 ∓ 0.065	0.325 ∓ 0.019	0.31493
14	0.56 ∓ 0.21	0.413 ∓ 0.056	0.454 ∓ 0.008	0.44833
15	0.73 ∓ 0.20	0.564 ∓ 0.037	0.587 ∓ 0.013	0.59030
16	0.89 ∓ 0.18	0.739 ∓ 0.034	0.734 ∓ 0.022	0.74538
17	1.05 ∓ 0.16	0.945 ∓ 0.057	0.907 ∓ 0.018	0.92098
18	1.21 ∓ 0.15	1.186 ∓ 0.075	1.131 ∓ 0.012	1.13095
19	1.37 ∓ 0.17	$1.468 = 0.067$	1.433 ∓ 0.032	1.40760
20	1.53 ∓ 0.35	$1.796 = 0.082$	1.856 ∓ 0.016	1.86748

Table 4.7 Bounds for $\mathbb{E}Z_{r:20}$ from Theorem 4.43 for $r = 11(1)19$ and $K = 1$ for the case of standard Cauchy distribution along with the corresponding exact values

r	Bounds	Exact value
11	0.0851 ∓ 0.0019	0.0850
12	0.2600 ∓ 0.0050	0.2599
13	0.4506 ∓ 0.0060	0.4506
14	0.6718 ∓ 0.0049	0.6720
15	0.9480 ∓ 0.0033	0.9484
16	1.3259 ∓ 0.0060	1.3268
17	1.9128 ∓ 0.0118	1.9140
18	3.0287 ∓ 0.0162	3.0293
19	6.2705 ∓ 0.0324	6.2648

monotonically with increasing values of K, revealing once again the accuracy of the orthogonal expansion bounds.

In addition to being useful in obtaining bounds as presented in Table 4.6, the Fourier coefficients α_{2k+1} in Table 4.5 are also useful in approximating the sum of squares of normal scores, viz., $\sum_{r=1}^{n} \mathbb{E}Z_{r:n}^2$, as discussed by Saw and Chow (1966) and Balakrishnan (1984).

Next, in the case of distributions for which the population mean does not exist, we can use Theorems 4.43–4.46 to obtain bounds for $\mathbb{E}Z_{r:n}$ and $\mathbb{E}Z_{r:n}^2$. For example, in the case of standard Cauchy distribution for which $\mathbb{E}Z_{n:n}$ and $\mathbb{E}Z_{1:n}$ do not exist, we can use Theorem 4.43 to obtain improved bounds for $\mathbb{E}Z_{r+m:n+2m}$ for $1 \le r \le n$, as the considered distribution is symmetric. With the choice of $n = 18$ and $m = 1$, the two-term bounds (i.e., with $K = 1$) obtained from (4.593) for $\mathbb{E}Z_{r:20}$, for $r = 11(1)19$, are presented in Table 4.7, along with the corresponding exact values taken from Barnett (1966).

From Table 4.7, it is evident that even a two-term approximation (i.e., with $K = 1$) is quite accurate for $\mathbb{E}Z_{r:n}$ for the case of standard Cauchy distribution.

The bounds presented in Theorems 4.43–4.46 can also be used for distributions like normal for which the population mean does exist. However, it should be noted that they can not be used to obtain bounds for the moments of extreme order statistics, unlike the bounds presented in Theorems 4.39–4.42.

4.10.6 Imputation-Type Bounds

The orthogonal expansion bounds presented in Subsects. 4.10.1–4.10.3 are unconditional-type bounds in the sense that they hold for any continuous distribution and only require the knowledge of a few population moments such as $\mathbb{E}X$ and $\mathbb{E}X^2$. However, for some distributions like normal, moments of order statistics have been numerically determined for some selected large sample sizes n; see, for example, the tables provided by Ruben (1954), Harter (1961), Borenius (1966), Yamauti (1972), Tietjen et al. (1977), Parrish (1992), and Harter and Balakrishnan (1996). Then, a natural question that arises is whether the knowledge of these moments or order statistics from some large sample sizes may be utilized to improve the bounds present in Subsects. 4.10.1–4.10.3. Of course, such improved bounds will be conditional imputation-type bounds in the sense that they will hold only for that continuous distribution from which we have used the known moments of order statistics from some large sample sizes. The first development of this form is due to Joshi and Balakrishnan (1983), which is as follows.

Theorem 4.47 *Let m and p be two integers for which $\mathbb{E}Z_{m:m}$ and $\mathbb{E}Z_{p:p}$ are known. Then, for $n = 2, 3, \ldots$ and $K = 0, 1, 2, \ldots$,*

$$\begin{aligned}
&\left|\mathbb{E}Z_{n:n} - c\mathbb{E}Z_{m:m} - d\mathbb{E}Z_{p:p} - \sum_{k=1}^{K} \alpha_k \beta_k\right| \\
&\le \left\{1 - \sum_{k=0}^{K} \alpha_k^2\right\}^{1/2} \left\{\frac{n^2}{2n-1} + c^2\frac{m^2}{2m-1} + d^2\frac{p^2}{2p-1} - 2c\frac{mn}{m+n-1}\right. \\
&\qquad \left. -2d\frac{pn}{p+n-1} + 2cd\frac{mp}{m+p-1} - \sum_{k=0}^{K} \beta_k^2\right\}^{1/2}, \qquad (4.605)
\end{aligned}$$

where α_k and β_k are the Fourier coefficients corresponding to g and h below, the imputational weights c and d are given by

$$c = \frac{R_{n,p}R_{m,p} - R_{n,m}R_{p,p}}{R_{m,p}^2 - R_{m,m}R_{p,p}}, \qquad d = \frac{R_{n,m}R_{m,p} - R_{n,p}R_{m,m}}{R_{m,p}^2 - R_{m,m}R_{p,p}} \qquad (4.606)$$

with

$$R_{i,j} = \sum_{k=0}^{K} \phi_{k,i}\phi_{k,j} - \frac{ij}{i+j-1} \tag{4.607}$$

and $\phi_{k,i} = \int_0^1 iu^{i-1}\phi_k(u)du$; *the equality in* (4.605) *holds iff* $F^{-1}(u)$, $nu^{n-1} - cmu^{m-1} - dpu^{p-1}$, $\phi_0(u), \ldots, \phi_K(u)$ *are linearly dependent.*

Proof In Lemma 2.2, upon choosing the functions $g, h \in L^2((0, 1), dx)$ to be

$$g(u) = F^{-1}(u) \text{ and } h(u) = nu^{n-1} - cmu^{m-1} - dpu^{p-1}, u \in (0, 1),$$

we observe the following:

$$\langle g, h\rangle = \int_0^1 g(u)h(u)du = \mathbb{E}Z_{n:n} - c\,\mathbb{E}Z_{m:m} - d\,\mathbb{E}Z_{p:p},$$
$$\|g\|^2 = \int_0^1 g^2(u)du = \mathbb{E}Z^2 = 1$$

using (4.586), (4.585), respectively, and

$$\|h\|^2 = \int_0^1 h^2(u)du = \frac{n^2}{2n-1} + c^2\frac{m^2}{2m-1} + d^2\frac{p^2}{2p-1} - 2c\frac{mn}{m+n-1}$$
$$-2d\frac{pn}{p+n-1} + 2cd\frac{mp}{m+p-1}.$$

The Fourier coefficients α_k and β_k in (4.605) are, respectively,

$$\alpha_k = \int_0^1 F^{-1}(u)\phi_k(u)du \qquad k = 0, 1, \ldots, K,$$
$$\beta_k = \int_0^1 g(u)\phi_k(u)du$$
$$= \int_0^1 (nu^{n-1} - cmu^{m-1} - dpu^{p-1})\phi_k(u)du$$
$$= \phi_{k,n} - c\phi_{k,m} - d\phi_{k,p}, \qquad k = 0, 1, \ldots, K.$$

Upon substituting all these expressions in the inequality in (2.151), we obtain

$$\left|\mathbb{E}Z_{n:n} - c\,\mathbb{E}Z_{m:m} - d\,\mathbb{E}Z_{p:p} - \sum_{k=1}^{K}\alpha_k\beta_k\right|$$
$$\leq \left\{1-\sum_{k=0}^{K}\alpha_k^2\right\}^{1/2}\left\{\frac{n^2}{2n-1}+c^2\frac{m^2}{2m-1}+d^2\frac{p^2}{2p-1}-2c\frac{mn}{m+n-1}\right.$$
$$\left.-2d\frac{pn}{p+n-1}+2cd\frac{mp}{m+p-1}-\sum_{k=0}^{K}[\phi_{k,n}-c\phi_{k,m}-d\phi_{k,p}]^2\right\}^{1/2}. \tag{4.608}$$

By minimizing the right-hand side of (4.608) with respect to c and d, we get the two equations to be

$$c\frac{m^2}{2m-1}-\frac{mn}{m+n-1}+d\frac{mp}{m+p-1}+\sum_{k=0}^{K}[\phi_{k,n}-c\phi_{k,m}-d\phi_{k,p}]\phi_{k,m}=0,$$
$$d\frac{p^2}{2p-1}-\frac{pn}{p+n-1}+c\frac{mp}{m+p-1}+\sum_{k=0}^{K}[\phi_{k,n}-c\phi_{k,m}-d\phi_{k,p}]\phi_{k,p}=0,$$

or equivalently, the equations

$$cR_{m,m}+dR_{m,p}=R_{n,m},$$
$$cR_{m,p}+dR_{p,p}=R_{n,p},$$

where $R_{i,j}$ is as defined earlier in (4.607). Upon noting that the solution of the above two equations is as given in (4.606), the bound in (4.605) readily follows. □

Of course, the imputation-type bounds presented in Theorem 4.47 can be improved for symmetric continuous distributions with finite variance exactly as done earlier in Theorem 4.41; see Joshi and Balakrishnan (1983).

Theorem 4.48 *For any symmetric continuous distribution having finite variance* σ^2, *with the standardized variable* $Z=\frac{X-\mu}{\sigma}$ *having its quantile function satisfying* $F^{-1}(u)=-F^{-1}(1-u)$ *for all* $u\in(0,1)$, *and with* $\mathbb{E}Z_{m:m}$ *and* $\mathbb{E}Z_{p:p}$ *being* *known, we have for* $n = 2, 3, \ldots$ *and* $K = 0, 1, \ldots,$

$$\left| \mathbb{E}Z_{n:n} - c\,\mathbb{E}Z_{m:m} - d\,\mathbb{E}Z_{p:p} - \sum_{k=0}^{K} \alpha_{2k+1}\beta_{2k+1} \right|$$

$$\leq \left\{1 - \sum_{k=0}^{K} \alpha_{2k+1}^2\right\}^{1/2} \left[\frac{n^2}{2}\left\{\frac{1}{2n-1} - B(n,n)\right\}\right.$$

$$+c^2\frac{m^2}{2}\left\{\frac{1}{2m-1} - B(m,m)\right\} + d^2\frac{p^2}{2}\left\{\frac{1}{2p-1} - B(p,p)\right\}$$

$$-cmn\left\{\frac{1}{m+n-1} - B(m,n)\right\} - dpn\left\{\frac{1}{p+n-1} - B(p,n)\right\}$$

$$\left. +cdmp\left\{\frac{1}{m+p-1} - B(m,p)\right\} - \sum_{k=0}^{K} \beta_{2k+1}^2\right]^{1/2}, \tag{4.609}$$

where α_k *and* β_k *are the Fourier coefficients corresponding to* g *and* h *below, the imputational weights* c *and* d *are given by*

$$c = \frac{S_{n,p}S_{m,p} - S_{n,m}S_{p,p}}{S_{m,p}^2 - S_{m,m}S_{p,p}}, \qquad d = \frac{S_{n,m}S_{m,p} - S_{n,p}S_{m,m}}{S_{m,p}^2 - S_{m,m}S_{p,p}} \tag{4.610}$$

with

$$S_{i,j} = \sum_{k=0}^{K} \phi_{2k+1,i}\phi_{2k+1,j} - \frac{ij}{2}\left\{\frac{1}{i+j-1} - B(i,j)\right\}, \tag{4.611}$$

$\phi_{k,i} = \int_0^1 iu^{i-1}\phi_k(u)du$, *and* $B(i,j)$ *is the complete beta function, the equality in* (4.609) *holds iff* $F^{-1}(u)$, $n\{u^{n-1} - (1-u)^{n-1}\} - cm\{u^{m-1} - (1-u)^{m-1}\}$ $- dp\{u^{p-1} - (1-u)^{p-1}\}$, $\phi_1(u), \phi_3(u), \ldots, \phi_{2K+1}(u)$ *are linearly dependent.*

Proof In Lemma 2.2, upon choosing the functions $g, h \in L^2((0,1), dx)$ to be

$$g(u) = F^{-1}(u)$$

and

$$h(u) = \frac{1}{2}\Big[n\{u^{n-1} - (1-u)^{n-1}\} - cm\{u^{m-1} - (1-u)^{m-1}\}$$

$$-dp\{u^{p-1} - (1-u)^{p-1}\}\Big], \qquad u \in (0,1),$$

we readily observe the following:

$$\langle g, h\rangle = \int_0^1 g(u)h(u)du = \mathbb{E}Z_{n:n} - c\,\mathbb{E}Z_{m:m} - d\,\mathbb{E}Z_{p:p},$$

$$\|g\|^2 = \int_0^1 g^2(u)du = \mathbb{E}Z^2 = 1,$$

due to (4.586) and (4.585), respectively, and

$$\begin{aligned}
\|h\|^2 &= \int_0^1 h^2(u)du \\
&= \frac{1}{2}\Bigg[n^2\left\{\frac{1}{2n-1} - B(n,n)\right\} + c^2m^2\left\{\frac{1}{2m-1} - B(m,m)\right\} \\
&\quad + d^2p^2\left\{\frac{1}{2p-1} - B(p,p)\right\} - 2cmn\left\{\frac{1}{m+n-1} - B(m,n)\right\} \\
&\quad - 2dpn\left\{\frac{1}{p+n-1} - B(p,n)\right\} \\
&\quad + 2cdmp\left\{\frac{1}{m+p-1} - B(m,p)\right\}\Bigg].
\end{aligned}$$

The Fourier coefficients α_k and β_k in (4.609) are, respectively,

$$\begin{aligned}
\alpha_k &= \int_0^1 F^{-1}(u)\phi_k(u)du, \\
\beta_k &= \int_0^1 g(u)\phi_k(u)du \\
&= \frac{1}{2}\int_0^1 [n\{u^{n-1} - (1-u)^{n-1}\} - cm\{u^{m-1} - (1-u)^{m-1}\} \\
&\qquad - dp\{u^{p-1} - (1-u)^{p-1}\}]\phi_k(u)du
\end{aligned}$$

for $k = 0, 1, \ldots$. Using the facts that

$$F^{-1}(u) = -F^{-1}(1-u) \quad \text{and} \quad g(u) = -g(1-u), \qquad u \in (0, 1),$$

we readily find

$$\alpha_{2k} = 0 \qquad \text{and} \qquad \beta_{2k} = 0 \qquad k = 0, 1, \ldots .$$

Moreover, we have

$$\beta_{2k+1} = \int_0^1 g(u)\phi_{2k+1}(u)du = \phi_{2k+1,n} - c\phi_{2k+1,m} - d\phi_{2k+1,p}$$

for $k = 0, 1, \ldots, K$. Upon substituting all these expressions in the inequality in (2.151), we obtain

$$\begin{aligned}
&\left|\mathbb{E}Z_{n:n} - c\,\mathbb{E}Z_{m:m} - d\,\mathbb{E}Z_{p:p} - \sum_{k=0}^{K}\alpha_{2k+1}\beta_{2k+1}\right| \\
&\le \left\{1 - \sum_{k=0}^{K}\alpha_{2k+1}^2\right\}^{1/2}\left[\frac{n^2}{2}\left\{\frac{1}{2n-1} - B(n,n)\right\}\right. \\
&\quad + c^2\frac{m^2}{2}\left\{\frac{1}{2m-1} - B(m,m)\right\} + d^2\frac{p^2}{2}\left\{\frac{1}{2p-1} - B(p,p)\right\} \\
&\quad - cmn\left\{\frac{1}{m+n-1} - B(m,n)\right\} - dpn\left\{\frac{1}{p+n-1} - B(p,n)\right\} \\
&\quad + cdmp\left\{\frac{1}{m+p-1} - B(m,p)\right\} \\
&\quad \left. - \sum_{k=0}^{K}\left\{\phi_{2k+1,n} - c\phi_{2k+1,m} - d\phi_{2k+1,p}\right\}^2\right]^{1/2}. \qquad (4.612)
\end{aligned}$$

By minimizing the right-hand side of (4.612) with respect to c and d, we get the equations

$$cS_{m,m} + dS_{m,p} = S_{n,m},$$
$$cS_{m,p} + dS_{p,p} = S_{n,p},$$

where $S_{i,j}$ is as defined earlier in (4.611). Upon noting that the solution of the above two equations is as given in (4.610), the bound in (4.609) readily follows. □

To demonstrate the accuracy of the imputation-type bounds presented in Theorems 4.47 and 4.48, Joshi and Balakrishnan (1983) used $m = 400$ and $p = 1000$ and the corresponding values of $\mathbb{E}Z_{m,m} = 2.96818$ and $\mathbb{E}Z_{p,p} = 3.24144$ for the standard normal distribution (see Harter, 1961; Harter & Balakrishnan, 1996) to produce the following imputation-type bounds for $\mathbb{E}Z_{n:n}$ for $n = 100(100)1000$ with $K = 13$. These bounds are presented in Table 4.8 along with the corresponding bounds obtained from Theorem 4.41 and the exact values taken from Harter and Balakrishnan (1996).

Table 4.8 Imputation-type bounds for $\mathbb{E}Z_{n:n}$ from Theorem 4.48 the standard normal distribution with $m = 400$, $p = 1000$ and $K = 13$

n	Bound in (4.591)	Bound in (4.609)	Exact value
100	2.50759 ∓ 0.00002	2.50760 ∓ 0.00001	2.50759
200	2.74531 ∓ 0.00146	2.74608 ∓ 0.00032	2.74604
300	2.87328 ∓ 0.00709	2.87777 ∓ 0.00054	2.87778
400	2.95666 ∓ 0.01623	2.96818 ∓ 0	2.96818
500	3.01587 ∓ 0.02733	3.03680 ∓ 0.00100	3.03670
600	3.06021 ∓ 0.03913	3.09195 ∓ 0.00195	3.09170
700	3.09468 ∓ 0.05157	3.13792 ∓ 0.00247	3.13755
800	3.12226 ∓ 0.06378	3.17719 ∓ 0.00235	3.17679
900	3.14483 ∓ 0.07576	3.21134 ∓ 0.00153	3.21105
1000	3.16364 ∓ 0.08742	3.24144 ∓ 0	3.24144

Remark 4.1 From Table 4.8, it is evident that the imputation-type approximations and bounds presented in Theorem 4.48 are considerably more precise than the bounds in Theorem 4.41 for the case of standard normal distribution. This improvement is natural as the former uses the extra information on $\mathbb{E}Z_{m:m}$ and $\mathbb{E}Z_{p:p}$ for imputational purpose!

Remark 4.2 Observe in Table 4.8 that the approximation for $\mathbb{E}Z_{m:m}$ and $\mathbb{E}Z_{p:p}$ are simply the corresponding exact values with the margin for bounds being 0. This is to be expected because these are the values that are made use of in the imputational process!

By proceeding along the lines of Subsects. 4.10.1 and 4.10.3, we can also present imputation-type bounds for $\mathbb{E}Z_{n:n}^2$ analogous to those given in Theorems 4.47 and 4.48 for $\mathbb{E}Z_{n:n}$. For example, for the case of symmetric distribution with $F^{-1}(u) = -F^{-1}(1-u)$ for all $u \in (0, 1)$, we have the following result analogous to Theorem 4.48. As its proof is similar to that of Theorem 4.48, we refrain from presenting it here.

Theorem 4.49 *For any symmetric and continuous distribution having finite fourth moment, with the standardized variable* $Z = \frac{x-\mu}{\sigma}$ *having its quantile function satisfying* $F^{-1}(u) = -F^{-1}(1-u)$ *for all* $u \in (0, 1)$*, and with* $\mathbb{E}Z_{m:m}^2$ *and* $\mathbb{E}_{p:p}^2$ *being known, we have for* $n = 2, 3, \ldots$ *and* $K = 0, 1, \ldots,$

$$\left|\mathbb{E}Z_{n:n}^2 - c\,\mathbb{E}Z_{m:m}^2 - d\,\mathbb{E}Z_{p:p}^2 - \sum_{k=0}^{K} \alpha_{2k}^{(2)} \beta_{2k}\right|$$

$$\leq \left\{\mathbb{E}Z^4 - \sum_{k=0}^{k} \alpha_{2k}^{(2)^2}\right\}^{1/2}$$

$$
\begin{aligned}
&\times\left[\frac{n^2}{2}\left\{\frac{1}{2n-1}+B(n,n)\right\}+c^2\frac{m^2}{2}\left\{\frac{1}{2m-1}+B(m,m)\right\}\right.\\
&+d^2\frac{p^2}{2}\left\{\frac{1}{2p-1}+B(p,p)\right\}-cmn\left\{\frac{1}{m+n-1}+B(m,n)\right\}\\
&-dpn\left\{\frac{1}{p+n-1}+B(p,n)\right\}+cdmp\left\{\frac{1}{m+p-1}+B(m,p)\right\}\\
&\left.-\sum_{k=0}^{K}\beta_{2k}^2\right]^{1/2},
\end{aligned}
\tag{4.613}
$$

where $\alpha_k^{(2)}$ and β_k are the Fourier coefficients corresponding to

$$g(u) = [F^{-1}(u)]^2$$

and

$$
\begin{aligned}
h(u) = n\left\{u^{n-1}+(1-u)^{n-1}\right\} - cm\left\{u^{m-1}+(1-u)^{m-1}\right\}\\
-dp\left\{u^{p-1}+(1-u)^{p-1}\right\},
\end{aligned}
$$

the imputational weights c and d are given by

$$c = \frac{T_{n,p}T_{m,p}-T_{n,m}T_{p,p}}{T_{m,p}^2-T_{m,m}T_{p,p}}, \qquad d = \frac{T_{n,m}T_{m,p}-T_{n,p}T_{m,n}}{T_{m,p}^2-T_{m,m}T_{p,p}} \tag{4.614}$$

with

$$T_{i,j} = \sum_{k=0}^{K}\phi_{2k,i}\phi_{2k,j} - \frac{ij}{2}\left\{\frac{1}{i+j-1}+B(i,j)\right\}, \tag{4.615}$$

$\phi_{k,i} = \int_0^1 iu^{i-1}\phi_k(u)du$, and $B(i,j)$ is the complete beta function; the equality in (4.613) *holds iff $[F^{-1}(u)]^2$, $n\{u^{n-1}+(1-u)^{n-1}\}+cm\{u^{m-1}+(1-u)^{m-1}\}$ $+dp\{u^{p-1}+(1-u)^{p-1}\}$, $\phi_0(u), \phi_2(u), \ldots, \phi_{2K}(u)$ are linearly dependent.*

Furthermore, in the case when the population mean does not exist, we can proceed along the lines of Subsect. 4.10.4 to develop imputation-type bounds for moments of order statistics other than the extremes, as done by Balakrishnan and Joshi (1985). For example, we have the following result for $\mathbb{E}Z_{n+1:r+2}$, with the given values of $\mathbb{E}Z_{m+1:m+2}$ and $\mathbb{E}Z_{p+1:p+2}$, for a symmetric continuous distribution with $\mathbb{E}Z_{3:5}^2$ being finite. We leave out the proof of this result as it is quite similar to that of Theorem 4.45.

Theorem 4.50 *For any symmetric and continuous distribution with finite* $\mathbb{E}Z_{3:5}^2$*, with the standardized variable* $Z = \frac{x-\mu}{\sigma}$ *having its quantile function satisfying* $F^{-1}(u) = -F^{-1}(1-u)$ *for all* $u \in (0, 1)$*, and with* $\mathbb{E}Z_{m+1:m+2}$ *and* $\mathbb{E}Z_{p+1:p+2}$ *being known, we have for* $n = 2, 3, \ldots$ *and* $K = 0, 1, \ldots$*,*

$$\left| \frac{n}{(n+1)(n+2)} \mathbb{E}Z_{n+1:n+2} - c \frac{m}{(m+1)(m+2)} \mathbb{E}Z_{m+1:m+2} \right.$$
$$\left. - d \frac{p}{(p+1)(p+2)} \mathbb{E}Z_{p+1:p+2} - \sum_{k=0}^{K} \alpha'_{2k+1} \beta_{2k+1} \right|$$
$$\leq \left\{ \frac{1}{30} \mathbb{E}Z_{3:5}^2 - \sum_{k=0}^{K} \alpha'^2_{2k+1} \right\}^{1/2}$$
$$\times \left[\frac{n^2}{2} \left\{ \frac{1}{2n-1} - B(n,n) \right\} + c^2 \frac{m^2}{2} \left\{ \frac{1}{2m-1} - B(m,m) \right\} \right.$$
$$+ d^2 \frac{p^2}{2} \left\{ \frac{1}{2p-1} - B(p,p) \right\} - cmn \left\{ \frac{1}{m+n-1} - B(m,n) \right\}$$
$$- dpn \left\{ \frac{1}{p+n-1} - B(p,n) \right\} + cdmp \left\{ \frac{1}{m+p-1} - B(m,p) \right\}$$
$$\left. - \sum_{k=0}^{K} \beta_{2k+1}^2 \right]^{1/2}, \tag{4.616}$$

where α'_k *and* β_k *are the Fourier coefficients corresponding to*

$$g(u) = F^{-1}(u)u(1-u),$$
$$h(u) = n\left\{u^{n-1} - (1-u)^{n-1}\right\} - cm\left\{u^{m-1} - (1-u)^{m-1}\right\}$$
$$-dp\left\{u^{p-1} - (1-u)^{p-1}\right\}, \tag{4.617}$$

and the imputational weights c *and* d *are as given in* (4.610) *with* $S_{i,j}$ *as defined in* (4.611)*; the equality in* (4.616) *holds iff* $F^{-1}(u)u(1-u)$, $n\{u^{n-1} - (1-u)^{n-1}\} - cm\{u^{m-1} - (1-u)^{m-1}\} - dp\{u^{p-1} - (1-u)^{p-1}\}$, $\phi_1(u), \phi_3(u), \ldots, \phi_{2K+1}(u)$ *are linearly dependent.*

It is important to note that the choice of the functions $g(u)$ and $h(u)$ in (4.617) is not the only choice for deriving the imputation-type bounds for $\mathbb{E}Z_{n+1:n+2}$. As pointed out by Balakrishnan and Joshi (1985), another convenient choice for the functions $g(u)$ and $h(u)$ is

$$\begin{aligned} g(u) &= F^{-1}(u), \\ h(u) &= n(n-1)\left\{u^{n-2}(1-u) - u(1-u)^{n-2}\right\} \\ &\quad -cm(m-1)\left\{u^{m-2}(1-u) - u(1-u)^{m-2}\right\} \\ &\quad -dp(p-1)\left\{u^{p-2}(1-u) - u(1-u)^{p-2}\right\}. \end{aligned} \tag{4.618}$$

This choice yields the following result for imputation-type bounds for $\mathbb{E}Z_{n-1:n}$, with the given values of $\mathbb{E}Z_{m-1:m}$ and $\mathbb{E}Z_{p-1:p}$, for a symmetric continuous distribution with finite second moment.

Theorem 4.51 *For any symmetric continuous distribution with finite variance σ^2, with the standardized variable $Z = \frac{X-\mu}{\sigma}$ having its quantile function satisfying $F^{-1}(u) = -F^{-1}(1-u)$ for all $u \in (0, 1)$, and with $\mathbb{E}Z_{m-1:m}$ and $\mathbb{E}Z_{p-1:p}$ being known, we have for $n = 4, 5, \ldots$ and $K = 0, 1, \ldots$,*

$$\begin{aligned} &\left|\mathbb{E}Z_{n-1:n} - c\,\mathbb{E}Z_{m-1:m} - d\,\mathbb{E}Z_{p-1:p} - \sum_{k=0}^{K} \alpha_{2k+1}\beta'_{2k+1}\right| \\ &\leq \left\{1 - \sum_{k=0}^{K} \alpha_{2k+1}^2\right\}^{1/2} \\ &\times \left[n^2(n-1)^2\left\{\frac{1}{(2n-1)(2n-2)(2n-3)} - \frac{1}{2}B(n,n)\right\}\right. \\ &+ c^2m^2(m-1)^2\left\{\frac{1}{(2m-1)(2m-2)(2m-3)} - \frac{1}{2}B(m,m)\right\} \\ &+ d^2p^2(p-1)^2\left\{\frac{1}{(2p-1)(2p-2)(2p-3)} - \frac{1}{2}B(p,p)\right\} \\ &- 2cm(m-1)n(n-1)\left\{\frac{1}{(m+n-1)(m+n-2)(m+n-3)} - \frac{1}{2}B(m,n)\right\} \\ &- 2dp(p-1)n(n-1)\left\{\frac{1}{(p+n-1)(p+n-2)(p+n-3)} - \frac{1}{2}B(p,n)\right\} \\ &+ 2cdm(m-1)p(p-1)\left\{\frac{1}{(m+p-1)(m+p-2)(m+p-3)} - \frac{1}{2}B(m,p)\right\} \\ &\left. - \sum_{k=0}^{K} \beta'^2_{2k+1}\right]^{1/2}, \end{aligned} \tag{4.619}$$

where α_k *and* β_k' *are the Fourier coefficients corresponding to* $g(u)$ *and* $h(u)$ *as given in* (4.618), *and the imputational weights* c *and* d *are given by*

$$c = \frac{U_{n,p}U_{m,p} - U_{n,m}U_{p,p}}{U_{m,p}^2 - U_{m,m}U_{p,p}}, \quad d = \frac{U_{n,m}U_{m,p} - U_{n,p}U_{m,m}}{U_{m,p}^2 - U_{m,m}U_{p,p}}, \tag{4.620}$$

$\phi_{k,i} = \int_0^1 iu^{i-1}\phi_k(u)du$, *and* $B(i, j)$ *is the complete beta function; the equality in* (4.619) *holds iff* $F^{-1}(u)$, $n(n-1)\{u^{n-2}(1-u) - u(1-u)^{n-2}\} - cm(m-1)\{u^{m-2}(1-u) - u(1-u)^{m-2}\} - dp(p-1)\{u^{p-1}(1-u) - u(1-u)^{p-2}\}$, $\phi_1(u), \phi_3(u), \ldots, \phi_{2K+1}(u)$ *are linearly dependent.*

The imputation-type bound for $\mathbb{E}Z_{n+1:n+2}$ in Theorem 4.50 simply reduces to the bound for $\mathbb{E}Z_{n+1:n+2}$ due to Joshi (1969) presented earlier in Theorem 4.43 in the special case when we take $c = d = 0$.

Remark 4.3 It should be noted that the imputation-type bounds for $\mathbb{E}Z_{n-1:n}$ presented in Theorems 4.50 and 4.51 are not directly comparable. The reason is that the former is applicable when the population mean does not exist and uses the finite value of $\mathbb{E}Z_{3:5}^2$, while the latter is applicable when the population mean exists and uses the value of $\mathbb{E}Z^2 = 1$.

While using Theorem 4.50, we need the values of the Fourier coefficients α_k'. With the choice of k-th order Legendre polynomial in (4.597) for $\phi_k(u)$, we then find from (4.617) that

$$\begin{aligned}
\alpha_k' &= \int_0^1 F^{-1}(u)u(1-u)\phi_k(u)du \\
&= \sqrt{2k+1}\sum_{\ell=0}^{k}(-1)^{k-\ell}\binom{k}{\ell}\binom{k+\ell}{\ell}\int_0^1 F^{-1}(u)u^{\ell+1}(1-u)du \\
&= \sqrt{2k+1}\sum_{\ell=0}^{k}(-1)^{k-\ell}\binom{k}{\ell}\binom{k+\ell}{\ell}B(\ell+2, 2)\mathbb{E}Z_{\ell+2:\ell+3} \\
&= \sqrt{2k+1}\sum_{\ell=0}^{k}(-1)^{k-\ell}\binom{k}{\ell}\binom{k+\ell}{\ell}\frac{\mathbb{E}Z_{\ell+2:\ell+3}}{(\ell+2)(\ell+3)}
\end{aligned} \tag{4.621}$$

for $k = 1, 3, \ldots$ (keep in mind that $\alpha_{2k}' = 0$ for $k = 0, 1, \ldots$ for symmetric distributions).

For the case of standard normal distribution, using the expected values of second largest order statistics in small samples (see, for example, Harter & Balakrishnan, 1996), the Fourier coefficients α_{2k+1}' have been computed from its expression in (4.621). These are presented in Table 4.9.

Next, by using the Fourier coefficients α_{2k+1} and α_{2k+1}' presented, for example, in Tables 4.5 and 4.9, respectively, the orthogonal expansion bounds presented in

Table 4.9 Values of the Fourier coefficients α'_{2k+1} for some selected values of k for the standard normal distribution

k	α'_{2k+1}
0	0.08574
1	−0.04678
2	−0.00446
3	−0.00120
4	−0.00047

Table 4.10 Bounds for $\mathbb{E}Z_{n-1:n}$ from (4.591), (4.593), (4.616) and (4.619) for $K = 8$ for the case of standard normal distribution along with the corresponding exact values

n	(4.591)	(4.593)	(4.616)	(4.619)	Exact value
50	1.85508 $\mp$ 0.00037	1.85488 $\mp$0.00000	1.85487 $\mp$ 0.00000	1.85494 $\mp$ 0.00026	1.85487
100	2.15506 $\mp$ 0.00982	2.14867 $\mp$ 0.00061	2.14816 $\mp$ 0.00015	2.14957 $\mp$ 0.00463	2.14814
150	2.32588 $\mp$ 0.02420	2.31000 $\mp$ 0.00380	2.30642 $\mp$ 0.00043	2.30843 $\mp$ 0.00627	2.30638
200	2.44407 $\mp$ 0.0.03521	2.42408 $\mp$ 0.01062	2.41369 $\mp$ 0.00039	2.41485 $\mp$ 0.00352	2.41365
250	2.53174 $\mp$ 0.04244	2.51531 $\mp$ 0.02115	2.49431 $\mp$ 0	2.49431 $\mp$ 0	2.49431
300	2.59948 $\mp$ 0.04730	2.59365 $\mp$ 0.03514	2.55862 $\mp$ 0.00046	2.55795 $\mp$ 0.00209	2.55867
350	2.65338 $\mp$ 0.05109	2.66398 $\mp$ 0.05231	2.61201 $\mp$ 0.00058	2.61136 $\mp$ 0.00206	2.61207
400	2.69723 $\mp$ 0.05468	2.72899 $\mp$ 0.07239	2.65761 $\mp$ 0	2.65761 $\mp$ 0	2.65761

Theorems 4.41 and 4.43 for $\mathbb{E}Z_{n-1:n}$ can be readily computed. Moreover, with the use of $\mathbb{E}Z_{249:250} = 2.49431$ and $\mathbb{E}Z_{399:400} = 2.65761$, the imputation-type bounds presented in Theorems 4.50 and 4.51 can also be computed for $\mathbb{E}Z_{n-1:n}$. All these bounds are presented in Table 4.10 for $n = 50(50)400$, along with the corresponding exact values taken from Harter and Balakrishnan (1996).

For the case of standard normal distribution, since the population mean exists (is in fact 0), all the bounds presented for $\mathbb{E}Z_{n-1:n}$ are valid. However they are not directly comparable as pointed out earlier in Remark 4.3.

From Table 4.10, it is evident that the imputation-type approximations and bounds in (4.616) are considerably more precise than those in (4.593) for the case of standard normal distribution. As mentioned earlier in Remark 4.1, this improvement is natural as the former uses the extra information on $\mathbb{E}Z_{m-1:m}$ and $\mathbb{E}Z_{p-1:p}$ for imputational purpose!

It is evident in Table 4.10 that the approximations in (4.616) and (4.619) for $\mathbb{E}Z_{m-1:m}$ and $\mathbb{E}Z_{p-1:p}$ are simply the corresponding exact values with the margin for bounds being 0. As mentioned earlier in Remark 4.2, this is to be expected because these are the values that are made use of in the imputational process!

It is important to note from Table 4.10 that the orthogonal expansion bounds in (4.591) and (4.593) both become less precise when n increases. This is indeed a general feature of orthogonal expansion bounds and one way to improve the precision will be to increase the choice of K accordingly.

4.11 Variance Bounds

4.11.1 Single Order Statistics

Here we determine the sharp bounds on the variances of order statistics based on iid samples. They are expressed in the scale units equal to the variances of the original random variables. We obviously assume that they are positive and finite. The results are based on Papadatos (1995). We first consider the extreme order statistics for which the evaluations are relatively simple.

Theorem 4.52 *If $X_1, \ldots, X_n$ are nondegenerate iid with a finite second moment, the following bounds are sharp.*

$$0 \leq \frac{\mathbb{V}ar\, X_{n:n}}{\mathbb{V}ar\, X_1} \leq n. \tag{4.622}$$

The lower (upper) one is attained in the limit by the two-point distributions when the probability of the smaller point tends to 0 *(*1*, respectively).*

The bounds identical to (4.622) hold to the sample minimum. The lower (upper) one is attained in the limit by the two-point distributions when the probability of the smaller point tends to 1 *(*0*, respectively).*

Proof We consider the sample maximum case. For this purpose we use the Hoeffding identity

$$\mathbb{C}ov(X, Y) = \iint_{\mathbb{R}^2} [H(x, y) - F(x)G(y)]\, dxdy, \tag{4.623}$$

for X, Y having the joint distribution function $H(x, y)$ and the marginals F and G, respectively (cf. e.g., Lehmann, 1966, Lemma 2). In particular, if $X = Y = X_1$, we get

$$\begin{aligned}
\mathbb{V}ar\, X_1 &= \iint_{\mathbb{R}^2} [F(\min\{x, y\}) - F(x)F(y)]\, dxdy \\
&= 2 \iint_{\{x \leq y\}} F(x)[1 - F(y)]\, dxdy \\
&= 2 \iint_{\{0 < F(x) \leq F(y) < 1\}} F(x)[1 - F(y)]\, dxdy,
\end{aligned} \tag{4.624}$$

with F denoting the distribution function of X_1. Since $X_{n:n}$ has the distribution function $F_{n:n} \circ F(x) = F^n(x)$, we can also write

$$\mathbb{V}ar\, X_{n:n} = 2 \iint_{\{0<F(x)\leq F(y)<1\}} F_{n:n} \circ F(x)[1 - F_{n:n} \circ F(y)]\, dxdy. \tag{4.625}$$

Accordingly,

$$\begin{aligned} \frac{\mathbb{V}ar\, X_{n:n}}{\mathbb{V}ar\, X_1} &= \frac{\iint_{\{0<F(x)\leq F(y)<1\}} \frac{F^n(x)[1-F^n(y)]}{F(x)[1-F(y)]} F(x)[1 - F(y)]\, dxdy}{\iint_{\{0<F(x)\leq F(y)<1\}} F(x)[1 - F(y)]\, dxdy} \\ &\leq \sup_{0<F(x)\leq F(y)<1} \frac{F^n(x)[1 - F^n(y)]}{F(x)[1 - F(y)]} \\ &\leq \sup_{0<\alpha\leq\beta<1} \frac{\alpha^n(1 - \beta^n)}{\alpha(1 - \beta)} = \sup_{0<\alpha\leq\beta<1} \alpha^{n-1} \sum_{i=0}^{n-1} \beta^i = n, \end{aligned} \tag{4.626}$$

which is attained when $\alpha \leq \beta$ tend to 1.

Suppose now that $X_1, \ldots, X_n$ have the common two-point distribution

$$\mathbb{P}_\gamma(X_1 = a) = \gamma = 1 - \mathbb{P}_\gamma(X_1 = b), \qquad 0 < \gamma < 1, \tag{4.627}$$

for some $a < b$. Then $\mathbb{V}ar_\gamma X_1 = (b - a)^2\gamma(1 - \gamma)$. The sample maximum has the same support points with the following probability masses

$$\mathbb{P}_\gamma(X_{n:n} = a) = \mathbb{P}_\gamma(X_i = a,\ i = 1, \ldots, n) = \gamma^n = 1 - \mathbb{P}_\gamma(X_{n:n} = b). \tag{4.628}$$

The corresponding variance amounts to $\mathbb{V}ar_\gamma X_{n:n} = (b-a)^2\gamma^n(1-\gamma^n)$. Therefore

$$\frac{\mathbb{V}ar_\gamma\, X_{n:n}}{\mathbb{V}ar_\gamma\, X_1} = \frac{\gamma^n(1 - \gamma^n)}{\gamma(1 - \gamma)} = \sum_{i=n-1}^{2n-2} \gamma^i, \tag{4.629}$$

and this tends to n as $\gamma \to 1$, which proves sharpness of the upper bound. Observe also that for $\gamma \to 0$

$$\frac{\mathbb{V}ar_\gamma\, X_{n:n}}{\mathbb{V}ar_\gamma\, X_1} = \sum_{i=n-1}^{2n-2} \gamma^i \to 0, \tag{4.630}$$

which is the trivial lower bound that cannot be improved.

The reasoning for the sample minima is similar. □

The remaining order statistics are studied below.

Theorem 4.53 *Under the assumptions of Theorem 4.52 for every* $j = 2, \ldots, n-1$ *there exists a unique* $0 < \gamma_* = \gamma_*(j,n) < 1$ *solving the equation*

$$(1-2\gamma)F_{j:n}(\gamma)[1-F_{j:n}(\gamma)] = \gamma(1-\gamma)f_{j:n}(\gamma)[1-2F_{j:n}(\gamma)] \tag{4.631}$$

and then

$$0 \le \frac{\mathbb{V}ar\, X_{j:n}}{\mathbb{V}ar\, X_1} \le \frac{F_{j:n}(\gamma_*)[1-F_{j:n}(\gamma_*)]}{\gamma_*(1-\gamma_*)}. \tag{4.632}$$

The upper bound is attained by the two-point distributions such that the probability of the smaller point amounts to $\gamma_* = \gamma_*(j,n)$. *The lower one is attained in the limit by any two-point distributions such that the probability of either point tends to* 0.

Proof We start with proving the optimality of the lower bound. The distribution functions of (4.627) are

$$F_\gamma(x) = \begin{cases} 0, \; 0 < x < a, \\ \gamma, \; a \le x < b, \\ 1, \; x \ge b, \end{cases} \qquad 0 < \gamma < 1, \tag{4.633}$$

and the distribution functions of $X_{j:n}$, $j = 2, \ldots, n-1$, from the respective samples have the forms

$$F_{j:n} \circ F_\gamma(x) = \begin{cases} 0, & 0 < x < a, \\ F_{j:n}(\gamma), & a \le x < b, \\ 1, & x \ge b, \end{cases} \qquad 0 < \gamma < 1. \tag{4.634}$$

Therefore

$$\frac{\mathbb{V}ar_\gamma\, X_{j:n}}{\mathbb{V}ar_\gamma\, X_1} = \frac{F_{j:n}(\gamma)[1-F_{j:n}(\gamma)]}{\gamma(1-\gamma)} \tag{4.635}$$

and

$$\lim_{\gamma\to 0} \frac{\mathbb{V}ar_\gamma\, X_{j:n}}{\mathbb{V}ar_\gamma\, X_1} = f_{j:n}(0), \tag{4.636}$$

$$\lim_{\gamma\to 1} \frac{\mathbb{V}ar_\gamma\, X_{j:n}}{\mathbb{V}ar_\gamma\, X_1} = f_{j:n}(1). \tag{4.637}$$

For every $j = 2, \ldots, n-1$ both the limits amount to 0.

For getting the upper bounds we mimic the arguments of the first part of the proof of Theorem 4.52 to obtain

$$\frac{\mathbb{V}ar\, X_{j:n}}{\mathbb{V}ar\, X_1} = \frac{\iint_{\{0<F(x)\le F(y)<1\}} \frac{F_{j:n}\circ F(x)[1-F_{j:n}\circ F(y)]}{F(x)[1-F(y)]} F(x)[1-F(y)]\,dxdy}{\iint_{\{0<F(x)\le F(y)<1\}} F(x)[1-F(y)]\,dxdy}$$
$$\le \sup_{0<\alpha\le\beta<1} \frac{F_{j:n}(\alpha)[1-F_{j:n}(\beta)]}{\alpha(1-\beta)} \tag{4.638}$$

Below we maximize the last fraction and show that it provides the optimal bound. Our first purpose is to prove that the above supremum is attained on the diagonal, i.e., for some $0<\alpha=\beta<1$. For brevity, we use the following notation

$$\underline{G}_{j:n}(\alpha) = \frac{F_{j:n}(\alpha)}{\alpha}, \qquad 0<\alpha\le 1, \tag{4.639}$$

$$\overline{G}_{j:n}(\beta) == \frac{F_{j:n}(\beta)}{1-\beta}, \qquad 0\le\beta<1, \tag{4.640}$$

$$\mathbf{G}_{j:n}(\alpha,\beta) = \underline{G}_{j:n}(\alpha)\,\overline{G}_{j:n}(\beta)$$
$$= \frac{F_{j:n}(\alpha)}{\alpha}\,\frac{1-F_{j:n}(\beta)}{1-\beta}, \qquad 0<\alpha\le\beta<1, \tag{4.641}$$

$$G_{j:n}(\gamma) = \mathbf{G}_{j:n}(\gamma,\gamma) = \frac{F_{j:n}(\gamma)}{\gamma}\,\frac{1-F_{j:n}(\gamma)}{1-\gamma}, \quad 0<\gamma<1. \tag{4.642}$$

We have already proved that $\underline{G}_{j:n}(\alpha)$ is increasing on $(0,\alpha_*)$ and decreasing on $(\alpha_*,1)$, where $\frac{j-1}{n-1}<\alpha_*=\alpha_*(j,n)<1$ is uniquely determined by the equation

$$f_{j:n}(\alpha) = \frac{F_{j:n}(\alpha)}{\alpha} \tag{4.643}$$

and $\overline{G}_{j:n}(\beta)$ is increasing on $(0,\beta_*)$ and decreasing on $(\beta_*,1)$, where $0<\beta_*=\beta_*(j,n)<\frac{j-1}{n-1}$ is determined by

$$f_{j:n}(\beta) = \frac{1-F_{j:n}(\beta)}{1-\beta} \tag{4.644}$$

(cf. formula (4.81) in Theorem 4.8). Therefore for all $0<\alpha<\beta_*$

$$\mathbf{G}_{j:n}(\alpha,\beta) = \underline{G}_{j:n}(\alpha)\,\overline{G}_{j:n}(\beta) \le \underline{G}_{j:n}(\min\{\beta,\beta_*\})\,\overline{G}_{j:n}(\beta), \tag{4.645}$$

and for all $\alpha_*<\beta<1$

$$\mathbf{G}_{j:n}(\alpha,\beta) = \underline{G}_{j:n}(\alpha)\,\overline{G}_{j:n}(\beta) \le \underline{G}_{j:n}(\alpha)\,\overline{G}_{j:n}(\max\{\alpha,\alpha_*\}), \tag{4.646}$$

because $\underline{G}_{j:n}(\alpha)$ is increasing on $(0, \beta_*)$, and $\overline{G}_{j:n}(\beta)$ is decreasing on $(\alpha_*, 1)$. Therefore it remains to consider the $\mathbf{G}_{j:n}$ for the arguments α and β from the triangle $\beta_* \leq \alpha \leq \beta \leq \alpha_*$. However, then we have

$$\mathbf{G}_{j:n}(\alpha, \beta) = \underline{G}_{j:n}(\alpha)\overline{G}_{j:n}(\beta) \leq \underline{G}_{j:n}(\alpha)\overline{G}_{j:n}(\alpha) = \mathbf{G}_{j:n}(\alpha, \alpha), \tag{4.647}$$

and

$$\mathbf{G}_{j:n}(\alpha, \beta) = \underline{G}_{j:n}(\alpha)\overline{G}_{j:n}(\beta) \leq \underline{G}_{j:n}(\beta)\overline{G}_{j:n}(\beta) = \mathbf{G}_{j:n}(\beta, \beta), \tag{4.648}$$

since $\underline{G}_{j:n}(\alpha)$ is increasing on (β_*, α_*), and $\overline{G}_{j:n}(\beta)$ is decreasing there.

Consequently, we can restrict ourselves to maximizing

$$G_{j:n}(\gamma) = \mathbf{G}_{j:n}(\gamma, \gamma) = \underline{G}_{j:n}(\gamma)\overline{G}_{j:n}(\gamma), \qquad 0 < \gamma < 1. \tag{4.649}$$

Moreover, since both $\underline{G}_{j:n}(\gamma)$ and $\overline{G}_{j:n}(\gamma)$ are increasing in $(0, \beta_*)$ and decreasing in $(\alpha_*, 1)$, we can focus on maximizing (4.649) over (β_*, α_*). Elementary calculations allow us to check the identity

$$(1-\gamma)\frac{d}{d\gamma}[nF_{j:n}(\gamma) - \gamma f_{j:n}(\gamma)] = (n-j)f_{j:n}(\gamma), \tag{4.650}$$

which is positive on $(0, 1)$. Since the expression in the square brackets vanishes at 0, we obtain

$$nF_{j:n}(\gamma) - \gamma f_{j:n}(\gamma) > 0, \qquad 0 < \gamma < 1. \tag{4.651}$$

The other useful identity is following

$$\gamma(1-\gamma)f'_{j:n}(\gamma) = [j - 1 - (n-1)\gamma]f_{j:n}(\gamma). \tag{4.652}$$

This immediately implies

$$\frac{d}{d\gamma}[\gamma(1-\gamma)f_{j:n}(\gamma) - (j - n\gamma)F_{j:n}(\gamma)] = nF_{j:n}(\gamma) - \gamma f_{j:n}(\gamma) > 0. \tag{4.653}$$

Again, vanishing of the expression in the brackets at 0, and positivity of the RHS imply that

$$\gamma(1-\gamma)f_{j:n}(\gamma) - (j - n\gamma)F_{j:n}(\gamma) > 0, \qquad 0 < \gamma < 1. \tag{4.654}$$

Applying (4.652) and (4.654) we obtain

$$\begin{aligned}
&(1-\gamma)[\gamma^2 f_{j:n}^2(\gamma) - \gamma f_{j:n}(\gamma)F_{j:n}(\gamma) - \gamma^2 f'_{j:n}(\gamma)F_{j:n}(\gamma)] \\
&= \gamma^2(1-\gamma)f_{j:n}^2(\gamma) - \gamma(1-\gamma)f_{j:n}(\gamma)F_{j:n}(\gamma) \\
&- [j-1-(n-1)\gamma]\gamma f_{j:n}(\gamma)F_{j:n}(\gamma) \\
&= \gamma f_{j:n}(\gamma)\,[\gamma(1-\gamma)f_{j:n}(\gamma) - (j-n\gamma)F_{j:n}(\gamma)] > 0. \qquad (4.655)
\end{aligned}$$

By the relation $F_{j:n}(\gamma) < \gamma f_{j:n}(\gamma)$ for $0 < \gamma < \alpha_*$, we further obtain

$$\begin{aligned}
0 &< \gamma^2 f_{j:n}^2(\gamma) - \gamma f_{j:n}(\gamma)F_{j:n}(\gamma) - \gamma^2 f'_{j:n}(\gamma)F_{j:n}(\gamma) \\
&< \gamma^2 f_{j:n}^2(\gamma) - F_{j:n}^2(\gamma) - \gamma^2 f'_{j:n}(\gamma)F_{j:n}(\gamma), \quad 0 < \gamma < \alpha_*. \qquad (4.656)
\end{aligned}$$

This implies that

$$\frac{d^2}{d\gamma^2}[\ln \underline{G}_{j:n}(\gamma)] = -\frac{\gamma^2 f_{j:n}^2(\gamma) - F_{j:n}^2(\gamma) - \gamma^2 f'_{j:n}(\gamma)F_{j:n}(\gamma)}{\gamma^2 F_{j:n}^2(\gamma)} \qquad (4.657)$$

is negative on $(0, \alpha_*)$. Since $\overline{G}_{j:n}(\gamma) = \underline{G}_{n+1-j:n}(1-\gamma)$ and $\alpha_*(j,n) = 1-\beta_*(n+1-j,n)$, we also have

$$\frac{d^2}{d\gamma^2}[\ln \overline{G}_{j:n}(\gamma)] < 0, \qquad \beta_* < \gamma < 1, \qquad (4.658)$$

and ultimately

$$\frac{d^2}{d\gamma^2}[\ln G_{j:n}(\gamma)] = \frac{d^2}{d\gamma^2}[\ln \underline{G}_{j:n}(\gamma)] + \frac{d^2}{d\gamma^2}[\ln \overline{G}_{j:n}(\gamma)] < 0, \qquad \beta_* < \gamma < \alpha_*. \qquad (4.659)$$

The smooth function $\ln G_{j:n}(\gamma)$, $0 < \gamma < 1$, is strictly increasing on $(0, \beta_*)$, strictly concave on (β_*, α_*), and strictly decreasing on $(\alpha_*, 1)$. Therefore it has a unique maximum located in (β_*, α_*), and the same obviously concerns $G_{j:n}(\gamma)$ itself. The maximum point γ_* is the unique point that solves the equation $G'_{j:n}(\gamma) = 0$, $0 < \gamma < 1$, which is equivalent to (4.631). This means that we can replace the last expression in (4.638) by $\frac{F_{j:n}(\gamma_*)[1-F_{j:n}(\gamma_*)]}{\gamma_*(1-\gamma_*)}$.

Note that the equality in (4.633) holds if the only value of $F(x)$ different from 0 and 1 is γ_*. This implies that the only parent distribution that attains the upper bound in (4.632) the two-point distribution with the probability of the smaller point equal to γ_*. The location of the support points does not matter here. □

Since

$$\begin{aligned}\mathbf{G}_{j:n}(\alpha,\beta) &= \underline{G}_{j:n}(\alpha)\overline{G}_{j:n}(\beta) = \overline{G}_{n+1-j:n}(1-\alpha)\underline{G}_{n+1-j:n}(1-\beta)\\ &= \mathbf{G}_{n+1-j:n}(1-\beta,1-\alpha),\end{aligned} \tag{4.660}$$

the variance bounds for $X_{j:n}$ and $X_{n+1-j:n}$ are identical. We can also see that $\gamma_*(j,n) = 1 - \gamma_*(n+1-j,n)$. This is obvious in view of the fact that under the transformation $Y_i = -X_i$, $i = 1,\ldots,n$, the order statistics $Y_{j:n}$ and $X_{n+1-j:n}$, $j = 1,\ldots,n$, have the same variances. Similar relations can be concluded from Theorem 4.52 are for the extreme order statistics.

Corollary 4.28 *For the odd sized iid samples we have*

$$\frac{\mathbb{V}ar\, X_{\frac{n+1}{2}:n}}{\mathbb{V}ar\, X_1} \le 1, \tag{4.661}$$

and the equality holds for the symmetric two-point distributions.

Proof We have

$$\begin{aligned}G_{\frac{n+1}{2}:n}(\gamma) &= \underline{G}_{\frac{n+1}{2}:n}(\gamma)\overline{G}_{\frac{n+1}{2}:n}(\gamma) = \overline{G}_{\frac{n+1}{2}:n}(1-\gamma)\underline{G}_{\frac{n+1}{2}:n}(1-\gamma)\\ &= G_{\frac{n+1}{2}:n}(1-\gamma).\end{aligned} \tag{4.662}$$

Since it is first increasing and then decreasing, its maximum point is necessarily located at $\frac{1}{2}$. Also, $F_{\frac{n+1}{2}:n}(\gamma) = 1 - F_{\frac{n+1}{2}:n}(1-\gamma)$ implies that $F_{\frac{n+1}{2}:n}\left(\frac{1}{2}\right) = \frac{1}{2}$, and in consequence $F^2_{\frac{n+1}{2}:n}\left(\frac{1}{2}\right)/\left(\frac{1}{2}\right)^2 = 1$. □

The inequality (4.661) was proved by Yang (1982). Lin and Huang (1989) verified its optimality.

4.11.2 Spacings

Now we proceed to the analogous inequalities for the variances of spacings $X_{j+1:n} - X_{j:n}$, $j = 1,\ldots,n-1$. They were presented in Kozyra and Rychlik (2017b). The most general result is following.

Theorem 4.54 *If $X_1,\ldots,X_n$ are iid with a positive and finite variance, then for every $j = 1,\ldots,n-1$ we have the following sharp bound*

$$\frac{\mathbb{V}ar\,(X_{j+1:n} - X_{j:n})}{\mathbb{V}ar\, X_1} \le \max_{0\le\alpha\le1} G_{j,j+1:n}(\alpha), \tag{4.663}$$

where

$$G_{j,j+1:n}(\alpha) = \binom{n}{j}\alpha^{j-1}(1-\alpha)^{n-j-1}\left[1-\binom{n}{j}\alpha^{j}(1-\alpha)^{n-j}\right]. \tag{4.664}$$

Let

$$\mathbb{P}_\alpha(X_1 = a) = \alpha = 1 - \mathbb{P}_\alpha(X_1 = b) \tag{4.665}$$

be arbitrary two-point distribution supported on $a < b$ *with an arbitrary probability* $0 < \alpha < 1$ *of the smaller point. If the maximum in (4.663) is attained at some* $0 < \alpha_* = \alpha_*(j,n) < 1$*, then the respective bound is attained by (4.665) with* $\alpha = \alpha_*$*. If* 0 *(*1*, respectively) is the global maximum point of (4.664), then the bound in (4.663) is attained in the limit by the marginal distributions (4.665) with* $\alpha \to 0$ *(*$\alpha \to 1$*, respectively).*

Proof We apply two identities

$$\mathbb{E}(X_{j+1:n} - X_{j:n}) = \binom{n}{j}\int_{\mathbb{R}} F^j(x)[1-F^{n-j}(x)]\,dx, \tag{4.666}$$

$$\mathbb{E}(X_{j+1:n} - X_{j:n})^2 = \binom{n}{j}\iint_{\mathbb{R}^2} F^j(\min\{x,y\})[1-F(\max\{x,y\})]^{n-j}\,dxdy, \tag{4.667}$$

due to Pearson (1902) and Irwin (1925), respectively. These and many other similar formulas may be also found in Jones and Balakrishnan (2002). Using (4.666) we can write

$$\begin{aligned}
[\mathbb{E}(X_{j+1:n}-X_{j:n})]^2 &= \binom{n}{j}^2\iint_{\mathbb{R}^2} F^i(x)[1-F(x)]^{n-j}F^j(y)[1-F(y)]^{n-j}dxdy\\
&= \binom{n}{j}^2\iint_{\mathbb{R}^2} F^j(\min\{x,y\})[1-F(\min\{x,y\})]^{n-j}\\
&\quad\times F^j(\max\{x,y\})[1-F(\max\{x,y\})]^{n-j}\,dxdy
\end{aligned} \tag{4.668}$$

Combining this with (4.667) we obtain

$$\begin{aligned}
\mathbb{V}ar(X_{j+1:n} - X_{j:n}) = \iint_{\mathbb{R}^2}\Bigg[&\binom{n}{j}F^j(\min\{x,y\})[1-F(\max\{x,y\})]^{n-j}\\
&-\binom{n}{j}^2F^j(\min\{x,y\})[1-F(\min\{x,y\})]^{n-j}\\
&\times F^j(\max\{x,y\})[1-F(\max\{x,y\})]^{n-j}\Bigg]dxdy
\end{aligned}$$

$$
\begin{aligned}
&= \iint_{\mathbb{R}^2} \binom{n}{i} F^{j-1}(\min\{x, y\})[1 - F(\max\{x, y\})]^{n-j-1} \\
&\times \left[1 - \binom{n}{j} F^{j}(\min\{x, y\})[1 - F(\max\{x, y\})]^{n-j}\right] \\
&\times F(\min\{x, y\})[1 - F(\max\{x, y\})] dxdy \\
&\leq \max_{0 \leq \alpha = F(\min\{x,y\}) \leq \beta = F(\max\{x,y\}) \leq 1} \mathbf{G}_{j,j+1:n}(\alpha, \beta) \\
&\times \iint_{\mathbb{R}^2} F(\min\{x, y\})[1 - F(\max\{x, y\})]\, dxdy \\
&\leq \max_{0 \leq \alpha \leq \beta \leq 1} \mathbf{G}_{j,j+1:n}(\alpha, \beta)\, \mathbb{V}ar\, X_1, \qquad (4.669)
\end{aligned}
$$

where

$$
\mathbf{G}_{j,j+1:n}(\alpha, \beta) = \binom{n}{j} \alpha^{j-1}(1-\beta)^{n-j-1}\left[1 - \binom{n}{j}\beta^{j}(1-\alpha)^{n-j}\right],
\quad 0 \leq \alpha \leq \beta \leq 1. \qquad (4.670)
$$

Since the expression in the square brackets as well as the factor preceding it are decreasing with respect to $\alpha \leq \beta \leq 1$, we obtain

$$
\max_{0 \leq \alpha \leq \beta \leq 1} \mathbf{G}_{j,j+1:n}(\alpha, \beta) = \max_{0 \leq \alpha \leq 1} G_{j,j+1:n}(\alpha), \qquad (4.671)
$$

which together with (4.669) complete the proof of the inequality (4.663).

Now we establish the conditions of its attainability. Suppose first that the maximum in (4.663) is attained at some $0 < \alpha_* < 1$. Then we get the equality in (4.669) if either $F(\min\{x, y\})[1 - F(\max\{x, y\})] = 0$ or $F(\min\{x, y\}) = F(\max\{x, y\}) = \alpha_*$ almost surely with respect to the Lebesgue measure on $\mathbb{R}^2$. This is equivalent to saying that 0, α_* and 1 are the only values of the marginal distribution function F. This defines (4.665) with $\alpha = \alpha_*$.

Assume now that (4.664) is maximized either at 0 or at 1. For the iid random variables with the parent distribution (4.665) we have

$$
\mathbb{V}ar_\alpha X_1 = (b-a)^2 \alpha(1-\alpha). \qquad (4.672)
$$

The spacing $X_{j+1:n} - X_{j:n}$ has then the two-point distribution

$$
\begin{aligned}
\mathbb{P}_\alpha(X_{j+1:n} - X_{j:n} = b - a) &= \mathbb{P}_\alpha(X_{1:n} = X_{j:n} = a, X_{j+1:n} = X_{n:n} = b) \\
&= \binom{n}{j} \alpha^{j}(1-\alpha)^{n-j} = B_{j,n}(\alpha), \\
\mathbb{P}_\alpha(X_{j+1:n} - X_{j:n} = 0) &= 1 - B_{j,n}(\alpha), \qquad (4.673)
\end{aligned}
$$

with the variance

$$\mathbb{V}ar_\alpha(X_{j+1:n} - X_{j:n}) = (b-a)^2 B_{j,n}(\alpha)[1 - B_{j,n}(\alpha)] \tag{4.674}$$

Combining (4.672) with (4.674) yields

$$\frac{\mathbb{V}ar_\alpha(X_{j+1:n} - X_{j:n})}{\mathbb{V}ar_\alpha X_1} = G_{j,j+1:n}(\alpha). \tag{4.675}$$

If 0 (1, respectively) maximizes $G_{j,j+1:n}(\alpha)$, $0 \le \alpha \le 1$, letting $\alpha \to 0$ ($\alpha \to 1$, respectively), we reach the bound (4.663) in the limit. □

The lower bounds are treated below.

Theorem 4.55 *If $X_1, \ldots, X_n$ are iid, and $0 < \mathbb{V}ar\, X_1 < \infty$, then for all $1 \le j \le n-1 \ge 2$ the following inequality*

$$\frac{\mathbb{V}ar\,(X_{j+1:n} - X_{j:n})}{\mathbb{V}ar\, X_1} \ge 0 \tag{4.676}$$

cannot be improved. If $j \ge 2$ ($j \le n-2$, respectively), it is attained in the limit by the parent distribution functions (4.665) with $\alpha \to 0$ ($\alpha \to 1$, respectively).

Proof Similarly to (4.669) we can get

$$\frac{\mathbb{V}ar\,(X_{j+1:n} - X_{j:n})}{\mathbb{V}ar\, X_1} \ge \min_{0 \le \alpha \le \beta \le 1} \mathbf{G}_{j,j+1:n}(\alpha, \beta) \ge 0. \tag{4.677}$$

It suffices to notice that

$$\begin{aligned} \lim_{\alpha,\beta \to 0} \mathbf{G}_{j,j+1:n}(\alpha, \beta) &= \lim_{\alpha \to 0} G_{j,j+1:n}(\alpha) = 0, \qquad j \ge 2, \\ \lim_{\alpha,\beta \to 1} \mathbf{G}_{j,j+1:n}(\alpha, \beta) &= \lim_{\alpha \to 1} G_{j,j+1:n}(\alpha) = 0, \qquad j \le n-2, \end{aligned} \tag{4.678}$$

(comp. (4.675)). □

Theorem 4.55 asserts that the only spacing with a possibly non-zero lower bound on the corresponding variance is $X_{2:2} - X_{1:2}$. This is treated in the proposition below.

Proposition 4.6 *Under the assumption of Theorem 4.54 with $n = 2$*

$$\frac{2}{3} \le \frac{\mathbb{V}ar(X_{2:2} - X_{1:2})}{\mathbb{V}ar\, X_1} \le 2. \tag{4.679}$$

The lower bound is attained by the uniform distributions, and the upper one is attained by (4.665) with either $\alpha \to 0$ or $\alpha \to 1$.

Proof The proof of the upper bound immediately follows from Theorem 4.54. In this case

$$G_{1,2:2}(\alpha) = 2[1 - 2\alpha(1-\alpha)] \le 2 = G_{1,2:2}(0) = G_{1,2:2}(1). \tag{4.680}$$

For getting the lower one, we first notice that

$$\mathbb{E}(X_{2:2} - X_{1:2})^2 = \mathbb{E}[(X_1 - \mu) - (X_2 - \mu)]^2 = 2\,\mathbb{V}ar\, X_1 = 2\sigma^2, \tag{4.681}$$

say, and so

$$\frac{\mathbb{V}ar(X_{2:2} - X_{1:2})}{\mathbb{V}ar\, X_1} = 2 - \frac{[\mathbb{E}(X_{2:2} - X_{1:2})]^2}{\sigma^2} \tag{4.682}$$

is minimized if $\frac{\mathbb{E}(X_{2:2}-X_{1:2})}{\sigma}$ is maximized. The maximum of the latter in concluded from Theorem 4.2 providing in the special case $n = 2$

$$\frac{\mathbb{E}(X_{2:2} - X_{1:2})}{\sigma} \le \frac{2\sqrt{3}}{3}, \tag{4.683}$$

which combined with (4.683) gives the LHS inequality in (4.679). The condition for getting the equality in (4.683), and in the LHS of (4.679) in consequence is

$$\frac{F^{-1}(x) - \mu}{\sigma} = \sqrt{3}(2x - 1), \qquad 0 < x < 1. \tag{4.684}$$

This ends the proof. □

Here we describe the precise upper bounds on variances of the extreme spacings when $n \ge 3$. Remind that all the lower bounds are equal to 0 then.

Corollary 4.29 *Under the assumptions of Theorem 4.54 for $n \ge 3$ we have*

$$\frac{\mathbb{V}ar(X_{2:n} - X_{1:n})}{\mathbb{V}ar\, X_1} \le n, \tag{4.685}$$

$$\frac{\mathbb{V}ar(X_{n:n} - X_{n-1:n})}{\mathbb{V}ar\, X_1} \le n. \tag{4.686}$$

The equalities in the former and latter inequalities are attained in the limit by the parent distributions (4.665) as $\alpha \to 0$ and $\alpha \to 1$, respectively.

Proof By Theorem 4.54, it suffices to verify that

$$G_{1,2:n}(\alpha) = n(1-\alpha)^{n-2}[1 - n\alpha(1-\alpha)^{n-1}], \qquad 0 \le \alpha \le 1, \tag{4.687}$$

and

$$G_{n-1,n:n}(\alpha) = n\alpha^{n-2}[1 - n\alpha^{n-1}(1-\alpha)], \qquad 0 \le \alpha \le 1, \tag{4.688}$$

are maximized at 0 and 1, respectively. We analyze the first function It has the derivative

$$G'_{1,2:n}(\alpha) = n(1-\alpha)^{n-3} g_{1,2:n}(\alpha), \tag{4.689}$$

where

$$g_{1,2:n}(\alpha) = n(1-\alpha)^{n-1}[2(n-1)\alpha - 1] - n + 2. \tag{4.690}$$

Observe that $g_{1,2:n}(0) = -2(n-1) < 0$ and $g_{1,2:n}(1) = -(n-2) < 0$. Moreover,

$$g'_{1,2:n}(\alpha) = n(n-1)(1-\alpha)^{n-2}(3 - 2n\alpha), \tag{4.691}$$

which implies that $g_{1,2:n}(\alpha)$ is increasing on $\left[0, \frac{3}{2n}\right]$ and decreasing on $\left[\frac{3}{2n}, 1\right]$. We show that its maximal value satisfies

$$g_{1,2:n}\left(\frac{3}{2n}\right) = 2n\left(1 - \frac{3}{2n}\right)^n - (n-2) < 0, \qquad n \ge 3. \tag{4.692}$$

This inequality is equivalent to

$$\frac{2n-3}{n-2} < \left(\frac{2n}{2n-3}\right)^{n-1}, \qquad n \ge 3. \tag{4.693}$$

By the Bernoulli inequality (see Theorem 2.46)

$$\left(\frac{2n}{2n-3}\right)^{n-1} > 1 + \frac{3(n-1)}{2n-3} = \frac{5n-6}{2n-3}, \qquad n \ge 2, \tag{4.694}$$

(see (2.291)). It remains to notice that $\frac{5n-6}{2n-3} \ge \frac{2n-3}{n-2}$, which is equivalent to $(n-1)(n-3) \ge 0$, $n \ge 3$, and verifies our claim (4.692). We proved that $g_{1,2:n}(\alpha) < 0$, and so $G'_{1,2:n}(\alpha) < 0$ for all $0 < \alpha < 1$. Therefore $G_{1,2:n}(\alpha)$ is maximized at $\alpha = 0$, and the respective value is $G_{1,2:n}(0) = n$.

For getting the maximum in (4.688) it suffices to notice that $G_{n-1,n:n}(\alpha) = G_{1,2:n}(1-\alpha)$. □

This corollary provides the upper bounds for all spacings of the samples of the size $n = 3$. Together with the following one they deliver all the bounds for the spacings in the case $n = 4$ as well.

Corollary 4.30 *Under the above assumptions*

$$\frac{\mathbb{V}ar(X_{3:4} - X_{2:4})}{\mathbb{V}ar\, X_1} \leq \frac{2\sqrt{2}}{3}, \tag{4.695}$$

and the equality holds for the marginal two-point distributions if the probabilities of the support point amount to $\frac{1}{2} - \frac{\sqrt{3}-\sqrt{6}}{6}$ *and* $\frac{1}{2} + \frac{\sqrt{3}-\sqrt{6}}{6}$.

Proof In the case $n = 4$, $j = 2$ we have

$$G_{2,3:n}(\alpha) = 6\alpha(1-\alpha)[1 - 6\alpha^2(1-\alpha)^2]. \tag{4.696}$$

This has the derivative

$$\begin{aligned} G'_{2,3:4}(\alpha) &= 6(2\alpha - 1)[18\alpha^2(1-\alpha)^2 - 1] \\ &= 6(2\alpha - 1)[3\sqrt{2}\alpha(1-\alpha) - 1][3\sqrt{2}\alpha(1-\alpha) + 1] \\ &= 108\left(\alpha - \frac{1}{2} + \frac{\sqrt{6}+\sqrt{3}}{6}\right)\left(\alpha - \frac{1}{2} + \frac{\sqrt{6}-\sqrt{3}}{6}\right)\left(\alpha - \frac{1}{2}\right) \\ &\times \left(\alpha - \frac{1}{2} - \frac{\sqrt{6}-\sqrt{3}}{6}\right)\left(\alpha - \frac{1}{2} - \frac{\sqrt{6}+\sqrt{3}}{6}\right). \end{aligned} \tag{4.697}$$

Therefore (4.696) extended to $\mathbb{R}$ has three local minima at $\frac{1}{2} - \frac{\sqrt{3}+\sqrt{6}}{6} < 0$, $\frac{1}{2}$ and $\frac{1}{2} + \frac{\sqrt{3}+\sqrt{6}}{6} > 1$, and two local maxima belonging to $[0, 1]$ at $\frac{1}{2} - \frac{\sqrt{3}-\sqrt{6}}{6}$ and $\frac{1}{2} + \frac{\sqrt{3}-\sqrt{6}}{6}$. The restriction to the interval $[0, 1]$ is increasing on $\left[0, \frac{1}{2} - \frac{\sqrt{3}-\sqrt{6}}{6}\right]$, decreasing on $\left[\frac{1}{2} - \frac{\sqrt{3}-\sqrt{6}}{6}, \frac{1}{2}\right]$, increasing on $\left[\frac{1}{2}, \frac{1}{2} + \frac{\sqrt{3}-\sqrt{6}}{6}\right]$, and finally decreasing on $\left[\frac{1}{2} + \frac{\sqrt{3}-\sqrt{6}}{6}, 1\right]$. Since (4.696) is symmetric about $\frac{1}{2}$, its local maximum values coincide, and are the global ones. Accordingly,

$$\begin{aligned} \max_{\alpha \in [0,1]} G_{2,3:4}(u) &= G_{2,3:4}\left(\frac{1}{2} - \frac{\sqrt{6}-\sqrt{3}}{6}\right) \\ &= G_{2,3:4}\left(\frac{1}{2} + \frac{\sqrt{6}-\sqrt{3}}{6}\right) = \frac{2\sqrt{2}}{3}. \end{aligned} \tag{4.698}$$

Referring to Theorem 4.54 we obtain the desired claim. □

We established less precise descriptions of the upper bounds on the variances of other spacings $X_{j+1:n} - X_{j:n}$ with $2 \leq j \leq n-2$ and $n \geq 5$. They are presented in Corollary 4.31. The exceptions are the central spacings $X_{\frac{n}{2}+1:n} - X_{\frac{n}{2}:n}$ for even $n \geq 6$ for which explicit formulae are presented in Corollary 4.32.

Corollary 4.31 *Let $n \geq 5$ and $2 \leq j \leq n-2$. Then*

$$\frac{\mathbb{V}ar(X_{j+1:n} - X_{j:n})}{\mathbb{V}ar\, X_1} \leq G_{j,j+1:n}(\alpha_*) \tag{4.699}$$

where $\alpha_* = \alpha_*(j,n)$ is the global maximum of (4.664) over $\alpha \in [0,1]$. This is determined as either the unique or one of three zeros (more precisely, either the first or the last one) of the function

$$h_{j,j+1:n}(\alpha) = [2(n-1)\alpha - 2j + 1]B_{j,n}(\alpha) - (n-2)\alpha + j - 1. \tag{4.700}$$

Moreover, $0 < \alpha_*(j,n) < \frac{1}{2}$ if $j < \frac{n}{2}$, and $1 > \alpha_*(j,n) > \frac{1}{2}$ if $j > \frac{n}{2}$.

The equality in (4.699) is attained by (4.665) with $\alpha = \alpha_*$.

Proof Under the assumption $2 \leq j \leq n-2$ function (4.664) vanishes at 0 and 1, and is positive on $(0,1)$. Therefore the possibly non-unique global maximum point is located inside the interval. The function has the derivative

$$\begin{aligned} G'_{j,j+1:n}(\alpha) &= \binom{n}{j}\alpha^{j-2}(1-\alpha)^{n-2-j}h_{j,j+1:n}(\alpha) \\ &= \binom{n}{j}\frac{\alpha^{j-2}(1-\alpha)^{n-2-j}}{n+1}\sum_{i=0}^{n+1} a_{i,n+1}B_{i,n+1}(\alpha) \end{aligned} \tag{4.701}$$

(cf. (4.700)), where

$$a_{i,n+1} = \begin{cases} -2(n-j)j, & \text{if } i = j, \\ 2(n-j)j, & \text{if } i = j+1, \\ (j-1)(n+1) - (n-2)i, & \text{otherwise.} \end{cases} \tag{4.702}$$

The arithmetic sequence $\tilde{a}_{i,n+1} = (j-1)(n+1)-(n-2)i$, $i = 0, \ldots n+1$, decreases from $\tilde{a}_{0,n+1} = (j-1)(n+1) > 0$ to $\tilde{a}_{n+1,n+1} = -(n+1)(n-1-j) < 0$. If we replace any pair $\tilde{a}_{j,n+1}$, $\tilde{a}_{j+1,n+1}$ by arbitrary $a < 0$ and $b > 0$, we obtain another sequence with consecutive signs $+-+-$ (we suppressed here multiple pluses and minuses, and dropped a possible zero at $j = \frac{(i-1)(n+1)}{n-2}$). This holds true for (4.702), in particular. By the variation diminishing property of the Bernstein polynomials (see Lemma 4.1), $G_{j,j+1:n}(\alpha)$, $0 \leq \alpha \leq 1$, is either first increasing and then decreasing or it is consecutively increasing, decreasing, increasing and ultimately decreasing. In the first case, the unique zero of (4.700) coincides with the global maximum point. In the latter one, the first and third zeros determine the local maxima of (4.664), and one at which the function takes on the greater value solves the global maximization problem.

Now we verify the claim about the location of the global maximum point. For $n \geq 5$ we have

$$G_{j,j+1:n}(\alpha) = \binom{n}{j}\alpha^{j-1}(1-\alpha)^{n-1-j}$$

$$\times \left[\sum_{\substack{i=0 \\ j\neq i\neq n-j}}^{n} \binom{n}{i}\alpha^i(1-\alpha)^{n-i} + \binom{n}{j}\alpha^{n-j}(1-\alpha)^j \right]$$

$$= \binom{n}{j}[\alpha(1-\alpha)]^{j-1}(1-\alpha)^{n-2j} \sum_{\substack{i=0 \\ j\neq i\neq n-j}}^{n} \binom{n}{i}\alpha^i(1-\alpha)^{n-i}$$

$$+ \binom{n}{j}^2 [\alpha(1-\alpha)]^{n-1}, \tag{4.703}$$

and

$$G_{j,j+1:n}(1-\alpha) = \binom{n}{j}[\alpha(1-\alpha)]^{j-1}\alpha^{n-2j} \sum_{\substack{i=0 \\ j\neq i\neq n-j}}^{n} \binom{n}{i}\alpha^i(1-\alpha)^{n-i}$$

$$+ \binom{n}{j}^2 [\alpha(1-\alpha)]^{n-1}. \tag{4.704}$$

Subtracting (4.704) from (4.703) we obtain

$$G_{j,j+1:n}(\alpha) - G_{j,j+1n}(1-\alpha) = \binom{n}{j}[\alpha(1-\alpha)]^{j-1} \sum_{\substack{i=0 \\ j\neq i\neq n-j}}^{n} \binom{n}{i}\alpha^i(1-\alpha)^{n-i}$$

$$\times [(1-\alpha)^{n-2j} - \alpha^{n-2j}]. \tag{4.705}$$

The sign of the difference is identical with that of the expression in square brackets. Therefore for $j < \frac{n}{2}$ this difference is positive on $(0, \frac{1}{2})$ and negative on $(\frac{1}{2}, 1)$. If $j > \frac{n}{2}$, the signs are reversed. This immediately implies our claims. The rest of the proof is obvious in view of Theorem 4.54. □

Corollary 4.32 *For even $n \geq 6$*

$$\frac{\mathbb{V}ar(X_{\frac{n}{2}+1:n} - X_{\frac{n}{2}:n})}{\mathbb{V}ar\, X_1} \leq \binom{n}{\frac{n}{2}}\frac{1}{2^{n-2}}\left[1 - \binom{n}{\frac{n}{2}}\frac{1}{2^n}\right], \tag{4.706}$$

and the equality is attained by the symmetric two-point distributions.

Proof Due to Theorem 4.54 it suffices to show that

$$G_{\frac{n}{2},\frac{n}{2}+1:n}(\alpha) = \binom{n}{\frac{n}{2}}\alpha^{\frac{n}{2}-1}(1-\alpha)^{\frac{n}{2}-1}\left[1-\binom{n}{\frac{n}{2}}\alpha^{\frac{n}{2}}(1-\alpha)^{\frac{n}{2}}\right],\ 0 \le \alpha \le 1, \tag{4.707}$$

has the global maximum at $\frac{1}{2}$. Observe that (4.707) is symmetric about $\frac{1}{2}$, and has a local extremum at $\frac{1}{2}$. From the proof of the previous corollary we know that (4.707) is either increasing-decreasing or increasing-decreasing-increasing-decreasing. In the first case $\frac{1}{2}$ is the global maximum, and in the other one $\frac{1}{2}$ is a local minimum and there are two global maxima located symmetrically about $\frac{1}{2}$. We show that the first case is the only possible proving that

$$G''_{\frac{n}{2},\frac{n}{2}+1:n}\left(\frac{1}{2}\right) < 0, \qquad n \ge 6. \tag{4.708}$$

In the proof we use $m = \frac{n}{2}$ for simplifying the notation. Then we have

$$\begin{aligned} G''_{m,m+1:2m}(\alpha) = {} & \frac{(2m)!\alpha^{m-3}(1-\alpha)^{m-3}}{(m!)^4} \\ & \times \Big[(m!)^2(m-1)(4m\alpha^2 - 6\alpha^2 - 4m\alpha + 6\alpha + m - 2) \\ & - 2\alpha^m(1-\alpha)^m(2m)!(2m-1) \\ & \times (4m\alpha^2 - 3\alpha^2 - 4m\alpha + 3\alpha + m - 1)\Big], \end{aligned} \tag{4.709}$$

$$G''_{m,m+1:2m}\left(\frac{1}{2}\right) = \frac{(2m)!}{2^{2m-1}(m!)^4}h_{m,m+1,:2m}(m), \tag{4.710}$$

where

$$h_{m,m+1:2m}(m) = \frac{1}{4^m}(2m)!(2m-1) - (m!)^2(m-1) \tag{4.711}$$

determines the sign of $G''_{m,m+1:2m}\left(\frac{1}{2}\right)$. We prove that $h_{m,m+1:2m}(m) < 0$ for $m \ge 3$ by induction. We check that $h_{m,m+1:2m}(3) = -\frac{63}{4}$ and assume that $h_{m,m+1:2m}(m) < 0$ for some $m \ge 3$ which is equivalent to the inequality $\binom{2m}{m}\frac{1}{4^m}\left(2 + \frac{1}{m-1}\right) < 1$. We show that the relation holds for $m+1$ as well. Indeed,

$$\binom{2m+2}{m+1}\frac{1}{4^{m+1}}\left(2+\frac{1}{m}\right) = \binom{2m}{m}\frac{1}{4^m}\left(2+\frac{1}{m}\right)\frac{2m+1}{2(m+1)}$$

$$< \binom{2m}{m} \frac{1}{4^m} \left(2 + \frac{1}{m-1}\right) \frac{2m+1}{2m+2}$$

$$< \binom{2m}{m} \frac{1}{4^m} \left(2 + \frac{1}{m-1}\right) < 1, \tag{4.712}$$

by the inductive assumption. This ends the proof. □

4.11.3 General L-Statistics

With a similar method, we can try to find the bounds for variances of arbitrary L-statistics. The respective results were presented in Kozyra (2017). Let

$$B_{i,j,n}(\alpha, \beta) = \frac{n!}{i!(j-i)!(n-j)!} \alpha^i (\beta - \alpha)^{j-i} (1-\beta)^{n-j} \tag{4.713}$$

for $0 \le i < j \le n$ and $0 \le \alpha \le \beta \le 1$. For a given non-zero $\tilde{c} = (c_1, \dots, c_n) \in \mathbb{R}^n$ and $0 \le \alpha \le \beta \le 1$ we define

$$\begin{aligned} \mathbf{G}_{\tilde{c}:n}(\alpha, \beta) = & \left[\sum_{i=0}^{n-1} \frac{n}{i+1} \left(\sum_{k=1}^{i+1} c_k \right) B_{i,n-1}(\alpha) \right] \\ & \times \left[\sum_{j=0}^{n-1} \frac{n}{n-j} \left(\sum_{m=j+1}^{n} c_m \right) B_{j,n-1}(\beta) \right] \\ & - \sum_{i=0}^{n-2} \sum_{j=i}^{n-2} \frac{n(n-1)}{(i+1)(n-1-j)} \left(\sum_{k=1}^{i+1} c_k \right) \left(\sum_{m=j+2}^{n} c_m \right) B_{i,j,n-2}(\alpha, \beta), \end{aligned} \tag{4.714}$$

and

$$\begin{aligned} G_{\tilde{c}:n}(\alpha) = \mathbf{G}_{\tilde{c}:n}(\alpha, \alpha) = & \left[\sum_{i=0}^{n-1} \frac{n}{i+1} \left(\sum_{k=1}^{i+1} c_k \right) B_{i,n-1}(\alpha) \right] \\ & \times \left[\sum_{j=0}^{n-1} \frac{n}{n-j} \left(\sum_{m=j+1}^{n} c_m \right) B_{j,n-1}(\alpha) \right] \\ & - \sum_{i=0}^{n-2} \frac{n(n-1)}{(i+1)(n-1-i)} \left(\sum_{k=1}^{i+1} c_k \right) \left(\sum_{m=i+2}^{n} c_m \right) B_{i,n-2}(\alpha). \end{aligned} \tag{4.715}$$

Theorem 4.56 *Let $X_1, \ldots, X_n$ be iid with a finite and positive variance. Then for arbitrary non-zero $\tilde{c} = (c_1, \ldots, c_n) \in \mathbb{R}^n$ we have*

$$\frac{\mathbb{V}ar(\sum_{i=1}^{n} c_i X_{i:n})}{\mathbb{V}ar\, X_1} \leq \max_{0\leq\alpha\leq\beta\leq 1} \mathbf{G}_{\tilde{c}:n}(\alpha, \beta). \tag{4.716}$$

If

$$\max_{0\leq\alpha\leq\beta\leq 1} \mathbf{G}_{\tilde{c}:n}(\alpha, \beta) = \max_{0\leq\alpha\leq 1} G_{\tilde{c}:n}(\alpha), \tag{4.717}$$

then the bound (4.716) is sharp. Moreover, under the notation

$$\mathbb{P}_\alpha(X_1 = a) = \alpha = 1 - \mathbb{P}_\alpha(X_1 = b), \qquad 0 < \alpha < 1, \tag{4.718}$$

for arbitrary $a < b$, the following yields.

(i) If $\max_{0\leq\alpha\leq 1} G_{\tilde{c}:n}(\alpha) = G_{\tilde{c}:n}(\alpha_*)$ for some $0 < \alpha_* < 1$, then the bound is attained by (4.718) with $\alpha = \alpha_*$.
(ii) If $\max_{0\leq\alpha\leq 1} G_{\tilde{c}:n}(\alpha) = G_{\tilde{c}:n}(0)$, then the bound (4.716) is attained in the limit for the parent two-point distribution (4.718) with $\alpha \to 0$.
(iii) If $\max_{0\leq\alpha\leq 1} G_{\tilde{c}:n}(\alpha) = G_{\tilde{c}:n}(1)$, then (4.716) becomes equality in the limit for (4.718) as $\alpha \to 1$.

Proof Let F denote the common distribution function of $X_1, \ldots, X_n$. Then for $1 \leq i < j \leq n$ and $x < y$

$$\begin{aligned}
&\mathbb{P}(X_{i:n} \leq x, X_{i+1:n}, X_{j:n} \in (x, y], X_{j+1:n} > y) \\
&= \frac{n!}{i!(j-i)!(n-j)!} F^i(x)[F(y) - F(x)]^{j-i}[1 - F(y)]^{n-j} \\
&= B_{i,j,n}(F(x), F(y)).
\end{aligned} \tag{4.719}$$

Therefore, if $1 \leq k < m \leq n$ and $x < y$, then

$$\mathbb{P}(X_{k:n} \leq x, X_{m:n} > y) = \sum_{j=k}^{m-1} \sum_{i=k}^{j} B_{i,j,n}(F(x), F(y)) = \tilde{F}_{k,m:n}(F(x), F(y)), \tag{4.720}$$

say. Obviously, for $k < m$

$$\mathbb{P}(X_{k:n} \le x, X_{m:n} \le y) = \begin{cases} \mathbb{P}(X_{k:n} \le x) - \mathbb{P}(X_{k:n} \le x, X_{m:n} > y), & x < y, \\ \mathbb{P}(X_{m:n} \le y), & x \ge y, \end{cases}$$
$$= \begin{cases} F_{k:n}(F(x)) - \tilde{F}_{k,m:n}(F(x), F(y)), & x < y, \\ F_{m:n}(F(y)), & x \ge y. \end{cases} \tag{4.721}$$

Consequently,

$$\begin{aligned}
&\mathbb{C}ov(X_{k:n}, X_{m:n}) \\
&= \iint_{\mathbb{R}^2} [\mathbb{P}(X_{k:n} \le x, X_{m:n} \le y) - \mathbb{P}(X_{k:n} \le x)\mathbb{P}(X_{m:n} \le y)]\, dxdy \\
&= \iint_{x<y} [F_{k:n}(F(x)) - \tilde{F}_{k,m:n}(F(x), F(y)) - F_{k:n}(F(x))F_{m:n}(F(y))]\, dxdy \\
&+ \iint_{x \ge y} [F_{m:n}(F(y)) - F_{k:n}(F(x))F_{m:n}(F(y))]\, dxdy \\
&= \iint_{x \le y} \Big\{ F_{k:n}\big(F(x)\big)\big[1 - F_{m:n}\big(F(y)\big)\big] \\
&+ F_{m:n}\big(F(x)\big)\big[1 - F_{k:n}\big(F(y)\big)\big] - \tilde{F}_{k,m:n}\big(F(x), F(y)\big)\Big\}\, dxdy \\
&= \iint_{0<F(x)\le F(y)<1} \left\{ \left[\sum_{i=k}^{n} B_{i,n}\big(F(x)\big)\right]\left[\sum_{i=0}^{m-1} B_{i,n}\big(F(y)\big)\right] \right. \\
&\left. + \left[\sum_{i=m}^{n} B_{i,n}\big(F(x)\big)\right]\left[\sum_{i=0}^{k-1} B_{i,n}\big(F(y)\big)\right] - \sum_{j=k}^{m-1}\sum_{i=k}^{j} B_{i,j,n}\big(F(x), F(y)\big)\right\} dxdy
\end{aligned} \tag{4.722}$$

(cf. (4.623)).

Applying the above establishments we can calculate

$$\begin{aligned}
&\mathbb{V}ar\left(\sum_{i=1}^{n} c_i X_{i:n}\right) \\
&= \sum_{k=1}^{n} c_k^2\, \mathbb{V}ar\, X_{k:n} + 2 \sum_{1 \le k < m \le 1} c_k c_m\, \mathbb{C}ov(X_{k:n}, X_{m:n})
\end{aligned}$$

$$= 2 \iint_{0<F(x)\le F(y)<1} \left\{ \sum_{k=1}^{n} c_k^2 F_{k:n}\big(F(x)\big)\big[1 - F_{k:n}\big(F(y)\big)\big] \right.$$

$$+ \sum_{k=1}^{n-1} \sum_{m=k+1}^{n} c_k c_m \Big[F_{k:n}\big(F(x)\big)\big[1 - F_{m:n}\big(F(y)\big)\big]$$

$$\left. + F_{m:n}\big(F(x)\big)\big[1 - F_{k:n}\big(F(y)\big)\big] - \tilde{F}_{k,m:n}\big(F(x), F(y)\big)\Big] \right\} dxdy$$

$$= 2 \iint_{0<F(x)\le F(y)<1} \left\{ \left[\sum_{k=1}^{n} c_k F_{k:n}\big(F(x)\big) \right] \right.$$

$$\times \left[\sum_{m=1}^{n} c_m [1 - F_{m:n}(F(y))] \right]$$

$$\left. - \sum_{k=1}^{n-1} \sum_{m=k+1}^{n} c_k c_m \tilde{F}_{k,m:n}(F(x), F(y)) \right\} dxdy$$

$$= 2 \iint_{0<F(x)\le F(y)<1} \left\{ \left[\sum_{i=1}^{n} \left(\sum_{k=1}^{i} c_k \right) B_{i,n}\big(F(x)\big) \right] \right.$$

$$\times \left[\sum_{j=0}^{n-1} \left(\sum_{m=j+1}^{n} c_m \right) B_{j,n}\big(F(y)\big) \right]$$

$$\left. - \sum_{i=1}^{n-1} \sum_{j=i}^{n-1} \left(\sum_{k=1}^{i} c_k \right) \left(\sum_{m=j+1}^{n} c_m \right) B_{i,j,n}\big(F(x), F(y)\big) \right\} dxdy$$

$$= 2 \iint_{0<F(x)\le F(y)<1} \mathbf{G}_{\tilde{c}:n}\big(F(x), F(y)\big) F(x)[1 - F(y)] dxdy$$

$$\le \max_{0\le\alpha\le\beta\le1} \mathbf{G}_{\tilde{c}:n}(\alpha, \beta)\, \mathbb{V}ar\, X_1, \tag{4.723}$$

which completes the proof of inequality (4.716).

The above calculations show that for the two-point distributions (4.718), we obtain

$$\mathbb{V}ar_\alpha \left(\sum_{i=1}^{n} c_i X_{i:n} \right) = G_{\tilde{c}:n}(\alpha)\, \mathbb{V}ar_\alpha\, X_1, \tag{4.724}$$

because except for 0 and 1, α is the unique value of the distribution function of (4.718), and it can replace $F(x)$ and $F(y)$ in (4.723). Accordingly, under (4.717) and condition (i) of Theorem 4.56, the upper bound is attained if the baseline distribution is (4.718) with $\alpha = \alpha_*$. Attainability of the bound in cases (ii) and (iii) follows from (4.724) and continuity of (4.715) on [0, 1]. □

We do not have analytic tools for calculating the maxima of (4.714) and (4.715) and comparing them when $\tilde{c} \in \mathbb{R}^n$ is arbitrary. If the maxima differ, we do not know either what is the optimal bound nor how much the bound (4.716) differs from the optimal one nor what is the way of determining the parent distribution which attains the bound (possibly in the limit) this sharp bound.

The following theorem asserts that nontrivial lower bounds are possible only if both c_1 and c_n are nonzero.

Theorem 4.57 *Under assumptions of Theorem 4.56, if $c_1 c_n = 0$, then the bound*

$$\frac{\mathbb{V}ar(\sum_{i=1}^n c_i X_{i:n})}{\mathbb{V}ar\, X_1} \geq 0 \tag{4.725}$$

is sharp. If $c_1 = 0$ ($c_n = 0$, respectively), then the equality is attained under conditions of Theorem 4.56(ii) ((iii), respectively).

Proof We first notice that $G_{\tilde{c}:n}(0) = nc_1^2$ and $G_{\tilde{c}:n}(1) = nc_n^2$. Therefore, if $c_1 = 0$, owing to (4.724) we obtain

$$\lim_{\alpha \to 0} \frac{\mathbb{V}ar_\alpha(\sum_{i=1}^n c_i X_{i:n})}{\mathbb{V}ar_\alpha\, X_1} = \lim_{\alpha \to 0} G_{\tilde{c}:n}(\alpha) = G_{\tilde{c}:n}(0) = 0. \tag{4.726}$$

Similarly, we treat the case $c_n = 0$. □

4.11.4 Symmetric Populations

We finally describe the bounds on variances of order statistics coming from iid samples with a symmetric marginal distribution. Notice first that the bounds for the jth smallest and jth greatest order statistics are identical, and they are attained for the same parent distributions. This is the consequence of the fact that distribution of an iid sample $X_1, \dots, X_n$ symmetrically distributed about $\mu \in \mathbb{R}$ is identical with the joint distribution of $2\mu - X_1, \dots, 2\mu - X_n$. Therefore $\mathbb{V}ar X_{j:n} = \mathbb{V}ar(2\mu - X_{n+1-j:n}) = \mathbb{V}ar X_{n+1-j:n}$. This allows us to concentrate on evaluating $\mathbb{V}ar X_{j:n}$ either for $1 \leq j \leq \frac{n}{2}$ or for $\frac{n}{2} + 1 \leq j \leq n$ only, and immediately apply the results for the other indices. Also, observe that the sharp upper bounds for the sample medians $X_{\frac{n+1}{2}:n}$ for the general iid odd-sized samples determined in Corollary 4.28 are obviously optimal under the restriction to symmetric parent distributions as well.

We first present the results of Papadatos (1997). They concern the nonextreme order statistics. The sample maxima and minina are treated separately later on.

Theorem 4.58 *Assume that $X_1, \ldots, X_n$ are iid and have a symmetric parent distribution with mean $\mu \in \mathbb{R}$ and variance $0 < \sigma^2 < \infty$. For $\frac{n}{2} + 1 \le j \le n - 1$ define the function*

$$S_{j:n}(\alpha) = \frac{[F_{j:n}(\alpha) + F_{j:n}(1-\alpha)][1 - F_{j:n}(\alpha) - F_{j:n}(1-\alpha)] + 2F_{j:n}(1-\alpha)}{2(1-\alpha)},$$

$$\frac{1}{2} \le \alpha < 1. \tag{4.727}$$

The function attains its maximal value in $\left[\frac{1}{2}, 1\right)$. If $\alpha_* = \alpha_*(j, n)$ is a (possibly non-unique) global maximum of (4.727), then

$$0 \le \frac{\mathbb{V}ar\, X_{j:n}}{\mathbb{V}ar\, X_1} \le S_{j:n}(\alpha_*). \tag{4.728}$$

The equality in the RHS inequality is attained if

$$\mathbb{P}_\alpha\left(X_1 = \mu \mp \frac{\sigma}{\sqrt{2(1-\alpha)}}\right) = 1 - \alpha,$$
$$\mathbb{P}_\alpha(X_1 = \mu) = 2\alpha - 1, \tag{4.729}$$

with $\alpha = \alpha_*$. The equality in the LHS of (4.728) is attained in the limit by (4.729) as $\alpha \to 1$. The same bounds with the same attainability conditions hold for $X_{n+1-j:n}$, $1 \le n + 1 - j \le \frac{n}{2}$.

Proof For simplicity of calculations we assume that $\mu = 0$. Then the symmetry assumption means that the parent distribution function F satisfies $F(x) = 1 - F(-x-)$, $x \in \mathbb{R}$. Note that the set $\{x \in \mathbb{R} : F(x) \neq 1 - F(-x)\}$ has zero Lebesgue measure. Therefore we can write

$$\iint_{x \le y \le 0} F(x)[1 - F(y)]\,dxdy = \iint_{0 \le x \le y} F(x)[1 - F(y)]\,dxdy, \tag{4.730}$$

$$\iint_{0 \le -x \le y} F(x)[1 - F(y)]\,dxdy = \iint_{0 \le y \le -x} F(x)[1 - F(y)]\,dxdy$$
$$= \iint_{0 \le x \le y} [1 - F(x)][1 - F(y)]\,dxdy. \tag{4.731}$$

The union of the sets $\{0 \le x \le y\}$, $\{x \le y \le 0\}$, $\{0 \le -x \le y\}$, and $\{0 \le y \le -x\}$ amounts to $\{x \le y\}$, and the intersection of either pair has the Lebesgue measure on $\mathbb{R}^2$ equal to 0. Accordingly,

$$\mathbb{V}ar\, X_1 = 2 \iint_{x \le y} F(x)[1 - F(y)]\, dxdy = 4 \iint_{0 \le x \le y} [1 - F(y)]\, dxdy. \tag{4.732}$$

Similarly, using

$$\iint_{x \le y \le 0} F_{j:n}(F(x))[1 - F_{j:n}(F(y))]\, dxdy$$
$$= \iint_{0 \le x \le y} [1 - F_{j:n}((1 - F(x))]F_{j:n}(1 - F(y))]\, dxdy, \tag{4.733}$$

$$\iint_{0 \le -x \le y} F_{j:n}(F(x))[1 - F_{j:n}(F(y))]\, dxdy$$
$$= \iint_{0 \le x \le y} F_{j:n}((1 - F(x))[1 - F_{j:n}(F(y))]\, dxdy, \tag{4.734}$$

$$\iint_{0 \le y \le -x} F_{j:n}(F(x))[1 - F_{j:n}(F(y))]\, dxdy$$
$$= \iint_{0 \le x \le y} [1 - F_{j:n}(F(x))]F_{j:n}(1 - F(y))\, dxdy, \tag{4.735}$$

we obtain

$$\mathbb{V}ar\, X_{j:n} = 2 \iint_{x \le y} F_{j:n}(F(x))[1 - F_{j:n}(F(y))]\, dxdy$$
$$= 2 \iint_{0 \le x \le y} \{[F_{j:n}(F(x)) + F_{j:n}(1 - F(x))][1 - F_{j:n}(F(y)) - F_{j:n}(1 - F(y))]$$
$$+ 2F_{j:n}(1 - F(x))]\}dxdy$$
$$= 2 \iint_{\substack{0 \le x \le y, \\ F(y) < 1}} \left\{ \frac{[F_{j:n}(F(x)) + F_{j:n}(1 - F(x))][1 - F_{j:n}(F(y)) - F_{j:n}(1 - F(y))]}{2(1 - F(y))} \right.$$
$$\left. + \frac{2F_{j:n}(1 - F(x))}{2(1 - F(y))} \right\} 2[1 - F(y)]\, dxdy$$
$$\le \sup_{\frac{1}{2} \le \alpha \le \beta < 1} \mathbf{S}_{j:n}(\alpha, \beta)\, 4 \iint_{0 \le x \le y} [1 - F(y)]\, dxdy \tag{4.736}$$

where

$$\mathbf{S}_{j:n}(\alpha,\beta)=\frac{[F_{j:n}(\alpha)+F_{j:n}(1-\alpha)][1-F_{j:n}(\beta)-F_{j:n}(1-\beta)]+2F_{j:n}(1-\beta)}{2(1-\beta)},$$

$$\frac{1}{2}\le\alpha\le\beta<1. \tag{4.737}$$

If $\frac{1}{2}\le\alpha\le\beta<1$, then

$$\frac{\partial}{\partial\alpha}\mathbf{S}_{j:n}(\alpha,\beta)=\frac{[f_{j:n}(\alpha)-f_{j:n}(1-\alpha)][1-F_{j:n}(\beta)-F_{j:n}(1-\beta)]}{2(1-\beta)}>0, \tag{4.738}$$

because both the factors in the numerator are positive for $j\ge\frac{n}{2}+1$ and $\alpha,\beta\in\left(\frac{1}{2},1\right)$. This implies that

$$\mathbf{S}_{j:n}(\alpha,\beta)\le\mathbf{S}_{j:n}(\beta,\beta)=S_{j:n}(\beta). \tag{4.739}$$

Since

$$1-F_{j:n}(\alpha)-F_{j:n}(1-\alpha)=F_{n+1-j:n}(\alpha)-F_{j:n}(\alpha)=\sum_{i=n+1-j}^{j-1}B_{i,n}(\alpha), \tag{4.740}$$

(4.727) is positive on $\left[\frac{1}{2},1\right)$. Moreover, by the de l'Hospital rule

$$\lim_{\alpha\to1}S_{j:n}(\alpha)=\lim_{\alpha\to1}-\frac{1}{2}\{f_{j:n}(\alpha)-3f_{j:n}(1-\alpha)$$
$$-2[F_{j:n}(\alpha)+F_{j:n}(1-\alpha)][f_{j:n}(\alpha)-f_{j:n}(1-\alpha)]\}=0. \tag{4.741}$$

because $f_{j:n}(0)=f_{j:n}(1)=0$ for $2\le j\le n-1$. Therefore we can rewrite (4.736) as

$$\mathbb{V}ar\,X_{j:n}\le\max_{\frac{1}{2}\le\alpha<1}S_{j:n}(\alpha)\,\mathbb{V}ar\,X_1. \tag{4.742}$$

Since the function (4.727) is continuous, it attains its positive maximum at some (possibly non-unique) $\alpha_*=\alpha_*(j,n)$ located in $\left[\frac{1}{2},1\right)$. This proves the RHS inequality in (4.728).

We obtain the equality in the RHS inequality of (4.728) if the only value of $F(x)$ and $F(y)$ in $\left[\frac{1}{2},1\right)$ is just α_*. This defines the unique symmetric distribution (4.729) with $\alpha=\alpha_*$, which attains the equality in (4.736), and the upper bound in (4.728).

Now we focus on the lower bound. Mimicking the above arguments we can show that if α and 1 are the only values of $F(x)$ and $F(y)$ for $\frac{1}{2} \leq x \leq y$ (i.e., the parent distribution is (4.729)), we get

$$\frac{\mathbb{V}ar_{\alpha} X_{j:n}}{\mathbb{V}ar_{\alpha} X_1} = S_{j:n}(\alpha). \tag{4.743}$$

Letting $\alpha \to 1$ we get the trivial bound

$$\lim_{\alpha \to 1} \frac{\mathbb{V}ar_{\alpha} X_{j:n}}{\mathbb{V}ar_{\alpha} X_1} = \lim_{\alpha \to 1} S_{j:n}(\alpha) = 0, \tag{4.744}$$

which cannot be improved.

The last statement of Theorem 4.58 is evident in view of the remarks preceding the theorem. □

Notice that for $j = \frac{n+1}{2}$ we have $F_{\frac{n+1}{2}:n}(\alpha) = 1 - F_{\frac{n+1}{2}:n}(1-\alpha)$ and (4.727) simplifies to

$$S_{\frac{n+1}{2}:n}(\alpha) = \frac{1 - F_{\frac{n+1}{2}:n}(\alpha)}{1-\alpha}. \tag{4.745}$$

If $\alpha \to 1$, then (4.745) tends to $f_{\frac{n+1}{2}:n}(1) = 0$ as well. Function (4.745) is maximized over $\left[\frac{1}{2}, 1\right)$ at $\frac{1}{2}$, because $F_{\frac{n+1}{2}:n}$ is concave on this interval, and the maximal slope of the line connecting $(\alpha, F_{\frac{n+1}{2}:n}(\alpha))$ with $(1, F_{\frac{n+1}{2}:n}(1)) = (1, 1)$ is attained at the minimal point, and amounts to 1. This confirms the result of Corollary 4.28.

Jasiński and Rychlik (2013) delivered a more precise description of the upper bounds and their attainability conditions than those of Papadatos (1997) presented on Theorem 4.58. Due to Theorem 4.58, in the problem of determining the variance bounds it suffices to take into account only two- and three point symmetric distributions. Jasiński and Rychlik (2013) found it more convenient to consider the order statistics $X_{j:n}$ with low indices $2 \leq j \leq \frac{n}{2}$. They also used a different parametrization $\gamma = 1 - \alpha$, $0 < \gamma \leq \frac{1}{2}$. Since the problem of establishing the variance bounds is location and scale invariant, they analyzed the following simple distributions

$$\mathbb{P}_{\gamma}(X_1 = \pm 1) = \gamma, \qquad \mathbb{P}_{\gamma}(X_1 = 0) = 1 - 2\gamma, \qquad 0 < \gamma \leq \frac{1}{2}, \tag{4.746}$$

with $\mathbb{E}_\gamma X_1 = 0$ and $\mathbb{V}ar_\gamma X_1 = 2\gamma$. The order statistics have the following three-point distributions

$$\begin{aligned}\mathbb{P}_\gamma(X_{j:n} = -1) &= F_{j:n}(\gamma),\\ \mathbb{P}_\gamma(X_{j:n} = 0) &= F_{j:n}(1-\gamma) - F_{j:n}(\gamma),\\ \mathbb{P}_\gamma(X_{j:n} = 1) &= 1 - F_{j:n}(1-\gamma).\end{aligned} \tag{4.747}$$

Using the formula $F_{n+1-j:n}(\gamma) = 1 - F_{j:n}(1-\gamma)$, we can write

$$\mathbb{E}_\gamma X_{j:n} = \mathbb{E}_\gamma X_{j:n}^2 = F_{n+1-j:n}(\gamma) - F_{j:n}(\gamma), \tag{4.748}$$

and

$$\mathbb{V}ar_\gamma X_{j:n} = F_{j:n}(\gamma) + F_{n+1-j:n}(\gamma) - [F_{j:n}(\gamma) - F_{n+1-j:n}(\gamma)]^2 = V_{j:n}(\gamma), \tag{4.749}$$

say. Accordingly, our problem is to determine the supremum of

$$S_{j:n}(\gamma) = \frac{V_{j:n}(\gamma)}{2\gamma} = \frac{\mathbb{V}ar_\gamma X_{j:n}}{\mathbb{V}ar_\gamma X_1} \tag{4.750}$$

over the set of parameters $0 < \gamma \leq \frac{1}{2}$. Using (4.727) and (4.749) with (4.750), one can easily check that $S_{j:n}(\alpha) = S_{j:n}(1-\gamma)$.

Theorem 4.59 *For each $n \geq 4$ and $2 \leq j \leq \frac{n}{2}$, the equation*

$$V_{j:n}(\gamma) = \gamma V'_{j:n}(\gamma) \tag{4.751}$$

(see (4.749)) has either 0 or 2 solutions in $\left(0, \frac{1}{2}\right)$. In the first case

$$\frac{\mathbb{V}ar\, X_{j:n}}{\mathbb{V}ar\, X_1} \leq S_{j:n}\left(\frac{1}{2}\right) = V_{j:n}\left(\frac{1}{2}\right) = \frac{1}{4^{n-1}} \sum_{i=0}^{j-1} \binom{n}{i} \left[2^n - \sum_{i=0}^{j-1} \binom{n}{i}\right]. \tag{4.752}$$

Otherwise, with $\gamma_* = \gamma_*(j,n)$ denoting the smaller of the solutions to (4.751), we have

$$\frac{\mathbb{V}ar\, X_{j:n}}{\mathbb{V}ar\, X_1} \leq \max\left\{S_{j:n}(\gamma_*),\, S_{j:n}\left(\frac{1}{2}\right)\right\}. \tag{4.753}$$

If the bound equals to $S_{k:n}\left(\frac{1}{2}\right)$, then it is attained by the two-point symmetric distributions. Otherwise it is attained by the three-point symmetric distributions *such that the probabilities of the side points are* γ_**, and that of the central one is* $1 - 2\gamma_*$.

The proof is based on a thorough analysis of variability of the functions (4.749). These are polynomials of degree $2n$ and they can be represented as linear combinations of Bernstein polynomials. The same concerns respective first and second derivatives. Knowing the signs of the coefficients of the corresponding linear combinations of the Bernstein polynomials, and applying the variation diminishing property we conclude some shape properties of (4.749). The signs of the combination coefficients in the representations of derivatives in the Bernstein bases are presented in the following lemma.

Lemma 4.4 *Let*

$$V'_{j:n}(\gamma) = \sum_{i=0}^{2n-1} \alpha_{j:n}(i) B_{i,2n-1}(\gamma), \tag{4.754}$$

$$V''_{j:n}(\gamma) = \sum_{i=0}^{2n-2} \beta_{j:n}(i) B_{i,2n-2}(\gamma). \tag{4.755}$$

Then, under the convention $\operatorname{sgn}(0) = 0$*, we have*

$$\operatorname{sgn}(\alpha_{j:n}(i)) = \begin{cases} 0, & i = 0, \dots, j-2, \\ +1, & i = j-1, \dots, 2j-2, \\ -1, & i = 2j-1, \dots, n-1, \\ +1, & i = n, \dots, 2n-j, \\ 0, & i = 2n-j+1, \dots, 2n-1. \end{cases} \tag{4.756}$$

and

$$\operatorname{sgn}(\beta_{j:n}(i)) = \begin{cases} 0, & i = 0, \dots, j-3, \\ +1, & i = j-2, \dots, 2j-3, \\ -1, & i = 2j-2, \\ +1, & i = 2j-1, \dots, n-1, \\ -1, & i = 2n-2j, \dots, 2n-j, \\ 0, & i = 2n-j+1, \dots, 2n-2. \end{cases} \tag{4.757}$$

If $j \le \frac{n-1}{2}$*, then* $\operatorname{sgn}(\beta_{j:n}(n)) = +1$ *and the sequence* $(\operatorname{sgn}(\beta_{j:n}(i)))_{i=n}^{2n-2j-1}$ *is non-increasing.*

The proof of Lemma 4.4 requires elaborate calculations, and it is moved to the end of this subsection.

Proof of Theorem 4.59 We first analyze (4.749). We easily see that $V_{j:n}(0) = 0$ and $V_{j:n}(1) = 2$. Moreover, $V_{j:n}(\gamma) > 0$, $0 < \gamma \le \frac{1}{2}$, because these are the variances of non-degenerate distributions. We show that (4.749) is increasing and so positive for $\frac{1}{2} < \gamma < 1$. Consider

$$V'_{j:n}(\gamma) = f_{j:n}(\gamma) + f_{n+1-j:n}(\gamma) - 2[F_{j:n}(\gamma) - F_{n+1-j:n}(\gamma)] \times [f_{j:n}(\gamma) - f_{n+1-j:n}(\gamma)]. \quad (4.758)$$

Clearly, $V'_{j:n}(0) = V'_{j:n}(1) = 0$, and $V'_{j:n}\left(\frac{1}{2}\right) = 2f_{j:n}\left(\frac{1}{2}\right) > 0$. Also $F_{j:n}(\gamma) - F_{n+1-j:n}(\gamma) = \sum_{i=j}^{n-j} B_{i,n}(\gamma) > 0$, $0 < \gamma < 1$, and $f_{j:n}(\gamma) < f_{n+1-j:n}(\gamma)$, $\frac{1}{2} < \gamma < 1$, imply positivity of (4.758) on $\left(\frac{1}{2}, 1\right)$, as claimed. By the variation diminishing property and Lemma 4.4 (see (4.756)), (4.749) is either increasing on the whole (0, 1), or consecutively increasing, decreasing and increasing there. In the latter case, it is increasing-decreasing-increasing in $\left(0, \frac{1}{2}\right)$, and increasing in $\left(\frac{1}{2}, 1\right)$.

Now we proceed to

$$V''_{j:n}(\gamma) = f'_{j:n}(\gamma) + f'_{n+1-j:n}(\gamma) - 2[f_{j:n}(\gamma) - f_{n+1-j:n}(\gamma)]^2 - 2[F_{j:n}(\gamma) - F_{n+1-j:n}(\gamma)][f'_{j:n}(\gamma) - f'_{n+1-j:n}(\gamma)] \quad (4.759)$$

with

$$f'_{j:n}(\gamma) = -f'_{n+1-j:n}(1-\gamma) = n(n-1)[B_{j-2,n-2}(\gamma) - B_{j-1,n-2}(\gamma)]. \quad (4.760)$$

It follows that

$$V''_{j:n}\left(\frac{1}{2}\right) = -4\left[F_{j:n}\left(\frac{1}{2}\right) - F_{n+1-j:n}\left(\frac{1}{2}\right)\right] f'_{j:n}\left(\frac{1}{2}\right) > 0, \quad (4.761)$$

i.e. (4.749) is convex in a neighborhood of $\frac{1}{2}$, because the difference in the brackets is positive and $f'_{j:n}(\gamma) < 0$ for $\gamma \in \left(\frac{j-1}{n-1}, 1\right) \ni \frac{1}{2}$. By the formula (4.757) of Lemma 4.4, (4.749) is either first convex and then concave, or consecutively convex, concave, convex, and finally concave in (0, 1).

We arrived to the following conclusions. If (4.749) is increasing-decreasing-increasing, then it is by turns convex increasing, concave increasing, concave decreasing, convex decreasing, and convex increasing in $\left(0, \frac{1}{2}\right)$, and convex increasing, concave increasing in $\left(\frac{1}{2}, 1\right)$. If (4.749) is convex-concave on (0, 1), then it is convex increasing in $\left(0, \frac{1}{2}\right)$, and first convex increasing and eventually concave increasing in $\left(\frac{1}{2}, 1\right)$. Finally, there is a possibility that (4.749) is increasing in the whole unit interval, and subsequently convex, concave, convex, and concave there. We do not arbitrate here if $\frac{1}{2}$ belongs to the first or second convexity regions.

We are now ready to tackle with the basic problem of maximizing function (4.750). Consider

$$S'_{:jn}(\gamma) = -\frac{T_{j:n}(\gamma)}{2\gamma^2}, \qquad 0 < \gamma < 1, \tag{4.762}$$

with

$$T_{j:n}(\gamma) = V_{j:n}(\gamma) - \gamma V'_{j:n}(\gamma). \tag{4.763}$$

The sign of the derivative (4.762) depends on the sign of (4.763). Define the straight line tangent to the graph of (4.749) at γ as

$$L_{j:n}(\gamma; x) = V_{j:n}(\gamma) + V'_{j:n}(\gamma)(x - \gamma). \tag{4.764}$$

Note that $T_{j:n}(\gamma) = L_{j:n}(\gamma; 0)$ can be interpreted as the value of the tangent line at 0. Function (4.750) attains its local maximum at some γ, if the values of the tangents at 0 change their signs from minus to plus there, and it reaches a local minimum, if the sign order is reversed. Therefore we concentrate on studying the sign changes of (4.763).

Note first that (4.763) cannot vanish at any point γ of decrease of (4.749), because then $T_{j:n}(\gamma) > V_{j:n}(\gamma) > 0$. The relation

$$T'_{j:n}(\gamma) = -\gamma V''_{j:n}(\gamma) \tag{4.765}$$

implies that (4.763) strictly increases on the interval of strict concavity of (4.749), and it may have at most one sign change (from minus to plus) there. Similarly, in each interval of strict convexity and increase of (4.749), function (4.750) has at most one local minimum. Function (4.763) has no zeros in the first convexity region, because $T_{j:n}(\gamma) = L_{j:n}(\gamma; 0) < V_{j:n}(0) = 0$ there. It follows that in the case that (4.749) is convex-concave, the function (4.750) has at most one local extremum, which is a minimum located in $\left(\frac{1}{2}, 1\right)$. If (4.749) is convex, concave, convex, and ultimately concave, then (4.750) has at most three extremes: one in each of the regions: the first and second concavity intervals, and the second convexity interval.

Below we show that the total number of zeros of (4.763) is odd, and there is an even number of zeros in $\left(0, \frac{1}{2}\right)$. The first statement is immediate, because $\lim_{\epsilon \searrow 0} T_{j:n}(\epsilon) < 0$, as claimed above, and $T_{j:n}(1) = 2 > 0$. To get the latter, it suffices to verify that

$$T_{j:n}\left(\frac{1}{2}\right) = 4F_{n+1-j:n}\left(\frac{1}{2}\right) - f_{j:n}\left(\frac{1}{2}\right) - 4F^2_{n+1-j:n}\left(\frac{1}{2}\right) < 0. \tag{4.766}$$

We show that the difference of the first two components is negative, which is equivalent to saying that

$$C(j,n) = j\binom{n}{j} - 2\sum_{i=0}^{j-1}\binom{n}{i} > 0, \qquad 2 \le j \le \frac{n}{2}. \tag{4.767}$$

We prove the relations by induction with respect to j. Firstly, $C(2,n) = n^2 - 3n - 2 > 0$ for all $n \ge 4$, because the function is increasing for $n \ge \frac{3}{2}$, and $C(2,4) = 2$. Moreover,

$$C(j,n) - C(j-1,n) = j\binom{n}{j} - (j+1)\binom{n}{j-1} = (n-2j)\binom{n}{j-1} \ge 0 \tag{4.768}$$

completes the proof of (4.767) and (4.766).

The final conclusions are following. If the function in (4.749) is increasing and convex-concave, then (4.763) is negative in $\left(0, \frac{1}{2}\right)$, and changes the sign from minus to plus in $\left(\frac{1}{2}, 1\right)$. So (4.750) is increasing in $\left(0, \frac{1}{2}\right)$, and attains its maximum at $\frac{1}{2}$. If (4.749) is increasing, decreasing, and increasing in $(0, \delta_*)$, (δ_*, δ_{**}), and $(\delta_{**}, 1)$, respectively, for some $0 < \delta_* < \delta_{**} < \frac{1}{2}$, then $T_{j:n}(\delta_*) = V_{j:n}(\delta_*) > 0$ and $T_{j:n}(\delta_{**}) = V_{j:n}(\delta_{**}) > 0$. Therefore there exist $0 < \gamma_* < \delta_*$ and $\delta_{**} < \gamma_{**} < \frac{1}{2}$ such that (4.763) changes the sign from minus to plus at γ_*, and in the opposite direction at γ_{**}. The points γ_* and γ_{**} provide the local maximum and minimum of (4.750), respectively. The function increases in $(0, \gamma_*)$ and $\left(\gamma_{**}, \frac{1}{2}\right)$, decreases in (γ_*, γ_{**}), and its global maximum amounts to $\max\left\{S_{j:n}(\gamma_*), S_{j:n}\left(\frac{1}{2}\right)\right\}$. The last possibility is that (4.749) is increasing, and has two regions of convexity as well as of concavity. Then (4.763) may have no zeros in $\left(0, \frac{1}{2}\right)$, and either one or three zeros in $\left(\frac{1}{2}, 1\right)$. Then (4.750) is increasing in the former interval, and has the maximum at $\frac{1}{2}$. Also, (4.749) may have two zeros in $\left(0, \frac{1}{2}\right)$, and the third one in $\left(\frac{1}{2}, 1\right)$. Then γ_* belongs to a concavity region (we do not attribute which one of two), provides a local maximum of (4.750), and the global one is either of $S_{j:n}(\gamma_*)$ and $S_{j:n}\left(\frac{1}{2}\right)$. This ends the proof. □

It remains to treat the extreme order statistics. The following result is due to Moriguti (1951).

Theorem 4.60 *If $X_1, \ldots, X_n$ are iid symmetrically distributed with a finite positive variance, then*

$$\lambda_* \le \frac{\mathbb{V}ar\, X_{n:n}}{\mathbb{V}ar\, X_1} \le \frac{n}{2}, \tag{4.769}$$

where $\lambda_* = \lambda_*(n)$ is the only solution to

$$\int_{\frac{1}{2}}^{1} \frac{n^2[x^{n-1} - (1-x)^{n-1}]^2}{n[x^{n-1} + (1-x)^{n-1}] - 2\lambda}\, dx = 1 \tag{4.770}$$

in the interval $\left(0, \frac{n}{2^{n-1}}\right)$. The upper bound is attained in the limit by the symmetric three-point distributions (4.729) with $\alpha \to 1$.

The bounds for the variances of sample minima $X_{1:n}$ are identical with (4.769) as well as their conditions of attainability.

It looks surprising that the conditions for attaining the upper bounds on variances of sample extremes are identical with the conditions for getting the lower bounds on the variances of other order statistics (comp. Theorem 4.58).

Proof Since the problem is location invariant, we assume $\mu = 0$ for simplicity. We first verify the upper bounds. Since

$$\begin{aligned}\int_0^{\frac{1}{2}} [F^{-1}(x)]^2 x^{n-1}\, dx &= \int_{\frac{1}{2}}^{1} [F^{-1}(1-x)]^2 (1-x)^{n-1}\, dx \\ &= \int_{\frac{1}{2}}^{1} [F^{-1}(x)]^2 (1-x)^{n-1}\, dx,\end{aligned} \tag{4.771}$$

we have

$$\mathbb{E}X_{n:n}^2 = n\int_0^1 [F^{-1}(x)]^2 x^{n-1}\, dx = n\int_{\frac{1}{2}}^{1} [F^{-1}(x)]^2 [x^{n-1} + (1-x)^{n-1}]\, dx. \tag{4.772}$$

and

$$\mathbb{V}ar\, X_1 = \mathbb{E}X_1^2 = \mathbb{E}X_{1:1}^2 = 2\int_{\frac{1}{2}}^{1} [F^{-1}(x)]^2\, dx \tag{4.773}$$

in particular. Therefore

$$\begin{aligned}\mathbb{V}ar\, X_{n:n} \le \mathbb{E}X_{n:n}^2 &= n\int_{\frac{1}{2}}^{1} [F^{-1}(x)]^2 [x^{n-1} + (1-x)^{n-1}]\, dx \\ &\le n\int_{\frac{1}{2}}^{1} [F^{-1}(x)]^2\, dx = \frac{n}{2}\, \mathbb{V}ar\, X_1.\end{aligned} \tag{4.774}$$

For proving attainability of the bound, we consider the marginal distributions

$$\mathbb{P}_\alpha(X_1 = \mp 1) = \alpha,$$
$$\mathbb{P}_\alpha(X_1 = 0) = 1 - 2\alpha, \qquad 0 < \alpha < \frac{1}{2}, \tag{4.775}$$

with the variances

$$\mathbb{V}ar_\alpha X_1 = \mathbb{E}_\alpha X_1^2 = 2\alpha. \tag{4.776}$$

The respective sample maxima have the distributions

$$\mathbb{P}_\alpha(X_{n:n} = -1) = \alpha^n = F_{n:n}(\alpha),$$
$$\mathbb{P}_\alpha(X_{n:n} = 0) = (1-\alpha)^n - \alpha^n = F_{n:n}(1-\alpha) - F_{n:n}(\alpha),$$
$$\mathbb{P}_\alpha(X_{n:n} = 1) = 1 - (1-\alpha)^n = 1 - F_{n:n}(1-\alpha), \tag{4.777}$$

and the first two moments

$$\mathbb{E}_\alpha X_{n:n} = 1 - (1-\alpha)^n - \alpha^n = 1 - F_{n:n}(1-\alpha) - F_{n:n}(\alpha), \tag{4.778}$$
$$\mathbb{E}_\alpha X_{n:n}^2 = 1 - (1-\alpha)^n - \alpha^n = 1 - F_{n:n}(1-\alpha) + F_{n:n}(\alpha). \tag{4.779}$$

Since

$$\lim_{\alpha\to 0} \mathbb{E}_\alpha X_{n:n} = \lim_{\alpha\to 0} [1 - F_{n:n}(1-\alpha) - F_{n:n}(\alpha)] = 0,$$
$$\lim_{\alpha\to 0} \frac{\mathbb{E}_\alpha X_{n:n}}{2\alpha} = \lim_{\alpha\to 0} \frac{1-F_{n:n}(1-\alpha)-F_{n:n}(\alpha)}{2\alpha} = \frac{f_{n:n}(1)-f_{n:n}(0)}{2} = \frac{n}{2},$$
$$\lim_{\alpha\to 0} \frac{[\mathbb{E}_\alpha X_{n:n}]^2}{2\alpha} = 0,$$
$$\lim_{\alpha\to 0} \frac{\mathbb{E}_\alpha X_{n:n}^2}{2\alpha} = \lim_{\alpha\to 0} \frac{1-F_{n:n}(1-\alpha)+F_{n:n}(\alpha)}{2\alpha} = \frac{f_{n:n}(1)+f_{n:n}(0)}{2} = \frac{n}{2},$$

we conclude that

$$\lim_{\alpha\to 0} \frac{\mathbb{V}ar_\alpha X_{n:n}}{\mathbb{V}ar_\alpha X_1} = \lim_{\alpha\to 0} \left[\frac{\mathbb{E}_\alpha X_{n:n}^2}{2\alpha} - \frac{[\mathbb{E}_\alpha X_{n:n}]^2}{2\alpha} \right] = \frac{n}{2}. \tag{4.780}$$

An alternative method of proving the upper bounds and respective attainability conditions is to maximize $S_{n:n}$ (see (4.749) and (4.750)) and calculate its limit at 0. The above approach looks more simple, though.

For establishing the lower bound, we first check that (4.770) has actually a unique solution in $\left(0, \frac{n}{2^{n-1}}\right)$. For this purpose we introduce the function

$$M_n(\lambda) = \int_{\frac{1}{2}}^{1} \frac{n^2[x^{n-1} - (1-x)^{n-1}]^2}{n[x^{n-1} + (1-x)^{n-1}] - 2\lambda}\, dx. \tag{4.781}$$

Note that for $0 < \lambda < \frac{n}{2^{n-1}}$ and $\frac{1}{2} \leq x < 1$ both the numerator and denominator are positive. Therefore (4.781) is increasing with respect to $0 < \lambda < \frac{n}{2^{n-1}}$. Applying the identity

$$\int_{\frac{1}{2}}^{1} n[x^{n-1} + (1-x)^{n-1}]\, dx = 1, \tag{4.782}$$

we can write

$$\begin{aligned}
1 - M_n(\lambda) &= \int_{\frac{1}{2}}^{1} \frac{n[x^{n-1} + (1-x)^{n-1}] - 2\lambda}{n[x^{n-1} + (1-x)^{n-1}] - 2\lambda}\, n[x^{n-1} + (1-x)^{n-1}]\, dx \\
&\quad - \int_{\frac{1}{2}}^{1} \frac{n^2[x^{n-1} - (1-x)^{n-1}]^2}{n[x^{n-1} + (1-x)^{n-1}] - 2\lambda}\, dx \\
&= \int_{\frac{1}{2}}^{1} \frac{n^2[x^{n-1} + (1-x)^{n-1}]^2 - n^2[x^{n-1} - (1-x)^{n-1}]^2}{n[x^{n-1} + (1-x)^{n-1}] - 2\lambda}\, dx \\
&\quad - \int_{\frac{1}{2}}^{1} \frac{2n\lambda[x^{n-1} + (1-x)^{n-1}]}{n[x^{n-1} + (1-x)^{n-1}] - 2\lambda}\, dx \\
&= \int_{\frac{1}{2}}^{1} \frac{4n^2 x^{n-1}(1-x)^{n-1} - 2n\lambda[x^{n-1} + (1-x)^{n-1}]}{n[x^{n-1} + (1-x)^{n-1}] - 2\lambda}\, dx
\end{aligned} \tag{4.783}$$

For $\lambda = 0$ the integrand is positive which implies that $M_n(0) < 1$. Note that

$$\begin{aligned}
x^{n-1}(1-x)^{n-1} &< \frac{1}{2^{2n-2}}, \\
x^{n-1} + (1-x)^{n-1} &> \frac{1}{2^{n-2}}, \qquad \frac{1}{2} < x < 1.
\end{aligned} \tag{4.784}$$

This implies that for $\lambda = \frac{n}{2^{n-1}}$ the numerator in last integral of (4.783) is always negative, and the respective denominator is always positive. This allows us to conclude that $M_n(0) < 1 < M_n\left(\frac{n}{2^{n-1}}\right)$, and by strict monotonicity of $M_n(\lambda)$, in $\left(0, \frac{n}{2^{n-1}}\right)$, there exists a unique $0 < \lambda_* = \lambda_*(n) < \frac{n}{2^{n-1}}$ satisfying $M_n(\lambda_*) = 1$.

Consider now two functions

$$f(x) = F^{-1}(x)\{n[x^{n-1} + (1-x)^{n-1} - 2\lambda_*\}^{1/2},$$
$$g(x) = \frac{n[x^{n-1} - (1-x)^{n-1}]}{\{n[x^{n-1} + (1-x)^{n-1}] - 2\lambda_*\}^{1/2}}, \qquad \frac{1}{2} < x < 1, \tag{4.785}$$

where $F^{-1}(x)$ denotes the quantile function of a symmetric distribution about 0, and apply the Schwarz inequality to them. Then we obtain

$$\begin{aligned}
\left(\int_{\frac{1}{2}}^{1} f(x)g(x)\,dx\right)^2 &= \left(\int_{\frac{1}{2}}^{1} F^{-1}(x)\,n[x^{n-1} - (1-x)^{n-1}]\,dx\right)^2 = (\mathbb{E}X_{n:n})^2 \\
&\le \int_{\frac{1}{2}}^{1} f^2(x)\,dx \int_{\frac{1}{2}}^{1} g^2(x)\,dx \\
&= \int_{\frac{1}{2}}^{1} [F^{-1}(x)]^2\{n[x^{n-1} - (1-x)^{n-1}] - 2\lambda_*\}\,dx \\
&\times \int_{\frac{1}{2}}^{1} \frac{n^2[x^{n-1} - (1-x)^{n-1}]^2}{n[x^{n-1} + (1-x)^{n-1}] - 2\lambda_*}\,dx \\
&= (\mathbb{E}X_{n:n}^2 - \lambda_*\mathbb{E}X_1^2)M_n(\lambda_*) \\
&= \mathbb{V}ar\, X_{n:n} + (\mathbb{E}X_{n:n})^2 - \lambda_*\mathbb{V}ar\, X_1,
\end{aligned} \tag{4.786}$$

which is obviously nonnegative and gives the LHS inequality in (4.769), and completes the proof. □

Moriguti (1951) claimed that $\lambda_* = \lambda_*(n)$ provides the optimal lower bound for the variance. The claim was repeated in several references of various authors (see, e.g., David & Nagaraja, 2003, p. 86). The attainability condition allegedly follows from the equality condition in the Schwarz inequality. In this case it has the form

$$F^{-1}(x)\{n[x^{n-1} + (1-x)^{n-1} - 2\lambda_*\}^{1/2} = C\,\frac{n[x^{n-1} - (1-x)^{n-1}]}{\{n[x^{n-1} + (1-x)^{n-1}] - 2\lambda_*\}^{1/2}} \tag{4.787}$$

with a positive C almost everywhere on $\left(\frac{1}{2}, 1\right)$, i.e.,

$$F^{-1}(x) = \begin{cases} -C\frac{n[(1-x)^{n-1} - x^{n-1}]}{n[x^{n-1} + (1-x)^{n-1}] - 2\lambda_*}, & 0 < x < \frac{1}{2}, \\ C\frac{n[x^{n-1} - (1-x)^{n-1}]}{n[x^{n-1} + (1-x)^{n-1}] - 2\lambda_*}, & \frac{1}{2} \le x < 1. \end{cases} \tag{4.788}$$

Note that for $n = 2$ the function takes on a linear form

$$F^{-1}(x) = \frac{C}{2}\frac{2x-1}{1-\lambda_*}, \qquad 0 < x < 1. \tag{4.789}$$

Also, (4.770) has the simple solution $\lambda_*(2) = \frac{2}{3}$ then. This gives the following conclusion.

Corollary 4.33 *Under the assumptions of Theorem* 4.60 *the lower bound*

$$\frac{\mathbb{V}ar\, X_{2:2}}{\mathbb{V}ar\, X_1} \geq \frac{2}{3}, \tag{4.790}$$

and the bound is attained for the uniform parent distributions.

The same statement holds for $X_{1:2}$ *as well.*

However, for $n \geq 3$ and $\frac{1}{2} < x < 1$

$$\frac{d}{dx}F^{-1}(x) = 2(n-1)C\,\frac{x^{n-2}(1-x)^{n-2} - \frac{\lambda_*}{n}[x^{n-2}+(1-x)^{n-2}]}{[x^{n-1}+(1-x)^{n-1}-\frac{2}{n}\lambda_*]^2}, \tag{4.791}$$

which continuously tends to $-\frac{2(n-1)C\lambda_*}{n[1-\frac{2}{n}\lambda_*]^2} < 0$ as $x \to 1$. This contradict the fact that (4.788) defines an antisymmetric about $\frac{1}{2}$ quantile function which has to be nondecreasing on the whole (0, 1). To the best of our knowledge, the mistake was not detected before.

Proof of Lemma 4.4 We start with verifying (4.756). Write

$$\begin{aligned} V'_{j:n}(\gamma) &= f_{j:n}(\gamma)[1 - 2F_{j:n}(\gamma) + 2F_{n+1-j:n}(\gamma)] \\ &\quad + f_{n+1-j:n}(\gamma)[1 + 2F_{j:n}(\gamma) - 2F_{n+1-j:n}(\gamma)] \\ &= n\binom{n-1}{j-1}[U_{j:n}(\gamma) + W_{j:n}(\gamma)], \end{aligned} \tag{4.792}$$

where

$$\begin{aligned} U_{j:n}(\gamma) &= \gamma^{j-1}(1-\gamma)^{n-j}\left[\sum_{i=0}^{j-1} - \sum_{i=j}^{n-j} + \sum_{i=n-j+1}^{n}\right] B_{i,n}(\gamma) \\ &= \left[\sum_{i=j-1}^{2j-2} - \sum_{i=2j-1}^{n-1} + \sum_{i=n}^{n+j-1}\right]\binom{n}{i-j+1}\gamma^i(1-\gamma)^{2n-1-i}, \end{aligned} \tag{4.793}$$

$$
\begin{aligned}
W_{j:n}(\gamma) &= \gamma^{n-j}(1-\gamma)^{j-1}\left[\sum_{i=0}^{j-1} +3\sum_{i=j}^{n-j} + \sum_{i=n-j+1}^{n}\right] B_{i,n}(\gamma) \\
&= \left[\sum_{i=n-j}^{n-1} +3\sum_{i=n}^{2n-2j} + \sum_{i=2n-2j+1}^{2n-j}\right]\binom{n}{i+j-n}\gamma^i(1-\gamma)^{2n-1-i},
\end{aligned}
\tag{4.794}
$$

with the convention

$$
\left[a\sum_{i=j}^{k} +b\sum_{i=\ell}^{m}\right] c_i = a\sum_{i=j}^{k} c_i + b\sum_{i=\ell}^{m} c_i.
$$

Put

$$
U_{j:n}(\gamma) = \sum_{i=0}^{2n-1} u_{j:n}(i)\gamma^i(1-\gamma)^{2n-1-i}, \tag{4.795}
$$

$$
W_{j:n}(\gamma) = \sum_{i=0}^{2n-1} w_{j:n}(i)\gamma^i(1-\gamma)^{2n-1-i}, \tag{4.796}
$$

for

$$
u_{j:n}(i) = \begin{cases} 0, & i = 0, \ldots, j-2, n+j, \ldots, 2n-1, \\ \binom{n}{i-j+1}, & i = j-1, \ldots, 2j-2, n, \ldots, n+j-1 \\ -\binom{n}{i-j+1}, & i = 2j-1, \ldots, n-1, \end{cases} \tag{4.797}
$$

$$
w_{j:n}(i) = \begin{cases} 0, & i = 0, \ldots, n-j-1, 2n-j+1, \ldots, 2n-1, \\ \binom{n}{i+j-n}, & i = n-j, \ldots, n-1, 2n-2j+1, \ldots, 2n-j, \\ 3\binom{n}{i+j-n}, & i = n, \ldots, 2n-2j, \end{cases} \tag{4.798}
$$

and define $a_{j:n}(i) = u_{j:n}(i) + w_{j:n}(i)$, $i = 0, \ldots, 2n-1$. Since $\binom{2n-1}{i}\alpha_{j:n}(i) = n\binom{n-1}{j-1}a_{j:n}(i)$, $i = 0, \ldots, 2n-1$, we analyze the signs of the latter ones. Condition $j \le \frac{n}{2}$ implies $j-2 < n-j-1$ and $n+j < 2n-j-1$. Therefore $a_{j:n}(i) = u_{j:n}(i) = w_{j:n}(i) = 0$, $i = 0, \ldots, j-2, 2n-j+1, \ldots, 2n-1$. For each $i = j-1, \ldots, 2j-2, n, \ldots, 2n-j$ the inequality $u_{j:n}(i)+w_{j:n}(i) > 0$ holds because both the summands are non-negative, and at least one of them is positive. If $j < \frac{n+1}{3}$, and so $2j-1 < n-j$, then

$$a_{j:n}(i) = \begin{cases} -\binom{n}{i-j+1}, & i = 2j-1, \dots, n-j-1, \\ \binom{n}{i+j-n} - \binom{n}{i-j+1}, & i = n-j, \dots, n-1. \end{cases} \tag{4.799}$$

Otherwise the latter formula holds for $i = 2j-1, \dots, n-1$. In order to obtain the final claim $\alpha_{j:n}(i) < 0$, $i = 2j-1, \dots, n-1$, it suffices to show that

$$\binom{n}{i+j-n} < \binom{n}{i-j+1}, \qquad i = \max\{2j-1, n-j\}, \dots, n-1. \tag{4.800}$$

Note that the sequence $\left(\binom{n}{i}\right)_{i=0}^{n}$ is first increasing, and then decreasing, and symmetric about $\frac{n}{2}$. Accordingly, since $j-1 < \frac{n}{2}$, the sequence $\left(\binom{n}{i+j-n}\right)_{i=n-j}^{n-1}$ $= \left(\binom{n}{k}\right)_{k=0}^{j-1}$ is increasing and has the maximal value $\binom{n}{j-1}$. The subsequence $\left(\binom{n}{i+j-n}\right)_{i=\max\{2j-1,n-j\}}^{n-1}$ has the same properties. The sequence $\left(\binom{n}{i-j+1}\right)_{i=2j-1}^{n-1}$, identical with $\left(\binom{n}{k}\right)_{k=j}^{n-j}$ is increasing-decreasing, symmetric about $\frac{n}{2}$, and its minimal value is $\binom{n}{j} = \binom{n}{n-j}$. Also, this provides the minimum of the sequence $\left(\binom{n}{i-j+1}\right)_{i=\max\{2j-1,n-j\}}^{n-1}$. So we showed that

$$\binom{n}{i+j-n} \le \binom{n}{j-1} < \binom{n}{j} \le \binom{n}{i-j+1} \tag{4.801}$$

for all $i = \max\{2j-1, n-j\}, \dots, n-1$, which completes the proof of (4.756).

For checking (4.757), we introduce further auxiliary notation. Write

$$V''_{j:n}(\gamma) = n\binom{n-1}{j-1}[U'_{j:n}(\gamma) + V'_{j:n}(\gamma)]. \tag{4.802}$$

Here

$$\begin{aligned} U'_{j:n}(\gamma) &= \sum_{i=j-1}^{n+j-1} u_{j:n}(i)[i\gamma^{i-1}(1-\gamma)^{2n-1-i} - (2n-1-i)\gamma^i(1-\gamma)^{2n-2-i}] \\ &= \sum_{i=0}^{2n-2} t_{j:n}(i)\gamma^i(1-\gamma)^{2n-2-i} \end{aligned} \tag{4.803}$$

with

$$t_{j:n}(i) = \begin{cases} 0, & i = 0, \ldots, j-3, \\ (j-1)u_{j:n}(j-1), & i = j-2, \\ (i+1)u_{j:n}(i+1) - (2n-1-i)u_{j:n}(i), & i = j-1, \ldots, n+j-2, \\ -(n-j)u_{j:n}(n+j-1), & i = n+j-1, \\ 0, & i = n+j, \ldots, 2n-2, \end{cases}$$

$$= \begin{cases} 0, & i = 0, \ldots, j-3, \\ j-1, & i = j-2, \\ t^*_{j:n}(i), & i = j-1, \ldots, 2j-3, \\ -(2j-1)\binom{n}{j} - (2n-2j+1)\binom{n}{j-1}, & i = 2j-2, \\ -t^*_{j:n}(i), & i = 2j-1, \ldots, n-2, \\ n\left[\binom{n}{j} + \binom{n}{j-1}\right], & i = n-1, \\ t^*_{j:n}(i), & i = n, \ldots, n+j-2, \\ -n+j, & i = n+j-1, \\ 0, & i = n+j, \ldots, 2n-2, \end{cases} \tag{4.804}$$

and

$$t^*_{j:n}(i) = (i+1)\binom{n}{i-j+2} - (2n-1-i)\binom{n}{i-j+1}, \quad i = j-1, \ldots, n-j-2. \tag{4.805}$$

Since

$$(i-j+2)\binom{n}{i-j+2} - (n-i+j-1)\binom{n}{i-j+1} = 0, \tag{4.806}$$

we simplify (4.805) to

$$t^*_{j:n}(i) = (j-1)\binom{n}{i-j+2} - (n-j)\binom{n}{i-j+1} = \binom{n+1}{i-j+2}\frac{(j-1)(n-i+j-1) - (n-j)(i-j+2)}{n+1}. \tag{4.807}$$

Hence the condition $t^*_{j:n}(i) > 0$ is equivalent to $i < 2j-3+2\frac{j-1}{n-1}$. Since $2 \le j \le \frac{n}{2}$ implies $0 < 2\frac{j-1}{n-1} < 1$, we have $t^*_{j:n}(i) > 0$ for $i = j-1, \ldots, 2j-3$. In the similar way we get $t^*_{j:n}(i) < 0$ for $i = 2j-2, \ldots, n+j-2$. Therefore

$$\operatorname{sgn}(t_{j:n}(i)) = \begin{cases} 0, \quad i = 0, \ldots, j-3, n+j, \ldots, 2n-2, \\ +1, i = j-2, \ldots, 2j-3, 2j-1, \ldots, n-1, \\ -1, i = 2j-2, n, \ldots, n+j-1. \end{cases} \tag{4.808}$$

Likewise

$$W'_{j:n}(\gamma) = \sum_{i=0}^{2n-2} z_{j:n}(i)\gamma^i(1-\gamma)^{2n-2-i} \tag{4.809}$$

with

$$z_{j:n}(i) = \begin{cases} 0, & i = 0, \ldots, n-j-2, \\ n-j, & i = n-j-1, \\ z^*_{j:n}(i), & i = n-j, \ldots, n-2, \\ n\left[3\binom{n}{j} - \binom{n}{j-1}\right], & i = n-1, \\ 3z^*_{j:n}(i), & i = n, \ldots, 2n-2j-1, \\ (2n-2j+1)\binom{n}{j-1} - (6j-3)\binom{n}{j}, & i = 2n-2j, \\ z^*_{j:n}(i), & i = 2n-2j+1, \ldots, 2n-j-1, \\ -j+1, & i = 2n-j, \\ 0, & i = 2n-j+1, \ldots, 2n-2, \end{cases} \tag{4.810}$$

and

$$\begin{aligned} z^*_{j:n}(i) &= (i+1)\binom{n}{i+j-n+1} - (2n-1-i)\binom{n}{i+j-n} \\ &= (n-j)\binom{n}{i+j-n+1} - (j-1)\binom{n}{i+j-n} \\ &= \binom{n+1}{i+j-n+1}\frac{(n-j)(2n-i-j) - (j-1)(i+j-n+1)}{n+1} \end{aligned} \tag{4.811}$$

for $i = n-j, \ldots, 2n-j-1$. It is easy to show that $z^*_{j:n}(i) > 0$ for $i = n-j, \ldots, 2n-2j$, and $z^*_{j:n}(i) < 0$ for $i = 2n-2j+1, \ldots, 2n-j-1$. We can also check that $z_{j:n}(2n-2j) < 0$, which corresponds to the inequality

$$4j^2 - 4(n+2)j + 3(n+1) < 0.$$

Indeed, the left-hand side amounts to $-5n + 3 < 0$ at $j = 2$, and decreases when $j < \frac{n}{2} + 1$. Summing up, we have

$$\operatorname{sgn}(z_{j:n}(i)) = \begin{cases} 0, & i = 0, \dots, n-j-2, 2n-j+1, \dots, 2n-2, \\ +1, & i = n-j-1, \dots, 2n-2j-1, \\ -1, & i = 2n-2j, \dots, 2n-j. \end{cases} \tag{4.812}$$

Our aim is to determine the signs of sums $b_{j:n}(i) = t_{j:n}(i) + z_{j:n}(i)$, which are identical with those of $\beta_{j:n}(i)$, $i = 0, \dots, 2n-2$. The summands take on different forms in various domains of variable i. We have

$$\begin{aligned} 0 &\le j-2 < \min\{2j-1, n-j\} \le \max\{2j-1, n-j\} < n-1 \\ &\le \min\{n+j-1, 2n-2j\} \le \max\{n+j-1, 2n-2j\} < 2n-j \le 2n-2. \end{aligned}$$

The relations $j < \frac{n+1}{3}$ (=, >, respectively) imply that $2j-1 < n-j$ (=, >, respectively), and $n+j-1 < 2n-2j$ (=, >, respectively). We first notice that

$$\operatorname{sgn}(b_{j:n}(i)) = \begin{cases} 0, & i = 0, \dots, j-3, 2n-j+1, \dots, 2n-2, \\ +1, & i = j-2, \dots, 2j-3, 2j-1, \dots, n-1, \\ -1, & i = 2n-2j, \dots, 2n-j, \end{cases} \tag{4.813}$$

because then both the summands either have identical signs or one of them vanishes, and the other determines the sign of the sum.

In order to get (4.757), we need to check that $b_{j:n}(2j-2) < 0$. If $j < \frac{n+1}{3}$, then $b_{j:n}(2j-2) = t_{j:n}(2j-2) < 0$ (cf. (4.808)). If $j = \frac{n+1}{3}$, then

$$b_{j:n}(2j-2) = -(2j-1)\binom{n}{j} - (2n-2j+1)\binom{n}{j-1} + n - j < 0 \tag{4.814}$$

because $2n-2j+1 > n-j$, $\binom{n}{j-1} > 1$ and so $(2n-2j+1)\binom{n}{j-1} > n-j$. Finally, for $j > \frac{n+1}{3}$ we have

$$\begin{aligned} b_{j:n}(2j-2) = &-(2j-1)\binom{n}{j} - (2n-2j+1)\binom{n}{j-1} \\ &+ (n-j)\binom{n}{3j-n-1} - (j-1)\binom{n}{3j-n-2}. \end{aligned} \tag{4.815}$$

Then $0 < 3j-n-1 \le j-1 < \frac{n}{2}$ implies $\binom{n}{3j-n-1} \le \binom{n}{j-1}$, and furthermore $(2n-2j+1)\binom{n}{j-1} > (n-j)\binom{n}{3j-n-1}$, and eventually negativity of (4.815).

From now on we assume that $j \le \frac{n-1}{2}$. We first show that $b_{j:n}(n) = t^*_{j:n}(n) + 3z^*_{j:n}(n) > 0$ for all $2 \le j \le \frac{n-1}{2}$. Since $t^*_{j:n}(n) < 0 < z^*_{j:n}(n)$, it suffices to prove that

$$
\begin{aligned}
c_{j:n}(n) &= \frac{z^*_{j:n}(n)}{-t^*_{j:n}(n)} = \frac{(n^2-2jn+1)(n-j+1)(n-j+2)}{(n^2-2jn+2n-1)j(j+1)} \\
&= \left[1 - \frac{2(n-1)}{n^2+2n-1-2jn}\right] \frac{n-j+1}{j} \frac{n-j+2}{j+1} > \frac{1}{3}. \qquad (4.816)
\end{aligned}
$$

All the factors of the product in the second line are positive and decreasing with respect to j. Hence setting $j = \frac{n-1}{2}$ we obtain

$$
c_{j:n}(n) \ge \frac{(n+3)(n+5)}{(3n-1)(n-1)}. \qquad (4.817)
$$

The fraction of the right-hand side is greater than $\frac{1}{3}$ iff $7n + 11 > 0$, which is true for all n, and completes the proof of the statement.

Now we proceed to proving non-increase of $(\text{sgn}(\beta_{j:n}(i)))_{i=n}^{2n-2j-1}$. We first check that this is true for $i = n, \ldots, \min\{n+j-2, 2n-2j-1\}$ for which $b_{j:n}(i) = t^*_{j:n}(i) + 3z^*_{j:n}(i)$ with $t^*_{j:n}(i) < 0 < z^*_{j:n}(i)$. Since $b_{j:n}(n) > 0$, we need to verify that either the sequence is positive for all i, or it changes the sign from plus to minus. In fact, we prove that the positive sequence

$$
\begin{aligned}
c_{j:n}(i) &= \frac{z^*_{j:n}(i)}{-t^*_{j:n}(i)} = \frac{2n^2-2jn-n+1-(n-1)i}{(n-1)i-2jn+3n-1} \frac{(i-j+2)!}{(i+j-n+1)!} \\
&\times \frac{(n+j-i-1)!}{(2n-i-j)!}, \qquad i=n, \ldots, \min\{n+j-2, 2n-2j-1\}, \qquad (4.818)
\end{aligned}
$$

is either increasing or decreasing or first increasing and then decreasing. Since $c_{j:n}(n) > 1$, this would imply that (4.818) is first greater than $\frac{1}{3}$, and then possibly less than $\frac{1}{3}$. And this would finally mean that the original sequence $(b_{j:n}(i))_{i=n}^{\min\{n+j-2, 2n-2j-1\}}$ is first positive and then possibly negative. Actually, it suffices to check that the positive sequence

$$
\begin{aligned}
\frac{c_{j:n}(n+k+1)}{c_{j:n}(n+k)} &= \frac{[n^2-2jn-n+2-(n-1)k][(n-1)k+n^2-2jn+2n-1]}{[n^2-2jn+1-(n-1)k][(n-1)k+n^2-2jn+3n-2]} \\
&\times \frac{(n-j+3+k)(n-j-k)}{(j+2+k)(j-1-k)}, \qquad k = 0, \ldots, \min\{j-3, n-2j-2\}, \qquad (4.819)
\end{aligned}
$$

is either greater than 1, or less than 1, or first greater and then less than 1. With $\nu = n - 1$ and $\zeta = n^2 - 2nj + 1$ introduced for abridging the notation, we can rewrite the numerator of (4.819) as

$$\begin{aligned} N_{j:n}(k) &= (\zeta - \nu - \nu k)(\zeta + 2\nu + \nu k)(n - j + 3 + k)(n - j - k) \\ &= \nu^2 k^4 + 6\nu^2 k^3 - [(n-j+3)(n-j)\nu^2 + (\zeta - \nu)(\zeta + 2\nu) - 9\nu^2]k^2 \\ &- 3[(\zeta - \nu)(\zeta + 2\nu) + \nu^2(n - j + 3)(n - j)]k \\ &+ (\zeta - \nu)(\zeta + 2\nu)(n - j + 3)(n - j). \end{aligned} \tag{4.820}$$

Likewise we represent the denominator as

$$\begin{aligned} M_{j:n}(k) &= (\zeta - \nu k)(\zeta + 3\nu + \nu k)(j + 2 + k)(j - 1 - k) \\ &= \nu^2 k^4 + 6\nu^2 k^3 - [(j+2)(j-1)\nu^2 + \zeta(\zeta + 3\nu) - 9\nu^2]k^2 \\ &- 3[\zeta(\zeta + 3\nu) + \nu^2(j + 2)(j - 1)]k \\ &+ \zeta(\zeta + 3\nu)(j + 2)(j - 1). \end{aligned} \tag{4.821}$$

We get

$$\frac{c_{j:n}(n + k + 1)}{c_{j:n}(n + k)} = \frac{N_{j:n}(k)}{M_{j:n}(k)} > 1 \tag{4.822}$$

iff

$$\begin{aligned} &N_{j:n}(k) - M_{j:n}(k) = (k^2 + 3k) \\ &\times [\nu^2(j+2)(j-1) - \nu^2(n-j+3)(n-j) + \zeta(\zeta+3\nu) - (\zeta-\nu)(\zeta+2\nu)] \\ &+ (\zeta-\nu)(\zeta+2\nu)(n-j+3)(n-j) - \zeta(\zeta+3\nu)(j+2)(j-1) > 0. \end{aligned} \tag{4.823}$$

Simple algebra shows that the expression in the square brackets

$$\begin{aligned} &\nu^2(j + 2)(j - 1) - \nu^2(n - j + 3)(n - j) + \zeta(\zeta + 3\nu) - (\zeta - \nu)(\zeta + 2\nu) \\ &= -(n - 1)^2(n + 2)(n + 1 - 2j) + 2(n - 1)(n^2 - 2jn + n) \\ &= (n - 1)(n + 1 - 2j)[2n - (n - 1)(n + 2)] \\ &= -(n - 2)(n - 1)(n + 1)(n + 1 - 2j) \end{aligned} \tag{4.824}$$

is negative. Therefore the quadratic function of (4.823) is concave and decreasing for $k > -\frac{3}{2}$. This is either non-negative or non-positive for all $k = 0, \ldots, \min\{n + j - 2, 2n - 2j - 1\}$ or it changes the sign from plus to minus. This means that (4.819) has the desired properties. Note that for $j \geq \frac{n+1}{3}$ we have $2n - 2j - 1 \leq n + j - 2$, and the proof is complete.

Suppose now that $j < \frac{n+1}{3}$ and so $n + j - 2 < 2n - 2j - 1$. Observe that $t_{j:n}(i) = 0, i = n+j, \ldots, 2n-2$, and $z_{j:n}(i) > 0, i = n-j-1, \ldots, 2n-2j-1$. If $n + j \le 2n - 2j - 1$ as well, we have $b_{j:n}(i) > 0, i = n + j, \ldots, 2n - 2j - 1$. Under the assumption, the final claim of Lemma 4.4 is equivalent with saying that $b_{j:n}(i) > 0, i = n + 1, \ldots, 2n - 2j - 1$. By the conclusion of the previous part of the proof it remains only to verify that $b_{j:n}(i) > 0, i = n + j - 2, n + j - 1$ when $2 \le j < \frac{n+1}{3}$. Notice that the condition forces $n \ge 6$.

We first evaluate

$$c_{j:n}(n + j - 2) = \frac{n^2 + 2n - 1 - (3n - 1)j}{n^2 + 1 - (n + 1)j} \frac{n!}{(2j - 1)!(n + 2 - 2j)!}. \tag{4.825}$$

The first fraction has the negative derivative $-\frac{2n(n-1)^2}{[n^2+1-(n+1)j]^2}$ with respect to j, and so is greater than its value at $j = \frac{n+1}{3}$ which amounts to $\frac{2n-1}{n^2-n+1}$. The assumption $2 \le j < \frac{n+1}{3}$ implies that $3 \le 2j-1 < n - \frac{n+1}{3} < n-2$ and $n-2 \ge n+2-2j > \frac{n+1}{3} + 1 > 3$, and so we have

$$\frac{n!}{(2j-1)!(n+2-2j)!} = \frac{1}{n+1}\binom{n+1}{2j-1} \ge \frac{1}{n+1}\binom{n+1}{3} = \frac{n(n-1)}{6} \ge n-1. \tag{4.826}$$

Therefore

$$c_{j:n}(n + j - 2) > \frac{(2n - 1)(n - 1)}{n^2 - n + 1} \ge 1, \tag{4.827}$$

and the latter inequality follows from the fact that $n(n - 2) \ge 0$ for all $n \ge 2$. In consequence, $b_{j:n}(n + j - 2) = t^*_{j:n}(n + j - 2) + 3z^*_{j:n}(n + j - 2) > 0$, as desired.

In a similar way, we evaluate

$$\begin{aligned} b_{j:n}(n + j - 1) &= -n + j + 3(n - j)\binom{n}{2j} - 3(j - 1)\binom{n}{2j - 1} \\ &= 3\binom{n + 1}{2j} \frac{(n - j)(n - 2j + 1) - 2j(j - 1)}{n + 1} - n + j. \end{aligned} \tag{4.828}$$

Since $2j \ge 4$ and $n + 1 - 2j \ge 3$, we have

$$\frac{1}{n + 1}\binom{n + 1}{2j} \ge \frac{n(n - 1)}{6} \ge n - 1 > n - j.$$

For getting positivity of (4.828), we only need to show that

$$(n - j)(n - 2j + 1) - 2j(j - 1) \geq 1.$$

Indeed, the left-hand side is decreasing with respect to j, and so is greater than

$$\left(n - \frac{n+1}{3}\right)\frac{n+1}{3} - 2\frac{n+1}{3}\left(\frac{n+1}{3} - 1\right) = \frac{n+1}{3} \geq \frac{7}{3} > 1. \tag{4.829}$$

This ends the proof. □

Chapter 5
Sampling from Finite Populations

In this chapter we consider bounds on the expectations of order statistics and their linear combinations for various drawing schemes from finite populations of a given size N. We also determine bounds on variances of order statistics based on sampling from finite populations. In Sect. 5.1, sharp positive upper bounds and negative lower ones for the properly centered L-statistics, expressed in various scale units, based on the standard model of drawing with replacement are presented. They were determined in Rychlik (2004). These results were completed by respective non-positive upper and non-negative lower bounds in various scale units coming from Goroncy and Rychlik (2006b) and Goroncy and Rychlik (2009). We finally present sharp upper bounds on the variances of order statistics from the drawing with replacement models. The results were established by Jasiński and Rychlik (2016). This drawing scheme provides an example of iid sample from a specific restricted family of distributions which are discrete and uniformly distributed on their domains.

A generalization to fixed nonuniform distributions on arbitrary N atoms is presented in Sect. 5.2. The general bounds for the expectations of order statistics based on iid samples with arbitrary distributions on arbitrary N points of the real axis are also presented. This is mainly based on the results of López-Blazquez (1998), López-Blazquez (2000), and López-Blazquez and Rychlik (2008).

In Sect. 5.3, we consider analogous problems for the random samples arising from the drawing without replacement models. They were studied by Balakrishnan et al. (2003) and Papadatos and Rychlik (2004) who determined the sharp positive upper mean-variance bounds on the expectations of single order statistics and L-statistics, respectively. The lower bounds on the expectations of positive L-statistics are based on the results of Goroncy and Rychlik (2008). We also present bounds of variances of order statistics from the model which were determined by Jasiński and Rychlik (2016). The drawing with and without replacement schemes from finite populations are of great interest due to numerous practical applications.

N. Balakrishnan, T. Rychlik, *Sharp Inequalities for Ordered Random Variables in Statistics and Reliability*, Frontiers in Probability and the Statistical Sciences,
https://doi.org/10.1007/978-3-031-61347-0_5

The last section is devoted to numerical evaluations of bounds of expectations of order statistics coming from sample drawings from finite populations determined in this chapter. Various sample sizes are considered for a fixed population size $N = 20$. These bounds are compared with the bounds for order statistics coming from iid samples with arbitrary marginals presented earlier in Chap. 4.

5.1 Classic With Replacement Model

5.1.1 Positive Upper and Negative Lower Bounds for General L-Statistics

Consider a population of N objects, labeled by real numbers $x_1, \ldots, x_N$. We perform an experiment that consists in drawing n objects from the population when the drawn element is either returned to the population or replaced by an identical one. Let X_i, $1 \le i \le n$, denote the random value of the numerical label assigned to the object derived in the ith drawing. Then, $X_1, \ldots, X_n$ are independent and have the identical distribution

$$\mathbb{P}(X_i = x_k) = \frac{1}{N}\#\{r : x_r = x_k\}, \qquad k = 1, \ldots, N. \tag{5.1}$$

The expectation and central absolute moments of X_i, $1 \le i \le n$, coincide with the respective population moments

$$\mu = \mathbb{E}X_i = \bar{x} = \frac{1}{N}\sum_{k=1}^{N} x_k, \tag{5.2}$$

$$\sigma_p^p = \mathbb{E}|X_i - \mu|^p = s_p^p = \frac{1}{N}\sum_{k=1}^{N} |x_k - \bar{x}|^p, \qquad 1 \le p < +\infty. \tag{5.3}$$

Also,

$$\sigma_\infty = \operatorname{esssup}|X_i - \mu| = s_\infty = \max\{\bar{x} - x_{1:N}, x_{N:N} - \bar{x}\} \tag{5.4}$$

is well defined and finite.

Here we provide optimal positive upper bounds and negative lower bounds on the expectations of arbitrary L-statistics based on the sample $X_1, \ldots, X_n$ in terms of location parameter (5.2) and various scale parameters expressed in terms of (5.3) and (5.4). The bounds are specified for single order statistics, and differences of order statistics. The main results come from Rychlik (2004). In the ensuring analysis, it is convenient to introduce a standard population of first N positive integers $\{1, \ldots, N\}$, and the random iid sample $V_1, \ldots, V_n$ drawn with replacement from the population. Since

$$\mathbb{P}(V_i \leq k) = \frac{k}{N}, \qquad i = 1, \ldots, n, \ k = 1, \ldots, N, \tag{5.5}$$

we have

$$\mathbb{P}(V_{i:n} \leq k) = \sum_{r=i}^{n} \binom{n}{r} \left(\frac{k}{N}\right)^r \left(1 - \frac{k}{N}\right)^{n-r} = F_{i:n}\left(\frac{k}{N}\right) \tag{5.6}$$

(see (4.2)), where $F_{i:n}(x)$, $0 \leq x \leq 1$, denotes the distribution function of the ith order statistic of the i.i.d. sample from the standard uniform distribution on the interval [0, 1]. Therefore, the probability mass function of each $V_{i:n}$ can be simply written as

$$\mathbb{P}(V_{i:n} = k) = F_{i:n}\left(\frac{k}{N}\right) - F_{i:n}\left(\frac{k-1}{N}\right), \quad i = 1, \ldots, n, \ k = 1, \ldots, N. \tag{5.7}$$

An alternative representation, based on multinomial arguments,

$$\mathbb{P}(V_{i:n} = k) = \sum_{n=0}^{i} \sum_{s=0}^{n-i} \frac{n!(k-1)^{i-1-r}(N-k)^{n-i-s}}{(i-1-r)!(n-i-s)!(r+s+1)!\,N^n} \tag{5.8}$$

follows from the fact that the event $\{V_{i:n} = k\}$ is realized in $i(n+1-i)$ ways so that $i-1-r$, $r+s+1$ and $n-i-s$ observations are less than, equal to and greater than k, respectively (see Arnold et al., 1992, p. 42).

For a given population represented by the sequence of numbers $\tilde{x} = (x_1, \ldots, x_N) \in \mathbb{R}^N$, we define a nondecreasing sequence

$$g_{\tilde{x}}(k) = x_{k:N}, \qquad k = 1, \ldots, N, \tag{5.9}$$

so that

$$(g_{\tilde{x}}(V_{1:n}), \ldots, g_{\tilde{x}}(V_{n:n})) \stackrel{d}{=} (X_{1:n}, \ldots, X_{n:n}) \tag{5.10}$$

holds. The relation provides a nice and useful representation for the expectation of as arbitrary L-statistic from the general with replacement models as

$$\begin{aligned}
\mathbb{E} \sum_{i=1}^{n} c_i X_{i:n} &= \mathbb{E} \sum_{i=1}^{n} c_i g_{\tilde{x}}(V_{i:n}) = \sum_{i=1}^{n} c_i \sum_{k=1}^{N} g_{\tilde{x}}(k) \mathbb{P}(V_{i:n} = k) \\
&= \sum_{k=1}^{n} g_{\tilde{x}}(k) \sum_{i=1}^{n} c_i \mathbb{P}(V_{i:n} = k)
\end{aligned}$$

$$= \sum_{k=1}^{N} x_{k:N} \sum_{i=1}^{n} c_i \left[F_{i:n}\left(\frac{k}{N}\right) - F_{i:n}\left(\frac{k-1}{N}\right) \right]$$

$$= \sum_{k=1}^{N} x_{k:N} \left[F_{\tilde{c}:n}\left(\frac{k}{N}\right) - F_{\tilde{c}:n}\left(\frac{k-1}{N}\right) \right], \tag{5.11}$$

where

$$F_{\tilde{c}:n}(x) = \sum_{i=1}^{n} c_i F_{i:n}(x)$$

for any $\tilde{c} \in \mathbb{R}^n$. For a given $\tilde{c}$, define

$$C_k = F_{\tilde{c}:n}\left(\frac{k}{N}\right) - F_{\tilde{c}:n}\left(\frac{k-1}{N}\right), \qquad k = 1, \ldots, N, \tag{5.12}$$

and note that

$$\sum_{k=1}^{N} C_k = \sum_{k=1}^{N} \sum_{i=1}^{n} c_i \mathbb{P}(V_{i:n} = k) = \sum_{i=1}^{n} c_i \sum_{k=1}^{N} \mathbb{P}(V_{i:n} = k) = \sum_{i=1}^{n} c_i. \tag{5.13}$$

Consequently, by (5.2)–(5.4) and (5.13), we have

$$\mathbb{E} \sum_{i=1}^{n} c_i \frac{X_{i:n} - \mu}{\sigma_p} = \sum_{k=1}^{N} C_k \frac{x_{k:N} - \bar{x}}{s_p}, \qquad 1 \le p \le +\infty. \tag{5.14}$$

It follows that the problem of determining the sharp upper bounds for the expectations of normalized L-statistics from the with replacement model is equivalent to establishing the upper bounds for the linear combination of the elements of normalized nondecreasing sequences $(x_{1:N}, \ldots, x_{N:N}) \in \mathbb{R}^N$ with fixed, not necessarily nondecreasing coefficients $\tilde{C} = (C_1, \ldots, C_N) \in \mathbb{R}^N$ defined in (5.12). To this end, we may apply the methods of Sect. 3.4. The crucial point is here determining the projection $P^{\nearrow}\tilde{C} = \underline{\tilde{C}} = (\underline{C}_1, \ldots, \underline{C}_N) \in \mathbb{R}^N$ of $\tilde{C}$ onto the convex cone $\mathcal{C}^{\nearrow}$ of nondecreasing vectors in the Euclidean space of dimension N. Using algorithms (2.78) and (2.79), we obtain

$$\underline{C}_k = \frac{F_{\tilde{c}:n}\left(\frac{j_m}{N}\right) - F_{\tilde{c}:n}\left(\frac{j_{m-1}}{N}\right)}{j_m - j_{m-1}}, \tag{5.15}$$

with $k = j_{m-1} + 1, \dots, j_m$, $m = 1, \dots, M$, for $j_0 = 0$, and

$$j_{m+1} = \min \left\{ j > j_m : \frac{F_{\tilde{c}:n}\left(\frac{j}{N}\right) - F_{\tilde{c}:n}\left(\frac{j_m}{N}\right)}{j - j_m} < \frac{F_{\tilde{c}:n}\left(\frac{\ell}{N}\right) - F_{\tilde{c}:n}\left(\frac{j_m}{N}\right)}{\ell - j_m}, \right.$$

$$\left. \ell = j_m + 1, \dots, N \right\} \tag{5.16}$$

and $j_M = N$ for some $1 \leq M \leq N$ determined by (5.16). By Theorem 2.13 and the Hölder inequality for $1 \leq p \leq +\infty$, we can write

$$\begin{aligned}
\mathbb{E} \sum_{i=1}^{n} c_i (X_{i:n} - \mu) &= \sum_{k=1}^{N} C_k (x_{k:N} - \bar{x}) \leq \sum_{k=1}^{N} \underline{C}_k (x_{k:N} - \bar{x}) \\
&= \sum_{k=1}^{N} (\underline{C}_k - \bar{\underline{C}}_q)(x_{k:N} - \bar{x}) \leq \|\tilde{\underline{C}} - \bar{\underline{C}}_q \tilde{\mathbf{1}}\|_q \|\tilde{x} - \bar{x}\tilde{\mathbf{1}}\|_p \\
&= N^{1/p} \|\tilde{\underline{C}} - \bar{\underline{C}}_q \tilde{\mathbf{1}}\|_q \sigma_p.
\end{aligned} \tag{5.17}$$

Here $N^{1/\infty} = \lim_{p\to\infty} N^{1/p} = 1$, by convention, and $\bar{\underline{C}}_q \tilde{\mathbf{1}}$ denotes the projection of $\tilde{\underline{C}}$ onto the family of constant sequences in $\mathbb{R}^N$ in the ℓ^q-norm. Actually, (5.17) remains true if we replace $\bar{\underline{C}}_q$ by any real constant, but this particular choice provides the optimal evaluation.

If $\tilde{\underline{C}}$ is a constant sequence with

$$\underline{C}_k = \frac{1}{N} \sum_{n=1}^{N} C_r = \frac{1}{N} \sum_{i=1}^{n} c_i, \qquad k = 1, \dots, N,$$

which occurs when

$$\frac{1}{k} \sum_{r=1}^{k} C_r \geq \frac{1}{N} \sum_{r=1}^{N} C_r, \qquad k = 1, \dots, N-1, \tag{5.18}$$

or equivalently,

$$\frac{N}{k} F_{\tilde{c}:n}\left(\frac{k}{N}\right) \geq \sum_{i=1}^{n} c_i, \qquad k = 1, \dots, N-1, \tag{5.19}$$

then $\bar{\underline{C}}_q = \frac{1}{N} \sum_{i=1}^{n} c_i$ and the RHS of (5.17) provides the zero bounds which is only attained by identically labeled objects with $\sigma_p = 0$, $1 \leq p \leq +\infty$. Later on, we determine possibly negative bounds for nonconstant $\tilde{x}$ under assumptions (5.18) and (5.19). However, we now concentrate on the L-statistics for which neither of (5.18) and (5.19) holds, i.e., we have

$$\exists\, 1 \le k \le n-1 \qquad \frac{N}{k} F_{\tilde{c}:n}\left(\frac{k}{N}\right) < \sum_{i=1}^{n} c_i. \tag{5.20}$$

Then $P^{\nearrow}\tilde{C} = \underline{\tilde{C}}$ is nonconstant which implies $\underline{C}_1 < \underline{C}_N$, $M \ge 2$ and $j_1 < N$. Furthermore, bounds (5.17) are strictly positive and attainable by populations with positive σ_p, $1 \le p \le +\infty$. Below, we present detailed results for the cases $p = 1$, $1 < p < +\infty$ and $p = +\infty$. Case $p = 2$ is also distinguished due to its importance and relative simplicity. The proofs are merely outlined. We now assume that $X_1, \ldots, X_n$ are the random values drawn with replacement from population $\{x_1, \ldots, x_N\}$ with given mean μ and scale parameter σ_p for some $1 \le p \le +\infty$, and $\tilde{c} \in \mathbb{R}^n$ is such that (5.19) does not hold.

Theorem 5.1 *For $\tilde{c} \in \mathbb{R}^n$ satisfying* (5.20)*, the bound*

$$\mathbb{E}\sum_{i=1}^{n} c_i \frac{X_{i:n}-\mu}{\sigma_1} \le \frac{N}{2}\left[\frac{\sum_{i=1}^{n} c_i - F_{\tilde{c}:n}\left(\frac{j_{m-1}}{N}\right)}{N-j_{M-1}} - \frac{F_{\tilde{c}:n}\left(\frac{j_1}{N}\right)}{j_1}\right] \tag{5.21}$$

is sharp and becomes an equality if

$$x_{k:n} = \begin{cases} \mu - \sigma_1 \frac{N}{2j_1}\,, & k = 1, \ldots, j_1, \\ \mu\,, & k = j_1+1, \ldots, j_{M-1}\,, \\ \mu + \sigma_1 \frac{N}{2(N-j_{M-1})}\,, & k = j_{M-1}+1, \ldots, N. \end{cases} \tag{5.22}$$

Proof The proof is similar to that of Theorem 3.8. Writing (5.17) with $p = 1$, we obtain

$$\mathbb{E}\sum_{i=1}^{n} c_i (X_{i:n}-\mu) \le \sum_{k=1}^{N} (C_k - \underline{\bar{C}}_\infty)(x_{k:N} - \bar{x}) \le N\|\underline{\tilde{C}} - \underline{\bar{C}}_\infty \tilde{\mathbf{1}}\|_\infty \sigma_1. \tag{5.23}$$

The extreme elements of $\underline{\tilde{C}}$ are

$$\underline{C}_1 = \frac{1}{j_1}\sum_{k=1}^{j_1}\left[F_{\tilde{c}:n}\left(\frac{k}{N}\right) - F_{\tilde{c}:n}\left(\frac{k-1}{N}\right)\right] = \frac{1}{j_1} F_{\tilde{c}:n}\left(\frac{j_1}{N}\right), \tag{5.24}$$

$$\begin{aligned} \underline{C}_N &= \frac{1}{N-j_{M-1}}\left[F_{\tilde{c}:n}(1) - F_{\tilde{c}:n}\left(\frac{j_{M-1}}{N}\right)\right] \\ &= \frac{1}{N-j_{M-1}}\left[\sum_{i=1}^{n} c_i - F_{\tilde{c}:n}\left(\frac{j_{M-1}}{N}\right)\right]. \end{aligned} \tag{5.25}$$

Therefore,

$$\underline{\bar{C}}_\infty = \frac{1}{2}(\underline{C}_1 + \underline{C}_N) = \frac{1}{2}\left[\frac{F_{\tilde{c}:n}\left(\frac{j_1}{N}\right)}{j_1} + \frac{\sum_{i=1}^n c_i - F_{\tilde{c}:n}\left(\frac{j_{M-1}}{N}\right)}{N - j_{M-1}}\right] \tag{5.26}$$

and

$$\|\underline{\tilde{C}} - \underline{\bar{C}}_\infty \tilde{\mathbf{1}}\|_\infty = \frac{1}{2}(\underline{C}_N - \underline{C}_1) = \frac{\sum_{i=1}^n c_i - F_{\tilde{c}:n}\left(\frac{j_{M-1}}{N}\right)}{2(N - j_{M-1})} - \frac{F_{\tilde{c}:n}\left(\frac{j_1}{N}\right)}{2j}, \tag{5.27}$$

which proves (5.21). The second inequality in (5.23) becomes an equality if

$$x_{k:N} - \mu \begin{cases} \le 0, & k = 1, \ldots, j_1, \\ = 0, & k = j_1 + 1, \ldots, j_{M-1}, \\ \ge 0, & k = j_{M-1} + 1, \ldots, N, \end{cases} \tag{5.28}$$

(cf. (2.180)). The condition for an equality in the first one is

$$x_{j_{m-1}+1:N} = \ldots = x_{j_m:N}, \qquad m = 1, \ldots, M. \tag{5.29}$$

Both (5.28) and (5.29) imply that

$$x_{1:N} = x_{j_1:N} \le \mu = x_{j_1+1:N} = x_{j_{M-1}:N} \le x_{j_{M-1}+1:N} = x_{N:N}. \tag{5.30}$$

This together with the moment conditions imply (5.22). □

Theorem 5.2 *For every $\tilde{c} \in \mathbb{R}^n$ satisfying* (5.20), *define*

$$B_m(c) = \frac{F_{\tilde{c}:n}\left(\frac{j_m}{N}\right) - F_{\tilde{c}:n}\left(\frac{j_{m-1}}{N}\right)}{j_m - j_{m-1}} - c, \quad m = 1, \ldots, M, \ c \in \mathbb{R}. \tag{5.31}$$

Then, for every $1 < p < \infty$, we have

$$\mathbb{E}\sum_{i=1}^n c_i \frac{X_{i:n} - \mu}{\sigma_p} \le N^{1/p}\left(\sum_{m=1}^M |B_m(\underline{\bar{C}}_q)|^q (j_m - j_{m-1})\right)^{1/q}, \tag{5.32}$$

where

$$\frac{1}{j_1}F_{\tilde{c}:n}\left(\frac{j_1}{N}\right) \le \underline{\bar{C}}_q \le \frac{1}{N - j_{M-1}}\left[\sum_{i=1}^n c_i - F_{\tilde{c}:n}\left(\frac{j_{M-1}}{N}\right)\right] \tag{5.33}$$

is the unique solution to the equation

$$\sum_{m=1}^{M} |B_m(c)|^{q/p} sgn\{B_m(c)\}(j_m - j_{m-1}) = 0. \tag{5.34}$$

The equality in (5.32) *holds if*

$$x_{k:N} = \mu + \sigma_p \frac{N^{1/p}|B_m(\underline{\bar{C}}_q)|^{q/p} sgn\{B_m(\underline{\bar{C}}_q)\}}{\left(\sum_{r=1}^{M} |B_r(\underline{\bar{C}}_q)|^q (j_r - j_{r-1})\right)^{1/p}} \tag{5.35}$$

for $k = j_{m-1} + 1, \ldots, j_m$, $m = 1, \ldots, M$.

Proof We refer to (5.17) again. The constant $\underline{\bar{C}}_q$ minimizing the convex function

$$c \mapsto \|\underline{\tilde{C}} - c\tilde{\mathbf{1}}\|_q^q = \sum_{i=1}^{N} |\underline{C}_k - c|^q = \sum_{m=1}^{M} |B_m(c)|^q (j_m - j_{m-1}) \tag{5.36}$$

lies obviously between the extreme values of $\underline{\tilde{C}}$ (see (5.24), (5.25))) and is uniquely determined by (5.34). Therefore, (5.32) holds. The equality in the Hölder inequality holds if

$$\begin{aligned} x_{k:N} - \mu &= \alpha|\underline{C}_k - \underline{\bar{C}}_q|^{q/p}\mathrm{sgn}\{\underline{C}_k - \underline{\bar{C}}_q\} \\ &= \alpha|B_m(\underline{\bar{C}}_q)|^{q/p}\mathrm{sgn}\{B_m(\underline{\bar{C}}_q)\} \end{aligned} \tag{5.37}$$

for $k = j_{m-1} + 1, \ldots, j_m$, $m = 1, \ldots, M$, and some $\alpha > 0$. This satisfies (5.29) so that (5.17) becomes an equality under (5.37). If, moreover,

$$\alpha = \sigma_p \frac{N^{1/p}}{\left(\sum_{m=1}^{M} |B_m(\underline{\bar{C}}_q)|^q (j_m - j_{m-1})\right)^{1/p}}, \tag{5.38}$$

then the moment conditions are satisfied as well. □

Theorem 5.3 *In particular, we have an inequality*

$$\mathbb{E}\sum_{i=1}^{n} c_i \frac{X_{i:n} - \mu}{\sigma_2} \leq \left[N \sum_{m=1}^{M} \frac{\left[F_{\tilde{c}:n}\left(\frac{j_m}{N}\right) - F_{\tilde{c}:n}\left(\frac{j_{m-1}}{N}\right)\right]^2}{j_m - j_{m-1}} - \left(\sum_{i=1}^{n} c_i\right)^2 \right]^{1/2} \tag{5.39}$$

and this becomes the equality if

$$x_{k:N} = \mu + \sigma_2 \frac{N^{1/2}\left(\frac{F_{\tilde{c}:n}\left(\frac{j_m}{N}\right) - F_{\tilde{c}:n}\left(\frac{j_{m-1}}{N}\right)}{j_m - j_{m-1}} - \sum_{i=1}^{n} c_i\right)}{\left[\sum_{m=1}^{M} \frac{\left[F_{\tilde{c}:n}\left(\frac{j_m}{N}\right) - F_{\tilde{c}:n}\left(\frac{j_{m-1}}{N}\right)\right]^2}{j_m - j_{m-1}} - \frac{1}{N}\left(\sum_{i=1}^{n} c_i\right)^2\right]^{1/2}} \tag{5.40}$$

for $k = j_{m-1} + 1, \ldots, j_m$, $m = 1, \ldots, M$.

Proof A simplification in the special case $p = 2$ follows from the fact that

$$\bar{\underline{C}}_2 = \frac{1}{N}\sum_{k=1}^{N} \underline{C}_k = \frac{1}{N}\sum_{k=1}^{N} C_k = \frac{1}{N}\sum_{i=1}^{n} c_i. \tag{5.41}$$

Then,

$$\begin{aligned} \|\underline{\tilde{C}} - \bar{\underline{C}}_2\|_2^2 &= \sum_{k=1}^{N} \underline{C}_k^2 - N\left(\frac{1}{N}\sum_{k=1}^{N} \underline{C}_k\right)^2 \\ &= \sum_{m=1}^{N} \frac{\left[F_{\tilde{c}:n}\left(\frac{j_m}{N}\right) - F_{\tilde{c}:n}\left(\frac{j_{m-1}}{N}\right)\right]^2}{j_m - j_{m-1}} - \frac{1}{N}\left(\sum_{i=1}^{n} c_i\right)^2 \end{aligned} \tag{5.42}$$

and so (5.39) holds. Also, these allow us to transform (5.35) into (5.40). □

In the case $p = +\infty$, the optimal constant $\bar{\underline{C}}_1$ is the median of sequence $\underline{\tilde{C}}$:

$$\bar{\underline{C}}_1 = \mathrm{MED}_N(\underline{\tilde{C}}) = \begin{cases} \underline{C}_{(N-1)/2}, & \text{if } N \text{ is odd}, \\ \frac{1}{2}(\underline{C}_{N/2} + \underline{C}_{N/2+1}), & \text{if } N \text{ is even}. \end{cases} \tag{5.43}$$

Therefore, we have the following result.

Theorem 5.4 *Let* $0 \le m_* \le M_* \le M$ *be defined by*

$$j_{m_*} = \#\{k : \underline{C}_k < \bar{\underline{C}}_1\}, \tag{5.44}$$

$$j_{M_*} = \#\{k : \underline{C}_k \le \bar{\underline{C}}_1\}, \tag{5.45}$$

with $j_0 = 0$ *and* j_m, $m = 1, \ldots, M$, *defined in* (5.16). *Then,*

$$\mathbb{E}\sum_{i=1}^{n} c_i \frac{X_{i:n}-\mu}{\sigma_\infty} \leq \sum_{i=1}^{n} c_i - F_{\tilde{c}:n}\left(\frac{j_{M_*}}{N}\right) - F_{\tilde{c}:n}\left(\frac{j_{m_*}}{N}\right) - (N - j_{m_*} - j_{M_*})\underline{\bar{C}}_1, \tag{5.46}$$

and the equality holds if

$$x_{k:N} = \begin{cases} \mu - \sigma_\infty, & k = 1, \ldots, j_{m_*}, \\ \mu + \sigma_\infty \frac{j_{m_*}+j_{M_*}-N}{j_{M_*}-j_{m_*}}, & k = j_{m_*}+1, \ldots, j_{M_*}, \\ \mu + \sigma_\infty, & k = j_{M_*}+1, \ldots, N. \end{cases} \tag{5.47}$$

Proof We have

$$\mathbb{E}\sum_{i=1}^{n} c_i (X_{i:n} - \mu) \leq \sum_{k=1}^{N} (\underline{C}_k - \underline{\bar{C}}_1)(x_{k:N} - \bar{x}) \leq \|\underline{\tilde{C}} - \underline{\bar{C}}_1\|_1 \sigma_\infty, \tag{5.48}$$

where

$$\begin{aligned} \|\underline{\tilde{C}} - \underline{\bar{C}}_1\|_1 &= \sum_{\underline{C}_k < \underline{\bar{C}}_1} (\underline{\bar{C}}_1 - \underline{C}_k) + \sum_{\underline{C}_k > \underline{\bar{C}}_1} (\underline{C}_k - \underline{\bar{C}}_1) \\ &= j_{m_*}\underline{\bar{C}}_1 - \sum_{k=1}^{j_{m_*}} \underline{C}_k + \sum_{k=j_{M_*}+1}^{N} \underline{C}_k - (N - j_{M_*})\underline{\bar{C}}_1, \end{aligned} \tag{5.49}$$

which is identical with the RHS of (5.46). The equality in the first inequality of (5.48) holds under condition (5.29), and for (5.47) in particular. By (2.182) and definitions (5.44) and (5.45), the other inequality becomes an equality if $x_{j_{m_*}:N}$ and $x_{j_{M_*}:N}$ attain the minimal and maximal possible values, respectively. Relation $m_* = M_*$ implies that n is even and $j_{m_*} = j_{M_*} = N/2$. Then, (5.47) take only two extreme values and sum up to $N\mu$. Otherwise,

$$0 \leq j_{m_*} < N/2 < j_{M_*} \leq N \tag{5.50}$$

which implies

$$-1 < \frac{j_{m_*} + j_{M_*} - N}{j_{M_*} - j_{m_*}} < +1, \tag{5.51}$$

and so $\|\tilde{x} - \mu\tilde{\mathbf{1}}\|_\infty = \sigma_\infty$. The first moment condition is immediately verified as well. □

The above results provide tools for establishing the best lower bounds. The method is standard: we replace the original sample $(X_1, \dots, X_n)$ drawn with replacement from population $\{x_1, \dots, x_N\}$ by $(Y_1, \dots, Y_n)$ drawn analogously from $\{2\mu - x_1, \dots, 2\mu - x_N\}$ with $\mu = \frac{1}{N}\sum_{i=1}^{N} x_i$. Then, Y_i and $2\mu - X_i$ $i = 1, \dots, n$, have identical joint distributions, and moments μ and σ_p, $1 \le p \le +\infty$, identical with those of the original sample. Since, moreover, the distributions of $Y_{i:n}$ and $2\mu - X_{n+1-i:n}$ $i = 1, \dots, n$, coincide, we have

$$\mathbb{E}\sum_{i=1}^{n} c_i \frac{Y_{i:n} - \mu}{\sigma_p} = -\mathbb{E}\sum_{i=1}^{n} c_i \frac{X_{n+1-i:n} - \mu}{\sigma_p} = -\mathbb{E}\sum_{i=1}^{n} c_{n+1-i} \frac{X_{i:n} - \mu}{\sigma_p}. \tag{5.52}$$

In particular, the minimal value of the first expectation is the negative of the maximal value of the last one. It then follows that the lower bound for the L-statistic with coefficients $c_1, \dots, c_n$ is equal to the negative of the upper bound for the L-statistic with the reversed sequence of coefficients $c_n, \dots, c_1$. For instance, the lower bound for the jth smallest order statistic $X_{j:n}$ is the negative of the upper bound for the jth greatest one $X_{n+1-j:n}$.

5.1.2 Upper Bounds for Single Order Statistics

Detailed evaluations for

$$\mathbb{E}X_{j:n} = \sum_{k=1}^{N} x_{k:N}\left[F_{j:n}\left(\frac{k}{N}\right) - F_{j:n}\left(\frac{k-1}{N}\right)\right], \quad 1 \le j \le n, \tag{5.53}$$

are presented below. The results are concluded from properties of the sequence

$$C_k = C_k(j, n, N) = F_{j:n}\left(\frac{k}{N}\right) - F_{j:n}\left(\frac{k-1}{N}\right), \quad k = 1, \dots, N. \tag{5.54}$$

Lemma 5.1 *If $j = 1$ and $j = n$, then (5.54) is decreasing and increasing, respectively. Otherwise, this is unimodal (i.e., either increasing or decreasing or first increasing and then decreasing) with the maximum at either one or two neighboring points $\lfloor \alpha N \rfloor$, $\lceil \alpha N \rceil$ and $\lceil \alpha N \rceil + 1$ with $\alpha = \alpha(j, n) = \frac{j-1}{n-1}$.*

Proof By the mean value theorem and (5.54),

$$C_k = \frac{1}{N} f_{j:n}(\zeta_k) \text{ for some } \zeta_k \in \left(\frac{k-1}{N}, \frac{k}{N}\right). \tag{5.55}$$

Therefore, it is essential to recall properties of functions $f_{j:n}(x)$, $0 < x < 1$, $1 \le j \le n$, here. Functions $f_{1:n}(x)$ and $f_{n:n}(x)$ are decreasing and increasing, respectively. This, together with (5.55) and relations $\zeta_1 < \cdots < \zeta_N$imply the first statements of the lemma. If $2 \le j \le n-1$, then $f_{j:n}(x)$ is increasing on $(0, \alpha)$ and decreasing on $(\alpha, 1)$. If $k \le \lfloor \alpha N \rfloor$, then $\alpha > \zeta_k > \zeta_{k-1}$, and, by (5.55) $C_k > C_{k-1}$. If $k \ge \lceil \alpha N \rceil + 1$, then $\alpha < \zeta_k < \zeta_{k+1}$, and $C_k < C_{k+1}$. Hence, C_k, $k = 1, \ldots, \lfloor \alpha N \rfloor$, is strictly increasing and C_k, $k = \lceil \alpha N \rceil + 1, \ldots, N$ is strictly decreasing. If $\alpha N = \frac{j-1}{n-1} N$ is integer, the lemma follows. Otherwise, it suffices to show that

$$C_{\lceil \alpha N \rceil} > \min \left\{ C_{\lfloor \alpha N \rfloor}, C_{\lceil \alpha N \rceil + 1} \right\}. \tag{5.56}$$

Function $f_{j:n}(x)$, restricted to interval $\left[\frac{\lfloor \alpha N \rfloor}{N}, \frac{\lceil \alpha N \rceil}{N}\right]$, is increasing for $x < \alpha$ and decreasing for $x > \alpha$ and so attains its minimum at either of the end points. Therefore, by (5.55),

$$\begin{aligned} C_{\lceil \alpha N \rceil} &= \frac{1}{N} f_{j:n}(\zeta_{\lceil \alpha N \rceil}) = \int_{\lfloor \alpha N \rfloor}^{\lceil \alpha N \rceil / N} f_{j:n}(x) dx \\ &> \frac{1}{N} \min \left\{ f_{j:n}\left(\frac{\lfloor \alpha N \rfloor}{N}\right), f_{j:n}\left(\frac{\lceil \alpha N \rceil}{N}\right) \right\} \\ &> \frac{1}{N} \min \left\{ f_{j:n}\left(\zeta_{\lfloor \alpha N \rfloor}\right), f_{j:n}\left(\zeta_{\lceil \alpha N \rceil + 1}\right) \right\} \\ &= \min \left\{ C_{\lfloor \alpha N \rfloor}, C_{\lceil \alpha N \rceil + 1} \right\}, \end{aligned} \tag{5.57}$$

which completes the proof. Existence of two neighboring maxima in sequence (5.55) is very unlikely, but it cannot be entirely excluded. Conditions $j = 1$ and $j = n$ are not necessary for monotonicity of the sequence. If N is not large enough, then $\alpha = \frac{j-1}{n-1}$ for some $2 \le j \le n-1$ may belong to either of the extreme intervals $(0, \frac{1}{N})$ and $\left(1 - \frac{1}{N}, 1\right)$, and (5.55) may become maximal at $k = 1$ or $k = N$, respectively. □

Using Lemma 5.1, we can easily establish the increasing-constant form of projection $\underline{\tilde{C}} = P^{\nearrow} \tilde{C}$ of $\tilde{C}$ onto the family of nondecreasing sequences in $\mathbb{R}^N$. Due to specific form (5.54) of the sequence, we are able to derive more precise conclusions.

Proposition 5.1 *Let $\tilde{C}$ be defined by (5.54) for some $1 \le j \le n$. If $j = 1$, then $P^{\nearrow} \tilde{C} = \frac{1}{N} \sum_{i=1}^{n} c_i \mathbf{1}$. If $j = n$, then $P^{\nearrow} \tilde{C} = \tilde{C}$. Otherwise, for*

$$H_{j:n}(x) = \frac{1 - F_{j:n}(x)}{1 - x}, \quad 0 \le x < 1, \tag{5.58}$$

there exists a unique $0 < \alpha_* < \alpha = \frac{j-1}{n-1}$ *satisfying the equation*

$$f_{j:n}(\alpha) = H_{j:n}(\alpha) \tag{5.59}$$

so that for

$$k_* = k_*(j, n, N) = \begin{cases} \lfloor \alpha_* N \rfloor \text{ if } H_{j:n}\left(\frac{\lfloor \alpha_* N \rfloor}{N}\right) \geq H_{j:n}\left(\frac{\lceil \alpha_* N \rceil}{N}\right) \\ \lceil \alpha_* N \rceil \text{ otherwise.} \end{cases} \tag{5.60}$$

projection $\underline{\tilde{C}} = P^{\nearrow}\tilde{C}$ *has the coordinates*

$$\underline{C}_k = \begin{cases} F_{j:n}\left(\frac{k}{N}\right) - F_{j:n}\left(\frac{k-1}{N}\right), & k = 1, \ldots, k_*, \\ \frac{1-F_{j:n}\left(\frac{k_*}{N}\right)}{N-k_*}, & k = k_* + 1, \ldots, N. \end{cases} \tag{5.61}$$

Proof The first two statements are immediate consequences of Lemma 5.1. For $2 \leq j \leq n-1$, when $\tilde{C}$ is increasing decreasing, from Theorem 2.29 we deduce that the first elements of $\underline{\tilde{C}}$ are identical with those of $\tilde{C}$, and the last ones are equal to the maximal mean of the last elements of $\tilde{C}$. Since

$$\frac{1}{N-k} \sum_{r=k+1}^{N} C_r = N H_{j:n}\left(\frac{k}{N}\right), \quad k = 0, \ldots, N-2, \tag{5.62}$$

it is useful to examine properties of (5.58) first. We claim that this is increasing on $(0, \alpha_*)$ and decreasing on $(\alpha_*, 1)$ for α_* defined by (5.59). Since $F_{j:n}(x)$ is a distribution function supported on $[0, 1]$ with the density function $f_{j:n}(x)$, function (5.58) is positive and satisfies $H_{j:n}(0) = 1$ and $H_{j:n}(1-) = f_{j:n}(1) = 0$. Also,

$$H'_{j:n}(x) = \frac{1 - F_{j:n}(x) - (1-x) f_{j:n}(x)}{(1-x)^2}. \tag{5.63}$$

We have $H'_{j:n}(0) = 1$ and $H'_{j:n}(1-) = \frac{1}{2} f'_{j:n}(1) \leq 0$, by the de l'Hopital rule. The numerator of (5.63) has the derivative $-f'_{j:n}(x)(1-x)$ which is negative for $x < \alpha = \frac{j-1}{n-1}$ and positive for $x > \alpha$. Therefore, (5.63) is first positive decreasing on some $(0, \alpha_*)$ with α_* satisfying (5.59), then negative decreasing on (α_*, α), and eventually negative increasing on $(\alpha, 1)$. It then follows that the desired claim holds. Therefore (5.60), and the projection has representation (5.61). □

Observe that $k_* = 0$ and the constant projection $\underline{\tilde{C}} = \frac{1}{N}\tilde{1}$ is possible for $j \geq 2$ in the case when the population size N is small with respect to the sample size n so that most of the probability mass of $f_{j:n}$ is concentrated in $\left[0, \frac{1}{N}\right]$. Precisely, this holds if $H_{j:n}\left(\frac{1}{N}\right) \leq H_{j:n}(0) = 1$. For instance, relation $N \leq \frac{n-1}{j-1}$ implies $\alpha_* < \alpha \leq \frac{1}{N}$, $k_* = 0$, and, consequently $\mathbb{E}X_{j:n} - \mu \leq 0$. If $H_{j:n}\left(\frac{1}{N}\right) > 1$, then $k_* \geq 1$, and the bounds in all the scale units σ_p, $1 \leq p \leq \infty$, are strictly positive and best possible. Tacitly assuming that $k_* \geq 1$, we use Proposition 5.1 in order to specify the results of Theorems 5.1–5.4 for single order statistics, omitting formal proofs. For completeness, we also cover the case of sample maximum $X_{n:n}$, defining $k_* = k_*(n, n, N) = N - 1$.

Corollary 5.1 *With* $1 \leq k_* \leq N - 1$ *defined in* (5.60)*, the inequality*

$$\mathbb{E}\frac{X_{j:n} - \mu}{\sigma_1} \leq \frac{N}{2}\left[\frac{1 - F_{j:n}\left(\frac{k_*}{N}\right)}{N - k_*} - F_{j:n}\left(\frac{1}{N}\right)\right] \tag{5.64}$$

holds and becomes the equality under condition

$$x_{k:N} = \begin{cases} \mu - \sigma_1 \frac{N}{2}, & \text{for } k = 1, \\ \mu, & \text{for } k = 2, \ldots, k_*, \\ \mu + \sigma_1 \frac{N}{2(N-k_*)}, & \text{for } k = k_* + 1, \ldots, N. \end{cases} \tag{5.65}$$

Corollary 5.2 *For* $1 < p < \infty$ *and* $1 \leq k_* \leq N - 1$ *defined in* (5.60)*, there is a unique*

$$F_{j:n}\left(\frac{1}{N}\right) < \underline{\bar{c}}_q < \frac{1 - F_{j:n}\left(\frac{k_*}{N}\right)}{N - k_*} \tag{5.66}$$

solving the equation

$$\begin{aligned} &\sum_{k=1}^{k_*}\left|F_{j:n}\left(\frac{k}{N}\right) - F_{j:n}\left(\frac{k-1}{N}\right) - c\right|^{q/p} sgn\left\{F_{j:n}\left(\frac{k}{N}\right) - F_{j:n}\left(\frac{k-1}{N}\right) - c\right\} \\ &+(N - k_*)\left[\frac{1 - F_{j:n}\left(\frac{k_*}{N}\right)}{N - k_*} - c\right]^{q/p} = 0 \end{aligned} \tag{5.67}$$

such that

$$\mathbb{E}\frac{X_{j:n} - \mu}{\sigma_p} \leq N^{1/p} B \tag{5.68}$$

for

$$B^q = \sum_{k=1}^{k_*} \left| F_{j:n}\left(\frac{k}{N}\right) - F_{j:n}\left(\frac{k-1}{N}\right) - \bar{c}_q \right|^q + (N-k_*) \left[\frac{1 - F_{j:n}\left(\frac{k_*}{N}\right)}{N - k_*} - \bar{c}_q \right]^q . \tag{5.69}$$

The equality in (5.68) *holds if*

$$x_{k:N} = \begin{cases} \mu + \sigma_p \frac{N^{1/p}}{B^{q/p}} \left| F_{j:n}\left(\frac{k}{N}\right) - F_{j:n}\left(\frac{k-1}{N}\right) - \bar{c}_q \right|^{q/p} & \\ \qquad \times sgn \left\{ F_{j:n}\left(\frac{k}{N}\right) - F_{j:n}\left(\frac{k-1}{N}\right) - \bar{c}_q \right\} & k = 1, \ldots, k_*, \\ \mu + \sigma_p \frac{N^{1/p}}{B^{q/p}} \left[\frac{1 - F_{j:n}\left(\frac{k_*}{N}\right)}{N - k_*} - \bar{c}_q \right]^{q/p}, & k = k_* + 1, \ldots, N. \end{cases} \tag{5.70}$$

Corollary 5.3 *In particular,*

$$\mathbb{E}\frac{X_{j:n} - \mu}{\sigma_2} \le N^{1/2} B \tag{5.71}$$

for

$$B^2 = \sum_{k=1}^{k_*} \left[F_{j:n}\left(\frac{k}{N}\right) - F_{j:n}\left(\frac{k-1}{N}\right) \right]^2 + \frac{\left[1 - F_{j:n}\left(\frac{k_*}{N}\right)\right]^2}{N - k_*} - \frac{1}{N}. \tag{5.72}$$

The equality in (5.71) *is attained if*

$$x_{k:n} = \begin{cases} \mu + \sigma_2 \frac{N^{1/2}}{B} \left[F_{j:n}\left(\frac{k}{N}\right) - F_{j:n}\left(\frac{k-1}{N}\right) - \frac{1}{N} \right], & k = 1, \ldots, k_*, \\ \mu + \sigma_2 \frac{N^{1/2}}{B} \left[\frac{1 - F_{j:n}\left(\frac{k_*}{N}\right)}{N - k_*} - \frac{1}{N} \right], & k = k_* + 1, \ldots, N. \end{cases} \tag{5.73}$$

Corollary 5.4 *Under definitions* (5.58) *and* (5.60) *and condition*

$$H_{j:n}\left(\left\lfloor \frac{N-1}{2} \right\rfloor \Big/ N \right) \ge H_{j:n}\left(\left\lfloor \frac{N+1}{2} \right\rfloor \Big/ N \right), \tag{5.74}$$

we have

$$\mathbb{E}\frac{X_{j:n}-\mu}{\sigma_\infty} \le \frac{N}{N-k_*}\left[\frac{k_*}{N} - F_{j:n}\left(\frac{k_*}{N}\right)\right], \tag{5.75}$$

which becomes an equality for

$$x_{k:N} = \begin{cases} \mu - \sigma_\infty, & k = 1, \dots, k_*, \\ \mu + \sigma_\infty \frac{k_*}{N-k_*}, & k = k_* + 1, \dots, N. \end{cases} \tag{5.76}$$

If (5.74) *does not hold, then*

$$\mathbb{E}\frac{X_{j:n}-\mu}{\sigma_\infty} \le \begin{cases} 1 - 2F_{j:n}(\frac{1}{2}), & \text{if } N \text{ is even}, \\ 1 - F_{j:n}\left(\frac{N+1}{2N}\right) - F_{j:n}\left(\frac{N-1}{2N}\right), & \text{if } N \text{ is odd}. \end{cases} \tag{5.77}$$

Equation (5.77) *becomes the equality for*

$$x_{k:N} = \begin{cases} \mu - \sigma_\infty, \; k = 1, \dots, \frac{N}{2}, \\ \mu + \sigma_\infty, \; k = \frac{N}{2} + 1, \dots, N, \end{cases} \tag{5.78}$$

if N is even, and for

$$x_{k:N} = \begin{cases} \mu - \sigma_\infty, \; k = 1, \dots, \frac{N-1}{2}, \\ \mu, \qquad\quad k = \frac{N+1}{2}, \\ \mu + \sigma_\infty, \; k = \frac{N+3}{2}, \dots, N, \end{cases} \tag{5.79}$$

if N is odd.

We obtain the conclusions of Corollaries 5.1–5.3 immediately, once we observe that $M = k_* + 1$, and $j_m = m$, $m = 1, \dots, k_*$. Only the statements in the case $p = \infty$ need some explanation. Relation (5.74) implies that $k_* \le \lfloor \frac{N-1}{2} \rfloor$, and $\mathrm{MED}_N(\underset{\sim}{C}) = \underline{C}_N = \frac{1-F_{j:n}\left(\frac{k_*}{N}\right)}{N-k_*}$. Then $\frac{k_*}{N-k_*} < 1$, and population (5.76) has the maximal absolute derivation δ_∞. Otherwise, $\mathrm{MED}_N(\underset{\sim}{C}) = \mathrm{MED}_N(\tilde{C})$. Note that, by definition,

$$\left(\frac{k}{N}, F_{j:n}\left(\frac{k}{N}\right)\right), \qquad k = 0, 1, \dots, k_*, N, \tag{5.80}$$

are the breaking points of a piecewise linear convex function that is the greatest convex minorant of pairs (5.80) for all $0 \le k \le N$. Therefore,

$$F_{j:n}\left(\frac{k}{N}\right) \le \frac{k}{N}, \qquad k = 0, 1, \dots, k_*, N, \tag{5.81}$$

and the inequality is strict except for the extreme $k = 0$ and N. Hence, bounds (5.75) and (5.77) are actually positive. Moreover, if N is even, then (5.77) coincides with the optimal bound for the general populations.

In the same way, we can prove more general results for trimmed means $T_{j,k:n}$. It suffices to replace in the above formulae the distribution functions $F_{j:n}(x)$ by the mixtures

$$F_{i,j:n}(x) = \frac{1}{j+1-i} \sum_{r=i}^{j} F_{r:n}(x). \tag{5.82}$$

The conclusions for $T_{1,j:n}$ and $T_{i,n:n}$ coincide with those for the sample maximum and minimum, respectively. Otherwise, the projection of the vector with coordinates

$$C_k = \frac{1}{j+1-i} \sum_{r=i}^{j} \left[F_{r:n}\left(\frac{k}{N}\right) - F_{r:n}\left(\frac{k-1}{N}\right) \right] \tag{5.83}$$

is determined in the form analogous to (5.61) which allows to rewrite Corollaries 5.1–5.4 in a more general form. The details can be found in Rychlik (2004).

Below we specify the bounds of Corollaries 5.1, 5.3, and 5.4 for the sample maximum. Then $k_* = N - 1$, and accordingly,

$$\mathbb{E}X_{n:n} - \mu \le \begin{cases} \frac{N}{2}\left[1 - F_{n:n}\left(\frac{N-1}{N}\right) - F_{n:n}\left(\frac{1}{N}\right)\right]\sigma_1, \\ \left\{ N \sum_{k=1}^{N} \left[F_{n:n}\left(\frac{k}{N}\right) - F_{n:n}\left(\frac{k-1}{N}\right)\right]^2 - 1 \right\} c_2, \\ \left[1 - F_{n:n}\left(\lfloor \frac{N}{2} \rfloor \Big/ N \right) - F_{n:n}\left(\lceil \frac{N}{2} \rceil \Big/ N \right) \right] \sigma_\infty, \end{cases}$$

$$\to \begin{cases} \frac{1}{2}[f_{n:n}(1) - f_{n:n}(0)]\sigma_1, \\ \left[\int_0^1 f_{n:n}^2(x)dx - 1 \right]^{1/2} \sigma_2, \\ \left[1 - 2F_{n:n}\left(\frac{1}{2}\right)\right] \sigma_\infty, \end{cases} = \begin{cases} \frac{n}{2}\sigma_1, \\ \frac{n-1}{\sqrt{2n-1}}\sigma_2, \\ \left(1 - \frac{1}{2^{n-1}}\right)\sigma_\infty. \end{cases} \tag{5.84}$$

as $N \to \infty$. The limits are identical with the general bounds of Sect. 4.2. Writing more complicated formulae, we obtain the same conclusions for $2 \neq p \in (1, +\infty)$. Later on, we provide arguments that the limiting property is extended to more general cases.

5.1.3 Upper Bounds for Differences of Order Statistics

Now we focus on the bounds for the expectations of differences of order statistics

$$\mathbb{E}(X_{j:n} - X_{i:n}) = \sum_{k=1}^{N} C_k x_{k:N}, \quad 1 \le i < j \le n, \tag{5.85}$$

with

$$C_k = F_{i,j:n}\left(\frac{k}{N}\right) - F_{i,j:n}\left(\frac{k-1}{N}\right) = f_{i,j:n}(\xi_k), \tag{5.86}$$

for some $\xi_k \in \left(\frac{k-1}{N}, \frac{k}{N}\right)$, and all $k = 1, \ldots, N$, where

$$f_{i,j:n}(x) = f_{j:n}(x) - f_{i:n}(x), \qquad 0 < x < 1, \tag{5.87}$$

and $F_{i,j:n}(x)$ stand for their antiderivatives.

Here, properties of (5.87) are crucial for evaluating (5.85). We have already proven (see (4.106)) that (5.87) is first negative and then positive. It has zero at

$$\beta = \beta(i, j, n) = 1 \Bigg/ \left\{1 + \left[\binom{n-1}{j-1} \Bigg/ \binom{n-1}{i-1}\right]^{1/(j-i)}\right\} \tag{5.88}$$

which becomes $\beta(i, i+1, n) = i/n$ in the case of spacings. Moreover, (5.87) vanishes at zero and is first decreasing (except for the case $i = 1$), then increasing negative and positive (in all the cases), and ultimately decreasing to zero at 1 (except for the case $j = n$). There are no analytic representations of the minimum and maximum points $0 < \alpha < \beta$ and $\beta < \gamma < 1$, respectively, except for the cases of spacings, when

$$\alpha = \frac{i}{n} - \sqrt{\frac{i(n-i)}{n^2(n-1)}}, \tag{5.89}$$

$$\gamma = \frac{i}{n} + \sqrt{\frac{i(n-i)}{n^2(n-1)}}. \tag{5.90}$$

Applying representation (5.86) and arguments of Lemma 5.1, we prove the following results.

Lemma 5.2 *Let* $0 \le \alpha < \beta < \gamma \le 1$ *denote the minimum, zero and maximum points of* (5.87)*, respectively. Then, sequence* (5.86) *is negative for* $k \le \lfloor \beta N \rfloor$ *and positive for* $k \ge \lceil \beta N \rceil + 1$*. Moreover, this is possibly first decreasing, then increasing and possibly ultimately decreasing. The global minimum is attained at* *one or two neighboring points among* $\lfloor \alpha N \rfloor$, $\lceil \alpha N \rceil$ *and* $\lceil \alpha N \rceil + 1$. *The global maximum is attained at one or two neighboring points among* $\lfloor \gamma N \rfloor$, $\lceil \gamma N \rceil$ *and* $\lceil \gamma N \rceil + 1$.

More precisely, it may happen that $\alpha < \frac{1}{N}$, and the minimal element is C_1, and there is not a decreasing subsequence of first elements. This is always true if $i = 1$ and $\alpha = 0$. It is also possible (for instance, for $j = n$) that $\gamma > 1 - \frac{1}{N}$ and C_n is maximal. In particular, there are differences $X_{j:n} - X_{i:n}$, with large j and small i (that are not necessarily the sample ranges) and small population sizes N such that (5.86) is strictly increasing. We also note that the relations between $C_{\lceil \beta N \rceil}$ and 0 may be arbitrary. If ℓ_* and L_* denote the indices of the last negative and first positive elements of (5.86), then $\lfloor \beta N \rfloor \le \ell_* < L_* \le \lceil \beta N \rceil + 1$. The projection of (5.86) onto the family of nondecreasing sequences is described below.

Proposition 5.2 *Define* $k_* = 1$ *for* $i = 1$ *and* $K_* = N - 1$ *for* $j = n$. *For* $i > 1$,

$$G_{i,j:n}(x) = \frac{F_{i,j:n}(x)}{x}, \qquad 0 < x \le 1, \tag{5.91}$$

and a unique $\alpha < \alpha_* < \beta$ *solving the equation*

$$f_{i,j:n}(x) = G_{i,j:n}(x), \tag{5.92}$$

put

$$k_* = \begin{cases} \lceil \alpha_* N \rceil, \text{ if } G_{i,j:n}\left(\frac{\lceil \alpha_* N \rceil}{N}\right) \le G_{i,j:n}\left(\frac{\lfloor \alpha_* N \rfloor}{N}\right), \\ \lfloor \alpha_* N \rfloor, \text{ otherwise.} \end{cases} \tag{5.93}$$

For $j < n$,

$$H_{i,j:n}(x) = -\frac{F_{i,j:n}(x)}{1-x}, \qquad 0 \le x < 1, \tag{5.94}$$

and a unique $\beta < \gamma_* < \gamma$ *solving the equation*

$$f_{i,j:n}(x) = H_{i,j:n}(x), \tag{5.95}$$

put

$$K_* = \begin{cases} \lfloor \gamma_* N \rfloor \text{ if } H_{i,j:n}\left(\frac{\lfloor \gamma_* N \rfloor}{N}\right) \ge H_{i,j:n}\left(\frac{\lceil \gamma_* N \rceil}{N}\right), \\ \lceil \gamma_* N \rceil, \text{ otherwise.} \end{cases} \tag{5.96}$$

Then,

$$\underline{C}_k = \begin{cases} \frac{F_{i,j:n}\left(\frac{k_*}{N}\right)}{k_*}, & k = 1, \ldots, k_*, \\ F_{i,j:n}\left(\frac{k}{N}\right) - F_{i,j:n}\left(\frac{k-1}{N}\right), & k = k_* + 1, \ldots, K_*, \\ -\frac{F_{i,j:n}\left(\frac{K_*}{N}\right)}{N-K_*}, & k = K_* + 1, \ldots, N. \end{cases} \tag{5.97}$$

Proof Define

$$k_* = \max\left\{1 \le k \le N : \frac{1}{k}\sum_{r=1}^{k} C_r = \min_{\ell=1,\ldots,N} \frac{1}{\ell}\sum_{r=1}^{\ell} C_r\right\}, \tag{5.98}$$

$$K_* = \min\left\{1 \le k \le N : \frac{1}{N-k}\sum_{r=N+1-k}^{N} C_r = \max_{\ell=1,\ldots,N} \frac{1}{N-\ell}\sum_{r=n+1-\ell}^{N} C_r\right\}. \tag{5.99}$$

Then $1 \le k_* \le K_* \le N$, and, by (2.177) and (2.178),

$$\underline{C}_1 = \ldots = \underline{C}_{k_*} = \frac{1}{k_*}\sum_{r=1}^{k_*} C_r = NG_{i,j:n}\left(\frac{k_*}{N}\right), \tag{5.100}$$

$$\underline{C}_{K_*+1} = \ldots = \underline{C}_N = \frac{1}{N-K_*}\sum_{r=N+1-K_*}^{N} C_r = NH_{i,j:n}\left(\frac{K_*}{N}\right). \tag{5.101}$$

Since $\tilde{C}$ is first negative and then positive,

$$\underline{C}_{k_*} < 0 < \underline{C}_{K_*+1}, \tag{5.102}$$

and

$$k_* \le \ell_* < L_* \le K_* + 1. \tag{5.103}$$

Let m_* denote either the unique minimum point or the greater of two minimum points of (5.86). Since sequence C_k, $k = 1, \ldots, m_*$, is nonincreasing, so is the sequence of means $\frac{1}{\ell}\sum_{r=1}^{\ell} C_r$, $\ell = 1, \ldots, m_*$. This implies that $k_* \ge m_*$. Similarly, we conclude that $K_* \le M_*$, where M_* is either the single or the smaller maximum of (5.86).

If $k_* = K_*$, then (5.97) describes the projection, but the middle line is redundant then. We show that otherwise $\underline{C}_k = C_k$, $k = k_* + 1, \ldots, K_*$, and so (5.97) is true as well. To this end, it suffices to verify that

$$\underline{C}_{k_*} < C_{k_*+1} < \cdots < C_{K_*} < \underline{C}_{K_*+1}. \tag{5.104}$$

The first inequality follows from the fact that

$$\frac{1}{k_*}\sum_{r=1}^{k_*} C_r < \frac{1}{k_*+1}\sum_{r=1}^{k_*+1} C_r, \tag{5.105}$$

which is a consequence of definition (5.88). Similarly, (5.99) implies the last inequality in (5.104). In order to deduce the remaining ones, we combine $C_{m_*} < C_{m_*+1} < \cdots < C_{M_*}$ with $m_* \le k_* \le K_* < M_*$. This completes the proof of (5.104).

It remains to show that definitions (5.93), (5.96) are identical with (5.98), (5.99), respectively. For the first identity, it suffices to check that the sequence

$$G_{i,j:n}\left(\frac{k}{N}\right), \qquad k = 1, \ldots, N, \tag{5.106}$$

attains its minimum (either unique or greater of two) at k_*. We have already proven (see Theorem 4.9 and its proof) that (5.91) is negative, and first decreasing (except for the case $i = 1$) and then increasing, and the unique minimum point is defined by (5.92). If $i = 1$, then (5.106) is minimized at $k_* = 1$. Otherwise, the minimum is attained at either of points $\lfloor \alpha_* N \rfloor$ and $\lceil \alpha_* N \rceil$. Definition (5.93) determines either the unique or the greater of two minima. Likewise, we prove an identity between (5.96) and (5.99). □

We are now in a position to present evaluations of bounds of expectations of differences of order statistics in various scale units. Observe that there are no additional restrictions on $1 \le i < j \le n$ to obtain positive sharp bounds. Using the same methods, we can extend the results to differences of trimmed means

$$T_{\ell,m:n} - T_{i,j:n} = \frac{1}{m+1-\ell}\sum_{r=\ell}^{m} X_{r:n} - \frac{1}{j+1-i}\sum_{r=i}^{j} X_{r:n},$$
$$1 \le i \le j,\ \ell \le m \le n. \tag{5.107}$$

For details, we refer to Rychlik (2004).

Corollary 5.5 *The inequality*

$$\mathbb{E}\frac{X_{j:n}-X_{i:n}}{\sigma_1}\le\frac{N}{2}\left[-\frac{F_{i,j:n}\left(\frac{K_*}{N}\right)}{N-K_*}-\frac{F_{i,j:n}\left(\frac{k_*}{N}\right)}{k_*}\right] \tag{5.108}$$

holds, and becomes an equality for

$$x_{k:N}=\begin{cases}\mu-\sigma_1\frac{N}{2k_*}, & k=1,\ldots,k_*,\\ \mu, & k=k_*+1,\ldots,K_*,\\ \mu+\sigma_1\frac{N}{2(N-K_*)}, & k=K_*+1,\ldots,N.\end{cases} \tag{5.109}$$

Corollary 5.6 *For* $1<p<\infty$ *and a unique*

$$\frac{F_{i,j:n}\left(\frac{k_*}{N}\right)}{k_*}<\underline{\bar{c}}_q<-\frac{F_{i,j:n}\left(\frac{K_*}{N}\right)}{N-K_*} \tag{5.110}$$

that solves the equation

$$\begin{aligned}0=&\sum_{k=k_*+1}^{K_*}\left|F_{i,j:n}\left(\frac{k}{N}\right)-F_{i,j:n}\left(\frac{k-1}{N}\right)-c\right|^{q/p}\\ &\times sgn\left\{F_{i,j:n}\left(\frac{k}{N}\right)-F_{i,j:n}\left(\frac{k-1}{N}\right)-c\right\}\\ &+k_*\left[c-\frac{F_{i,j:n}\left(\frac{k_*}{N}\right)}{k_*}\right]^{q/p}+(N-K_*)\left[-\frac{F_{i,j:n}\left(\frac{K_*}{N}\right)}{N-K_*}-c\right]^{q/p},\end{aligned} \tag{5.111}$$

we have

$$\mathbb{E}\frac{X_{j:n}-X_{i:n}}{\sigma_p}\le N^{1/p}B \tag{5.112}$$

with

$$\begin{aligned}B^q=&k_*\left[\underline{\bar{c}}_q-\frac{F_{i,j:n}\left(\frac{k_*}{N}\right)}{k_*}\right]^q+(N-K_*)\left[-\frac{F_{i,j:n}\left(\frac{K_*}{N}\right)}{N-K_*}-\underline{\bar{c}}_q\right]^q\\ &+\sum_{k=k_*+1}^{K_*}\left|F_{i,j:n}\left(\frac{k}{N}\right)-F_{i,j:n}\left(\frac{k-1}{N}\right)-\underline{\bar{c}}_q\right|^q.\end{aligned} \tag{5.113}$$

The equality in (5.112) *holds if*

$$x_{k:N} = \begin{cases} \mu - \sigma_p \frac{N^{1/p}}{B^{q/p}} \left[\underline{c}_q - \frac{F_{i,j:n}\left(\frac{k_*}{N}\right)}{k_*} \right]^{q/p}, & k = 1, \ldots, k_*, \\ \mu + \sigma_p \frac{N^{1/p}}{B^{q/p}} \left| F_{i,j:n}\left(\frac{k}{N}\right) - F_{i,j:n}\left(\frac{k-1}{N}\right) - \underline{c}_q \right|^{q/p} & \\ \quad \times sgn\left\{ F_{i,j:n}\left(\frac{k}{N}\right) - F_{i,j:n}\left(\frac{k-1}{N}\right) - \bar{c}_q \right\}, & k = k_* + 1, \ldots, K_*, \\ \mu + \sigma_p \frac{N^{1/p}}{B^{q/p}} \left[-\frac{F_{i,j:n}\left(\frac{K_*}{N}\right)}{N - K_*} - \underline{c}_q \right]^{q/p}, & K = K_* + 1, \ldots, N. \end{cases} \tag{5.114}$$

Corollary 5.7 *In particular,*

$$\mathbb{E} \frac{X_{j:n} - X_{i:n}}{\sigma_2} \le N^{1/2} B$$

for

$$B^2 = \frac{F^2_{i,j:n}\left(\frac{k_*}{N}\right)}{k_*} + \frac{F^2_{i,j:n}\left(\frac{K_*}{N}\right)}{N - K_*} + \sum_{k=k_*+1}^{K_*} \left[F_{i,j:n}\left(\frac{k}{N}\right) - F_{i,j:n}\left(\frac{k-1}{N}\right) \right]^2, \tag{5.115}$$

and this is attained by

$$x_{k:N} = \begin{cases} \mu + \sigma_2 \frac{N^{1/2}}{B} F_{i,j:n}\left(\frac{k_*}{N}\right), & k = 1, \ldots, k_*, \\ \mu + \sigma_2 \frac{N^{1/2}}{B} \left[F_{i,j:n}\left(\frac{k}{N}\right) - F_{i,j:n}\left(\frac{k-1}{N}\right) \right], & k = k_* + 1, \ldots, K_*, \\ \mu - \sigma_2 \frac{N^{1/2}}{B} \frac{F_{i,j:n}\left(\frac{K_*}{N}\right)}{N - K_*}, & k = K_* + 1, \ldots, N. \end{cases} \tag{5.116}$$

Corollary 5.8 *Under notation* (5.91) *and* (5.93), *if*

$$G_{i,j:n}\left(\left\lceil \frac{N+1}{2} \right\rceil \Big/ N \right) \le G_{i,j:N}\left(\left\lceil \frac{N-1}{2} \right\rceil \Big/ N \right) < 0 \tag{5.117}$$

then the inequality

$$\mathbb{E} \frac{X_{j:n} - X_{i:n}}{\sigma_\infty} \le -\frac{N}{k_*} F_{i,j:n}\left(\frac{k_*}{N}\right) \tag{5.118}$$

holds and becomes the equality if

$$x_{k:N} = \begin{cases} \mu - \sigma_\infty \frac{N-k_*}{k_*}, & k = 1, \ldots, k_*, \\ \mu + \sigma_\infty, & k = k_* + 1, \ldots, N. \end{cases} \tag{5.119}$$

Under notation (5.94) *and* (5.97), *if*

$$0 < H_{i,j:n}\left(\left\lfloor \frac{N+3}{2} \right\rfloor \Big/ N\right) \leq H_{i,j:n}\left(\left\lfloor \frac{N+1}{2} \right\rfloor \Big/ N\right), \tag{5.120}$$

then inequality

$$\mathbb{E}\frac{X_{j:n} - X_{i:n}}{\sigma_\infty} \leq -\frac{N}{N - K_*} F_{i,j:n}\left(\frac{K_*}{N}\right) \tag{5.121}$$

is tight, and becomes an equality for

$$x_{k:N} = \begin{cases} \mu - \sigma_\infty, & k = 1, \ldots, K_*, \\ \mu + \sigma_\infty \frac{K}{N-K_*}, & k = K_* + 1, \ldots, N. \end{cases} \tag{5.122}$$

Otherwise,

$$\mathbb{E}\frac{X_{j:n} - X_{i:n}}{\sigma_\infty} \leq \begin{cases} -2F_{i,j:n}\left(\frac{1}{2}\right), & \text{if } N \text{ is even}, \\ -F_{i,j:n}\left(\frac{N+1}{2}\right) - F_{i,j:n}\left(\frac{N-1}{2}\right), & \text{if } N \text{ is odd}, \end{cases} \tag{5.123}$$

holds, and becomes an equality under conditions (5.78) *and* (5.79), *respectively.*

Conditions (5.117) and (5.120) are equivalent to $\text{MED}_N(\underline{\tilde{C}}) = \underline{C}_{k_*}$ and $\underline{C}_{K_*+1}$, respectively. Otherwise, $\text{MED}_N(\underline{\tilde{C}}) = \text{MED}_N(\tilde{C})$. The first bound in (5.123) is optimal for the general bounded random populations. For the sample range, we have in particular

$$\mathbb{E}(X_{n:n} - X_{1:n}) \leq \begin{cases} \frac{N}{2}\left[-F_{1,n:n}\left(\frac{N-1}{N}\right) - F_{1,n:n}\left(\frac{1}{N}\right)\right]\sigma_1, \\ \left\{\sum_{k=1}^N N\left[F_{1,n:n}\left(\frac{k}{N}\right) - F_{1,n:n}\left(\frac{k-1}{N}\right)\right]^2\right\}\sigma_2, \\ \left[-F_{1,n:n}\left(\lfloor\frac{N}{2}\rfloor \Big/ N\right) - F_{1,n:n}\left(\lceil\frac{N}{2}\rceil \Big/ N\right)\right]\sigma_\infty, \end{cases}$$

$$\to \begin{cases} \frac{1}{2}[f_{1,n:n}(1) - f_{1,n:n}(0)]\sigma_1, \\ \left[\int_0^1 f_{1,n:n}^2(x)dx\right]^{1/2}\sigma_2, \\ \left[-2F_{1,n:n}\left(\frac{1}{2}\right)\right]\sigma_\infty, \end{cases}$$

$$= \begin{cases} n\,\sigma_1, \\ \frac{\sqrt{2}n}{\sqrt{2n-1}}\left[1-\frac{1}{\binom{2n-2}{n-1}}\right]^{1/2}\sigma_2, \\ 2\left(1-\frac{1}{2^{n-1}}\right)\sigma_\infty, \end{cases} \tag{5.124}$$

as $N \to \infty$. The limiting values coincide with the optimal bounds for general populations.

5.1.4 Negative Upper and Positive Lower Bounds for General L-Statistics

Now we consider L-statistics based on the with replacement model whose coefficient vectors $\tilde{c} \in \mathbb{R}^n$ satisfy equivalent conditions (5.18) and (5.19). The conditions do not admit positive upper bounds on $\sum_{i=1}^n c_i \frac{\mathbb{E}X_{i:n}-\mu}{\sigma_p}$ for $1 \le p \le +\infty$. Under these assumptions, we present non-positive sharp upper bounds. The results are mostly based on the paper by Goroncy and Rychlik (2006b). It is convenient to introduce the following function:

$$L_{\tilde{c}:n}(x) = F_{\tilde{c}:n}(x) - \sum_{i=1}^n c_i x. \tag{5.125}$$

Note that (5.19) can be expressed as

$$L_{\tilde{c}:n}\left(\frac{k}{N}\right) \ge 0, \qquad k = 1, \ldots, N-1.$$

Theorem 5.5 *Let $p \in [1, \infty)$ and $\tilde{c} \in \mathbb{R}^n$ be a vector satisfying condition* (5.19) *with $L_{\tilde{c}:n}$ defined in (5.125). Put*

$$U_{p,\tilde{c}:n}(x) = \frac{L_{\tilde{c}:n}(x)}{[x^p(1-x)+x(1-x)^p]^{1/p}}. \tag{5.126}$$

Then,

$$\mathbb{E}\left(\sum_{i=1}^n c_i \frac{X_{i:n}-\mu}{\sigma_p}\right) \le -\min_{j=1,\ldots,N-1} U_{p,\tilde{c}:n}\left(\frac{j}{N}\right), \tag{5.127}$$

with the equality attained for the following ordered population:

$$x_{k:N}^{(j,p)} = \begin{cases} \mu - \sigma_p(N-j)\left[\frac{1}{\frac{j}{N}(1-\frac{j}{N})^p+(\frac{j}{N})^p(1-\frac{j}{N})}\right]^{1/p}, & k=1,\ldots,j, \\ \mu + \sigma_p j\left[\frac{1}{\frac{j}{N}(1-\frac{j}{N})^p+(\frac{j}{N})^p(1-\frac{j}{N})}\right]^{1/p}, & k=j+1,\ldots,N, \end{cases} \tag{5.128}$$

where $j = j_*$ minimizes $U_{p,\bar{c}:n}(\frac{j}{N})$, (see (5.126)*) $j = 1,\ldots,N-1$.*

Proof By (5.11) with (5.12), we obtain

$$\mathbb{E}\left(\sum_{i=1}^{n} c_i \frac{X_{i:n}-\mu}{\sigma_p}\right) = \sum_{k=1}^{N} C_k \frac{x_{k:N}-\mu}{\sigma_p} = \sum_{k=1}^{N}(C_k-\bar{C})\frac{x_{k:N}-\mu}{\sigma_p},$$

where constant $\bar{C} = \frac{1}{N}\sum_{k=1}^{N} C_k = \frac{1}{N}\sum_{i=1}^{n} c_i = \frac{n}{N}\bar{c}$ can be subtracted, because $x_{k:N} - \mu$, $k = 1,\ldots,n$, sum up to 0 (cf. (5.2)). Notice that

$$\sum_{k=1}^{j}(C_k-\bar{C}) = F_{\bar{c}:n}\left(\frac{j}{N}\right) - \frac{j}{N}\sum_{i=1}^{n} c_i = L_{\bar{c}:n}\left(\frac{j}{N}\right) \geq 0, \quad j=1,\ldots,N-1, \tag{5.129}$$

and μ and σ_p^p coincide with the sample mean and the p-th absolute central moment of the deterministic sample $\{x_1,\ldots,x_N\}$. If all the elements of (5.129) are positive, Theorem 3.17 shows that under condition (5.129), the following deterministic bounds hold:

$$\sum_{k=1}^{N}(C_k-\bar{C})\frac{x_{k:N}-\mu}{\sigma_p} \leq -\min_{j=1,\ldots,N-1}\frac{\sum_{k=1}^{j}(C_k-\bar{C})}{\left[(\frac{j}{N})^p(1-\frac{j}{N})+\frac{j}{N}(1-\frac{j}{N})^p\right]^{1/p}}, \tag{5.130}$$

and the equality is attained by the deterministic sample (5.128) of size N with $j = j_*$ providing the minimum of the right-hand side of (5.130). Plugging (5.129) into (5.130), we obtain the desired inequality. Population (5.128) with properly chosen $j = j_*$ provides the equality there. If either of (5.129) amounts to zero, then the respective numerator in (5.130) is zero as well, and it certainly provides the minimum. Note that then the conclusion is equivalent to that of Theorem 3.15. □

Using similar arguments, we apply Theorem 3.18 for proving the following result for $p = \infty$.

Theorem 5.6 *Let $p = \infty$, and $\tilde{c} \in \mathbb{R}^n$ be a vector satisfying condition (5.19). Put*

$$U_{\infty,\tilde{c}:n}(x) = \frac{L_{\tilde{c}:n}(x)}{\max\{x, 1-x\}}. \tag{5.131}$$

Then,

$$\mathbb{E}\left(\sum_{i=1}^{n} c_i \frac{x_{i:n} - \bar{x}}{\sigma_\infty}\right) \leq - \min_{j=1,\ldots,N-1} U_{\infty,\tilde{c}:n}\left(\frac{j}{N}\right), \tag{5.132}$$

with the equality attained for the following ordered population:

$$x_{k:N}^{(j,\infty)} = \begin{cases} \mu - \sigma_\infty, & k = 1, \ldots, j, \\ \mu + \sigma_\infty \frac{j}{N-j}, & k = j+1, \ldots, N, \end{cases} \quad \text{if } 1 \leq j \leq \frac{N}{2}, \tag{5.133}$$

$$x_{k:N}^{(j,\infty)} = \begin{cases} \mu - \sigma_\infty \frac{N-j}{j}, & k = 1, \ldots, j, \\ \mu + \sigma_\infty, & k = j+1, \ldots, N, \end{cases} \quad \text{if } \frac{N}{2} \leq j \leq N-1, \tag{5.134}$$

with $j = j_*$ minimizing $U_{\infty,\tilde{c}:n}(\frac{j}{N})$ (comp. (5.131), $j = 1, \ldots, N-1$.

Applying standard arguments (see (5.52)), we immediately conclude corresponding nonnegative upper bounds.

Corollary 5.9 *Let $p \in [1, \infty)$ and $\tilde{c}$ be a vector of coefficients satisfying*

$$\forall\, j = 1, \ldots, N-1 \quad \sum_{i=1}^{n} c_{n+1-i} F_{i:n}\left(\frac{j}{N}\right) \geq \frac{j}{N} \sum_{i=1}^{n} c_i. \tag{5.135}$$

Put

$$U_{p,\tilde{c}':n}(x) = \frac{\sum_{i=1}^{n} c_{n+1-i}[F_{i:n}(x) - x]}{[x^p(1-x) + x(1-x)^p]^{1/p}}. \tag{5.136}$$

Then,

$$\mathbb{E}\left(\sum_{i=1}^{n} c_i \frac{X_{i:n} - \mu}{\sigma_p}\right) \geq \min_{j=1,\ldots,N-1} U_{p,\tilde{c}':n}\left(\frac{j}{N}\right), \tag{5.137}$$

and the equality is attained for population (5.128) with $j = j_*$ minimizing $U_{p,\tilde{c}':n}(\frac{j}{N})$, $j = 1, \ldots, N-1$.

Corollary 5.10 *For $p = \infty$ and $\tilde{c} \in \mathbb{R}^n$ satisfying* (5.135), *put*

$$U_{\infty,\tilde{c}':n}(x) = \frac{\sum_{i=1}^{n} c_{n+1-i}[F_{i:n}(x) - x]}{\max\{x, 1-x\}}. \tag{5.138}$$

Then,

$$\mathbb{E}\left(\sum_{i=1}^{n} c_i \frac{X_{i:n} - \bar{x}}{\sigma_\infty}\right) \geq \min_{j=1,\ldots,N-1} U_{\infty,\tilde{c}':n}\left(\frac{j}{N}\right), \tag{5.139}$$

with the equality attained for the populations defined in (5.133) *and* (5.134) *with $j = j_*$ minimizing $U_{\infty,\tilde{c}':n}(\frac{j}{N})$, $j = 1, \ldots, N-1$.*

Note that $U_{p,\tilde{c}':n}(x) = U_{p,\tilde{c}:n}(1-x)$ for all $1 \leq p \leq \infty$. Indeed, if we write

$$U_{p,\tilde{c}:n}(x) = \frac{L_{\tilde{c}:n}(x)}{M_p(x)}, \tag{5.140}$$

where the numerator is defined in (5.125), and

$$M_p(x) = \begin{cases} [x^p(1-x) + x(1-x)^p]^{1/p}, & 1 \leq p < \infty. \\ \max\{x, 1-x\}, & p = \infty, \end{cases} \tag{5.141}$$

we observe that $L_{\tilde{c}':n}(x) = L_{\tilde{c}:n}(1-x)$ and $M_p(x) = M_p(1-x)$. Accordingly, if $U_{p,\tilde{c}:n}$ restricted to arguments $\frac{1}{N}, \ldots, \frac{N-1}{n}$ attains its minimum at some $\frac{j_*}{N}$, then $U_{p,\tilde{c}':n}$ attains its minimum at $1 - \frac{j_*}{N}$. Obviously, under symmetry of $L_{\tilde{c}:n}$ (and $L_{\tilde{c}:n}$), the respective minima are symmetrically located about $\frac{1}{2}$.

Below we show that all the bounds tend to zero as the population size N increases to infinity, with a possible exception of the case $p = 1$. This confirms the results of Sect. 4.5 which assert there are no sharp negative upper and positive lower bounds for the expectations of centered L-statistics based on general populations except for ones evaluated in the mean absolute deviation from the mean σ_1 units.

Corollary 5.11

(i) *For arbitrary $1 < p \leq \infty$ and $\tilde{c} \in \mathbb{R}^n$ satisfying* (5.19) *for all N sufficiently large, the non-positive upper bounds of* (5.127) and (5.132) *converge to zero, as $N \to \infty$;*
(ii) *For $p = 1$ and $\tilde{c} \in \mathbb{R}^n$ satisfying* (5.19) *for all N sufficiently large, the bounds of* (5.127) *tend to zero as $N \to \infty$ iff either $c_1 = \bar{c}$ or $c_n = \bar{c}$ or $L_{\tilde{c}:n}(x_*) = 0$ for some $0 < x_* < 1$.*

Proof

(i) Take $p \in (1, \infty]$, $x \in (0, 1)$ and recall formula (5.140). Notice that the numerator of (5.140) defined in (5.125) is a polynomial of degree n, and satisfies $L_{\tilde{c}:n}(0) = L_{\tilde{c}:n}(1) = 0$. By (5.125), (4.2), and (4.6),

$$\lim_{x \searrow 0} \frac{L_{\tilde{c}:n}(x)}{x} = \lim_{x \searrow 0} \sum_{i=1}^{n} c_i \left[\frac{F_{i:n}(x)}{x} - 1 \right] = \sum_{i=1}^{n} c_i \left[f_{i:n}(0) - 1 \right] = nc_1 - \sum_{i=1}^{n} c_i$$

and

$$\begin{aligned} \lim_{x \nearrow 1} \frac{L_{\tilde{c}:n}(x)}{1-x} &= \lim_{x \nearrow 1} \sum_{i=1}^{n} c_i \left[\frac{F_{i:n}(x)}{1-x} - \frac{x}{1-x} \right] \\ &= \sum_{i=1}^{n} c_i \left[f_{i:n}(1) - 1 \right] = nc_n - \sum_{i=1}^{n} c_i \end{aligned}$$

are finite. On the other hand,

$$\begin{aligned} \lim_{x \searrow 0} \frac{M_p(x)}{x} &= \lim_{x \searrow 0} x^{1/p-1} = +\infty, \\ \lim_{x \nearrow 1} \frac{M_p(x)}{1-x} &= \lim_{x \nearrow 1} (1-x)^{1/p-1} = +\infty, \end{aligned}$$

under convention $\frac{1}{\infty} = 0$. Therefore,

$$\lim_{x \searrow 0} \frac{L_{\tilde{c}:n}(x)}{M_p(x)} = \lim_{x \nearrow 1} \frac{L_{\tilde{c}:n}(x)}{M_p(x)} = 0,$$

and in particular

$$\lim_{N \to \infty} U_{p,\tilde{c}:n} \left(\frac{1}{N} \right) = \lim_{N \to \infty} U_{p,\tilde{c}:n} \left(\frac{N-1}{N} \right) = 0.$$

(ii) For $p = 1$, we have

$$\lim_{x \searrow 0} U_{1,\tilde{c}:n}(x) = \lim_{x \searrow 0} \frac{L_{\tilde{c}:n}(x)}{2x} = \frac{1}{2} \left(nc_1 - \sum_{i=1}^{n} c_i \right)$$

and

$$\lim_{x \nearrow 1} U_{1,\tilde{c}:n}(x) = \lim_{x \nearrow 1} \frac{L_{\tilde{c}:n}(x)}{2(1-x)} = \frac{1}{2} \left(\sum_{i=1}^{n} c_i - nc_n \right).$$

Therefore, under conditions $c_1 = \bar{c}$ and $c_n = \bar{c}$, we have

$$\lim_{N\to\infty} U_{1,\tilde{c}:n}(\frac{1}{N}) = 0$$

and

$$\lim_{N\to\infty} U_{1,\tilde{c}:n}(\frac{N-1}{N}) = 0,$$

respectively. If $L_{\tilde{c}:n}(x_*) = 0$, then

$$\lim_{j/N\longrightarrow x_*} U_{1,\tilde{c}:n}(j/N) = 0,$$

which completes the proof.

□

Assume that we observe n random variables drawn in the with replacement scheme from a N-valued population. In order to determine upper bounds for a specific L-statistic with a fixed coefficient vector $\tilde{c} = (c_1, \ldots, c_n)$ in the scale units σ_p with specified $1 \le p \le \infty$, we need to evaluate functions (5.140), with the numerator and denominators defined in (5.125) and (5.141), respectively, at a finite number of points $x = \frac{1}{N}, \ldots, \frac{N-1}{N}$, and choose the minimal value. In the lower bound problem, we merely replace the original vector $\tilde{c}$ by the reversed one $\tilde{c}' = (c_n, \ldots, c_1)$.

If we would like to consider specific types of L-statistics for various n and N, we should rather analyze (5.140) over the continuous domain $0 < x < 1$ first. Numerator (5.125) is a polynomial dependent on the coefficients $c_1, \ldots, c_n$ with the degree equal to the sample size n, and this is independent of p. If the type of L-statistic is specified (e.g., the sample minimum, range, etc.), (5.125) actually depends on the sample size n. In any case, this is nonnegative on (0, 1) under (5.19), and vanishes at the end-points. On the other hand, denominator (5.141) is identical for all the L-statistics, and depends only on the order p of scale units. This is positive on (0, 1), and symmetric about $\frac{1}{2}$. We also have $M_p(0) = M_p(1) = 0$ except for $p = \infty$. For $1 \le p \le 3$, the function is first increasing and then decreasing with the maximum at $\frac{1}{2}$. If $3 < p < \infty$, (5.141) is first increasing, then decreasing, increasing and ultimately decreasing. It has a local minimum at $\frac{1}{2}$, and two maxima at points located symmetrically about $\frac{1}{2}$. The maximum points move from the center $\frac{1}{2}$ to the ends 0 and 1, as p increases. The limiting broken line M_∞ has the maxima $M_\infty(0) = M_\infty(1) = 1$ at the ends and the minimum $M_\infty\left(\frac{1}{2}\right) = \frac{1}{2}$ at the center.

We first consider the upper bounds for the sample minima. The respective coefficient vector has the form $\tilde{c} = (1, 0, \ldots, 0)$ and its coordinates sum up to 1. Therefore (5.140) can be rewritten as

$$U_{p,n}^{(m)}(x) = \frac{L_n^{(m)}(x)}{M_p(x)},$$

where

$$L_n^{(m)}(x) = F_{1:n}(x) - x = 1 - x - (1 - x)^n,$$

and $M_p(x)$, $1 \le p \le \infty$, are defined in (5.141).

Corollary 5.12

(i) *For arbitrary $p \in [1, \infty)$, the sharp upper negative bound on the expectation of the sample minimum is*

$$\mathbb{E}\left(\frac{X_{1:n} - \mu}{\sigma_p}\right) \le -\frac{N^{1/p}[1 - N^{1-n}]}{[(N-1) + (N-1)^p]^{1/p}}, \tag{5.142}$$

and the equality holds for (5.128) *with $j = N - 1$;*

(ii) *Let $p = \infty$. The upper negative bound for the sample minimum is*

$$\mathbb{E}\left(\frac{X_{1:n} - \mu}{\sigma_\infty}\right) \le -\frac{1 - N^{1-n}}{N - 1}, \tag{5.143}$$

and the equality holds for (5.134) *with $j = N - 1$.*

Also, for $n = 2$, bounds (5.142), (5.143) *are attained by* (5.128) *and* (5.133), *respectively, with $j = 1$ as well, and in the case $n = 2$ with $p = 1$, the equality in* (5.142) *is attained by* (5.128) *with arbitrary $j = 1, \ldots, N - 1$.*

Proof

(i) For $p = 1$, we have

$$U_{1,n}^{(m)}(x) = \frac{1 - x - (1 - x)^n}{2x(1 - x)} = \frac{1 - (1 - x)^{n-1}}{2x} = \frac{1}{2}\sum_{i=0}^{n-2}(1 - x)^i,$$

which is constant for $n = 2$ and decreasing for $n \ge 3$. Hence, the respective bounds are $U_{1,2}^{(m)}\left(\frac{j}{N}\right) = \frac{1}{2}$, $j = 1, \ldots, N - 1$, and $U_{1,n}^{(m)}\left(\frac{N-1}{N}\right) = \frac{1}{2}$, $n \ge 3$.

For $1 < p < \infty$, we define

$$V_{p,n}^{(m)}(x) = [U_{p,n}^{(m)}(x)]^p = \frac{[1 - x - (1 - x)^n]^p}{x^p(1 - x) + x(1 - x)^p}. \tag{5.144}$$

It has the derivative

$$V_{p,n}^{(m)'}(x) = \frac{[1 - x - (1 - x)^n]^{p-1}}{[x^p(1 - x) + x(1 - x)^p]^2} K(x),$$

where

$$K(x) = (1 - p)x^p(1 - x) - (1 - x)^{p+1} - px^{p-1}(1 - x)^2 + (1 - x)^{p+n}$$
$$+ (np - 1)x^p(1 - x)^n + p(n - 1)x(1 - x)^{n+p-1} + px^{p-1}(1 - x)^{n+1}.$$

It is obvious that the signs of $V_{p,n}^{(m)'}(x)$ and $K(x)$ coincide. Note that $K(x)$ is a linear combination of products of x and $(1 - x)$ raised to powers that sum up to either $p + 1$ or $n + p$. We multiply the former ones by

$$1^{n-1} = \sum_{i=0}^{n-1} \binom{n-1}{i} x^i (1 - x)^{n-i-1},$$

in order to obtain a homogenous representation

$$K(x) = \sum_{i=0}^{n-1} \binom{n-1}{i} (1 - p)x^{i+p}(1 - x)^{n-i} - \sum_{i=0}^{n-1} \binom{n-1}{i} x^i (1 - x)^{n-i+p}$$
$$- \sum_{i=0}^{n-1} \binom{n-1}{i} px^{i+p-1}(1 - x)^{n-i-1} + p(n - 1)x(1 - x)^{n+p-1}$$
$$+ (np - 1)x^p(1 - x)^n + (1 - x)^{p+n} + px^{p-1}(1 - x)^{n+1}.$$

Dividing it by $(1 - x)^{n+p}$, and changing the variable $y = \frac{x}{1-x} \in (0, \infty)$, we get

$$\begin{aligned} R(y) &= (n - 1)(p - 1)y - \sum_{i=2}^{n-1} \binom{n-1}{i} y^i \\ &\quad + \sum_{i=1}^{n-2} \left[\binom{n-1}{i} (1 - p) - \binom{n-1}{i+1} p \right] y^{i+p} + (1 - p)y^{n+p-1} \\ &= \sum_{i=1}^{n-1} A_i y^i + \sum_{i=1}^{n-1} B_i y^{i+p}, \end{aligned} \tag{5.145}$$

with

$$\begin{aligned}
A_1 &= (n-1)(p-1) > 0, \\
A_i &= -\binom{n-1}{i} < 0, && i = 2, \dots, n-1, \\
B_i &= \binom{n-1}{i}(1-p) - \binom{n-1}{i+1}p < 0, && i = 1, \dots, n-2, \\
B_{n-1} &= 1 - p < 0.
\end{aligned}$$

Now we recall the Descartes rule of sign (see Theorem 2.46) asserting that the number of sign changes of a linear combination of power functions with increasing exponents does not exceed the number of sign changes of the coefficients, and the first and last signs coincide with the signs of the respective first and last coefficients (see also, Karlin & Studden, 1966, Chapter I, Corollary 4.4). This allows us to conclude that (5.145) is first positive and then negative. Accordingly, (5.144) is first increasing and then decreasing, and its restriction to $x = \frac{1}{N}, \dots, \frac{N-1}{N}$ attains the minimum at either of the end-points $\frac{1}{N}, \frac{N-1}{N}$. To compare the values of (5.144) there, it suffices to compare the numerators, because the denominators are symmetric about $\frac{1}{2}$. Define

$$L_s(x) = L_n^{(m)}(x) - L_n^{(m)}(1-x) = 1 - x - (1-x)^n - x - x^n. \tag{5.146}$$

Notice that $L_s(x) = 0$ if $n = 2$. Consequently, $L_2^{(m)}(\frac{1}{N}) = L_2^{(m)}(\frac{N-1}{N})$, and we obtain

$$\begin{aligned}
\mathbb{E}\left(\frac{X_{1:2} - \mu}{\sigma_p}\right) &\leq -U_{p,2}^{(m)}\left(\frac{1}{N}\right) = -U_{p,2}^{(m)}\left(\frac{N-1}{N}\right) \\
&= -(N-1)\left[\frac{N^{1-p}}{(N-1) + (N-1)^p}\right]^{1/p}.
\end{aligned}$$

In the other cases when $n \geq 3$, we have

$$\begin{aligned}
L_s(0) &= L_s(1) = L_s\left(\frac{1}{2}\right) = 0, \\
L_s''(x) &= n(n-1)[x^{n-2} - (1-x)^{n-2}].
\end{aligned} \tag{5.147}$$

It follows that (5.146) is strictly concave for $0 < x < \frac{1}{2}$ and strictly convex for $\frac{1}{2} < x < 1$. This, combined with (5.147), implies that $L_s(x) > 0$, and

$$L_n^{(m)}(x) > L_n^{(m)}(1-x), \quad 0 < x < \frac{1}{2}, \tag{5.148}$$

and $L_n^{(m)}(\frac{1}{N}) > L_n^{(m)}(\frac{N-1}{N})$ in particular. It follows that

$$U_{p,n}^{(m)}\left(\frac{N-1}{N}\right) < U_{p,n}^{(m)}\left(\frac{1}{N}\right),$$

and we get

$$\mathbb{E}\left(\frac{X_{1:n}-\mu}{\sigma_p}\right) \le -U_{p,n}^{(m)}\left(\frac{N-1}{N}\right) = -\frac{N^{1/p}[1-N^{1-n}]}{[(N-1)+(N-1)^p]^{1/p}}.$$

The equality holds here for (5.128) with $j = N-1$.

(ii) Consider

$$U_{\infty,n}^{(m)}(x) = \frac{L_n^{(m)}(x)}{M_\infty(x)} = \frac{1-x-(1-x)^n}{\max\{x, 1-x\}}, \qquad 0 < x < 1. \tag{5.149}$$

By (5.148), it suffices to minimize (5.149) over $\frac{1}{2} < x < 1$ only. We have

$$U_{\infty,n}^{(m)}(x) = \frac{1-x-(1-x)^n}{x} = \sum_{i=1}^{n-1}(1-x)^i,$$

which is obviously decreasing and attains its minimum over the lattice $\frac{k}{N}$, $k = 1, \dots, N-1$, at $\frac{N-1}{N}$. Therefore, we get the following negative upper bound

$$\mathbb{E}\left(\frac{X_{1:n}-\mu}{\sigma_\infty}\right) \le -U_{\infty,n}^{(m)}\left(\frac{N-1}{N}\right) = -\frac{1-N^{1-n}}{N-1}.$$

The equality holds there for (5.134) with $j = N-1$.

□

In the special cases $p = 1$ and $p = 2$, we obtain

$$\mathbb{E}\left(\frac{X_{1:n}-\mu}{\sigma_1}\right) \le -\frac{N[1-N^{1-n}]}{2(N-1)}, \tag{5.150}$$

$$\mathbb{E}\left(\frac{X_{1:n}-\mu}{\sigma_2}\right) \le -\frac{1-N^{1-n}}{\sqrt{N-1}}, \tag{5.151}$$

respectively. If N increases, then, due to Corollary 5.11, the bounds of (5.142) and (5.143) tend to 0 except for the case $p = 1$, where the limiting value is -1. If $n \to \infty$, we derive limits of (5.142) and (5.143), once we replace $1 - N^{1-n}$ by 1 there.

Application of Corollaries 5.9 and 5.10 gives immediately the following results.

Corollary 5.13 *The positive sharp lower bounds on the expectations of the sample maxima $X_{n:n}$ are identical with the negatives of the negative upper bounds for the sample minima $X_{1:n}$ described in Corollary* 5.12.

We now proceed to determining lower bounds on the expectations of Gini mean differences

$$\mathrm{GMD}_n(\tilde{X}) = \frac{1}{n(n-1)} \sum_{i,j=1}^{n} |X_i - X_j| = \frac{2}{n(n-1)} \sum_{i=1}^{n} (2i-n-1) X_{i:n} \tag{5.152}$$

which are obviously nonnegative. The reversed coefficient vector $\tilde{c}'$ of the right-hand side representation has the coordinates

$$c_i' = c_{n+1-i} = \frac{2(n+1-2i)}{n(n-1)} = -c_i.$$

Obviously $\bar{c} = 0$. Corresponding function (5.125) has the representation

$$\begin{aligned}
L^{(g)}(x) &= \sum_{i=1}^{n} c_i' F_{i:n}(x) = \sum_{i=1}^{n} \frac{2(n+1-2i)}{n(n-1)} \sum_{r=i}^{n} \binom{n}{r} x^r (1-x)^{n-r} \\
&= \frac{2}{n(n-1)} \sum_{i=1}^{n} \binom{n}{i} x^i (1-x)^{n-i} \sum_{r=1}^{i} (n+1-2r) \\
&= \frac{2}{n(n-1)} \sum_{i=1}^{n} i(n-i) \binom{n}{i} x^i (1-x)^{n-i} \\
&= 2x(1-x) \sum_{i=0}^{n-2} \binom{n-2}{i} x^i (1-x)^{n-i-2} = 2x(1-x), \quad 0 < x < 1.
\end{aligned}$$

Notice that this is symmetric about $\frac{1}{2}$, and does not depend on the sample size n. So, we drop the index n from the notation. In order to determine lower bounds on the expectations of Gini mean differences, we analyze

$$U_p^{(g)}(x) = \frac{L^{(g)}(x)}{M_p(x)} = U_p^{(g)}(1-x). \tag{5.153}$$

Corollary 5.14

(i) *Let $p = 1$. The lower positive bound on the expectation of Gini mean differences in the mean absolute deviation units σ_1 is*

$$\mathbb{E}\left(\frac{1}{n(n-1)}\sum_{i,j=1}^{n}\frac{|X_i - X_j|}{\sigma_1}\right) \geq 1,$$

and the equality holds for population (5.128) *with arbitrary $j = 1, \ldots N-1$.*

(ii) *Let $1 < p < \infty$. Then,*

$$\mathbb{E}\left(\frac{1}{n(n-1)}\sum_{i,j=1}^{n}\frac{|X_i - X_j|}{\sigma_p}\right) \geq \frac{2(N-1)^{1-1/p}}{N^{1-1/p}[1+(N-1)^{p-1}]^{1/p}},$$

and the equality holds for (5.128) *with $j = 1$ and $j = N-1$.*

(iii) *Let $p = \infty$. The lower positive bound on expectation of Gini mean differences given in the σ_∞ units is*

$$\mathbb{E}\left(\frac{1}{n(n-1)}\sum_{i,j=1}^{n}\frac{|X_i - X_j|}{\sigma_\infty}\right) \geq \frac{2}{N},$$

and the equality holds for (5.133) *and* (5.134) *with $j = 1$ and $N-1$, respectively.*

Proof The proofs of cases (i) and (iii) are immediate. More elaborate arguments are needed for proving (ii). Consider

$$Z_p^{(g)}(x) = \frac{2^p}{[U_p^{(g)}(x)]^p} = (1-x)^{1-p} + x^{1-p} = Z_p^{(g)}(1-x),$$

with the derivative

$$Z_p^{(g)'}(x) = (1-p)[x^{-p} - (1-x)^{-p}] = -Z_p^{(g)'}(1-x).$$

Notice that the latter is negative for $x < \frac{1}{2}$ and positive for $x > \frac{1}{2}$, and the former is first decreasing and then increasing. Therefore, symmetric function (5.153) is increasing on $(0, \frac{1}{2})$, and decreasing elsewhere. The restriction to the discrete points $\frac{1}{N}, \ldots, \frac{N-1}{N}$ attains its minimum at the extreme points of the domain, at which it amounts to

$$U_p^{(g)}\left(\frac{1}{N}\right) = U_p^{(g)}\left(\frac{N-1}{N}\right) = \frac{2(N-1)^{1-1/p}}{N^{1-1/p}[1+(N-1)^{p-1}]^{1/p}}.$$

Application of Corollary 5.9 then yields the desired statement. □

In the special case $p = 2$, we have

$$\mathbb{E}\left(\frac{1}{n(n-1)}\sum_{i,j=1}^{n}\frac{|X_i - X_j|}{\sigma_2}\right) \geq \frac{2\sqrt{N-1}}{N}.$$

Observe that all the bounds do not depend on the sample size n, and they vanish for $p > 1$, as the population size N increases. The upper bounds for properly standardized Gini mean difference can be determined with the use of Theorems 5.1–5.3. Although we have some simplification due to the fact that $\underline{\tilde{C}} = \tilde{C}$, the respective formulae have still quite complicated forms as

$$\mathbb{E}\mathrm{GMD}_n(\tilde{X}) \leq \begin{cases} -\frac{N}{2}\left[F_{\tilde{c}:n}\left(\frac{N-1}{N}\right) + F_{\tilde{c}:n}\left(\frac{1}{N}\right)\right]\sigma_1, & \\ \left[N\sum_{k=1}^{N}\left[F_{\tilde{c}:n}\left(\frac{k}{N}\right) - F_{\tilde{c}:n}\left(\frac{k-1}{N}\right)\right]^2\right]^{1/2}\sigma_2, & \\ -\left[F_{\tilde{c}:n}\left(\frac{N+1}{2N}\right) + F_{\tilde{c}:n}\left(\frac{N-1}{2N}\right)\right]\sigma_\infty, & N \text{ odd}, \\ -\left[\frac{3}{2}F_{\tilde{c}:n}\left(\frac{N+2}{2N}\right) + F_{\tilde{c}:n}\left(\frac{N}{2N}\right) - \frac{1}{2}F_{\tilde{c}:n}\left(\frac{N-2}{2N}\right)\right]\sigma_\infty, & N \text{ even}, \end{cases}$$

where

$$F_{\tilde{c}:n}(x) = \sum_{i=1}^{n}\frac{2(2i-n-1)}{n(n-1)}F_{i:n}(x).$$

For $1 < p \neq 2$, we do not have closed analytic formulae.

5.1.5 Lower Bounds for Differences of Order Statistics

We first treat the sharp lower bounds for the sample ranges $X_{n:n} - X_{1:n}$. It follows from Corollaries 5.9 and 5.10 that in order to find them, we should minimize the symmetric function

$$U_{p,n}^{(r)}(x) = \frac{1 - x^n - (1-x)^n}{M_p(x)} = U_{p,n}^{(r)}(1-x) \tag{5.154}$$

over $x = \frac{1}{N}, \ldots, \frac{N-1}{N}$, for arbitrarily fixed n, $N \geq 1$ and $1 \leq p \leq \infty$. In contrast to the previous examples, there are no general partitions of the two-values populations (5.128), (5.133), or (5.134) that provide the minimum for all the choices of N, n, and p. The following corollary shows that for any population size N and $1 \leq j_* \leq N-1$, we can find sample size n and parameter p of the scale units σ_p such that the minimal expectation of the sample range is attained by a population with two subsets of identical elements of sizes j_* and $N - j_*$.

Corollary 5.15 *For every $\frac{j_*}{N} \in (0,1)$, there exist $3 \le p < \infty$ and $n \ge 2$ such that*

$$\min_{j=1,\dots,N-1} U^{(r)}_{p,n}\left(\frac{j}{N}\right) = U^{(r)}_{p,n}\left(\frac{j_*}{N}\right) = U^{(r)}_{p,n}\left(\frac{N-j_*}{N}\right).$$

Proof We start with recalling properties of (5.141) described above. The function is symmetric, has a single local minimum at $\frac{1}{2}$ for $p \le 3$, and two symmetrically located maxima for $p > 3$. By continuity of (5.141) in both x and p, the arguments of maxima change continuously with respect to p, and the limiting values are 0 and 1. Therefore for every fixed $0 < \frac{j_*}{N} < 1$, there exists $p \ge 3$ such that (5.141) attains its maximum at $\frac{j_*}{N}$ and $1 - \frac{j_*}{N}$. Then, we have

$$\frac{1}{M_p(\frac{j_*}{N})} = \frac{1}{M_p(\frac{N-j_*}{N})} < \min_{j \ne j_*, N-j*} \frac{1}{M_p(\frac{j}{N})} - \epsilon,$$

for some positive ϵ. Notice that the numerator of (5.154) tends to 1 as $n \to \infty$ and $x \in (0,1)$ is fixed. Therefore, we can find n_0 sufficiently large such that for all $n \ge n_0$, we have

$$\min_{j \ne j_*, N-j_*} U^{(r)}_{p,n}\left(\frac{j}{N}\right) = \min_{j \ne j_*, N-j_*} \frac{L_n\left(\frac{j}{N}\right)}{M_p\left(\frac{j}{N}\right)} > U^{(r)}_{p,n}\left(\frac{j_*}{N}\right) = U^{(r)}_{p,n}\left(\frac{N-j_*}{N}\right).$$

This proves the statement. □

Therefore, we focus only on the most interesting special cases of bounds for $p = 1, \infty$, and 2.

Corollary 5.16 *Let $p = 1$. For $n = 2, 3$ and arbitrary N, we have*

$$\mathbb{E}\frac{X_{2:2} - X_{1:2}}{\sigma_1} \le 1, \tag{5.155}$$

$$\mathbb{E}\frac{X_{3:3} - X_{1:3}}{\sigma_1} \le \frac{3}{2}, \tag{5.156}$$

and both the bounds are attained if the population is two-valued.

Suppose now that $n \ge 4$. If N is even, we have the following sharp lower bound:

$$\mathbb{E}\left(\frac{X_{n:n} - X_{1:n}}{\sigma_1}\right) \ge 2 - \frac{1}{2^{n-2}}, \tag{5.157}$$

and the equality holds for (5.128) *with* $j = \frac{N}{2}$. *If N is odd, we have*

$$\mathbb{E}\left(\frac{X_{n:n} - X_{1:n}}{\sigma_1}\right) \geq \frac{N^2-1}{2N^2}\left[1 - \left(\frac{N+1}{2N}\right)^n - \left(\frac{N-1}{2N}\right)^n\right], \tag{5.158}$$

with the equality for (5.128) *with* $j = \frac{N-1}{2}$ *and* $j = \frac{N-1}{2}$.

Proof We aim at minimizing the following symmetric function:

$$U^{(r)}_{1,n}(x) = \frac{L^{(r)}_{1,n}(x)}{M_1(x)} = \frac{1 - x^n - (1-x)^n}{2x(1-x)}, \qquad 0 < x < 1. \tag{5.159}$$

over the discrete set $\frac{1}{N}, \ldots, \frac{N-1}{N}$. Note that $U^{(r)}_{1,2}(x) = 1$ and $U^{(r)}_{1,3}(x) = \frac{3}{2}$ are constant, and so the first statement holds.

For $n \geq 4$, we have

$$L^{(r)}_{1,n}(x) = F_{1:n}(x) - F_{n:n}(x) = \sum_{i=1}^{n-1} B_{i,n}(x) = \sum_{i=1}^{n-1} \binom{n}{i} x^i (1-x)^{n-i}.$$

Therefore,

$$U^{(r)}_{1,n}(x) = \frac{1}{2}\sum_{i=1}^{n-1}\binom{n}{i} x^{i-1}(1-x)^{n-i-1} = \frac{1}{2}\sum_{i=0}^{n-2}\binom{n}{i+1} x^i (1-x)^{n-i-2}.$$

and

$$U^{(r)\prime}_{1,n}(x) = \frac{1}{2}\sum_{i=0}^{n-3} a_i x^i (1-x)^{n-i-3} \tag{5.160}$$

with

$$a_i = \frac{n!(i+1)}{(i+2)!(n-i-2)!} - \frac{n!(n-2-i)}{(i+1)!(n-i-1)!}, \qquad i = 0, \ldots n-3.$$

We have $a_i > 0$ iff $i > \frac{n-3}{2}$. Using the variation diminishing property of the Bernstein polynomials (see Lemma 4.1), we conclude that (5.160) is first negative and then positive. Therefore, (5.159) is first decreasing and then increasing. By symmetry, it attains its minimum in the middle point $x = \frac{1}{2}$.

If N is even, we can simply take $x = \frac{j}{N} = \frac{1}{2}$, and calculate the lower bound

$$\mathbb{E}\left(\frac{X_{n:n} - X_{1:n}}{\sigma_1}\right) \geq U_{1,n}^{(r)}\left(\frac{1}{2}\right) = 2 - \frac{1}{2^{n-2}}.$$

If N is odd, we take the nearest neighbors $x = \frac{j}{N} = \frac{N-1}{2N}$, and $\frac{N+1}{2N}$ of the global minimum point $\frac{1}{2}$, and get

$$\mathbb{E}\left(\frac{X_{n:n} - X_{1:n}}{\sigma_1}\right) \geq U_{1,n}^{(r)}\left(\frac{N-1}{2N}\right) = U_{1,n}^{(r)}\left(\frac{N+1}{2N}\right)$$
$$= \frac{N^2-1}{2N^2}\left[1 - \left(\frac{N+1}{2N}\right)^n - \left(\frac{N-1}{2N}\right)^n\right].$$

□

The bounds (5.157) are identical for all even-sized populations. They together with (5.155) and (5.156) coincide with the sharp lower bounds for the sample ranges coming from iid samples with arbitrary marginal distributions (see Corollary 4.19). The RHS of (5.158) is greater than the RHS of (5.157), but the difference becomes negligible for large N. It can be concluded from the above proof, and confirmed by direct calculations.

Corollary 5.17 *Let $p = \infty$. The lower positive bound on expectation of the sample range is*

$$\mathbb{E}\left(\frac{X_{n:n} - X_{1:n}}{\sigma_\infty}\right) \geq \frac{N^n - 1 - (N-1)^n}{N^{n-1}(N-1)}.$$

The equality holds above for population (5.133) *with $j = 1$ and population* (5.134) *with $j = N - 1$.*

Proof We first need to analyze

$$U_{\infty,n}^{(r)}(x) = \frac{L_n^{(r)}(x)}{M_\infty(x)} = \frac{1 - x^n - (1-x)^n}{\max\{x, 1-x\}} = U_{\infty,n}^{(r)}(1-x), \qquad 0 < x < 1. \tag{5.161}$$

Due to symmetry, it suffices to consider the function over $\left(\frac{1}{2}, 1\right)$, and drop $1 - x$ from the denominator, because $x > 1 - x$ then. We have

$$U_{\infty,n}^{(r)\prime}(x) = \frac{L_n^{(r)\prime}(x)}{x} - \frac{L_n^{(r)}(x)}{x^2}, \quad \frac{1}{2} < x < 1,$$

which is negative, because $L_n^{(r)}(x) > 0$ for $0 < x < 1$, and $L_n^{(r)'}(x) < 0$ for $\frac{1}{2} < x < 1$. Therefore, (5.161) is decreasing for $x \in (\frac{1}{2}, 1)$ and attains its minimum at $x = \frac{N-1}{N}$. By symmetry, another minimum point of (5.161) is $x = \frac{1}{N}$. Summing up, we obtain

$$\mathbb{E}\left(\frac{X_{n:n} - X_{1:n}}{\sigma_\infty}\right) \geq U_{\infty,n}^{(r)}\left(\frac{N-1}{N}\right) = U_{\infty,n}^{(r)}\left(\frac{1}{N}\right) = \frac{N^n - 1 - (N-1)^n}{N^{n-1}(N-1)}.$$

□

Corollary 5.18 *Let $p = 2$.*

(i) *For $n = 2, \ldots, 5$, we have*

$$\mathbb{E}\left(\frac{X_{n:n} - X_{1:n}}{\sigma_2}\right) \geq \frac{N}{\sqrt{N-1}}\left[1 - \frac{1}{N^n} - \left(\frac{N-1}{N}\right)^n\right]$$

and the equality holds for population (5.128) *with $j = 1$ and $j = N - 1$.*

(ii) *If $n \geq 6$ and N is even, then*

$$\mathbb{E}\left(\frac{X_{n:n} - X_{1:n}}{\sigma_2}\right) \geq \min\left\{\frac{N}{\sqrt{N-1}}\left[1 - \frac{1}{N^n} - \left(\frac{N-1}{N}\right)^n\right], 2 - \frac{1}{2^{n-2}}\right\}.$$

If the minimum of the RHS is attained at the first argument, the equality holds for (5.128) *with $j = 1$ and $N - 1$. Otherwise, the bound is attained by* (5.128) *with $j = \frac{N}{2}$.*

(iii) *If $n \geq 6$ and N is odd, we have*

$$\mathbb{E}\left(\frac{X_{n:n} - X_{1:n}}{\sigma_2}\right) \geq \min\left\{\frac{N}{\sqrt{N-1}}\left[1 - \frac{1}{N^n} - \left(\frac{N-1}{N}\right)^n\right]\right.$$
$$\left.\frac{2N}{\sqrt{N^2-1}}\left[1 - \left(\frac{N-1}{2N}\right)^n - \left(\frac{N+1}{2N}\right)^n\right]\right\},$$

If the minimum is attained at the first argument, the equality holds for (5.128) *with $j = 1$ and $N - 1$. The equality is attained by (5.128) and with $j = \frac{N-1}{2}$ and $\frac{N+1}{2}$ in the opposite case.*

Observe that

$$U_{2,n}^{(r)}\left(\frac{1}{N}\right) = U_{2,n}^{(r)}\left(\frac{N-1}{N}\right) = \frac{N}{\sqrt{N-1}}\left[1 - \frac{1}{N^n} - \left(\frac{N-1}{N}\right)^n\right]$$
$$\to \frac{N}{\sqrt{N-1}},$$

$$U_{2,n}^{(r)}\left(\frac{N/2}{N}\right) = 2 - \frac{1}{2^{n-2}} \to 2,$$

$$U_{2,n}^{(r)}\left(\frac{N-1}{2N}\right) = U_{2,n}^{(r)}\left(\frac{N+1}{2N}\right) = \frac{2N}{\sqrt{N^2-1}}\left[1 - \left(\frac{N-1}{2N}\right)^n - \left(\frac{N+1}{2N}\right)^n\right]$$
$$\to \frac{2}{\sqrt{1-\frac{1}{N^2}}},$$

as N is fixed and $n \to \infty$, and

$$U_{2,n}^{(r)}\left(\frac{1}{N}\right) = U_{2,n}^{(r)}\left(\frac{N-1}{N}\right) = \frac{N}{\sqrt{N-1}}\left[1 - \frac{1}{N^n} - \left(\frac{N-1}{N}\right)^n\right] \to 0,$$
$$U_{2,n}^{(r)}\left(\frac{N-1}{2N}\right) = U_{2,n}^{(r)}\left(\frac{N+1}{2N}\right) = \frac{2N}{\sqrt{N^2-1}}\left[1 - \left(\frac{N-1}{2N}\right)^n - \left(\frac{N+1}{2N}\right)^n\right]$$
$$\to U_{2,n}^{(r)}\left(\frac{N/2}{N}\right) = 2 - \frac{1}{2^{n-2}},$$

as n is fixed and $N \to \infty$. Hence, if n is large, then $U_{2,n}^{(r)}(x)$ attains its minimum in the middle and if N is large, then it attains its minimum at the ends of the lattice domain. This means that in the former case, the bound is attained by the equally partitioned population (and almost equally, in the odd case). In the latter case, the extreme population has only one element different from the others.

Proof of Corollary 5.18 We consider

$$U_{2,n}^{(r)}(x) = \frac{1 - x^n - (1-x)^n}{\sqrt{x(1-x)}} = U_{2,n}^{(r)}(1-x), \qquad 0 < x < 1. \tag{5.162}$$

It is more convenient to study

$$V_{2,n}^{(r)}(x) = [U_{2,n}^{(r)}(x)]^2 = \frac{[1 - x^n - (1-x)^n]^2}{x(1-x)} = \frac{L^2(x)}{M(x)},$$

say, with the derivative

$$V_{2,n}^{(r)'}(x) = \frac{L(x)}{M^2(x)}\left[2L'(x)M(x) - L(x)M'(x)\right] = \frac{L(x)}{M^2(x)}K(x),$$

where the fraction is positive, and

$$K(x) = (2n-1)x(1-x)^n - (2n-1)x^n(1-x) + (1-x)^{n+1} - x^{n+1} + x - (1-x),$$

say, has the sign coinciding with the sign of the derivative. We can rewrite it as

$$K(x) = (2n-1)x(1-x)^n - (2n-1)x^n(1-x) + (1-x)^{n+1} - x^{n+1}$$
$$+ \sum_{i=0}^{n} \binom{n}{i} x^{i+1}(1-x)^{n-i} - \sum_{i=0}^{n} \binom{n}{i} x^i (1-x)^{n+1-i}$$
$$= \sum_{i=0}^{n+1} A_i x^i (1-x)^{n+1-i}, \tag{5.163}$$

where

$$A_0 = 0,$$
$$A_1 = n > 0,$$
$$A_i = \binom{n}{i-1} - \binom{n}{i} \begin{cases} < 0, \text{ if } & i < \frac{n+1}{2}, \\ = 0, \text{ if } & i = \frac{n+1}{2}, \\ > 0, \text{ if } & i > \frac{n+1}{2}, \end{cases} \qquad i = 2, \ldots, n-1,$$
$$A_n = -n < 0,$$
$$A_{n+1} = 0.$$

Using the variation diminishing property of (5.163), we deduce that it is either positive, negative, again positive and negative or first positive, and then negative. Therefore, (5.162) is either increasing, decreasing, then increasing again and decreasing or first increasing and then decreasing. Since K is symmetric and has a local extremum at $\frac{1}{2}$, we have $K'\left(\frac{1}{2}\right) > 0$ in the former case, and $K'\left(\frac{1}{2}\right) \leq 0$ in the latter one. Note that

$$K'\left(\frac{1}{2}\right) = 2\left(1 - \frac{n(n-1)+1}{2^{n-1}}\right)$$

is non-positive iff $2^{n-1} - n \leq (n-1)^2$ which is true for $n = 2, \ldots, 5$ only. Then, (5.162) is increasing on $\left(0, \frac{1}{2}\right)$ and decreasing on $\left(\frac{1}{2}, 1\right)$, and by Theorem 5.5, we have

$$\mathbb{E}\left(\frac{X_{n:n} - X_{1:n}}{\sigma_2}\right) \geq U_{2,n}^{(r)}\left(\frac{1}{N}\right) = U_{2,n}^{(r)}\left(\frac{N-1}{N}\right)$$
$$= \frac{N}{\sqrt{N-1}}\left[1 - \frac{1}{N^n} - \left(\frac{N-1}{N}\right)^n\right]$$

which verifies the validity of (i).

For $n \geq 6$, there is a local minimum of (5.162) at $\frac{1}{2}$, and the minimum over the grid $\frac{1}{N}, \ldots, \frac{N-1}{N}$ can be attained either at the center or at the extremes. We consider the cases of population size N even and odd separately. If N is even, we need to compare

$$U_{2,n}^{(r)}\left(\frac{1}{N}\right) = U_{2,n}^{(r)}\left(\frac{N-1}{N}\right) = \frac{N}{\sqrt{N-1}}\left[1 - \frac{1}{N^n} - \left(\frac{N-1}{N}\right)^n\right] \tag{5.164}$$

with

$$U_{2,n}^{(r)}\left(\frac{1}{2}\right) = 2 - \frac{1}{2^{n-2}}. \tag{5.165}$$

If (5.164) is less than (5.165), then this is the desired bound, and it is attained by (5.128) with $j = 1$ and $N - 1$. Otherwise, the bound amounts to (5.165), and is attained by (5.128) with $j = \frac{N}{2}$. In the case of odd sized populations, we repeat the arguments, replacing merely (5.165) by

$$U_{2,n}^{(r)}\left(\frac{N-1}{2N}\right) = U_{2,n}^{(r)}\left(\frac{N+1}{2N}\right) = \frac{2N}{\sqrt{N^2-1}}\left[1 - \left(\frac{N-1}{2N}\right)^n - \left(\frac{N+1}{2N}\right)^n\right].$$

□

We finally present lower bounds for spacings determined in Goroncy and Rychlik (2009).

Corollary 5.19 *Suppose that $X_1, \ldots, X_n$ is a sample drawn with replacement from a numerical population of size N such that it contains at least two different elements. If $1 \leq k \leq \frac{n}{2}$, then we have*

$$\mathbb{E}\frac{X_{k+1:n} - X_{k:n}}{\sigma_p} \geq \begin{cases} \binom{n}{k}\frac{(N-1)^k}{N^{n-1-1/p}[(N-1)+(N-1)^p]^{1/p}}, & 1 \leq p < \infty, \\ \binom{n}{k}\frac{(N-1)^{k-1}}{N^{n-1}}, & p = \infty. \end{cases}$$

The equality holds for (5.128) *in the case $1 \leq p < \infty$ and for* (5.133) *in the case $p = \infty$, with $j = 1$ in both the cases.*

If $\frac{n}{2} \leq k \leq n - 1$, then

$$\mathbb{E}\frac{X_{k+1:n} - X_{k:n}}{\sigma_p} \geq \begin{cases} \binom{n}{k}\frac{(N-1)^{n-k}}{N^{n-1-1/p}[(N-1)+(N-1)^p]^{1/p}}, & 1 \leq p < \infty, \\ \binom{n}{k}\frac{(N-1)^{n-k-1}}{N^{n-1}}, & p = \infty. \end{cases}$$

Then, the equality holds for (5.128) *in the case $1 \leq p < \infty$ and for* (5.134) *in the case $p = \infty$, with $j = N - 1$ in both the cases.*

Proof Fix k and note that the reversed coefficient vector $\tilde{c}'$ of the L-statistic $X_{k+1:n} - X_{k:n}$ has 1 and -1 at the positions $n-k$ and $n-k+1$, and zeros elsewhere. Therefore, function *(5.125)* has the particular form

$$L_{\tilde{c},n}(x) = L_n^{(k)}(x) = F_{n-k:n}(x) - F_{n+1-k:n}(x) = \binom{n}{k} x^{n-k}(1-x)^k \qquad (5.166)$$

According to Corollaries 5.9 and 5.10, we should consider functions

$$U_{p,n}^{(k)}(x) = \frac{L_n^{(k)}(x)}{M_p(x)}, \qquad (5.167)$$

and determine their minima restricted to the points $\frac{j}{N}$, $j = 1, \dots, N-1$. First, we focus on the case $p = 1$ for which we have

$$U_{1,n}^{(k)}(x) = \frac{1}{2}\binom{n}{k} x^{n-k-1}(1-x)^{k-1}, \quad 0 < x < 1. \qquad (5.168)$$

Notice that the derivative of (5.168) is given by

$$[U_{1,n}^{(k)}(x)]' = \frac{1}{2}\binom{n}{k} x^{n-k-2}(1-x)^{k-2}[(n-k-1)(1-x) - (k-1)x].$$

We deduce that function (5.168) is increasing for $k = 1$, decreasing for $k = n-1$ and first increasing and then decreasing in the other cases. Therefore, after restricting to the arguments $\frac{1}{N}, \dots, \frac{N-1}{N}$, its minimal value may be one of the following numbers:

$$U_{1,n}^{(k)}\left(\frac{1}{N}\right) = \frac{1}{2}\binom{n}{k}\frac{(N-1)^{k-1}}{N^{n-2}},$$

$$U_{1,n}^{(k)}\left(\frac{N-1}{N}\right) = \frac{1}{2}\binom{n}{k}\frac{(N-1)^{n-k-1}}{N^{n-2}}.$$

It is easy to check that $U_{1,n}^{(k)}\left(\frac{1}{N}\right) \le U_{1,n}^{(k)}\left(\frac{N-1}{N}\right)$ if $k \le \frac{n}{2}$, and the inequality is reversed if $k \ge \frac{n}{2}$.

Assume now that $1 < p < \infty$. In this case, function (5.167) is given by

$$U_{p,n}^{(k)}(x) = \binom{n}{k}\frac{x^{n-k}(1-x)^k}{[x^p(1-x) + x(1-x)^p]^{1/p}} = 2\frac{U_{1,n}^{(k)}(x)}{[g_p(x)]^{1/p}}, \qquad (5.169)$$

where the numerator is defined in (5.168) and

$$g_p(x) = \frac{x^p(1-x) + x(1-x)^p}{x^p(1-x)^p} = x^{1-p} + (1-x)^{1-p}. \qquad (5.170)$$

Notice that

$$g_p'(x) = (p-1)[(1-x)^{-p} - x^{-p}]$$

is negative on $(0, 1/2)$ and positive on $(1/2, 1)$. Accordingly, the symmetric function (5.170) is first decreasing, and then increasing, and so the sequence $1/[g_p(\frac{i}{N})]^{1/p}$, $i = 1, \ldots, N-1$, attains its minimal values at $x = \frac{1}{N}$ and $\frac{N-1}{N}$. If both the positive factors of (5.169), restricted to arguments $\frac{1}{N}, \ldots, \frac{N-1}{N}$, attain simultaneously the respective minima either at $\frac{1}{N}$ or at $\frac{N-1}{N}$, then the product is minimized at the same arguments. By Corollary 5.9, the lower bound for the kth spacing is the smaller of the following two numbers:

$$U_{p,n}^{(k)}\left(\frac{1}{N}\right) = \binom{n}{k} \frac{(N-1)^k}{N^{n-1-1/p}[(N-1)+(N-1)^p]^{1/p}},$$

$$U_{p,n}^{(k)}\left(\frac{N-1}{N}\right) = \binom{n}{k} \frac{(N-1)^{n-k}}{N^{n-1-1/p}[(N-1)+(N-1)^p]^{1/p}}.$$

Evidently, $U_{p,n}^{(k)}\left(\frac{1}{N}\right) \le U_{p,n}^{(k)}\left(\frac{N-1}{N}\right)$ for $k \le \frac{n}{2}$, and $U_{p,n}^{(k)}\left(\frac{1}{N}\right) \ge U_{p,n}^{(k)}\left(\frac{N-1}{N}\right)$ for $k \ge \frac{n}{2}$.

The remaining case is $p = \infty$. It is easy to see that (5.166) is first increasing, and then decreasing, and it is minimized at one of the endpoints of the discrete domain $\frac{1}{N}, \ldots, \frac{N-1}{N}$. Taking into account the fact that for $p = \infty$ the symmetric function (5.141) is maximized by the endpoints as well, and repeating arguments of the previous paragraph, we conclude that the respective sequence of values (5.167) attains its minimum either at $x = \frac{1}{N}$ or at $x = \frac{N-1}{N}$. The corresponding values are

$$U_{\infty,n}^{(k)}\left(\frac{1}{N}\right) = \binom{n}{k} \frac{(N-1)^{k-1}}{N^{n-1}},$$

$$U_{\infty,n}^{(k)}\left(\frac{N-1}{N}\right) = \binom{n}{k} \frac{(N-1)^{n-k-1}}{N^{n-1}}.$$

We easily verify that $U_{\infty,n}^{(k)}\left(\frac{1}{N}\right) \le U_{\infty,n}^{(k)}\left(\frac{N-1}{N}\right)$ for $k \le \frac{n}{2}$, and the inequality is opposite for $k \ge \frac{n}{2}$. The minimal values of $U_{p,n}^{(k)}\left(\frac{j}{N}\right)$, $j = 1, \ldots, N-1$, determined above for various $1 \le p \le \infty$ provide the lower bounds for accordingly standardized expectations of spacings. The attainability conditions can be immediately concluded. □

A natural question that arises here is that what are the sharp lower bounds for the differences of non-adjacent order statistics. A partial answer was given in Corollaries 5.15–5.18 where the sample ranges were treated. Especially, Corollary 5.15 shows that for various p and large sample sizes n, the bounds can be attained by two-valued populations with arbitrary proportions of elements in each group. This implies in particular that the respective bounds may look different. We can expect similar properties for the differences of other non-adjacent, especially remote, order statistics. However, it is worth pointing out that for any for fixed p, n and N, properly chosen vector $\tilde{c} \in \mathbb{R}^n$, and properly constructed $\overline{\tilde{C}} \in \mathbb{R}^N$, one can numerically determine $N-1$ values of either of functions (5.136) and (5.138) at $\frac{k}{N}$, $k = 1, \ldots, N-1$, and choose the minimal one for the value of bound. This would enable to determine the lower bounds for the expectations of the differences of arbitrary order statistics with arbitrary population and sample sizes.

5.1.6 Bounds on Variances of Order Statistics

We finally evaluate variances of order statistics based on random data coming from the drawing with replacement model. With n and N denoting the sample and population sizes, respectively, we define discrete analogs of functions (4.639)–(4.642):

$$\underline{G}_{k:n}\left(\frac{i}{N}\right) = \frac{N}{i} F_{k:n}\left(\frac{i}{N}\right), \qquad 1 \le i \le N-1, \tag{5.171}$$

$$\overline{G}_{k:n}\left(\frac{i}{N}\right) = \frac{N}{N-i}\left[1 - F_{k:n}\left(\frac{i}{N}\right)\right], \qquad 1 \le i \le N-1, \tag{5.172}$$

$$\begin{aligned}\mathbf{G}_{k:n}\left(\frac{i}{N}, \frac{j}{N}\right) &= \underline{G}_{k:n}\left(\frac{i}{N}\right)\overline{G}_{k:n}\left(\frac{j}{N}\right) = \frac{N^2}{i(N-j)} F_{k:n}\left(\frac{i}{N}\right) \\ &\quad \times \left[1 - F_{k:n}\left(\frac{j}{N}\right)\right], \qquad 1 \le i \le j \le N-1,\end{aligned} \tag{5.173}$$

$$G_{k:n}\left(\frac{i}{N}\right) = \mathbf{G}_{k:n}\left(\frac{i}{N}, \frac{i}{N}\right) = \frac{N^2}{i(N-i)} F_{k:n}\left(\frac{i}{N}\right)\left[1 - F_{k:n}\left(\frac{i}{N}\right)\right], \quad 1 \le i \le N-1. \tag{5.174}$$

Lemma 5.3 *For $k = 1$, sequences (5.172) are decreasing, and for $k = n$ they are increasing. For every $2 \le k \le n-1$, $n \ge 3$, the sequence $\overline{G}_{k:n}$ is either decreasing, or increasing or it is first increasing and then decreasing, and its maximal value amounts to*

$$\max_{1 \le i \le N-1} \overline{G}_{k:n}\left(\frac{i}{N}\right) = \begin{cases} \overline{G}_{k:n}\left(\frac{1}{N}\right), & if\ u_1 \le \frac{1}{N}, \\ \overline{G}_{k:n}\left(\frac{N-1}{N}\right), & if\ u_1 \ge \frac{N-1}{N}, \\ \max\left\{\overline{G}_{k:n}\left(\frac{\lfloor Nu_1 \rfloor}{N}\right), \overline{G}_{k:n}\left(\frac{\lceil Nu_1 \rceil}{N}\right)\right\}, & otherwise, \end{cases} \tag{5.175}$$

where $0 < u_1 = u_1(k,n) < \frac{k-1}{n-1}$ is the unique solution to Eq. (4.81).

The statements are immediate consequences of the facts proven in Theorem 4.8 that functions $\overline{G}_{k:n}(u)$, $0 < u < 1$, defined in (4.640) are decreasing for $k = 1$, increasing for $k = n$, and first increasing and then decreasing with the maxima at $u_1(k, n)$ for $2 \le k \le n$. The following lemma is concluded from the above lemma and the fact that $F_{k:n}(x) = 1 - F_{n+1-k:n}(1 - x)$, $1 \le k \le n$.

Lemma 5.4 *For $k = 1$ and $k = n$, the sequences* (5.171) *are decreasing and increasing, respectively. For every $2 \le k \le n - 1$, $n \ge 3$, the sequence $\underline{G}_{k:n}$ is either decreasing, or increasing or it is first increasing and then decreasing, and its maximal value amounts to*

$$\max_{1 \le i \le N-1} \underline{G}_{k:n}\left(\frac{i}{N}\right) = \begin{cases} \underline{G}_{k:n}\left(\frac{1}{N}\right), & if\ u_2 \le \frac{1}{N}, \\ \overline{G}_{k:n}\left(\frac{N-1}{N}\right), & if\ u_2 \ge \frac{N-1}{N}, \\ \max\left\{\underline{G}_{k:n}\left(\frac{\lfloor Nu_2 \rfloor}{N}\right), \underline{G}_{k:n}\left(\frac{\lceil Nu_2 \rceil}{N}\right)\right\}, & otherwise, \end{cases} \tag{5.176}$$

where $\frac{k-1}{n-1} < u_2 = u_2(k, n) < 1$ is the unique solution to the equation $xf_{k:n}(x) = F_{k:n}(x)$.

We denote the numbers of the elements maximizing sequences (5.172) by $i_1 = i_1(k, n, N)$. In the very unlikely but possible cases when $1 < Nu_1(k, n) < N - 1$ and the maximum is attained by two integer points, we shall denote them by $\underline{i}_1$ and $\overline{i}_1 = \underline{i}_1 + 1$. However, if there is no need for distinguishing them, we simply write i_1. On the other hand, it may be sometimes convenient to write either $\underline{i}_1$ or $\overline{i}_1$ even if the maximum point of (5.172) is unique. The same convention shall be used if (5.171) has two maxima, and then we denote by i_2, $\underline{i}_2$ and $\overline{i}_2$ the arguments maximizing it. Note that the relation $u_1 < u_2$ implies $\underline{i}_1 \le \underline{i}_2$ and $\overline{i}_1 \le \overline{i}_2$.

Lemma 5.5 *For every $1 \le k \le n \ge 2$ and $1 < i \le j < N - 1$, there exists $1 \le \ell \le N - 1$ such that*

$$\mathbf{G}_{k:n}\left(\frac{i}{N}, \frac{j}{N}\right) \le G_{k:n}\left(\frac{\ell}{N}\right). \tag{5.177}$$

Moreover, vectors $G_{1:n}\left(\frac{i}{N}\right)$, $i = 1, \ldots, N - 1$, are decreasing for $n \ge 2$, and $G_{n:n}\left(\frac{i}{N}\right)$, $i = 1, \ldots, N - 1$, $n \ge 2$, are increasing. For each $2 \le k \le n - 1$, $n \ge 3$, $G_{k:n}\left(\frac{i}{N}\right)$, $i = 1, \ldots, N - 1$, is unimodal with the maximal value

$$\max_{1 \le i \le N-1} G_{k:n}\left(\frac{i}{N}\right) = \begin{cases} G_{k:n}\left(\frac{1}{N}\right), & if\ u_0 \le \frac{1}{N}, \\ G_{k:n}\left(\frac{N-1}{N}\right), & if\ u_0 \ge \frac{N-1}{N}, \\ \max\left\{G_{k:n}\left(\frac{\lfloor Nu_0 \rfloor}{N}\right), G_{k:n}\left(\frac{\lceil Nu_0 \rceil}{N}\right)\right\}, & otherwise, \end{cases} \tag{5.178}$$

for some $u_1(k,n) < u_0 = u_0(k,n) < u_2(k,n)$ *being the unique solution to Eq.* (4.631).

Proof The proof is similar to the proof of Theorem 4.53. Consider first the case $k = 1$. The triangular array $\mathbf{G}_{1:n}\left(\frac{i}{N}, \frac{j}{N}\right)$, $1 \le i \le j \le N-1$, is the product of two positive decreasing sequences. Therefore, (5.177) holds then with $\ell = 1$. It is also obvious that the diagonal elements $G_{1:n}\left(\frac{i}{N}\right)$, $i = 1, \ldots, N-1$, of the array decrease. Similarly, we obtain the conclusions for $k = n$.

The remaining cases need more elaborate arguments. We admit the most sophisticated possibility that both $\underline{G}_{k:n}\left(\frac{1}{N}\right)$ and $\overline{G}_{k:n}\left(\frac{1}{N}\right)$, $1 \le i \le N-1$, have two maximal values at the points $\underline{i}_2 < \overline{i}_2$ and $\underline{i}_1 < \overline{i}_1$, respectively. Otherwise, we adhere to the convention that $\underline{i}_2 = \overline{i}_2$ and/or $\underline{i}_1 = \overline{i}_1$. If $j > i \ge \underline{i}_1$, then $\overline{G}_{k:n}\left(\frac{j}{N}\right) \le \overline{G}_{k:n}\left(\frac{i}{N}\right)$ and so

$$\mathbf{G}_{k:n}\left(\frac{i}{N}, \frac{j}{N}\right) = \underline{G}_{k:n}\left(\frac{i}{N}\right)\overline{G}_{k:n}\left(\frac{j}{N}\right) \le \underline{G}_{k:n}\left(\frac{i}{N}\right)\overline{G}_{k:n}\left(\frac{i}{N}\right) = G_{k:n}\left(\frac{i}{N}\right). \tag{5.179}$$

If $i < j \le \overline{i}_2$, then $\underline{G}_{k:n}\left(\frac{i}{N}\right) \le \underline{G}_{k:n}\left(\frac{j}{N}\right)$ and

$$\mathbf{G}_{k:n}\left(\frac{i}{N}, \frac{j}{N}\right) = \underline{G}_{k:n}\left(\frac{i}{N}\right)\overline{G}_{k:n}\left(\frac{j}{N}\right) \le \underline{G}_{k:n}\left(\frac{j}{N}\right)\overline{G}_{k:n}\left(\frac{j}{N}\right) = G_{k:n}\left(\frac{j}{N}\right). \tag{5.180}$$

The remaining case is that $i < \underline{i}_1 \le \overline{i}_2 < j$. Then, $\underline{G}_{k:n}\left(\frac{i}{N}\right) < \underline{G}_{k:n}\left(\frac{\underline{i}_1}{N}\right)$ and $\overline{G}_{k:n}\left(\frac{j}{N}\right) < \overline{G}_{k:n}\left(\frac{\overline{i}_2}{N}\right)$, and consequently

$$\mathbf{G}_{k:n}\left(\frac{i}{N}, \frac{j}{N}\right) = \underline{G}_{k:n}\left(\frac{i}{N}\right)\overline{G}_{k:n}\left(\frac{j}{N}\right) \le \underline{G}_{k:n}\left(\frac{\underline{i}_1}{N}\right)\overline{G}_{k:n}\left(\frac{\overline{i}_2}{N}\right) = \mathbf{G}_{k:n}\left(\frac{\underline{i}_1}{N}, \frac{\overline{i}_2}{N}\right). \tag{5.181}$$

By the previous establishments, the last expression is not greater than either of $G_{k:n}\left(\frac{\underline{i}_1}{N}\right)$ and $G_{k:n}\left(\frac{\overline{i}_2}{N}\right)$. It follows that the maximal value of the arrays $\mathbf{G}_{k:n}\left(\frac{i}{N}, \frac{j}{N}\right)$, $1 \le i \le j \le N-1$, $2 \le k \le n-1$, are attained on the diagonal with $1 \le i = j \le N-1$. We showed in the proof of Theorem 4.53 that the functions $G_{k:n}(u)$, $0 < u < 1$, $2 \le k \le n-1$, defined in (4.642) are increasing-decreasing with the maxima attained by the points $u_0(k,n)$ determined by Eq. (4.631). It follows that the functions with the domain restricted to the lattice $\frac{i}{N}$, $i = 1, \ldots, N-1$, have the maxima described by (5.178). □

If $\overline{i}_1 \leq \underline{i}_2$, then

$$G_{k:n}(i) = \underline{G}_{k:n}(i)\overline{G}_{k:n}(i), \qquad i \leq \overline{i}_1,$$

is increasing, because so is the first factor, and the other is nondecreasing, and

$$G_{k:n}(i) = \underline{G}_{k:n}(i)\overline{G}_{k:n}(i), \qquad i \geq \underline{i}_2,$$

is decreasing, because the first factor is nondecreasing then, and the latter one is decreasing. It follows that the maximal value of $G_{k:n}$ is attained among $\frac{\overline{i}_1}{N}, \frac{\overline{i}_1+1}{N}, \ldots, \frac{\underline{i}_2}{N}$. In particular, if $\overline{i}_1 = \underline{i}_2$, this uniquely determines the maximum point. A thorough analysis shows that under the condition $\overline{i}_1 \leq \underline{i}_2$, the array $\mathbf{G}_{1:n}\left(\frac{i}{N}, \frac{j}{N}\right)$, $1 \leq i \leq j \leq N-1$, cannot attain the maximal value beyond the diagonal. Though it is possible when $\underline{i}_1 = \underline{i}_2 < \overline{i}_1 = \overline{i}_2$ for which $\max_{1 \leq i \leq j \leq N-1} \mathbf{G}_{1:n}\left(\frac{i}{N}, \frac{j}{N}\right)$ is attained for (i, j) equal to $(\underline{i}_1, \underline{i}_1)$, $(\underline{i}_1, \overline{i}_1)$ and $(\overline{i}_1, \overline{i}_1)$. This is only possible when n is relatively large with respect to N so that u_1 and u_2 are very close to each other so that they lie in one interval of the form $\left[\frac{i}{N}, \frac{i+1}{N}\right]$, and each $\underline{G}_{k:n}$ and $\overline{G}_{k:n}$ takes on identical values at the interval end-points. The last condition seems to be impossible to meet.

Theorem 5.7 *Let $X_1, \ldots, X_n$, $n \geq 2$, denote the random results of consecutive drawings with replacement from an arbitrary non-constant population $x_1 \leq \ldots \leq x_N > x_1$ for some $N \geq 2$. With the notation of* (4.642), (5.174), *and $u_0(k,n)$ defined in* (4.631), *we have the following:*

(i) If $u_0(k,n) \leq \frac{1}{N}$, then

$$\frac{\mathbb{V}ar\, X_{k:n}}{\mathbb{V}ar\, X_1} \leq G_{k:n}\left(\frac{1}{N}\right), \tag{5.182}$$

and the equality is attained for $x_1 < x_2 = \ldots = x_N$.

(ii) If $u_0(k,n) \geq 1 - \frac{1}{N}$, then

$$\frac{\mathbb{V}ar\, X_{k:n}}{\mathbb{V}ar\, X_1} \leq G_{k:n}\left(1 - \frac{1}{N}\right), \tag{5.183}$$

and the equality is attained for $x_1 = \ldots = x_{N-1} < x_N$.

(iii) If $\frac{1}{N} < u_0(k,n) < 1 - \frac{1}{N}$, then

$$\frac{\mathbb{V}ar\ X_{k:n}}{\mathbb{V}ar\ X_1} \le \max\left\{G_{k:n}\left(\frac{\lfloor u_0 N\rfloor}{N}\right), G_{k:n}\left(\frac{\lceil u_0 N\rceil}{N}\right)\right\} = G_{k:n}\left(\frac{i_0}{N}\right), \tag{5.184}$$

say, for some $i_0 = i_0(k, n, N)$, *and the equality is attained if* $x_1 = \ldots = x_{i_0} < x_{i_0+1} = \ldots = x_N$.

Case (i) of the theorem occurs always for $k = 1$, and is also possible for other small k, when n is large with respect to N. Inequality $u_0(k, n) \ge 1 - \frac{1}{N}$ holds for $k = n$ and possibly for other large k, when n is large and N is small. Theorem 5.7 can be summarized as follows:

$$\frac{\mathbb{V}ar\ X_{k:n}}{\mathbb{V}ar\ X_1} \le \max_{1\le i\le N-1} G_{k:n}\left(\frac{i}{N}\right). \tag{5.185}$$

If the maximum is attained at $\frac{i_0}{N}$, then the bound is attained by $x_1 = \ldots = x_{i_0} < x_{i_0+1} = \ldots = x_N$. Relation (5.185) together with

$$G_{n+1-k:n}\left(\frac{i}{N}\right) = G_{k:n}\left(\frac{N-i}{N}\right) \tag{5.186}$$

imply that the bounds for the variances of kth smallest and kth largest order statistics are identical. We observe that this property is shared by the order statistics from arbitrary iid samples. It is implied by the following general argument. The simple transformation of variables $Y_k = -X_k$, $k = 1, \ldots, n$, changes the ordering of order statistics $Y_{k:n} = -X_{n+1-k:n}$, $k = 1, \ldots, n$, but preserves the joint distribution in the family under consideration, and the variance values $\mathbb{V}ar\ Y_k = \mathbb{V}ar\ X_k$, $\mathbb{V}ar\ Y_{k:n} = \mathbb{V}ar\ X_{n+1-k:n}$, $k = 1, \ldots, n$. We notice the same in the drawing without replacement model and for dependent identically distributed samples.

Proof of Theorem 5.7 We recall the Hoeffding (1940) formula for the variance of a random variable Z with a distribution function H and a finite second moment

$$\mathbb{V}ar\ Z = \iint_{\mathbb{R}^2} H(\min\{x, y\})[1 - H(\max\{x, y\})]\, dx\, dy.$$

It is easy to verify that for Z having a discrete distribution function $H(x) = \sum_{k=1}^{N} \alpha_k \mathbf{1}_{[x_k,\infty)}(x)$ for some $\alpha_k > 0$, $k = 1, \ldots, N$, $\sum_{k=1}^{N} \alpha_k = 1$ and $x_1 < x_2 < \ldots < x_N$, the Hoeffding formula takes on the form

$$\mathbb{V}ar\ Z = \sum_{i,j=1}^{N-1} H(\min\{x_i, x_j\})[1 - H(\max\{x_i, x_j\})](x_{i+1} - x_i)(x_{j-1} - x_j). \tag{5.187}$$

By the continuity argument, the above formula can be extended to the case $x_1 \le \ldots \le x_N$. Although the extension looks artificial in the first glance, because a natural interpretation $\alpha_k = \mathbb{P}(Z = x_k)$ loses its meaning, it is useful in our urn model. By (5.187), we can write here

$$\mathbb{V}ar\, X_1 = \sum_{i,j=1}^{N-1} \frac{\min\{i,j\}}{N}\left[1-\frac{\max\{i,j\}}{N}\right](x_{i+1}-x_i)(x_{j+1}-x_j). \tag{5.188}$$

For $X_{k:n}$ with distribution function $F_{k:n} \circ F$, we similarly obtain

$$\mathbb{V}ar\, X_{k:n} = \sum_{i,j=1}^{N-1} F_{k:n}\left(\frac{\min\{i,j\}}{N}\right) \times \left[1-F_{k:n}\left(\frac{\max\{i,j\}}{N}\right)\right](x_{i+1}-x_i)(x_{j+1}-x_j). \tag{5.189}$$

Therefore,

$$\begin{aligned}
\mathbb{V}ar\, X_{k:n} &= \sum_{i,j=1}^{N-1} \frac{F_{k:n}\left(\frac{\min\{i,j\}}{N}\right)\left[1-F_{k:n}\left(\frac{\max\{i,j\}}{N}\right)\right]}{\frac{\min\{i,j\}}{N}\left[1-\frac{\max\{i,j\}}{N}\right]} \\
&\quad\times \frac{\min\{i,j\}}{N}\left[1-\frac{\max\{i,j\}}{N}\right](x_{i+1}-x_i)(x_{j+1}-x_j) \\
&\le \max_{1\le i\le j\le N-1} \frac{F_{k:n}\left(\frac{i}{N}\right)\left[1-F_{k:n}\left(\frac{j}{N}\right)\right]}{\frac{i}{N}\left[1-\frac{j}{N}\right]} \\
&\quad\times \sum_{i,j=1}^{N-1} \frac{\min\{i,j\}}{N}\left[1-\frac{\max\{i,j\}}{N}\right](x_{i+1}-x_i)(x_{j+1}-x_j) \\
&= \max_{1\le i\le j\le N-1} \mathbf{G}_{k:n}\left(\frac{i}{N},\frac{j}{N}\right)\mathbb{V}ar\, X_1 \\
&= \max_{1\le i\le N-1} G_{k:n}\left(\frac{i}{N}\right)\mathbb{V}ar\, X_1,
\end{aligned} \tag{5.190}$$

where the last equality follows from Lemma 5.5. Using (5.178), we specify the bounds.

Now we proceed to establishing the bound attainability conditions. The inequality in (5.190), becomes equality iff for every $1 \le i \le j \le N-1$ either

$$\mathbf{G}_{k:n}\left(\frac{i}{N},\frac{j}{N}\right) = \max_{1\le i\le j\le N-1} \mathbf{G}_{k:n}\left(\frac{i}{N},\frac{j}{N}\right)$$

or

$$(x_{i+1} - x_i)(x_{j+1} - x_j) = 0.$$

Note that the latter condition for all $1 \leq i \leq j \leq N-1$, i.e. $x_{i+1} - x_i = 0$, $1 \leq i \leq N-1$, contradicts our assumption $x_1 < x_N$, and $\mathbb{V}ar\, X_1 > 0$ in consequence. If $(x_{i+1} - x_i)(x_{j+1} - x_j) > 0$ for some $1 \leq i < j \leq N-1$, then in order to attain the equality in (5.190), the non-diagonal element $\mathbf{G}_{k:n}\left(\frac{i}{N}, \frac{j}{N}\right)$ of the triangular array has to be necessarily the maximum point of the array. This seems to be very unlikely in view of the comments preceding Theorem 5.7. However, the attainability condition is satisfied if for some $1 \leq i_0 \leq N-1$, we have $x_1 = \ldots = x_{i_0} < x_{i_0+1} = \ldots = x_N$ and $G_{k:n}\left(\frac{i_0}{N}\right) = \mathbf{G}_{k:n}\left(\frac{i_0}{N}, \frac{i_0}{N}\right) = \max_{1 \leq i \leq j \leq N-1} \mathbf{G}_{k:n}\left(\frac{i}{N}, \frac{j}{N}\right)$.

Summing up, the variance bound is attained by a two-valued population with precisely described numbers $i_0 = i_0(k, n, N)$ smaller and $N - i_0(k, n, N)$ greater-valued elements of the population, where $i_0(k, n, N)$ is an argument maximizing the sequence $G_{k:n}\left(\frac{i}{N}\right)$, $1 \leq i \leq N-1$. It is unique if the sequence has a unique maximum. If it has two neighboring maxima, then the maximizing variance populations take on the forms $x_1 = \ldots = x_{\lfloor u_0 N \rfloor} < x_{\lceil u_0 N \rceil} = \ldots = x_N$ and $x_1 = \ldots = x_{\lceil u_0 N \rceil} < x_{\lceil u_0 N \rceil + 1} = \ldots = x_N$. □

Below we specify the results of Theorem 5.7 for the sample extremes and the sample medians.

Corollary 5.20 *For the sample minimum, we have*

$$\frac{\mathbb{V}ar\, X_{1:n}}{\mathbb{V}ar\, X_1} \leq N\left(\frac{N-1}{N}\right)^{n-1}\left[1 - \left(\frac{N-1}{N}\right)^n\right], \tag{5.191}$$

and the equality is attained if $x_1 < x_2 = \ldots = x_N$. The ratio of the variances of the sample maximum and sample element satisfies inequality (5.191) *as well, and the attainability condition is $x_1 = \ldots = x_{N-1} < x_N$.*

Corollary 5.21 *Suppose n is odd and $k = \frac{n+1}{2}$.*

(i) If N is even, then

$$\frac{\mathbb{V}ar\, X_{\frac{n+1}{2}:n}}{\mathbb{V}ar\, X_1} \leq G_{\frac{n+1}{2}:n}\left(\frac{1}{2}\right) = 1, \tag{5.192}$$

and the inequality becomes equality iff $x_1 = \ldots x_{\frac{N}{2}} < x_{\frac{N}{2}+1} = \ldots = x_N$.

(ii) If N is odd, then

$$\begin{aligned}\frac{\mathbb{V}ar\, X_{\frac{n+1}{2}:n}}{\mathbb{V}ar\, X_1} &\le G_{\frac{n+1}{2}:n}\left(\frac{N-1}{2N}\right) = G_{\frac{n+1}{2}:n}\left(\frac{N+1}{2N}\right) \\ &= \frac{4N^2}{N^2-1} F_{\frac{n+1}{2}:n}\left(\frac{N-1}{2N}\right) F_{\frac{n+1}{2}:n}\left(\frac{N+1}{2N}\right) < 1,\end{aligned} \tag{5.193}$$

with the equality in the first line attained iff either $x_1 = \ldots x_{\frac{N-1}{2}} < x_{\frac{N+1}{2}} = \ldots = x_N$ *or* $x_1 = \ldots x_{\frac{N+1}{2}} < x_{\frac{N+3}{2}} = \ldots = x_N$.

Proof For $k = \frac{n+1}{2}$, function $G_{\frac{n+1}{2}:n}(x)$ is symmetric about $\frac{1}{2}$. Since it is increasing-decreasing (cf. Lemma 5.5), its maximum point is $\frac{1}{2}$. If N is even, then

$$\max_{1\le i\le N-1} G_{\frac{n+1}{2}:n}\left(\frac{i}{N}\right) = G_{\frac{n+1}{2}:n}\left(\frac{1}{2}\right) = 1,$$

because $f_{\frac{n+1}{2}:n}(x)$ is symmetric about $\frac{1}{2}$ and so $F_{\frac{n+1}{2}:n}\left(\frac{1}{2}\right) = 1 - F_{\frac{n+1}{2}:n}\left(\frac{1}{2}\right) = \frac{1}{2}$. If N is odd, the multiplicities of $\frac{1}{N}$ closest to $\frac{1}{2}$ are $\frac{N-1}{2N}$ and $\frac{N+1}{2N}$, mutually symmetric about $\frac{1}{2}$. Therefore,

$$\max_{1\le i\le N-1} G_{\frac{n+1}{2}:n}\left(\frac{i}{N}\right) = G_{\frac{n+1}{2}:n}\left(\frac{N-1}{2N}\right) = G_{\frac{n+1}{2}:n}\left(\frac{N+1}{2N}\right),$$

and the respective value is presented in (5.52). This is less than 1, because $G_{\frac{n+1}{2}:n}\left(\frac{N-1}{2N}\right) = G_{\frac{n+1}{2}:n}\left(\frac{N+1}{2N}\right) < G_{\frac{n+1}{2}:n}\left(\frac{1}{2}\right)$. □

Corollary 5.21 shows that it is possible that $G_{k:n}\left(\frac{\lfloor u_0 N\rfloor}{N}\right) = G_{k:n}\left(\frac{\lceil u_0 N\rceil}{N}\right)$, and consequently some variance bound is attained by two-valued populations with different numbers of subpopulation elements. However, functions $\underline{G}_{\frac{n+1}{2}:n}$ and $\overline{G}_{\frac{n+1}{2}:n}$ are not symmetric about $\frac{1}{2}$, which makes it impossible to attain the variance bounds for the sample medians by three-valued populations for any n and N by the method described in the comment following the proof of Lemma 5.5.

For fixed $n \ge 2$,

$$\lim_{N\to\infty} N\left(\frac{N-1}{N}\right)^{n-1}\left[1-\left(\frac{N-1}{N}\right)^n\right] = n, \tag{5.194}$$

which means that the variance bounds for the extreme order statistics in the with replacement model tend to respective bounds in the general iid model (comp. Theorem 4.52) as the population size increases. The same happens for the other order statistics. Indeed, for given $2 \leq k \leq n$ and sufficiently large N, we have $\frac{1}{N} < u_0(k,n) < \frac{N-1}{N}$, and then

$$\lim_{N\to\infty} \max\left\{G_{k:n}\left(\frac{\lfloor Nu_0 \rfloor}{N}\right), G_{k:n}\left(\frac{\lceil Nu_0 \rceil}{N}\right)\right\} = G_{k:n}(u_0) \tag{5.195}$$

by continuity of $G_{k:n}$. The RHS represents the variance bounds for $X_{k:n}$ coming from arbitrary iid sample (comp. Theorem 4.53). Clearly, the limit majorizes all the elements of the sequence. However, the convergence is not necessarily monotone which shows the example of sample medians from Corollary 5.21. The bound subsequence elements (5.192) for even N coincide with the limit equal to 1, and the subsequence in (5.193) with odd N is actually less than 1 and increasing.

5.2 With Replacement Model for General Distributions

5.2.1 Fixed Drawing Distribution

As before, we assume that a finite population of N items is labeled by numbers $x_{1:N} \leq \cdots \leq x_{N:N}$, which are ordered in ascending order for convenience. However, now we admit nonuniform distributions on the population. We consider iid random variables $X_1, \ldots, X_n$ with a common distribution

$$\mathbb{P}(X_i = x_{k:N}) = \sum_{\{r: x_{r:N} = x_{k:N}\}} p_r, \qquad 1 \leq k \leq N, \tag{5.196}$$

for a given $\tilde{p} = (p_1, \ldots, p_N) \in \mathbb{R}^N$ such that $p_k > 0$, $k = 1, \ldots, N$, and $\sum_{k=1}^N p_k = 1$. Therefore,

$$\mu = \mathbb{E}X_1 = \sum_{k=1}^{N} x_{k:N}\, p_k \tag{5.197}$$

and

$$\sigma_p^p = \mathbb{E}|X_1 - \mu|^p = \sum_{k=1}^{N} |x_{k:N} - \mu|^p\, p_k, \qquad 1 \leq p < \infty, \tag{5.198}$$

(definition of σ_∞ is identical with (5.4)).

5.2.1.1 General L-Statistics

Here we study the problem of optimal bounds for the expectations of standardized L-statistics $\mathbb{E}\sum_{i=1}^{n} c_i \frac{X_{i:n}-\mu}{\sigma_p}$, $1 < p < \infty$, where the vectors of coefficients $\tilde{c} = (c_1, \ldots, c_n) \in \mathbb{R}^n$ and probabilities $\tilde{p} = (p_1, \ldots, p_N) \in \mathbb{R}^N$ are fixed, and various populations $\tilde{x} = (x_{1:N}, \ldots, x_{N:N})$ are considered. Note that this is another version of the problems studied in Sects. 4.6, 4.7, and later on in Chap. 6: we evaluate L-statistics of iid samples from a restricted family of distributions such that the respective distribution functions take on their values in a fixed finite set $Q = \{q_k : k = 0, \ldots, N\}$ with

$$\begin{aligned} q_0 &= 0, \\ q_k &= \sum_{r=1}^{k} p_r, \qquad k = 1, \ldots, N-1, \\ q_N &= 1. \end{aligned} \tag{5.199}$$

Admitting that some elements of population are equal implies that the definition (5.196) of probability distribution is slightly awkward. It does not have a satisfactory justification except for the case $p_1 = \cdots = p_N = \frac{1}{N}$ which is naturally motivated by the standard drawing with replacement model. Moreover, this ensures existence of solutions, because the set of nondecreasing vectors $\tilde{x} \in \mathbb{R}^N$ is closed. Also, this can be approximated by the sequences of strictly increasing vectors. These results can be applied for solving a more general and natural problem of evaluating L-statistics arising from the iid sampling from arbitrary populations $\tilde{x}$ of size N (at most) with arbitrary marginal distributions $\tilde{p}$.

One cannot expect nontrivial solutions to the problem in the cases $p = 1$ and $p = \infty$, because the general bounds of Sect. 4.4 for $p = 1$ and $p = \infty$ are attained by discrete distributions on either two or three points. Therefore these cases are not considered here, and we concentrate on $1 < p < \infty$, for which the distributions attaining the general bounds have absolutely continuous components. Below we determine optimal upper bounds on the expectations of arbitrarily fixed L-statistics arising from fixed distributions $\tilde{p}$ over various populations $\tilde{x}$ by applying projections of expectation functionals onto the family of nondecreasing sequences in $\ell^2(\mathbb{R}^N, \tilde{p})$. Then we examine closer the special cases of single order statistics and their differences, and the sample maximum and range in particular.

For a given vector of probability masses $\tilde{p} = (p_1, \ldots, p_N)$, it is convenient to introduce $\tilde{p}$-distributed iid random variables Y_i, $i = 1, \ldots, n$, on the standard increasing population $\{1, \ldots, N\}$ such that

$$\mathbb{P}(Y_i = k) = p_k, \qquad k = 1, \ldots, N. \tag{5.200}$$

Then,

$$\mathbb{P}(Y_i \le k) = \sum_{r=1}^{k} p_r = q_k, \tag{5.201}$$

$$\mathbb{P}(Y_{i:n} \le k) = \sum_{r=i}^{n} \binom{n}{r} q_k^r (1-q_k)^{n-r} = F_{i:n}(q_k), \tag{5.202}$$

$$\mathbb{P}(Y_{i:n} = k) = F_{i:n}(q_k) - F_{i:n}(q_{k-1}), \qquad k = 1, \ldots, N. \tag{5.203}$$

Applying transformation (5.9), dependent on population $\tilde{x}$, we obtain a sequence $g_{\tilde{x}}(Y_{1:n}), \ldots, g_{\tilde{x}}(Y_{n:n})$ with the same distribution as that of the original vector $X_{1:n}, \ldots, X_{n:n}$. Therefore,

$$\mathbb{E}X_{i:n} = \mathbb{E}g_{\tilde{x}}(Y_{i:n}) = \sum_{k=1}^{N} x_{k:N}[F_{i:n}(q_k) - F_{i:n}(q_{k-1})] \tag{5.204}$$

and

$$\begin{aligned} \mathbb{E}\sum_{i=1}^{n} c_i X_{i:n} &= \sum_{i=1}^{n} c_i \sum_{k=1}^{N} x_{k:N}[F_{i:n}(q_k) - F_{i:n}(q_{k-1})] \\ &= \sum_{k=1}^{N} x_{k:N} \sum_{i=1}^{n} c_i [F_{i:n}(q_k) - F_{i:n}(q_{k-1})] \\ &= \sum_{k=1}^{N} x_{k:N}[F_{\tilde{c}:n}(q_k) - F_{\tilde{c}:n}(q_{k-1})] \\ &= \sum_{k=1}^{N} C_k x_{k:N} p_k, \end{aligned} \tag{5.205}$$

where

$$C_k = \frac{F_{\tilde{c}:n}(q_k) - F_{\tilde{c}:n}(q_{k-1})}{p_k}, \qquad k = 1, \ldots, N, \tag{5.206}$$

are dependent on $\tilde{c} \in \mathbb{R}^n$ and $\tilde{p} \in \mathbb{R}^N$ being known, and $\tilde{x} \in \mathbb{R}^N$ is nondecreasing and unknown. Observe that

$$\sum_{k=1}^{N} C_k p_k = \sum_{k=1}^{N} [F_{\tilde{c}:n}(q_k) - F_{\tilde{c}:n}(q_{k-1})] = F_{\tilde{c}:n}(1) = \sum_{i=1}^{n} c_i. \tag{5.207}$$

The reason for using a weighted linear combination of the last sum in (5.205) instead of a simpler non-weighted form of the penultimate expression is that the population moments (5.197) and (5.198) are expressed in terms of weighted combinations. Accordingly, by (5.205) and (5.207), we have

$$\mathbb{E}\sum_{i=1}^{n} c_i(X_{i:n}-\mu) = \sum_{k=1}^{N} C_k(x_{k:N}-\mu)p_k = \sum_{k=1}^{N}(C_k-c)(x_{k:N}-\mu)p_k \tag{5.208}$$

for arbitrary real c.

Now we use a weighted version of the Moriguti inequality presented in Theorem 2.14 for determining the projection $\underline{\tilde{C}} = P^{\nearrow}\tilde{C}$ of $\tilde{C}$ defined in (5.206) onto the family of nondecreasing sequences in ℓ^2-norm with weights $(p_1, \ldots, p_N)$. Set $j_0 = 0$ and

$$\begin{aligned} j_m = \min\Bigg\{ j_{m-1}+1 \le j \le N : \frac{F_{\tilde{c}:n}(q_j) - F_{\tilde{c}:n}(q_{j_{m-1}})}{q_j - q_{j_{m-1}}} \\ = \min_{j_{m-1}+1\le k\le N} \frac{F_{\tilde{c}:n}(q_k) - F_{\tilde{c}:n}(q_{j_{m-1}})}{q_k - q_{j_{m-1}}} \Bigg\} \end{aligned} \tag{5.209}$$

for $m = 1, \ldots, M$, with $1 \le M \le N$ such that $j_M = N$. Then, the coordinates of projection have the form

$$\underline{C}_k = \frac{F_{\tilde{c}:n}(q_{j_m}) - F_{\tilde{c}:n}(q_{j_{m-1}})}{q_{j_m} - q_{j_{m-1}}}, \qquad k = j_{m-1}+1, \ldots, j_m,\ m = 1, \ldots, M. \tag{5.210}$$

Alternatively, each $\underline{C}_k$ is defined as the slope on the interval (q_{k-1}, q_k) of the greatest convex function minorizing the points $(q_k, F_{\tilde{c}:n}(q_k))$, $k = 0, \ldots, N$. We also see that

$$\begin{aligned} \sum_{k=1}^{N} \underline{C}_k p_k &= \sum_{m=1}^{M} \underline{C}_{j_m} \sum_{k=j_{m-1}+1}^{j_m} p_k = \sum_{m=1}^{M} [F_{\tilde{c}:n}(q_{j_m}) - F_{\tilde{c}:n}(q_{j_{m-1}})] \\ &= F_{\tilde{c}:n}(1) = \sum_{i=1}^{n} c_i. \end{aligned} \tag{5.211}$$

Projection $\underline{\tilde{C}}$ is constant with coordinates $\underline{C}_k = \sum_{i=1}^{n} c_i, i = 1, \ldots, N$, if $M = 1$ or equivalently,

$$F_{\tilde{c}:n}(1) = \sum_{i=1}^{n} c_i \le F_{\tilde{c}:n}(q_k)/q_k, \qquad k = 1, \ldots, N-1. \tag{5.212}$$

Concavity of $F_{\tilde{c}:n}(x)$ implies nonincrease of $F'_{\tilde{c}:n}(x) = f_{\tilde{c}:n}(x)$, and (5.212) in particular, but this is not necessary for getting (5.212). Similarly, using representation

$$C_k = \frac{F_{\tilde{c}:n}(q_k) - F_{\tilde{c}:n}(q_{k-1})}{q_{k-1} - q_k} = f_{\tilde{c}:n}(\xi_k), \quad \xi_k \in (q_{k-1}, q_k), \ k = 1, \ldots, N, \tag{5.213}$$

we deduce that nondecrease of $f_{\tilde{c}:n}(x)$ implies that of $\tilde{C}$, and $\underline{\tilde{C}} = P^{\nearrow}\tilde{C} = \tilde{C}$ as a consequence, but this is not a necessary condition for that. Now we state the main results of this section.

Theorem 5.8 *For given* $1 < p < \infty$, $\tilde{c} \in \mathbb{R}^n$, *and* $\tilde{p} \in S^N$ *with all positive coordinates, with the notation*

$$B_m(c) = \frac{F_{\tilde{c}:n}(q_{j_m}) - F_{\tilde{c}:n}(q_{j_{m-1}})}{q_{j_m} - q_{j_{m-1}}} - c, \quad m = 1, \ldots, M, \ c \in \mathbb{R}, \tag{5.214}$$

there exists a unique

$$\frac{F_{\tilde{c}:n}(q_{j_1})}{q_{j_1}} < \underline{c}_q < \frac{\sum_{i=1}^n c_i - F_{\tilde{c}:n}(q_{j_{M-1}})}{1 - q_{j_{M-1}}} \tag{5.215}$$

solving the equation

$$\sum_{m=1}^{M} |B_m(c)|^{q/p} sgn\{B_m(c)\}(q_{j_m} - q_{j_{m-1}}) = 0 \tag{5.216}$$

such that

$$\mathbb{E}\sum_{i=1}^{n} c_i \frac{X_{i:n} - \mu}{\sigma_p} \le B = \left(\sum_{m=1}^{M}(q_{j_m} - q_{j_{m-1}})|B_m(\underline{c}_q)|^q\right)^{1/q}. \tag{5.217}$$

If

$$\sum_{i=1}^{n} c_i > \frac{F_{\tilde{c}:n}(q_k)}{q_k} \text{ for some } k = 1, \ldots, N-1, \tag{5.218}$$

then the RHS bound of (5.217) *is strictly positive and is attained by*

$$x_{k:N} = \mu + \sigma_p \frac{|B_m(\underline{c}_q)|^{q/p}}{B^{q/p}} sgn\{B_m(\underline{c}_q)\}, \tag{5.219}$$

for $k = j_{m-1} + 1, \ldots, j_m$, *and* $m = 1, \ldots, M$.

Proof Applying (5.208), Theorem 2.14, and the weighted Hölder inequality, we obtain

$$
\begin{aligned}
\mathbb{E}\sum_{i=1}^{n} c_i(X_{i:n}-\mu) &= \sum_{k=1}^{N}(C_k - c)(x_{k:N}-\mu)p_k \le \sum_{k=1}^{N}(\underline{C}_k - c)(x_{k:N}-\mu)p_k \\
&\le \left(\sum_{k=1}^{N}|\underline{C}_k - c|^q p_k\right)^{1/q}\left(\sum_{k=1}^{N}|x_{k:N}-\mu|^p p_k\right)^{1/p} \\
&= \left(\sum_{m=1}^{M}(q_{j_m}-q_{j_{m-1}})|B_m(c)|^q\right)^{1/q}\sigma_p. \qquad (5.220)
\end{aligned}
$$

If (5.212) holds, then $M = 1$, $\bar{\underline{c}}_q = \sum_{i=1}^{n} c_i$ and $B_1(\bar{\underline{c}}_q) = 0$. The zero bound is attained if all x_k, $k = 1, \dots, N$, are equal which forces $\sigma_p = 0$. In the opposite case (5.218), a strictly convex function

$$
c \to \|\tilde{\underline{C}} - c\tilde{\mathbf{1}}\|_{q,\tilde{p}}^{q} = \sum_{m=1}^{M}|B_m(c)|^q(q_{j_m}-q_{j_{m-1}}) \qquad (5.221)
$$

has a unique minimum at $\bar{\underline{c}}_q \in (\underline{C}_1, \underline{C}_N)$ determined by (5.216). Therefore bound (5.220) with $c = \bar{\underline{c}}_q$ is the most sharp one.

We prove that this is attained by (5.219). For obtaining an equality in the second inequality of (5.220) with $c = \bar{\underline{c}}_q$, we need that $x_{k:N} - \mu$ are proportional to $|\underline{C}_k - \bar{\underline{c}}_q|^{q/p}\mathrm{sgn}\{\underline{C}_k - \bar{\underline{c}}_q\}$. This suffices for getting an equality in the former inequality of (5.220) as well. Also, we easily check that population (5.219) satisfies the equality conditions as well as both the moment conditions. □

The case $p = 2$ is of special interest. Then (5.216) has the form

$$
\sum_{m=1}^{M}(q_{j_m}-g_{j_{m-1}})B_m(c) = \sum_{i=1}^{n} c_i - c = 0, \qquad (5.222)
$$

and with $\bar{\underline{c}}_2 = \sum_{i=1}^{n} c_i$, we obtain

$$
\begin{aligned}
B^2 = \sum_{m=1}^{M}(q_{j_m}-g_{j_{m-1}})B_m^2\left(\sum_{i=1}^{n} c_i\right) &= \sum_{m=1}^{M}\frac{[F_{\tilde{c}:n}(q_{j_m}) - F_{\tilde{c}:n}(q_{j_{m-1}})]^2}{q_{j_m}-g_{j_{m-1}}} \\
- 2\sum_{i=1}^{n} c_i \sum_{m=1}^{M}[F_{\tilde{c}:n}(q_{j_m}) - F_{\tilde{c}:n}(q_{j_{m-1}})] &+ \left(\sum_{i=1}^{n} c_i\right)\sum_{m=1}^{M}(q_{j_m}-g_{j_{m-1}})
\end{aligned}
$$

$$= \sum_{m=1}^{M} \frac{[F_{\tilde{c}:n}(q_{j_m}) - F_{\tilde{c}:n}(q_{j_{m-1}})]^2}{q_{j_m} - g_{j_{m-1}}} - \left(\sum_{i=1}^{n} c_i\right)^2. \tag{5.223}$$

Hence, the following statement holds.

Theorem 5.9 *For* $p = 2$,

$$\mathbb{E}\sum_{i=1}^{n} c_i \frac{X_{i:n} - \mu}{\sigma_2} \leq \left[\sum_{m=1}^{M} \frac{[F_{\tilde{c}:n}(q_{j_m}) - F_{\tilde{c}:n}(q_{j_{m-1}})]^2}{q_{j_m} - g_{j_{m-1}}} - \left(\sum_{i=1}^{n} c_i\right)^2\right]^{1/2} \tag{5.224}$$

holds. Moreover, if (5.218) *holds, then* (5.223) *is positive and* (5.224) *becomes an equality under the conditions*

$$x_{k:N} = \mu + \sigma_2 \frac{\frac{F_{\tilde{c}:n}(q_{j_m}) - F_{\tilde{c}:n}(q_{j_{m-1}})}{q_{j_m} - g_{j_{m-1}}} - \sum_{i=1}^{n} c_i}{\left[\sum_{m=1}^{M} \frac{[F_{\tilde{c}:n}(q_{j_m}) - F_{\tilde{c}:n}(q_{j_{m-1}})]^2}{q_{j_m} - g_{j_{m-1}}} - \left(\sum_{i=1}^{n} c_i\right)^2\right]^{1/2}} \tag{5.225}$$

for $k = j_{m-1} + 1, \ldots, j_m$, *and* $m = 1, \ldots, M$.

We complete this subsubsection with some general remarks. Let $\tilde{c} \in \mathbb{R}^n$ and $\tilde{p} \in \boldsymbol{S}^N$ be fixed, and $\tilde{q}$ be defined by (5.199). Then,

$$f_{\tilde{c}:n,\tilde{q}}(x) = \sum_{k=1}^{N} \frac{F_{\tilde{c}:n}(q_k) - F_{\tilde{c}:n}(q_{k-1})}{q_{k-1} - q_k} \mathbf{1}_{[q_{k-1},q_k)}(x) = \sum_{k=1}^{N} C_k \mathbf{1}_{[q_{k-1},q_k)}(x) \tag{5.226}$$

is the projection of the polynomial

$$f_{\tilde{c}:n}(x) = \sum_{i=1}^{n} c_i f_{i:n}(x) \tag{5.227}$$

onto the linear subspace of functions which are constant on each of intervals $[q_{k-1}, q_k)$, $k = 1, \ldots, N$,

$$\mathcal{L}_{\tilde{q}} = \left\{ g : g(x) = \sum_{k=1}^{N} A_k \mathbf{1}_{[q_{k-1},q_k)}(x),\ A_k \in \mathbb{R},\ k = 1, \ldots, N \right\} \tag{5.228}$$

of the Hilbert space $L^2([0,1),dx)$. The norms of the elements of the subspace coincide with the norms of the coefficient vectors $\tilde{A} = (A_1, \ldots, A_N)$ in the space $\ell^2(\mathbb{R}^N, \tilde{p})$, because

$$\|g\| = \int_0^1 \left[\sum_{k=1}^N A_k \mathbf{1}_{[q_{k-1}, g_k)}(x)\right]^2 dx = \sum_{k=1}^N \int_{q_{k-1}}^{q_k} A_k^2\, dx = \sum_{k=1}^N A_k^2\, p_k. \tag{5.229}$$

Applying the greatest convex minorant method, we conclude that

$$\begin{aligned} \underline{f}_{\tilde{c}:n,\tilde{q}}(x) &= \sum_{m=1}^M \frac{F_{\tilde{c}:n}(q_{j_m}) - F_{\tilde{c}:n}(q_{j_{m-1}})}{q_{j_m} - q_{j_{m-1}}} \mathbf{1}_{[q_{j_{m-1}}, q_{j_m})}(x) \\ &= \sum_{k=1}^N \underline{C}_k \mathbf{1}_{[q_{k-1}, q_k)}(x) \end{aligned} \tag{5.230}$$

is the projection of (5.226) onto the family of nondecreasing functions $\mathcal{C}^{\nearrow}$ in $L^2([0,1),dx)$. It belongs to (5.228) as well, and has the norm

$$\|\underline{f}_{\tilde{c}:n,\tilde{q}}\| = \left\{\sum_{m=1}^M \frac{[F_{\tilde{i}:n}(q_{j_m}) - F_{\tilde{i}:n}(q_{j_{m-1}})]^2}{q_{j_m} - q_{j_{m-1}}}\right\}^{1/2} = \left(\sum_{k=1}^N \underline{C}_k^2\, p_k\right)^{1/2}. \tag{5.231}$$

Also,

$$\left\|\underline{f}_{\tilde{c}:n,\tilde{q}} - \sum_{i=1}^n c_i \mathbf{1}\right\| = \left[\|\underline{f}_{\tilde{c}:n,\tilde{q}}\|^2 - \left(\sum_{i=1}^n c_i\right)^2\right]^{1/2} \tag{5.232}$$

coincides with the upper bound determined in Theorem 5.9. If the probability vectors $\tilde{p}_N = (p_{1,N}, \ldots, p_{N,N}) \in \mathbb{S}^N$, $N = 1, 2, \ldots$, satisfy

$$\lim_{N\to\infty} \max_{1\le k\le N} p_{k,N} = 0, \tag{5.233}$$

then

$$f_{\tilde{c}:n,\tilde{q}} \to f_{\tilde{c}:n}, \text{ as } N \to \infty. \tag{5.234}$$

Since $\mathcal{C}^{\nearrow}$ is convex, the projection operator $P^{\nearrow}$ is continuous, and (5.234) holds, we have

$$P^{\nearrow} f_{\tilde{c}:n,\tilde{q}} \to \underline{f}_{\tilde{c}:n,\tilde{q}} \Leftrightarrow P^{\nearrow} f_{\tilde{c}:n} = \underline{f}_{\tilde{c}:n} \tag{5.235}$$

By the continuity of the norm functional

$$\left\| \underline{f}_{\tilde{c}:n,\tilde{q}} - \sum_{i=1}^{n} c_i \mathbf{1} \right\| \to \left[\|\underline{f}_{\tilde{c}:n}\|^2 - \left(\sum_{i=1}^{n} c_i \right)^2 \right]^{1/2}. \tag{5.236}$$

That means that under condition (5.233), which is satisfied in particular by the classic drawing with replacement models, the upper bounds on the expectations of arbitrary normalized L-statistics coming from iid samples drawn from finite populations tend to the general bounds on respective L-statistics coming from arbitrary populations. In fact, we verified the claim in the case $p = 2$, but for $p \neq 2$ the conclusions are analogous. These follow from the fact that the convergence in (5.234) is uniform, but more subtle arguments should be used then.

5.2.1.2 Single Order Statistics and Their Differences

In the special case of single order statistics, we have

$$\mathbb{E}X_{j:n} = \sum_{k=1}^{N} C_k \, x_{k:N} \, p_k \tag{5.237}$$

with

$$C_k = \frac{F_{j:n}(q_k) - F_{j:n}(q_{k-1})}{q_k - g_{k-1}} = f_{j:n}(\xi_k), \tag{5.238}$$

for some $\xi_k \in (q_{k-1}, q_k)$, and $k = 1, \dots, N$ (cf. (5.55)). Using arguments of the previous subsection, we determine the projection $\underline{\tilde{C}} = (\underline{C}_1, \dots, \underline{C}_N)$ of $\tilde{C} = (C_1, \dots, C_N)$ onto the family of non decreasing vectors in $\ell^2(\mathbb{R}^n, \tilde{p})$. The only difference is that instead of considering equally distant points $q_k = \frac{k}{N}$, $k = 0, 1, \dots, N$, we cope with general increasing sequences partitioning interval $[0, 1]$. From (5.238), we deduce that $\tilde{C}$ is decreasing for $j = 1$, and then $\underline{\tilde{C}} = \sum_{i=1}^{n} c_i \tilde{\mathbf{1}}$ is constant. Also, $\tilde{C}$ is increasing for $j = n$, and $\underline{\tilde{C}} = \tilde{C}$ then. For $2 \le j \le n-1$ and $1 \le \ell \le N$ such that $\alpha = \frac{j-1}{n-1} \in (q_{\ell-1}, q_\ell]$, sequence (5.238) is first increasing, and then decreasing, with the maximum at either one or two neighboring points among $C_{\ell-1}$, C_ℓ, and $C_{\ell+1}$. (Note that assumption $2 \le j \le n-1$ does not exclude monotonicity of (5.238).) Therefore $\underline{C}_k$ are first identical with C_k, and ultimately constant. The change point k_* is such that the sequence $H_{j:n}(q_k)$, $k = 1, \dots, N$, for function $H_{j:n}(x)$ defined in (5.58) attains its maximum at k_*. Since the function is first increasing and then decreasing with the maximum at α_* defined by (5.59), the maximum of the sequence is attained at one of the end points of the interval containing α_*. Precisely, we can write

$$k_* = \begin{cases} \ell & \text{if } H_{j:n}(q_\ell) \geq H_{j:n}(q_{\ell-1}), \\ \ell - 1 & \text{if } H_{j:n}(q_\ell) < H_{j:n}(q_{\ell-1}), \end{cases} \tag{5.239}$$

for $1 \leq \ell \leq N$ such that $\alpha_* \in (q_{\ell-1}, q_\ell]$, and then

$$\underline{C}_k = \begin{cases} \frac{F_{j:n}(q_k) - F_{j:n}(q_{k-1})}{p_k}, & k = 1, \ldots, k_*, \\ \frac{1 - F_{j:n}(q_{k_*})}{1 - q_{k_*}}, & k = k_* + 1, \ldots, N. \end{cases} \tag{5.240}$$

By definition, $k_* = 0$ and (5.215) is constant equal to 1 iff

$$H_{j:n}(q_1) = H_{j:n}(p_1) \geq H_{j:n}(0) = 1. \tag{5.241}$$

Then $\mathbb{E}X_{j:n} \leq \mu$. Below we omit this case. If

$$H_{j:n}(q_{N-2}) \leq H_{j:n}(q_{N-1}), \tag{5.242}$$

then $k_* = N - 1$, and $\underline{\tilde{C}}$ is identical with $\tilde{C}$. This holds for $j = n$ whatever $\tilde{p} = (p_1, \ldots, p_N) \in \mathbf{S}^N$ is. Therefore, we extend the definition of k_* to $k_* = N - 1$ in this case. Now we present positive bounds for single order statistics under the assumption that (5.241) does not hold.

Corollary 5.22 *Assume $H_{j:n}(p_1) < 1$. Then for a given $1 < p < \infty$, there exists a unique*

$$F_{j:n}(q_1) < \underline{\bar{c}}_q < \frac{1 - F_{j:n}(q_{k_*})}{1 - q_{k_*}} \tag{5.243}$$

determined by the equation

$$\sum_{k=1}^{k_*} p_k \left| \frac{F_{j:n}(q_k) - F_{j:n}(q_{k-1})}{p_k} - c \right|^{q/p} sgn \left\{ \frac{F_{j:n}(q_k) - F_{j:n}(q_{k-1})}{p_k} - c \right\}$$
$$+ (1 - q_{k_*}) \left[\frac{1 - F_{j:n}(q_{k_*})}{1 - q_{k_*}} - c \right]^{q/p} = 0 \tag{5.244}$$

such that

$$\mathbb{E}\frac{X_{j:n} - \mu}{\sigma_p} \leq B \tag{5.245}$$

with

$$B^q = \sum_{k=1}^{k_*} p_k \left| \frac{F_{j:n}(q_k) - F_{j:n}(q_{k-1})}{p_k} - \bar{c}_q \right|^q + (1 - q_{k_*}) \left[\frac{1 - F_{j:n}(q_{k_*})}{1 - q_{k_*}} - \bar{c}_q \right]^q. \tag{5.246}$$

Moreover, (5.245) *is the equality if*

$$x_{k:N} = \begin{cases} \mu + \frac{\sigma_p}{B^{q/p}} \left| \frac{F_{j:n}(q_k) - F_{j:n}(q_{k-1})}{p_k} - \bar{c}_q \right|^{q/p} \\ \qquad \times sgn \left\{ \frac{F_{j:n}(q_k) - F_{j:n}(q_{k-1})}{p_k} - \bar{c}_q \right\}, & k = 1, \dots, k_*, \\ \mu + \frac{\sigma_p}{B^{q/p}} \left[\frac{1 - F_{j:n}(q_{k_*})}{1 - q_{k_*}} - \bar{c}_q \right]^{q/p}, & k = k_* + 1, \dots, N. \end{cases} \tag{5.247}$$

The following special case $p = 2$ will be studied further.

Corollary 5.23 *The inequality*

$$\mathbb{E}\frac{X_{j:n} - \mu}{\sigma_2} \le B = \left\{ \sum_{k=1}^{k_*} \frac{[F_{j:n}(q_k) - F_{j:n}(q_{k-1})]^2}{p_k} + \frac{[1 - F_{j:n}(q_{k_*})]^2}{1 - q_{k_*}} - 1 \right\}^{1/2} \tag{5.248}$$

is sharp and becomes an equality for

$$x_{k:N} = \begin{cases} \mu + \frac{\sigma_2}{B} \left[\frac{F_{j:n}(q_k) - F_{j:n}(q_{k-1})}{p_k} - 1 \right], & k = 1, \dots, k_*, \\ \mu + \frac{\sigma_2}{B} \left[\frac{1 - F_{j:n}(q_{k_*})}{1 - q_{k_*}} - 1 \right], & k = k_* + 1, \dots, N. \end{cases} \tag{5.249}$$

For the sample maximum, (5.248) takes on the form

$$\mathbb{E}\frac{X_{n:n} - \mu}{\sigma_2} \le B = \left\{ \sum_{k=1}^{N} \frac{[F_{n:n}(q_k) - F_{n:n}(q_{k-1})]^2}{p_k} - 1 \right\}^{1/2} = \left\{ \sum_{k=1}^{N} \frac{[q_k^n - q_{k-1}^n]^2}{q_k - q_{k-1}} - 1 \right\}^{1/2} \tag{5.250}$$

for $0 = q_0 < q_1 < \cdots < q_N = 1$, and the equality conditions are

$$x_{k:n} = \mu + \frac{\sigma_2}{B} \left[\frac{q_k^n - q_{k-1}^n}{q_k - q_{k-1}} - 1 \right], \quad k = 1, \dots, N. \tag{5.251}$$

In order to evaluate the expectations of differences of order statistics

$$\mathbb{E}(X_{j:n} - X_{i:n}) = \sum_{k=1}^{N} C_k \, x_{k:N} \, p_k, \tag{5.252}$$

we analyze the respective coefficients

$$C_k = \frac{F_{i,j:n}(q_k) - F_{i,j:n}(q_{k-1})}{p_k} = f_{j:n}(\xi_k) - f_{i:n}(\xi_k), \tag{5.253}$$

for some $\xi_k \in (q_{k-1}, q_k)$, and $k = 1, \ldots, N$. Slightly modifying respective arguments from the previous case, we conclude that the projection of $\tilde{C} = (C_1, \ldots, C_N)$ onto the family of nondecreasing sequences with $\tilde{p}$-weighted ℓ^2-norm has the form

$$\underline{C}_k = \begin{cases} \frac{F_{i,j:n}(q_{k_*})}{q_{k_*}}, & k = 1, \ldots, k_*, \\ \frac{F_{i,j:n}(q_k) - F_{i,j:n}(q_{k-1})}{p_k}, & k = k_* + 1, \ldots, K_*, \\ \frac{1 - F_{i,j:n}(q_{K_*})}{1 - q_{K_*}}, & k = K_* + 1, \ldots, N, \end{cases} \tag{5.254}$$

for $1 \le k_* \le K_* < N$. Here k_* is the index minimizing

$$G_{i,j:n}(q_k), \qquad k = 1, \ldots, N, \tag{5.255}$$

and $K_* + 1$ maximizes

$$H_{i,j:n}(q_k), \qquad k = 1, \ldots, N, \tag{5.256}$$

where functions $G_{i,j:n}(x)$ and $H_{i,j:n}(x)$ are defined in (5.91) and (5.94), respectively. Since the former is first decreasing and then increasing with the minimum at α_* determined from (5.92), sequence (5.255) is minimized at either of the neighbors of α_* from the set $\{q_k : k = 0, \ldots, N\}$. Since (5.94) is increasing on $(0, \gamma_*)$ and decreasing on $(\gamma_*, 1)$ for γ_* defined by (5.95), argument q_{K_*+1} is either the left or the right closest neighbor of γ_*. We also notice that (5.254) takes at least two different values and $\underline{C}_{k_*} < 0 < \underline{C}_{K_*+1}$. Combining Theorem 5.8 with (5.254), we obtain the following strictly positive bounds.

Corollary 5.24 *If* $1 < p < \infty$, *and*

$$\frac{F_{i,j:n}(q_{k_*})}{q_{k_*}} < \bar{\underline{c}}_q < -\frac{F_{i,j}(q_{K_*})}{1 - q_{K_*}} \tag{5.257}$$

is uniquely determined as the root of the equation

$$\sum_{k=k_*+1}^{K_*} p_k \left| \frac{F_{i,j:n}(q_k) - F_{i,j:n}(q_{k-1})}{p_k} - c \right|^{\frac{q}{p}} sgn \left\{ \frac{F_{i,j:n}(q_k) - F_{i,j:n}(q_{k-1})}{p_k} - c \right\}$$
$$+ q_{k_*} \left[c - \frac{F_{i,j:n}(q_{k_*})}{q_{k_*}} \right]^{\frac{q}{p}} + (1 - q_{K_*}) \left[-\frac{F_{i,j:n}(q_{K_*})}{1 - q_{K_*}} - c \right]^{\frac{q}{p}} = 0, \tag{5.258}$$

then

$$\mathbb{E} \frac{X_{j:n} - X_{i:n}}{\sigma_p} \le B \tag{5.259}$$

for

$$B^q = q_{k_*} \left[\bar{c}_q - \frac{F_{i,j:n}(q_{k_*})}{q_{k_*}} \right]^q + (1 - q_{K_*}) \left[-\frac{F_{i,j:n}(q_{K_*})}{1 - q_{K_*}} - \bar{c}_q \right]^q$$
$$+ \sum_{k=k_*+1}^{K_*} p_k \left| \frac{F_{i,j:n}(q_k) - F_{i,j:n}(q_{k-1})}{p_k} - \bar{c}_q \right|^q. \tag{5.260}$$

Inequality (5.259) *becomes the equality if*

$$x_{k:N} = \begin{cases} \mu - \frac{\sigma_p}{B^{q/p}} \left[\bar{c}_q - \frac{F_{i,j:n}(q_{k_*})}{q_{k_*}} \right]^{q/p}, & k = 1, \ldots, k_*, \\ \mu + \frac{\sigma_p}{B^{q/p}} \left| \frac{F_{i,j:n}(q_k) - F_{i,j:n}(q_{k-1})}{p_k} - \bar{c}_q \right|^{q/p} & \\ \qquad sgn \left\{ \frac{F_{i,j:n}(q_k) - F_{i,j:n}(q_{k-1})}{p_k} - \bar{c}_q \right\}, & k = k_* + 1, \ldots, K_*, \\ \mu + \frac{\sigma_p}{B^{q/p}} \left[-\frac{F_{i,j:n}(q_{k_*})}{1 - q_{k_*}} - \bar{c}_q \right]^{q/p}, & k = K_* + 1, \ldots, N. \end{cases} \tag{5.261}$$

Since $\bar{c}_2 = 0$, we can write the following.

Corollary 5.25 *We have*

$$\mathbb{E} \frac{X_{j:n} - X_{i:n}}{\sigma_2} \le B \tag{5.262}$$

for

$$B^2 = \frac{F_{i,j:n}^2(q_{k_*})}{q_{k_*}} + \frac{F_{i,j:n}^2(q_{K_*})}{1 - q_{K_*}} + \sum_{k=k_*+1}^{K_*} \frac{[F_{i,j:n}(q_k) - F_{i,j:n}(q_{k-1})]^2}{p_k} \tag{5.263}$$

with the equality conditions

$$x_{k:N} = \begin{cases} \mu + \frac{\sigma_2}{B}\frac{F_{i,j:n}(q_{k*})}{q_{k*}}, & k = 1, \ldots, k_*, \\ \mu + \frac{\sigma_2}{B}\frac{F_{i,j:n}(q_{k*}) - F_{i,j:n}(q_{k-1})}{p_k}, & k = k_* + 1, \ldots, K_*, \\ \mu - \frac{\sigma_2}{B}\frac{F_{i,j:n}(q_{k*})}{1-q_{k*}}, & k = K_* + 1, \ldots, N. \end{cases} \tag{5.264}$$

In the case of sample range, we have $k_* = 1$, $K_* = N - 1$, and (5.262) with (5.263) are replaced by

$$\mathbb{E}\frac{X_{n:n} - X_{1:n}}{\sigma_2} \leq B = \left\{\sum_{k=1}^{N} \frac{[F_{1,n:n}(q_k) - F_{1,n:n}(q_{k-1})]^2}{p_k}\right\}^{1/2}$$
$$= \sum_{k=1}^{N} \frac{[q_k^n - q_{k-1}^n + (1 - q_{k-1})^n - (1 - q_k)^n]^2}{q_k - q_{k-1}} \tag{5.265}$$

and

$$x_{k:N} = \mu + \frac{\sigma_2}{B} \frac{q_k^n - q_{k-1}^n + (1 - q_{k-1})^n - (1 - q_k)^n}{q_k - q_{k-1}}, \quad k = 1, \ldots, N, \tag{5.266}$$

respectively. Observe that the sample ranges are not necessarily the only differences of order statistics for which $k_* = 1$, $K_* = N - 1$ and $\tilde{C} = \underline{\tilde{C}}$.

5.2.2 Arbitrary Drawing Distribution

Below we discuss the problem of deriving optimal bounds for expectations of L-statistics from finite populations with arbitrary distribution $\tilde{p} \in \mathbb{S}^N$, maximizing the norms in (5.231) and (5.232) with respect to $\tilde{p} \in S^N$. It is difficult to derive analytic solutions to the problem, and only few particular cases have been treated in the literature. We concentrate on the mean-variance bounds with $p = 2$. López-Blazquez (1998) provided a recursively constructed solution for the sample maxima. Recall that for the samples of n iid random variables distributed at N points with a marginal distribution function taking on values $q_0 = 0 < q_1 < \ldots < q_{N-1} < q_N = 1$, the following bound

$$\frac{\mathbb{E}X_{n:n} - \mu}{\sigma_2} \leq B(n, N, \tilde{q}) = \left[\sum_{k=1}^{N} \frac{(q_k^n - q_{k-1}^n)^2}{q_k - q_{k-1}} - 1\right]^{1/2} \tag{5.267}$$

(cf. (5.250)) with $\tilde{q}$ denoting $(q_1, \ldots, q_{N-1})$ is attained if the support points satisfy (5.251). Our purpose is to determine

$$B(n, N) = \sup_{0<q_1<\ldots<q_{N-1}<1} B(n, N, \tilde{q}). \tag{5.268}$$

Theorem 5.10 *Let $n \geq 2$ be fixed. Then, $B^2(n, N)$ are determined recursively for $N = 1, 2, \ldots$ as follows:*

$$B^2(n, 1) = 0,$$
$$B^2(n, N+1) = [1 + B^2(n, N)]q_{n,N}^{2n-1} + \frac{(1 - q_{n,N}^n)^2}{1 - q_{n,N}} - 1. \tag{5.269}$$

where $q_{n,N}$ is the unique solution in $(0, 1)$ to the equation

$$(2n-1)[B^2(n, N) + 1]q^{2n-2} + \left(\frac{1 - q^n}{1 - q}\right)^2 - 2nq^{n-1}\frac{1 - q^n}{1 - q} = 0. \tag{5.270}$$

Moreover, $B(n, N) = B(n, N, \tilde{q}^*)$ for

$$q_i^* = \prod_{k=i}^{N-1} q_{n,\kappa}, \qquad i = 1, \ldots, N-1. \tag{5.271}$$

Proof Relation $B^2(n, 1) = 0$ is obvious, because $X_{n:n} - \mu = 0$ for degenerate one-point distributions. For $N = 2$, we can determine optimal $q = q_1$ by maximizing directly

$$h(n, 2, q) = \frac{(1 - q^n)^2}{1 - q} + q^{2n-1} - 1 = \sum_{i=1}^{n-1} q^i - \sum_{i=n}^{2n-2} q^i, \tag{5.272}$$

which has derivative

$$h'(n, 2, q) = \sum_{i=1}^{n-1} iq^{i-1} - \sum_{i=n}^{2n-2} iq^{i-1}. \tag{5.273}$$

(note that this is equal to (5.270) with $N = 1$). By the variation diminishing property of power functions, (5.273) has one change of sign in $(0, +\infty)$, and the change is from plus to minus. Since $h'(n, 2, 1) = -n^2+1 < 0$, the sign change point $q_{n,2}$, say, that solves (5.273) belongs to $(0, 1)$. Accordingly, we have $B(n, 2) = B(n, 2, q_{n,2})$.

Take now $N \geq 2$. For increasing $\tilde{q} = (q_1, \ldots, q_N) \in (0, 1)^N$, define increasing $\tilde{\rho} = (\rho_1, \ldots, \rho_{N-1}) \in (0, 1)^{N-1}$ taking $\rho_i = \frac{q_i}{q_N}$, $i = 1, \ldots, N-1$. Then we can write

$$\begin{aligned}
B^2(n, N+1, \tilde{q}) &= \sum_{k=1}^{N} \frac{(q_k^n - q_{k-1}^n)^2}{q_k - q_{k-1}} + \frac{(1 - q_N^n)^2}{1 - q_N} - 1 \\
&= q_N^{2n-1} \sum_{k=1}^{N} \frac{(\rho_k^n - \rho_{k-1}^n)^2}{\rho_k - \rho_{k-1}} + \frac{(1 - q_N^n)^2}{1 - q_N} - 1 \\
&= q_N^{2n-1} [B^2(n, N, \tilde{\rho}) + 1] + \frac{(1 - q_N^n)^2}{1 - q_N} - 1 \\
&\leq q_N^{2n-1} [B^2(n, N) + 1] + \frac{(1 - q_N^n)^2}{1 - q_N} - 1 \\
&\leq \sup_{0<q<1} \left\{ q^{2n-1} [B^2(n, N) + 1] + \frac{(1-q^n)^2}{1-q} - 1 \right\}. \qquad (5.274)
\end{aligned}$$

The expression inside the curly brackets is a polynomial of degree $2n - 1$ in q. We denote it by $h(n, N, q)$, and prove that it has a unique maximum in $(0, 1)$ at the point $q_{n,N}$ solving the equation

$$h'(n, N, q) = (2n-1)[B^2(n, N) + 1] q^{2n-2} + \left(\frac{1 - q^n}{1 - q} \right)^2 - 2nq^{n-1} \frac{1 - q^n}{1 - q} = 0 \qquad (5.275)$$

(see (5.270)). First we notice that $h'(n, N, 0) = 1 > 0$, and

$$h'(n, N, 1) = (2n-1)[B^2(n, N) + 1] - n^2 < 0, \qquad (5.276)$$

because $B(n, N) < \frac{n-1}{\sqrt{2n-1}}$, and the RHS is the sharp mean-variance bound for the expectation of the maximum of n iid random variables with an arbitrary marginal distribution, and it is attained by a continuous distribution (see Theorem 4.1). Therefore Eq. (5.270) has at least one solution in $(0, 1)$. We aim at showing that it has only one solution there, and then it provides the maximum of $h(n, N, q)$.

Dividing (5.270) by q^{2n-1} and changing the variable $(0, 1) \ni q \to s = \frac{1}{q} \in (1, +\infty)$, we obtain the equation

$$\left(\frac{s^n - 1}{s - 1} \right)^2 - 2n \frac{s^n - 1}{s - 1} + (2n-1)[B^2(n, N) + 1] = 0. \qquad (5.277)$$

By another transformation $(1, +\infty) \ni s \to t = \frac{s^n-1}{s-1} = 1 + s + \ldots + s^{n-1} \in (n, +\infty)$, we get

$$t^2 - 2nt + (2n-1)[B^2(n, N) + 1] = 0. \tag{5.278}$$

The discriminant $\Delta = 4n^2 - 4(2n-1)[B^2(n, N) + 1] = -4h'(n, N\ 1)$ of this quadratic equation is positive (comp. (5.276)), which implies that (5.278) has two solutions $t_{1,2} = n \mp \sqrt{n^2 - (2n-1)[B^2(n, N) + 1]}$, and only the greater one belongs to $(n, +\infty)$. This implies that (5.277) has a single solution in $(1, +\infty)$, and consequently (5.270) has a unique zero in $(0, 1)$ which attains the supremum in (5.274).

Observe that $B^2(n, 2) = B^2(n, 2, q_{n,2})$ for uniquely determined $0 < q_{n,2} < 1$, and for $N \geq 2$

$$B^2(n, N+1, (q_1, \ldots, q_N)) \leq \max_{0<q<1} h(n, N, q) \tag{5.279}$$

which becomes equality when $q = q_{n,N}$ determined by (5.270), and

$$B^2\left(n, N, \left(\frac{q_1}{q_{n,N}}, \ldots \frac{q_{N-1}}{q_{n,N}}\right)\right) = B^2(n, N).$$

This shows that $B(n, N+1)$ coincides with the RHS of (5.279), and the equality is attained by $(q_1^*, \ldots, q_N^*)$ determined recursively by (5.271) for $N = 2, 3, \ldots$. □

The solution is especially nice for $n = 2$. Then the bound is attained by a uniform distribution on N points with equal distances among the neighbors.

Corollary 5.26 *If X_1, X_2 are iid supported on N points, and have mean μ and variance σ_2^2, then*

$$\frac{\mathbb{E}X_{2:2} - \mu}{\sigma_2} \leq \frac{\sqrt{N^2-1}}{\sqrt{3}\,N}. \tag{5.280}$$

The equality holds in (5.280) *iff X_1 is uniformly distributed on the points $x_{k:N} = \mu + \sqrt{3}\sigma_2 \frac{2k-1-N}{\sqrt{N^2-1}}$, $k = 1, \ldots, N$.*

Proof For $n = N = 2$, (5.272) has the form $h(2, 2, q) = q - q^2, 0 < q < 1$. Its maximal value $B^2(2, 2) = \frac{1}{4}$ is attained at $q = q_{2,2} = \frac{1}{2}$. For $N \geq 2$, we use the induction arguments. Put $B^2(2, N) = \frac{N^2-1}{3N^2}$. Then

$$h'(2, N, q) = 3\frac{4N^2-1}{3N^2} q^2 + (1+q)^2 - 4q(1+q) = \frac{N^2-1}{N^2} q^2 - 2q + 1 = 0 \tag{5.281}$$

(cf. (5.275)) has one solution in (0, 1) equal to $q_{2,N} = \frac{N}{N+1}$. Plugging $B^2(2, N)$ and $q_{2,N}$ into the RHS of (5.269), we obtain $B^2(2, N+1) = \frac{(N+1)^2-1}{3(N+1)^2}$, as desired. Using (5.271) we calculate $q_i^* = \prod_{k=i}^{N-1} \frac{k}{k+1} = \frac{i}{N}$, $i = 1, \ldots, N-1$, which means that the distribution attaining (5.280) is uniform. Using ultimately (5.251), we determine the support points. □

López-Blazquez (2000) studied an analogous problem for the sample ranges. He showed that for all n and N, (5.265) is strictly concave in $0 < q_1 < \ldots < q_{N-1} < 1$, and there exist unique $0 < q_1^* < \ldots < q_{N-1}^* < 1$ maximizing it. It follows that one can use standard numerical procedures for determining the bounds and distribution functions attaining them for arbitrary sample and population sizes. He also managed to determine analytic formulae in the cases $n = 2$ and $n = 3$ as

$$\frac{\mathbb{E}(X_{2:2} - X_{1:2})}{\sigma_2} \le 2\frac{\sqrt{N^2-1}}{\sqrt{3}\,N},$$

$$\frac{\mathbb{E}(X_{3:3} - X_{1:3})}{\sigma_2} \le \frac{\sqrt{3(N^2-1)}}{N}.$$

Both these bounds are attained by uniform discrete distributions.

López-Blazquez and Rychlik (2008) investigated the problem of determining positive sharp upper bounds on $\frac{\mathbb{E}X_{j:n}-\mu}{\sigma_2}$ over all iid samples with at most N-valued marginals for $j = 2, \ldots, n-1$. The case $j = n$ was treated above, whereas for $j = 1$ one cannot expect positive bounds. We first note that for fixed parent distribution function values $0 = q_0 < q_1 \le \ldots \le q_{N-1} < q_N = 1$, the problem is equivalent to finding the projection of $f_{j:n}$ (see (4.6)) onto the convex cone

$$\mathcal{C}_{\bar{q}}^{\nearrow} = \left\{ s(x) = \sum_{k=1}^{N} s_i \mathbf{1}_{[q_{k-1},q_k)}(x) : s_1 \le \ldots \le s_N \right\} \tag{5.282}$$

in $L^2((0, 1), dx)$. So we can formulate our problem as the problem of selecting the best choice of knots $0 = q_0 < q_1 \le \cdots \le q_{N-1} < q_N = 1$ for the best nondecreasing and at most N-valued approximation of $f_{j:n}$. Recall that the projection of $f_{j:n}$ onto the family $\mathcal{C}^{\nearrow}$ of all nondecreasing functions in $L^2((0, 1), dx)$ is

$$\underline{f}_{j:n}(x) = f_{j:n}(\min\{x, \alpha_{j:n}\}) \tag{5.283}$$

with $\alpha_{j:n}$ defined in formula (4.81) of Theorem 4.8.

We start by describing the best approximation of $\underline{f}_{j:n}$ by at most N-valued functions with arbitrary breaking points in the L^2-norm. The problem of choosing the split points for the best L^2-approximation of a given function is a classic and still open problem in approximation theory. We start with a simple lemma formulated for a general function g.

Lemma 5.6 *For $g : (0, 1) \to \mathbb{R}$ with the antiderivative $G(x) = \int_0^x g(t)\,dt$ and fixed $0 = q_0 < q_1 < \ldots < q_{M-1} < q_M = 1$, the closest element of the linear subspace*

$$\mathcal{L}_{\tilde{q}} = \left\{ s(x) = \sum_{k=1}^{M} s_k \mathbf{1}_{[q_{k-1}, q_k)}(x) : s_k \in \mathbb{R},\ k = 1, \ldots, M \right\} \tag{5.284}$$

of $L^2((0, 1), dx)$ has the form $s^*(x) = \sum_{k=1}^{M} s_k^* \mathbf{1}_{[q_{k-1}, q_k)}(x)$ with

$$s_k^* = \frac{G(q_k) - G(q_{k-1})}{q_k - q_{k-1}}, \qquad k = 1, \ldots, M, \tag{5.285}$$

and then for all $s \in \mathcal{L}_{\tilde{q}}$,

$$\|g - s\|^2 \geq \|g - s^*\|^2 = \|g\|^2 - \sum_{k=1}^{M} \frac{[G(q_k) - G(q_{k-1})]^2}{q_k - q_{k-1}}. \tag{5.286}$$

Proof We have

$$||g - s||^2 = \int_0^1 g^2(x)dx - \sum_{k=1}^{M} \{2s_k[G(q_k) - G(q_{k-1})] + s_k^2(q_k - q_{k-1})\},$$

$$\frac{\partial ||g - s||^2}{\partial s_k} = -2[G(q_k) - G(q_{k-1})] + 2s_k(q_k - q_{k-1}),$$

$$\frac{\partial^2 ||g - s||^2}{\partial s_k \partial s_\ell} = \begin{cases} 2(q_k - q_{k-1}) > 0, & k = \ell, \\ 0, & k \neq \ell. \end{cases}$$

It follows that $||g - s||^2$ is minimized when

$$s_k = \frac{G(q_k) - G(q_{k-1})}{q_k - q_{k-1}}, \qquad k = 1, \ldots, M, \tag{5.287}$$

and the minimum takes on the form (5.286). □

Note that nondecrease of g implies nondecrease of s^*. This concerns $f_{i:n}$ in particular. Another observation in case $g = f_{j:n}$ is that if we have a possibility of approximating the function by one with N values at most, we should choose all of them distinct. Suppose that we chose $M < N$ different values. Adding $q \in (0, \min\{q_1, \alpha_{i:n}\})$, we get the best constant approximations $\frac{E_{j:n}(q)}{q}$ and $\frac{E_{j:n}(q_1) - E_{j:n}(q)}{q_1 - q}$, on $(0, q)$ and (q, q_1), respectively. We have $\frac{E_{j:n}(q)}{q} <$

$\frac{\underline{F}_{j:n}(q_1)-\underline{F}_{j:n}(q)}{q_1-q}$, because all the values of $\underline{f}_{j:n}$ on $(0, q)$ are strictly less than those for the arguments from (q, q_1). This pair better approximates $\underline{f}_{j:n}$ on $(0, q_1)$ than the single constant $\frac{\underline{F}_{j:n}(q_1)}{q_1}$. This shows that it is better to use all $N-1$ split points.

The next lemma shows that we should choose split points from $(0, \alpha_{j:n}]$.

Lemma 5.7 *Suppose that* $g \in L^2(0,1)$ *is a constant on a non-empty interval* $(a, b) \subset (0, 1)$. *If* $s \in \mathcal{L}_{\tilde{q}}$ *has split points in* (a, b), *then it is possible to construct* $s' \in \mathcal{L}_{\tilde{q}'}$ *such that its split points are not contained there and*

$$||g - s'||^2 \leq ||g - s||^2.$$

Proof Suppose that $g(t) = c$ for all $t \in (a, b)$. Assume also that the interval is maximal in the sense that $a = \inf\{u \in (0,1) : g(t) = c \text{ for all } t \in (u, b)\}$ and $b = \sup\{u \in (0,1) : g(t) = c \text{ for all } t \in (a, u)\}$. Moreover, we assume that $0 < a < b < 1$. (The cases $a = 0$ and/or $b = 1$ can be discussed similarly). Suppose that $s(x) = \sum_{k=1}^{N} s_k \mathbf{1}_{[q_{k-1}, q_k)}(x)$ is a step function with some of its split points in the interval (a, b). Let

$$k_0 = \min\{k : q_k \in [a, b]\} \quad \text{and} \quad k_1 = \max\{k : q_k \in [a, b]\}.$$

We consider two cases.

The first case is when $k_0 = k_1$, and there is only one split point u_{k_0} in the interval (a, b). In this case, we have

$$\begin{aligned}
\int_{q_{k_0}-1}^{q_{k_0}+1} [g(x) - s(x)]^2 dx &= \int_{q_{k_0}-1}^{a} [g(x) - s_{k_0}]^2 dx + \int_{a}^{u_{k_0}} [c - s_{k_0}]^2 dx \\
&+ \int_{q_{k_0}}^{b} [c - s_{k_0+1}]^2 dx + \int_{b}^{q_{k_0}+1} [g(x) - s_{k_0+1}]^2 dx \\
&= \int_{q_{k_0}-1}^{a} [g(x) - s_{k_0}]^2 dx + \int_{b}^{q_{k_0}+1} [g(x) - s_{k_0+1}]^2 dx + b(c - s_{k_0+1})^2 \\
&- a(cs_{k_0})^2 + q_{k_0}\left(s_{k_0+1} - s_{k_0}\right)\left(2c - s_{k_0} - s_{k_0+1}\right). \qquad (5.288)
\end{aligned}$$

Note that (5.288) is a linear function of q_{k_0}. As $q_{k_0} \in [a, b]$, this is minimized when q_{k_0} is chosen to be equal to a or b. So a new step function s' can be constructed by removing the split point q_{k_0} from the original function s and substituting it by either a or b. As new levels, we can consider the mean values in the newly formed intervals.

The second case occurs when $k_1 - k_0 \geq 1$, and there is more than one split point in (a, b). We will show that this case can be reduced to the previous one. Construct a new step function

$$s^{'}(x) = \begin{cases} s(x), & \text{either } t \leq q_{k_0-1} \text{ or } x \leq q_{k_1+1}, \\ m_1, & x \in \left(q_{k_0-1}, a\right], \\ c, & x \in (a, b], \\ m_2, & x \in \left(b, q_{k_1+1}\right], \end{cases}$$

with m_1 and m_2 being the mean values of g in the intervals $\left(q_{k_0-1}, a\right]$ and $\left(b, q_{k_1+1}\right]$, respectively. It can be easily checked that $||g - s'||^2 \leq ||g - s||^2$. Note that we removed the split points $q_{k_0}, \ldots, q_{k_1}$ from the original step function, and added two new split points: a and b. Accordingly, the number of split points of $s^{'}$ in $(0, 1)$ is at most $(N-1)-(k_1-k_0+1)+2 = N-(k_1-k_0) \leq N-1$. It follows that s' has no more than $N-1$ split points and it does not have them in the interior of the flatness interval (a, b). We can add new split points beyond (a, b), and further improve the approximation. □

Lemma 5.8 *For every* $s(x) = \sum_{k=1}^{N} \frac{G(q_k)-G(q_{k-1})}{q_k-q_{k-1}} \mathbf{1}_{[q_{k-1},q_k)}(x)$ *with* $0 < q_1 < \ldots < q_{N-1} = \alpha_{j:n}$, *there exists* $s'(x) = \sum_{k=1}^{N} \frac{G(q_k')-G(q_{k-1}')}{q_k'-q_{k-1}'} \mathbf{1}_{[q_{k-1}', q_k')}(x)$ *with* $0 < q_1' < \ldots < q_{N-1}' < \alpha_{j:n}$ *such that* $||\underline{f}_{j:n} - s'||^2 < ||\underline{f}_{j:n} - s||^2$.

Proof By (5.286) and $0 < q_1 < \ldots < q_{N-1} = \alpha_{j:n}$ we have

$$\begin{aligned} ||\underline{f}_{j:n} - s||^2 &= ||\underline{f}_{j:n}||^2 - \sum_{k=1}^{M} \frac{[\underline{F}_{j:n}(q_k) - \underline{F}_{j:n}(q_{k-1})]^2}{q_k - q_{k-1}} \\ &= ||\underline{f}_{j:n}||^2 - \sum_{k=1}^{M} \frac{[F_{j:n}(q_k) - F_{j:n}(q_{k-1})]^2}{q_k - q_{k-1}}. \end{aligned} \tag{5.289}$$

Then

$$\begin{aligned} \frac{\partial ||\underline{f}_{j:n} - s||^2}{\partial q_{N-1}} &= \left(2f_{j:n}(q_{N-1}) - \frac{1-F_{j:n}(q_{N-1})}{1-q_{N-1}} - \frac{F_{j:n}(q_{N-1}) - F_{j:n}(q_{N-2})}{q_{N-1} - q_{N-2}}\right) \\ &\times \left(\frac{1-F_{j:n}(q_{N-1})}{1-q_{N-1}} - \frac{F_{j:n}(q_{N-1}) - F_{j:n}(q_{N-2})}{q_{N-1} - q_{N-2}}\right). \end{aligned} \tag{5.290}$$

Since

$$f_{j:n}(\alpha_{j:n}) - \frac{1 - F_{j:n}(\alpha_{j:n})}{1 - \alpha_{j:n}},$$

we have

$$\frac{\partial \|\underline{f}_{j:n} - s\|^2}{\partial q_{N-1}} = \left(f_{j:n}(q_{N-1}) - \frac{F_{j:n}(q_{N-1}) - F_{j:n}(q_{N-2})}{q_{N-1} - q_{N-2}} \right)^2 > 0. \tag{5.291}$$

This means that fixing $q_1, \ldots, q_{N-2}$ and moving q_{N-1} to the left, we decrease the distance from $\underline{f}_{j:n}$ to s. □

Summing up, we conclude that

$$\inf_{s\in\mathcal{L}_N} ||\underline{f}_{j:n} - s||^2 = \inf_{s\in\mathcal{L}_N^{\nearrow}} ||\underline{f}_{j:n} - s||^2 = \inf_{s\in\mathcal{L}_N^*} ||\underline{f}_{j:n} - s||^2, \tag{5.292}$$

where

$$\mathcal{L}_N = \left\{ \sum_{k=1}^{N} s_k \mathbf{1}_{[q_{k-1},q_k)}(x) : s_k \in \mathbb{R},\ 0 \le q_1 \le \ldots \le q_{N-1} \le 1 \right\}, \tag{5.293}$$

$$\mathcal{L}_N^{\nearrow} = \left\{ \sum_{k=1}^{N} s_k \mathbf{1}_{[q_{k-1},q_k)}(x) : s_1 \le \ldots \le s_{N-1}, \right.$$
$$\left. 0 \le q_1 \le \ldots \le q_{N-1} \le 1 \right\} \subset \mathcal{L}_N, \tag{5.294}$$

$$\mathcal{L}_N^* = \left\{ \sum_{k=1}^{N} \frac{F_{j:n}(q_k) - F_{j:n}(q_{k-1})}{q_k - q_{k-1}} \mathbf{1}_{[q_{k-1},q_k)}(x) : \right.$$
$$\left. 0 < q_1 < \ldots < q_{N-1} \le \alpha_{j:n} \right\} \subset \mathcal{L}_N^{\nearrow}. \tag{5.295}$$

Below we show that the problem of maximizing $\frac{\mathbb{E}X_{j:n}-\mu}{\sigma_2}$ over iid samples with at most N-valued marginals, which is equivalent to determining $\inf_{s\in\mathcal{L}_N^{\nearrow}} ||f_{j:n} - s||^2$, is also equivalent to finding the projection of $\underline{f}_{j:n}$ onto the family $\mathcal{L}_N$ of at most N-valued functions.

Proposition 5.3 *We have*

$$\forall\, s \in \mathcal{L}_N^{\nearrow} \qquad ||f_{j:n} - s||^2 \ge ||f_{j:n} - \underline{f}_{j:n}||^2 + ||\underline{f}_{j:n} - s||^2, \tag{5.296}$$

$$\forall\, s \in \mathcal{L}_N^{*} \qquad ||f_{j:n} - s||^2 = ||f_{j:n} - \underline{f}_{j:n}||^2 + ||\underline{f}_{j:n} - s||^2. \tag{5.297}$$

Proof Since $\underline{f}_{j:n}$ is the projection of $f_{j:n}$ onto the convex cone $\mathcal{C}^{\nearrow}$ of nondecreasing functions, therefore for every $s \in \mathcal{L}_N^{\nearrow} \subset \mathcal{C}^{\nearrow}$, we have

$$\langle f_{j:n} - \underline{f}_{j:n}, \underline{f}_{j:n} - s\rangle \geq 0$$

and consequently,

$$\begin{aligned}||f_{j:n} - s||^2 &= ||f_{j:n} - \underline{f}_{j:n}||^2 + 2\langle f_{j:n} - \underline{f}_{j:n}, \underline{f}_{j:n} - s\rangle + ||\underline{f}_{j:n} - s||^2 \\ &\geq ||f_{j:n} - \underline{f}_{j:n}||^2 + ||\underline{f}_{j:n} - s||^2.\end{aligned}$$

(comp. (2.136)). Function $\underline{f}_{j:n}$ is identical with $f_{j:n}$ on the interval $[0, \alpha_{j:n}]$, and is equal to the constant $f_{j:n}(\alpha_{j:n}) = \frac{1-F_{j:n}(\alpha_{j:n}}{1-\alpha_{j:n}}$ on $[\alpha_{j:n}, 1]$. Therefore for every $s \in \mathcal{L}_N^*$ which is constant on $[\alpha_{j:n}, 1]$, we have

$$\begin{aligned}&\langle f_{j:n} - \underline{f}_{j:n}, \underline{f}_{j:n} - s\rangle \\ &= \int_{\alpha_{j:n}}^1 [f_{j:n}(x) - f_{j:n}(\alpha_{j:n})] \left(f_{j:n}(\alpha_{j:n}) - \frac{1 - F_{j:n}(q_{N-1})}{1 - q_{N-1}} \right) dx \\ &= [1 - F_{j:n}(\alpha_{j:n}) - (1 - \alpha_{j:n}) f_{j:n}(\alpha_{j:n})] \left(f_{j:n}(\alpha_{j:n}) - \frac{1 - F_{j:n}(q_{N-1})}{1 - q_{N-1}} \right) = 0.\end{aligned}$$

□

We are not able to provide an analytic method of determining optimal $0 < q_1^* < \ldots < q_{N-1}^* < \alpha_{j:n}$ that minimizes the last expression in (5.292). We can only present an algorithm tending to local minima of the problems. Since for various starting points the procedure leads to the same points, we guess that these are the global minimum points as well. For constructing the algorithm, we need the following lemma.

Lemma 5.9 *Let* $0 < s_1 < \ldots < s_N < f_{j:n}(\alpha_{j:n})$. *Then the best approximation of* $\underline{f}_{j:n}$ *by non-decreasing step functions* $s = \sum_{k=1}^N s_k \mathbf{1}_{q_{k-1}, q_k}$ *with* N *fixed levels* $s_1, \ldots, s_N$ *in the space* $L^2((0,1), dx)$ *is the step function with jumps at the points*

$$q_k = f_{i:n}^{-1}\left(\frac{s_k + s_{k+1}}{2}\right), \quad k = 1, \ldots, N-1, \tag{5.298}$$

where $f_{i:n}^{-1}$ *denotes the inverse of the increasing part of* $f_{j:n}$ *restricted to the interval* $\left[0, \frac{j-1}{n-1}\right] \supset [0, \alpha_{j:n}]$.

Proof We consider

$$||\underline{f}_{j:n} - s||^2 = \int_0^1 \left[\underline{f}_{i:n}(x) - \sum_{k=1}^{N} s_k \mathbf{1}_{q_{k-1},q_k}(x) \right]^2 dx$$
$$= \sum_{k=1}^{N} \int_{q_{k-1}}^{q_k} \left[\underline{f}_{i:n}(x) - s_k \right]^2 dx. \tag{5.299}$$

The partial derivatives of the first and second orders have the respective forms

$$\frac{\partial ||\underline{f}_{j:n} - s||^2}{\partial q_k} = (s_{k+1} - s_k)(2\underline{f}_{j:n}(q_k) - s_k - s_{k+1}), \tag{5.300}$$

$$\frac{\partial^2 ||\underline{f}_{j:n} - s||^2}{\partial q_k \partial q_\ell} = \begin{cases} 2(s_{k+1} - s_k)\underline{f}'_{j:n}(q_k), & \ell = k, \\ 0, & \ell \neq k. \end{cases} \tag{5.301}$$

Gradient (5.300) vanishes if

$$\underline{f}_{j:n}(q_k) = \frac{s_k + s_{k+1}}{2} \in (0, f_{j:n}(\alpha_{j:n})), \quad k = 1, \ldots, N-1,$$

where $\underline{f}_{j:n}$ coincides with $f_{j:n}$, and is strictly increasing. So, the optimal split points are uniquely defined by (5.298). Moreover, they satisfy $0 < q_1 < \ldots < q_{N-1} < \alpha_{i:n}$, i.e., they all belong to interval of increase of $\underline{f}_{j:n}$. The Hessian matrix with entries (5.301) evaluated at q_k, $k = 1, \ldots, N-1$, defined in (5.298) has positive elements on the diagonal and zeros elsewhere, and so is positive definite. This implies that (5.298) provide the desired minimum of (5.299). □

The algorithm is as follows:

1. We choose arbitrary initial split points $0 < q_1 < \ldots < q_{N-1} < \alpha_{j:n}$, and set $H = 0$;
2. Then, we determine optimal levels

$$s_k = \frac{F_{j:n}(q_k) - F_{j:n}(q_{k-1})}{q_k - q_{k-1}}, \qquad k = 1, \ldots, N.$$

and calculate $H = \sum_{k=1}^{N} s^2(q_k - q_{k-1})$. If it is less than the value determined in the previous iteration plus a small number $\varepsilon > 0$ prescribed in advance, then we stop. Otherwise, we pass to the next point;

3. For the levels determined above, we calculate new split points

$$q_k = f_{i:n}^{-1}\left(\frac{s_k + s_{k+1}}{2}\right), \quad k = 1, \dots, N-1,$$

and go back to point 2.

If we accept H after a number of iterations, we assume that $\sqrt{H-1}$ is an accurate approximation of the bound on $\frac{\mathbb{E}X_{j:n}-\mu}{\sigma_2}$ taken from iid samples from N-valued populations, and the bound is attained if

$$\mathbb{P}\left(X_i = \mu + \frac{\sigma}{\sqrt{H-1}}(s_k - 1)\right) = q_k - q_{k-1}, \qquad k = 1, \dots, N.$$

It seems that analogous procedures can be applied for other L-statistics. For instance, the projections of the differences $f_{j:n} - f_{i:n}$ onto the family of nondecreasing functions are constant on the neighborhoods of 0 and 1, and strictly increasing in between. It is reasonable to choose the split points from the respective increase interval if we try to evaluate $\frac{\mathbb{E}X_{j:n}-X_{i:n}}{\sigma_2}$. The maximal bounds certainly tend to the general bounds in the iid case as we admit population size N increasing to infinity. However, we can deduce from the single order statistics case that condition (5.233) is not generally satisfied for the probability vectors providing the maximal bounds.

5.3 Without Replacement Model

5.3.1 *Positive Upper and Negative Lower Bounds for General L-Statistics*

Here we study L-statistics based on samples $X_1, \dots, X_n$ that arise by the drawing without replacement scheme from a finite population of numbers $\tilde{x} = (x_1, \dots, x_N) \in \mathbb{R}^N$. Observations X_i, $i = 1, \dots, n$, have the same marginal distribution

$$\mathbb{P}(X_i = x_k) = \frac{1}{N}\#\{r : x_r = x_k\}, \qquad i = 1, \dots, n, \ k = 1, \dots, N, \tag{5.302}$$

and the same moments

$$\mu = \mathbb{E}X_i = \bar{x} = \frac{1}{N}\sum_{k=1}^{n} x_k, \tag{5.303}$$

$$\sigma_p^p = \mathbb{E}|X_i - \mu|^p = \frac{1}{N}\sum_{k=1}^{N} |x_k - \mu|^p, \qquad 1 \le p < \infty, \tag{5.304}$$

like the observations in the classic with replacement model. However, the without replacement sampling implies that the observations are dependent. Another difference is the restriction that the sample size n cannot exceed the population size N. When $n = N$, the consecutive observations $X_1, \ldots, X_n$ are random, but the corresponding order statistics are deterministic, and we can apply the results of Sect. 3.4. Again, it is highly useful to consider the standard population $\{1, \ldots, N\}$ of size N and nondecreasing transformations

$$g_{\tilde{x}}(k) = x_{k:N}, \qquad k = 1, \ldots, N, \quad \tilde{x} \in \mathbb{R}^N. \tag{5.305}$$

If $W_1, \ldots, W_n$ denote the random observations drawn without replacement from the standard population, we have

$$(X_{1:n}, \ldots, X_{n:n}) \stackrel{d}{=} (g_{\tilde{x}}(W_{1:n}), \ldots, g_{\tilde{x}}(W_{n:n})) \tag{5.306}$$

and so

$$\mathbb{E}X_{i:n} = \mathbb{E}g_{\tilde{x}}(W_{i:n}) = \sum_{k=1}^{N} x_{k:N}\mathbb{P}(W_{i:n} = k). \tag{5.307}$$

The distribution of the ith order statistic based on the standard without replacement model has the form

$$p_{i:n}(k) = \mathbb{P}(W_{i:n} = k) = \frac{\binom{k-1}{i-1}\binom{N-k}{n-i}}{\binom{N}{n}}, \qquad k = 1, \ldots, N. \tag{5.308}$$

This follows from the fact that the event $\{W_{i:n} = k\}$ happens when the number k is drawn together with $i - 1$ observations from the set $\{1, \ldots, k - 1\}$ and $n - i$ observations from the set $\{k + 1, \ldots, N\}$. The denominator represents the number of all possible subsets of size n of the set with N elements. In (5.308) and later on, we adopt the convention that $\binom{k}{\ell} = 0$ for $k < \ell$ which means that

$$\mathbb{P}(W_{i:n} = k) = 0, \qquad k = 1, \ldots, i - 1, N - n + i + 1, \ldots, N. \tag{5.309}$$

This is obvious, because the ith smallest number in any subset of $\{1, \ldots, N\}$ cannot be less than i. Also, the greatest possible value of $W_{i:n}$ is $N - n + i$, which happens when $N - n + i, N - n + i + 1, \ldots, N$ are the $n + 1 - i$ greatest elements of a subsample of size n.

Using (5.307) and (5.308), for arbitrary fixed $\tilde{c} \in \mathbb{R}^n$, we can write

$$\mathbb{E}\sum_{i=1}^{n} c_i X_{i:n} = \sum_{i=1}^{n} c_i \sum_{k=1}^{N} x_{k:N} p_{i:n}(k)$$
$$= \sum_{k=1}^{N} x_{k:N} \sum_{i=1}^{n} c_i p_{i:n}(k) = \sum_{k=1}^{N} C_k x_{k:N}, \tag{5.310}$$

where

$$C_k = \sum_{i=1}^{n} c_i p_{i:n}(k) = \sum_{i=1}^{n} c_i \frac{\binom{k-1}{i-1}\binom{N-k}{n-i}}{\binom{N}{n}}, \qquad k = 1, \ldots, N \tag{5.311}$$

are known coefficients. Representation (5.310) allows us to implement the method applied before for the with replacement models for establishing various bounds on the expectations of L-statistics. This consist in replacing the vector of coefficients $\tilde{C} = (C_1, \ldots, C_N)$ defined in (5.311) by its projection $\underline{\tilde{C}} = (\underline{C}_1, \ldots, \underline{C}_N)$ onto the convex cone of nondecreasing sequences and ultimate application of norm inequalities. We have

$$\underline{C}_k = \frac{1}{j_m - j_{m-1}} \sum_{r=j_{m-1}+1}^{j_m} C_r, \quad k = j_{m-1}+1, \ldots, j_m, \; m = 1, \ldots, M, \tag{5.312}$$

for $j_0 = 0$ and

$$j_m = \min\left\{ j_{m-1} + 1 \le j \le N : \frac{1}{j - j_{m-1}} \sum_{r=j_{m-1}+1}^{j} C_r \right.$$
$$\left. = \min_{j_{m-1}+1 \le k \le N} \frac{1}{k - j_{m-1}} \sum_{r=j_{m-1}+1}^{k} C_r \right\}, \tag{5.313}$$

for $m = 1, \ldots, M$ with some $1 \le M \le N$ such that $j_M = N$. Definitions (5.311) and (5.312) imply that

$$\sum_{k=1}^{N} \underline{C}_k = \sum_{k=1}^{N} C_k = \sum_{i=1}^{n} c_i \sum_{k=1}^{N} p_{i:n}(k) = \sum_{i=1}^{n} c_i. \tag{5.314}$$

Then we have

$$\sum_{k=1}^{N} C_k\, x_{k:N} \leq \sum_{k=1}^{N} \underline{C}_k\, x_{k:N} \tag{5.315}$$

for every nondecreasing sequence $(x_{1:N}, \ldots, x_{N:N}) \in \mathbb{R}^N$. Therefore, combining (5.310), (5.314), (5.315) with the Hölder inequality, we get

$$\begin{aligned}
\mathbb{E}\sum_{i=1}^{n} c_i(X_{i:n}-\mu) &= \sum_{k=1}^{N} C_k(x_{k:N}-\mu) \\
\leq \sum_{k=1}^{N} \underline{C}_k(x_{k:N}-\mu) &= \sum_{k=1}^{N}(\underline{C}_k - c)(x_{k:N}-\mu) \\
\leq \|\tilde{\underline{C}} - c\tilde{\mathbf{1}}\|_q \|\tilde{x} - \mu\tilde{\mathbf{1}}\|_p &= N^{1/p}\|\tilde{\underline{C}} - c\tilde{\mathbf{1}}\|_q \sigma_p
\end{aligned} \tag{5.316}$$

for arbitrary $c \in \mathbb{R}$. This should be chosen so as to minimize $\|\tilde{\underline{C}} - c\tilde{\mathbf{1}}\|_q$ in order to provide the best bounds in (5.316). These are positive and attainable for populations with arbitrary $\mu \in \mathbb{R}$ and $\sigma_p > 0$ iff $\tilde{\underline{C}} \neq \frac{1}{N}\sum_{i=1}^{n} c_i \tilde{\mathbf{1}}$, which is equivalent to the condition

$$\begin{aligned}
\exists\, 1 \leq k \leq N-1 \qquad \frac{1}{k}\sum_{r=1}^{k} C_r &= \frac{1}{k\binom{N}{n}}\sum_{r=1}^{k}\sum_{i=1}^{n} c_i \binom{r-1}{i-1}\binom{N-r}{n-i} \\
&< \frac{1}{N}\sum_{r=1}^{N} C_r = \frac{1}{N}\sum_{i=1}^{n} c_i .
\end{aligned} \tag{5.317}$$

Otherwise, an equality occurs in (5.316) only when $x_{1:N} = x_{N:N}$ and $\sigma_p = 0$. The bounds (5.316) specified for particular choices of $1 \leq p \leq \infty$, with the corresponding conditions of attainability, are presented below.

Theorem 5.11 *Inequality*

$$\mathbb{E}\sum_{i=1}^{n} c_i \frac{X_{i:n}-\mu}{\sigma_1} \leq \frac{N}{2}\left[\frac{1}{N-j_{M-1}}\sum_{k=j_{M-1}+1}^{N} C_k - \frac{1}{j_1}\sum_{k=1}^{j_1} C_k\right] \tag{5.318}$$

holds, and under condition (5.317) *becomes the equality if*

$$x_{k:N} = \begin{cases} \mu - \sigma_1 \frac{N}{2j_1}, & k = 1, \ldots, j_1 \\ \mu, & k = j_1+1, \ldots, j_{M-1}, \\ \mu + \sigma_1 \frac{N}{2(N-j_{M-1})}, & k = j_{M-1}+1, \ldots, N. \end{cases} \tag{5.319}$$

Theorem 5.12 *For a given* $1 < p < \infty$, *let*

$$\frac{1}{j_1}\sum_{k=1}^{j_1} C_k < \underline{\bar{c}}_q < \frac{1}{N-j_{M-1}}\sum_{k=j_{M-1}+1}^{N} C_k$$

be the unique solution to

$$\sum_{m=1}^{M}(j_m - j_{m-1})|\underline{C}_{j_m} - c|^{q/p} sgn\{\underline{C}_{j_m} - c\} = 0. \tag{5.320}$$

Then the inequality

$$\mathbb{E}\sum_{i=1}^{n} c_i \frac{X_{i:n}-\mu}{\sigma_p} \le N^{1/p}\left(\sum_{m=1}^{M}(j_m - j_{m-1})|\underline{C}_{j_m} - \underline{\bar{c}}_q|^q\right)^{1/q} \tag{5.321}$$

holds, and becomes an equality if conditions (5.317) *and*

$$x_{k:N} = \mu + \sigma_p \frac{N^{1/p}\,\underline{C}_{j_m} - \underline{\bar{c}}_q|^{q/p} sgn\left\{\underline{C}_{j_m} - \underline{\bar{c}}_q\right\}}{\left(\sum_{r=1}^{m}(j_r - j_{r-1})|\underline{C}_{j_r} - \underline{\bar{c}}_q|^q\right)^{1/p}}, \tag{5.322}$$

for $k = j_{m-1}+1, \ldots, j_m$, $m = 1, \ldots, M$, *are satisfied.*

Theorem 5.13 *In particular, we have*

$$\mathbb{E}\sum_{i=1}^{n} c_i \frac{X_{i:n}-\mu}{\sigma_2} \le B = \left[N\sum_{m=1}^{M}\frac{[\sum_{r=j_{m-1}}^{j_m} C_r]^2}{j_m - j_{m-1}} - \left(\sum_{i=1}^{n} c_i\right)^2\right]^{1/2} \tag{5.323}$$

which under condition (5.317) *becomes an equality for*

$$x_{k:N} = \mu + \sigma_2 \frac{N^{1/2}\left[\frac{1}{j_m - j_{m-1}}\sum_{r=j_{m-1}+1}^{j_m} C_r - \sum_{i=1}^{n} c_i\right]}{\left[\sum_{m=1}^{M}\frac{[\sum_{r=j_{m-1}}^{j_m} C_r]^2}{j_m - j_{m-1}} - \frac{1}{N}\left(\sum_{i=1}^{n} c_i\right)^2\right]^{1/2}} \tag{5.324}$$

for $k = j_{m-1}+1, \ldots, j_m$, *and* $m = 1, \ldots, M$.

Theorem 5.14 *With $\underline{\bar{C}}_1 = \mathrm{MED}_N(\underline{\tilde{C}})$ being the median of $\underline{\tilde{C}}$ (cf.* (5.43)*), and j_{m_*} and j_{M_*} denoting the numbers of elements of $\underline{\tilde{C}}$ which are less and not greater than $\underline{\bar{C}}_1$, respectively (cf.* (5.44) *and* (5.45)*), we have*

$$\mathbb{E}\sum_{i=1}^{n} c_i \frac{X_{i:n}-\mu}{\sigma_\infty} \leq \frac{1}{N-j_{M_*}} \sum_{k=j_{M_*}+1}^{N} C_k - \frac{1}{j_{m_*}} \sum_{k=1}^{j_{m_*}} C_k \\ - (N - j_{m_*} - j_{M_*})\underline{\bar{C}}_1. \tag{5.325}$$

Under (5.317)*, inequality* (5.325) *becomes the equality for*

$$x_{k:N} = \begin{cases} \mu - \sigma_\infty, & k = 1, \ldots, j_{m_*}, \\ \mu + \sigma_\infty \frac{j_{m_*}+j_{M_*}-N}{j_{M_*}-j_{m_*}}, & k = j_{m_*}+1, \ldots, j_{M_*}, \\ \mu + \sigma_\infty, & k = j_{M_*+1}, \ldots, N. \end{cases}$$

The proofs of Theorems 5.11–5.14 are analogous to their counterparts for the with replacement case (see Theorems 5.1–5.4, respectively). Also, the method of determining the lower bounds for $\sum_{i=1}^{n} c_i X_{i:n}$ is the same: we calculate the upper bounds for $\sum_{i=1}^{n} c_{n+1-i} X_{i:n}$ and change the sign. The only difference is that vectors $\tilde{C}$ are defined differently. For the drawing with replacement model, C_k are defined as increments of a fixed polynomial $F_{\tilde{c}:n}$ among the lattice points $\frac{k}{N}$, $k = 0, \ldots, N$ (see (5.12)). This makes it possible to simplify some expressions in statements of results, and what is more important, allows us to conclude essential properties of the sequences from the properties of the polynomial. A similar approach is not possible for (5.311) in the without replacement model, which are expressible by means of combinatorial notions only. This causes that the statements of Theorems 5.11–5.14 are by no means simpler than in the case of replacing (5.311) by arbitrary real numbers $C_1, \ldots, C_k$. Another consequence is that the analysis of properties of $\tilde{C}$ corresponding to specific types of L-statistics is by far more complex than for the corresponding problems in the with replacement model.

We can observe that the limiting values of the bounds in the drawing without replacement model are identical with the bounds for the general populations.

Theorem 5.15 *For given $\tilde{c} \in \mathbb{R}^n$ and $1 \leq p \leq \infty$, there exists a sequence of populations $\Pi_N = \{x_{1,N}, \ldots, x_{N,N}\}$, $N = n, n+1, \ldots,$, with a given mean and pth central absolute moment σ_p^p (or absolute mean deviation σ_∞) such that for the samples $X_{1,N}, \ldots, X_{n,N}$ drawn without replacement from populations π_N, $N = n, n+1, \ldots$, with the respective order statistics denoted by $X_{1:n}(N), \ldots, X_{n:n}(N)$, we have*

$$\lim_{N\to\infty} \mathbb{E}\sum_{i=1}^{n} c_i[X_{i:n}(N) - \mu] = \|\underline{f}_{\tilde{c}:n} - \underline{\bar{c}}_q \mathbf{1}\|_q \sigma_p, \tag{5.326}$$

which is the sharp upper bound for the respective L-statistic from the iid samples of general populations.

Proof For fixed parameters $\tilde{c}$, p, μ and σ_p, there exists a distribution function F such that for $X_1, \ldots, X_n$ that are iid with the common distribution function F, the following holds:

$$\mathbb{E}\sum_{i=1}^{n} c_i(X_{i:n} - \mu) = \|\underline{f}_{\tilde{c}:n} - \underline{\tilde{c}}_q \mathbf{1}\|_q \sigma_p. \tag{5.327}$$

Using arguments similar to those of Lemma 5.10, we conclude that there exists a sequence of populations π_N, $N = n, n+1, \ldots$, with given μ and σ_p such that the samples drawn without replacement from Π_N tend weakly to $(X_1, \ldots, X_n)$. Note that each F attaining the particular bound (5.327) has a bounded support. Continuous functions

$$h_{i:n}(X_1, \ldots, X_n) = X_{i:n}, \qquad 1 \le i \le n, \tag{5.328}$$

although generally unbounded, can be truncated so to contain the whole finite support of a given F and all populations Π_N for sufficiently large N. Therefore, by (5.353), (5.327) and (5.328), we obtain

$$\mathbb{E}\sum_{i=1}^{n} c_i[X_{i:n}(N) - \mu] = \mathbb{E}\sum_{i=1}^{n} c_i[h_{i:n}(X_{1,N}, \ldots, X_{n,N}) - \mu]$$

$$\to \mathbb{E}\sum_{i=1}^{n} c_i[h_{i:n}(X_1, \ldots, X_n) - \mu] = \|\underline{f}_{\tilde{c}:n} - \underline{\tilde{c}}_q \mathbf{1}\|_q \sigma_p. \tag{5.329}$$

□

Similar conclusions can be deduced for the L-statistics with coefficient vectors $\tilde{c}$ satisfying assumptions (5.405) that assure nonpositive upper and nonnegative bounds (see e.g., Theorem 4.19 and Corollary 5.36).

On the other hand, for the minimal population size $N = n$ (that means in the exhaustive drawing without replacement model), we can observe that all the bounds established in this subsection are identical with the respective deterministic bounds for L-statistics of Sect. 3.4. Indeed, in the case of exhaustive drawing, the order statistics have deterministic values, and the sample and population moments are identical. In Volume 2, we show that the exhaustive drawing models provide the optimal bounds on the expectations of L-statistics coming from arbitrarily dependent identically distributed samples. This allows us to formulate the following conclusions. In both the drawing with and without replacement schemes, it is possible to approximate the general bounds on the expectations of L-statistics of iid general samples if the population size is large. In the first case, the expectations of L-statistics never exceed the general bounds. This is not necessarily true once we consider the without replacement model.

We also observe that the bounds for N close to n are greater than the bounds for large N. The former are close to the bounds in the more general possibly dependent identically distributed model (and they are equal when $N = n$), and the latter approximate the bounds in the more restricted iid model.

Below we consider the special cases of single order statistics and differences of order statistics.

5.3.2 Upper Bounds for Single Order Statistics

The problem for the single order statistics $X_{j:n}$, $1 \le j \le n$, is relatively easy. We apply (5.310) with

$$C_k = p_{j:n}(k) = \frac{\binom{k-1}{j-1}\binom{N-k}{n-j}}{\binom{N}{n}}, \quad k = 1, \ldots, N, \tag{5.330}$$

which are positive iff $k = j, \ldots, N - n + j$, and sum up to 1. Calculating

$$\frac{C_{k+1}}{C_k} = \frac{\binom{k}{j-1}\binom{N-k-1}{n-j}}{\binom{k-1}{j-1}\binom{N-k}{n-j}} = \frac{k(N-n+j-k)}{(k+1-j)(N-k)} \tag{5.331}$$

for $k = j, \ldots, N-n+j-1$, we check that $C_{k+1} \ge C_k$ for $k \le \frac{j-1}{n-1}N$ and $C_{k+1} \le C_k$ for $k \ge \frac{j-1}{n-1}N$. This means that the elements of sequence (5.330) are first equal to zero for $k = 1, \ldots, j-1$, then positive and increase for $k = j, \ldots, \lceil \frac{j-1}{n-1}N \rceil$, next they are positive and decrease for $k = \lceil \frac{j-1}{n-1}N \rceil + 1, \ldots, N-n+j$, and finally they are equal to zero for $k = N-n+j+1, \ldots, N$. Two maxima are possible at $\frac{j-1}{n-1}N$ and $\frac{j-1}{n-1}N + 1$ if they are integer. If $j = 1$, then (5.330) is nonincreasing: positive decreasing for $k = 1, \ldots, N-n+1$, and ultimately vanishing. If $j = n$, then this has first $n-1$ elements equal to zero, and then it is positive increasing. Otherwise, (5.330) is a nonnegative sequence that is nondecreasing for the small arguments and nonincreasing for the large ones. Hence, we have the following.

Proposition 5.4 *If $j = 1$, then $\underline{\tilde{C}} = P^{\nearrow}\tilde{C} = \frac{1}{N}\tilde{\mathbf{1}}$. If $j = n$, then $\underline{\tilde{C}} = P^{\nearrow}\tilde{C} = \tilde{C}$. If $2 \le j \le n-1$, then there exists $j-1 \le k_* < \lceil \frac{j-1}{n-1}N \rceil$ defined by*

$$k_* = \max\left\{k : C_k < \frac{1}{N-k}\sum_{r=k+1}^{N} C_r\right\} \tag{5.332}$$

such that

$$\underline{C}_k = \begin{cases} C_k, & k = 1, \ldots, k_*, \\ \frac{1}{N-k_*} \sum_{r=k_*+1}^{N} C_k, & k = k_* + 1, \ldots, N. \end{cases} \tag{5.333}$$

The first two conclusions are immediate. It only suffices to note that $k_* < k_* = \lceil \frac{j-1}{n-1} N \rceil$ which is either the unique or the smaller point maximizing (5.330). Also, $k_* \geq j - 1$, because the maximal mean of the last elements is positive and greater than $C_1 = C_{j-1} = 0$. Accordingly, for $j = 1$, the projection method provides the apparent relation

$$\mathbb{E}X_{1:n} - \mu = \mathbb{E}(X_{1:n} - X_1) \leq 0. \tag{5.334}$$

For the other cases, it is possible to determine nontrivial positive bounds. In order to include case $j = n$, it is convenient to define $k_* = N - 1$ then.

Corollary 5.27 *For* $2 \leq j \leq n \leq N$, *with* C_k, $k = 1, \ldots, N$, *and* k_* *defined in* (5.330) *and* (5.332), *respectively, we have*

$$\mathbb{E}\frac{X_{j:n} - \mu}{\sigma_1} \leq \frac{N}{2(N-k_*)} \sum_{k=k_*+1}^{N} C_k = \frac{N}{2(N-k_*)} \mathbb{P}(W_{j:n} > k_*), \tag{5.335}$$

(see (5.308)*) with the attainability conditions*

$$x_{k:N} = \begin{cases} \mu - \sigma_1 \frac{N}{2i}, & k = 1, \ldots, i, \\ \mu, & k = i + 1, \ldots, k_*, \\ \mu + \sigma_1 \frac{N}{2(N-k_*)}, & k = k_* + 1, \ldots, N, \end{cases} \tag{5.336}$$

for some $1 \leq i \leq j - 1$.

Corollary 5.28 *Under the above assumptions, with* $1 < p < \infty$ *and a unique*

$$0 < \underline{\bar{c}}_q < \frac{1}{N-k_*} \sum_{k=k_*+1}^{N} C_k \tag{5.337}$$

solving the equation

$$\sum_{k=1}^{k_*} |C_k - c|^{q/p} sgn\{C_k - c\} + (N - k_*) \left(\frac{1}{N-k_*} \sum_{k=k_*+1}^{N} C_k - c \right)^{q/p} = 0, \tag{5.338}$$

we have

$$\mathbb{E}\frac{X_{j:n}-\mu}{\sigma_p} \le N^{1/p}B \tag{5.339}$$

for

$$B^q = \sum_{k=1}^{k_*} |C_k - \bar{\underline{c}}_q|^q + (N-k_*)\left(\frac{1}{N-k_*}\sum_{k=k_*+1}^{N} C_k - \bar{\underline{c}}_q\right)^q. \tag{5.340}$$

The equality in (5.339) *occurs when*

$$x_{k:N} = \begin{cases} \mu + \sigma_p \frac{N^{1/p}}{B^{q/p}} |C_k - \bar{\underline{c}}_q|^{q/p} sgn\{C_k - \bar{\underline{c}}_q\}, & k = 1, \ldots, k_*, \\ \mu + \sigma_p \frac{N^{1/p}}{B^{q/p}} \left[\frac{1}{N-k_*}\sum_{k=k_*+1}^{N} C_k - \bar{\underline{c}}_q\right]^{q/p}, & k = k_*+1, \ldots, N. \end{cases} \tag{5.341}$$

Corollary 5.29 *In particular, the inequality*

$$\mathbb{E}\frac{X_{j:n}-\mu}{\sigma_2} \le N^{1/2}B \tag{5.342}$$

for

$$B^2 = \sum_{k=1}^{k_*} C_k^2 + \frac{1}{N-k_*}\left[\sum_{k=k_*+1}^{N} C_k\right]^2 - \frac{1}{N} \tag{5.343}$$

becomes the equality if

$$x_{k:N} = \begin{cases} \mu + \sigma_2 \frac{N^{1/2}}{B}(C_k - \frac{1}{N}), & k = 1, \ldots, k_*, \\ \mu + \sigma_2 \frac{N^{1/2}}{B}\left(\frac{1}{N-k_*}\sum_{k=k_*+1}^{N} C_k - \frac{1}{N}\right), & k = k_*+1, \ldots, N. \end{cases} \tag{5.344}$$

For getting the forms of bounds in the case $p = \infty$, it is essential to determine the median of sequence $\underline{\tilde{C}} = P^{\nearrow}\tilde{C}$. It is either equal to $\underline{C}_{k_*+1} = \underline{C}_N = \frac{1}{N-k_*}\sum_{k=k_*+1}^{N} C_k$ if $k_* \le \lfloor \frac{N}{2} \rfloor$ or to the median of the original $\tilde{C}$ otherwise. Condition $k_* < \lfloor \frac{N}{2} \rfloor$ is equivalent to

$$\frac{1}{\lceil N/2 \rceil + 1}\sum_{k=\lfloor N/2 \rfloor+2}^{N} C_k \ge \frac{1}{\lceil N/2 \rceil + 2}\sum_{k=\lfloor N/2 \rfloor+3}^{N} C_k \tag{5.345}$$

which can be simplified to

$$C_{\lfloor N/2\rfloor+1} \geq \frac{1}{\lceil N/2\rceil+2} \sum_{k=\lfloor N/2\rfloor+3}^{N} C_k. \tag{5.346}$$

Then, with the notation of Theorem 5.14, we have $j_{m_*} = k_*$ and $j_{M_*} = N$. Therefore, the following holds.

Corollary 5.30 *Under condition* (5.346),

$$\mathbb{E}\frac{X_{j:n}-\mu}{\sigma_\infty} \leq \frac{N}{N-k_*}\left(\frac{k_*}{N} - \sum_{k=1}^{k_*} C_k\right) = \frac{N}{N-k_*}\left(\frac{k_*}{N} - \mathbb{P}(W_{j:n} \leq k_*)\right), \tag{5.347}$$

(cf. (5.308)*) which becomes the equality for*

$$x_{k:N} = \begin{cases} \mu - \sigma_\infty, & k = 1, \ldots, k_*, \\ \mu + \sigma_\infty \frac{k_*}{N-k_*}, & k = k_* + 1, \ldots, N. \end{cases} \tag{5.348}$$

Otherwise,

$$\begin{aligned} \mathbb{E}\frac{X_{j:n}-\mu}{\sigma_\infty} &\leq \begin{cases} 1 - 2\sum_{k=1}^{N/2} C_k, & \text{if } N \text{ is even}, \\ 1 - 2\sum_{k=1}^{\frac{N-1}{2}} C_k - C_{\frac{N+1}{2}}, & \text{if } N \text{ is odd}, \end{cases} \\ &= \mathbb{P}\left(W_{j:n} > \frac{N+1}{2}\right) - \mathbb{P}\left(W_{j:n} < \frac{N+1}{2}\right) \end{aligned} \tag{5.349}$$

is a tight bound that is attained for

$$x_{k:N} = \begin{cases} \mu - \sigma_\infty, & k \leq \frac{N}{2}, \\ \mu, & k = \frac{N+1}{2} \text{ (if } N \text{ is even)}, \\ \mu + \sigma_\infty, & k \geq \frac{N}{2} + 1, \ldots, N. \end{cases} \tag{5.350}$$

For the sample maximum in particular,

$$\mathbb{E}(X_{n:n} - \mu) \leq \begin{cases} \frac{n}{2}\sigma_1, \\ \left[N\binom{N}{n}^{-2}\sum_{k=n}^{N}\binom{k-1}{n-1}^2 - 1\right]^{1/2}\sigma_2, \\ \left[\mathbb{P}\left(W_{n:n} > \frac{N+1}{2}\right) - \mathbb{P}\left(W_{n:n} < \frac{N+1}{2}\right)\right]\sigma_\infty. \end{cases} \tag{5.351}$$

Balakrishnan et al. (2003, Lemma 5.1) proved

$$\lim_{N\to\infty} N\binom{N}{n}^{-2}\sum_{k=n}^{N}\binom{k-1}{n-1}^2 = \lim_{N\to\infty}\frac{n^2}{N}\sum_{k=n}^{N}\prod_{r=1}^{n-1}\left(\frac{k-r}{N-r}\right)^2$$
$$= n^2\int_0^1 x^{2n-2}dx = \frac{n^2}{n-1} \tag{5.352}$$

which implies that the middle term in (5.351) tends to $\frac{n-1}{\sqrt{2n-1}}\sigma_2$. We also present the following lemma.

Lemma 5.10 (Balakrishnan et al. (2003, Lemma 6.1)) *Let $X_1, \ldots, X_n$ be an iid sample with a given distribution function F that has finite mean μ and variance σ^2. Then there exists a sequence of numerical populations $\Pi_N = \{x_{1,N}, \ldots, x_{N,N}\}$, $N = n, n+1, \ldots$, such that the samples $X_{1,N}, \ldots, X_{n,N}$, $N = n, n+1, \ldots$, of size n drawn without replacement from the populations Π_N have the mean μ and variance σ^2, and satisfy*

$$(X_{1,N}, \ldots, X_{n,N}) \xrightarrow{d} (X_1, \ldots, X_n), \qquad \text{as } N \to \infty. \tag{5.353}$$

The proof is constructive. Firstly, it is shown that for $(W_{1,N}, \ldots, W_{n,N})$, $N = n, n+1, \ldots$, denoting the sequences of random variables drawn without replacement from the standard populations $\{1, \ldots, N\}$ of size N, we have

$$\left(\frac{W_{1,N}}{N+1}, \ldots, \frac{W_{n,N}}{N+1}\right) \xrightarrow{d} (U_1, \ldots, U_n), \text{ as } N \to \infty \tag{5.354}$$

where the RHS denotes the iid sample from the standard uniform distribution on [0, 1]. Then the Skorokhod theorem (see, for example Billingsley, 1986, Theorem 23.6) is used to prove that

$$X_{i,N} = a_N F^{-1}\left(\frac{W_{i:N}}{N+1}\right) + b_N, \qquad i = 1, \ldots, n, \; N = n, n+1, \ldots, \tag{5.355}$$

for $a_N \to 1$ and $b_N \to 0$ chosen so as to obey the moment conditions satisfy (5.353). Using (5.354), we conclude that

$$\lim_{N\to\infty} \mathbb{P}\left(W_{n:n} > \frac{N+1}{2}\right) - \mathbb{P}\left(W_{n:n} < \frac{N+1}{2}\right)$$
$$= 1 - 2F_{n:n}\left(\frac{1}{2}\right) = 1 - \frac{1}{2^{n-1}}. \tag{5.356}$$

This shows that all the bounds in (5.351) tend to their counterparts for the general samples as N becomes large.

Papadatos and Rychlik (2004) proved generalizations of Corollaries 5.27–5.30 to the expectations of trimmed means

$$\mathbb{E}T_{i,j:n} = \sum_{k=1}^{N} C_k x_{k:N} \tag{5.357}$$

with

$$\begin{aligned} C_k &= \frac{1}{j+1-i} \sum_{r=i}^{j} p_{r:n}(k) \\ &= \frac{1}{(j+1-i)\binom{N}{n}} \sum_{r=i}^{j} \binom{k-1}{r-1}\binom{N-k}{n-r}, \quad k = 1, \ldots, N. \end{aligned} \tag{5.358}$$

It can be shown that this is nonincreasing for $i = 1$ and nondecreasing for $j = n$, and first nondecreasing and then nonincreasing in all the other cases $2 \le i < j \le n-1$. Therefore, the lower trivial bounds (5.334) are extended to all the lower selection differentials. For the other cases, the statement of Corollaries 5.27–5.30 hold true under the condition that we replace (5.330) by (5.358). Certainly, the analysis of the latter one is far more complicated.

5.3.3 *Upper Bounds for Differences of Order Statistics*

Similar problems occur when we treat the expectations of differences of order statistics

$$\mathbb{E}(X_{j:n} - X_{i:N}) = \sum_{k=1}^{N} C_k x_{k:N}, \qquad 1 \le i < j \le n, \tag{5.359}$$

with

$$C_k = p_{j:n}(k) - p_{i:n}(k) = \frac{\binom{k-1}{j-1}\binom{N-k}{n-j} - \binom{k-1}{i-1}\binom{N-k}{n-i}}{\binom{N}{n}}, \quad k = 1, \ldots, N. \tag{5.360}$$

Lemma 5.11 *Sequence (5.360) is first nonpositive, and then nonnegative, and sums up to zero. Furthermore, this is first nonincreasing (except for the case $i = 1$), then nondecreasing and ultimately nonincreasing (except for the case $j = n$).*

Proof The easiest fact to prove is

$$\sum_{k=1}^{N} C_k = \sum_{k=1}^{N} \mathbb{P}(W_{j:n} = k) - \sum_{k=1}^{N} \mathbb{P}(W_{i:n} = k) = 0. \tag{5.361}$$

Sequences $p_{i:n}(k)$ and $p_{j:n}(k)$, $k = 1, \ldots, N$, vanish for $k \in \{1, \ldots, i-1\} \cup \{N - n + i + 1, \ldots, N\}$ and $k \in \{1, \ldots, j-1\} \cup \{N - n + j + 1, \ldots, N\}$, respectively. If $j > N - n + i$, then

$$C_k = \begin{cases} 0, & k = 0, \ldots, i-1, \\ -p_{i:n}(k), & k = 1, \ldots, N - n + i, \\ 0, & k = N - n + i + 1, \ldots, j - 1, \\ p_{j:n}(k), & k = j, \ldots, N - n + j, \\ 0, & k = N - n + j + 1, \ldots, N, \end{cases} \tag{5.362}$$

and the statements immediately follow from nonnegativity and unimodality of $p_{i:n}(k)$ and $p_{j:n}(k)$.

Assume now that $j \le N - n + i$ and take $j \le k \le N - n + i$. We have

$$\frac{p_{j:n}(k)}{p_{i:n}(k)} = \frac{(i-1)!(n-i)!}{(j-1)!(n-j)!} \frac{(k-j+1)\ldots(k-i)}{(N-n+i+1-k)\ldots(N-n+j-k)}. \tag{5.363}$$

The numerator is positive increasing, and the denominator is positive decreasing. Therefore (5.363) is increasing, and so

$$C_k = p_{j:n}(k) - p_{i:n}(k), \qquad j \le k \le N - n + i, \tag{5.364}$$

is either negative, or positive, or changes the sign from $-$ to $+$. This together with the fact that

$$C_k = \begin{cases} 0, & k = 0, \ldots, i-1, \\ -p_{i:n}(k) < 0, & k = i-1, \ldots, j-1, \\ p_{j:n}(k) > 0, & k = N - n + i + 1, \ldots, N - n + j, \\ 0, & k = N - n + j + 1, \ldots, N, \end{cases} \tag{5.365}$$

imply that (5.360) is first nonpositive and then nonnegative.

Now we consider the differences

$$\begin{aligned} \Delta_{i,j:n}(k) &= C_{k+1} - C_k = [p_{j:n}(k+1) - p_{j:n}(k)] - [p_{i:n}(k+1) - p_{i:n}(k)] \\ &= \Delta[p_{j:n}(k)] - \Delta[p_{i:n}(k)], \quad k = 1, \ldots, N-1. \end{aligned} \tag{5.366}$$

We aim at showing that (5.366) are first nonpositive, then nonnegative and ultimately nonpositive. We readily see that

$$\Delta[p_{j:n}(k)] = 0 \quad \text{iff either } k < j-1 \text{ or } k = \frac{j-1}{n-1}N \text{ or } k > N-n+j, \tag{5.367}$$

$$\Delta[p_{j:n}(k)] > 0 \quad \text{iff } j-1 \le k < \frac{j-1}{n-1}N, \tag{5.368}$$

$$\Delta[p_{j:n}(k)] < 0 \quad \text{iff } \frac{j-1}{n-1}N < k \le N-n+j, \tag{5.369}$$

and similar relations hold for $\Delta[p_{i:n}(k)]$. Therefore, $\Delta_{i,j:n}(k) > 0$ for

$$\begin{aligned} k \in \underline{H}_m := \{k : \Delta[p_{j:n}(k)] > 0 \text{ and } \Delta[p_{i:n}(k)] \le 0\} \\ \cup \{k : \Delta[p_{j:n}(k)] \ge 0 \text{ and } \Delta[p_{i:n}(k)] < 0\}. \end{aligned} \tag{5.370}$$

Applying (5.367)–(5.369), we check that

$$\left\{k_1 := \frac{i-1}{n-1}N < k < k_2 := \frac{j-1}{n-1}N\right\} \subset \underline{H}_m \subset \overline{\underline{H}}_m = \{k_1 \le k \le k_2\}, \tag{5.371}$$

where $\overline{\underline{H}}_m$ denotes the closure of $\underline{H}_m$. If k_1 is integer, then $\Delta[p_{i:n}(k_1)] = 0$ and $\Delta_{i,j:n}(k_1) = \Delta[p_{j:n}(k_1)] \ge 0$. Likewise, for k_2 integer, $\Delta[p_{j:n}(k_2)] = 0$ and $\Delta_{i,j:n}(k_2) = -\Delta[p_{i:n}(k_2)] \ge 0$. Accordingly,

$$\Delta_{i,j:n}(k) \ge 0 \text{ for } k \in \overline{\underline{H}}_m. \tag{5.372}$$

Also,

$$\Delta_{i,j:n}(k) \le 0 \text{ for } k \in \underline{H}_\ell \cup \underline{H}_r, \tag{5.373}$$

where

$$\underline{H}_\ell := \{k : 1 \le k \le \min\{j-2, k_1\}\}, \tag{5.374}$$

$$\underline{H}_r := \{k : \max\{N-n+i+1, k_2\} \le k \le N\}, \tag{5.375}$$

because

$$\underline{H}_\ell \cup \underline{H}_r = \{k : \Delta[p_{j:n}(k)] \le 0 \text{ and } \Delta[p_{i:n}(k)] \ge 0\}. \tag{5.376}$$

It remains to analyze (5.366) in the remaining sets

$$\underline{H}^+ := \{k : \min\{\Delta[p_{j:n}(k)], \Delta[p_{i:n}(k)]\} > 0\} = \{k : j-1 \le k < k_1\} \quad (5.377)$$

and

$$\underline{H}^- := \{k : \max\{\Delta[p_{j:n}(k)], \Delta[p_{i:n}(k)]\} < 0\} = \{k : k_2 < k \le N-n+i\}. \quad (5.378)$$

Note that we have

$$\underline{H}_\ell \cup \underline{H}^+ \cup \overline{\underline{H}}_m \cup \underline{H}^- \cup \underline{H}_r = \{1, \ldots, N-1\}, \quad (5.379)$$

where all the subsets of the LHS are disjoint, and they are written so that each of them (if nonempty) lies on the right to the preceding one. By (5.372) and (5.373), it suffices to show that (5.366) changes the sign in $\underline{H}^+$ and $\underline{H}^-$ at most once, from $-$ to $+$ and from $+$ to $-$, respectively. To this end, for $k \in \underline{H}^+ \cup \underline{H}^-$, we consider

$$\begin{aligned}\rho_{i,j:n}(k) = \frac{\Delta[p_{j:n}(k)]}{\Delta[p_{i:n}(k)]} &= \frac{(i-1)!(n-i)!}{(j-1)!(n-j)!} \\ &\times \frac{(k-j-2)\cdots(k+1-i)}{(N-n+i+1-k)\cdots(N-n+j-k)}\,\frac{k_2-k}{k_1-k}. \end{aligned} \quad (5.380)$$

The first of the three fractions is a positive constant, and the second one is positive increasing. If $k \in \underline{H}^+$, and so $k < k_1 < k_2$, then the last factor is positive increasing, and so is the whole product (5.380). Since

$$\Delta_{i,j:n}(k) = [\rho_{i,j:n}(k) - 1]\Delta[p_{i:n}(k)] \quad (5.381)$$

and $\Delta[p_{i:n}(k)] > 0$, we see that (5.381) is either $-$ or $+$ or $-+$ in $\underline{H}^+$.

Similarly, for $k \in \underline{H}^-$ satisfying $k > k_2 > k_1$, the last fraction is positive increasing as well. Since $\Delta[p_{i:n}(k)] < 0$ now, (5.381) may change the sign from $+$ to $-$ only. This completes the proof. □

We can observe that sequence (5.360) representing the expectation functional for differences of order statistics from the without replacement model behaves analogously to its counterpart (5.86) for the with replacement problem. Therefore, it is not surprising that its projection $\underline{\tilde{C}} = P^{\nearrow}\tilde{C}$ has a similar form: it is first negative constant, then strictly increasing and finally positive constant. Let z_* and Z_* denote the indices of the last negative and first positive elements of (5.360). Let m_* and

M_* be the maximal and minimal arguments attaining the minimum and maximum of (5.360), respectively. Then, we have the following.

Proposition 5.5 *With notation* (5.360), *let*

$$k_* = \min\left\{m_* \le k \le z_* : \frac{1}{k}\sum_{r=1}^{k} C_r < C_{k+1}\right\}, \tag{5.382}$$

$$K_* = \max\left\{Z_* - 1 \le k \le M_* - 1 : C_k < \frac{1}{N-k}\sum_{r=k+1}^{N} C_r\right\}. \tag{5.383}$$

Then the projection of (5.360) *onto the set of nondecreasing sequences in* $\ell^2(\mathbb{R}^N)$ *has the form*

$$\underline{C}_k = \begin{cases} \frac{1}{k_*}\sum_{r=1}^{k_*} C_r, & k = 1, \dots, k_*, \\ C_k, & k = k_* + 1, \dots, K_*, \\ \frac{1}{N-K_*}\sum_{r=K_*+1}^{N} C_r, & k = K_* + 1, \dots, N. \end{cases} \tag{5.384}$$

This allows us to apply the general results for arbitrary L-statistics, presented in Theorems 5.11–5.14 to the special case of differences.

Corollary 5.31 *The inequality*

$$\mathbb{E}\frac{X_{j:n} - X_{i:n}}{\sigma_1} \le \frac{N}{2(N-k_*)}\sum_{k=K_*+1}^{N} C_k - \frac{N}{2k_*}\sum_{k=1}^{k_*} C_k \tag{5.385}$$

is sharp and becomes the equality for

$$x_{k:N} = \begin{cases} \mu - \sigma_1 \frac{N}{2k_*}, & k = 1, \dots, k_*, \\ \mu, & k = k_* + 1, \dots, K_*, \\ \mu + \sigma_2 \frac{N}{2(N-K_*)}, & k = K_* + 1, \dots, N. \end{cases} \tag{5.386}$$

Corollary 5.32 *If* $1 < p < \infty$ *and*

$$\frac{1}{k_*}\sum_{k=1}^{k_*} C_k < \bar{\underline{c}}_q < \frac{1}{N-K_*}\sum_{k=k_*+1}^{K} C_k \tag{5.387}$$

solves the equation

$$k_* \left[c - \frac{1}{k_*} \sum_{k=1}^{k_*} C_k \right]^{q/p} + (N - K_*) \left[\frac{1}{N - K_*} \sum_{k=K_*+1}^{N} C_k - c \right]^{q/p}$$

$$+ \sum_{k=K_*+1}^{K_*} |C_k - c|^{q/p} sgn\{C_k - c\} = 0, \tag{5.388}$$

then

$$\mathbb{E} \frac{X_{j:n} - X_{i:n}}{\sigma_p} \le N^{1/p} B, \tag{5.389}$$

where

$$B^q = k_* \left[\bar{\underline{c}}_q - \frac{1}{k_*} \sum_{k=1}^{k_*} C_k \right]^q + (N - K_*) \left[\frac{1}{N - K_*} \sum_{k=K_*+1}^{N} C_k - \bar{\underline{c}}_q \right]^q$$

$$+ \sum_{k=k_*+1}^{K_*} |C_k - \bar{\underline{c}}_q|^q. \tag{5.390}$$

Relation (5.389) becomes the equality if

$$x_{k:N} = \begin{cases} \mu - \sigma_p \frac{N^{1/p}}{B^{q/p}} \left[\bar{\underline{c}}_q - \frac{1}{k_*} \sum_{k=1}^{k_*} C_k \right]^{q/p}, & k = 1, \dots, k_*, \\ \mu + \sigma_p \frac{N^{1/p}}{B^{q/p}} |C_k - \bar{\underline{c}}_q|^{q/p} sgn\{C_k - \bar{\underline{c}}_q\}, & k = k_*+1, \dots, K_*, \\ \mu + \sigma_p \frac{N^{1/p}}{B^{q/p}} \left[\frac{1}{N-K_*} \sum_{k=K_*+1}^{N} C_k - \bar{\underline{c}}_q \right]^{q/p}, & k = K_*+1, \dots, N. \end{cases} \tag{5.391}$$

The following special case has been presented in Papadatos and Rychlik (2004).

Corollary 5.33 *For $p = 2$,*

$$\mathbb{E} \frac{X_{j:n} - X_{i:n}}{\sigma_2} \le N^{1/2} B, \tag{5.392}$$

with

$$B^2 = \frac{1}{k_*} \left(\sum_{k=1}^{k_*} C_k \right)^2 + \frac{1}{N - K_*} \left(\sum_{k=K_*+1}^{N} C_k \right)^2 + \sum_{k=k_*+1}^{K_*} C_k^2. \tag{5.393}$$

The equality in (5.392) *is obtained if*

$$X_{k:N} = \begin{cases} \mu + \sigma_2 \frac{N^{1/2}}{Bk_*} \sum_{k=1}^{k_*} C_r, & k = 1, \dots, k_*, \\ \mu + \sigma_2 \frac{N^{1/2}}{B} C_k, & k = k_* + 1, \dots, K_*, \\ \mu + \sigma_2 \frac{N^{1/2}}{B(N-K_*)} \sum_{k=K_*+1}^{N} C_r, & k = K_* + 1, \dots, N. \end{cases} \tag{5.394}$$

The bounds for $p = \infty$ depend on the value of the median of $\underline{\tilde{C}}$. This can be equal either to $\underline{C}_1 = \frac{1}{k_*} \sum_{k=1}^{k_*} C_k$ or to $\underline{C}_N = \frac{1}{N-K_*} \sum_{k=K_*+1}^{N} C_k$ or lie in between. The first case occurs when $k_* \geq \lceil \frac{N+1}{2} \rceil$, or equivalently,

$$0 > \frac{1}{\lceil \frac{N-1}{2} \rceil} \sum_{k=1}^{\lceil \frac{N-1}{2} \rceil} C_k \geq C_{\lceil \frac{N+1}{2} \rceil}. \tag{5.395}$$

The equivalent conditions for the second case are $K_* \leq \lfloor \frac{N-1}{2} \rfloor$ and

$$0 < \frac{1}{\lfloor \frac{N}{2} \rfloor} \sum_{k=\lfloor \frac{N+3}{2} \rfloor}^{N} C_k \leq C_{\lfloor \frac{N+1}{2} \rfloor}. \tag{5.396}$$

Corollary 5.34 *If* (5.395) *holds, then*

$$\mathbb{E} \frac{X_{j:n} - X_{i:n}}{\sigma_\infty} \leq \frac{N}{k_*} \sum_{k=k_*+1}^{N} C_k \tag{5.397}$$

is true, and becomes the equality for

$$x_{k:N} = \begin{cases} \mu - \sigma_\infty \frac{N-k_*}{k_*}, & k = 1, \dots, k_*, \\ \mu + \sigma_\infty, & k = k_* + 1, \dots, N. \end{cases} \tag{5.398}$$

Under (5.396), *we have*

$$\mathbb{E} \frac{X_{j:n} - X_{i:n}}{\sigma_\infty} \leq \frac{N}{N - K_*} \sum_{k=K_*+1}^{N} C_k \tag{5.399}$$

which becomes the equality for

$$x_{k:N} = \begin{cases} \mu - \sigma_\infty, & k = 1, \dots, K_*, \\ \mu + \sigma_\infty \frac{K_*}{N-K_*}, & k = K_* + 1, \dots, N. \end{cases} \tag{5.400}$$

Otherwise,

$$\begin{aligned} \mathbb{E}\frac{X_{j:n} - X_{i:n}}{\sigma_\infty} &\leq \begin{cases} 1 - 2\sum_{k=1}^{N/2} C_k, & \text{if } N \text{ is even}, \\ 1 - 2\sum_{k=1}^{\frac{N-1}{2}} C_k - C_{\frac{N+1}{2}}, & \text{if } N \text{ is odd}, \end{cases} \\ &= \mathbb{P}\left(W_{j:n} > \frac{N+1}{2}\right) - \mathbb{P}\left(W_{j:n} < \frac{N+1}{2}\right) \\ &- \mathbb{P}\left(W_{i:n} > \frac{N+1}{2}\right) + \mathbb{P}\left(W_{i:n} < \frac{N+1}{2}\right) \end{aligned} \tag{5.401}$$

is sharp and becomes the equality under conditions (5.350).

In the special case of sample range, for which $\underline{\tilde{C}} = \tilde{C}$, $k_* = 1$ and $K_* = N - 1$, we have

$$\mathbb{E}(X_{n:n} - X_{1:n}) \leq \begin{cases} n\,\sigma_1, \\ \frac{\sqrt{2N}}{\binom{N}{n}} \left[\sum_{k=n}^{N} \binom{k-1}{n-1}^2 - \binom{N}{2n-1}\right]^{1/2} \sigma_2, \\ \Big[\mathbb{P}\left(W_{n:n} > \frac{N+1}{2}\right) - \mathbb{P}\left(W_{n:n} < \frac{N+1}{2}\right) \\ \quad -\mathbb{P}\left(W_{1:n} > \frac{N+1}{2}\right) - \mathbb{P}\left(W_{1:n} < \frac{N+1}{2}\right)\Big]\sigma_\infty, \end{cases} \tag{5.402}$$

$$\to \begin{cases} n\sigma_1, \\ \frac{\sqrt{2}n}{\sqrt{2n-1}} \left[1 - \binom{2n-2}{n-1}^{-1}\right]^{1/2} \sigma_2, & \text{as } N \to \infty. \\ 2[1 - 2^{1-n}]\sigma_\infty, \end{cases} \tag{5.403}$$

The concise form of the middle term in the first line follows from Balakrishnan et al. (2003, Lemma 2.2). The limiting value is concluded from (5.352) and

$$\lim_{N\to\infty} \frac{N}{\binom{N}{n}^2} \binom{N}{2n-1} = \frac{2n^2}{2n-1} \binom{2n-2}{n-1}^{-1}. \tag{5.404}$$

The limit of the third expression follows from (5.354).

5.3.4 Negative Upper and Positive Lower Bounds for General L-Statistics

Now we concentrate on L-statistics whose coefficients do not satisfy condition (5.317), and assume that

$$\frac{1}{k}\sum_{r=1}^{k} C_r = \frac{1}{k\binom{N}{n}}\sum_{r=1}^{k}\sum_{i=1}^{n} c_i \binom{r-1}{i-1}\binom{N-r}{n-i}$$
$$\geq \frac{1}{N}\sum_{r=1}^{N} C_r = \frac{1}{N}\sum_{i=1}^{n} c_i, \qquad k = 1, \dots, N-1. \tag{5.405}$$

We also restrict ourselves to nontrivial populations satisfying $x_{1:N} < x_{N:N}$, and consequently, $\sigma_p > 0$ for all $1 \leq p \leq \infty$.

Theorem 5.16 *Let $1 \leq p < \infty$ and $\bar{c} \in \mathbb{R}^n$ be a vector satisfying (5.405). Define*

$$U_{p,\bar{c}}(j) = \left[\frac{N^{p+1}}{j^p(N-j) + j(N-j)^p}\right]^{1/p} \sum_{i=1}^{j}(C_i - \bar{C}), \qquad j = 1, \dots, N-1. \tag{5.406}$$

Then for $X_1, \dots, X_n$ drawn with replacement from any nonconstant population of size N, we have

$$\mathbb{E}\left(\sum_{i=1}^{n} c_i \frac{X_{i:n} - \mu}{\sigma_p}\right) \leq - \min_{j=1,\dots,N-1} U_{p,\bar{c}}(j), \tag{5.407}$$

with the equality attained for the following ordered population

$$x_{k:N}^{(j,p)} = \begin{cases} \mu - \sigma_p(N-j)\left[\frac{N^{p+1}}{j(N-j)^p + j^p(N-j)}\right]^{1/p}, & k = 1, \dots, j, \\ \mu + \sigma_p j\left[\frac{N^{p+1}}{j(N-j)^p + j^p(N-j)}\right]^{1/p}, & k = j+1, \dots, N, \end{cases} \tag{5.408}$$

with $j = j_*$, where j_* minimizes (5.406).

Theorem 5.17 *Under the assumptions of Theorem* 5.16 *with* $p = \infty$, *put*

$$U_{\infty,\bar{c}}(j) = \frac{N}{\max\{j, N-j\}} \sum_{i=1}^{j} (C_i - \bar{C}), \qquad j = 1, \ldots, N-1. \tag{5.409}$$

Then

$$\mathbb{E}\left(\sum_{i=1}^{n} c_i \frac{X_{i:n} - \mu}{\sigma_\infty}\right) \le - \min_{j=1,\ldots,N-1} U_{\infty,\bar{c}}(j), \tag{5.410}$$

with the equality attained for the ordered population

$$x_{k:N}^{(j,\infty)} = \begin{cases} \mu - \sigma_\infty, & k = 1, \ldots, j, \\ \mu + \sigma_\infty \frac{j}{N-j}, & k = j+1, \ldots, N, \end{cases} \quad \text{if} \quad 1 \le j \le \frac{N}{2}, \tag{5.411}$$

$$x_{k:N}^{(j,\infty)} = \begin{cases} \mu - \sigma_\infty \frac{N-j}{j}, & k = 1, \ldots, j, \\ \mu + \sigma_\infty, & k = j+1, \ldots, N, \end{cases} \quad \text{if} \quad \frac{N}{2} \le j \le N-1, \tag{5.412}$$

with $j = j_*$, *where* j_* *minimizes* (5.409).

The proofs of these two theorems are similar to the proofs of Theorems 5.5 and 5.6, respectively. The crucial step is the representation of the standardized expectation of the linear combination of order statistics as a standardized linear combination of ordered deterministic numbers

$$\mathbb{E}\left(\sum_{i=1}^{n} c_i \frac{X_{i:n} - \mu}{\sigma_p}\right) = \sum_{k=1}^{N} C_k \frac{x_{k:N} - \mu}{\sigma_p} = \sum_{k=1}^{N} (C_k - \bar{C}) \frac{x_{k:N} - \mu}{\sigma_p}.$$

Then we can apply the sharp evaluations of Theorems 3.17 and 3.18 for deterministic combinations. The only difference is that instead of

$$C_k = \sum_{i=1}^{n} c_i \left[F_{i:n}\left(\frac{k}{N}\right) - F_{i:n}\left(\frac{k-1}{N}\right) \right], \qquad k = 1, \ldots, N,$$

(see (5.12)) we consider

$$C_k = \sum_{i=1}^{n} c_i \frac{\binom{k-1}{i-1}\binom{N-k}{n-i}}{\binom{N}{n}}, \qquad k = 1, \ldots, N \tag{5.413}$$

(see (5.311)). Note that the attainability conditions in (5.128), (5.133), and (5.134) are identical with those of (5.408), (5.411), and (5.412), respectively. In a standard way we conclude non-negative lower bounds for L-statistics with coefficients satisfying

$$\frac{1}{k}\sum_{r=1}^{k} C_r = \frac{1}{k\binom{N}{n}}\sum_{r=1}^{k}\sum_{i=1}^{n} c_i \binom{r-1}{i-1}\binom{N-r}{n-i}$$
$$\leq \frac{1}{N}\sum_{r=1}^{N} C_r = \frac{1}{N}\sum_{i=1}^{n} c_i, \qquad k = 1, \ldots, N-1. \tag{5.414}$$

Corollary 5.35 *Under assumption (5.414), with the standard notation* $\tilde{c}' = (c_n, \ldots, c_1)$, *for every* $1 \leq p \leq \infty$, *we have*

$$\mathbb{E}\left(\sum_{i=1}^{n} c_i \frac{X_{i:n} - \mu}{\sigma_p}\right) \geq \min_{j=1,\ldots,N-1} U_{p,\tilde{c}'}(j)$$

(cf. (5.407) *and* (5.410)*), and the equality is attained for the ordered population* $x_{k:N}^{(N-j'_*,p)}$, $k = 1, \ldots, N$, *(cf.* (5.408), (5.411) *and* (5.412)*) where* j'_* *minimizes* $U_{p,\tilde{c}'}(j)$, $j = 1, \ldots, N-1$.

When N increases to infinity, the bounds of Theorems 5.16 and 5.17 (and Corollary 5.35) behave similarly to those for the with replacement model (comp. Corollary 5.11)

Corollary 5.36 *For arbitrary* $\tilde{c} \in \mathbb{R}^n$ *satisfying* (5.405), *for every* $1 < p \leq \infty$ *and for* $p = 1$ *when either* $c_1 = \bar{c}$ *or* $c_n = \bar{c}$, *bounds* (5.407) *and* (5.410) *tend to* 0 *as* $N \to \infty$.

Proof The statement is obvious when (5.405) is the equality for some $k = 1, \ldots, N-1$. We show that the statement is true even if all inequalities in (5.405) are sharp. By (5.311),

$$C_1 = \sum_{i=1}^{n} p_{i:n}(1) = \sum_{i=1}^{n} \frac{\binom{0}{i-1}\binom{N-1}{n-i}}{\binom{N}{n}} = c_1 \frac{n}{N},$$

whereas $\bar{C} = \frac{n}{N}\,\bar{c}$ (see (5.314)). Therefore,

$$\lim_{N\to\infty} U_{p,\tilde{c}}(1) = \lim_{N\to\infty} \begin{cases} \left[\frac{N^{p+1}}{N-1+(N-1)^p}\right]^{1/p} \frac{n}{N}(c_1 - \bar{c}), & 1 \leq p < \infty, \\ \frac{N}{N-1}\,\frac{n}{N}(c_1 - \bar{c}), & p = \infty, \end{cases}$$
$$= \begin{cases} \frac{n}{2}(c_1 - \bar{c}), & p = 1, \\ 0, & 1 < p \leq \infty. \end{cases} \tag{5.415}$$

Similarly, we obtain $C_N = c_n \frac{n}{N}$ and

$$\sum_{k=1}^{N-1}(C_k - \bar{C}) = -(C_N - \bar{C}) = \frac{n}{N}(\bar{c} - c_n).$$

It follows that

$$\lim_{N\to\infty} U_{p,\tilde{c}}(N-1) = \lim_{N\to\infty} \begin{cases} \left[\frac{N^{p+1}}{N-1+(N-1)^p}\right]^{1/p} \frac{n}{N}(\bar{c} - c_N), & 1 \le p < \infty, \\ \frac{N}{N-1}\,\frac{n}{N}(\bar{c} - c_N), & p = \infty, \end{cases}$$
$$= \begin{cases} \frac{n}{2}\,(\bar{c} - c_N), & p = 1, \\ 0, & 1 < p \le \infty. \end{cases}$$

Under the assumption of the corollary, $\min_{1\le k\le N-1} U_{p,\tilde{c}}(k) \to 0$ when N goes to infinity. □

5.3.5 Special Cases

Below we describe optimal inequalities for some L-statistics which are of special interest. In order to find the optimal bounds on expectations of specific L-statistics, we need to analyze

$$U_{\tilde{c},p}(j) = A_{\tilde{c}}(j)\, B_p(j), \quad 1 \le j \le N-1, \tag{5.416}$$

where

$$A_{\tilde{c}}(j) = \sum_{k=1}^{j}(C_k - \bar{C}), \tag{5.417}$$

with $\tilde{C} = (C_1, \dots, C_N)$ defined in (5.311) (see also (5.413)), dependent on the coefficients $c_1, \dots, c_n$, and

$$B_p(j) = M_p\left(\frac{j}{N}\right) = \begin{cases} \left[\frac{N^{1+p}}{j^p(N-j)+j(N-j)^p}\right]^{1/p}, & 1 \le p < \infty, \\ \frac{N}{\max\{j, N-j\}}, & p = \infty, \end{cases} \tag{5.418}$$

(cf. (5.141)) depending only on the range p of the scale unit σ_p and being the same for various L-statistics. For specified types of L-statistics, (5.417) and (5.418) depend on n, N and j as well.

We start with an analysis of the upper bounds for the sample minima. For fixed n, the coefficient vector $\tilde{c} = (1, 0, \dots, 0)$ has the mean $\bar{c} = \frac{1}{n}$. Therefore for a given $N > n$,

$$C_k = \mathbb{P}(W_{1:n} = k) = \frac{\binom{N-k}{n-1}}{\binom{N}{n}}, \qquad k = 1, \dots, N,$$

and $\bar{C} = \frac{1}{N}$. We have

$$\begin{aligned} A_{\tilde{c}}(j) &= \sum_{k=1}^{j}(C_k - \bar{C}) = \sum_{k=1}^{j} \mathbb{P}(W_{1:n} = k) - \frac{j}{N} \\ &= 1 - \mathbb{P}(W_{1:n} > j) - \frac{j}{N} = \frac{N-j}{N} - \frac{\binom{N-j}{n}}{\binom{N}{n}} \\ &= \begin{cases} \frac{N-j}{N} - \frac{(N-j)(N-j-1)\dots(N-j-n+1)}{N(N-1)\dots(N-n+1)}, & j \le N-n, \\ \frac{N-j}{N}, & j > N-n, \end{cases} \\ &= A_n^m(j), \end{aligned} \tag{5.419}$$

say. The equality in the second line follows from the fact that the event $\{W_{1:n} > j\}$ is equivalent to drawing n numbers from the subset $\{j+1, \dots, N\}$

Corollary 5.37 *We have*

$$\mathbb{E}\left(\frac{X_{1:n} - \mu}{\sigma_p}\right) \le - \begin{cases} \left[\frac{N}{(N-1)+(N-1)^p}\right]^{1/p}, & 1 \le p < \infty, \\ \frac{1}{N-1}, & p = \infty. \end{cases} \tag{5.420}$$

If $1 \le p < \infty$ *(*$p = \infty$*, respectively), then the equality in* (5.420) *holds for* (5.408) *((*5.412*), respectively) with* $j = N - 1$*. Moreover, if* $n = 2$ *and* $1 < p < \infty$ *(*$p = \infty$*, respectively), then the equality holds for* (5.408) *((*5.411*), respectively) with* $j = 1$ *as well. In the case* $n = 2$ *and* $p = 1$*, the equality is attained by* (5.408) *with arbitrary* $j = 1, \dots, N - 1$.

Proof We aim at analyzing (5.416) with (5.417) specified in (5.419) and (5.418).

Fix first $n = 2$. Then

$$A_2^m(j) = \frac{j(N-j)}{N(N-1)}.$$

Under the notation $U_{p,n}^m(j) = A_n^m(j) B_p(j)$, we have

$$U_{p,2}^m(j) = \begin{cases} \frac{N}{N-1} \cdot \frac{\frac{j}{N}(1-\frac{j}{N})}{[\frac{j}{N}(1-\frac{j}{N})^p + (\frac{j}{N})^p(1-\frac{j}{N})]^{1/p}}, & 1 \le p < \infty, \\ \frac{\min\{j, N-j\}}{N-1}, & p = \infty. \end{cases} \tag{5.421}$$

If $p = 1$, then

$$U^m_{1,2}(j) = \frac{N}{2(N-1)}$$

is a constant and so is minimized by every $j = 1, \ldots, N-1$. Let $1 < p < \infty$ now. We have

$$U^m_{p,2}(j) = \frac{N}{N-1} \cdot \frac{1}{[g_p(\frac{j}{N})]^{1/p}}, \qquad 1 \le j \le N-1,$$

where

$$g_p(x) = x^{1-p} + (1-x)^{1-p}, \qquad 0 < x < 1,$$

is a function symmetric about $1/2$ that is first decreasing and then increasing. Therefore

$$\min_{1 \le j \le N-1} U^m_{p,2}(j) = U^m_{p,2}(1) = U^m_{p,2}(N-1) = \left[\frac{N}{(N-1)+(N-1)^p}\right]^{1/p}. \tag{5.422}$$

From (5.421), we also easily conclude that

$$\min_{1 \le j \le N-1} U^m_{\infty,2}(j) = U^m_{\infty,2}(1) = U^m_{\infty,2}(N-1) = \frac{1}{N-1}.$$

Now consider $n > 2$. Note that for fixed j and increasing n, (5.419) is first increasing for $2 \le n \le N+1-j$, and ultimately a constant for $N+1-j \le n \le N$ with the maximal value $\frac{N-j}{N}$. Accordingly, we have

$$A^m_n(j) \ge A^m_2(j), \quad 1 \le j \le N-1, \quad 2 \le n \le N. \tag{5.423}$$

In particular, $A^m_n(1)$ is strictly increasing in n, and $A^m_n(N-1) = 1/N$ is a constant for $2 \le n \le N$. This implies

$$U^m_{p,n}(j) \ge U^m_{p,2}(j) \ge U^m_{p,2}(N-1) = U^m_{p,n}(N-1) \tag{5.424}$$

for arbitrary $1 \le p \le \infty$, $2 \le n \le N$ and $1 \le j \le N-1$. We get the first inequality once we multiply (5.423) by a positive term (5.418), independent of n. The other one has been established in (5.422). The equality follows from the fact that both the factors in $U^m_{p,n}(N-1)$ are independent of n. The last expression amounts to (5.420).

It suffices to show that at least one inequality in (5.424) is strict for $j \leq N-2$ and $n \geq 3$. If $p > 1$, then $U^m_{p,2}(j) > U^m_{p,2}(N-1)$ for $j = 2, \ldots, N-2$, and $A^m_n(1) > A^m_2(1)$ implies

$$U^m_{p,n}(1) > U^m_{p,2}(1) = U^m_{p,2}(N-1) = U^m_{p,n}(N-1).$$

If $p = 1$, the relation $U^m_{1,n}(j) = U^m_{1,n-1}(j)$ for some $j \leq N-2$, yields $A^m_n(j) = \frac{N-j}{N}$, and so

$$U^m_{1,n}(j) = \frac{N}{2j} > U^m_{1,n}(N-1) = \frac{N}{2(N-1)}.$$

This completes the proof. □

Bounds (5.420) are independent of the sample size n. As the population size N increases, they tend to 0 except for the case $p = 1$, with the limiting value $\frac{1}{2}$. Corollary 5.12 shows that respective evaluations for the sampling-with-replacement scheme are $1 - 1/N^{n-1}$ times smaller, and are attained by the same populations. For $p = 1$ and $p = 2$, we obtain simple forms

$$\mathbb{E}\left(\frac{X_{1:n} - \mu}{\sigma_p}\right) \leq \begin{cases} -\frac{N}{2(N-1)}, & p = 1, \\ -\frac{1}{\sqrt{N-1}}, & p = 2. \end{cases} \tag{5.425}$$

The latter inequality was first established by Balakrishnan et al. (2003). Their proof was different, and the main tool they used was non-negativity of the correlation of the maximum and minimum in the sample of size 2 drawn without replacement. This was concluded from Takahasi and Futatsuya (1998, Theorem 4.1) who proved that for arbitrary order statistics $X_{i:n}$ and $X_{j:n}$, $1 \leq i, j \leq n \leq N$, based on the samples drawn without replacement from any population of size N are nonnegatively correlated. The proof is based on the notion of positive likelihood ratio dependence.

Corollaries 5.35 and 5.37 allow us to get the following conclusion

Corollary 5.38 *In the without replacement model,*

$$\mathbb{E}\left(\frac{X_{n:n} - \mu}{\sigma_p}\right) \geq \begin{cases} \left[\frac{N}{(N-1)+(N-1)^p}\right]^{1/p}, & 1 \leq p < \infty, \\ \frac{1}{N-1}, & p = \infty. \end{cases}$$

The equality is attained if one element of the population is less than all the other identical ones. If $n = 2$ in particular, it suffices to assume that the population has exactly $N-1$ identical elements. When $n = 2$ and $p = 1$, the lower bounds are attained by arbitrary two-valued populations.

Now we present lower bounds for Gini mean differences

$$\mathrm{GMD}_n(\tilde{X}) = \frac{1}{n(n-1)} \sum_{i,j=1}^{n} |X_i - X_j| = \frac{2}{n(n-1)} \sum_{i=1}^{n} (2i-n-1) X_{i:n} \qquad (5.426)$$

(see also (5.152). To this end, we use the reversed coefficient vector of the L-statistic representation, which has the components

$$c_i' = c_{n+1-i} = \frac{2(n+1-2i)}{n(n-1)} = -c_i, \qquad i = 1, \ldots, n.$$

This immediately implies that $\bar{c'} = 0$, and as a consequence $\bar{C} = \frac{n}{N}\bar{c}' = 0$. Next, we determine

$$\begin{aligned} C_k &= \sum_{r=1}^{n} c_r' \, \mathbb{P}(W_{r:n} = k) = \sum_{r=1}^{n} \frac{2(n+1-2r)}{n(n-1)} \cdot \frac{\binom{k-1}{r-1}\binom{N-k}{n-r}}{\binom{N}{n}} \\ &= \sum_{s=0}^{n-1} \frac{2(n-1-2s)}{n(n-1)} \cdot \frac{\binom{k-1}{s}\binom{N-k}{n-1-s}}{\binom{N}{n}} \\ &= \sum_{s=0}^{n-1} \frac{2(n-1-2s)}{N(n-1)} \cdot \frac{\binom{k-1}{s}\binom{N-k}{n-1-s}}{\binom{N-1}{n-1}}, \qquad k = 1, \ldots, N. \end{aligned}$$

Note that these are the expectations of linear functions $\frac{2(n-1-2S_k)}{N(n-1)}$ of random variables S_k with the hypergeometric distributions $\mathcal{HG}(N-1, k-1, n-1)$, $k = 1, \ldots, N$, where S_k represents the number of white balls in a sample of size $n-1$ drawn without replacement from an urn with $k-1$ white and $N-k$ black balls. Therefore

$$C_k = \mathbb{E}\left(\frac{2(n-1-2S_k)}{N(n-1)}\right) = \frac{2(N+1-2k)}{N(N-1)}, \qquad k = 1, \ldots, N,$$

and so

$$\begin{aligned} A_{\bar{c}}(j) &= A^g(j) = \sum_{k=1}^{j} C_k = \sum_{k=1}^{j} \frac{2(N+1-2k)}{N(N-1)} = \frac{2j(N-j)}{N(N-1)}, \\ U_{\bar{c},p}(j) &= U_p^g(j) = A^g(j) B_p(j), \qquad j = 1, \ldots, N-1, \end{aligned} \qquad (5.427)$$

which do not depend on n. We can note that functions (5.427) that we minimize are identical with those of the deterministic model (see (3.309) and (3.310)) with $n = N$ considered in Corollary 3.13. Hence, we immediately obtain the following results.

Corollary 5.39

(i) If $1 \le p < \infty$, *then*

$$\mathbb{E}\left(\frac{\mathrm{GMD}_n(\tilde{X})}{\sigma_p}\right) \ge 2\left[\frac{N}{(N-1)+(N-1)^p}\right]^{1/p},$$

and the equality holds for (5.408) *with* $j = 1$ *and with* $j = N-1$ *for* $p > 1$, *and with arbitrary* $j = 1, \dots, N-1$ *for* $p = 1$.

(ii) If $p = \infty$, *then*

$$\mathbb{E}\left(\frac{\mathrm{GMD}_n(\tilde{X})}{\sigma_p}\right) \ge \frac{2}{N-1},$$

and the equality holds for (5.411) *with* $j = 1$ *and for* (5.412) *with* $j = N-1$.

We also observe that the bounds of Corollary 5.39 are $\frac{N}{N-1}$ times greater than the respective bounds in the drawing with replacement model presented in Corollary 5.14, and twice as much as those for the sample maxima determined in Corollary 5.38. The upper bounds are obviously positive as well and have the forms

$$\mathbb{E}\mathrm{GMD}_n(\tilde{X}) \le \begin{cases} 2\sigma_1, & \\ \sqrt{\frac{4(N+1)}{3(N-1)}}\,\sigma_2, & \\ \frac{N}{N-1}\sigma_\infty, & N \text{ even}, \\ \frac{N-1}{n}\sigma_\infty, & N \text{ odd}. \end{cases}$$

They are identical with the respective bounds in exhaustive drawing model from N-valued population (cf. (3.66), (3.175), and (3.197)).

We finally discuss the lower bounds in the case of sample ranges. The coefficient vector of the L-statistic $X_{n:n} - X_{1:n}$ is $\tilde{c} = (-1, 0, \dots, 0, 1)$. In order to find the lower positive bound, we use the reversed coefficient vector $\tilde{c}' = (1, 0, \dots, 0, -1)$. Obviously $\bar{c}' = 0$ which implies $\bar{C} = 0$, whereas

$$C_k = \mathbb{P}(W_{1:n} = k) - \mathbb{P}(W_{n:n} = k) = \frac{\binom{N-k}{n-1}}{\binom{N}{n}} - \frac{\binom{k-1}{n-1}}{\binom{N}{n}}, \quad k = 1, \dots, N. \tag{5.428}$$

Accordingly,

$$\begin{aligned} A_{\tilde{c}}(j) &= \sum_{k=1}^{j} C_k = 1 - \mathbb{P}(W_{1:n} > j) - \mathbb{P}(W_{n:n} \le j) \\ &= \frac{1}{\binom{N}{n}}\left[\binom{N}{n} - \binom{N-j}{n} - \binom{j}{n}\right] = A_n^r(j). \end{aligned} \tag{5.429}$$

Our purpose is to minimize the symmetric function

$$U_{p,n}^r(j) = A_n^r(j) \cdot B_p(j), \tag{5.430}$$

where $B_p(j)$ is defined in (5.418). In contrast to the previous examples, there are no general partitions of the two-valued populations (5.408), (5.411), or (5.412) that provides here the minimum (5.407) and (5.410) for all choices of N, n, p. It has been shown in Corollary 3.12 that for $n = N$ (which is the exhaustive drawing without replacement) and $p > 3$, optimal partitions j_* and $N - j_*$ of two-valued populations, depending on particular choices of p, may take arbitrary values in the set $\{1, \dots, N-1\}$. Corollary 5.15 shows that similar conclusions hold for the drawing with replacement model. We can expect the same for the without replacement case which for large population sizes N gives insignificantly different results. Numerical examples also confirm the hypothesis that for the drawing without replacement scheme, there are no general rules of partitioning two-valued populations (see (5.408), (5.411), and (5.412)) which provide the minimal values of the standardized expectation of the sample ranges. However, the formal verification of the hypothesis in this case is much more difficult. The reason is that (5.430) is the product of two expressions of different types, and the first one (5.429) is a combinatorial factor, which cannot be transformed into an analytic form, which would allow to use standard analytic tools of the optimization problems.

Therefore we focus only on the most interesting special cases of bounds for $p = 1, 2, \infty$. For $p = 1$ and $p = \infty$, we have some general conclusions.

Corollary 5.40 *For $p = 1$, we have*

$$\mathbb{E}\left(\frac{X_{n:n} - X_{1:n}}{\sigma_1}\right) \geq \begin{cases} \frac{N}{N-1}, & 2 = n \leq N, \\ \frac{3N}{2(N-1)}, & 3 = n \leq N, \\ 2 - 4\frac{\binom{N/2}{n}}{\binom{N}{n}}, & n \geq 4,\ N - even, \\ \frac{2N^2}{N^2-1}\left[1 - \frac{\binom{\frac{N-1}{2}}{n}}{\binom{N}{n}} - \frac{\binom{\frac{N+1}{2}}{n}}{\binom{N}{n}}\right] & n \geq 4,\ N - odd. \end{cases} \tag{5.431}$$

In the first two cases, the equality holds for (5.408) with $p = 1$ for every $j = 1, \dots, N-1$. In the last two cases, the equality is attained by (5.408) *with $p = 1$ when $j = N/2$ and $j = \frac{N-1}{2}, \frac{N+1}{2}$, respectively.*

Proof We minimize

$$U_{1,n}^r(j) = \frac{N^2}{2j(N-j)}\left[1 - \frac{\binom{N-j}{n}}{\binom{N}{n}} - \frac{\binom{j}{n}}{\binom{N}{n}}\right] = U_{1,n}^r(N-j), \quad j = 1, \dots, N-1. \tag{5.432}$$

First notice that for $n = 2$ and 3, functions

$$U^r_{1,2}(j) = \frac{N}{N-1}, \tag{5.433}$$

$$U^r_{1,3}(j) = \frac{3N}{2(N-1)}, \tag{5.434}$$

are constant and minimized by every $j = 1, \ldots, N-1$. For $n = 4$, we have

$$U^r_{1,4}(j) = \frac{N(j^2 - jN + 2N^2 - 9N + 11)}{(N-1)(N-2)(N-3)}, \tag{5.435}$$

which is a quadratic convex function symmetric about $N/2$. Therefore, it attains the minimum at $j = \frac{N}{2}$ if N is even, and at $j = \frac{N-1}{2}$ and $j = \frac{N+1}{2}$, if N is odd.

Cases $n \geq 5$ need more elaborate arguments. Let N be fixed. Under the notation

$$b(j) = \frac{N^2}{2j(N-j)},$$

$$c(j,n) = \frac{N^2\binom{j}{n}}{2j(N-j)\binom{N}{n}},$$

$$a(j,n) = c(j,n) + c(N-j,n),$$

we have

$$U^r_{1,n}(j) = b(j) - a(j,n) = b(j) - c(j,n) - c(N-j,n). \tag{5.436}$$

We consider the increments of the difference sequence $a(j,n) - a(j+1,n)$, when n increases given by

$$\begin{aligned}\Delta(j,n) &= [a(j,n) - a(j+1,n)] - [a(j,n+1) - a(j+1,n+1)] \\ &= c(j,n) + c(N-j,n) - c(j+1,n) - c(N-j-1,n) - c(j,n+1) \\ &\quad - c(N-j,n+1) + c(j+1,n+1) + c(N-j-1,n+1).\end{aligned}$$

Observing that

$$c(j,n+1) = \frac{j-n}{N-n}c(j,n),$$

$$c(N-j,n+1) = \frac{N-j-n}{N-n}c(N-j,n),$$

$$c(j+1, n+1) = \frac{j-n+1}{N-n}\, c(j+1, n),$$

$$c(N-j-1, n+1) = \frac{N-j-n-1}{N-n}\, c(N-j-1, n),$$

and writing $c_k = c(k, n)$ for brevity, we get

$$\Delta(j, n) = \frac{1}{N-n}\left[(N-j)c_j + j\, c_{N-j} - (N-j-1)c_{j+1} - (j+1)c_{N-j-1}\right].$$

Since

$$c_{j+1} = \frac{j(N-j)}{(N-j-1)(j-n+1)}\, c_j,$$

$$c_{N-j-1} = \frac{j(N-j-n)}{(j+1)(N-j-1)}\, c_{N-j},$$

we finally obtain

$$\begin{aligned}\Delta(j, n) &= \frac{1}{N-n}\left[-\frac{(n-1)(N-j)}{j-n+1}\, c_j + \frac{j(n-1)}{N-j-1}\, c_{N-j}\right] \\ &= \frac{N(n-1)}{2(N-1)\ldots(N-n)} \\ &\quad \times [(N-j-2)\ldots(N-j-n+1) - (j-1)\ldots(j-n+2)].\end{aligned}$$

The fraction in front of the brackets is positive. The expression inside the brackets is the difference of two products of consecutive $n-2$ integers arranged in the decreasing order. This is positive iff $N-j-2 > j-1$ and $N-j-n+1 > 0$, i.e., for $j < \frac{N-1}{2}$ and $j \leq N-n$. Likewise, this is negative iff $j > \frac{N-1}{2}$ and $j \geq n-1$, and amounts to zero in all the remaining cases. Accordingly, we can write

$$\Delta(j, n) \geq 0, \qquad j < \frac{N-1}{2},$$

and, equivalently,

$$a(j, n) - a(j+1, n) \geq a(j, n+1) - a(j+1, n+1), \qquad j < \frac{N-1}{2}. \tag{5.437}$$

Note that the differences

$$b(j) - b(j+1) = \frac{N^2(N-2j+1)}{2j(N-j)(j+1)(N-j-1)}$$

are independent of n and are positive iff $j < \frac{N-1}{2}$.

Now we are in a position to prove

$$U^r_{1,n}(j) > U^r_{1,n}(j+1), \qquad j < \frac{N-1}{2}, \tag{5.438}$$

for all $4 \leq n \leq N$. We apply the induction argument with respect to n. Using (5.435), we observe that the statement holds true for $n = 4$. Assume the same for some $n \geq 4$. By (5.436), it is equivalent to the claim

$$b(j) - b(j+1) > a(j,n) - a(j+1,n), \qquad j < \frac{N-1}{2}.$$

This combined with (5.437) yield

$$b(j) - b(j+1) > a(j,n+1) - a(j+1,n+1),$$

which coincides with

$$U^r_{1,n+1}(j) > U^r_{1,n+1}(j+1), \qquad j < \frac{N-1}{2},$$

and establishes (5.438) for all $n \geq 4$. This claim, together with the symmetry of (5.432) with respect to $\frac{N}{2}$, completes the proof of the fact that (5.432) is minimized either by $j = \frac{N}{2}$ or $j = \frac{N-1}{2}$ and $\frac{N+1}{2}$ in the cases of N being even and odd, respectively. □

We observe that for $n > \frac{N}{2}$ with even N, and for $n > \frac{N+1}{2}$ with N odd, the third and fourth bounds in (5.431) simplify to 2 and $\frac{2N^2}{N^2-1}$, respectively. They are attained by populations with two values, and the numbers of both types of elements are either equal or almost equal. The respective bounds in the maximal absolute mean deviation units σ_∞ are attained by the populations with only one element different from the others. The same populations have minimal expectations of ranges for the drawing with replacement models.

Corollary 5.41 *For* $p = \infty$,

$$\mathbb{E}\left(\frac{X_{n:n} - X_{1:n}}{\sigma_\infty}\right) \geq \frac{n}{N-1},$$

and the equality holds for (5.411) *when* $j = 1$ *and for* (5.412) *when* $j = N - 1$.

Proof We analyze

$$U^r_{\infty,n}(j) = A^r_n(j)\, B_\infty(j) = A^r_n(j) \frac{N}{\max\{j, N-j\}}, \qquad j = 1, \ldots, N-1,$$

where $A_n^r(j)$ is defined in (5.429). It is easy to see that sequence (5.428) is non-increasing, and so

$$2A_n^r(j) = 2\sum_{i=1}^{j} C_i \geq \sum_{i=1}^{j-1} C_i + \sum_{i=1}^{j+1} C_i = A_n^r(j-1) + A_n^r(j+1), \quad 2 \leq j \leq N-2,$$

which implies that $A_n^r(j) \geq \min\{A_n^r(j-1), A_n^r(j+1)\}$. Hence

$$\min_{1\leq j\leq N-1} A_n^r(j) = \min\{A_n^r(1), A_n^r(N-1)\} = A_n^r(1) = A_n^r(N-1) > 0, \qquad (5.439)$$

by symmetry. Also,

$$\min_{1\leq j\leq N-1} B_\infty(j) = B_\infty(1) = B_\infty(N-1) < \min_{2\leq j\leq N-2} B_\infty(j). \qquad (5.440)$$

Therefore

$$\begin{aligned}\min_{1\leq j\leq N-1} A_n^r(j)B_\infty(j) &\geq \min_{1\leq j\leq N-1} A_n^r(j) \min_{1\leq j\leq N-1} B_\infty(j)\\ &= A_n^r(1)B_\infty(1) = A_n^r(N-1)B_\infty(N-1) = \frac{n}{N-1}.\end{aligned}$$

By (5.439) and (5.440), these are the unique minimum points. □

Finally, we discuss the case $p = 2$. We have

$$\mathbb{E}\left(\frac{X_{n:n} - X_{1:n}}{\sigma_2}\right) \geq \min_{1\leq j\leq N-1} U_{2,n}^r(j),$$

where

$$U_{2,n}^r(j) = \frac{N}{\sqrt{j(N-j)}\binom{N}{n}}\left[\binom{N}{n} - \binom{j}{n} - \binom{N-j}{n}\right]. \qquad (5.441)$$

Numerical evaluations for various n and N show that (5.441) is minimized by $j = 1, N-1$ if n is small, and by $j = \frac{N}{2}$ (or $j = \frac{N\pm1}{2}$ for odd-sized populations) if n is large, and there are no other cases. The same statement has been proved in Corollary 5.18 for the sampling with replacement model. The formal proof in the without replacement case is an open problem. When replacing is allowed, the combinatorial part is replaced by a polynomial of degree n in variable $\frac{j}{N}$, which makes the analysis easier. Our conjecture for the without replacement scheme can be supported by partial results. For instance, using

$$U_{2,n}^r(j) = 2\frac{\sqrt{j(N-j)}}{N} U_{1,n}^r(j),$$

by constancy of $U^r_{1,2}(j)$ and $U^r_{1,3}(j)$ (cf. (5.433) and (5.434)), we obtain

$$\mathbb{E}\left(\frac{X_{2:2}-X_{1:2}}{\sigma_2}\right) \geq \min_{1\leq j\leq N-1} 2\frac{\sqrt{j(N-j)}}{N}\frac{N}{N-1} = \frac{2}{\sqrt{N-1}},$$
$$\mathbb{E}\left(\frac{X_{3:3}-X_{1:3}}{\sigma_2}\right) \geq \min_{1\leq j\leq N-1} 2\frac{\sqrt{j(N-j)}}{N}\frac{3N}{2(N-1)} = \frac{3}{\sqrt{N-1}}.$$

The first inequality was derived by Balakrishnan et al. (2003). Likewise, we show

$$\mathbb{E}\left(\frac{X_{2:2}-X_{1:2}}{\sigma_p}\right) \geq 2\left[\frac{N}{(N-1)^p+N-1}\right]^{1/p},$$
$$\mathbb{E}\left(\frac{X_{3:3}-X_{1:3}}{\sigma_p}\right) \geq 3\left[\frac{N}{(N-1)^p+N-1}\right]^{1/p},$$

for arbitrary $p > 1$. These bounds are attained by the populations where only one element differs from the others. We can finally recall the results of Corollary 3.12 which represent the bounds in the exhaustive drawing without replacement model with $N = n$.

5.3.6 Bounds on Variances of Order Statistics

Now we proceed to the bounds on variances of order statistics in the drawing without replacement model. We assume that we have a population with N real valued elements $x_{1:N} \leq \ldots \leq x_{N:N} > x_{1:N}$. We randomly draw without replacement $2 \leq n < N$ elements, and denote them by $X_1, \ldots X_n$. Note that unlike in the drawing with replacement scheme, admitting $n > N$ does not make sense. For $n = N$, the order statistics take on deterministic values, and evaluating their variances does not make sense, either. If all $x_{1:N}, \ldots, x_{N:N}$ are distinct, then

$$\mathbb{P}(X_{k:n} = x_{i:N}) = p_{k:n}(i) = \frac{\binom{i-1}{k-1}\binom{N-i}{n-k}}{\binom{N}{n}}, \qquad i = 1, \ldots, N, \tag{5.442}$$

under the convention that $\binom{a}{b} = 0$ for $a < b$ (comp. (5.308)). This means that $p_{k:n}(i) = 0$ for $i = 1, \ldots, k-1, N-n+k+1, \ldots, N$, which confirms the intuition that the kth minimal and $(n+1-k)$th maximal element of the sample cannot be less than kth minimal element and greater than $(n+1-k)$th maximal element of the population, respectively. Then we also have

$$\mathbb{P}(X_{k:n} \leq x_{i:N}) = P_{k:n}(i) = \sum_{j=1}^{i} p_{k:n}(j), \qquad i = 1, \ldots, N \tag{5.443}$$

If some repetitions in the population are admitted, then the distribution functions of X_1 and $X_{k:n}$ still have the forms

$$\mathbb{P}(X_1 \le x) = \frac{1}{N}\sum_{i=1}^{N} \mathbf{1}_{[x_i,\infty)}(x), \tag{5.444}$$

$$\begin{aligned}\mathbb{P}(X_{k:n} \le x) &= P_{k:n}\left(\sum_{i=1}^{N} \mathbf{1}_{[x_i,\infty)}(x)\right) = \sum_{i=1}^{N} p_{k:n}(i)\mathbf{1}_{[x_i,\infty)}(x) \\ &= \sum_{i=1}^{N} P_{k:n}(i)\mathbf{1}_{[x_i,x_{i+1})}(x),\end{aligned} \tag{5.445}$$

with the convention $x_{N+1} = +\infty$.

Lemma 5.12

(i) Sequence $p_{1:n}(i)$ is positive decreasing for $i = 1, \ldots, N - n + 1$, and equals zero for $i = N - n + 2, \ldots, N$.
(ii) Sequence $p_{n:n}(i) = 0$ for $i = 1, \ldots, n - 1$, and is positive increasing for $i = n, \ldots, N$.
(iii) For $2 \le k \le n - 1$, $p_{k:n}(i)$ is positive, first increasing and then decreasing for $i = k, \ldots, N - n + k$, and is equal zero otherwise. If $N\frac{k-1}{n-1}$ is not integer, the $p_{k:n}(i)$ attains its unique maximum at $M = \lceil N\frac{k-1}{n-1}\rceil$. Otherwise, it has two neighboring maximizing arguments $\underline{M} = N\frac{k-1}{n-1}$ and $\overline{M} = N\frac{k-1}{n-1} + 1$.

The conclusions are immediate consequences of the relations of the ratios

$$\frac{p_{k:n}(i+1)}{p_{k:n}(i)} = \frac{i(N-n+k-i)}{(i+1-k)(N-i)}, \qquad i = k, N-n+k-1,$$

with 1, which allows to compare the values of $p_{k:n}(i)$ and $p_{k:n}(i+1)$. For convenience, we also use the notation $\underline{M} = \overline{M} = M$ if $N\frac{k-1}{n-1}$ is not integer. We further define

$$\underline{Q}_{k:n}(i) = \frac{P_{k:n}(i)}{i}, \qquad 1 \le i \le N-1, \tag{5.446}$$

$$\overline{Q}_{k:n}(i) = \frac{1 - P_{k:n}(i)}{N-i}, \qquad 1 \le i \le N-1, \tag{5.447}$$

$$\mathbf{Q}_{k:n}(i,j) = \underline{Q}_{k:n}(i)\,\overline{Q}_{k:n}(j), \qquad 1 \le i \le j \le N-1, \tag{5.448}$$

$$Q_{k:n}(i) = \mathbf{Q}_{k:n}(i,i), \qquad 1 \le i \le N-1 \tag{5.449}$$

(comp. (5.171) to (5.174)). The following three lemmas play the roles analogous to Lemmas 5.3–5.5 in the drawing with replacement case.

Lemma 5.13

(i) *Sequence* $\underline{Q}_{1:n}(i)$, $i = 1, \ldots, N-1$, *is positive decreasing.*
(ii) *Sequence* $\underline{Q}_{n:n}(i) = 0$ *for* $i = 1, \ldots, n-1$, *and is positive increasing for* $i = n, \ldots, N-1$.
(iii) *If* $2 \le k \le n-1$, *then* $\underline{Q}_{k:n}(i)$, $i = 1, \ldots, N-1$, *is first nondecreasing and then nonincreasing. Precisely, it amounts to zero for* $i = 1, \ldots, k-1$, *is positive increasing for* $i = k, \ldots, \overline{M}$, *and positive decreasing for* $i = N - n + k, \ldots, N-1$. *For* $i = \overline{M}, \ldots, N-n+k$ *it is unimodal, and attains its (global) maximum either at a single point* $i_2 = i_2(k, n, N)$ *or two neighboring points* $\underline{i}_2 = \underline{i}_2(k, n, N)$ *and* $\overline{i}_2 = \underline{i}_2 + 1$.

Unimodality means here that the sequence is either monotone or increasing-decreasing. For convenience, we write $\underline{i}_2 = \overline{i}_2 = i_2$ in the case of unique maximum.

Proof It is based on one simple observation. If we construct a sequence of averages $\bar{a}_i = \frac{1}{i}\sum_{j=1}^{i} a_j$, $i = 1, 2, \ldots$, by including new elements of a_i, $i = 1, 2, \ldots$, into the averages, then due to the relations

$$\bar{a}_{i+1} = \frac{i}{i+1}\bar{a}_i + \frac{1}{i+1}a_{i+1},$$

we observe that we increase (decrease or just preserve) the value of the average iff the new element is greater than (less than or equal to, respectively) the hitherto average.

(i) Note that (5.446) are constructed as the averages of consecutive elements of (5.442). For $k = 1$, the first $N - n + 1$ elements of (5.442) are positive decreasing, which means that each partial average of them is greater than the succeeding element. Therefore the partial averages decrease then. For $i \ge N - n + 1$, we have $0 < \bar{a}_i > a_{i+1} = 0$, which allows us to notice that the whole $\underline{Q}_{1:n}(i)$, $i = 1, \ldots, N-1$, is strictly decreasing.
(ii) In the case $k = n$, the first $n - 1$ values of $p_{k:n}(i)$ are equal to zero, and so are their means $\underline{Q}_{k:n}(i)$, $i = 1, \ldots, n-1$. For $i \ge n$, we have $p_{k:n}(i) > \max\{p_{k:n}(1), \ldots, p_{k:n}(i-1)\}$, which implies $\underline{Q}_{k:n}(i) > \underline{Q}_{k:n}(i-1)$.
(iii) By similar arguments, we deduce that for $2 \le k \le n-1$, the subsequence $\underline{Q}_{k:n}(i)$, $i = 1, \ldots, \overline{M}$, is first equal to 0 and then positive increasing. We certainly have $\underline{Q}_{k:n}(\overline{M}) < p_{k:n}(\overline{M})$, and so there can be any relation between $\underline{Q}_{k:n}(\overline{M})$ and $p_{k:n}(\overline{M}+1)$. The same we can say about the mutual relation between $\underline{Q}_{k:n}(\overline{M})$ and $\underline{Q}_{k:n}(\overline{M}+1)$. We observe further increase of $\underline{Q}_{k:n}(i)$ as long as it is still less than $p_{k:n}(i+1)$. However, if $\underline{Q}_{k:n}(i) \ge p_{k:n}(i+1)$, and so $\underline{Q}_{k:n}(i) \ge \underline{Q}_{k:n}(i+1)$, then we also obtain $\underline{Q}_{k:n}(i+1) > p_{k:n}(i+2)$ and $\underline{Q}_{k:n}(i+1) > \underline{Q}_{k:n}(i+2)$. The sequence $\underline{Q}_{k:n}(i)$ is definitely decreasing for $i = N-n+k, \ldots, N-1$, because then $0 < \underline{Q}_{k:n}(i) > p_{k:n}(i+1) = 0$.

□

By similar arguments, we obtain the following lemma.

Lemma 5.14

(i) *Sequence* $\overline{Q}_{1:n}(i)$, *is positive decreasing for* $i = 1, \dots, N-n$, *and vanishes for* $i = N-n+1, \dots, N-1$.
(ii) *Sequence* $\overline{Q}_{n:n}(i)$, $i = 1, \dots, N-1$, *is positive increasing.*
(iii) *If* $2 \le k \le n-1$, *then* $\overline{Q}_{k:n}(i)$, $i = 1, \dots, N-1$, *is first nondecreasing and then nonincreasing. Precisely, it is positive increasing for* $i = 1, \dots, k$, *positive decreasing for* $i = \underline{M} \dots, N-n+k$, *and equal to zero for* $i = N-n+k+1, \dots, N-1$. *For* $i = k, \dots, \underline{M}$, *it is unimodal, and attains its (global) maximum either at a single point* $i_1 = i_1(k,n,N)$ *or two neighboring points* $\underline{i}_1 = \underline{i}_1(k,n,N)$ *and* $\overline{i}_1 = \underline{i}_1 + 1$.

If the maximum point is unique, we write $\overline{i}_1 = i_1$. Note that $k \le \overline{i}_1(k,n,N) \le \underline{M}(k,n,N) \le \overline{M}(k,n,N) \le \underline{i}_2(k,n,N) \le N-n+k$.

Lemma 5.15

(i) *The triangular array* $\mathbf{Q}_{1:n}(i,j)$, $1 \le i \le j \le N-1$, *attains its unique maximum at the vertical point* $(1,1)$.
(ii) $\mathbf{Q}_{n:n}(i,j)$, $1 \le i \le j \le N-1$, *attains its unique maximum at* $(N-1, N-1)$.
(iii) *If* $2 \le k \le n-1$, $\mathbf{Q}_{k:n}(i,j)$, $1 \le i \le j \le N-1$, *attains its maximum* $\mathbf{Q}_{k:n}(i,i) = Q_{k:n}(i)$ *at a diagonal point* (i,i) *for* $\overline{i}_1(k,n,N) \le i \le \underline{i}_2(k,n,N)$.

Proof The idea of proof is analogous to the proof of Lemma 5.5. The elements of the array $\mathbf{Q}_{k:n}(i,j)$, $1 \le i \le j \le N-1$, are the products of the nonnegative elements $\underline{Q}_{k:n}(i)$ and $\overline{Q}_{k:n}(j)$ of the sequences depending on different arguments i and j. For $k = 1$ both the sequences are nonincreasing, and hence their product is maximized at $(i,j) = (1,1)$. If $k = n$, the sequences are nondecreasing, and the pair $(N-1, N-1)$ provides the maximum of the product. For $2 \le k \le n-1$ both sequences $\underline{Q}_{k:n}(i)$ and $\overline{Q}_{k:n}(j)$ are unimodal. The respective maximum arguments $\underline{i}_2 \le \overline{i}_2$ (possibly identical), and $\underline{i}_1 \le \overline{i}_1$ (possibly identical as well) satisfy $\overline{i}_1 \le \underline{M} \le \overline{M} \le \underline{i}_2$. If $i < \underline{i}_1$ and $j > \overline{i}_2$, then $\mathbf{Q}_{k:n}(i,j) \le \mathbf{Q}_{k:n}(\underline{i}_1, \overline{i}_2)$ (comp. (5.181)). If $\underline{i} \le i < j$, then $\mathbf{Q}_{k:n}(i,j) \le Q_{k:n}(i)$ (comp. (5.179)). If $i < j \le \overline{i}_2$, then $\mathbf{Q}_{k:n}(i,j) \le Q_{k:n}(j)$ (see (5.180)). This shows that the array has the maximum at some diagonal point. Note finally that both $\underline{Q}_{k:n}(i)$ and $\overline{Q}_{k:n}(i)$ are nondecreasing for $i \le \overline{i}_1$ and nonincreasing for $i \ge \underline{i}_2$ which imply that while looking for the array maxima it suffices to restrict ourselves to the arguments $\overline{i}_1 \le i \le \underline{i}_2$. □

Numerical examples show that the diagonal sequences $Q_{k:n}(i)$, $i = 1, \dots, N-1$, are unimodal for $2 \le k \le n-1$, and attain their maxima either at a single point $i_0(k,n,N)$ (very often) or two neighboring integers (very seldom). However, we do not have a formal proof of this hypothesis. Theorem 5.18 asserts that the maximal values of the arrays $\mathbf{Q}_{k:n}$ provide optimal upper bounds for the variances of respective order statistics $X_{k:n}$ from the drawing without replacement model.

Theorem 5.18 *Let $X_1, \ldots, X_n$, $n \geq 2$, denote the random outcomes of consecutive drawings without replacement from a non-constant population $x_{1:N} \leq \ldots \leq x_{N:N} > x_{1:N}$ for some $N \geq n+1$.*

(i) For $k = 1$, we have

$$\frac{\mathbb{V}ar\, X_{1:n}}{\mathbb{V}ar\, X_1} \leq N^2\, Q_{1:n}\,(1) = n\,\frac{N-n}{N-1}, \tag{5.450}$$

with the equality for $x_{1:N} < x_{2:N} = \ldots = x_{N:N}$.

(ii) For $k = n$,

$$\frac{\mathbb{V}ar\, X_{n:n}}{\mathbb{V}ar\, X_1} \leq N^2\, Q_{n:n}\,(N-1) = n\,\frac{N-n}{N-1}, \tag{5.451}$$

with the equality for $x_{1:N} = \ldots = x_{N-1:N-1} < x_{N:N}$.

(iii) For $2 \leq k \leq n-1$,

$$\frac{\mathbb{V}ar\, X_{k:n}}{\mathbb{V}ar\, X_1} \leq N^2 \max_{\bar{i}_1(k,n,N) \leq i \leq \underline{i}_2(k,n,N)} Q_{k:n}\,(i) = Q_{k:n}\,(i_0) \tag{5.452}$$

say, for some $i_0 = i_0(k, n, N)$, with the equality for $x_{1:N} = \ldots = x_{i_0:N} < x_{i_0+1:N} = \ldots = x_{N:N}$.

Assertions (i) and (ii) can be found in Čiginas and Pumputis (2014). In contrast with the drawing with replacement case, the populations with $N-1$ identical values and a single different value from the rest attain the variance bounds only for the sample extremes. Since $\mathbf{Q}_{n+1-k:n}(i, j) = \mathbf{Q}_{k:n}(N-j, N-i)$, we easily deduce that the bounds are identical for the kth smallest and largest order statistics. The claim is apparently confirmed by the statements (i) and (ii) of the theorem.

Proof Due to (5.187) and (5.445),

$$\begin{aligned} \mathbb{V}ar\, X_{k:n} = & \sum_{i,j=1}^{N-1} P_{k:n}\,(\min\{i, j\})\,[1 - P_{k:n}\,(\max\{i, j\})] \\ & \times (x_{i+1:N} - x_{i:N})(x_{j+1:N} - x_{j:N}). \end{aligned} \tag{5.453}$$

This can be evaluated as follows:

$$\begin{aligned} \mathbb{V}ar\, X_{k:n} = & \sum_{i,j=1}^{N-1} \frac{P_{k:n}\,(\min\{i, j\})\,[1 - P_{k:n}\,(\max\{i, j\})}{\frac{\min\{i,j\}}{N}\left[1 - \frac{\max\{i,j\}}{N}\right]} \\ & \times \frac{\min\{i, j\}}{N}\left[1 - \frac{\max\{i, j\}}{N}\right](x_{i+1:N} - x_{i:N})(x_{j+1:N} - x_{j:N}) \end{aligned}$$

$$= N^2 \sum_{i,j=1}^{N-1} \underline{Q}_{k:n}(\min\{i,j\})\overline{Q}_{k:n}(\max\{i,j\})$$

$$\times \frac{\min\{i,j\}}{N}\left[1-\frac{\max\{i,j\}}{N}\right](x_{i+1:N}-x_{i:N})(x_{j+1:N}-x_{j:N})$$

$$\leq N^2 \max_{1\leq i\leq j\leq N-1} \underline{Q}_{k:n}(i)\overline{Q}_{k:n}(j)$$

$$\times \sum_{i,j=1}^{N-1} \frac{\min\{i,j\}}{N}\left[1-\frac{\max\{i,j\}}{N}\right](x_{i+1:N}-x_{i:N})(x_{j+1:N}-x_{j:N})$$

$$= N^2 \max_{1\leq i\leq j\leq N-1} \mathbf{Q}_{k:n}(i,j)\, \mathbb{V}ar\, X_1 \tag{5.454}$$

(cf. (5.188)). The above maxima for fixed k, n, and N has been described in Lemma 5.15. They coincide with $\max_{1\leq i\leq N-1} Q_{k:n}(i)$. For the extreme order statistics, the maximal values can be determined explicitly.

The equality in (5.454) is attained iff for every $1 \leq i \leq j \leq N-1$,

$$\text{either } \mathbf{Q}_{k:n}(i,j) = \max_{1\leq i\leq j\leq N-1} \mathbf{Q}_{k:n}(i,j) \ \text{ or } (x_{i+1:N}-x_{i:N})(x_{j+1:N}-x_{j:N}) = 0. \tag{5.455}$$

Take $i_0 = i_0(k,n,N)$ satisfying

$$Q_{k:n}(i_0) = \mathbf{Q}_{k:n}(i_0,i_0) = \max_{1\leq i\leq N-1} Q_{k:n}(i) = \max_{1\leq i\leq j\leq N-1} \mathbf{Q}_{k:n}(i,j) \tag{5.456}$$

and arbitrary two different numbers $x_{1:N} < x_{N:N}$. Construct the population $x_{1:N} = \ldots = x_{i_0:N} < x_{i_0+1:N} = \ldots = x_{N:N}$. We immediately check that i_0 and the population satisfy the conditions (5.455). Consequently, the drawing without replacement scheme from this two-valued population attains the variance bound (5.454). □

We finally present a simple explicit variance estimate for the sample medians.

Theorem 5.19 *Under the assumptions of Theorem* 5.18, *with n odd, we have*

$$\frac{\mathbb{V}ar\, Y_{\frac{n+1}{2}:n}}{\mathbb{V}ar\, Y_1} \leq 1. \tag{5.457}$$

If N is even, the equality is attained by drawing without replacement from arbitrary population satisfying $x_{1:N} = \ldots = x_{N/2:N} < x_{N/2+1:N} = \ldots = x_{N:N}$.

The proof is based on Theorem 1 of Lin and Huang (1989) presented below.

Lemma 5.16 *Suppose that F is an arbitrary distribution function, and* G_1, G_2 *are two continuous distribution functions on* $[0,1]$, *symmetric about* $\frac{1}{2}$ *and such that* *the ratio* $\frac{G_2}{G_1}$ *is nondecreasing on* $\left(0, \frac{1}{2}\right)$. *If random variables* Z_i *have distribution functions* $G_i \circ F$, $i = 1, 2$, *respectively and finite variances, then* $\mathbb{V}ar\, Z_1 \geq \mathbb{V}ar\, Z_2$.

We do not rewrite the proof of the lemma. It is based on discrete approximations of distribution functions $G_i \circ F$ and their variances. The latter are represented as the double finite sums, and the inequality of Yang (1982) is then used for comparing them (see also Lin & Huang, 1989, Lemma 2).

Proof of Theorem 5.19 We fix

$$F(x) = \frac{1}{N} \sum_{k=1}^{N} \mathbf{1}_{[x_k, \infty)}(x), \qquad x \in \mathbb{R}, \tag{5.458}$$

$$G_1(u) = u, \qquad 0 < u < 1, \tag{5.459}$$

$$G_2(u) = N p_{\frac{n+1}{2}:n}(i+1)\left(u - \frac{i}{N}\right) + P_{\frac{n+1}{2}:n}(i),$$
$$\frac{i}{N} \leq u \leq \frac{i+1}{N}, \qquad i = 1, \ldots, N-1. \tag{5.460}$$

Certainly G_1 satisfies the assumptions of Lemma 5.16, and $G_1 \circ F = F$ is the distribution function of X_1. Also,

$$G_2 \circ F(x) = \sum_{k=1}^{N} p_{\frac{n+1}{2}:n}(i) \mathbf{1}_{[x_k, \infty)}(x)$$

is the distribution function of $X_{\frac{n+1}{2}:n}$. Since $p_{\frac{n+1}{2}:n}(i) = p_{\frac{n+1}{2}:n}(N-i)$, we also have $P_{\frac{n+1}{2}:n}(i) = 1 - P_{\frac{n+1}{2}:n}(N-i)$, $i = 1, \ldots, N-1$. Accordingly, the pairs of points $\left(\frac{i}{N}, P_{\frac{n+1}{2}:n}(i)\right)$ and $\left(1 - \frac{i}{N}, P_{\frac{n+1}{2}:n}(N-i)\right)$ are located symmetrically about $\left(\frac{1}{2}, \frac{1}{2}\right)$. Function G_2 is the linear interpolation of points $\left(\frac{i}{N}, P_{\frac{n+1}{2}:n}(i)\right)$, $i = 0, \ldots, N$. Hence it is a continuous distribution function on $[0, 1]$, symmetric about $\frac{1}{2}$.

It remains to check that

$$\frac{G_2(u)}{G_1(u)} = N p_{\frac{n+1}{2}:n}(i+1) + \frac{i}{u}\left[Q_{\frac{n+1}{2}:n}(i) - p_{\frac{n+1}{2}:n}(i+1)\right], \quad \frac{i}{N} \leq u \leq \frac{i+1}{N}, \tag{5.461}$$

is nondecreasing for $0 < u < \frac{1}{2}$. Observe that sequence $p_{\frac{n+1}{2}:n}(i)$ is unimodal, with either one maximum at $\frac{N}{2}$ if N is even, or two maxima at $\frac{N-1}{2}$ and $\frac{N-1}{2}$ if N is odd. Therefore $Q_{\frac{n+1}{2}:n}(i) \leq p_{\frac{n+1}{2}:n}(i+1)$ for $i = 1, \ldots, \frac{N}{2}-1$ and $i = 1, \ldots, \frac{N-1}{2}$ in the even and odd cases, respectively. It follows that the expression inside the brackets in (5.461) is either zero or negative, and the ratio $\frac{G_2(u)}{G_1(u)}$ is nondecreasing for $u < \frac{1}{2}$ and $u < \frac{N+1}{2N}$ in the even and odd cases, respectively. Lemma 5.16 implies desired inequality (5.457).

If N is even and $x_1 = \ldots = x_{\frac{N}{2}} < x_{\frac{N}{2}+1} = \ldots = x_N$, then $P_{\frac{n+1}{2}:n}\left(\frac{N}{2}\right) = 1 - P_{\frac{n+1}{2}:n}\left(\frac{N}{2}\right) = \frac{1}{2}$. We have

$$N^2 \max_{1\le i\le j\le N-1} \mathbf{Q}_{\frac{n+1}{2}:n}(i, j) \le 1 = N^2\mathbf{Q}_{\frac{n+1}{2}:n}\left(\frac{N}{2}, \frac{N}{2}\right)$$

by Theorem 5.18 and (5.457). We see that the unit bound is sharp and it is attained by the two-valued populations with equal numbers of elements of subpopulations.

□

Numerical examples suggest the following bounds for the odd-sized populations

$$\frac{\mathbb{V}ar\ X_{\frac{n+1}{2}:n}}{\mathbb{V}ar\ X_1} \le N^2\, Q_{\frac{n+1}{2}}\left(\frac{N-1}{2}\right) = N^2\, Q_{\frac{n+1}{2}}\left(\frac{N+1}{2}\right), \tag{5.462}$$

but the claim needs a formal proof.

As the sample size n remains fixed and the population size N increases, the joint distributions of the samples coming from the with and without replacement models become more and more similar to each other. One can expect that the sharp bounds on the variances of the order statistics become close as well. Since those from the with replacement model tend to the general bounds in the iid case, we can suspect the same in the without replacement model as well. The hypothesis is informally confirmed by the examples of sample extremes and medians, for which the bounds in the without replacement scheme are presented explicitly in Theorems 5.18 and 5.19, respectively.

5.4 Numerical Evaluations

In this section, we present exemplary numerical values of sharp upper bounds for expected order statistics coming from finite populations. We consider the models of sampling with and without replacement. In both the cases, we assume that the population size is $N = 20$, and consider all the order statistics based on the samples of sizes $n = 5, 10, 15$, and 20. We evaluate the differences $\mathbb{E}X_{j:n} - \mu$, $j = 1, \ldots, n$, in three scale units σ_p, $p = 1, 2, \infty$, being the mean absolute deviation from the mean, standard deviation, and maximal distance from the support ends to the mean, respectively. In all the cases, we present numerical approximations of the bounds with the accuracy rounded up to 6 decimals.

Table 5.1 Sharp upper bounds for $\mathbb{E}\frac{X_{j:n}-\mu}{\sigma_p}$, $p = 1$, for the drawing with replacement models with population size $N = 20$, and sample sizes $n = 5, 10, 15, 20$, and for $\mathbb{E}\frac{X_{j:20}-\mu}{\sigma_p}$, $p = 1$, for iid samples with arbitrary marginal distribution

j	$n = 5$	$n = 10$	$n = 15$	$n = 20$	A, $n = 20$
1	−0.526313	−0.526316	−0.526316	−0.526316	−0.500000
2	0.288500	−0.380404	−0.526316	−0.526316	0.500682
3	0.586219	0.405226	0.145261	−0.268249	0.506603
4	0.828500	0.548556	0.470019	0.358931	0.517952
5	2.500000	0.613945	0.545791	0.505830	0.533726
6		0.696177	0.586315	0.546011	0.553620
7		0.828124	0.628885	0.574993	0.577782
8		1.054847	0.682128	0.604882	0.606694
9		1.560476	0.754127	0.639381	0.641171
10		5.000000	0.849121	0.680027	0.682429
11			0.978403	0.728333	0.732238
12			1.181862	0.790214	0.793208
13			1.527824	0.868412	0.869296
14			2.27138	0.96674	0.966774
15			7.500000	1.093001	1.096225
16				1.270820	1.277098
17				1.549688	1.549836
18				1.984788	2.016589
19				3.041265	3.048067
20				10.000000	10.000000

We first present the results for the drawing with replacement scheme. They are accompanied by the analogous bounds for order statistics $X_{j:20}$ coming from iid samples with arbitrary parent distribution. These are contained in the last columns of the respective tables. The letter A standing there is the abbreviation for the parent distributions with an arbitrary shape of the iid sample. They merely need to satisfy moment conditions: finiteness of the first and second moments in Tables 5.1 and 5.2, respectively, and boundedness of support for Table 5.3. These general bounds coincide with the limiting values of the bounds from the with replacement and without replacement models as the population size N tends to $+\infty$. Tables 5.1, 5.2, and 5.3 contain the bounds gauged in units σ_1, σ_2 and σ_∞, respectively.

In the first one, the bounds for $j = 1$ are concluded from Corollary 5.12 (see (5.150)). The corresponding bound for general population is based on Corollary 4.15. Negative bounds for $j = 2$ with $n = 10, 15, 20$ and $j = 3$ with $n = 20$ are directly based on Theorem 5.5. Positive bounds are evaluated with use of Corollary 5.1. The explicit bounds for sample maxima are presented in formula (5.84). The positive bounds in general populations come from Corollary 4.12.

Table 5.2 Sharp upper bounds for $\mathbb{E}\frac{X_{j:n}-\mu}{\sigma_p}$, $p = 2$, for the drawing with replacement models with population size $N = 20$, and the samples sizes $n = 5, 10, 15, 20$, and for $\mathbb{E}\frac{X_{j:20}-\mu}{\sigma_p}$, $p = 2$, for iid samples with arbitrary marginal distribution

j	$n = 5$	$n = 10$	$n = 15$	$n = 20$	A, $n = 20$
1	−0.229414	−0.229416	−0.229416	−0.229416	0.000000
2	0.125754	−0.165814	−0.229416	−0.229416	0.030013
3	0.374704	0.176634	0.063318	−0.116927	0.101396
4	0.682296	0.303746	0.204593	0.156455	0.172095
5	1.333333	0.431450	0.294431	0.224365	0.239363
6		0.569969	0.381462	0.297714	0.304444
7		0.732325	0.469848	0.364942	0.368713
8		0.942619	0.561069	0.430460	0.433409
9		1.265402	0.661801	0.497104	0.499702
10		2.064742	0.774209	0.566328	0.568805
11			0.907258	0.639447	0.642094
12			1.073396	0.718213	0.721266
13			1.299328	0.805513	0.808559
14			1.660310	0.904430	0.907144
15			2.599735	1.019192	1.021830
16				1.156497	1.160531
17				1.333364	1.337737
18				1.577955	1.584500
19				1.976687	1.988132
20				3.042435	3.042435

The positive bounds in the standard deviation units presented in Table 5.2 are deduced from Corollary 5.3, and the explicit formula for the sample maxima are presented in (5.84). The formula (5.151) provides the negative bounds for the sample minima. The remaining negative bounds are determined with the use of Theorem 5.5. For the general population with finite variances, the zero bounds for the sample maxima are derived in Theorem 4.19. For the sample maxima, we use the classic Gumbel and Hartley-David formula (4.20). The remaining values are based on Theorem 4.8.

The positive bounds for the order statistics in the drawing with replacement model in Table 5.3 are established with use of (5.84) for the maxima and Corollary 5.4 otherwise. The values from the first row are based on (5.143) in Corollary 5.12. The other negative bounds are calculated on the basis of Theorem 5.6. The corresponding evaluations in the general case come from Theorem 4.19 for $j = 1$, and Corollary 4.14 otherwise.

As we can expect, the bounds are increasing with respect to j, and decreasing with respect to n. They also decrease as p increases. However, the strict monotonicity is not observed if we take into account the approximations of the bounds up to

Table 5.3 Sharp upper bounds for $\mathbb{E}\frac{X_{j:n}-\mu}{\sigma_p}$, $p=\infty$, for the drawing with replacement models with population size $N=20$, and the samples sizes $n=5, 10, 15, 20$, and for $\mathbb{E}\frac{X_{j:20}-\mu}{\sigma_p}$, $p=\infty$, for iid samples with arbitrary marginal distribution

j	$n=5$	$n=10$	$n=15$	$n=20$	A, $n=20$
1	−0.052631	−0.052632	−0.052632	−0.052632	0.000000
2	0.028850	−0.038040	−0.052632	−0.052632	0.001364
3	0.195600	0.040523	0.014526	−0.026825	0.013207
4	0.625000	0.117683	0.049383	0.035891	0.035904
5	0.937500	0.229164	0.103876	0.063139	0.067452
6		0.392409	0.173686	0.098608	0.107240
7		0.656250	0.257840	0.150665	0.155563
8		0.890625	0.364257	0.209822	0.213389
9		0.978516	0.508254	0.278767	0.282343
10		0.998047	0.698242	0.360054	0.364858
11			0.881531	0.456667	0.464476
12			0.964844	0.580427	0.586417
13			0.992615	0.736824	0.738592
14			0.999023	0.884682	0.884682
15			0.999970	0.958611	0.958611
16				0.988182	0.988182
17				0.997423	0.997423
18				0.999598	0.999598
19				0.999960	0.999960
20				0.999999	0.999999

six decimals presented here. For instance, for the sample minima with $N=20$, we have

$$\mathbb{E}\frac{X_{1:n}-\mu}{\sigma_p} \le B(1,n,20,p) = \begin{cases} -\frac{10}{19}[1-20^{1-n}] \approx \frac{10}{19}, & p=1, \\ -\frac{1}{\sqrt{19}}[1-20^{1-n}] \approx \frac{1}{\sqrt{19}}, & p=2, \\ -\frac{1}{19}[1-20^{1-n}] \approx \frac{1}{19}, & p=\infty \end{cases} \tag{5.463}$$

(cf. (5.142) and (5.143) in Corollary 5.12). We see that they decrease with respect to sample size in very slow rates. In particular, the decrease of the first sequence is

$$B(1,n,20,1) - B(1,n+1,20,1) = \frac{10}{20^n}. \tag{5.464}$$

Also, we do not see the differences between the negative bounds for $X_{1:n}$ and $X_{2:n}$ when $n=15$ and $n=20$. The reason is that the bounds in both these cases are attained by the samples containing $N-1=19$ elements with identical numerical values, and one with the greater value (see (5.128) in Theorem 5.5 and (5.134) in Theorem 5.6). For $p = 1$, the bounds have the forms

$$B(j, n, 20, 1) = \frac{200}{19}\left[F_{j:n}\left(\frac{19}{20}\right) - \frac{19}{20}\right], \qquad j = 1, 2. \tag{5.465}$$

Since $F_{1:n}(x) - F_{2:n}(x) = nx(1-x)^{n-1}$, the difference of bounds amounts to

$$B(2, n, 20, 1) - B(1, n, 20, 1) = n20^{2-n}. \tag{5.466}$$

This is not observed even if we take 16 decimals approximations. Similar conclusions concern the cases $p = 2$ and $p = \infty$.

The differences between the bounds for order statistics $X_{j:20}$, $j = 2, \ldots, 20$, coming from the with replacement model with $N = 20$, and the general iid model decrease as j increases. Obviously, the latter are greater. For $p = 1$ and the sample maximum, the bound equal to 10 (see (4.178) in Corollary 4.12) is precise for arbitrary parent distributions, whereas for the with replacement model with $N = n = 20$, we have

$$\mathbb{E}\frac{X_{20:20} - \mu}{\sigma_1} \le 10\left[1 - \left(\frac{19}{20}\right)^{20} - \left(\frac{1}{20}\right)^{20}\right] \tag{5.467}$$

(see (5.84)), which is insignificantly less than 10. Analogous differences are unnoticeable for $p = 2$ and $p = \infty$ as well.

The counterparts of numerical bounds in Tables 5.1, 5.2 and 5.3 for the drawing without replacement model are presented in Tables 5.4, 5.5 and 5.6, respectively. The bounds for the sample minima in these tables are concluded from Corollary 5.37, see (5.425) and (5.420). Analogous results for the sample maxima are derived with the use of (5.351). The remaining numerical bounds in Tables 5.4, 5.5 and 5.6 are based on Corollaries 5.27, 5.29 and 5.30, respectively.

The bounds in Tables 5.4, 5.5 and 5.6 are increasing with respect to the number of order statistic j, nonincreasing with respect to sample size n, and decreasing as p increases. They are strictly greater than their counterparts in Tables 5.1, 5.2 and 5.3. In particular, all the bounds for $j \ge 2$ are positive here which is not always true in the with replacement model.

Moreover, the bounds for the without replacement scheme with $2 \le j \le n = N = 20$ are greater than the respective bounds in the general iid case (see the last columns in Tables 5.1, 5.2 and 5.3). The only exception is the case $j = n$, $p = 1$, for which the bounds are equal to $\frac{n}{2}$ in both the without replacement model and iid with an arbitrary marginal case. However, the bound in the without replacement scheme is attained by given populations of the fixed size $N = 20$, and that in the iid case is attained in the limit as the parent distribution tends to a degenerate one. The justification of this phenomenon as follows. If $n = N = 20$, they are the best deterministic bounds for the order statistics from the samples of size 20. It will

Table 5.4 Sharp upper bounds for $\mathbb{E}\frac{X_{j:n}-\mu}{\sigma_p}$, $p = 1$, for the drawing without replacement models with population size $N = 20$ and the samples sizes $n = 5, 10, 15, 20$

j	$n = 5$	$n = 10$	$n = 15$	$n = 20$
1	−0.526316	−0.526316	−0.526316	−0.526316
2	0.526316	0.526316	0.526316	0.526316
3	0.620485	0.555556	0.555556	0.555556
4	0.866873	0.597910	0.588235	0.588235
5	2.500000	0.663976	0.625000	0.625000
6		0.762582	0.666667	0.666667
7		0.910552	0.714286	0.714286
8		1.178682	0.790764	0.769231
9		1.772458	0.882001	0.833333
10		5.000000	1.002752	0.909091
11			1.177761	1.000000
12			1.447798	1.111111
13			1.878225	1.250000
14			2.865497	1.428571
15			7.500000	1.666667
16				2.000000
17				2.500000
18				3.333333
19				5.000000
20				10.000000

Table 5.5 Sharp upper bounds for $\mathbb{E}\frac{X_{j:n}-\mu}{\sigma_p}$, $p = 2$, for the drawing without replacement models with population size $N = 20$ and the samples sizes $n = 5, 10, 15, 20$

j	$n = 5$	$n = 10$	$n = 15$	$n = 20$
1	−0.229416	−0.229416	−0.229416	−0.229416
2	0.229416	0.229416	0.229415	0.229416
3	0.736762	0.333333	0.333333	0.333333
4	1.361821	0.536259	0.420084	0.420084
5	1.446337	0.701829	0.500000	0.500000
6		0.861016	0.577350	0.577350
7		1.036087	0.654654	0.654654
8		1.442856	0.847855	0.733799
9		2.136208	0.914298	0.816466
10		2.413574	1.1660[illegible]9	0.904534
11			1.245252	1.000000
12			1.597109	1.105542
13			2.166151	1.224749
14			2.432404	1.362770
15			3.334086	1.527525
16				1.732051
17				2.000000
18				2.380476
19				3.000000
20				4.358990

Table 5.6 Sharp upper bounds for $\mathbb{E}\frac{X_{j:n}-\mu}{\sigma_p}$, $p=\infty$, for the drawing without replacement models with population size $N=20$ and the samples sizes $n=5, 10, 15, 20$

j	$n=5$	$n=10$	$n=15$	$n=20$
1	−0.052632	−0.052632	−0.052632	−0.052632
2	0.052632	0.052632	0.052632	0.052632
3	0.240970	0.111111	0.111111	0.111111
4	0.696594	0.195820	0.176471	0.176471
5	0.967492	0.327952	0.250000	0.250000
6		0.525165	0.333333	0.333333
7		0.821105	0.428571	0.428571
8		0.976986	0.581527	0.538462
9		0.998907	0.764002	0.666667
10		0.999989	0.967492	0.818182
11			1.000000	1.000000
12			1.000000	1.000000
13			1.000000	1.000000
14			1.000000	1.000000
15			1.000000	1.000000
16				1.000000
17				1.000000
18				1.000000
19				1.000000
20				1.000000

be shown in Vol. 2 that these bounds are also sharp upper bounds for the order statistics coming from arbitrarily dependent identically distributed random samples. The dependent identically distributed model is obviously wider than the iid case, and consequently, the upper bounds in it are greater than those in the narrower iid model.

Given $p=1, 2, \infty$, for some j the bounds are identical for various n. This is not an effect of rounding. For instance, for $j=1$ the bounds

$$\mathbb{E}\frac{X_{1:n}-\mu}{\sigma_p} \le \begin{cases} -\frac{N}{2(N-1)}, & p=1, \\ -\frac{1}{\sqrt{N-1}}, & p=2, \\ -\frac{1}{N-1}, & p=\infty, \end{cases} \tag{5.468}$$

(cf. Corollary 5.37) depend merely on the population size $N = 20$, and are independent of the sample size n. The same phenomenon happens for $j=2$ with $n=5, 10, 15, 20$, $j=3$ with $n=10, 15, 20$, and $j=4, \ldots, 7$ with $n=15, 20$. The reason is as follows. For order statistics with relatively small ranks, the positive subsequence

$$C_k = C_k(j, n, N) = \frac{\binom{k-1}{j-1}\binom{N-k}{n-j}}{\binom{N}{n}}, \qquad k=j, \ldots, N-n+j, \tag{5.469}$$

of the sequence (5.330) is either decreasing or first slightly increasing and then decreasing. It follows that k_* defined in (5.332) is equal to $j-1$ then, and the projection (5.333) of (5.330) onto the family of nondecreasing sequences has the simple two-valued form

$$\underline{C}_k = \begin{cases} 0, & k = 1, \ldots, j-1, \\ \frac{1}{N+1-j}, & k = j, \ldots, N. \end{cases} \tag{5.470}$$

It depends on the population size N and the number j of the order statistic, but it does not depend explicitly on the sample size n. Since pth bounds on the expected order statistics are the qth norms of optimally centered vectors $\tilde{\underline{C}} = (\underline{C}_1, \ldots, \underline{C}_N)$ multiplied by the pth root of N, they do not depend on the particular value of n either.

It looks surprising at first glance that the bounds for $j = 1$ and $j = 2$ have the same absolute values and only different signs. It has the following justification. The bound for the sample minimum is attained by the population containing $N-1$ identical elements and one greater than the others (see Corollary 5.37). On the other hand, the projection $\tilde{\underline{C}} = (\underline{C}_1, \ldots, \underline{C}_N)$ of the vector (5.330) for the second minimum has the form (5.470) with $j = 2$, i.e., $N-1$ identical elements and one smaller than the others. To fix the ideas, we assume that $p = 1$ $\mu = 0$ and $\sigma_1 = 1$. The sample providing the bound for $X_{1:n}$ and satisfying the conditions has $N-1$ balls with value $-\frac{N}{2(N-1)}$ and one ball with value $\frac{N}{2}$. Since for $n \geq 2$, we have $\mathbb{P}\left(X_{1:n} = -\frac{N}{2(N-1)}\right) = 1$, the following holds: $\mathbb{E}X_{1:n} = \mathbb{E}\frac{X_{1:n}-\mu}{\sigma_1} = -\frac{N}{2(N-1)}$. The standardized version of $\tilde{\underline{C}}$ for $j = 2$ contains one number equal to $-\frac{N}{2}$, and $N-1$ numbers equal to $\frac{N}{2(N-1)}$. Then, $\mathbb{E}X_{2:n} = \frac{N}{2(N-1)}$ for $n \geq 2$. Similar arguments can be used for $p = 2$ and $p = \infty$.

Chapter 6
IID Samples from Shape Restricted Families

This chapter is devoted to evaluations of bounds for the expectations of order statistics and L-statistics based on iid samples whose parent distribution functions satisfy some shape restrictions. We mainly focus on the distributions with monotone density, failure rate and reversed failure rate functions. The bounds of Chap. 4 on the expected L-statistics centered about the population mean are expressed in terms of the scale units based on absolute central moments of various orders of the parent distribution. The positive upper bounds presented here are determined with use of the method of projection in Hilbert spaces (see Sect. 2.4.2), and they are expressed in the standard deviation units of the baseline distribution. The negative ones are established by means of the norm maximization method (see Sect. 2.4.7). They are trivially equal to zero except for ones measured in the mean absolute deviation from the mean units.

In Sect. 6.1, we consider samples with decreasing density and failure rate and reversed failure rate functions (DD, DFR, and DRFR, for short). We formulate general results for arbitrary L-statistics, and specify them for single order statistics and spacings. This section is mainly based on the papers by Danielak (2003) and Danielak and Rychlik (2004) (positive bounds for DD and DFR distributions), Rychlik (2009a), Rychlik (2009b) (respective negative bounds), and Goroncy and Rychlik (2024) (DRFR distributions). In Sect. 6.2, we present bounds for the expectations of order statistics from symmetric unimodal populations established in Gajek and Rychlik (1998) and Rychlik (2009c). Section 6.3 contains analogs of the results of Sect. 6.1 for order statistics from the populations with increasing density (ID), failure rate (IFR) and reversed failure rate (IRFR) functions. The results are based on Goroncy and Rychlik (2015), Goroncy and Rychlik (2016), Goroncy and Rychlik (2024), and Rychlik (2014b). Parent distributions with monotone density and failure rate on the average functions are treated in Sect. 6.4. The positive upper bounds for expectations of order statistics and spacings are based on Rychlik (2002) and Danielak and Rychlik (2003b). Negative upper bounds for order statistics with

N. Balakrishnan, T. Rychlik, *Sharp Inequalities for Ordered Random Variables in Statistics and Reliability*, Frontiers in Probability and the Statistical Sciences,
https://doi.org/10.1007/978-3-031-61347-0_6

small ranks are by Rychlik (2014a) and Goroncy and Rychlik (2018). Theoretical results of Sects. 6.1–6.4 are illustrated by some numerical evaluations in Sect. 6.5.

Inequalities of Sect. 6.6 have different forms. There are for L-statistics from iid populations with different baseline distribution functions F and W. If $F^{-1} \circ W$ is either convex or concave, the Jensen and Steffensen-Jensen inequalities (see Sect. 2.6) allow us to compare the expectation of L-statistics from one sample with the expected value of the same L-statistic from the other sample transformed by $F^{-1} \circ W$. When we properly fix W, we obtain evaluations of expected L-statistics from monotone density and failure rate function distributions by quantiles of the parent distribution of fixed orders. Analogous results can be obtained for L-statistics from populations with monotone generalized failure rates and reversed failure rates, and for symmetric baseline distributions with some shape restrictions. The greatest contributions to the topic are due to van Zwet (1964) and Rychlik (1998).

6.1 DD, DFR and DRFR Distributions

6.1.1 *Projection Bounds*

The most general statement of this section describes the sharp upper bounds on the expectations of L-statistics based on iid samples with a marginal distribution function F succeeding a fixed distribution function W in the convex transform order (see Sect. 2.8.1), which means that the composition $F^{-1} \circ W$ is convex. We assume that F and W have positive and finite variances, and W is absolutely continuous and supported on an interval (a, d), where $-\infty \le a < d \le +\infty$. Our aim is to bound the expressions $\mathbb{E} \sum_{i=1}^n c_i \frac{X_{i:n}-\mu}{\sigma}$ for arbitrary $\tilde{c} = (c_1, \ldots, c_n) \in \mathbb{R}^n$, where μ and σ^2 denote the expectation and variance, respectively, of X_1 with the distribution function $F \succeq_c W$. We notice that

$$0 = \int_0^1 [F^{-1}(x) - \mu]dx = \int_a^d [F^{-1} \circ W(x)]W(dx) = \langle F^{-1} \circ W, \mathbf{1} \rangle_W, \tag{6.1}$$

$$\sigma^2 = \int_0^1 [F^{-1}(x) - \mu]^2 dx = \int_a^d [F^{-1} \circ W(x)]^2 W(dx) = ||F^{-1} \circ W||_W^2, \tag{6.2}$$

where $\langle \cdot, \cdot \rangle_W$ and $\| \cdot \|_W$ denote the inner product and norm of the Hilbert space $L^2((a, d), W(dx))$, and $\mathbf{1}$ is the constant function on (a, d) equal to 1. We further observe that the set of functions $F^{-1} \circ W - \mu$, $F \succeq_c W$, constitutes the closed convex cone

$$C^0_{\succeq_c W} = \{g \in L^2((a, d), W(dx)) : g \text{ is nondecreasing, convex and orthogonal to } \mathbf{1}\} \tag{6.3}$$

in $L^2((a,d), W(dx))$. Then for $\bar{c} = (c_1, \ldots, c_n) \in \mathbb{R}^n$, with the notation

$$f_{\bar{c}:n}(x) = \sum_{i=1}^{n} c_i f_{i:n}(x) = n \sum_{i=1}^{n} c_i \binom{n-1}{i-1} x^{i-1}(1-x)^{n-i}, \qquad 0 < x < 1, \tag{6.4}$$

we have

$$\begin{aligned}
\mathbb{E} \sum_{i=1}^{n} c_i \frac{X_{i:n} - \mu}{\sigma} &= \frac{\int_0^1 [F^{-1}(x) - \mu] f_{\bar{c}:n}(x)dx}{(\int_0^1 [F^{-1}(x) - \mu]^2 dx)^{1/2}} \\
&= \frac{\int_a^d [F^{-1} \circ W(x) - \mu] f_{\bar{c}:n} \circ W(x) W(dx)}{(\int_a^d [F^{-1} \circ W(x) - \mu]^2 W(dx))^{1/2}} \\
&= \frac{\int_a^d [F^{-1} \circ W(x) - \mu] \left[f_{\bar{c}:n} \circ W(x) - \sum_{i=1}^n c_i \right] W(dx)}{(\int_a^d [F^{-1} \circ W(x) - \mu]^2 W(dx))^{1/2}} \\
&= \frac{\langle F^{-1} \circ W - \mu \mathbf{1}, f_{\bar{c}:n} \circ W - \sum_{i=1}^n c_i \mathbf{1} \rangle_W}{\| F^{-1} \circ W - \mu \mathbf{1} \|_W}
\end{aligned} \tag{6.5}$$

to be a normed linear functional on (6.3), represented by $f_{\bar{c}:n} \circ W(x) - \sum_{i=1}^n c_i \mathbf{1} \in L^2((a,d), W(dx))$. The passage from the second line to the third line follows from (6.1).

Let $P^0_{\succeq_c W}$ denote the projection operator onto (6.3). The following general result allows us to determine bounds on expected L-statistics of samples for parent distributions restricted by the relation $F \succeq_c W$.

Theorem 6.1 *Let $X_1, \ldots, X_n$ be an iid sample with a common distribution function F that has a finite variance and satisfies $F \succeq_c W$ for a given distribution function W. Then for arbitrary fixed $\bar{c} \in \mathbb{R}^n$, we have*

$$\mathbb{E} \sum_{i=1}^{n} c_i \frac{X_{i:n} - \mu}{\sigma} \le \left\| P^0_{\succeq_c W} \left(f_{\bar{c}:n} \circ W - \sum_{i=1}^{n} c_i \mathbf{1} \right) \right\|_W \tag{6.6}$$

which becomes equality iff

$$\frac{F^{-1} \circ W(x) - \mu}{\sigma} = \frac{P^0_{\succeq_c W} \left(f_{\bar{c}:n} \circ W - \sum_{i=1}^n c_i \mathbf{1} \right)(x)}{\| P^0_{\succeq_c W} \left(f_{\bar{c}:n} \circ W - \sum_{i=1}^n c_i \mathbf{1} \right) \|_W} \tag{6.7}$$

provided that the RHS of (6.6) is positive.

The proof immediately follows by combining (6.5) and (6.3) with (2.144) and (2.145). As we see, the problem is in determining the sharp upper bound on the normed functional (6.5) over the convex cone (6.3). We have

$$\begin{aligned}\mathbb{E}\sum_{i=1}^{n} c_i \frac{X_{i:n}-\mu}{\sigma} &= \frac{\langle F^{-1}\circ W-\mu\mathbf{1},\, f_{\tilde{c}:n}\circ W-\sum_{i=1}^{n} c_i\mathbf{1}\rangle_W}{\|F^{-1}\circ W-\mu\mathbf{1}\|_W} \\ &\le \frac{\left(F^{-1}\circ W-\mu\mathbf{1},\, P^0_{\succeq_c W}\left(f_{\tilde{c}:n}\circ W-\sum_{i=1}^{n} c_i\mathbf{1}\right)_W\right.}{\|F^{-1}\circ W-\mu\mathbf{1}\|_W} \\ &\le \left\| P^0_{\succeq_c W}\left(f_{\tilde{c}:n}\circ W-\sum_{i=1}^{n} c_i\mathbf{1}\right)\right\|_W . \end{aligned} \tag{6.8}$$

We obtain the equalities in both inequalities if

$$F^{-1}\circ W-\mu\mathbf{1}=\alpha P^0_{\succeq_c W}\left(f_{\tilde{c}:n}\circ W-\sum_{i=1}^{n} c_i\mathbf{1}\right)$$

for some positive α. The only α assuring the value of variance of F is equal to $\frac{\sigma}{\|P^0_{\succeq_c W}(f_{\tilde{c}:n}\circ W-\sum_{i=1}^{n} c_i\mathbf{1})\|_W}$.

The problem of determining projections $P^0_{\succeq_c W}\left(f_{\tilde{c}:n}\circ W-\sum_{i=1}^{n} c_i\mathbf{1}\right)$ for various vectors $\tilde{c}\in\mathbb{R}^n$ and distribution functions W, crucial in our approach, is very difficult in general. Firstly, function $f_{\tilde{c}:n}$ (see (6.4)) is an arbitrary positive polynomial of degree $n-1$ with abundant possibilities of mutual locations of increase, decrease, convexity and concavity regions. Even in the problem of projecting onto monotone functions, it is impossible to get explicit formulae valid for all polynomials. Adding the convexity condition makes the problem by far more complicated. There are no known explicit, recursive nor any other general methods of constructing projections onto convex cones of convex functions. Moreover, composing $f_{\tilde{c}:n}$ with a nondecreasing W does not change much monotonicity properties, but may strongly disturb convexity. Therefore, the sharp bounds presented in (6.6) were effectively determined only for a very few specific choices of L-statistics and distribution functions W. We mainly focus here on the expectations of single order statistics coming from iid samples with decreasing density, failure rate, and reversed failure rate parent distributions.

These are nonparametric families of absolutely continuous distributions with important interpretations in survival analysis. We say that F with corresponding density functions f has a decreasing density, failure rate, and reversed failure rate, respectively, if the functions $f(x)$, $\frac{f(x)}{1-F(x)}$, and $\frac{f(x)}{F(x)}$, respectively, are nonincreasing on the support of F. Note that the above conditions imply that $F(x)$, $\ln(1-F(x))$ and $\ln F(x)$ are concave, respectively, whereas their inverses $F^{-1}(x)$, $F^{-1}(1-e^{-x})$, and $F^{-1}(e^x)$ are convex, respectively. In other words, F succeeds the standard uniform distribution function $U(x)=x$, $0<x<1$, exponential distribution function $Z(x) = 1 - e^{-x}$, $x > 0$, and negative exponential one $V(x) = e^x$, $x < 0$, respectively, in the convex transform order.

Convex and nondecreasing compositions $F^{-1} \circ W$ may be constant on some neighborhoods of the left-end point of the support of W. This may imply that the $F \succeq_c W$ which attains the positive bound in (6.6), and defined in (6.7) has an atom at its left-end of the support. However, the respective quantile function F^{-1} can be approximated with an arbitrary accuracy by the quantile functions of absolutely continuous distributions succeeding a fixed W in the convex transform order, with the means and variances identical with the mean and variances of the optimal discontinuous distribution. As a consequence, the bounds for the families succeeding U, Z and V can be attained in the limit by the absolutely continuous distributions with decreasing density, failure rate and reversed failure rate functions, respectively. Moreover, one can even choose the elements of the sequences attaining the bounds in the limit which have strictly decreasing density, failure rate and reversed failure rate functions. For convenience, we present below bounds and the limiting, possible discontinuous members of the classes of distribution functions succeeding U, Z and V in the convex transform order which attain the bounds. We also call the families of distributions functions F with decreasing density, failure rate and reversed failure rate (and use the acronyms DD, DFR, and DRFR, respectively) if F succeeds the uniform, exponential, and negative exponential distributions, respectively, in the convex transform order.

Now we present a simple lemma which allows us to simplify the solutions of projection problems.

Lemma 6.1 *We consider a closed convex cone $\mathcal{C}$ in a real Hilbert space $\mathcal{H}$. Suppose that Ph is a projection of $h \in \mathcal{H}$ onto $\mathcal{C}$, and g_* is a nonzero element of $\mathcal{C}$.*

(i) The projection of $h + g_*$ onto $\mathcal{C}$ amounts to $Ph + g_*$.
(ii) We have $\langle Ph, g_* \rangle = \langle h, g_* \rangle$.

Proof

(i) We first observe that $\mathcal{C} = \{g - g_* : g \in \mathcal{C}\}$. Therefore,

$$\|(P(h + g_*) - (h + g_*)\| = \inf_{g \in \mathcal{C}} \|g - h - g_*\| = \inf_{g + g_* \in \mathcal{C}} \|g + g_* - h - g_*\|$$
$$= \inf_{g \in \mathcal{C}} \|g - h\| = \|Ph - h\| = \|Ph + g_* - (h + g_*)\|. \tag{6.9}$$

(ii) For every $g \in \mathcal{C}$, we minimize the function

$$\|h - g - cg_*\|^2 = \|h - g\| - 2c\langle h - g, g_* \rangle + c^2\|g_*\|^2 \tag{6.10}$$

with respect to $c \in \mathbb{R}$. Obviously, it is minimized at $c = c_g = \frac{\langle h - g, g_* \rangle}{\|g_*\|^2}$. In particular, for $g = Ph$, we have $c_{Ph} = 0$, which implies that $\langle h, g_* \rangle = \langle Ph, g_* \rangle$.

□

Define the convex cone

$$\mathcal{C}_{\succeq_c W} = \{g \in L^2((a,d), W(dx)) : g \text{ is nondecreasing, convex}\}. \tag{6.11}$$

Since the constant functions belong to the cone, by Lemma 6.1(ii), for every $h \in L^2((a,d), W(dx))$, we have

$$\langle P_{\succeq_c W} h, \mathbf{1}\rangle_W = \int_a^d P_{\succeq_c W} h(x) W(dx) = (h, \mathbf{1})_W = \int_a^d h(x) W(dx) \tag{6.12}$$

Consider the functional

$$\begin{aligned} &\left\langle F^{-1} \circ W - \mu \mathbf{1}, f_{\tilde{c}:n} \circ W - \sum_{i=1}^n c_i \mathbf{1} \right\rangle_W \\ &= \int_a^d \left[F^{-1} \circ W(x) - \mu\right] \left[f_{\tilde{c}:n} \circ W(x) - \sum_{i=1}^n c_i\right] W(dx) \end{aligned} \tag{6.13}$$

acting on (6.11). The constant function $g_* = \mathbf{1}$ belongs to (6.11) and satisfies $\left\langle f_{\tilde{c}:n} \circ W - \sum_{i=1}^n c_i \mathbf{1}, \mathbf{1}\right\rangle_W = 0$. Lemma 6.1(ii) implies

$$P^0_{\succeq_c W}\left(f_{\tilde{c}:n} \circ W - \sum_{i=1}^n c_i \mathbf{1}\right) = P_{\succeq_c W}\left(f_{\tilde{c}:n} \circ W - \sum_{i=1}^n c_i \mathbf{1}\right), \tag{6.14}$$

where $P_{\succeq_c W}$ denotes the projection onto (6.11). Projecting functions onto $\mathcal{C}_{\succeq_c W}$ is easier than projecting onto $\mathcal{C}^0_{\succeq_c W}$, because then one does not need to bother about obeying a troublesome orthogonality condition. By Lemma 6.1(i), we also have

$$P_{\succeq_c W}\left(f_{\tilde{c}:n} \circ W - \sum_{i=1}^n c_i \mathbf{1}\right) = P_{\succeq_c W}\left(f_{\tilde{c}:n} \circ W\right) - \sum_{i=1}^n c_i \mathbf{1}, \tag{6.15}$$

because $\alpha \mathbf{1} \in \mathcal{C}_{\succeq_c W}$ for any $\alpha \in \mathbb{R}$. Note that

$$\begin{aligned} &\left\| P_{\succeq_c W}\left(f_{\tilde{c}:n} \circ W\right) - \sum_{i=1}^n c_i \mathbf{1}\right\|_W^2 \\ &= \left\| P_{\succeq_c W}\left(f_{\tilde{c}:n} \circ W\right)\right\|_W^2 - 2\sum_{i=1}^n c_i \left\langle P_{\succeq_c W}\left(f_{\tilde{c}:n} \circ W\right), \mathbf{1}\right\rangle_W + \left(\sum_{i=1}^n c_i\right)^2 \end{aligned}$$

$$
\begin{aligned}
&= \left\| P_{\succeq_c W} \left(f_{\bar{c}:n} \circ W \right) \right\|_W^2 - 2 \sum_{i=1}^{n} c_i \left\langle f_{\bar{c}:n} \circ W, \mathbf{1} \right\rangle_W + \left(\sum_{i=1}^{n} c_i \right)^2 \\
&= \left\| P_{\succeq_c W} \left(f_{\bar{c}:n} \circ W \right) \right\|_W^2 - \left(\sum_{i=1}^{n} c_i \right)^2 . \qquad (6.16)
\end{aligned}
$$

Accordingly, we have the following.

Corollary 6.1 *Under the assumptions of Theorem* 6.1, *we have*

$$
\mathbb{E} \sum_{i=1}^{n} c_i \frac{X_{i:n} - \mu}{\sigma} \le \sqrt{ \| P_{\succeq_c W} f_{\bar{c}:n} \circ W \|_W^2 - \left(\sum_{i=1}^{n} c_i \right)^2 } \qquad (6.17)
$$

which becomes equality iff

$$
\frac{F^{-1} \circ W(x) - \mu}{\sigma} = \frac{P_{\succeq_c W} f_{\bar{c}:n} \circ W(x) - \sum_{i=1}^{n} c_i}{\sqrt{ \| P_{\succeq_c W} f_{\bar{c}:n} \circ W \|_W^2 - \left(\sum_{i=1}^{n} c_i \right)^2 }} \qquad (6.18)
$$

provided that the RHS of (6.17) is positive.

6.1.2 Positive Upper Bounds for Single Order Statistics

In the case of single order statistic, we obviously obtain

$$
\mathbb{E} \frac{X_{j:n} - \mu}{\sigma} \le \sqrt{ \| P_{\succeq_c W} f_{j:n} \circ W \|_W^2 - 1 } \qquad (6.19)
$$

which becomes equality iff

$$
\frac{F^{-1} \circ W(x) - \mu}{\sigma} = \frac{P_{\succeq_c W} f_{j:n} \circ W(x) - 1}{\sqrt{ \| P_{\succeq_c W} f_{j:n} \circ W \|_W^2 - 1 }}. \qquad (6.20)
$$

Below we determine the projections $P_{\succeq_c W} f_{j:n} \circ W$ for W being the standard uniform, exponential, and negative exponential distribution functions in order to calculate the bounds on the expectations of single order statistics coming from simple samples with the DD, DFR, and DRFR marginal distributions. First we analyze shape properties of the compositions $f_{j:n} \circ W$ with $W = U$, Z and V. We focus on establishing the monotonicity and convexity/concavity intervals of the functions.

Lemma 6.2 *Consider the functions* $h_{j:n,U} = f_{j:n} \circ U = f_{j:n}$, $0 < x < 1$, $1 \le j \le n$, $n \ge 2$.

(i) Function $f_{1:n}$ *is convex decreasing on* $(0, 1)$.

(ii) For $j = 2$, *the function is concave increasing on* $\left(0, \frac{1}{n-1}\right)$, *concave decreasing on* $\left(\frac{1}{n-1}, \frac{1}{n-2}\right)$, *and convex decreasing on* $\left(\frac{1}{n-2}, 1\right)$.

(iii) For $3 \le j \le n-2$, *define*

$$u_0 = u_0(j, n) = \frac{j-1}{n-1}, \tag{6.21}$$

$$u_1 = u_1(j, n) = \frac{j-1}{n-1} - \sqrt{\frac{(j-1)(n-j)}{(n-1)^2(n-2)}}, \tag{6.22}$$

$$u_2 = u_2(j, n) = \frac{j-1}{n-1} + \sqrt{\frac{(j-1)(n-j)}{(n-1)^2(n-2)}}. \tag{6.23}$$

Then, $f_{j:n}$ *is convex increasing on* $(0, u_1)$, *concave increasing on* (u_1, u_0), *concave decreasing on* (u_0, u_2), *and convex decreasing on* $(u_2, 1)$.

(iv) Function $f_{n-1:n}$ *is convex increasing on* $\left(0, \frac{n-3}{n-1}\right)$, *concave increasing on* $\left(\frac{n-3}{n-1}, \frac{n-2}{n-1}\right)$, *and concave decreasing on* $\left(\frac{n-2}{n-1}, 1\right)$.

(v) For $j = n$, *the density function* $f_{n:n}$ *is convex increasing on* $(0, 1)$.

Proof The first derivatives of $f_{j:n}$ have the forms

$$\begin{aligned} f'_{j:n}(x) &= n[B_{j-2,n-2}(x) - B_{j-1,n-2}(x)] \\ &= \frac{n!}{(j-1)!(n-j)!} x^{j-2}(1-x)^{n-j-1}[(j-1) - (n-1)x] \end{aligned} \tag{6.24}$$

The first and last summands in the first line vanish if $j = 1$ and $j = n$, respectively. This implies decrease and increase of the function, respectively. The latter line concerns only the cases $2 \le j \le n-1$. It follows that the derivative (6.24) is positive on $\left(0, \frac{j-1}{n-1}\right)$, and negative on $\left(\frac{j-1}{n-1}, 1\right)$, respectively.

The second derivative amounts to

$$f''_{j:n}(x) = n(n-1)[B_{j-3,n-3}(x) - 2B_{j-2,n-3}(x) + B_{j-1,n-3}(x)]. \tag{6.25}$$

The first two summands disappear for $j = 1$, which implies that $f_{1:n}$ is convex. The same conclusion holds for $f_{n:n}$, because in this case the last two summands vanish. For $j = 2$ and $j = n-1$, the first and last terms do not appear, respectively. In the first case, the function is first negative and then positive on $(0, 1)$, and in the latter one the sign order is reversed (cf. the VDP of Bernstein polynomials, see Lemma 4.1). Simple calculations show that the respective sign change points are equal to $\frac{1}{n-2}$ and $\frac{n-3}{n-2}$. For $3 \le j \le n-2$, (6.25) can be rewritten as

$$f''_{j:n}(x) = \frac{n!}{(j-1)!(n-j)!} x^{j-3}(1-x)^{n-j-2}$$
$$\times [(n-1)(n-2)x^2 - 2(j-1)(n-2)x + (j-1)(j-2)]. \quad (6.26)$$

The sign of the second derivative is equal to the sign of the convex quadratic expression in the brackets. Simple calculations show it is positive at 0 and 1, and has a positive determinant, which implies that its zeros belong to (0, 1). The precise value of zeros are $u_i(j,n)$, $i = 1, 2$, presented in (6.22) and (6.23). These observations allow us to complete the proof. □

Lemma 6.3 *Take* $h_{j:n,Z} = f_{j:n} \circ Z(x) = f_{j:n}(1-e^{-x})$, $1 \le j \le n$, $n \ge 2$.

(i) Function $h_{1:n,Z}$ *is convex decreasing on* $(0, \infty)$.

(ii) For $j = 2$, *the function is concave increasing on* $\left(0, \ln \frac{n-1}{n-2}\right)$, *concave decreasing on* $\left(\ln \frac{n-1}{n-2}, 2\ln \frac{n-1}{n-2}\right)$, *and convex decreasing on* $\left(2\ln \frac{n-1}{n-2}, \infty\right)$.

(iii) For $3 \le j \le n-1$, *set*

$$z_0 = z_0(j,n) = -\ln \frac{n-j}{n-1}, \quad (6.27)$$

$$z_1 = z_1(j,n) = -\ln\left(\frac{n-j}{n-1} + \frac{j-1+\sqrt{(j-1)^2+4(n-1)(j-1)(n-j)}}{2(n-1)^2}\right), \quad (6.28)$$

$$z_2 = z_2(j,n) = -\ln\left(\frac{n-j}{n-1} - \frac{j-1-\sqrt{(j-1)^2+4(n-1)(j-1)(n-j)}}{2(n-1)^2}\right), \quad (6.29)$$

Then $h_{j:n,Z}$ *is convex increasing on* $(0, z_1)$, *concave increasing on* (z_1, z_0), *concave decreasing on* (z_0, z_2), *and convex decreasing on* (z_2, ∞).

(iv) The density function $h_{n:n,Z}$ *is convex increasing on* $(0, \ln(n-1))$, *and concave increasing on* $(\ln(n-1), \infty)$.

Proof Note that $h_{j:n,Z}$ is the composition of $f_{j:n}$ with a strictly increasing function $Z(x) = 1 - e^{-x}$ acting on $\mathbb{R}_+$. Therefore it is decreasing for $j = 1$, increasing for $j = n$, and increasing-decreasing with the maximum point $z_0(j,n) = Z^{-1}(u_0(j,n)) = -\ln \frac{n-j}{n-1}$ otherwise. For $1 \le j \le n-1$, its second derivative has the form

$$h''_{j:n,Z}(x) = n[(n+1-j)(n-j)B_{j-3,n-1}(1-e^{-x})$$
$$- (n+1-j)(2n-2j+1)B_{j-2,n-1}(1-e^{-x})$$
$$+ (n-j)^2 B_{j-1,n-1}(1-e^{-x})]. \quad (6.30)$$

The first two terms vanish for $j = 1$, and then the function $h_{1:n,Z}$ is convex. The first one disappears for $j = 2$. For convenience of analysis, we change the variable $\mathbb{R}_+ \ni x \mapsto y = 1 - e^{-x} \in (0, 1)$. For $j = 2$, we obtain a linear combination of two consecutive Bernstein polynomials. This can be represented as a product of a positive polynomial and a linear increasing function which has zero at $y_0(j, n) = \frac{2n-3}{(n-1)^2} \in (0, 1)$. Putting $x_0(2, n) = Z^{-1}(y_0(2, n))$, we obtain $2 \ln \frac{n-1}{n-2}$. For $3 \le j \le n - 1$, the above variable change provides a linear combination of three consecutive Bernstein polynomials, which similarly can be represented as the product of a positive polynomial and a convex quadratic function with two zeros in $(0, 1)$. Determining the precise values of zeros, and transforming them by Z^{-1}, we finally obtain $z_i(j, n)$, $i = 1, 2$, presented in (6.28) and (6.29). Note that these representations allow us to verify easily the relations $z_1(j, n) < z_0(j, n) < z_2(j, n)$.

In the sample maximum case, the second derivative has a slightly different form

$$h''_{n:n,Z}(x) = n[2B_{n-3,n-1}(1 - e^{-x}) - B_{n-2,n-1}(1 - e^{-x})]. \tag{6.31}$$

Elementary calculations show that $(0, -\ln(n - 1))$ and $(\ln(n - 1), \infty)$ are the intervals of convexity and concavity of $h_{n:n,Z}$. This completes the proof. □

Lemma 6.4 *Put* $h_{j:n,V} = f_{j:n} \circ V(x) = f_{j:n}(e^x)$, $x < 0$, $1 \le j \le n$, $n \ge 2$.

(i) *The density function* $h_{1:n,V}$ *is concave decreasing on* $(-\infty, -\ln(n - 1))$, *and convex decreasing on* $(-\ln(n - 1), 0)$.

(ii) *For* $2 \le j \le n - 2$, *define*

$$v_0 = v_0(j, n) = \ln \frac{j-1}{n-1}, \tag{6.32}$$

$$v_1 = v_1(j, n) = \ln \left(\frac{j-1}{n-1} - \frac{n - j - \sqrt{(n-j)^2 + 4(n-1)(j-1)(n-j)}}{2(n-1)^2} \right), \tag{6.33}$$

$$v_2 = v_2(j, n) = \ln \left(\frac{j-1}{n-1} + \frac{n - j + \sqrt{(n-j)^2 + 4(n-1)(j-1)(n-j)}}{2(n-1)^2} \right). \tag{6.34}$$

Then $h_{j:n,V}$ *is convex increasing on* $(-\infty, v_1)$, *concave increasing on* (v_1, v_0), *concave decreasing on* (v_0, v_2), *and convex decreasing on* $(v_2, 0)$.

(iii) *For* $j = n - 1$, *the function is convex increasing on* $\left(-\infty, 2 \ln \frac{n-2}{n-1}\right)$, *concave increasing on* $\left(2 \ln \frac{n-2}{n-1}, \ln \frac{n-2}{n-1}\right)$, *and concave decreasing on* $\left(\ln \frac{n-2}{n-1}, 0\right)$.

(iv) *Function* $h_{n:n,Z}$ *is convex increasing on* $(-\infty, 0)$.

The inequalities

$$f_{j:n}(x) = f_{n+1-j:n}(1 - x), \qquad 0 < x < 1, \quad j = 1, \dots, n, \tag{6.35}$$

imply that

$$u_0(j,n) = 1 - u_0(n+1-j,n), \tag{6.36}$$

$$u_1(j,n) = 1 - u_2(n+1-j,n), \tag{6.37}$$

$$u_2(j,n) = 1 - u_1(n+1-j,n) \tag{6.38}$$

(cf. (6.21)–(6.23)). Further, (6.35) implies the relations

$$\begin{aligned} h_{j:n,V}(-x) &= f_{j:n}(e^{-x}) = f_{j:n} \circ V(-x) = f_{n+1-j:n}(1-e^{-x}) \\ &= f_{n+1-j:n} \circ Z(x), \qquad x > 0, \quad j = 1, \ldots, n, \end{aligned} \tag{6.39}$$

and

$$v_0(j,n) = -z_0(n+1-j,n), \tag{6.40}$$

$$v_1(j,n) = -z_2(n+1-j,n), \tag{6.41}$$

$$v_2(j,n) = -z_1(n+1-j,n) \tag{6.42}$$

(cf. (6.27)–(6.29) and (6.32)–(6.34)); consequently. If $h_{j:n,Z}$ is increasing (decreasing) on some interval $(a,b) \subset \mathbb{R}_+$, then $h_{n+1-j,n,V}$ is decreasing (increasing, respectively) on $(-b,-a)$. If $h_{j:n,Z}$ is convex (concave) on (a,b), then so is $h_{n+1-j:n,V}$ on $(-b,-a)$. Therefore Lemma 6.4 can be concluded from the statements of Lemma 6.3. We finally note that by writing convex and concave in Lemmas 6.2–6.4, we meant that the functions studied there are strictly convex and strictly concave.

In Assumptions (A1) presented below, we describe the shape conditions which are satisfied by majority of functions treated in Lemmas 6.2–6.4.

(A1) For some $-\infty \le a \le b < c \le d \le \infty$ and distribution function W with a positive probability density w on (a,d) and finite second moment, a function $h \in L^2((a,d), W(dx))$ is nonnegative, bounded, twice differentiable, satisfies $h(a) = 0$,

$$\int_a^d h(x)W(dx) = 1, \tag{6.43}$$

and it is strictly increasing and strictly convex on (a,b), strictly increasing and strictly concave on (b,c) and strictly decreasing on (c,d).

We admit here that intervals (a,b) and (c,d) are empty. Note that the assumptions are satisfied by $h_{j:n,W}$, $W = U, Z, V$, $j = 2,\ldots,n-1$, and $h_{n:n,V}$. Under (A1), we describe below the projection $P_{\succeq_c W}h$ of h onto the convex cone (6.11). Taking particular forms of $h_{j:n,W}$ and applying Corollary 6.1, we obtain tools for calculating the bounds on expected order statistics from selected families of distributions. The projections of the remaining functions $h_{1:n,W}$, $W = U, Z, V$ and $h_{n:n,W}$, $W = U, V$, can be calculated with not so elaborate tools.

Theorem 6.2 *Suppose that functions h and W satisfy assumptions* (A1). *Define the functions*

$$\lambda_*(\beta) = \frac{\int_\beta^d (x-\beta)[h(x)-h(\beta)]W(dx)}{\int_\beta^d (x-\beta)^2 W(dx)}, \tag{6.44}$$

$$K(\beta) = \lambda_*(\beta) - h'(\beta), \tag{6.45}$$

$$L(\beta) = \int_\beta^d h(x)W(dx) - \int_\beta^d [\lambda_*(\beta)(x-\beta) + h(\beta)]W(dx) \tag{6.46}$$

for $a \le \beta \le b$, and the set

$$\mathcal{K} = \{a < \beta \le b : K(\beta) \ge 0 \text{ and } L(\beta) = 0\}. \tag{6.47}$$

If $\mathcal{K}$ is nonempty and $\beta_* = \sup \mathcal{K}$, then

$$P_{\succeq_c W} h(x) = \begin{cases} h(x), & a \le x \le \beta_*, \\ \lambda_*(\beta_*)(x-\beta_*) + h(\beta_*), & \beta_* \le x < d. \end{cases} \tag{6.48}$$

If $\mathcal{K} = \emptyset$, take

$$\lambda_* = \frac{\int_a^d (x-a)h(x)W(dx) - \int_a^d (x-a)W(dx)}{\int_a^d (x-a)^2 W(dx) - [\int_a^d (x-a)W(dx)]^2}, \tag{6.49}$$

$$\gamma_* = \frac{\int_a^d (x-a)^2 W(dx) - \int_a^d (x-a)h(x)W(dx) \int_a^d (x-a)W(dx)}{\int_a^d (x-a)^2 W(dx) - [\int_a^d (x-a)W(dx)]^2}. \tag{6.50}$$

If $\lambda_* > 0$ and $a > -\infty$, then

$$P_{\succeq_c W} h(x) = \lambda_*(x-a) + \gamma_*. \tag{6.51}$$

If $\mathcal{K} = \emptyset$ and either $a = -\infty$ or $\lambda_* \le 0$, then

$$P_{\succeq_c W} h(x) = 1. \tag{6.52}$$

The formal proof of Theorem 6.2 is very sophisticated. It is performed in two steps. The first step consists in describing of possible shapes of the projection, and it is based on geometric arguments. This makes it possible to focus on some parametric families of candidates for projection. The latter step consists in determining the optimal parameters, and analytic tools are used then. We postpone the detailed proof to the Appendix located at the end of this book (see Theorem A.1, and also the papers by Danielak (2003) and Goroncy and Rychlik (2024)).

Here we present some informal intuitive arguments. Obviously, one obtains an optimal approximation of the function on some interval if the approximation coincides with the projected function on the interval. In our projection problem, it is only possible for some subinterval $(a, \beta] \subset (a, b]$. If we took $\beta = b$, then the optimal approximation on the remaining interval (b, d) would be the line tangent to h at b, but this could not be the desired part of projection, because the line runs above h on the interval, which contradicts condition (6.12). Therefore, we should select a change point on the left to the inflection point b. Function (6.44) represents the slope on the best approximation of h on the interval (β, d) by a linear function starting at β at the level $h(\beta)$. Gluing it with the original function h on (a, β), we obtain a convex function if (6.45) is nonnegative. The second condition $L(\beta) = 0$ (see (6.46)) is forced by (6.12). Then $\lambda_*(\beta)(x - \beta) + h(\beta)$ is also the best linear approximation of h on (β, d) with no border restrictions. If the set (6.47) contains more than one elements, we choose the maximal one β_* for constructing the projection, because this coincides with h on the longest interval (a, β_*), and its linear part is the best linear approximation of h on (β_*, d).

It may happen that the interval (a, b) of convex increase of h is so short that it is better to take the optimal linear approximation with the shape (6.49) and intercept (6.50) rather than to take a part of h and follow the restrictions on the slope of the linear part. If the best linear approximation appears to be decreasing, it is replaced by the constant function, which is equal to 1 due to condition (6.12). Note that in the case $a = -\infty$, a strictly increasing linear projection is excluded, because it is negative for the arguments close to $-\infty$, and so it can be replaced at least by its lower truncation at the level 0.

We also present a proposition which simplifies determination of the projection of the form (6.48). Its proof is based on some arguments of the proof of Theorem 6.2, and is postponed to the Appendix. It can be also found in Gajek and Rychlik (1998).

Proposition 6.1 *Suppose that the set of zeros of* (6.46) *does not contain intervals and*

$$\mathcal{K}^+ = \{a < \beta < b : K(\beta) > 0\} \tag{6.53}$$

is nonempty. Then each subinterval of (6.53) *contains at most one zero of* (6.46) *where* (6.46) *changes the sign from minus to plus. In particular, if* (6.53) *is an interval and* (6.47) *is nonempty, then it consists of a single point* β_*, *which is the shape changing point of the projection* $P_{\succeq_c W} h$ *of the form* (6.48).

Now we are in a position to present the positive bounds in the DD, DFR, and DRFR cases.

Theorem 6.3 (DD Distributions) *Assume that $X_1, \ldots, X_n$ are iid with a common concave distribution function, finite mean μ and positive variance σ^2.*

If $j \le \frac{n+1}{2}$, then $\mathbb{E}X_{j:n} \le \mu$.

If $\frac{n+1}{2} < j \le \frac{2}{3}(n+1)$, then

$$\mathbb{E}\frac{X_{j:n}-\mu}{\sigma} \le \sqrt{3}\left(\frac{2j}{n+1}-1\right), \tag{6.54}$$

and the equality holds if all X_i are uniformly distributed on $[\mu-\sqrt{3}\sigma, \mu+\sqrt{3}\sigma]$.

If $\frac{2}{3}(n+1) < j \le n-1$, then

$$\mathbb{E}\frac{X_{j:n}-\mu}{\sigma} \le B \tag{6.55}$$

for

$$B^2 = \frac{(n!)^2}{(2n-1)!}\binom{2j-2}{j-1}\binom{2(n-j)}{n-j}F_{2j-1:2n-1}(\beta_*) - 1$$
$$+(1-\beta_*)f_{j:n}^2(\beta_*) - \lambda_*(\beta_*)(1-\beta_*)^2 f_{j:n}(\beta_*) + \lambda_*^2(\beta_*)\frac{(1-\beta_*)^3}{3}, \tag{6.56}$$

where

$$\lambda_* = \lambda_{*,j:n,U}(\beta_*) = \frac{3}{(1-\beta_*)^3}\left\{\frac{j}{n+1}[1-F_{j+1:n+1}(\beta_*)]\right.$$
$$\left. -\beta_*[1-F_{j:n}(\beta_*)] - \frac{3}{2(1-\beta_*)}f_{j:n}(\beta_*)\right\} \tag{6.57}$$

and β_* is the smallest positive zero of the polynomial

$$L_{j:n,U}(\beta) = \sum_{i=1}^{j-1} 2(2n+1-3j+i)f_{i:n+2}(\beta) - (n-j)(n-j-1)f_{j:n+2}(\beta). \tag{6.58}$$

Bound (6.55) *is attained if*

$$F(x) = \begin{cases} 0, & \frac{x-\mu}{\sigma} < a_1, \\ f_{j:n}^{-1}\left(\frac{x-\mu}{\sigma}B+1\right), & a_1 \le \frac{x-\mu}{\sigma} < a_2, \\ \beta_* + \frac{1-f_{j:n}(\beta_*)}{\lambda_*} + \frac{B}{\lambda_*}\frac{x-\mu}{\sigma}, & a_2 \le \frac{x-\mu}{\sigma} \ge a_3, \\ 1, & \frac{x-\mu}{\sigma} \ge a_3, \end{cases} \tag{6.59}$$

where $a_1 = -\frac{1}{B}$, $a_2 = \frac{f_{j:n}(\beta_*)-1}{B}$ and $a_3 = \frac{\lambda_*(1-\beta_*)+f_{j:n}(\beta_*)-1}{B}$.

The distribution function is absolutely continuous and has a finite support. It is the inverse of an increasing and convex part of a Bernstein polynomial (up to a linear transformation) on the left and is linear on the right.

Proof We determine the bounds by means of projection methods, applying Theorem 6.2 for the case $h = f_{j:n} = nB_{j-1,n-1} \in L^2((0,1), dx)$, $1 \leq j \leq n-1$.

Density functions $f_{1:n}$ are decreasing, and so their projections onto the set of nondecreasing functions is the constant function equal to 1. This is a member of the cone of nondecreasing convex functions as well. Therefore $P_{\succeq_c W} f_{1:n} = \mathbf{1}$. Since $f_{2:n}$ does not have convex increasing parts, $P_{\succeq_c U} f_{2:n}$ are linear. For $3 \leq j \leq n-1$, we need to specify functions (6.44), (6.45) and (6.46). It is obvious that they can be expressed by means of Bernstein polynomials. After some tedious calculations, we obtain

$$\lambda_*(\beta) = \lambda_{*,j:n,U}(\beta) = \frac{\sum_{i=0}^{j-1} \lambda_i B_{i,n+1}(\beta)}{2(n+1)(1-\beta)^3} \tag{6.60}$$

with

$$\lambda_i = \begin{cases} 6(j-i), & i = 0, \ldots, j-2, \\ 6-(n+1-j)(n+2-j), & i = j-1, \end{cases} \tag{6.61}$$

$$K(\beta) = K_{j:n,U}(\beta) = \frac{\sum_{i=0}^{j-1} a_i B_{i,n+1}(\beta)}{2(n+1)(1-\beta)^3}, \tag{6.62}$$

where

$$a_i = \begin{cases} 6(j-i), & i = 0, \ldots, j-3, \\ 12 - 2\frac{(n+3-j)!}{(n-j)!}, & i = j-2, \\ 6 + \frac{(n+2-j)!}{(n-j)!}(2n-2j-3), & i = j-1, \end{cases} \tag{6.63}$$

and

$$L(\beta) = L_{j:n,U}(\beta) = \frac{\sum_{i=0}^{j-1} b_i B_{i,n+1}(\beta)}{4(n+1)(1-\beta)} \tag{6.64}$$

with

$$b_i = \begin{cases} 2(2n+2-3j+i), & i = 0, \ldots, j-2, \\ -(n-j)(n-1-j), & i = j-1. \end{cases} \tag{6.65}$$

Now we analyze the behavior of (6.62) and (6.64). Since the denominators are positive, we apply the variation diminishing property of Bernstein polynomials to examine the sign changes of the numerators. We have $a_i > 0$, $i = 0, \ldots, j-3$, $a_{j-2} < 0$ and $a_{j-1} = 0$ if $j = n-1$ and $a_{j-1} > 0$ for $j < n-1$. By VDP, (6.62) is first positive and then negative for $j = n-1$ (+, − for short), and first positive, then possibly negative and finally positive (either + − + or +) in other cases. However, $K(\beta)$ is negative at the inflection point b, because this is the difference between the slope $\lambda_*(b)$ of the optimal linear approximation $\lambda_*(b)(x-b)+h(b)$ of h on $[b,d)$ that certainly cuts the graph of h in $[b,d)$ and the slope $h'(b)$ of the tangent line $h'(b)(x-b)+h(b)$ which runs over h. Therefore, $K(\beta)$ does have a unique zero $\alpha \in (0,b)$ and $\mathcal{K}^+ = (0,\alpha)$.

Suppose that $j \le \frac{2}{3}(n+1)$. Then the first nonzero element of (6.65) is positive and $L(0+) > 0$. By Proposition 6.1, $L(\beta)$ is positive on the whole $(0,\alpha)$ and $D(\beta, \lambda_*(\beta))$ increases there. Therefore the projection is linear in this case. The optimal linear approximation of $f_{j:n}$ on (0, 1) has the form

$$\ell_0(x) = 6\left[\frac{2j}{n+1} - 1\right]x + 2\left(2 - \frac{3j}{n+1}\right). \tag{6.66}$$

If $\frac{n+1}{2} < j \le \frac{2}{3}(n+1)$, then (6.66) is increasing and hence becomes the projection. If $j \le \frac{n+1}{2}$, then the projection is the constant function equal to 1.

Assume now that $\frac{2}{3}(n+1) < j \le n-1$. Then $L(0+) < 0$, and we show that (6.64) has a unique zero in $\mathcal{K}^+ = (0,\alpha)$. Suppose, on the contrary, that this is not true. This implies that the projection has the form (6.66). This is, however, negative at a neighborhood of zero, and can be improved by cutting the negative part, which contradicts the assumption. Therefore there exists a unique $\beta_* \in (0,\alpha)$, where $L(\beta)$ vanishes. A thorough analysis of signs of (6.65) shows that $L(\beta)$ has another zero greater than α if $j < n-1$, and the zero is unique when $j = n-1$. Summing up, the solution of the projection problem is as follows. If $1 \le j \le \frac{n+1}{2}$, then $P_{\succeq_c U} f_{j:n} = \mathbf{1}$. If $\frac{n+1}{3} < j \le \frac{2}{3}(n+1)$, then it has linear form (6.66). If $\frac{2}{3}(n+1) < j \le n-1$, we have (6.48) with β_* and $\lambda_*(\beta_*)$ defined in (6.58) and (6.57), respectively. All that remains is to calculate bounds

$$B = \|P_{\succeq_c U} f_{j:n} - \mathbf{1}\|_U = \left(||P_{\succeq_c U} f_{j:n}\|_U^2 - 1\right)^{1/2} \tag{6.67}$$

and transform

$$\frac{F^{-1}(x) - \mu}{\sigma} = \frac{P_{\succeq_c U} f_{j:n}(x) - 1}{B} \tag{6.68}$$

(cf. (6.18)) in order to determine the distribution functions attaining the bounds in the respective cases. □

Analogous bounds for the DFR distributions are presented below. Denote the expectations of consecutive order statistics from the iid sample with the standard exponential distribution $Z(x) = 1 - \exp(-x)$, $x \ge 0$, by

$$S(j,n) = \mathbb{E}Z_{j:n} = \sum_{i=1}^{j} \frac{1}{n+1-i}, \qquad 1 \le j \le n. \tag{6.69}$$

Theorem 6.4 (DFR Distributions) *Let $X_1, \dots, X_n$ be iid with a common distribution function F whose cumulative failure rate Λ_F is concave. We also assume that F has a finite variance σ^2, and denote the respective mean by μ.*

If $S(j,n) \le 1$, then $\mathbb{E}X_{j:n} \le \mu$.

If $1 < S(j,n) \le 2$, then

$$\mathbb{E}\frac{X_{j:n}-\mu}{\sigma} \le S(j,n) - 1 = \sum_{i=1}^{j} \frac{1}{n+1-i} - 1, \tag{6.70}$$

and the equality holds for the exponential distribution with location and scale parameters $\mu - \sigma$ and σ, respectively.

If $S(j,n) > 2$, then

$$\mathbb{E}\frac{X_{j:n}-\mu}{\sigma} \le B \tag{6.71}$$

for

$$\begin{aligned} B^2 = {} & \frac{(n!)^2}{(2n-1)!}\binom{2j-2}{j-1}\binom{2n-2j}{n-j} F_{2j-1:2n-1}(\gamma_*) \\ & + (1-\gamma_*)[2\lambda_*^2 + 2\lambda_* f_{j:n}(\gamma_*) + f_{j:n}^2(\gamma_*)] - 1 \end{aligned} \tag{6.72}$$

with

$$\begin{aligned} \lambda_* &= \lambda_*(-\ln(1-\gamma_*)) \\ &= \frac{\sum_{i=1}^{j} S(j+1-i, n+1-i) f_{i:n+1}(\gamma_*) - (n+1-j) f_{j:n+1}(\gamma_*)}{2(n+1)(1-\gamma_*)} \end{aligned} \tag{6.73}$$

and γ_* being the smallest positive zero of

$$\begin{aligned} L_*(\gamma) = {} & \sum_{i=1}^{j-1} [2 - S(j-1-i, n+1-i)] f_{i:n+1}(\gamma) \\ & - [n-j-1+S(1,n+1-j)] f_{j:n+1}(\gamma). \end{aligned} \tag{6.74}$$

The equality in (6.71) is attained iff

$$F(x) = \begin{cases} 0, & \frac{x-\mu}{\sigma} < -\frac{1}{B}, \\ f_{j:n}^{-1}\left(\frac{x-\mu}{\sigma}B + 1\right), & -\frac{1}{B} \le \frac{x-\mu}{\sigma} < \frac{f_{j:n}(\gamma_*)-1}{B}, \\ 1 - (1-\gamma_*)\exp\left(-\frac{x-\mu}{\sigma}\frac{B}{\lambda_*} - \frac{1-f_{j:n}(\gamma_*)}{\lambda_*}\right) & \frac{x-\mu}{\sigma} \ge \frac{f_{j:n}(\gamma_*)-1}{B}. \end{cases} \tag{6.75}$$

Proof The bounds coincide with the norms of the projections of the compositions of the exponential distribution function with Bernstein polynomials

$$h(x) = f_{j:n} \circ Z(x) = nB_{j-1,n-1}(1 - e^{-x}) \tag{6.76}$$

onto the convex cone of nondecreasing convex and integrating to zero functions in $L^2([0, +\infty), e^{-x}dx)$. The form of projection and its parameters are determined by means of Theorem 6.2. The first step consists in specifying analytic forms of (6.44), (6.45) and (6.46) for h given in (6.76). The calculations are more elaborate than in the DD case, and the final formulae become quite complicated, but fortunately they can be expressed in terms of combinations of the exponential distribution function with the Bernstein polynomials as

$$\lambda_*(\gamma) = \lambda_{*,j:n,Z}(Z^{-1}(\gamma)) = \frac{\sum_{i=1}^{j-1} \lambda_i B_{i,n}(\gamma)}{2(1-\gamma)}, \tag{6.77}$$

$$K(\gamma) = K_{j:n,Z}(Z^{-1}(\gamma)) = \frac{\sum_{i=0}^{j-1} a_i B_{i,n}(\gamma)}{2(1-\gamma)}, \tag{6.78}$$

$$L(\gamma) = L_{j:n,Z}(Z^{-1}(\gamma)) = \frac{1}{2}\sum_{i=0}^{j-1} b_i B_{i,n}(\gamma), \tag{6.79}$$

with

$$\lambda_i = \begin{cases} S(j-i, n-i), & i = 1, \ldots, j-2, \\ \frac{1}{n-j} - (n+1-j), \; i = j-1, \end{cases} \tag{6.80}$$

$$a_i = \begin{cases} S(j-i, n-i), & i = 0, \ldots, j-3, \\ \frac{1}{n+1-j} + \frac{1}{n-j} - 2\frac{(n+2-j)!}{(n-j)!}, & i = j-2, \\ (2n-1-2j)(n+1-j) - \frac{1}{n+1-j}, & i = j-1, \end{cases} \tag{6.81}$$

$$b_i = \begin{cases} 2 - S(j-i, n-i), \; i = 0, \ldots, j-2, \\ -\frac{n-j}{n+1-j}, \quad i = j-1, \end{cases} \tag{6.82}$$

respectively, where

$$\gamma = 1 - \exp(-\beta), \qquad \beta > 0. \tag{6.83}$$

Therefore, the VDP of Bernstein polynomials can be used for analyzing variability of (6.78) and (6.79). The arguments are analogous to those of the previous proof and we only mention the main steps. Examining the signs of (6.81) and referring to negativity of $K(1 - e^{-\beta})$ at the inflection point, we conclude that (6.78) is positive on a right neighborhood of 0. If $b_0 \geq 0$, then projection is linear

$$P_{\succeq_c Z}(f_{j:n} \circ Z)(x) = [S(j,n) - 1]_+ x + 1 - [S(j,n) - 1]_+. \tag{6.84}$$

If $b_0 < 0$, then the projection cannot be linear and has the representation (6.48) for $\beta_* = -\ln(1-\gamma_*)$ with the slope defined in (6.73), where γ_* is determined from (6.74). At the final stage, we determine bounds by calculating the norms of projections and the distributions that provide the extreme expectations of order statistics. □

Note that although conditions on (6.69) for specific forms of bounds are not explicit, we can use the integral approximation

$$\ln \frac{n+1}{n+1-j} < S(j,n) < \ln \frac{n+1/2}{n+1/2-j} \tag{6.85}$$

to conclude that

$$S(j,n) \in \begin{cases} (0,1), & \text{if } j < (1-e^{-1})\left(n+\frac{1}{2}\right), \\ (1,2), & \text{if } (1-e^{-1})(n+1) < j < (1-e^{-2})\left(n+\frac{1}{2}\right), \\ (2,+\infty), & \text{if } j > (1-e^{-2})(n+1). \end{cases}$$

Therefore, we need to verify the original contributions directly for integers $j \in \left((1-e^{-1})\left(n+\frac{1}{2}\right), (1-e^{-1})(n+1)\right)$ and $j \in \left(1-e^{-2})\left(n+\frac{1}{2}\right), (1-e^{-2})(n+1)\right)$. Each of the intervals is shorter than $\frac{1}{2}$ and includes one integer at most

The bounds on the expectations of trimmed means $\frac{1}{k+1-j}\sum_{i=j}^{k} X_{:n}$, $1 \le j \le k \le n$, based on iid samples with parent DD and DFR distributions have been presented by Danielak (2003). It occurs that most of the corresponding functions $\frac{1}{k+1-j}\sum_{i=j}^{k} f_{i:n} \circ W$, $W = U, Z$ satisfy assumptions (A1), and Theorem 6.2 can be applied for determining the bounds. The respective results are similar to those presented in Theorems 6.3 and 6.4, but the formulae are more complicated. We present them here without proofs. Extensions of the above two theorems to arbitrary trimmed means have been presented in Danielak (2003). It can be proved that

$$\bar{f}_{j,k:n} \circ W(x) = \frac{1}{k+1-j}\sum_{i=j}^{k} f_{i:n} \circ W(x),$$
$$W = U, Z, \qquad 1 \le j \le k \le n, \tag{6.86}$$

satisfy assumptions (A1) unless either

$$j = 1,\ W = U, Z \qquad \text{or} \qquad j = k = n,\ W = U. \tag{6.87}$$

In the former case,

$$\frac{1}{k}\sum_{i=1}^{k} X_{i:n} \le \frac{1}{n}\sum_{i=1}^{n} X_i \tag{6.88}$$

and the trivial bounds

$$ET_{1,k:n} \le EX_1 = \mu \tag{6.89}$$

hold. In the latter, we apply the Gumbel-Hartley-David-like general bounds, obtained with straightforward use of the Schwarz inequality. For all the other cases, we directly apply the statements of Lemma A.1 and Theorem 6.2. The final results are similar to those for the single order statistics, though the formulae are more sophisticated here.

Theorem 6.5 (DD Distributions) *Let the assumptions of Theorem* 6.3 *hold.*
If $j + k \le n + 1$, then

$$\mathbb{E}T_{j,k:n} \le \mu. \tag{6.90}$$

If $n + 1 < j + k \le \frac{4}{3}(n + 1)$, then

$$\mathbb{E}\frac{T_{j,k:n} - \mu}{\sigma} \le \sqrt{3}\left(\frac{j + k}{n + 1} - 1\right) \tag{6.91}$$

with the equality for the uniform distribution $[\mu - \sqrt{3}\sigma, \mu + \sqrt{3}\sigma]$.
If $\frac{4}{3}(n + 1) < j + k \le 2n - 1$, then

$$\mathbb{E}\frac{T_{j,k:n} - \mu}{\sigma} \le B \tag{6.92}$$

for

$$\begin{aligned}B^2 = {} & \frac{(n!)^2}{(2n-1)!(k+1-j)^2}\left\{\sum_{i=j}^{k}\binom{2i-2}{i-1}\binom{2n-2i}{n-i}F_{2i-1:2n-1}(\beta_*)\right. \\ & \left. + 2\sum_{i=j+1}^{k}\sum_{m=j}^{i-1}\binom{i+m-2}{i-1}\binom{2n-i-m}{n-i}F_{i+m-1:2n-1}(\beta_*)\right\} \\ & + (1-\beta_*)\bar{f}^2_{j,k:n}(\beta_*) + \lambda_*(1-\beta_*)^2\bar{f}_{j,k:n}(\beta_*) + \frac{1}{3}(1-\beta_*)^3\lambda_*^2 - 1,\end{aligned} \tag{6.93}$$

with

$$\begin{aligned}\lambda_* = \lambda_*(\beta_*) = {} & \frac{3\left\{\sum_{i=j}^{k}\frac{i}{n+1}[1 - F_{i+1:n+1}(\beta_*)] - \beta_*[1 - F_{i:n}(\beta_*)]\right\}}{(k+1-j)(1-\beta_*)^3} \\ & - \frac{3}{2(k+1-j)(1-\beta_*)}\sum_{i=j}^{k} f_{i:n}(\beta_*),\end{aligned} \tag{6.94}$$

and β_* being the smallest positive zero of

$$L(\beta) = \sum_{i=1}^{i-1}(k+1-j)[4n-3k-3j+2i+2]f_{i:n+2}(\beta)$$
$$- \sum_{i=j}^{k}[n^2+3k^2-4nk+(n-k)(2i-1)]f_{i:n+2}(\beta). \qquad (6.95)$$

and the equality in (6.92) holds for (6.59) with $f_{j:n}$ replaced by $\bar{f}_{j,k:n}$.

Let

$$T(j,k,n) = \frac{1}{k+1-j}\sum_{i=j}^{k}\mathbb{E}Z_{i:n} = \frac{1}{k+1-j}\sum_{i=j}^{k}S(i,n)$$
$$= \frac{1}{k+1-j}\sum_{i=1}^{k}\frac{k+1-\max\{i,j\}}{n+1-i}, \qquad 1 \le j \le k \le n, \qquad (6.96)$$

denote the expectations of trimmed means from the standard exponential samples.

Theorem 6.6 (DFR Distributions) *Under the assumptions of Theorem 6.4, we have the following.*

If $T(j,k,n) \le 1$, then

$$\mathbb{E}T_{j,k:n} \le \mu.$$

If $1 < T(j,k,n) \le 2$, then

$$\mathbb{E}\frac{T_{j,k:n}-\mu}{\sigma} \le T(j,k,n) - 1 = \sum_{i=1}^{k}\frac{k+1-\max\{i,j\}}{(k+1-j)(n+1-i)} - 1, \qquad (6.97)$$

with the equality attained for the exponential distribution with location $\mu-\sigma$ and scale σ.

Otherwise,

$$\mathbb{E}\frac{T_{j,k:n}-\mu}{\sigma} \le B \qquad (6.98)$$

for

$$B^2 = \frac{(n!)^2}{(2n-1)!(k+1-j)^2} \left\{ \sum_{i=j}^{k} \binom{2i-2}{i-1} \binom{2n-2i}{n-i} F_{2i-1:2n-1}(\gamma_*) \right.$$

$$\left. + 2 \sum_{i=j+1}^{k} \sum_{m=j}^{i-1} \binom{i+m+2}{i-1} \binom{2n-i-m}{n-i} F_{i+m-1:2n-1}(\gamma_*) \right\}$$

$$+ (1-\gamma_*)[2\lambda_*^2 + 2\lambda_* \bar{f}_{j,k:n}(\gamma_*) + \bar{f}_{j,k:n}^2(\gamma_*)] - 1, \tag{6.99}$$

$$\lambda_* = \lambda_*(-\ln(1-\gamma_*))$$

$$= \frac{\sum_{i=j}^{k} \left\{ \sum_{m=1}^{i} S(i+1-m, n+1-m) f_{m:n+1}(\gamma_*) - (n+1-i) f_{i:n+1}(\gamma_*) \right\}}{2(k+1-j)(n+1)(1-\gamma_*)}, \tag{6.100}$$

and γ_* being the smallest positive zero of

$$L(\gamma) = (k+1-j) \sum_{i=1}^{j-1} [2 - T(j+1-i, k+1-i, n+1-i)] f_{i:n+1}(\gamma)$$

$$+ \{(k+1-j)[2 - T(1, k+1-j, n+1-j)] - n - 1 + j\} f_{j:n+1}(\gamma)$$

$$+ \sum_{i=j+1}^{k} \{(k+1-i)[2 - T(1, k+1-i, n+1-i)] - n - 1 + i\} f_{i:n+1}(\gamma). \tag{6.101}$$

Then the equality condition is (6.75) *with $f_{j:n}$ replaced by $\bar{f}_{j,k:n}$.*

Theorem 6.7 (DRFR Distributions) *Consider the iid sequence $X_1, \ldots, X_n$ of random variables based on the distribution function F with concave* $\ln F$*, expectation μ and standard deviation σ, and let $X_{1:n}, \ldots, X_{n:n}$ be respective order statistics. For fixed $2 \le j \le n-1$, we define*

$$\lambda_{*,j:n,V}(\beta) = \frac{1}{\beta^2 + 2\beta - 2e^\beta + 2}$$

$$\times \left[f_{j:n}(e^\beta)(1 - e^\beta + \beta) - \beta - \sum_{i=j}^{n} \frac{1}{i}(1 - F_{i:n}(e^\beta)) \right], \tag{6.102}$$

$$K_{j:n,V}(\beta) = j[f_{j+1:n}(e^{\beta}) - f_{j:n}(e^{\beta})] + \frac{1}{\beta^2 + 2\beta - 2e^{\beta} + 2}$$
$$\times \left[(\beta^2 + 3\beta - 3e^{\beta} + 3) f_{j:n}(e^{\beta}) - \beta - \sum_{i=j}^{n} \frac{1}{i}(1 - F_{i:n}(e^{\beta})) \right], \tag{6.103}$$

$$L_{j:n,V}(\beta) = 1 - F_{j:n}(e^{\beta}) - f_{j:n}(e^{\beta})(1 - e^{\beta}) - \frac{e^{\beta} - \beta - 1}{\beta^2 + 2\beta - 2e^{\beta} + 2}$$
$$\times \left[(1 - e^{\beta} + \beta) f_{j:n}(e^{\beta}) - \beta - \sum_{i=j}^{n} \frac{1}{i}(1 - F_{i:n}(e^{\beta})) \right]. \tag{6.104}$$

If either $j = 1$ or $2 \le j \le n-1$ and the set

$$\mathcal{K}_{j:n,V} = \{-\infty < \beta \le b = v_1(j,n) : K_{j:n,V}(\beta) \ge 0 \text{ and } L_{j:n,V}(\beta) = 0\} \tag{6.105}$$

(see (6.33) for the definition of v_1) is empty, then we have $\mathbb{E}X_{j:n} \le u$.

If the set (6.105) is nonempty and $\beta_* = \sup \mathcal{K}_{j:n,V}$, then we have the following bound:

$$\mathbb{E}\frac{X_{j:n} - \mu}{\sigma} \le B, \tag{6.106}$$

where

$$B^2 = 2n \frac{\binom{2j-2}{j-1}\binom{2n-2j}{n-j}}{\binom{2n}{n}} F_{2j-1:2n-1}(e^{\beta_*}) + \frac{1}{\beta_*^2 + 2\beta_* - 2e^{\beta_*} + 2}$$
$$\times \left[\left[(1 - e^{\beta_*})^2 - \beta_*^2 e^{\beta_*} \right] f_{j:n}^2(e^{\beta_*}) + \left(\beta + \sum_{i=j}^{n} \frac{1}{i}(1 - F_{i:n}(e^{\beta_*})) \right)^2 \right] - 1. \tag{6.107}$$

The equality in (6.106) holds for the absolutely continuous DRFR distribution function

$$F(x) = \begin{cases} f_{j:n}^{-1}\left(1 + \frac{x-\mu}{\sigma} B\right), & \frac{x-\mu}{\sigma} < \frac{[f_{j:n}(e^{\beta_*}) - 1]}{B}, \\ \exp\left(\frac{\frac{x-\mu}{\sigma} B + 1 - f_{j:n}(e^{\beta_*})}{\lambda_{*,j:n,V}(\beta_*)} + \beta_* \right), & \frac{[f_{j:n}(e^{\beta_*}) - 1]}{B} \le \frac{x-\mu}{\sigma} \\ & < \frac{[f_{j:n}(e^{\beta_*}) - \lambda_{*,j:n,V}(\beta_*)\beta_* - 1]}{B}, \\ 1, & \frac{x-\mu}{\sigma} \ge \frac{[f_{j:n}(e^{\beta_*}) - \lambda_{*,j:n,V}(\beta_*)\beta_* - 1]}{B}, \end{cases} \tag{6.108}$$

Proof For $j = 1$ the conclusion is trivial. For $2 \leq j \leq n-1$, we directly apply the conclusions of Theorem 6.2 with specific $W = V$ and $h = h_{j:n,V}$. Simple calculations show that (6.44)–(6.46) take then on the particular forms (6.102)–(6.104). If (6.105) is empty, in view of the fact that here $a = -\infty$, the only possible form of projection is $P_{\succeq_c V} h_{j:n,V} = \mathbf{1}$. This implies the zero bound (see (6.52)).

If (6.105) is nonempty, and β_* is its maximal point, then $P_{\succeq_c V} h_{j:n,V}$ has the form (6.48) with $h = h_{j:n,V}$ and $\lambda_* = \lambda_{*,j:n,V}$. Its squared norm amounts to

$$\|P_{\succeq_c V} h_{j:n,V}\|_V^2 = \int_{-\infty}^{0} [P_{\succeq_c V} h_{j:n,V}(x)]^2 e^x \, dx$$

$$= 2n \frac{\binom{2j-2}{j-1}\binom{2n-2j}{n-j}}{\binom{2n}{n}} F_{2j-1:2n-1}(e^{\beta_*}) + \lambda_{j:n,V}^2(\beta_*)(\beta_*^2 + 2\beta_* - 2e^{\beta_*} + 2)$$

$$+ 2\lambda_{j:n,V}(\beta_*) f_{j:n}(e^{\beta_*})(e^{\beta_*} - 1 - \beta_*) + f_{j:n}^2(e^{\beta_*})(1 - e^{\beta_*}) = B^2 + 1$$

(cf. (6.19) and (6.107)). Due to (6.19), the inequality (6.106) with B defined in (6.107) holds. The equality in (6.106) occurs when (6.20) is satisfied. The respective distribution function has the explicit form (6.108). □

Here we do not obtain more precise results like in Theorems 6.3 and 6.4, because we are unable to use the VDP for analyzing functions (6.102)–(6.104). Although it does not directly follow from the statement of Theorem 6.7, the set (6.105) is empty for small values of j, and then we obtain the zero bound, whereas it is nonempty for large j, and then we obtain a strictly positive bound defined in (6.107). We do not have precise descriptions of $j = j(n)$ at which the values of bounds switch from 0 to positive bounds. We can only say that such indices satisfy $j(n) \leq \frac{n+1}{2}$. This follows from the fact that the bounds in the DD case amount to 0 for $j \leq \frac{n+1}{2}$, and they are positive otherwise (see, e.g., Rychlik, 2009b). The relations $V \prec_c U \prec_c Z$ imply that DFR $\subset$ DD $\subset$ DRFR. This in particular means that the bounds for the expectations of given order statistics from decreasing reversed failure rate distributions cannot be less than the corresponding bounds for the decreasing density case.

The bounds for the sample maximum in the DD and DRFR cases

$$\mathbb{E}\frac{X_{n:n} - \mu}{\sigma} \leq \frac{n-1}{\sqrt{2n-1}} \tag{6.109}$$

coincide with the general one (see Theorem 4.1). It follows from the fact that the distribution function (4.21) attaining the bound has a decreasing density function

$$f(x) = \frac{\left[\frac{1}{n}\left(\frac{n-1}{\sqrt{2n-1}}\frac{x-\mu}{\sigma} + 1\right)\right]^{-\frac{n-2}{n-1}}}{n\sqrt{2n-1}\sigma}, \quad -\frac{\sqrt{2n-1}}{n-1} \leq \frac{x-\mu}{\sigma} \leq \sqrt{2n-1}, \tag{6.110}$$

on its support, and consequently it belongs to the DRFR family as well.

Some modifications of Theorems 6.3 and 6.4 have been presented by Gajek and Rychlik (1998) who have determined optimal evaluations of order statistics of nonnegative samples from DD and DFR distributions in terms of the square root of the second raw population moment. They are based on the following general result.

Theorem 6.8 *Suppose that $X_1, \ldots, X_n$ are iid nonnegative random variables with a common distribution function $F \succeq_c W$ and finite second raw moment $m_2^2 = \mathbb{E}X_1^2$. Then for arbitrary $\tilde{c} \in \mathbb{R}^n$, we have*

$$\mathbb{E}\sum_{i=1}^{n} c_i \frac{X_{i:n}}{m_2} \leq \| P^+_{\succeq_c W}(f_{\tilde{c}:n} \circ W)\|_W \tag{6.111}$$

with the equality condition

$$\frac{F^{-1} \circ W(x)}{m_2} = \frac{P^+_{\succeq_c W}(f_{\tilde{c}:n} \circ W)(x)}{\| P^+_{\succeq_c W}(f_{\tilde{c}:n} \circ W)\|_W} \tag{6.112}$$

with $P^+_{\succeq_c W}$ denoting the projection operator onto

$$\mathcal{C}^+_{\succeq_c W} = \{g \in L^2((a, d), W(dx)) : g(a) = 0, \quad g \text{ is nondecreasing and convex}\} \tag{6.113}$$

for some $a > -\infty$.

For functions satisfying assumptions (A1) with finite a, the projections onto (6.113) are described in the following theorem.

Theorem 6.9 *Under notation* (6.44)–(6.47), *if* (6.47) *is nonempty and β_* is its maximal element, then $P^+_{\succeq_c W}h$ has the form* (6.48). *Otherwise,*

$$P^+_{\succeq_c W}h(x) = \lambda_*(a)(x - a). \tag{6.114}$$

The first statement is obvious. If $P_{\succeq_c W}h \in \mathcal{C}^+_{\succeq_c W} \subset \mathcal{C}_{\succeq_c W}$, then $P_{\succeq_c W}h = P^+_{\succeq_c W}h$ as well. The other asserts that otherwise the projection is linear. It certainly starts from the zero level and has the optimal slope (6.44) which is positive by positivity of $h(x) - h(a) = h(x)$, $a < x < d$. The solution seems reasonable, but it cannot be justified in a few words. It has been derived by means of a geometric proof by careful analysis of various cases (see Gajek & Rychlik, 1998, Lemmas 1 and 2). Applications of Theorem 6.9 are presented below.

Theorem 6.10 (DD Distributions) *Let $X_1, \ldots, X_n$ be nonnegative iid random variables with a concave distribution function F and finite second moment m_2^2.*

If $j \le \min\{\frac{2}{3}(n+1), n-1\}$, *then*

$$\mathbb{E}\frac{X_{j:n}}{m_2} \le \sqrt{3}\frac{j}{n+1}, \tag{6.115}$$

which is the equality for the uniform distribution on the interval $[0, \sqrt{3}m_2]$.

If $\frac{2}{3}(n+1) < j \le n-1$, *then under notation* (6.56)–(6.58),

$$\mathbb{E}\frac{X_{j:n}}{m_2} \le B_1 = \sqrt{B^2+1} \tag{6.116}$$

holds true, and becomes equality for

$$F(x) = \begin{cases} 0, & x < 0, \\ f_{j:n}^{-1}\left(\frac{x}{m_2}B_1\right), & 0 \le \frac{x}{m_2} < \frac{f_{j:n}(B_*)}{B_1}, \\ \frac{x}{m_2}\frac{B_1}{\lambda_*} + \beta_* - \frac{f_{j:n}(\beta_*)}{\lambda_*}, & \frac{f_{j:n}(\beta_*}{B_1} \le \frac{x}{m_2} < \frac{f_{j:n}(\beta_*)+\lambda_*(1-\beta_*)}{B_1}, \\ 1, & \frac{X}{m_2} \ge \frac{f_{j:n}(\beta_*)+\lambda_*(1-\beta_*)}{B_1}. \end{cases} \tag{6.117}$$

For $j = n$, *we have*

$$\mathbb{E}\frac{X_{n:n}}{m_2} \le \frac{n}{\sqrt{2n-1}} \tag{6.118}$$

and the equality holds for the distribution function $F(x) = \left(\frac{x^2}{(2n-1)m_2^2}\right)^{\frac{1}{2(n-1)}}$, $0 < x < \sqrt{2n-1}m_2$.

Bound (6.118) for the sample maximum is sharp for arbitrary marginal distribution functions F.

Theorem 6.11 (DFR Distributions) *Suppose that* $X_1, \ldots, X_n$ *are nonnegative iid random variables with distribution function* F *that has concave cumulative failure rate* $\Lambda_F = -\ln(1-F)$ *and finite second moment* m_2^2.

If $S(j,n) \le 2$, *then*

$$\mathbb{E}\frac{X_{j:n}}{m_2} \le \frac{S(j,n)}{\sqrt{2}} \tag{6.119}$$

which becomes an equality for the exponential distribution with scale parameter $m_2/\sqrt{2}$.

Otherwise, with the notation (6.72)–(6.74),

$$\mathbb{E}\frac{X_{j:n}}{m_2} \le B_1 = \sqrt{B^2+1} \tag{6.120}$$

and the equality holds for

$$F(x) = \begin{cases} 0, & x < 0, \\ f_{j:n}^{-1}\left(\frac{x}{m_2}B_1\right), & 0 \le x < \frac{f_{j:n}(\gamma_*)}{B_1}, \\ 1-(1-\gamma_*)\exp\left(-\frac{x}{m_2\lambda_*} - \frac{f_{j:n}(\gamma_*)}{\lambda_*}\right), & x > \frac{f_{j:n}(\gamma_*)}{B_1}. \end{cases} \quad (6.121)$$

6.1.3 Positive Upper Bounds for Spacings

Now we consider the problem of establishing optimal bounds on the expectations of spacings from the DD and DFR populations considered in Danielak and Rychlik (2004). The spacings coming from the DRFR families are discussed in Sect. 6.3.7. Due to Theorem 6.1, the basic step is in determining the L^2-projections $P_{\succeq_c W} h$ of

$$h = f_{j,j+1:n} \circ W = f_{j+1:n} \circ W - f_{j:n} \circ W, \quad 1 \le j \le n-1, \quad (6.122)$$

for $W = U, Z$ onto (6.11). Observe that (6.122) are orthogonal to the constant functions in $L^2((a,d), W(dx))$ and by Lemma 6.1,

$$P^0_{\succeq_c W} h = P_{\succeq_c W} h. \quad (6.123)$$

Most of the functions (6.122) fulfill the following assumptions:

(A2) For some $0 < a < b < c \le d \le +\infty$ and a distribution function W with a positive density w on $(0,d)$ and finite second moment, function $h \in L^2((0,d), W(dx)$ is bounded, twice differentiable, and satisfies $h(0) = 0$, $\lim_{x \nearrow d} h(x) \ge 0$,

$$\int_0^d h(x)w(x)dx = 0, \quad (6.124)$$

and it is decreasing on $(0,a)$ and (c,d) (if the latter is not empty), strictly convex increasing on (a,b), and strictly concave increasing on (b,c).

Conditions (A2) imply that h has a unique zero, say θ, in $(0,d)$ located between a and c, but the mutual relation between θ and the inflection point b is not determined.

Lemma 6.5

(i) Let $W = U$. If $n = 2$, then $f_{1,2:2}$ is linear increasing. If $n \ge 3$, then for $j = 1$ the function $f_{1,2:n}$ is first concave increasing, then decreasing, $f_{j,j+1:n}$, $2 \le j \le n-2$, are consecutively decreasing, convex increasing, concave increasing and decreasing, and $f_{n-1,n:n}$ is first decreasing, then convex

increasing. Moreover, functions $f_{j,j+1:n}$ *have unique zeros in* $(0, 1)$ *located at* $\theta = \frac{j}{n}$.

(ii) *Let* $W = Z$. *If* $n = 2$, *then* $f_{1,2:2} \circ Z$ *is concave increasing. If* $n \geq 3$, *then* $f_{1,2:n} \circ Z$ *is first concave increasing, then decreasing, for* $2 \leq j \leq n-2$ *the compositions* $f_{j,j+1:n} \circ Z$ *are decreasing, convex increasing, concave increasing and decreasing, and finally* $f_{n-1,n:n} \circ Z$ *is first decreasing, then convex increasing and ultimately concave increasing. Functions* $f_{j,j+1:n} \circ Z$ *have unique zeros in* $(0, \infty)$ *located at* $\theta = -\ln\left(1 - \frac{j}{n}\right)$.

Proof

(i) Let $W = U$. For $n = 2$ we have $f_{1,2:2}(x) = 4x - 2$. Assuming $n \geq 3$, we get

$$f'_{j,j+1:n}(x) = n(n-1)\left[-B_{j-2,n-2}(x) + 2B_{j-1,n-2}(x) - B_{j,n-2}(x)\right]$$

$$f''_{j,j+1:n}(x) = n(n-1)(n-2) \times \left[-B_{j-3,n-3}(x) + 3B_{j-2,n-3}(x) - 3B_{j-1,n-3}(x) + B_{j,n-3}(x)\right].$$

If $2 \leq j \leq n-2$, the VDP asserts that $f'_{j,j+1:n}$ is either first negative, then positive and ultimately negative ($-+-$, for brevity) or negative everywhere in $(0, 1)$. The latter one is impossible because $f_{j,j+1:n}$ integrates to 0 in $(0, 1)$ and vanishes at 0 and 1. Thus, $f_{j,j+1:n}$ has first a minimum, then a maximum, and it is convex and concave about the minimum and maximum, respectively. This, combined with Lemma 4.1, implies that $f''_{j,j+1:n}$ is $-+-+$ and our claim then follows. Similar, but simplified considerations, apply to the remaining cases.

(ii) Assume that $W = Z$. For $n = 2$, the function $f_{1,2:2} \circ Z(x) = 2\left(1 - 2e^{-x}\right)$ is concave increasing on $[0, \infty)$. Take $n \geq 3$. Denoting $C_{j,m}(x) = B_{j,m} \circ Z(x)$, we get

$$f_{j,j+1:n} \circ Z(x) = n\left[-C_{j-1,n-1}(x) + C_{j,n-1}(x)\right]$$

and

$$\left(f_{j,j+1:n} \circ Z\right)'(x) = n(n-1)e^{-x}\left[-C_{j-2,n-2}(x) + 2C_{j-1,n-2}(x) - C_{j,n-2}(x)\right],$$

$$\begin{aligned}\left(f_{j,j+1:n} \circ Z\right)''(x) = n(n-1)(n-2)e^{-x}&\left[-(n-j+1)C_{j-3,n-2}(x)\right.\\ &+ (3n-3j+1)C_{j-2,n-2}(x) - (3n-3j-1)C_{j-1,n-2}(x)\\ &+ \left.(n-j-1)C_{j,n-2}(x)\right].\end{aligned}$$

Since each $C_{l,m}$ is a superposition of increasing function Z and a Bernstein polynomial, Lemma 4.1 holds for linear combinations of $C_{l,m}$ as well. Analyzing signs of $\left(f_{j,j+1:n} \circ Z\right)'$ and $\left(f_{j,j+1:n} \circ Z\right)''$, analogously to the proof of part (i), we easily obtain the desired conclusions.

□

It follows that $f_{j,j+1:n} \circ W$ satisfy $(A2)$ for $W = U$, $2 \le j \le n-2$ and $W = V$, $2 \le j \le n-1$, and $h = f_{n-1,n:n} \circ Z$ are the only functions which do not eventually decrease on the right. For establishing the bounds on expected spacings coming from DD and DFR populations, we determine the projections of respective $f_{j,j+1:n} \circ W$ onto the convex cone of nondecreasing convex functions (6.11) in $L^2((0,d), W(dx)$. Note that here 0 stands as the left-end point of the support of W, and a is used for denoting the minimum point of h. From now on, we assume that $h \in L^2((0,d), W(dx)$ satisfies $(A2)$.

Theorem 6.12 *Assume conditions* (A2) *and define*

$$\bar{h}(\alpha) = \frac{\int_0^\alpha h(x)W(dx)}{\int_0^\alpha W(dx)}, \qquad a < \alpha < d. \tag{6.125}$$

Then there exists a unique $\alpha_* \in (a, \theta)$ solving the equation

$$h(\alpha) = \bar{h}(\alpha). \tag{6.126}$$

Moreover, with the notation (6.44)–(6.46), if

$$\mathcal{K} = \{\beta \in (\alpha, b) : K(\beta) \ge 0,\ L(\beta) = 0\} \tag{6.127}$$

is nonempty, then

$$P_{\succeq_c W}h = \begin{cases} h(\alpha_*), & 0 \le x \le \alpha_*, \\ h(x), & \alpha_* \le x \le \beta_*, \\ h(\beta_*) + \lambda_*(\beta_*)(x - \beta_*), & \beta_* \le x < d, \end{cases} \tag{6.128}$$

for β_* being the maximal element of (6.127).

Otherwise,

$$P_{\succeq_c W}h(x) = \begin{cases} \bar{h}(\beta_*), & 0 \le x \le \beta_*, \\ \bar{h}(\beta_*)\left[1 - \frac{x-\beta_*}{\int_{\beta_*}^d (y-\beta_*)W(dy)}\right], & \beta_* \le x < d, \end{cases} \tag{6.129}$$

for β_* being the maximal element of $[0, \alpha_*]$ satisfying

$$\int_\beta^d h(x)W(dx)\left\{\int_\beta^d (x-\beta)^2 W(dx) - \left[\int_\beta^d (x-\beta)W(dx)\right]^2\right\}$$
$$= \int_0^\beta W(dx) \int_\beta^d (x-\beta)W(dx) \int_\beta^d (x-\beta)h(x)W(dx). \tag{6.130}$$

The proof of Theorem 6.12 is elaborate and it is postponed to the Appendix (see Theorem A.2). It can also be found in the paper of Danielak and Rychlik (2004).

Before we apply Theorem 6.12 for establishing the bounds for spacings $X_{j+1:n} - X_{j:n}$, $j = 2, \ldots, n-2$, in the DD case and $j = 2, \ldots, n-1$ for the DFR case, we consider the remaining cases for which the projections and bounds are determined much easier. For $j = 1$ and $n \geq 3$, the functions $f_{1,2:n}$ and $f_{1,2:n} \circ Z$ are first concave increasing and ultimately decreasing. It follows that they satisfy assumptions (A2) with $b = a = 0$ (up to a vertical translation), and their projections are linear. Therefore, we have

$$\mathbb{E}\frac{X_{2:n} - X_{1:n}}{\sigma} \leq \frac{2\sqrt{3}}{n+1} \tag{6.131}$$

for the DD distributions, with an equality attained by the uniform distributions supported on intervals of length $2\sqrt{3}\sigma$ and

$$\mathbb{E}\frac{X_{2:n} - X_{1:n}}{\sigma} \leq \frac{1}{n-1} \tag{6.132}$$

for the DFR ones, and these are attained by the exponential distributions with the scale parameter σ. Actually, bounds (6.131) and (6.132) can also be extended to $n = 2$. In the DD case, $f_{1,2:2}$ is linear increasing, and $P^{\nearrow}_{\succeq_c U} f_{1,2:2} = f_{1,2:2}$ provides the well-known Plackett bound for the range (see Theorem 4.2), attained by the uniform distribution. In the DFR case, $f_{1,2:2} \circ Z$ is concave increasing, so that it has a linear projection as well. This allows us to extend formula (6.132) to the case $n = 2$ as well.

If $j = n - 1 \geq 2$ and $W = U$, then $h = f_{n-1,n:n}$ are first decreasing and then convex increasing. Therefore, the projection $P^{\nearrow} f_{n-1,n:n}$ onto the family of nondecreasing functions is first constant and then convex increasing. This is an element of $C_{\succeq_c U}$, and the general bounds of Corollary 4.4 derived by direct application of the Schwarz inequality is also valid for the DD family of distributions.

Theorem 6.13 (DD Distributions) *Assume that $X_1, \ldots, X_n$ are iid, have a concave distribution function and finite variance. For $2 \leq j \leq n-2$, set*

$$\begin{aligned} K_*(\gamma) = & \sum_{i=1}^{j-2} f_{i:n+2}(\gamma) + \left[1 + \frac{(n+3-j)!}{3(n-j)!}\right] f_{j-1:n+2}(\gamma) \\ & + \left[1 + \frac{(n-j+2)!}{(n-j)!}\left(\frac{1}{2} - \frac{2}{3}(n-j)\right)\right] f_{j:n+2}(\gamma) \\ & + \left[1 + \frac{(n+1-j)!}{(n-1-j)!}\left(\frac{1}{3}(n-j) - \frac{5}{6}\right)\right] f_{j+1:n+2}(\gamma), \end{aligned} \tag{6.133}$$

$$L_*(\beta) = -\sum_{i=1}^{j-1} f_{i:n+2}(\beta) + \left[\frac{(n+2-j)!}{6(n-j)!} - 1\right] f_{j:n+2}(\beta)$$

$$-\left[\frac{1}{6}(n+1-j)(n-4-j)+1\right] f_{j+1:n+2}(\beta). \tag{6.134}$$

Let $\alpha_* = \frac{j-1}{n-1}$ *and* γ_* *be the smallest positive zero of (6.133). If*

$$K_*(\alpha_*) > 0 \qquad \text{and} \qquad L_*(\alpha_*) < 0 \le L_*(\gamma_*), \tag{6.135}$$

then

$$\mathbb{E}\frac{X_{j+1:n} - X_{j:n}}{\sigma} \le B \tag{6.136}$$

for

$$\begin{aligned}
B^2 &= \alpha_*[f_{j+1:n}(\alpha_*) - f_{j:n}(\alpha_*)]^2 \\
&+ \frac{(n!)^2}{(2n-1)!}\left\{\binom{2j}{j}\binom{2(n-j-1)}{n-j-1}[F_{2j+1:2n-1}(\beta_*) - F_{2j+1:2n-1}(\alpha_*)]\right. \\
&- 2\binom{2j-1}{j}\binom{2n-2j-1}{n-j}[F_{2j:2n-1}(\beta_*) - F_{2j:2n-1}(\alpha_*)] \\
&+ \left.\binom{2j-2}{j-1}\binom{2n-2j}{n-j}[F_{2j-1:2n-1}(\beta_*) - F_{2j-1:2n-1}(\alpha_*)]\right\} \\
&+ \frac{1}{3}\lambda_*^2(1-\beta_*)^3 + \lambda_*(1-\beta_*)^2[f_{j+1:n}(\beta_*) - f_{j:n}(\beta_*)] \\
&+ (1-\beta_*)[f_{j+1:n}(\beta_*) - f_{j:n}(\beta_*)]^2
\end{aligned} \tag{6.137}$$

with β_* *being the smallest positive zero of* (6.134) *and*

$$\begin{aligned}
\lambda_* = \lambda_*(\beta_*) &= \frac{3[1 - F_{j+1:n+1}(\beta_*)]}{(n+1)(1-\beta_*)^3} \\
&- \frac{3(n+1-j)[(n-j)f_{j+1:n+2}(\beta_*) - (n-j+2)f_{j:n+2}(\beta_*)]}{2(n+1)(n+2)(1-\beta_*)^3}.
\end{aligned} \tag{6.138}$$

The equality in (6.136) *is attained if*

$$F(x) = \begin{cases} 0, & \frac{x-\mu}{\sigma} < a_1, \\ f_{j,j+1:n}^{-1}\left(\frac{x-\mu}{\sigma}B\right), & a_1 \le \frac{x-\mu}{\sigma} < a_2, \\ \frac{x-\mu}{\sigma}\frac{B}{\lambda_*} - \frac{f_{j,j+1:n}(\beta_*)}{\lambda_*} + \beta_*, & a_2 \le \frac{x-\mu}{\sigma} < a_3, \\ 1, & \frac{x-\mu}{\sigma} \ge a_3, \end{cases} \tag{6.139}$$

for $a_1 = \frac{f_{j,j+1:n}(\alpha_*)}{B}$, $a_2 = \frac{f_{j,j+1:n}(\beta_*)}{B}$, *and* $a_3 = \frac{\lambda_*(1-\beta_*)+f_{j,j+1:n}(\beta_*)}{B}$.

If (6.135) *does not hold, then*

$$\mathbb{E}\frac{X_{j+1:n} - X_{j:n}}{\sigma} \le \frac{f_{j+1:n+1}(\beta_*)}{\beta_*(n+1)}\sqrt{\frac{1+3\beta_*}{3(1-\beta_*)}} \tag{6.140}$$

for $\beta_* \in (0, 1)$ *being the unique root of*

$$6\sum_{i=1}^{j+1} i f_{i+1:n+3}(\beta) = (n+1-j)[(n+2-j)f_{j+1:n+3}(\beta)+4(j+1)f_{j+2:n+3}(\beta)]. \tag{6.141}$$

The equality in (6.140) *holds for*

$$F(x) = \begin{cases} 0, & \frac{x-\mu}{\sigma} < a_4, \\ \beta_* + \frac{(1-\beta_*)^2}{2}\left(1 + \frac{x-\mu}{\sigma}\sqrt{\frac{1+3\beta_*}{3(1-\beta_*)}}\right), & a_4 \le \frac{x-\mu}{\sigma} < a_5, \\ 1, & \frac{x-\mu}{\sigma} \ge a_5, \end{cases} \tag{6.142}$$

for $a_4 = -\sqrt{\frac{3(1-\beta_*)}{1+3\beta_*}}$ *and* $a_5 = \frac{1+\beta_*}{1-\beta_*}\sqrt{\frac{3(1-\beta_*)}{1+3\beta_*}}$.

Proof Theorem 6.13 is deduced from Theorem 6.12 in the following way. For $h = f_{j+1:n} - f_{j:n}$, (6.126) has a simple solution $\alpha_* = \frac{j-1}{n-1}$. Functions (6.133) and (6.134) are identical with (6.45) and (6.46), respectively, up to some positive functional factors. They are combinations of Bernstein polynomials of degree $n+1$, and the VDP can be used for the analysis of their sign changes. This implies that (6.133) changes the sign once in $(0, b)$, and is $+-$ there. Furthermore, (6.134) is first negative and changes the sign from $-$ to $+$ once at most. The necessary and sufficient conditions for the projection of the form (6.128) is that the first zero γ_*, say, of (6.133) lies in (α_*, b), and (6.134) changes the sign from $-$ to $+$ in (α_*, γ_*), which is equivalent to (6.135). Otherwise, the projection is the broken line (6.129) with the breaking point determined by (6.130), which for $h = f_{j,j+1:n}$ simplifies to (6.141). The difference of the LHS and RHS is a combination of Bernstein polynomials which changes the sign once in (0, 1). Distributions functions (6.139) and (6.142) have finite supports. They have atoms of height $\alpha_* = \frac{j-1}{n-1}$ and $\beta_* < \alpha_*$, respectively, at the left-end support points, and uniform right tails. The former also has a decreasing density on the left which is the inverse of a positive polynomial on its domain of increase. □

Solutions to the bounds problems for the DFR class of distributions are represented in a similar way. The main difference is that the bounds are attained by distributions with exponential right tails.

Theorem 6.14 (DFR Distributions) *Assume that $X_1, \ldots, X_n$ are iid with a distribution function F that has a concave cumulative failure rate $\Lambda_F = -\ln(1-F)$ and finite variance. For $2 \le j \le n-1$, define*

$$K_*(\gamma) = \frac{1}{n-j}\sum_{i=1}^{j-2} f_{i:n+2}(\gamma) - \left[\frac{1}{n-j} + 2\frac{(n+2-j)!}{(n-j)!}\right] f_{j-1:n+2}(\gamma)$$
$$+ \left[\frac{1}{n-j} + n + 1 - j - 4\frac{(n+1-j)!}{(n-1-j)!}\right] f_{j:n+2}(\gamma)$$
$$+ \left[\frac{1}{n-j} + (n-j)(2n-3-2j)\right] f_{j+1:n+2}(\gamma) \tag{6.143}$$

and

$$L_*(\beta) = -\sum_{i=1}^{j-1} f_{i:n+1}(\beta) + \left[\frac{(n+1-j)!}{(n-1-j)!} - 1\right] f_{j:n+1}(\beta)$$
$$- [(n-j)(n-j-2)+1] f_{j+1:n+1}(\beta). \tag{6.144}$$

If for $\alpha_* = \frac{j-1}{n-1}$ and γ_* being the smallest positive zero of (6.143), we have

$$K_*(\alpha_*) > 0 \quad \text{and} \quad L_*(\alpha_*) < 0 \le L_*(\gamma_*), \tag{6.145}$$

then

$$\mathbb{E}\frac{X_{j+1:n} - X_{j:n}}{\sigma} \le B \tag{6.146}$$

for

$$B^2 = \alpha_* f_{j,j+1:n}^2(\alpha_*) + (1-\beta_*)[2\lambda_*^2 - 2\lambda_* f_{j,j+1:n}(\beta_*) + f_{j,j+1:n}^2(\beta_*)]$$
$$+ \frac{(n!)^2}{(2n-1)!}\left\{\binom{2j}{j}\binom{2(n-j-1)}{n-j-1}[F_{2j+1:2n-1}(\beta_*) - F_{2j+1:2n-1}(\alpha_*)]\right.$$
$$- 2\binom{2j-1}{j}\binom{2n-2j-1}{n-j}[F_{2j:2n-1}(\beta_*) - F_{2j:2n-1}(\alpha_*)]$$
$$\left.+ \binom{2j-2}{j-1}\binom{2n-2j}{n-j}[F_{2j-1:2n-1}(\beta_*) - F_{2j-1:2n-1}(\alpha_*)]\right\} \tag{6.147}$$

with β_* being the smallest positive zero to (6.144) and

$$\lambda_* = \lambda_*(-\ln(1-\beta_*)) = \frac{1}{2(1-\beta_*)}\Bigg[\frac{1}{n-j}\sum_{i=1}^{j+1} f_{i:n+1}(\beta_*)$$
$$-(n+1-j)f_{j:n+1}(\beta_*) - (n-j)f_{j+1:n+1}(\beta_*)\Bigg]. \quad (6.148)$$

The equality in (6.146) holds for

$$F(x) = \begin{cases} 0, & \frac{x-\mu}{\sigma} < a_1, \\ f^{-1}_{j:j+1:n}\left(\frac{x-\mu}{\sigma}B\right), & a_1 \le \frac{x-\mu}{\sigma} < a_2, \\ 1-(1-\beta_*)\exp\left(-\frac{x-\mu}{\sigma}\frac{B}{\lambda_*} + +\frac{f_{j,j+1:n}(\beta_*)}{B}\right), & \frac{x-\mu}{\sigma} \ge a_2. \end{cases} \quad (6.149)$$

with $a_1 = \frac{f_{j,j+1:n}(\alpha_*)}{B}$ and $a_2 = \frac{f_{j,j+1:n}(\beta_*)}{B}$.
Otherwise,

$$\mathbb{E}\frac{X_{j+1:n} - X_{j:n}}{\sigma} \le \frac{f_{j,j+1:n}(\beta_*)}{(n+1)\beta_*}\sqrt{\frac{1+\beta_*}{1-\beta_*}}, \quad (6.150)$$

where $\beta_* \in (0, 1)$ is the unique root of the equation

$$\frac{1}{n-j}\sum_{i=1}^{j+1} i f_{i:n+2}(\beta) = (n+1-j)f_{j+1:n+2}(\beta) + 2(j+1)f_{j+2:n+2}(\beta). \quad (6.151)$$

The equality in (6.150) is attained by

$$F(x) = \begin{cases} 0, & \frac{x-\mu}{\sigma} < -\sqrt{\frac{1-\beta_*}{1+\beta_*}} \\ 1-(1-\beta_*)\exp\left(-\frac{x-\mu}{\sigma}\sqrt{1-\beta_*^2} - 1 + \beta_*\right), & \frac{x-\mu}{\sigma} \ge -\sqrt{\frac{1-\beta_*}{1+\beta_*}}. \end{cases} \quad (6.152)$$

The proof is similar and is therefore not presented here.

6.1.4 Nonpositive Upper Bounds for Single Order Statistics

Now we focus on the small order statistics from the DD and DFR for which the projection bounds described above vanish. In contrast to the previous results, in this case, we are able to determine the bounds not only in the standard deviation units, but also in other scale units generated by pth absolute cental moments for any $p \ge 1$. The results are based on Ryuchlik (2009a,2009b). We first consider the family of parent distributions with decreasing density functions, and verify when zero upper bounds cannot be improved.

Theorem 6.15 (DD Distributions) *Let $X_1, \dots, X_n$ be iid with a common nondegenerate distribution function F that has a nonincreasing density function and expectation μ. If $\frac{j}{n+1} \le \frac{1}{2}$, then*

$$\frac{\mathbb{E}X_{j:n} - \mu}{\sigma_p} \le 0, \qquad 1 \le p \le \infty, \tag{6.153}$$

and the equality holds iff $\frac{j}{n+1} = \frac{1}{2}$ and F is a uniform distribution function.

Proof Define a density function on $[0, 1]$, symmetric about $\frac{1}{2}$ by

$$\psi(x) = \psi(1-x) = \begin{cases} f_{\frac{n+1}{2}:n}(x), & \text{if } n \text{ is odd}, \\ \frac{1}{2}\left[f_{\frac{n}{2}:n}(x) + f_{\frac{n}{2}+1:n}(x)\right], & \text{if } n \text{ is even}. \end{cases} \tag{6.154}$$

Then

$$\Psi(x) = \begin{cases} \sum_{k=\frac{n+1}{2}}^{n} \binom{n}{k} x^k (1-x)^{n-k}, & \text{if } n \text{ is odd}, \\ \frac{1}{2}\binom{n}{n/2}[x(1-x)]^{n/2} + \sum_{k=\frac{n}{2}+1}^{n} \binom{n}{k} x^k (1-x)^{n-k}, & \text{if } n \text{ is even}, \end{cases} \tag{6.155}$$

stands for the corresponding distribution function. Clearly $F_{j:n}(0) = \Psi(0) = 0$ and $F_{j:n}(1) = \Psi(1) = 1$, $j = 1, \dots, n$. Comparing further $F_{j:n}$ (see (4.2)) with (6.155), we immediately check that $F_{j:n}(x) > \Psi(x)$, $0 < x < 1$, $1 \le j < \frac{n+1}{2}$, and we have the equality $F_{\frac{n+1}{2}:n}(x) = \Psi(x)$, $0 < x < 1$, for odd-sized samples. One can say that $F_{j:n}$, $j \le \frac{n+1}{2}$, precede Ψ in the usual stochastic order. Also, we can easily show that (6.154) is increasing on $\left[0, \frac{1}{2}\right]$ and decreasing on $\left[\frac{1}{2}\ 1\right]$. It follows that the antiderivative (6.155) is first strictly convex and then strictly concave, and consequently,

$$\Psi(x) < x, \qquad 0 < x < \frac{1}{2}, \tag{6.156}$$

with the reversed relation for $\frac{1}{2} < x < 1$. We also have $\Psi(0) = 0$ and $\Psi\left(\frac{1}{2}\right) = \frac{1}{2}$, and so Ψ succeeds the identity function $\mathrm{id}(x) = x$ on $\left[0, \frac{1}{2}\right]$ in the stochastic order.

Using the above notions, for $j \le \frac{n+1}{2}$, we can write

$$\begin{aligned}
\mathbb{E}X_{j:n} - \mu &= \int_0^1 F^{-1}(x)F_{j:n}(dx) - \mu \\
&\le \int_0^1 F^{-1}(x)\Psi(dx) - \mu = \int_0^1 F^{-1}(x)[\psi(x) - 1]dx \\
&= \int_0^{1/2} [F^{-1}(x) + F^{-1}(1-x)][\psi(x) - 1]dx \\
&= \int_0^{1/2} [F^{-1}(x) + F^{-1}(1-x)](\Psi - \mathrm{id})(dx) \le 0. \qquad (6.157)
\end{aligned}$$

The subsequent relations are implied by the following arguments. The first inequality holds by the fact that $F_{j:n}$ precedes Ψ in the stochastic order. The passage from the second line to the third one is based on splitting the integral into two parts with the respective domains $\left[0, \frac{1}{2}\right]$ and $\left[\frac{1}{2}, 1\right]$, change of the variable in the latter, and symmetry of (6.154). The inequality in the last line follows from the facts that the function $\left(0, \frac{1}{2}\right) \ni x \to [F^{-1}(x) + F^{-1}(1-x)]$ is nonincreasing when F^{-1} is convex, and Ψ dominates the identity function in the stochastic order in $\left(0, \frac{1}{2}\right)$, and the functions are identical at the interval end-points (see also Theorem 2.9).

In order to establish the equality conditions, we analyze (6.157). The first inequality becomes equality there if either F^{-1} is a constant, which is excluded by assumption, or $F_{j:n}(x) = \Psi(x)$, $0 < x < 1$, which holds for $j = \frac{n+1}{2}$ only. Since (6.156) holds, the last integral vanishes if its integrand is a constant. By our assumption, this is absolutely continuous, and the respective quantile density $(F^{-1})'$ is symmetric about $\frac{1}{2}$ then. Another assumption asserts that $(F^{-1})'$ is nondecreasing on $[0, 1]$. Therefore $(F^{-1})'$ is a positive constant, and the quantile function F^{-1} is linear. It follows that the corresponding distribution is uniform. Combining both the equality conditions, we obtain the final claim of Theorem 6.15. □

Theorem 6.16 (DD Distributions) *Let the assumptions of Theorem* 6.15 *hold together with* $0 < \sigma_p < \infty$ *for some* $1 < p \le \infty$. *If* $\frac{j}{n+1} < \frac{1}{2}$, *then*

$$\mathbb{E}\frac{X_{i:n} - \mu}{\sigma_p} \le 0 \qquad (6.158)$$

is sharp. The equality is attained in limit for the mixtures of atoms at $\mu - \frac{\sigma_p}{M_p(\alpha)}\frac{(1-\alpha)^2}{2}$ *and uniform distributions on* $\left[\mu - \frac{\sigma_p}{M_p(\alpha)}\frac{(1-\alpha)^2}{2}, \mu + \frac{\sigma_p}{M_p(\alpha)}\frac{1-\alpha^2}{2}\right]$ *with probabilities* α *and* $1-\alpha$*, respectively, as* $\alpha \to 1$*, where*

$$M_p(\alpha) = \begin{cases} \left\{\frac{(1-\alpha)^{p+1}[2(p+1)\alpha(1-\alpha)^{p-1}+(1-\alpha)^{p+1}+(1+\alpha)^{p+1}]}{2^{p+1}(p+1)}\right\}^{1/p}, & p < \infty, \\ \frac{1-\alpha^2}{2}, & p = \infty. \end{cases} \tag{6.159}$$

Proof Inequality (6.158) follows from Theorem 6.15 (see also Theorem 6.3). Denote the distribution functions of the mixture distributions described in Theorem 6.16 by F_α (we assume that p is fixed, and we drop it in the notation). Note that all of them are concave, and so satisfy the shape conditions of Theorem 6.16. The respective quantile functions are

$$\begin{aligned} F_\alpha^{-1}(x) &= \begin{cases} \mu - \frac{\sigma_p}{M_p(\alpha)}\frac{(1-\alpha)^2}{2}, & 0 < x \le \alpha, \\ \mu + \frac{\sigma_p}{M_p(\alpha)}\left(x - \frac{1+\alpha^2}{2}\right), & \alpha \le x < 1, \end{cases} \\ &= \mu + \frac{\sigma_p}{M_p(\alpha)}\left((x-\alpha)\mathbf{1}_{[\alpha,1]}(x) - \frac{(1-\alpha)^2}{2}\right). \end{aligned} \tag{6.160}$$

Elementary calculations show all F_α, $0 < \alpha < 1$, satisfy the moment conditions as well. Indeed,

$$\begin{aligned} \mathbb{E}_\alpha X_1 - \mu &= \int_0^1 [F_\alpha^{-1}(x) - \mu]\,dx \\ &= \frac{\sigma_p}{M_p(\alpha)}\int_0^1 \left((x-\alpha)\mathbf{1}_{[\alpha,1]}(x) - \frac{(1-\alpha)^2}{2}\right)dx \\ &= \frac{\sigma_p}{M_p(\alpha)}\left[-\alpha\frac{(1-\alpha)^2}{2} + \frac{1-\alpha^2}{2} - (1-\alpha)\left(\alpha + \frac{(1-\alpha)^2}{2}\right)\right] = 0. \end{aligned} \tag{6.161}$$

For $p < \infty$, we have

$$\begin{aligned} \frac{\mathbb{E}_\alpha|X_1 - \mu|^p}{\sigma_p^p} &= \int_0^1 \left|\frac{F_\alpha^{-1}(x) - \mu}{\sigma_p}\right|^p dx \\ &= \frac{1}{M_p^p(\alpha)}\int_0^1 \left|(x-\alpha)\mathbf{1}_{[\alpha,1]}(x) - \frac{(1-\alpha)^2}{2}\right|^p dx \end{aligned}$$

$$= \frac{1}{M_p^p(\alpha)} \left[\alpha \frac{(1-\alpha)^{2p}}{2^p} + \int_\alpha^{(1+\alpha^2)/2} \left(\frac{(1-\alpha)^2}{2} + \alpha - x \right)^p dx \right.$$

$$\left. + \int_{(1+\alpha^2)/2}^1 \left(x - \alpha - \frac{(1-\alpha)^2}{2} \right)^p dx \right] = 1 \tag{6.162}$$

and

$$\frac{\mathbb{E}_\alpha X_{j:n} - \mu}{\sigma_p} = \int_0^1 \frac{F_\alpha^{-1}(x) - \mu}{\sigma_p} f_{j:n}(x)\, dx$$

$$= \frac{1}{M_p(\alpha)} \left[\int_\alpha^1 (x-\alpha) f_{j:n}(x)\, dx - \frac{(1-\alpha)^2}{2} \right]$$

$$= \frac{\sum_{k=0}^{j-1} \frac{(j-k)n!}{k!(n+1-k)!} \alpha^k (1-\alpha)^{n+1-k} - \frac{(1-\alpha)^2}{2}}{\left\{ \frac{(1-\alpha)^{p+1}[2(p+1)\alpha(1-\alpha)^{p-1} + (1-\alpha)^{p+1} + (1+\alpha)^{p+1}]}{2^{p+1}(p+1)} \right\}^{1/p}}$$

$$= -(1-\alpha)^{1-1/p} \frac{[2^{p+1}(p+1)]^{1/p} \left[\frac{1}{2} - \sum_{k=0}^{j-1} \frac{(j-k)n!}{k!(n+1-k)!} \alpha^k (1-\alpha)^{n-1-k} \right]}{[2(p+1)\alpha(1-\alpha)^{p-1} + (1-\alpha)^{p+1} + (1+\alpha)^{p+1}]^{1/p}}. \tag{6.163}$$

As $\alpha \to 1$, the fraction in the last line tends to $\frac{(p+1)^{1/p}}{2}$, whereas the preceding factor vanishes. This shows that the conclusion of Theorem 6.16 holds for every $1 < p < \infty$.

For $p = \infty$, we have

$$\text{ess sup} \frac{|X_1 - \mu|}{\sigma_\infty} = \frac{\max\{(1-\alpha)^2, 1-\alpha^2\}}{2M_\infty(\alpha)} = 1.$$

Moreover,

$$\frac{\mathbb{E}_\alpha X_{j:n} - \mu}{\sigma_\infty} = \int_0^1 \frac{F_\alpha^{-1}(x) - \mu}{\sigma_\infty} f_{j:n}(x)\, dx$$

$$= \frac{1}{M_\infty(\alpha)} \left[\int_\alpha^1 (x-\alpha) f_{j:n}(x)\, dx - \frac{(1-\alpha)^2}{2} \right]$$

$$= \frac{1-\alpha}{1+\alpha} \left[2 \sum_{k=0}^{j-1} \frac{(j-k)n!}{k!(n+1-k)!} \alpha^k (1-\alpha)^{n-1-k} - 1 \right], \tag{6.164}$$

which tends to 0 as α tends to 1. This completes the proof. □

Theorem 6.17 (DD Distributions) *If $X_1, \ldots, X_n$ are iid with concave parent distribution and finite first moment, then we have*

$$\mathbb{E}\frac{X_{j:n}-\mu}{\sigma_1} \le \begin{cases} -\frac{1}{2}, & \text{if } 0 < \frac{j}{n+1} \le \frac{3}{8}, \\ -2\left(1-\frac{2j}{n+1}\right), & \text{if } \frac{3}{8} \le \frac{j}{n+1} < \frac{1}{2}. \end{cases} \tag{6.165}$$

The former bound is attained in limit by the combinations of atoms at $\mu - \frac{2\sigma_1}{(1+\alpha)^2}$ and uniform distributions on $\left[\mu - \frac{2\sigma_1}{(1+\alpha)^2}, \mu + \frac{2\sigma_1}{1-\alpha^2}\right]$, as $\alpha \to 1$. The latter one becomes the equality for the uniform distribution on $[\mu - 2\sigma_1, \mu + 2\sigma_1]$.

Note that the distributions attaining the bounds in the limit coincide with those described in the statement of Theorem 6.16 with particular $M_1 = \frac{(1-\alpha)^2}{4}$.

Proof We use the representation

$$\mathbb{E}\frac{X_{j:n}-\mu}{\sigma_1} = \int_0^1 \frac{F^{-1}(x)-\mu}{\sigma_1} f_{j:n}(x)dx = \int_0^1 \frac{F^{-1}(x)-\mu}{\sigma_1}[f_{j:n}(x)-1]dx. \tag{6.166}$$

We apply the norm maximization method described in Sect. 2.4.7. Our aim is to maximize the functional

$$T_h(g) = \int_0^1 g(x)h(x)dx, \tag{6.167}$$

where

$$h(x) = f_{j:n}(x) - 1 \tag{6.168}$$

is fixed, and

$$g(x) = \frac{F^{-1}(x)-\mu}{||F^{-1}-\mu\mathbf{1}||_1}, \tag{6.169}$$

belongs to the set

$$\mathcal{S}_1 = \{g \in L^1((0,1), dx) : g\text{-nondecreasing convex}, \quad T_1(g) = 0, \quad ||g||_1 = 1\}, \tag{6.170}$$

where $T_1(g) = \int_0^1 g(x)dx$. Since $T_h(g) < 0$ for all $g \in \mathcal{S}_1$, our problem is equivalent to maximizing $||g||_1$ over the set

$$\mathcal{S}_2 = \{g \in L^1((0,1), dx) : g\text{-nondecreasing convex}, \quad T_1(g) = 0, \quad T_h(g) = -1\}. \tag{6.171}$$

Consider the family of broken lines

$$\tilde{g}_\alpha(x) = a(\alpha) + b(\alpha)(x-\alpha)\mathbf{1}_{[\alpha,1]}(x), \quad 0 \le \alpha < 1, \tag{6.172}$$

which are continuous, first constant on $[0, \alpha]$ and then linear increasing on $[\alpha, 1]$. They belong to $\mathcal{S}_2$ if $T_{\mathbf{1}}(\tilde{g}_\alpha) = 0$ and $T_h(\tilde{g}_\alpha) = -1$, which uniquely determine the intercepts and slopes

$$b(\alpha) = \left[\frac{(1-\alpha)^2}{2} - \sum_{k=0}^{j-1} \frac{(j-k)n!}{k!(n+1-k)!}\alpha^k(1-\alpha)^{n+1-k}\right]^{-1} > 0, \tag{6.173}$$

$$a(\alpha) = -b(\alpha)\frac{(1-\alpha)^2}{2} < 0, \qquad 0 < \alpha < 1. \tag{6.174}$$

It is easy to verify that every piecewise linear element of $\mathcal{S}_2$ is a convex combination of functions (6.172) with parameters (6.173) and (6.174). Every such a combination satisfies

$$\left|\left|\sum_{i=1}^{k} \gamma_i \tilde{g}_{\alpha_i}\right|\right|_1 < \sum_{i=1}^{k} \gamma_i ||\tilde{g}_{\alpha_i}||_1 \le \max_{1\le i\le k} ||\tilde{g}_{\alpha_i}||_1, \tag{6.175}$$

and the first inequality is strict, because $\tilde{g}_{\alpha_i}$ are positive and negative on different intervals for different α_i (cf. Theorem 2.37).

Take now any $g \in \mathcal{S}_2$, and approximate it by a piecewise linear convex interpolation g_k satisfying $g_k(\alpha_i) = g(\alpha_i)$ for some $\alpha_0 = 0 < \alpha_1 < \ldots < \alpha_k = 1$. If the sets $\{\alpha_1, \ldots, \alpha_k\}$, $k = 1, 2, \ldots$ are increasing and $\max_{0\le i\le k} |\alpha_{i+1} - \alpha_i| \to 0$, then g_k decrease uniformly to g, and so $||g - g_k|| \to 0$ as $k \to \infty$. As a consequence,

$$T_1(g_k) \to T_1(g) = 0, \tag{6.176}$$

$$T_h(g_k) \to T_h(g) = -1, \tag{6.177}$$

as $k \to \infty$. Note that g_k do not necessarily belong to $\mathcal{S}_2$, but the following modifications $\tilde{g}_k = \frac{g_k - T_1(g_k)}{|T_h(g_k)|}$, $k = 1, 2, \ldots$, do. They satisfy

$$||\tilde{g}_k - g||_1 = \left|\left|\frac{g_k}{|T_h(g_k)|} - g_k + g_k - g - \frac{T_1(g_k)}{|T_h(g_k)|}\mathbf{1}\right|\right|_1$$

$$\le \left|\frac{1}{|T_h(g_k)|} - 1\right| ||g_k||_1 + ||g_k - \tilde{g}||_1 + \left|\frac{T_1(g_k)}{T_h(g_k)}\right| ||\mathbf{1}||_1 \to 0, \tag{6.178}$$

because $||\tilde{g}_k||_1 \to ||\tilde{g}||_1$, the last norm amounts to 1, and the remaining three terms in the last line vanish as $k \to \infty$. It follows that $||\tilde{g}_k||_1 \to ||g||_1$ as well. Therefore,

$$||g||_1 \leftarrow ||\tilde{g}_k||_1 = \left\|\sum_{i=1}^{k} \gamma_i \tilde{g}_{\alpha_i}\right\|_1 < \max_{1\le i\le k} ||\tilde{g}_{\alpha_i}||_1$$

and finally

$$||g||_1 \le \sup_{0<\alpha<1} ||\tilde{g}_\alpha||_1, \qquad g \in \mathcal{S}_2,$$

which implies that it suffices to restrict ourselves to functions (6.172) in the maximization problem.

Since

$$\sup_{g\in\mathcal{S}_1} T_h(g) = -\frac{1}{\sup_{g\in\mathcal{S}_2} ||g||_1} = -\frac{1}{\sup_{0<\alpha<1} ||\tilde{g}_\alpha||_1},$$

it suffices to minimize the reciprocal of

$$||\tilde{g}_\alpha||_1 = b(\alpha)\int_0^1 \left|(x-\alpha)\mathbf{1}_{\alpha,1]}(x) - \frac{(1-\alpha)^2}{2}\right| dx = b(\alpha)\frac{(1-\alpha^2)^2}{4},$$

which has the form

$$\begin{aligned} U(\alpha) &= \frac{2(1-\alpha)^2 - 4\sum_{k=0}^{j-1}\frac{(j-k)n!}{k!(n+1-k)!}\alpha^k(1-\alpha)^{n-1-k}}{(1-\alpha^2)^2} \\ &= \frac{2}{(1+\alpha)^2}\left[1 - 2\sum_{k=1}^{j-1} c_k\alpha^k(1-\alpha)^{n-1-k}\right] \end{aligned} \qquad (6.179)$$

with $c_k = \frac{(j-k)n!}{k!(n+1-k)!}$, $k = 0, \ldots, j-1$. For $j \ge 2$, we analyze

$$\begin{aligned} U'(\alpha) = \frac{4}{(1+\alpha)^3}\Bigg\{ &-\sum_{k=0}^{j-1} c_k[k\alpha^{k-1}(1-\alpha)^{n-1-k} - (n-1-k)\alpha^k(1-\alpha)^{n-2-k}] \\ &\times (1-\alpha+2\alpha) - \left[1 - 2\sum_{k=0}^{j-1} c_k\alpha^k(1-\alpha)^{n+1-k}\right]\Bigg\} \end{aligned}$$

$$= \frac{4}{(1+\alpha)^3}\Bigg[-\sum_{k=0}^{j-2}(k+1)c_{k+1}\alpha^k(1-\alpha)^{n-1-k}$$

$$+\sum_{k=0}^{j-1}(n+1-3k)c_k\alpha^k(1-\alpha)^{n-1-k}$$

$$+\sum_{k=0}^{j}2(n-k)c_{k-1}\alpha^k(1-\alpha)^{n-1-k} - \sum_{k=0}^{n-1}\binom{n-1}{k}\alpha^k(1-\alpha)^{n-1-k}\Bigg]$$

$$= \frac{4}{(1+\alpha)^3}\sum_{k=0}^{n-1}a_k\alpha^k(1-\alpha)^{n-1-k}, \tag{6.180}$$

say. Our aim is to show that function (6.180) is first positive and then negative. To this end, it suffices to show

$$\begin{aligned} &a_0 = 0, \\ &a_k > 0, \quad k = 1, \ldots, j-1, \\ &a_k < 0, \quad k = j+1, \ldots, n-1, \end{aligned} \tag{6.181}$$

and refer to the variation diminishing property of Bernstein polynomials.

For $k = 0, \ldots, j-2$, we have

$$\begin{aligned} a_k &= -(k+1)c_{k+1} + (n+1-3k)c_k + 2(n-k)c_{k-1} - \binom{n-1}{k} \\ &= \frac{k(n-1)!}{k!(n+2-k)!}[k^2 - 3k + 3n^2 - 4jn + 3n + 2]. \end{aligned} \tag{6.182}$$

Therefore $a_0 = 0$, and the sign of (6.182) is identical to that of the expression inside the brackets for $k \geq 1$. The expression attains its positive minimum $6\left(\frac{n+1}{2} - j\right) + \frac{j}{3} - 1 > 2$ at $k = 1$ and 2. Also a_{j-1} is positive, because this arises by adding a positive term jc_j to the positive expression (6.182) evaluated at $k = j-1$. For proving (6.181), it suffices to note that $a_k = -\binom{n-1}{k} < 0$, $k = j+1, \ldots, n-1$.

We have thus concluded that (6.179) is first increasing and then decreasing. Therefore,

$$\inf_{0\leq\alpha<1} U(\alpha) = \min\{U(0), U(1)\} = \begin{cases} U(0) = 2\left(1 - \frac{2j}{n+1}\right), & \text{if } \frac{3}{8} \leq \frac{j}{n+1} < \frac{1}{2}, \\ U(1) = \frac{1}{2}, & \text{if } 0 < \frac{j}{n+1} \leq \frac{3}{8}. \end{cases} \tag{6.183}$$

For $j = 1$, the analysis is much simpler. We have

$$U(\alpha) = \frac{2}{(1+\alpha)^2}\left[1 - \frac{2}{n+1}(1-\alpha)^{n-1}\right],$$

$$U'(\alpha) = \frac{4}{(1+\alpha)^3}\left[(1-\alpha)^{n-1} - 2\frac{n-1}{n+1}\alpha(1-\alpha)^{n-2} - 1\right]$$

$$= -\frac{4}{(1+\alpha)^3}\left[\frac{(n-1)^2}{n+1}\alpha(1-\alpha)^{n-2} + \sum_{k=2}^{n-1}\binom{n-1}{k}\alpha^k(1-\alpha)^{n-k}\right] < 0,$$

and

$$\inf_{0\le\alpha<1} U(\alpha) = U(1) = \frac{1}{2}. \qquad (6.184)$$

Formulae (6.183) and (6.184) provide the desired bounds for the negative deviations of $\mathbb{E}X_{j:n}$ from the mean μ expressed in the mean deviation from the mean units σ_1. The conditions of attainability (possibly in limit) can be derived from (6.169) and (6.172). The functions represent the quantile functions of uniform distributions with atoms at left-end point of their supports. We omit the details here. □

Now we proceed to order statistics with small ranks from the DFR populations. Theorem 6.15 does not have a nontrivial counterpart here, because no partial sum $S(j,n) = \sum_{i=1}^{j}\frac{1}{n+1-j}$ of the harmonic series amounts to 1 except for $S(1,1)$. Then we obviously have $\mathbb{E}X_{1:1} = \mu$ when the parent distribution is exponential. The counterpart of Theorem 6.16 is as follows.

Theorem 6.18 (DFR Distributions) *For $X_1, \ldots, X_n$ being iid random variables with a concave cumulative failure rate, finite mean μ, positive finite pth absolute central moment σ_p^p for some fixed $p > 1$, and integer $1 \le j \le n$ satisfying $S(j,n) < 1$, the bound*

$$\mathbb{E}\frac{X_{j:n} - \mu}{\sigma_p} \le 0 \qquad (6.185)$$

is attained in the limit by a sequence of parent distribution functions F_β, $0 < \beta < 1$, which are mixtures of atoms at $\mu - \frac{\sigma_p}{N_p(\beta)}\beta$ and exponential distributions with location parameter $\mu - \frac{\sigma_p}{N_p(\beta)}\beta$ and scale $\frac{\sigma_p}{N_p(\beta)}$ with probabilities $1-\beta$ and β, respectively, as β tends to 0, where

$$N_p(\beta) = \left\{\beta^p - \beta^{p+1} + \beta e^{-\beta}\left[\int_0^\beta y^p e^y\, dy + \Gamma(p+1)\right]\right\}^{1/p}. \qquad (6.186)$$

Proof Inequality (6.185) follows from Theorem 6.4. It remains to prove that the parent distribution functions F_β attain the equality in the limit. We first notice that all of them satisfy $F_\beta \succeq_c Z$. We also have

$$F_\beta(x) = \begin{cases} 0, & x < \mu - \frac{\sigma_p}{N_p(\beta)}\beta, \\ 1 - \beta \exp\left(-\frac{x-\mu}{\sigma_p} N_p(\beta) - \beta\right), & x \geq \mu - \frac{\sigma_p}{N_p(\beta)}\beta, \end{cases}$$

and

$$F_\beta^{-1}(x) = \begin{cases} \mu - \frac{\sigma_p}{N_p(\beta)}\beta, & 0 < x < 1 - \beta, \\ \mu + \frac{\sigma_p}{N_p(\beta)}\left[-\ln(1-x) + \ln\beta - \beta\right], & 1 - \beta \leq x < 1. \end{cases}$$

Now we check the moment conditions. Indeed,

$$\begin{aligned} \mathbb{E}_\beta X_1 = \int_0^1 F_\beta^{-1}(x)\,dx &= \left(\mu - \frac{\sigma_p}{N_p(\beta)}\beta\right)(1-\beta) \\ &+ \left[\mu + \frac{\sigma_p}{N_p(\beta)}(\ln\beta - \beta)\right]\beta - \frac{\sigma_p}{N_p(\beta)} \int_{1-\beta}^1 \ln(1-x)\,dx = \mu \end{aligned}$$

and

$$\begin{aligned} \mathbb{E}_\beta |X_1 - \mu|^p = \int_0^1 |F_\beta^{-1}(x) - \mu|^p\,dx = \frac{\sigma_p^p}{N_p^p(\beta)}\Big[\beta^p(1-\beta) \\ + \int_{1-\beta}^{1-\beta e^{-\beta}} \left(\ln\frac{1-x}{\beta} + \beta\right)^p dx + \int_{1-\beta e^{-\beta}}^1 \left(-\ln\frac{1-x}{\beta} - \beta\right)^p dx\Big] \\ = \frac{\sigma_p^p}{N_p^p(\beta)}\left[\beta^p(1-\beta) + \beta e^{-\beta}\int_0^\beta y^p e^y\,dy + \beta e^{-\beta}\int_0^\infty y^p e^{-y}\,dy\right] = \sigma_p^p, \end{aligned}$$

by definition (6.186). Moreover, using the formula

$$\int_\alpha^\infty (x-\alpha) f_{j:n}(1-e^{-x})e^{-x}\,dx = \sum_{k=0}^{j-1} S(j-k, n-k)\binom{n}{k}(1-e^{-\alpha})^k e^{-(n-k)\alpha} \tag{6.187}$$

(cf. Gajek & Rychlik, 1998, pp. 176–177), we calculate

$$\mathbb{E}_\beta \frac{X_{j:n} - \mu}{\sigma_p} = \int_0^1 \frac{F_\beta^{-1}(x) - \mu}{\sigma_p} [f_{j:n}(x) - 1]\, dx$$

$$= \frac{1}{N_p(\beta)} \int_{1-\beta}^1 [-\ln(1-x) + \ln \beta][f_{j:n}(x) - 1]\, dx$$

$$= \frac{1}{N_p(\beta)} \int_{-\ln \beta}^\infty (y + \ln \beta)[f_{j:n}(1 - e^{-y}) - 1] e^{-y}\, dx$$

$$= \frac{1}{N_p(\beta)} \left[\int_{-\ln \beta}^\infty (y + \ln \beta) f_{j:n}(1 - e^{-y}) e^{-y}\, dx - \beta \int_0^\infty y e^{-y}\, dy \right]$$

$$= \frac{\sum_{k=0}^{j-1} S(j-k, n-k) \binom{n}{k} (1-\beta)^k \beta^{n-k} - \beta}{\left\{ \beta^p - \beta^{p+1} + \beta e^{-\beta} \left[\int_0^\beta y^p e^y\, dy + \Gamma(p+1) \right] \right\}^{1/p}}$$

$$= \beta^{1-1/p} \frac{\sum_{k=0}^{j-1} S(j-k, n-k) \binom{n}{k} (1-\beta)^k \beta^{n-k-1} - 1}{\left\{ \beta^{p-1} - \beta^p + e^{-\beta} \left[\int_0^\beta y^p e^y\, dy + \Gamma(p+1) \right] \right\}^{1/p}}$$

If $\beta \to 0$, then the fraction tends to a negative number $-\Gamma^{-1/p}(p+1)$, and the preceding term vanishes. This completes the proof. □

Strictly negative sharp upper bounds on the centered expectations of order statistics appear when we gauge them in the first absolute central moment σ_1.

Theorem 6.19 (DFR Distributions) *Let $X_1, \ldots, X_n$ be independent random variables with a common distribution function F which has a concave cumulative failure rate, finite mean μ, and positive mean absolute deviation σ_1. Suppose that $1 \le j \le n$ satisfies $S(j, n) < 1$. Then,*

$$\mathbb{E}_F \frac{X_{j:n} - \mu}{\sigma_1} \le \begin{cases} -\frac{1}{2}, & \text{if } 0 < S(j,n) < 1 - e^{-1}, \\ -\frac{e}{2}[1 - S(j,n)], & \text{if } 1 - e^{-1} < S(j,n) < 1. \end{cases} \tag{6.188}$$

In the first case, the equality is attained in limit by the mixtures of atoms at $\mu - \frac{\sigma_1 e^\beta}{2}$ and exponential distributions with location $\mu - \frac{\sigma_1 e^\beta}{2}$ and scale $\frac{\sigma_1 e^\beta}{2\beta}$ with probabilities $1 - \beta$ and β, as $\beta \to 0$. In the latter case, the inequality becomes equality for the exponential distribution with location $\mu - \frac{\sigma_1 e}{2}$ and scale $\frac{\sigma_1 e}{2}$.

Using (6.85), we can check that the first condition in (6.188) is satisfied by $0 < i \le i_1 = \lfloor (1 - e^{-1+e^{-1}})(n+1) \rfloor \approx \lfloor 0.46854(n+1) \rfloor$, and the latter holds for $i_1 < i \le i_2 = \lfloor (1 - e^{-1})(n+1) \rfloor \approx \lfloor 0.63212(n+1) \rfloor$, where $\lfloor x \rfloor$ denotes the greatest integer not exceeding x.

Proof The arguments are analogous to those of the proof of Theorem 6.17. We maximize the linear functional

$$\begin{aligned}\mathbb{E}\frac{X_{j:n}-\mu}{\sigma_1} &= \int_0^\infty \frac{F^{-1}(1-e^{-x})-\mu}{\sigma_1}[f_{j:n}(1-e^{-x})-1]e^{-x}dx \\ &= \int_0^\infty g(x)h(x)e^{-x}dx = T_h(g),\end{aligned} \tag{6.189}$$

with $h(x) = f_{j:n}(1-e^{-x})-1$, $x > 0$, and

$$g(x) = \frac{F^{-1}(1-e^{-x})-\mu}{\sigma_1}, \qquad x > 0, \tag{6.190}$$

being an element of the family

$$\begin{aligned}\mathcal{S}_3 = \{g \in L^1((0,\infty), e^{-x}dx) : g\text{-nondecreasing convex,}\\ T_{\mathbf{1}}(g) = 0, \quad ||g||_1 = 1\},\end{aligned} \tag{6.191}$$

where $T_{\mathbf{1}}(g) = \int_0^\infty g(x)e^{-x}dx$. We consider the dual problem of maximizing the norm $||g||_1$ over the set

$$\begin{aligned}\mathcal{S}_4 = \{g \in L^1((0,\infty), e^{-x}dx) : g\text{-nondecreasing convex,}\\ T_{\mathbf{1}}(g) = 0, \quad T_h(g) = -1\}.\end{aligned} \tag{6.192}$$

The following two-piece broken lines, first constant, and then increasing

$$\tilde{g}_\alpha(x) = a(\alpha) + b(\alpha)(x-\alpha)\mathbf{1}_{[\alpha,+\infty)}(x), \qquad 0 \le \alpha < \infty, \tag{6.193}$$

belong to (6.192) iff

$$b(\alpha) = \left[e^{-\alpha} - \sum_{k=0}^{j-1} S(j-k, n-k)\binom{n}{k}(1-e^{-\alpha})^k e^{-(n-k)\alpha}\right]^{-1} > 0 \tag{6.194}$$

and

$$a(\alpha) = -b(\alpha)e^{-\alpha} < 0. \tag{6.195}$$

One can show that every $g \in \mathcal{S}_4$ is a limit of convex combinations of functions (6.193) with parameters (6.194) and (6.195). Therefore the norm is maximized over the set of broken lines (6.193). Since for all $F \succeq_c Z$ with finite mean and j satisfying $S(j,n) < 1$, we have

$$\frac{\mathbb{E}X_{i:n} - \mu}{\sigma_1} \leq -\frac{1}{\sup_{0\leq\alpha<\infty} ||\tilde{g}_\alpha||_1},$$

we can directly solve the original problem minimizing

$$\begin{aligned}
\frac{1}{||\tilde{g}_\alpha||_1} &= \left[b(\alpha) \int_0^\infty \left| (x-\alpha)\mathbf{1}_{[\alpha,+\infty)}(x) - e^{-\alpha} \right| e^{-x}\, dx \right]^{-1} \\
&= \frac{1/b(\alpha)}{\left(\int_0^{\alpha+e^{-\alpha}} - \int_{\alpha+e^{-\alpha}}^\infty\right) e^{-x-\alpha} dx + \left(\int_{\alpha+e^{-\alpha}}^\infty - \int_\alpha^{\alpha+e^{-\alpha}}\right)(x-c)e^{-x} dx} \\
&= \frac{e^{-\alpha} - \sum_{k=0}^{j-1} S(j-k, n-k)\binom{n}{k}(1-e^{-\alpha})^k e^{-(n-k)\alpha}}{2\exp(-\alpha - e^{-\alpha})} \\
&= \frac{e^\beta}{2}\left[1 - \sum_{k=0}^{j-1} S(j-k, n-k)\binom{n}{k}(1-\beta)^k \beta^{n-k-1} \right] \\
&= \frac{e^\beta}{2}\left[1 - \sum_{k=0}^{j-1} c_k B_{n-k-1,n-1}(\beta) \right] = U(\beta),
\end{aligned} \tag{6.196}$$

say, where $\beta = e^{-\alpha} \in (0, 1]$, and $c_k = \frac{n}{n-k} S(j-k, n-k)$, $k = 0, \ldots, j-1$. Our purpose is to prove that function U is first increasing and then decreasing on the interval $[0, 1]$. So, we analyze

$$\begin{aligned}
2e^{-\beta}U'(\beta) &= 1 - \sum_{k=0}^{j-1} c_k \left[B_{n-k-1,n-1}(\beta) - B'_{n-k-1,n-1}(\beta) \right] \\
&= \sum_{k=0}^{n-1} B_{n-k-1,n-1}(\beta) - \sum_{k=0}^{j-1} c_k B_{n-k-1,n-1}(\beta) \\
&\quad - (n-1)(x+1-x) \sum_{k=0}^{j-1} c_k \left[B_{n-k-2,n-2}(\beta) - B_{n-k-1,n-2}(\beta) \right] \\
&= \sum_{k=0}^{n-1} d_k B_{n-k-1,n-1}(\beta),
\end{aligned} \tag{6.197}$$

say, using the variation diminishing property of the Bernstein polynomials. Taking into account the fact that

$$\sum_{k=0}^{n-1} d_k B_{n-k-1,n-1}(\beta) = \sum_{k=0}^{n-1} d_{n-k-1} B_{k,n-1}(\beta),$$

it suffices to verify that the coefficients $d_0, \ldots, d_{n-1}$ in (6.197) change the sign once at most, and the possible change is from minus to plus.

For $j = 1$, we obtain

$$\begin{aligned} d_0 &= 1 - nc_0 = 1 - nS(1, n) = 0, \\ d_1 &= 1 - c_0 = 1 - S(1, n) = 1 - \frac{1}{n} > 0, \\ d_k &= 1 > 0, \qquad\qquad k = 2, \ldots, n-1, \end{aligned}$$

which immediately gives the desired claim. Assume now that $j \geq 2$. Then

$$d_0 = (n-1)c_1 - nc_0 + 1 = nS(i-1, n-1) - nS(i, n) + 1 = 0$$

as well. For $1 \leq k \leq j-2$, we have

$$\begin{aligned} d_k &= (n-1-k)c_{k+1} - (n-2k)c_k - kc_{k-1} + 1 \\ &= n\,S(j-k-1, n-k-1) - \frac{n(n-2k)}{n+1-k} S(j-k, n-k) \\ &\quad - \frac{nk}{n-k} S(j-k+1, n-k+1) + 1 \\ &= n\left[1 - \frac{n(n-2k)}{n-k} - \frac{k}{n+1-k}\right] S(j-k+1, n-k+1) \\ &\quad + n\left[\frac{n-2k}{(n+1-k)(n-k)} - \frac{1}{n-k} - \frac{1}{n+1-k}\right] + 1 \\ &= \frac{nk}{(n+1-k)(n-k)}\left[S(j-k+1, n-k+1) - 2 - \frac{k-1}{n}\right] \\ &= -\frac{nk\left\{[1 - S(j, n)] + \left(1 - \frac{k-1}{n}\right) + S(k-1, n)\right\}}{(n+1-k)(n-k)} < 0, \end{aligned}$$

because all three summands inside the curly brackets are positive. Furthermore,

$$\begin{aligned}
d_{j-1} &= -(n+2-2j)c_{j-1} - (j-1)c_{j-2} + 1 \\
&= -\frac{(n+2-2j)n}{(n+1-j)^2} - \frac{(j-1)n}{n+2-j}\left(\frac{1}{n+1-j} + \frac{1}{n-j}\right) + 1 \\
&= \frac{(j-1)n}{(n+1-j)^2} - \frac{(j-1)n}{n+2-j}\left(\frac{1}{n+1-j} + \frac{1}{n-j}\right) - \frac{(j-1)n}{n+1-j} \\
&= -\frac{j-1}{n+1-j}\left[\frac{n}{(n+2-j)(n+1-j)(n-j)} + \frac{n}{(n+2-j)} + 1\right] < 0,
\end{aligned}$$

and clearly $d_k = 1 > 0$, $k = j+1, \ldots, n-1$, which is sufficient for the desired claim. The sign of $d_j = 1 - jc_{j-1} = 1 - \frac{jn}{(n+1-j)^2}$, which is positive for small j, and negative for the others, is immaterial here.

Summing up, we conclude that the sharp bounds in our problem have the forms

$$\begin{aligned}
\frac{\mathbb{E}X_{j:n} - \mu}{\sigma_1} &\le -\min\{U(0), U(1)\} \\
&= \begin{cases} -U(0) = -\frac{1}{2}, & \text{if } 0 < S(j,n) < 1 - e^{-1}, \\ -U(1) = -\frac{e}{2}[1 - S(j,n)], & \text{if } 1 - e^{-1} < S(j,n) < 1. \end{cases}
\end{aligned} \tag{6.198}$$

The attainability conditions follow from the relations

$$\frac{F_\beta^{-1}(1-e^{-x}) - \mu}{\sigma_1} = \frac{\tilde{g}_{-\ln\beta}(x)}{||\tilde{g}_{-\ln\beta}||_1} = \frac{(x+\ln\beta)\mathbf{1}_{[-\ln\beta,+\infty)}(x) - \beta}{2\beta e^{-\beta}}, \quad 0 < \beta \le 1, \tag{6.199}$$

If $1 - e^{-1} < S(j,n) < 1$, we choose $\beta = 1$ for which (6.199) determines the exponential distribution with desired moments μ and σ_1. For $\beta < 1$ relation (6.199) defines mixtures of exponential distributions with atoms at their left-end points, which satisfy the moment conditions as well. Decreasing the contribution of the exponential term in the case $S(j,n) < 1 - e^{-1}$, we attain $-U(0) = -\frac{1}{2}$ in the second line of (6.198). □

6.2 Symmetric Unimodal Distributions

6.2.1 Positive Upper Bounds for Single Order Statistics

Now we consider order statistics from iid samples with parent symmetric unimodal distributions. We also use the acronym SU while referring to this family of distributions. Applying the standard trick of folding for symmetric distributions, we can use Theorem 6.2 for establishing bounds on expectations of L-statistics of symmetric unimodal distributions. Then,

$$\mathbb{E}\sum_{i=1}^{n} c_i \frac{X_{i:n}-\mu}{\sigma} = \int_{1/2}^{1} [F^{-1}(x)-\mu] f_{\bar{c}^s:n}(x)dx \tag{6.200}$$

for

$$(F^{-1}-\mu\mathbf{1})_{|[1/2,1)} \in C_{\succeq_s U} = \left\{ g \in L^2\left(\left[\frac{1}{2},1\right),dx\right) : g\left(\frac{1}{2}\right)=0, \right.$$
$$\left. g \text{ is nondecreasing and convex} \right\} \tag{6.201}$$

and $c_i^s = c_i - c_{n+1-i}$, $i=1,\ldots,n$. Accordingly, the problem of evaluating (6.200) can be solved by determining the projection of $f_{\bar{c}^s:n}$ restricted by $[\frac{1}{2},1)$ onto (6.201). Notation $C_{\succeq_s U}$ is used here because the cone contains the upper halves of quantile functions of distribution functions which succeed the standard uniform one in s-order for symmetric distributions, introduced in van Zwet (1964). This means that $F^{-1}\circ U = F^{-1}$ is convex on $\left[\frac{1}{2},1\right)$ and obviously concave on $\left(0,\frac{1}{2}\right]$. We do not develop the idea of s-ordering later on. We present the results for the special case of single order statistics of symmetric unimodal populations established in Gajek and Rychlik (1998), with a simplification deduced from Danielak (2003). The following lemma, proved in Gajek and Rychlik (1998, Lemma 3), solves the question about which folded density functions

$$f_{j:n}^s(x) = f_{j:n}(x) - f_{n+1-j:n}(x), \quad x \in \left[\frac{1}{2},1\right), \tag{6.202}$$

satisfy assumptions (A1).

Lemma 6.6 *If* $\frac{n+1}{2} < j \le n-1$, *then* (6.202) *are positive, first increasing and then decreasing in* $(\frac{1}{2},1)$ *and become zero at the ends. Moreover, in the case* $j \le n-2$, *they are first concave and then convex iff* $j \le \frac{n+1+\sqrt{3n-5}}{2}$ *and consecutively convex, concave and convex otherwise. For* $j=n-1$, *they are concave for* $n \le 7$ *and first convex and then concave for* $n \ge 8$.

This has been proved with use of the variation diminishing property of the Bernstein polynomials. It follows that assumptions (A1) are satisfied for all $\frac{n+1}{2} < j \le n-1$. The lemma, together with Theorem 6.9, allow us to write the following result.

Theorem 6.20 (SU Distributions) *Let* $X_1,\ldots,X_n$ *be iid rv with a symmetric unimodal distribution, and finite mean* μ *and variance* σ^2.

For $\frac{n+1}{2} < j \le n-1$, *put*

$$L^s(\beta) = -6(2j-n-1)\sum_{i=1}^{n-j} f_{i:n+2}(\beta) + (j^2 - 11j + 6n + 6) f_{n+1-j:n+2}(\beta)$$
$$+ 2 \sum_{i=n+2-j}^{j-1} [2n - 3j + i + 1] f_{i:n+2}(\beta) - (n-j)(n+1-j) f_{j:n+2}(\beta). \tag{6.203}$$

If $L^s(\frac{1}{2}+) > 0$, *then*

$$\mathbb{E}\frac{X_{j:n} - \mu}{\sigma} \le \sqrt{3}\left(\frac{2j}{n+1} - 1\right) \tag{6.204}$$

is a sharp inequality attainable by the uniform distribution on $[\mu - \sqrt{3}\sigma, \mu + \sqrt{3}\sigma]$. *Otherwise,*

$$\mathbb{E}\frac{X_{j:n} - \mu}{\sigma} \le \frac{B}{\sqrt{2}}, \tag{6.205}$$

where

$$B^2 = \frac{(n!)^2}{(2n-1)!}\binom{2j-2}{j-1}\binom{2n-2j}{n-j}\Big[F_{2j-1:2n-1}(\beta_*) + F_{2n-2j+1:2n-1}(\beta_*)$$
$$- F_{2j-1:2n-1}\left(\frac{1}{2}\right) - F_{2n-2j+1:2n-1}\left(\frac{1}{2}\right)\Big]$$
$$- \frac{2(n!)^2}{(2n-1)!}\binom{n-1}{j-1}^2\left[F_{n:2n-1}(\beta_*) - \frac{1}{2}\right] + \frac{1}{3}\lambda_*^2(1-\beta_*)^3$$
$$+ (1-\beta_*)[f_{j:n}(\beta_*) - f_{n+1-j:n}(\beta_*)]^2 + (1-\beta_*)^2\lambda_*[f_{j:n}(\beta_*) - f_{n+1-j:n}(\beta_*)], \tag{6.206}$$

with β_* *being the smallest zero of* (6.203) *in* $[\frac{1}{2}, 1)$, *and*

$$\lambda_* = \lambda_*(\beta_*) = 3j\frac{F_{n+2-j:n+1}(\beta_*) - F_{j+1:n+1}(\beta_*)}{(n-1)(1-\beta_*)^3}$$
$$- 3\beta_*\frac{F_{n+1-j:n}(\beta_*) - F_{j:n}(\beta_*)}{(1-\beta_*)^3} - 3\frac{f_{j:n}(\beta_*) - f_{n+1-j:n}(\beta_*)}{2(1-\beta_*)}. \tag{6.207}$$

The equality in (6.205) *is attained by*

$$F(x) = \begin{cases} 0, & \frac{x-\mu}{\sigma} < -a_2, \\ \frac{x-\mu}{\sigma}\frac{\sqrt{2}B}{\lambda_*} + 1 - \beta_* + \frac{f^s_{j:n}(\beta_*)}{\lambda_*}, & -a_2 \le \frac{x-\mu}{\sigma} < -a_1, \\ (f^s_{j:n})^{-1}\left(\sqrt{2}B\frac{x-\mu}{\sigma}\right), & -a_1 \le \frac{x-\mu}{\sigma} < a_1, \\ \frac{x-\mu}{\sigma}\frac{\sqrt{2}B}{\lambda_*} + \beta_* - \frac{f^s_{j:n}(\beta_*)}{\lambda_*}, & a_1 \le \frac{x-\mu}{\sigma} < a_2, \\ 1, & \frac{x-\mu}{\sigma} > a_2, \end{cases} \tag{6.208}$$

for $a_1 = \frac{f^s_{j:n}(\beta_*)}{\sqrt{2}B}$, $a_2 = \frac{f^s_{j:n}(\beta_*)+\lambda_*(1-\beta_*)}{\sqrt{2}B}$.

The proof is based on Theorem 6.9. Functions (6.44)–(6.46) for the case of $h = f^s_{j:n} = f_{j:n} - f_{n+1-j:n}$ are modifications of (6.60), (6.62), and (6.64), respectively, which consist in subtracting from the original functions their versions with j replaced by $n+1-j$. It has been shown that the symmetrized modification of (6.62) is positive right to $\frac{1}{2}$, and changes the sign at some γ_* left to the inflection point of (6.202). The numerator of symmetrized (6.64) is given in (6.203). It is either always positive in $\left(\frac{1}{2}, \gamma_*\right)$ or changes the sign from $-$ to $+$ there. In the first case, the projection is linear, and it contains a left fragment of (6.202) otherwise. It is impossible to find explicit formulae in terms of j and n on positivity of $L^s(\frac{1}{2}+)$, which implies the uniform bounds (6.204). Analysis of examples shows that this occurs if j is relatively small with respect to n. It is also evident without calculating $L^s(\frac{1}{2}+)$ that (6.204) holds for $\frac{n+1}{2} < j \le \frac{n+1+\sqrt{3n-5}}{2}$, because (6.203) does not have any convex region in the right neighborhood of $\frac{1}{2}$. The same happens for $j = n-1 \le 6$.

It is much easier to find the projection bounds in the remaining cases $j \le \frac{n+1}{2}$ and $j = n$. If $j \le \frac{n+1}{2}$, then $f^s_{j:n}(x) \le 0$ for $x \in [\frac{1}{2}, 1)$, and its nearest neighbor in (6.201) is the zero function which provides the zero bound. Clearly, the RHS of (6.204) representing the standardized expectations of order statistics from uniform populations are nonpositive then. Finally, notice that the Moriguti-Schwarz bounds (4.325) for arbitrary symmetric distributions gives us the solution in the remaining case $j = n$. This is consistent with the fact that $f^s_{n:n}$ belong to (6.201).

In other words, symmetric unimodal distributions are the distributions which succeed the uniform distribution in the s-ordering (see van Zwet, 1964; Lawrence, 1975). It follows that the above results can be extended with use of the projection method to the families of distributions which succeed different symmetric distributions in the s-ordering. We do not present such generalizations here, because they do not seem to have useful applications.

6.2.2 Nonpositive Upper Bounds for Single Order Statistics

Now we concentrate on small rank order statistics based on iid samples with symmetric unimodal baseline distributions. Recall first the statement that for any nondegenerate symmetric parent distribution with finite mean μ, we have $\mathbb{E}X_{j:n} < \mu$ when $j < \frac{n+1}{2}$ and $\mathbb{E}X_{j:n} = \mu$ if n is odd and $j = \frac{n+1}{2}$ (see Sect. 4.6). We show that the trivial zero bound on $\frac{\mathbb{E}X_{j:n}-\mu}{\sigma_p}$ is attained by symmetric unimodal population except for the sample minimum $X_{1:n}$ and scale unit being the mean absolute deviation from the mean σ_1.

Theorem 6.21 (SU Distributions) *Let $X_1, \ldots, X_n$ be iid with parent symmetric unimodal distribution, some mode μ, and finite pth absolute central moment σ_p^p. If either $1 < j < \frac{n+1}{2}$ or $p > 1$, then*

$$\mathbb{E}\frac{X_{j:n} - \mu}{\sigma_p} \le 0, \tag{6.209}$$

and the equality is attained in the limit by the families of mixtures of an atom at μ and uniform distributions symmetric about μ, when the probability of the atom tends to 1.

Proof The inequality (6.209) is obvious. So, we only need to construct the family of symmetric unimodal distributions which attain the equality in (6.209) in the limit. For simplicity we assume that $\mu = 0$, and consider the family of symmetric about zero unimodal distributions which are the mixtures of the uniform distribution on the interval $[-\beta, \beta]$, and the atom at 0, with respective probabilities 2β and $1 - 2\beta$, where β varies over $(0, 1/2]$. The corresponding distribution and quantile functions have the forms

$$F_\beta(x) = \begin{cases} 0, & x \le -\beta, \\ x + \beta, & -\beta \le x < 0, \\ x + 1 - \beta, & 0 \le x \le \beta, \\ 1, & x \ge \beta, \end{cases} \tag{6.210}$$

and

$$F_\beta^{-1}(x) = \begin{cases} x - \beta, & 0 \le x \le \beta, \\ 0, & \beta \le x \le 1 - \beta, \\ x - 1 + \beta, & 1 - \beta \le x \le 1. \end{cases}$$

Clearly, $\mathbb{E}_\beta X_1 = 0$ and

$$\sigma_p^p = \mathbb{E}_\beta|X_1|^p = \int_0^1 |F_\beta^{-1}(x)|^p dx = 2\int_0^\beta (\beta - x)^p dx = \frac{2}{p+1}\beta^{p+1},$$

and, consequently,

$$\sigma_p(\beta) = \left(\frac{2}{p+1}\right)^{1/p} \beta^{1+1/p}. \tag{6.211}$$

Determine now

$$\begin{aligned}
\mathbb{E}_\beta X_{j:n} &= \int_0^1 F_\beta^{-1}(x) f_{j:n}(x)\, dx = \int_0^{\frac{1}{2}} F_\beta^{-1}(x)[f_{j:n}(x) - f_{j:n}(1-x)]\, dx \\
&= \int_0^\beta (x-\beta)[f_{j:n}(x) - f_{n+1-j:n}(x)]\, dx \\
&= \frac{j}{n+1} F_{j+1:n+1}(\beta) - \beta F_{j:n}(\beta) \\
&\quad - \frac{n+1-j}{n+1} F_{n+2-j:n+1}(\beta) + \beta F_{n+1-j:n}(\beta).
\end{aligned} \tag{6.212}$$

Note that $B_{j,n}$ is a polynomial of degree n with the first nonzero coefficient at the power of degree j. The property is inherited by $F_{j:n} = \sum_{k=j}^n B_{k,n}$. Therefore, (6.212) can be rewritten as

$$\mathbb{E}_\beta X_{j:n} = \beta^{j+1} \sum_{k=0}^{n-j} a_k \beta^k$$

with $a_0 = \frac{j}{n+1}\binom{n+1}{j+1} - \binom{n}{j} = -\frac{1}{j+1}\binom{n}{j} \neq 0$ and some real $a_1, \ldots, a_{n-j}$. This, combined with (6.211), imply

$$\frac{\mathbb{E}_\beta X_{j:n}}{\sigma_p} = \frac{\mathbb{E}_\beta X_{j:n} - \mu}{\sigma_p} = \left(\frac{p+1}{2}\right) \beta^{j-1/p} \sum_{k=0}^{n-j} a_k \beta^k,$$

which tends to 0 as $\beta \to 0$, when $j - \frac{1}{p} > 0$, i.e., either $j \geq 2$ or $p > 1$. □

Note that the same conclusion holds when $p = +\infty$, because $\sigma_\infty = \beta$ for distribution functions (6.210). The nontrivial bounds appear when $j = 1$ and $p = 1$.

Theorem 6.22 (SU Distributions) *If the parent distribution is symmetric unimodal, then*

$$\mathbb{E}\frac{X_{1:n} - \mu}{\sigma_1} \leq -2\frac{n-1}{n+1}, \tag{6.213}$$

and the equality is attained by the uniform distributions.

Proof Again, we assume that the symmetry center is equal to 0. If F is symmetric about 0, then the quantile function is antisymmetric about $\frac{1}{2}$. Therefore,

$$\begin{aligned}\mathbb{E}\frac{X_{1:n}}{\sigma_1} &= \frac{\int_0^1 F^{-1}(x) f_{1:n}(x)\, dx}{\int_0^1 |F^{-1}(y)|\, dy} \\ &= \int_{1/2}^1 \frac{F^{-1}(x)}{2\int_{1/2}^1 F^{-1}(y)\, dy}[f_{1:n}(x) - f_{n:n}(x)]\, dx. \qquad (6.214)\end{aligned}$$

Set

$$g(x) = \frac{F^{-1}(x)}{2\int_{1/2}^1 F^{-1}(y)\, dy}, \qquad \frac{1}{2} \le x < 1, \qquad (6.215)$$

$$G(x) = \int_{1/2}^x g(y)\, dy, \qquad \frac{1}{2} \le x < 1. \qquad (6.216)$$

By assumption, function (6.215) is nondecreasing convex, and vanishes at $\frac{1}{2}$. Furthermore, $G\left(\frac{1}{2}\right) = 0$ and $G(1) = \int_{\frac{1}{2}}^1 g(x)\, dx = \frac{1}{2}$. In particular, the conditions are satisfied by the unique linear function

$$g_0(x) = 4x - 2, \qquad \frac{1}{2} \le x < 1. \qquad (6.217)$$

Let g be another nondecreasing convex function on $\left(\frac{1}{2}, 1\right)$ which vanishes at $\frac{1}{2}$, and integrates to $\frac{1}{2}$ on the interval. It has to have another crossing point $\frac{1}{2} < x_0 < 1$ with g_0, and run below g_0 on $\left(\frac{1}{2}, x_0\right)$, and above it on $(x_0, 1)$. As a consequence,

$$G(x) = \int_{1/2}^x g(y)\, dy < \int_{1/2}^x g_0(y)\, dy = G_0(x), \qquad \frac{1}{2} < x \le x_0,$$

and

$$\frac{1}{2} - G(x) = \int_x^1 g(y)\, dy > \int_x^1 g_0(y)\, dy = \frac{1}{2} - G_0(x), \qquad x_0 \le x < 1,$$

which means that $G_0 < G$ on whole $\left(\frac{1}{2}, 1\right)$, and they are equal at the interval ends. Note, moreover, that $f_{1:n}(x) - f_{n:n}(x) = n[(1-x)^{n-1} - x^{n-1}]$ is decreasing. Referring to Theorem 2.9, we obtain

$$\begin{aligned} \mathbb{E}\frac{X_{1:n}}{\sigma_1} &= \int_{1/2}^{1} [f_{1:n}(x) - f_{n:n}(x)]\, G(dx) \\ &\leq \int_{1/2}^{1} [f_{1:n}(x) - f_{n:n}(x)]\, G_0(dx) \\ &= \int_{1/2}^{1} n[(1-x)^{n-1} - x^{n-1}](4x-2)\, dx = -2\frac{n-1}{n+1}. \end{aligned} \tag{6.218}$$

We conclude by observing that the relation

$$\frac{F^{-1}(x)}{2\int_{1/2}^{1} |F^{-1}(y)|\, dy} = 4x - 2, \quad 1/2 \leq x < 1,$$

assuring the upper bound in (6.218) uniquely determines the family of symmetric uniform distributions on intervals $[-2\sigma_1, 2\sigma_1]$, $\sigma_1 > 0$. Letting μ differ from 0, we obtain the uniform distributions on $[\mu - 2\sigma_1, \mu + 2\sigma_1]$. □

6.3 ID, IFR and IRFR Distributions

6.3.1 Positive Upper Bounds for Extreme Order Statistics from ID and IFR Families

The projection method can be used for establishing positive sharp upper mean-variance bounds on L-statistics based on iid samples with the parent distribution functions F preceding a fixed distribution function W in the convex transform order. Using the projection method, we establish the bounds for single order statistics from samples coming from the distributions preceding the uniform, exponential, and negative exponential distributions in the order. The families of the respective quantile functions F^{-1} belong to the closures of the families with increasing density, failure rate, and reversed failure rate functions, respectively. For easy reference, we call them the increasing density (ID), failure rate (IFR), and reversed failure rate (IRFR) families. We also study the spacings from the ID and IFR models.

We first formulate the most general result (cf. Theorem 6.1). For fixed $\bar{c} = (c_1, \ldots, c_n) \in \mathbb{R}^n$, we use the standard notation $f_{\bar{c}:n}$ for the linear combination

$\sum_{i=1}^n c_i f_{i:n}$ of density functions of order statistics from standard uniform samples. Let

$$C^0_{\preceq_c W} = \{g \in L^2((a_0, d), W(dx)) : g \text{ is nondecreasing, concave and orthogonal to } \mathbf{1}\}, \tag{6.219}$$

where (a_0, d) is the support interval of the probability distribution function W, and $P^0_{\preceq_c W} h$ denote the projection of $h \in L^2((a_0, d), W(dx))$ onto $C^0_{\preceq_c W}$.

Theorem 6.23 *Let $X_1, \ldots, X_n$ be an iid sample with a common distribution function F that has a finite mean μ and variance σ^2, and satisfies $F \preceq_c W$ for a given cdf W. Then for arbitrary fixed $\tilde{c} \in \mathbb{R}^n$, we have*

$$\mathbb{E} \sum_{i=1}^n c_i \frac{X_{i:n} - \mu}{\sigma} \le \left\| P^0_{\preceq_c W} \left(f_{\tilde{c}:n} \circ W - \sum_{i=1}^n c_i \mathbf{1} \right) \right\|_W \tag{6.220}$$

which becomes equality iff

$$\frac{F^{-1} \circ W(x) - \mu}{\sigma} = \frac{P^0_{\preceq_c W} \left(f_{\tilde{c}:n} \circ W - \sum_{i=1}^n c_i \mathbf{1} \right)(x)}{\left\| P^0_{\preceq_c W} \left(f_{\tilde{c}:n} \circ W - \sum_{i=1}^n c_i \mathbf{1} \right) \right\|_W} \tag{6.221}$$

under the condition that the RHS of (6.220) is positive.

The inequality is concluded from the following relations:

$$\begin{aligned}
\mathbb{E} \sum_{i=1}^n c_i \frac{X_{i:n} - \mu}{\sigma} &= \int_0^1 \frac{F^{-1}(x) - \mu}{\sigma} \sum_{i=1}^n c_i [f_{i:n}(x) - 1] dx \\
&= \int_0^d \frac{F^{-1} \circ W(x) - \mu}{\sigma} \sum_{i=1}^n c_i [f_{i:n} \circ W(x) - 1] W(dx) \\
&\le \int_0^d \frac{F^{-1} \circ W(x) - \mu}{\sigma} P^0_{\preceq_c W} \left(\sum_{i=1}^n c_i [f_{i:n} \circ W - 1] \right)(x) W(dx) \\
&\le \left(\int_0^d \left[\frac{F^{-1} \circ W(x) - \mu}{\sigma} \right]^2 W(dx) \right)^{\frac{1}{2}} \\
&\quad \times \left(\int_0^d \left[P^0_{\preceq_c W} \left(\sum_{i=1}^n c_i [f_{i:n} \circ W - 1] \right)(x) \right]^2 W(dx) \right)^{\frac{1}{2}}
\end{aligned}$$

$$= \left\| P^0_{\preceq_c W} \left(\sum_{i=1}^{n} c_i [f_{i:n} \circ W - 1] \right) \right\|_W \left(\int_0^1 \left[\frac{F^{-1}(x) - \mu}{\sigma} \right]^2 dx \right)^{\frac{1}{2}}$$

$$= \left\| P^0_{\preceq_c W} \left(\sum_{i=1}^{n} c_i [f_{i:n} \circ W - 1] \right) \right\|_W . \tag{6.222}$$

Both the inequalities in the series of relations (6.222) become equalities if $\frac{F^{-1} \circ W(x) - \mu}{\sigma}$ is proportional to the projection. The variance condition forces that this fraction has norm equal to 1, which implies (6.221). We subtract the constant $\sum_{i=1}^{n} c_i$ from the linear combination of density functions $\sum_{i=1}^{n} c_i f_{i:n}$ in the first line of (6.222) in order to simplify calculations of projections. Note that $\sum_{i=1}^{n} c_i [f_{i:n} \circ W(x) - 1]$ is orthogonal to the constant functions. Therefore, $P^0_{\succeq_c W} \left(f_{\tilde{c}:n} \circ W - \sum_{i=1}^{n} c_i \mathbf{1} \right)(x)$ coincides with $P_{\succeq_c W} \left(f_{\tilde{c}:n} \circ W - \sum_{i=1}^{n} c_i \mathbf{1} \right)(x)$ which is the projection of $f_{\tilde{c}:n} \circ W(x) - \sum_{i=1}^{n} c_i$ onto the convex cone

$$\mathcal{C}_{\preceq_c W} = \{ g \in L^2((a,d), W(dx)) : g \text{ nondecreasing, concave} \} \tag{6.223}$$

(cf. Lemma 6.1). Mimicking the arguments of Sect. 4.7, we also conclude the following (cf. Corollary 6.1).

Corollary 6.2 *Under the assumptions of Theorem* 6.23, *we have*

$$\mathbb{E} \sum_{i=1}^{n} c_i \frac{X_{i:n} - \mu}{\sigma} \leq \sqrt{ \| P_{\preceq_c W} f_{\tilde{c}:n} \circ W \|_W^2 - \left(\sum_{i=1}^{n} c_i \right)^2 } \tag{6.224}$$

which becomes equality iff

$$\frac{F^{-1} \circ W(x) - \mu}{\sigma} = \frac{P_{\preceq_c W} f_{\tilde{c}:n} \circ W(x) - \sum_{i=1}^{n} c_i}{\sqrt{ \| P_{\preceq_c W} f_{\tilde{c}:n} \circ W \|_W^2 - \left(\sum_{i=1}^{n} c_i \right)^2 }} \tag{6.225}$$

provided that the RHS of (6.224) *is positive.*

Obviously, in the case of single order statistics, we have

$$\mathbb{E} \frac{X_{j:n} - \mu}{\sigma} \leq \sqrt{ \| P_{\preceq_c W} f_{j:n} \circ W \|_W^2 - 1 } \tag{6.226}$$

with the equality condition

$$\frac{F^{-1} \circ W(x) - \mu}{\sigma} = \frac{P_{\preceq_c W} f_{j:n} \circ W(x) - 1}{\sqrt{ \| P_{\preceq_c W} f_{j:n} \circ W \|_W^2 - 1 }} . \tag{6.227}$$

Our first evaluations do not need a reference to projections onto (6.223).

Corollary 6.3 (ID and IFR Distributions) *Suppose that $X_1, \dots, X_n$ are iid with a convex distribution function (cumulative failure rate, respectively), finite mean and variance. If $n = 2$, we have*

$$\frac{\mathbb{E}X_{2:2} - \mu}{\sigma} \le \frac{1}{\sqrt{3}} \tag{6.228}$$

and the equality is attained by the uniform distribution on $[\mu - \sqrt{3}\sigma, \mu + \sqrt{3}\sigma]$.

For $n \ge 3$, we have

$$\frac{\mathbb{E}X_{2:n} - \mu}{\sigma} \le B = B(n),$$

where

$$B^2(n) = \frac{1}{(2n-1)(2n-3)}\left[n^2(n-1) - \frac{(n^2-2n)^{2n-1}}{(n-1)^{4n-5}}\right] - 1, \tag{6.229}$$

and becomes equality for the distribution function

$$F(x) = \begin{cases} 0, & \frac{x-\mu}{\sigma} < -\frac{1}{B}, \\ f_{2:n}^{-1}\left(\frac{x-\mu}{\sigma}B + 1\right), & -\frac{1}{B} \le \frac{x-\mu}{\sigma} < \frac{1}{B}\left[\frac{(n-2)^{n-2}n^{n-1}}{(n-1)^{2n-3}} - 1\right], \\ 1, & \frac{x-\mu}{\sigma} \ge \frac{1}{B}\left[\frac{(n-2)^{n-2}n^{n-1}}{(n-1)^{2n-3}} - 1\right]. \end{cases} \tag{6.230}$$

Proof Inequality (6.228) is a special case of the Gumbel-Hartley-David inequality for the sample maxima from general populations (see (4.20)). The equality is attained there by uniform distributions, which have nondecreasing density functions, and increasing failure rates consequently. In the latter case, we again obtain the general bound with the attainability condition established in Balakrishnan (1993) (cf. Corollary 4.2). In order to check that the bound is attained by an ID distribution, it suffices to notice that $f_{2:n}$ are concave on their intervals of increase $\left(0, \frac{1}{n-1}\right)$ (see Lemma 6.2). The point where $f_{2:n}$ changes from increasing to the constant function belongs to the interval, which implies that $f_{2:n}$ is concave increasing. Distribution function (6.230) coincides with its inverse up to a linear transformation of the argument. Obviously, ID distribution function (6.230) attaining the general bound is also IFR. □

Note that the above bounds are not valid in the IRFR case. Indeed, the uniform distribution has a decreasing reversed failure rate. Also, for $n \ge 3$, the functions $f_{2:n}(e^x)$ are convex in some neighborhoods of $-\infty$ (see Lemma 6.4) which contradicts the fact that (6.230) is an IRFR distribution.

Also, the problem of finding the bounds on sample maxima from ID populations seems to be relatively easy. Note that case $n = 2$ has already been solved in Corollary 6.3.

Theorem 6.24 (ID Distributions) *Let $X_1, \ldots, X_n$, $n \geq 3$, be iid random variables with a convex distribution function, mean μ, and variance $0 < \sigma < \infty$. Then,*

$$\frac{\mathbb{E}X_{n:n} - \mu}{\sigma} \leq \sqrt{3}\,\frac{n-1}{n+1}, \tag{6.231}$$

and the equality holds for the uniform parent distribution on the interval $[\mu - \sqrt{3}\sigma, \mu + \sqrt{3}\sigma]$.

Proof As we noted above, we first determine the projections $P_{\preceq_c U}(f_{n:n} - \mathbf{1})$ of the strictly increasing and convex functions $f_{n:n}(x) - 1 = nx^{n-1} - 1$ onto the family $\mathcal{C}_{\preceq_c U}$ of non-decreasing concave functions in the space $L^2((0, 1), dx)$. It is easy to note that the projection has to be linear (a formal verification is left as an exercise).

Minimizing the squared distance

$$D(\alpha, \beta) = \int_0^1 [f_{n:n}(x) - 1 - \alpha(x - \beta)]^2 dx$$

with respect to the parameters α, β of the linear function, we obtain $\alpha^*(x - \beta^*) = 6\,\frac{n-1}{n+1}\left(x - \frac{1}{2}\right)$. Since $\alpha^* > 0$, we have $P_{\preceq_c U}(f_{n:n} - \mathbf{1})(x) = \alpha^*(x - \beta^*)$. Its norm is

$$||P_{\preceq_c U}(f_{n:n} - \mathbf{1})||_U = 6\,\frac{n-1}{n+1}\left[\int_0^1 \left(x - \frac{1}{2}\right)^2 dx\right]^{1/2} = \sqrt{3}\,\frac{n-1}{n+1},$$

which proves (6.231). The equality condition

$$\frac{F^{-1}(x) - \mu}{\sigma} = \frac{P_U^0(f_{n:n} - \mathbf{1})(x)}{||P_U^0 f_{n:n} - \mathbf{1}||_U} = \sqrt{3}\,(2x - 1), \quad 0 < x < 1,$$

determines

$$F(x) = \frac{1}{2}\left(\frac{x - \mu}{\sqrt{3}\sigma} + 1\right), \qquad \mu - \sqrt{3}\sigma < x < \mu + \sqrt{3}\sigma,$$

which is the uniform distribution on $[\mu - \sqrt{3}\sigma, \mu + \sqrt{3}\sigma]$. □

The maximum case for the IFR distributions needs more effort. Compositions $f_{n:n} \circ Z$, $n \geq 3$, are continuous strictly increasing from 0 to n on $(0, \infty)$, strictly convex on $(0, \ln(n - 1)$, and strictly concave on $(\ln(n - 1, \infty)$. In the following lemma, we describe the shape of its projection onto the cone of nondecreasing concave functions.

Lemma 6.7 *Let W be a probability distribution function supported on $\mathbb{R}_+$, and let $h \in L^2(\mathbb{R}_+, W(dx))$ be bounded, differentiable, strictly increasing, first strictly* *convex and then concave function, with the inflection point $b > 0$ say. Then for every $g \in \mathcal{C}_{\preceq_c W}$, there exists $g_* \in \mathcal{C}_{\preceq_c W}$ of the form*

$$g_*(x) = g_*(\alpha, c; x) = \begin{cases} \alpha(x-c) + h(c), & x \le c, \\ h(x), & x \ge c, \end{cases} \tag{6.232}$$

with $c \ge b$ and $\alpha \ge g'(c)$ such that $||g - h_*||_W \le ||g - h||_W$.

The conclusion of the lemma is intuitive. The projection coincides with the original function on the right where the latter is increasing and concave. The best approximation of the left convex increasing part is linear. Obviously, the interval of linear increase of the projection cannot be smaller than the interval of convex increase of h. A formal proof of the claim is not immediate. We present it in Lemma A.5 of the Appendix. It can be also found in Rychlik (2014b).

Theorem 6.25 (IFR Distributions) *For $n \ge 3$, denote by $X_1, \ldots, X_n$ iid random variables with convex cumulative failure rate with mean μ and variance $0 < \sigma^2 < \infty$. Define the functions*

$$S_n(x) = \frac{\sum_{i=1}^n \frac{x^i}{i} - nx^n - (nx^{n-1} - 1)\ln(1-x)}{\ln^2(1-x) + 2\ln(1-x) + 2x}, \tag{6.233}$$

$$T_n(x) = \max\{S_n(x), f_{n-1:n}(x)\}, \tag{6.234}$$

and

$$\begin{aligned} D_n(x, a) &= a^2[\ln^2(1-x) + 2\ln(1-x) + 2x] \\ &\quad - 2a\left[\sum_{i=1}^n \frac{x^i}{i} - nx^n - (nx^{n-1} - 1)\ln(1-x)\right] + \frac{2n(n-1)^2}{2n-1}x^{2n-1} \end{aligned}$$

for $0 < x < 1$ and $a > 0$. Set

$$x_n = \arg \min_{\frac{n-2}{n-1} < x < 1} D_n(x, T_n(x)).$$

Then,

$$\frac{\mathbb{E}X_{n:n} - \mu}{\sigma} \le B = B(n), \tag{6.235}$$

where

$$B^2 = [T_n(x_n)\ln(1-x_n) + nx_n^{n-1}]^2 x_n + \frac{n^2}{2n-1}(1-x_n^{2n-1})$$
$$+ 2[T_n(x_n)\ln(1-x_n) + nx_n^{n-1}]T_n(x_n)[x_n + (1-x_n)\ln(1-x_n)]$$
$$+ T_n^2(x_n)\{2x_n + (1-x_n)[\ln(1-x_n) - \ln^2(1-x_n)]\} - 1. \tag{6.236}$$

The equality in (6.235) holds iff the parent distribution function is

$$F(x) = \begin{cases} 0, & \frac{x-\mu}{\sigma} < a_1, \\ 1-(1-x_n)\exp\left(-\frac{1}{T_n(x_n)}\left[\frac{x-\mu}{\sigma}B - nx_n^{n-1} + 1\right]\right), & a_1 \le \frac{x-\mu}{\sigma} < a_2, \\ \left[\frac{1}{n}\left(\frac{x-\mu}{\sigma}B + 1\right)\right]^{1/(n-1)}, & a_2 \le \frac{x-\mu}{\sigma} < a_3, \\ 1, & x \ge a_3, \end{cases} \tag{6.237}$$

where $a_1 = \frac{T_n(x_n)\ln(1-x_n)+nx_n^{n-1}-1}{B}$, $a_2 = \frac{nx_n^{n-1}-1}{B}$ *and* $a_3 = \frac{n-1}{B}$.

Proof The problem involves in determining the projection of $g(x) = f_{n:n} \circ Z(x) - 1 = n(1-e^{-x})^{n-1} - 1$ belonging to $L^2((0,\infty), e^{-x}dx)$ onto the convex cone of nondecreasing concave functions in the space. By Lemma 6.7, the projection has the form (6.232) with $g = f_{n:n} \circ Z - 1$ for some $c \ge \ln(n-1)$, and $\alpha \ge g'(c) = n(n-1)(1-e^{-c})^{n-2}e^{-c}$. We determine the optimal parameters c_* and α_* minimizing the squared distance

$$\Delta(c,\alpha) = \int_0^\infty [h(\alpha, c; x) - g(x)]^2 e^{-x}dx = \int_0^c [\alpha(x-c) + g(c) - g(x)]^2 e^{-x}dx$$
$$= \alpha^2 \int_0^c (x-c)^2 e^{-x}dx - 2\alpha \int_0^c (x-c)[g(x) - g(c)]e^{-x}dx$$
$$+ \int_0^c [g(x) - g(c)]^2 e^{-x}dx \tag{6.238}$$

under the above inequality constraints on the parameters. For a given $c > 0$, the global minimum of (6.238) with respect to real α is attained at

$$\alpha(c) = \frac{\int_0^c (x-c)[g(x) - g(c)]e^{-x}dx}{\int_0^c (x-c)^2 e^{-x}dx}. \tag{6.239}$$

It is positive, because so are the numerator and denominator. However, we cannot guarantee that $\alpha(c) \ge g'(c)$ which is necessary for concavity of (6.232). The constrained minimum of (6.238) is attained at $\alpha_*(c) = \max\{\alpha(c), g'(c)\}$. Tedious but routine calculations show that

$$\int_0^c [g(x) - g(c)]^2 e^{-x} dx = n^2 \int_0^c [(1 - e^{-x})^{n-1} - (1 - e^{-c})^{n-1}]^2 e^{-x} dx$$

$$= \frac{2n(n-1)^2}{2n-1}(1 - e^{-c})^{2n-1},$$

$$\int_0^c (x-c)[g(x)-g(c)]e^{-x} dx = n \int_0^c (x-c)[(1-e^{-x})^{n-1} - (1-e^{-c})^{n-1}]e^{-x} dx$$

$$= \sum_{i=1}^n \frac{(1-e^{-c})^i}{i} - n(1-e^{-c})^n$$

$$+ nc(1-e^{-c})^{n-1} - c,$$

$$\int_0^c (x-c)^2 e^{-x} dx = c^2 + 2c + 2(1 - e^{-c}).$$

The above formulae allow us to write $\Delta(c, \alpha_*(c))$ for $\ln(n-1) \le c < \infty$ in an analytic form. Changing the variable $x = 1 - e^{-c}$, $\frac{n-2}{n-1} \le x < 1$, we obtain

$$S_n(x) = \alpha(-\ln(1-x)) = \alpha(c),$$

$$f_{n-1:n}(x) = g'(-\ln(1-x)) = g'(c),$$

$$T_n(x) = \max\{\alpha(-\ln(1-x)), g'(-\ln(1-x))\}$$

$$= \max\{\alpha(c), g'(c)\} = \alpha_*(c),$$

$$D_n(x, T_n(x)) = \Delta(-\ln(1-x), \max\{\alpha(-\ln(1-x)), g'(-\ln(1-x))\})$$

$$= \Delta(c, \alpha_*(c)).$$

If $x_n = 1 - e^{-c_*}$ minimizes $D_n(x, T_n(x))$ over the interval $[\frac{n-2}{n-1}, 1) = [1 - e^{-b}, 1)$, then $P_{\preceq_c Z} f_{n:n} \circ Z(x) = h(\alpha_*(c_*), c_*; x)$ (see (6.232)). This implies that the optimal bound for the maximum of IFR random variables is

$$\frac{\mathbb{E}X_{n:n} - \mu}{\sigma} \le ||P_{\preceq_c Z} f_{n:n} \circ Z - 1||_Z$$

$$= \left[\int_0^\infty h^2(\alpha_n^*(c_n), c_n; x)e^{-x} dx - 1\right]^{1/2}$$

$$= \left\{\int_0^{c_n} [\alpha_n^*(c_n)(x - c_n) + n(1 - e^{-c_n})^{n-1}]^2 e^{-x} dx\right.$$

$$\left. + n^2 \int_{c_n}^\infty (1 - e^{-x})^{2n-2} e^{-x} dx - 1\right\}^{1/2}$$

$$= \left\{ \int_0^{x_n} \{T_n(x_n)[\ln(1-x_n) - \ln(1-x)] + nx_n^{n-1}\}^2 dx \right.$$

$$\left. + n^2 \int_{x_n}^1 x^{2n-2} dx - 1 \right\}^{1/2}$$

$$= \left\{ [T_n(x_n)\ln(1-x_n) + nx_n^{n-1}]^2 x_n \right.$$

$$- 2[T_n(x_n)\ln(1-x_n) + nx_n^{n-1}]T_n(x_n) \int_0^{x_n} \ln(1-x)dx$$

$$\left. + T_n^2(x_n) \int_0^{x_n} \ln^2(1-x)dx + \frac{n^2}{2n-1}(1-x_n^{2n-1}) - 1 \right\}^{1/2},$$

which is equal to the square root of (6.236).

The distribution function attaining the bound is characterized by the relation

$$\frac{F^{-1} \circ Z(x) - \mu}{\sigma} B = h(\alpha_*(c_*), c_*; x) - 1$$

$$= \begin{cases} \alpha_*(c_*)(x - c_*) + n(1 - e^{-c_*})^{n-1} - 1, & 0 < x < c_*, \\ n(1 - e^{-x})^{n-1} - 1, & x \geq c_*. \end{cases}$$

Since this is an increasing concave function, F has a convex cumulative failure rate. Changing the variable, we obtain

$$\frac{F^{-1}(x) - \mu}{\sigma} B = \begin{cases} T_n(x_n) \ln \frac{1-x_n}{1-x} + nx_n^{n-1} - 1, & 0 < x < x_n, \\ nx^{n-1} - 1, & x_n \leq x < 1. \end{cases}$$

We easily check that the explicit representation of the respective distribution function is (6.237). This completes the proof. □

6.3.2 *Positive Upper Bounds for Nonextreme Order Statistics from ID Family*

The problems of calculating optimal bounds for other order statistics are far more complicated. Below we present shape assumptions on the functions $f_{j:n} \circ W$, $2 \leq j \leq n-1$, $W = U, Z, V$, whose projections onto the cone (6.219) of nondecreasing concave functions enable us to determine the bounds. They are similar to assumptions (A2) of Sect. 6.1. The difference is merely in admitting that the support of W can start from arbitrary point a_0, being $-\infty$ in particular. The results are used for determining the bounds for expectations of order statistics and spacings based on populations with increasing density, failure rate, and reversed failure rate distributions. The assumptions are explicitly presented below:

(A3) For an absolutely continuous distribution function with a positive density function on the interval (a_0, d) and a finite second moment, let $h \in L^2((a_0, d), W(dx))$ be a bounded, twice differentiable function on (a_0, d) such that

$$\int_{a_0}^{d} h(x)W(dx) = 0.$$

We further assume that h is strictly decreasing on (a_0, a), strictly convex increasing on (a, b), strictly concave increasing on (b, c) with $h(c) > 0 \geq h(0)$, and strictly decreasing on (c, d) with $h(d) = h(a_0)$ for some $-\infty \leq a_0 \leq a < b < c < d \leq \infty$.

We present now the precise forms of projections of functions satisfying (A3). To this end, we introduce some auxiliary functions

$$M(\beta) = h(\beta)[1 - W(\beta)] - \int_{\beta}^{d} h(x)W(dx), \qquad a_0 \leq \beta \leq d. \tag{6.240}$$

It is relevant to notice that $M(a_0) = h(a_0)$, and $M(a) < M(d) = 0 < M(c)$. Since $M'(\beta) = h'(\beta)[1 - W(\beta)]$, function (6.240) is decreasing on (a_0, a), increasing on (a, c), and decreasing on (c, d). Let β_* denote the unique zero of (6.240) in (a, c). Put further

$$\lambda_*(\alpha) = \frac{\int_{a_0}^{\alpha} (x - \alpha)[h(x) - h(\alpha)]W(dx)}{\int_{a_0}^{\alpha} (x - \alpha)^2 W(dx)}, \tag{6.241}$$

$$K(\alpha) = \lambda_*(\alpha) - h'(\alpha), \tag{6.242}$$

$$L(\alpha) = \int_{a_0}^{\alpha} [h(x) - h(\alpha) - \lambda_*(\alpha)(x - \alpha)]W(dx), \quad a_0 \leq \alpha < d, \tag{6.243}$$

(cf. (6.44)–(6.46)).

Theorem 6.26 *Suppose that the zero $a < \beta_* < c$ of (6.240) belongs to (b, c), the set $\mathcal{K} = \{\alpha \in (b, \beta_*) : K(\alpha) \geq 0, L(\alpha) = 0\}$ is nonempty, and $\alpha_* = \inf\{\alpha \in \mathcal{K}\}$. Then,*

$$P_{\preceq_c W} h(x) = \begin{cases} h(\alpha_*) + \lambda_*(\alpha_*)(x - \alpha_*), & a_0 \leq x < \alpha_*, \\ h(x), & \alpha_* \leq x < \beta_*, \\ h(\beta_*), & \beta_* \leq x < d, \end{cases} \tag{6.244}$$

is the projection of h on (6.223). *Otherwise, we have*

$$P_{\preceq_c W}h(x) = \frac{\int_{\alpha_*}^{d} h(x)W(dx)}{1 - W(\alpha_*)} \left[\frac{(x-\alpha_*)\mathbf{1}_{(a_0,\alpha_*)}(x)}{-\int_{a_0}^{\alpha_*}(x-\alpha_*)W(dx)} + 1 \right] \tag{6.245}$$

for α_* *maximizing*

$$\|P_{\preceq_c W}h\|_W^2 = \frac{\left[\int_\alpha^d h(x)W(dx)\right]^2}{[1-W(\alpha)]^2\left[\int_{a_0}^{\alpha}(x-\alpha)W(dx)\right]^2} \times \left[\int_{a_0}^{\alpha}(x-\alpha)^2W(dx) - \left(\int_{a_0}^{\alpha}(x-\alpha)W(dx)\right)^2\right] \tag{6.246}$$

among $\alpha \geq \beta_*$ *satisfying*

$$\frac{\int_\alpha^d h(x)W(dx)}{1-W(\alpha)} = -\frac{\int_{a_0}^{\alpha}(x-\alpha)h(x)W(dx)\int_{a_0}^{\alpha}(x-\alpha)W(dx)}{\int_{a_0}^{\alpha}(x-\alpha)^2W(dx) - \left(\int_{a_0}^{\alpha}(x-\alpha)W(dx)\right)^2} > 0. \tag{6.247}$$

The proof is presented in the Appendix (see Theorem A.3, and also Goroncy & Rychlik, 2015). We only mention here two partial results derived in the proof which are directly applied in our further reasoning. They are as follows:

$$\frac{dD(\alpha, \lambda_*(\alpha))}{d\alpha} = \frac{d}{d\alpha}\left[\int_0^{\alpha}[\lambda_*(\alpha)(x-\alpha)+h(\alpha)-h(x)]^2W(dx)\right] = 2K(\alpha)L(\alpha) \tag{6.248}$$

and

$$L(\alpha) = 0 \quad \Rightarrow \quad \frac{d^2D(\alpha, \lambda_*(\alpha))}{d\alpha^2} > 0. \tag{6.249}$$

Now we use Theorem 6.26 for establishing accurate upper bounds on the ratios $\frac{\mathbb{E}X_{j:n}-\mu}{\sigma}$, $3 \leq j \leq n-1$, $n \geq 4$, for the iid samples with ID parent distribution. For this purpose, we need to project $h = h_{j:n} = f_{j:n} - 1$ onto the convex cone of nondecreasing concave elements of $L^2((0,1), dx)$. We see that $h_{j:n}$, $3 \leq j \leq n-1$, satisfy assumptions (A3) (cf. Lemma 6.2). Namely, they are convex increasing on $(0,b)$ for $b = b_{j:n} = \frac{j-1}{n-1} - \sqrt{\frac{(j-1)(n-j)}{(n-1)^2(n-2)}}$, concave increasing on (b,c) for

$c = c_{j:n} = \frac{j-1}{n-1}$, and decreasing on (c, d) for $d = 1$. Here $a_0 = a = 0$, i.e., the first interval of decrease is empty. Also $h_{j:n}$ integrates to 0 with weight $w(x) = 1$ on (0, 1). Owing to Theorem 6.26, the projection $S_{\preceq_c U} h_{j:n}$ may have one of the following forms: it is either subsequently linear increasing, equal to h, and constant (l-h-c for brevity), or merely first linear and ultimately constant (i.e., l-c, respectively).

We first check conditions for the l-h-c form. We start with specifying definitions of functions studied above for $h = h_{j:n} = f_{j:n} - 1$. Let

$$M_{j:n}(\beta) = f_{j:n}(\beta)(1-\beta) - 1 + F_{j:n}(\beta), \qquad 0 < \beta < 1, \tag{6.250}$$

(cf. (6.240)), where $1 - F_{j:n}(x) = \sum_{k=0}^{j-1} B_{k,n}(x)$ is the survival function of the jth order statistic from the standard uniform sample. Since $M'_{j:n}(\beta) = f'_{j:n}(\beta)(1-\beta)$, (6.250) increases from $M_{j:n}(0) = -1$ to $M_{j:n}(c_{j:n})$, and then decreases to $M_{j:n}(1) = 0$. So it has a single zero at $0 < \beta_* = \beta^*_{j:n} < c_{j:n}$, and a necessary condition for the l-h-c form is

$$M_{j:n}(b_{j:n}) < 0. \tag{6.251}$$

Referring to (6.241) and (6.242), we define further

$$\lambda^*_{j:n}(\alpha) = \frac{\int_0^\alpha (x-\alpha) f_{j:n}(x)dx - f_{j:n}(\alpha)\int_0^\alpha (x-\alpha)dx}{\int_0^\alpha (x-\alpha)^2 dx}$$

$$= \frac{3}{\alpha^3}\left[\frac{j}{n+1} F_{j+1:n+1}(\alpha) - \alpha F_{j:n}(\alpha) + \frac{\alpha^2}{2} f_{j:n}(\alpha)\right], \tag{6.252}$$

$$K_{j:n}(\alpha) = \lambda^*_{j:n}(\alpha) - f'_{j:n}(\alpha), \qquad 0 < \alpha < 1. \tag{6.253}$$

The following lemma is useful in what follows.

Lemma 6.8 *Function* (6.253) *changes its sign once from minus to plus in the interval* $(b_{j:n}, c_{j:n})$.

Proof Tedious elementary calculations show that function (6.252) has the following representation in the Bernstein basis:

$$\lambda^*_{j:n}(\alpha) = \frac{3n(n-1)(j+2)}{2j(j-1)} B_{j-2,n-2}(\alpha)$$
$$+ \sum_{k=j-1}^{n-2} \frac{3n(n-1)(j-k-3)}{(k+1)(k+2)(k+3)} B_{k,n-2}(\alpha).$$

Therefore, (6.253) can be written as

$$K_{j:n}(\alpha) = \sum_{k=j-2}^{n-2} b_k B_{k,n-2}(\alpha), \tag{6.254}$$

where

$$b_{j-2} = -\frac{n(n-1)}{2j(j+1)}(2j^2 - j - 6) < 0, \qquad \text{for } j \geq 3,$$

$$b_{j-1} = \frac{n(n-1)}{j(j+1)(j+2)}[j(j+1)(j+2) - 6] > 0, \qquad \text{for } j \geq 2,$$

$$b_k = \frac{3n(n-1)(j-k-3)}{(k+1)(k+2)(k+3)} < 0, \qquad k = j, \ldots, n-2.$$

The variation diminishing property of the Bernstein polynomials implies that (6.253) is either negative on the whole (0, 1), or first negative, then positive, and ultimately negative. The relations $K_{j:n}(b_{j:n}) < 0 < K_{j:n}(c_{j:n})$ exclude the first case, and allow us to conclude the statement of Lemma 6.8. □

Let $b_{j:n} < \gamma_* = \gamma^*_{j:n} < c_{j:n}$ be the change sign point of (6.253) in $(b_{j:n}, c_{j:n})$. Another necessary condition for the l-h-c-type projection is

$$K_{j:n}(\beta^*_{j:n}) > 0, \tag{6.255}$$

which implies that the interval $[\gamma^*_{j:n}, \beta^*_{j:n})$ is nonempty. The projection can have the l-h-c form only if the point $\alpha_* = \alpha^*_{j:n}$ of its change from the linear part into h itself belongs to the interval. Therefore, yet another necessary condition for the projection of this shape is that the function

$$L_{j:n}(\alpha) = \int_0^{\alpha} [f_{j:n}(x) - \lambda^*_{j:n}(\alpha)(x-\alpha) - f_{j:n}(\alpha)]dx$$

$$= F_{j:n}(\alpha) + \frac{\alpha^2}{2}\lambda^*_{j:n}(\alpha) - \alpha f_{j:n}(\alpha) \tag{6.256}$$

vanishes for some $\gamma_{j:n} < \alpha < \beta^*_{j:n}$.

Lemma 6.9 *If* $j < \frac{n+1}{3}$, *then* (6.256) *is negative in some neighborhoods of* 0 *and* 1, *and possibly positive in one interval contained in* (0, 1). *If* $j \geq \frac{n+1}{3}$, *then* (6.256) *is first positive and eventually negative in* (0, 1).

Proof It follows immediately from the equation

$$L_{j:n}(\alpha) = -\frac{(j-1)(j-2)}{4(j+1)} B_{j,n}(\alpha) + \sum_{k=j+1}^{n} \frac{3j-k-1}{2(k+1)} B_{k,n}(\alpha)$$

and VDP of Bernstein polynomials. □

We are finally in a position to present the following theorem on the bounds for expectations of order statistics in increasing density populations.

Theorem 6.27 (ID Distributions) *Assume that $j < \frac{n+1}{3}$, relations (6.251), and (6.255) hold, and the last sign change point $\alpha_* = \alpha^*_{j:n}$ of (6.256) belongs to $(\gamma^*_{j:n}, \beta^*_{j:n})$. Then, $h_{j:n} = f_{j:n} - 1$ has the projection of the form*

$$P_{\preceq_c U} h_{j:n}(x) = \begin{cases} \lambda_*(x-\alpha_*) + f_{j:n}(\alpha_*) - 1, & 0 \le x < \alpha_*, \\ f_{j:n}(x) - 1, & \alpha_* \le x < \beta_* \\ f_{j:n}(\beta_*) - 1, & \beta_* \le x < 1, \end{cases} \tag{6.257}$$

where $\alpha_* = \alpha^*_{j:n}$ and $\lambda_* = \lambda^*_{j:n}$ and $\beta_* = \beta^*_{j:n}$ are defined above. Accordingly,

$$\frac{\mathbb{E}X_{j:n} - \mu}{\sigma} \le B = B(j, n),$$

where

$$B^2 = \frac{1}{3}\lambda_*^2\alpha_*^3 + \lambda_*\alpha_*^2 f_{j:n}(\alpha_*) + (1 - \beta_* + \alpha_*) f^2_{j:n}(\alpha_*) - 1$$
$$+ n\frac{\binom{2j-2}{j-1}\binom{2n-2j}{n-j}}{\binom{2n-1}{n}}[F_{2j-1:2n-1}(\beta_*) - F_{2j-1:2n-1}(\alpha_*)], \tag{6.258}$$

and the equality holds if

$$F(x) = \begin{cases} 0, & \frac{x-\mu}{\sigma} < a_1, \\ \frac{1}{\lambda_*}\left(B\frac{x-\mu}{\sigma} - f_{j:n}(\alpha_*) + 1\right) + \alpha_*, & a_1 \frac{x-\mu}{\sigma} < a_2, \\ f^{-1}_{j:n}\left(B\frac{x-\mu}{\sigma} + 1\right), & a_2 \le \frac{x-\mu}{\sigma} < a_3, \\ 1, & \frac{x-\mu}{\sigma} \ge a_3. \end{cases} \tag{6.259}$$

for $a_1 = \frac{f_{j:n}(\alpha_*)-1-\lambda_*}{B}$, $a_2 = \frac{f_{j:n}(\alpha_*)-1}{B}$ and $a_3 = \frac{f_{j:n}(\beta_*)-1}{B}$.

Proof Let conditions (6.251) and (6.255) hold, and suppose that (6.256) vanishes at some $\gamma_{j:n} < \alpha < \beta^*_{j:n}$. Then, $M_{j:n}(\alpha) > 0$ and $\frac{dD(\alpha,\lambda^*_{j:n}(\alpha))}{d\alpha} = 2K_{j:n}(\alpha)L_{j:n}(\alpha)$ changes its sign from − to + (cf. (6.248) and (6.249)). This implies that (6.256) changes the sign from + to − there. Due to Lemma 6.9, it is only possible for the latter change sign point in the case $j < \frac{n+1}{3}$. If it happens that (6.256) amounts to 0 at $\gamma_{j:n}$, where (6.253) changes the sign from $-$ to $+$, (6.248) and (6.249) would force that $\gamma_{j:n}$ to be a local maximum of (6.256). We show that this is impossible. We observe that in this case (6.256) starts from 0 at 0, then it is negative, then positive, and finally negative with $L_{j:n}(1) = \frac{3j-n-1}{2(n+1)} < 0$. So it has at least two local extremes in (0, 1). However, its derivative

$$L'_{j:n}(\alpha) = -\frac{n(j-1)(j-2)}{4(j+1)} B_{j-1,n-1}(\alpha) + \frac{nj(j-1)(j+4)}{4(j+1)(j+2)} B_{j,n-1}(\alpha)$$
$$- \sum_{k=j+1}^{n-1} \frac{3nj}{2(k+1)(k+2)} B_{k,n-1}(\alpha)$$

has at most two sign changes in (0, 1). It follows that (6.256) itself has exactly two local extremes there: the first one is strictly negative, and the other is positive. It is hence impossible that $L_{j:n}(\alpha) = L'_{j:n}(\alpha) = 0$ for any $0 < \alpha < 1$. In the same way, we show that $K_{j:n}(\alpha) = K'_{j:n}(\alpha) = 0$ cannot occur for any $0 < \alpha < 1$, because relations $K_{j:n}(\alpha) = L_{j:n}(\alpha) = 0$ and $\frac{d^2 D(\alpha,\lambda^*_{j:n}(\alpha))}{d\alpha^2} > 0$ imply that just one of the functions changes its sign, and the other has a local extremum.

Summing up, the projection l-h-c is possible only when (6.256) changes its sign from $+$ to $-$ at some point $\alpha \in (\gamma^*_{j:n}, \beta^*_{j:n})$ where (6.253) is strictly positive. This is only possible for $j < \frac{n+1}{3}$ if (6.256) has two sign changes, and the latter belongs to $(\gamma^*_{j:n}, \beta^*_{j:n})$. This is the only point where the projection changes its form from linear one to $h_{j:n}$. The other transformation point is $\beta^*_{j:n}$, the unique zero of (6.250), at which the projection becomes a constant. If these conditions are satisfied, the remaining conclusions are immediate. Formula

$$B^2(j,n) = ||P_{\preceq_c U} h_{j:n}||^2_U = \int_0^{\alpha_*} [\lambda_*(x-\alpha_*) + f_{j:n}(\alpha_*)]^2 dx$$
$$+ \int_{\alpha_*}^{\beta_*} f^2_{j:n}(x)dx + f^2_{j:n}(\beta_*)(1-\beta_*) - 1 \qquad (6.260)$$

represents the squared L^2-norm of (6.257). Its explicit form is presented in (6.258) (cf. (6.222)). Moreover,

$$\frac{F^{-1}(x)-\mu}{\sigma} = \frac{P_{\preceq_c U} h_{j:n}(x)}{||P_{\preceq_c U} h_{j:n}||_U} \qquad (6.261)$$

describes the attainability condition (cf. (6.221)). The respective distribution function has the form (6.259). □

One may wonder if the conditions of Theorem 6.27 hold true for any j and n. Note that the projection may coincide with the original function $h_{j:n}$ only on a part of interval $(b_{j:n}, c_{j:n})$ of the concave increase of $f_{j:n}$. The length of the interval $c_{j:n} - b_{j:n} = \sqrt{\frac{(j-1)(n-j)}{(n-1)^2(n-2)}}$ is small, especially for large n and extreme j, whereas the values of $h_{j:n}(x)$, $b_{j:n} < x < c_{j:n}$, where $f_{j:n}$ is concave increasing are close to the maximal ones, and rather far from average ones. Condition (6.251) cannot be verified analytically for general j and n, because (6.250) is a polynomial of degree n with complicated coefficients, and $b_{j:n}$ has a complex form as well. However, numerical evaluations show that it is satisfied very rarely. Namely, it holds for $j = 3, \dots, n-1$ when $n = 4, \dots 8$, for $j = n-3, n-2, n-1$ when $n = 9, 10$ and for $j = n-2, n-1$ when $n = 11, \dots, 1000$. Neither of these pairs satisfy $j < \frac{n+1}{3}$ which is necessary for (6.257). Therefore, we are almost sure that the projection of $f_{j:n} - 1$ does not have an interval part identical with the original function for any $3 \le j \le n-1 < \infty$, and all the bounds are described in Theorem 6.28 below. We are unable to verify the hypothesis analytically, though.

So we suppose now that the assumptions of Theorem 6.27 do not hold. For $0 < \alpha < 1$, we define

$$\begin{aligned} A_{j:n}(\alpha) &= \int_0^\alpha (x-\alpha)[f_{j:n}(x) - 1]dx \\ &= \frac{j}{n+1} F_{j+1:n+1}(\alpha) - \alpha F_{j:n}(\alpha) + \frac{\alpha^2}{2}, \end{aligned} \tag{6.262}$$

$$\begin{aligned} C_{j:n}(\alpha) &= (\alpha^2 - \alpha^3) A_{j:n}(\alpha) + 2\left[\frac{\alpha^3}{3} - \frac{\alpha^4}{4}\right] \int_0^\alpha [f_{j:n}(x) - 1]dx \\ &= \frac{j\alpha^2(1-\alpha)}{n+1} F_{j+1:n+1}(\alpha) - \frac{\alpha^3(1-\alpha)}{3} F_{j:n}(\alpha) - \frac{\alpha^4}{6}[1 - F_{j:n}(\alpha). \end{aligned} \tag{6.263}$$

It will be seen below that vanishing of (6.263) is equivalent to the equality (6.247) in the special case $h(x) = f_{j:n}(x) - 1$ and $w(x) = 1$, $0 < x < 1$. Lemma 6.10 describes the behavior of functions (6.262) and (6.263).

Lemma 6.10

(i) *If* $3 \le j < \frac{n+1}{2}$, *then* (6.262) *changes the sign once from plus to minus. For* $\frac{n+1}{2} \le j \le n-1$, *it is nonnegative.*

(ii) *If* $3 \le j < \frac{n+1}{3}$, *function* (6.263) *is either nonpositive in* $(0, 1)$ *or consecutively nonpositive, nonnegative and ultimately nonpositive. Otherwise, it changes the sign once from minus to plus.*

Proof

(i) Function (6.262) has derivatives $A'_{j:n}(\alpha) = \alpha - F_{j:n}(\alpha)$ and $A''_{j:n}(\alpha) = 1 - f_{j:n}(\alpha)$. The latter is positive at some neighborhoods of 0 and 1, and negative in between. So the former, which vanishes at 0 and 1, is increasing near 0 and 1, and decreasing in the middle. This implies that it is first positive, and then negative in (0, 1). Therefore (6.262) first increases from $A_{j:n}(0) = 0$ to a positive maximum, and finally decreases to $A_{j:n}(1) = \frac{j}{n+1} - \frac{1}{2}$ which is negative for $j < \frac{n+1}{2}$, and nonnegative otherwise.

(ii) The Bernstein basis representation of (6.263) is

$$C_{j:n}(\alpha) = -\sum_{k=4}^{j+3} \frac{(k-3)(k-2)(k-1)k}{6(n+1)(n+2)(n+3)(n+4)} B_{k,n+4}(\alpha)$$

$$+ \sum_{k=j+3}^{n+3} \frac{(k-1)k(n+4-k)(3j+2-k)}{3(n+1)(n+2)(n+3)(n+4)} B_{k,n+4}(\alpha) = \sum_{k=4}^{n+3} c_k B_{k,n+4}(\alpha).$$

Clearly $c_k < 0$, $k = 4, \ldots, j+2$. The sign of c_{j+3} is immaterial for our analysis. The signs of c_k, $k = j+1, \ldots, n+3$, depend on values of $3j+2-k$. They are all nonnegative when $j \geq \frac{n+1}{3}$, and change their signs from + to − otherwise. Referring again to the VDP of Bernstein polynomials, we conclude that function (6.263) is first nonpositive and then nonnegative for $j \geq \frac{n+1}{3}$. In the opposite case, it is either always nonnegative or consecutively nonnegative, nonpositive and again nonnegative.

□

Now we arrive at the theorem which presents the sharp upper bounds for most of the order statistics coming from iid ID populations.

Theorem 6.28 (ID Distributions) *Assume that for some* $3 \leq j \leq n-1$, *assumptions of Theorem* 6.27 *do not hold. Then* (6.263) *has two sign changes for* $j < \frac{n+1}{3}$. *Let* $\alpha_* = \alpha^*_{j:n}$ *denote either the smaller of or the unique sign change point of* (6.263) *in* (0, 1). *Then,*

$$P_{\preceq_c U} h_{j:n}(x) = \lambda_*(x - \alpha_*)\mathbf{1}_{(0,\alpha_*)}(x) + \gamma_*,$$

where

$$\lambda_* = \lambda_*(\alpha_*) = \frac{12 A_{j:n}(\alpha_*)}{4\alpha_*^3 - 3\alpha_*^4},$$

$$\gamma_* = \gamma_*(\alpha_*, \lambda_*) = \lambda_* \frac{\alpha_*^2}{2}.$$

Consequently,

$$\frac{\mathbb{E}X_{j:n} - \mu}{\sigma} \leq B = B(j, n), \tag{6.264}$$

where

$$B^2(j,n) = ||P_{\preceq_c U} h_{j:n}||_U^2 = \frac{\gamma_*^3 - (\gamma_* - \lambda_* \alpha_*)^3}{3\lambda_*} + \gamma_*^2(1-\alpha_*).$$

The equality in (6.264) *holds if the parent distribution is a mixture of the uniform distribution on* $\left[\mu - \sigma\frac{\lambda_*\alpha_* - \gamma_*}{B(j,n)}, \mu + \sigma\frac{\gamma_*}{B(j,n)}\right]$ *and an atom at* $\mu + \sigma\frac{\gamma_*}{B(j,n)}$ *with respective probabilities* α_* *and* $1-\alpha_*$.

Proof We know that under the assumptions of Theorem 6.28 the projection has the form $\lambda_*(x-\alpha_*)\mathbf{1}_{[0,\alpha_*)}(x) + \gamma_*$, where

$$\gamma_* = \gamma_*(\alpha_*, \lambda_*) = \lambda_* \frac{\alpha_*^2}{2},$$

$$\lambda_* = \lambda_*(\alpha_*) = \frac{12\int_0^{\alpha_*}(x-\alpha_*)[f_{j:n}(x)-1]dx}{4\alpha_*^3 - 3\alpha_*^4} = \frac{12A_{j:n}(\alpha_*)}{4\alpha_*^3 - 3\alpha_*^4},$$

and $\alpha_* > \beta_{j:n}^*$ is a point where (6.263) changes the sign from − to + For $j < \frac{n+1}{3}$, α_* is just the unique sign change point of (6.263). If $j \geq \frac{n+1}{3}$, relation $C_{j:n}(\alpha) \leq 0$, $0 < \alpha < 1$, contradicts the existence of l-c projection. Under the alternative claim of Lemma 6.10(ii), α coincides with the smaller sign change point of (6.263). If the final form of projection is determined, we can easily calculate the bound in (6.264) which is just the norm of the projection, and provide the attainability condition with the use of (6.221). □

6.3.3 Positive Upper Bounds for Spacings from ID Family

Now we consider the problem of bounding the expectations of spacings $\frac{\mathbb{E}(X_{j+1:n} - X_{j:n})}{\sigma}$, $j = 1, \ldots, n-1$, where $X_1, \ldots X_n$ are independent and have a common ID distribution. For this purpose, we use the projection methods. The basic step here in determining the projection of the function $f_{j,j+1:n} = f_{j+1:n} - f_{j:n}$ onto the convex cone of nondecreasing concave functions in $L^2((0,1), dx)$. For $j = 1$ and $n-1$, the problems are relatively easy and we solve them at the end. Now we focus on spacings for $j = 2, \ldots, n-2$. Then $f_{j,j+1:n}$ satisfy assumptions (A3): they decrease from 0 to $a = a_{j,j+1:n} = \frac{j}{n} - \sqrt{\frac{j(n-j)}{n^2(n-1)}}$, then convexly increase from $a_{j,j+1:n}$ to some $b = b_{j,j+1:n}$, concavely increase to $c = c_{j,j+1:n} = \frac{j}{n} + \sqrt{\frac{j(n-j)}{n^2(n-1)}}$, and eventually decrease to $d_{j,j+1:n} = 1$. The inflection point $b_{j,j+1:n}$ does not have a simple analytic form. We can check that $b_{j,j+1:n} < (=,>)\frac{j}{n}$ when $j < (=,>)\frac{n}{2}$. Note that the functions in (6.240)–(6.243) can be treated as linear operators acting on functions $h \in L^2((0,d), w(x)dx)$.

The same property have (6.262) and (6.263) defined for specific $h_{j:n} = f_{j:n} - 1$ from $L^2((0,1), dx)$. This observation allows us to apply the results of the previous subsection to obtain bounds for expectations of spacings. For constructing the projections of $f_{j,j+1:n} = f_{j+1:n} - f_{j:n} = h_{j+1:n} - h_{j:n}$, we can use the following functions:

$$\begin{aligned} M_{j,j+1:n}(\beta) &= M_{j+1:n}(\beta) - M_{j:n}(\beta) \\ &= n[B_{j,n-1}(\beta) - B_{j-1,n-1}(\beta)](1-\beta) - B_{j,n}(\beta), \end{aligned} \tag{6.265}$$

$$\begin{aligned} \lambda^*_{j,j+1:n}(\alpha) &= \lambda^*_{j+1:n}(\alpha) - \lambda^*_{j:n}(\alpha) = \frac{3n}{2\alpha}[B_{j,n-1}(\alpha) - B_{j-1,n-1}(\alpha)] + \frac{3}{\alpha^3} \\ &\times \left[\frac{1}{n+1}F_{j+2:n+1}(\alpha) - \frac{j}{n+1}B_{j+1,n+1}(\alpha) + \alpha B_{j,n}(\alpha)\right], \end{aligned} \tag{6.266}$$

$$\begin{aligned} K_{j,j+1:n}(\alpha) &= K_{j+1:n}(\alpha) - K_{j:n}(\alpha) = \lambda^*_{j,j+1:n}(\alpha) \\ &+ n(n-1)[B_{j-2,n-2}(\alpha) - 2B_{j-1,n-2}(\alpha) + B_{j,n-2}(\alpha)], \end{aligned} \tag{6.267}$$

$$\begin{aligned} L_{j,j+1:n}(\alpha) &= L_{j+1:n}(\alpha) - L_{j:n}(\alpha) = -B_{j,n}(\alpha) + \frac{\alpha^2}{2}\lambda^*_{j,j+1:n}(\alpha) \\ &- n\alpha[B_{j,n-1}(\alpha) - B_{j-1,n-1}(\alpha)], \end{aligned} \tag{6.268}$$

$$\begin{aligned} A_{j,j+1:n}(\alpha) &= A_{j+1:n}(\alpha) - A_{j:n}(\alpha) \\ &= \frac{1}{n+1}[F_{j+2:n+1}(\alpha) - jB_{j+1,n+1}(\alpha)] + \alpha B_{j,n}(\alpha), \end{aligned} \tag{6.269}$$

$$\begin{aligned} C_{j,j+1:n}(\alpha) &= C_{j+1:n}(\alpha) - C_{j:n}(\alpha) = \frac{\alpha^2(1-\alpha)}{n+1}[F_{j+2:n+1}(\alpha) - jB_{j+1,n+1}(\alpha)] \\ &+ \left[\frac{\alpha^3(1-\alpha)}{3} - \frac{\alpha^4}{6}\right]B_{j,n}(\alpha). \end{aligned} \tag{6.270}$$

Some useful properties of them are described in the following lemma.

Lemma 6.11

(i) *Function* (6.265) *is decreasing on* $(0, a_{j,j+1:n})$ *and on* $(c_{j,j+1:n}, 1)$, *and increasing on* $(a_{j,j+1:n}, c_{j,j+1:n})$, *and vanishes at* 0 *and* 1. *Its unique zero* $\beta_* = \beta^*_{j,j+1:n} = \frac{j}{n-1} \in \left(\frac{j}{n}, c_{j,j+1:n}\right) \subset (a_{j,j+1:n}, c_{j,j+1:n})$.
(ii) *Function* (6.267) *changes the sign once from* $-$ *to* $+$ *in* $(b_{j,j+1:n}, c_{j,j+1:n})$ *at some* $\gamma_* = \gamma^*_{j,j+1:n}$, *say.*
(iii) *Function* (6.268) *is positive at some neighborhoods of* 0 *and* 1, *and there is possibly one subinterval of* (0, 1), *where it is nonpositive.*
(iv) *Function* (6.269) *is increasing from* 0 *at* 0 *to* $\frac{1}{n+1}$ *at* 1.
(v) *Function* (6.270) *is first nonpositive, and then nonnegative in* (0, 1).

Proof

(i) We have

$$M_{j,j+1:n}(\beta) = -(n-j+1)B_{j-1,n}(\beta) + (n-j-1)B_{j,n}(\beta)$$
$$= \frac{(n-1)\beta - j}{\beta} B_{j,n}(\beta),$$
$$\frac{M'_{j,j+1:n}(\beta)}{n} = -(n-j+1)B_{j-2,n-1}(\beta) + 2(n-j)B_{j-1,n-1}(\beta)$$
$$-(n-j-1)B_{j,n-1}(\beta).$$

So (6.265) has zeros at 0, $\frac{j}{n-1}$, and 1, and by the variation diminishing property of its derivative, it is first decreasing, then increasing and finally decreasing.

(ii) Applying (6.267) and (6.254), we obtain

$$K_{j,j+1:n}(\alpha) = \frac{n(n-1)(2j^2-j-6)}{2j(j+1)} B_{j-2,n-2}(\alpha)$$
$$- \frac{n(n-1)(4j^3+9j^2-j-12)}{2j(j+1)(j+2)} B_{j-1,n-2}(\alpha)$$
$$+ \frac{n(n-1)[(j+1)(j+2)(j+3)+3]}{(j+1)(j+2)(j+3)} B_{j,n-2}(\alpha)$$
$$+ \sum_{k=j+1}^{n-2} \frac{3n(n-1)}{(k+1)(k+2)(k+3)} B_{k,n-2}(\alpha).$$

Its sequence of signs is either $+$ or $+-+$. Conditions $K_{j,j+1:n}(b) < 0 < K_{j,j+1:n}(c)$ exclude the former case, and provide the conclusion.

(iii) The statement can be deduced from the equation

$$L_{j,j+1:n}(\alpha) = \frac{(j-1)(j-2)}{4(j+1)} B_{j,n}(\alpha) - \frac{(j-1)(j+4)}{4(j+2)} B_{j+1,n}(\alpha)$$
$$+ \sum_{k=j+2}^{n} \frac{3}{2(k+1)} B_{k,n}(\alpha),$$

and VDP of Bernstein polynomials.

(iv) Increase of (6.269) is implied by $A'_{j,j+1:n}(\alpha) = B_{j,n}(\alpha) > 0$, and its values at 0 and 1 are easily calculated

(v) The conclusion follows immediately from the representation

$$C_{j,j+1:n}(\alpha) = -\frac{(j+3)(j+2)(n+1-j)(2j-1)}{3(n+1)(n+2)(n+3)(n+4)} B_{j+3,n+4}(\alpha)$$
$$- \frac{(j+3)(j+4)[(j+1)(j+2)-6(n-j)]}{6(n+1)(n+2)(n+3)(n+4)} B_{j+4,n+4}(\alpha)$$
$$+ \sum_{k=j+5}^{n+3} \frac{(k-1)k(n+4-k)}{(n+1)(n+2)(n+3)(n+4)} B_{k,n+4}(\alpha).$$

□

The bounds for the case of spacings are presented in the following two theorems.

Theorem 6.29 (ID Distributions) *If $M_{j,j+1:n}(b_{j,j+1:n}) < 0 < K_{j,j+1:n}(\beta^*_{j,j+1:n})$, (6.268) has two sign changes in* (0, 1), *and the former sign change point $\alpha_{j,j+1:n}$ of (6.268) belongs to $(\gamma^*_{j,j+1:n}, \beta^*_{j,j+1:n})$. Then, the projection $P_{\preceq_c U} f_{j,j+1:n}$ of $f_{j,j+1:n} = f_{j+1:n} - f_{j:n}$ satisfies*

$$P_{\preceq_c U} f_{j,j+1:n}(x) = \begin{cases} \lambda_*(x-\alpha_*) + f_{j,j+1:n}(\alpha_*), & 0 \le x < \alpha_*, \\ f_{j,j+1:n}(x), & \alpha_* \le x < \beta_*, \\ f_{j,j+1:n}(\beta_*), & \beta_* \le x < 1, \end{cases}$$

for $\alpha_* = \alpha^*_{j,j+1:n}$ and $\lambda_* = \lambda^*_{j,j+1:n}(\alpha^*_{j,j+1:n})$ and $\beta_* = \beta^*_{j,j+1:n}$ defined in the assumptions, formula (6.266) and Lemma 6.11(i), *respectively. Consequently,*

$$\frac{\mathbb{E}(X_{j+1:n} - X_{j:n})}{\sigma} \le B = B(j, j+1, n),$$

where

$$B^2 = \frac{1}{3}\lambda_*^2\alpha_*^3 + \lambda_*\alpha_*^2 f_{j,j+1:n}(\alpha_*) + (1-\beta_*+\alpha_*) f^2_{j,j+1:n}(\alpha_*)$$
$$+ n\frac{\binom{2j-2}{j-1}\binom{2n-2j}{n-j}}{\binom{2n-1}{n}}[F_{2j-1,2n-1}(\beta_*) - F_{2j-1:2n-1}(\alpha_*)]$$
$$+ n\frac{\binom{2j}{j}\binom{2n-2j-2}{n-j-1}}{\binom{2n-1}{n}}[F_{2j+1,2n-1}(\beta_*) - F_{2j+1:2n-1}(\alpha_*)]$$
$$- 2n\frac{\binom{2j-1}{j}\binom{2n-2j-1}{n-j}}{\binom{2n-1}{n}}[F_{2j,2n-1}(\beta_*) - F_{j:2n-1}(\alpha_*)], \tag{6.271}$$

and the equality holds if

$$F(x) = \begin{cases} 0, & \frac{x-\mu}{\sigma} < a_1, \\ \frac{1}{\lambda_*}\left(B\frac{x-\mu}{\sigma} - f_{j,j+1n}(\alpha_*)\right) + \alpha_*, & a_1\frac{x-\mu}{\sigma} < a_2, \\ f^{-1}_{j,j+1:n}\left(B\frac{x-\mu}{\sigma}\right), & a_2 \leq \frac{x-\mu}{\sigma} < a_3, \\ 1, & \frac{x-\mu}{\sigma} \geq a_3, \end{cases} \tag{6.272}$$

for $a_1 = \frac{f_{j,j+1:n}(\alpha_*)-\lambda_*}{B}$, $a_2 = \frac{f_{j,j+1:n}(\alpha_*)}{B}$ *and* $a_3 = \frac{f_{j,j+1:n}(\beta_*)}{B}$.

Proof Our arguments are similar to those in the proof of Theorem 6.27. The l-h-c projection is possible if $M_{j,j+1:n}(b_{j,j+1:n}) < 0$, $K_{j,j+1:n}(\beta^*_{j,j+1:n}) > 0$, and for some $b_{j,j+1:n} \leq \alpha < \beta^*_{j,j+1:n}$ we have $K_{j,j+1:n}(\alpha) \geq 0$ and $L_{j,j+1\,n}(\alpha) = 0$. If $K_{j,j+1:n}(\alpha) > 0$ (which may occur when $\gamma^*_{j,j+1:n} < \alpha < \beta^*_{j,j+1:n}$), we also need that (6.268) should change its sign from $+$ to $-$. It happens if (6.268) has two sign change points, and the first of them belongs to $(\gamma^*_{j,j+1:n}, \beta^*_{j,j+1:n})$. If $K_{j,j+1:n}(\alpha) = L_{j,j+1:n}(\alpha) = 0$, we necessarily require $K'_{j,j+1:n}(\alpha)L'_{j,j+1:n}(\alpha) = 0$. Note that $L_{j,j+1:n}(0) = 0$, $L_{j,j+1:n}(1) = \frac{3}{2(n+1)} > 0$, and consider

$$\begin{aligned} \frac{L'_{j,j+1:n}(\alpha)}{n} &= \frac{(j-1)(j-2)}{4(j+1)} B_{j-1,n-1}(\alpha) \\ &- \left[\frac{(j-1)(j-2)}{4(j+1)} + \frac{(j-1)(j+4)}{4(j+2)}\right] B_{j\,n-1}(\alpha) \\ &+ \left[\frac{(j-1)(j+4)}{4(j+2)} + \frac{3}{2(j+3)}\right] B_{j+1,n-1}(\alpha) \\ &- \sum_{k=j+2}^{n-1} \frac{3}{2(k+1)(k+2)} B_{k,n-1}(\alpha). \end{aligned}$$

If (6.268) has two sign changes, then it has at least 2 local extremes in (0, 1). The above formula shows that it has two positive local maxima and a negative minimum. Neither of them amounts to 0. If it has no sign changes, and the derivative is first nonnegative and then nonpositive, there is only one positive maximum. If the derivative has two sign changes, it is theoretically possible (but extremely unlikely) that (6.268) has a local minimum in (0, 1) whose value is 0. Relation (6.249) implies that (6.267) should change its sign from $+$ to $-$ which cannot occur for $\alpha \in (b_{j,j+1:n}, c_{j,j+1:n})$. So we exclude the case $K_{j,j+1:n}(\alpha) = L_{j,j+1:n}(\alpha) = L'_{j,j+1:n}(\alpha) = 0$. We also show that $K_{j,j+1:n}(\alpha) = K'_{j,j+1:n}(\alpha) = L_{j,j+1:n}(\alpha) = 0$ is impossible as well. Note that (6.267) starts from 0 at 0, then is consecutively positive, negative and positive, and ends at $K_{j,j+1:n}(1) = \frac{3}{n+1} > 0$. Applying the variation diminishing property for its derivative, we deduce that its sequence of signs is $+ - +-$ which implies that (6.267) has two strictly positive maxima and one negative minimum.

So the only admissible $\alpha_* = \alpha^*_{j,j+1:n}$ is the first sign change point of (6.268) under the condition that $\gamma^*_{j,j+1:n} < \alpha^*_{j,j+1:n} < \beta^*_{j,j+1:n}$ so that $K_{j,j+1:n}(\alpha^*_{j,j+1:n}) > 0$. If these conditions are fulfilled, we are in a position to calculate the norm of the l-h-c projection, with unique determined parameters $\alpha^*_{j,j+1:n}$, $\beta^*_{j,j+1:n}$, and $\lambda^*_{j,j+1:n}(\alpha^*_{j,j+1:n})$. The norm

$$B^2(j, j+1, n) = ||P_{\preceq_c U} f_{j,j+1:n}||^2_U = \int_0^{\alpha_*} [\lambda_*(x-\alpha_*) + f_{j,j+1:n}(\alpha_*)]^2 dx$$
$$+ \int_{\alpha_*}^{\beta_*} f^2_{j,j+1:n}(x)dx + f^2_{j,j+1:n}(\beta_*)(1-\beta_*),$$

simplified to (6.271), provides the optimal bound in Theorem 6.29. The attainability condition

$$\frac{F^{-1}(x)-\mu}{\sigma} = \frac{P_{\preceq_c U} f_{j,j+1:n}(x)}{||P_{\preceq_c U} f_{j,j+1:n}||_U}$$

is used for describing the distribution (6.272) which attains the bound. □

The other possibility is as presented in the following theorem.

Theorem 6.30 (ID Distributions) *Suppose that the assumptions of Theorem* 6.29 *do not hold. Then*

$$P_{\preceq_c U} f_{j,j+1:n}(x) = \lambda_*(x-\alpha_*)\mathbf{1}_{(0,\alpha_*)}(x) + \gamma_*, \tag{6.273}$$

where $\alpha_* = \alpha^*_{j,j+1:n}$ *is the unique sign change point of* (6.270),

$$\lambda_* = \lambda_*(\alpha_*) = \frac{12A_{j,j+1:n}(\alpha_*)}{4\alpha_*^3 - 3\alpha_*^4},$$

$$\gamma_* = \gamma_*(\alpha_*, \lambda_*) = \lambda_* \frac{\alpha_*^2}{2}.$$

Consequently,

$$\frac{\mathbb{E}(X_{j+1:n} - X_{j:n})}{\sigma} \le B = B(j, j+1, n),$$

where

$$B^2 = \frac{\gamma_*^3 - (\gamma_* - \lambda_*\alpha_*)^3}{3\lambda_*} + \gamma_*^2(1-\alpha_*).$$

The bound is attained by the mixture of the uniform distribution on the interval $\left[\mu - \sigma\frac{\lambda_*\alpha_* - \gamma_*}{B}, \mu + \sigma\frac{\gamma_*}{B}\right]$ *with probability* α_*, *and an atom at* $\mu + \sigma\frac{\gamma_*}{B}$ *with probability* $1 - \alpha_*$.

Proof If the assumptions of Theorem 6.29 are not satisfied, the projection is a broken line (6.273). The only possible breaking point here is the unique zero α_* of (6.270), where the function changes its sign from $-$ to $+$. The other parameters $\lambda_*(\alpha_*)$ and $\gamma_*(\alpha_*, \lambda_*(\alpha_*))$ satisfy (6.245) with (6.247). If the projection is precisely determined, calculation of the bounds and description of the distribution that attains it become standard tasks. □

The remaining cases of the first and last spacings $X_{j+1:n} - X_{j:n}$, $j = 1, n-1$, are analyzed in the following theorem.

Theorem 6.31 (ID Distributions) *Let* $X_1, \ldots, X_n$ *be iid with a convex distribution function, and finite second moment. Then*

$$\frac{\mathbb{E}(X_{2:n} - X_{1:n})}{\sigma} \le B = B(1, 2, n), \tag{6.274}$$

where

$$\begin{aligned} B^2 &= \int_0^{1/(n-1)} f_{1,2:n}^2(x)dx + f_{1,2:n}^2\left(\frac{1}{n-1}\right)\frac{n-2}{n-1} \\ &= \frac{n^2(n-1)^2}{2n-3}\left[1 - \left(\frac{n-2}{n-1}\right)^{2n-3}\right] - n^3\left[1 - \left(\frac{n-2}{n-1}\right)^{2n-2}\right] \\ &\quad + \frac{n^4}{2n-1}\left[1 - \left(\frac{n-2}{n-1}\right)^{2n-1}\right] + \left(\frac{n}{n-1}\right)^2\left(\frac{n-2}{n-1}\right)^{2n-3}. \end{aligned}$$

The equality in (6.274) *holds iff*

$$\frac{F^{-1}(x) - \mu}{\sigma} = \begin{cases} \frac{n}{B}(1-x)^{n-2}(nx-1), & 0 < x < \frac{1}{n-1}, \\ \frac{n(n-2)^{n-2}}{(n-1)^{n-1}B}, & \frac{1}{n-1} < x < 1. \end{cases} \tag{6.275}$$

Also,

$$\frac{\mathbb{E}(X_{n:n} - X_{n-1:n})}{\sigma} \le B = B(n-1, n, n) = \frac{2\sqrt{3}}{n+1},$$

and the bound is attained by the uniform distribution on $[\mu - \sqrt{3}\sigma, \mu + \sqrt{3}\sigma]$.

Proof The ideas of proofs are identical with those of Corollary 6.3 and Theorem 6.24, and so we only outline them here. In the former case, the function $f_{1,2:n}(x) = n(1-x)^{n-2}(nx-1)$ is concave increasing on $\left(0, \frac{2}{n}\right)$, and then decreasing. Its projection

$$P_U^{\nearrow} f_{1,2:n}(x) = f_{1,2:n}\left(\min\left\{x, \frac{1}{n-1}\right\}\right)$$

onto the family $\mathcal{C}_U^{\nearrow}$ of nondecreasing functions in $L^2((0,1), dx)$ is concave. Hence it is the projection onto (6.223) as well. The bound in (6.274) is equal to its L^2-norm, and this is identical with the general bound for arbitrary parent distributions. Observe that formula (6.275) defines a distribution which has an increasing density on the interval $\left[\mu - \frac{n\sigma}{B}, \mu + \frac{n(n-2)^{n-2}\sigma}{(n-1)^{n-1}B}\right]$, and the jump of size $\frac{n-2}{n-1}$ at the right support end point.

In the latter case, $f_{n-1,n:n}(x) = f_{1,2:n}(1-x) = nx^{n-2}(nx-n+1)$ is decreasing on $\left(0, 1-\frac{2}{n}\right)$, and convex increasing on $\left(1-\frac{2}{n}, 1\right)$. Every concave nondecreasing function has at most two common values with $f_{n-1,n:n}$. A function that has less than two meeting points with $f_{n-1,n:n}$ can be replaced by one with just two contact points, which gives a better approximation. The latter one can be further improved if we replace it by the line passing through the crossing points. This allows us to conclude that the projection is linear. The rest of the proof is straightforward. □

6.3.4 Positive Upper Bounds for Nonextreme Order Statistics from IFR Family

Now we proceed to the problem of obtaining bounds for expectations of single order statistics $X_{j:n}$, $3 \le j \le n-1$, from iid sample with an IFR distribution with use of the projection method. Cases $j = 2$ and $j = n$ have been solved above in Corollary 6.3 and Theorem 6.25, respectively. Since $\mathbb{E}X_{1:n} < \mu$ for every nondegenerate parent distribution, case $j = 1$ shall be treated below by means of the norm maximization method. The basic tool for applying the projection method is in determining the projection of the function

$$h(x) = h_{j:n}(x) = f_{j:n} \circ Z(x) - 1 = f_{j:n}(1 - e^{-x}) - 1 \tag{6.276}$$

onto the convex cone

$$\mathcal{C}_{\preceq_c Z} = \{g \in L^2((0,\infty), e^{-x}dx) : g(x) \text{ is nondecreasing and concave}\}. \tag{6.277}$$

Observe that functions (6.276), for $3 \le j \le n-1$, satisfy assumptions (A3) with $a = 0$,

$$b = b_{j:n} = -\ln\left(1 - \frac{(j-1)(2n-3) - \sqrt{(j-1)(4n^2 - 4n - 1 + 5j - 4jn)}}{2(n-1)^2}\right),$$

$$c = c_{j:n} = -\ln\left(\frac{n-j}{n-1}\right),$$

and $d = +\infty$. Here the first interval of decrease of (6.276) is empty, which is acceptable. By Theorem 6.26, the projection of (6.276) onto (6.277) can be either of l-h-c or l-c type. In Theorem 6.32 below, we present the bounds corresponding to the first case. To this end, we specify the general functions (6.240)–(6.243) for particular $h = h_{j:n}$ defined in (6.276). Using auxiliary formulas

$$\int_0^\alpha (x-\alpha)e^{-x}dx = 1 - e^{-\alpha} - \alpha, \tag{6.278}$$

$$\int_0^\alpha (x-\alpha)^2 e^{-x}dx = \alpha^2 - 2\alpha + 2 - 2e^{-\alpha}, \tag{6.279}$$

$$\int_0^\alpha f_{j:n}(1-e^{-x})e^{-x}dx = F_{j:n}(1-e^{-\alpha}), \tag{6.280}$$

$$\int_0^\alpha (x-\alpha) f_{j:n}(1-e^{-x})e^{-x}dx = \sum_{k=1}^{j} \frac{F_{k:n}(1-e^{-\alpha})}{n-k+1} - \alpha, \tag{6.281}$$

we determine functions

$$M_{j:n}(\beta) = f_{j:n}(1-e^{-\beta})e^{-\beta} - 1 + F_{j:n}(1-e^{-\beta}), \tag{6.282}$$

$$\lambda_{j:n}(\alpha) = \frac{\sum_{k=1}^{j} \frac{F_{k:n}(1-e^{-\alpha})}{n-k+1} - f_{j:n}(1-e^{-\alpha})(1-e^{-\alpha}-\alpha) - \alpha}{\alpha^2 - 2\alpha + 2 - 2e^{-\alpha}}, \tag{6.283}$$

$$\begin{aligned} K_{j:n}(\alpha) = \lambda_{j:n}(\alpha) - n[(n-j+1)B_{j-2,n-1}(1-e^{-\alpha}) \\ - (n-j)B_{j-1,n-1}(1-e^{-\alpha})], \end{aligned} \tag{6.284}$$

$$\begin{aligned} L_{j:n}(\alpha) = F_{j:n}(1-e^{-\alpha}) - f_{j:n}(1-e^{-\alpha})(1-e^{-\alpha}) \\ - \lambda_{j:n}(\alpha)(1-e^{-\alpha}-\alpha), \end{aligned} \tag{6.285}$$

on the positive half-axis.

Theorem 6.32 (IFR Distributions) *Suppose that $M_{j:n}(b_{j:n}) < 0$ so that the unique zero $0 < \beta_{j:n} < c_{j:n}$ of (6.282) belongs to $(b_{j:n}, c_{j:n})$. Also, suppose that $\mathcal{K}_{j:n} = \{b_{j:n} < \alpha < \beta_{j:n} : K_{j:n} \geq 0$ and $L_{j:n} = 0\}$ is nonempty. Let $\alpha_{j:n}$ denote the smallest (possibly unique) element of $\mathcal{K}_{j:n}$, and $\lambda_{j:n} = \lambda_{j:n}(\alpha_{j:n})$. Then*

$$\frac{\mathbb{E}X_{j:n} - \mu}{\sigma} \leq B = B(j, n), \tag{6.286}$$

where

$$\begin{aligned} B^2 &= f_{j:n}^2(1 - e^{-\alpha_{j:n}})(1 - e^{-\alpha_{j:n}}) + 2\lambda_{j:n} f_{j:n}(1 - e^{-\alpha_{j:n}})(1 - e^{-\alpha_{j:n}} - \alpha_{j:n}) \\ &\quad + \lambda_{j:n}^2(\alpha_{j:n}^2 - 2\alpha_{j:n} + 2 - 2e^{-\alpha_{j:n}}) + f_{j:n}^2(1 - e^{-\beta_{j:n}})e^{-\beta_{j:n}} - 1 \\ &\quad + n\frac{\binom{2j-2}{j-1}\binom{2n-2j}{n-j}}{\binom{2n-1}{n}} \left[F_{2j-1:2n-1}(1 - e^{-\beta_{j:n}}) - F_{2j-1:2n-1}(1 - e^{-\alpha_{j:n}})\right]. \end{aligned}$$

The equality in (6.286) holds for the distribution function given by

$$F(y) = \begin{cases} 0, & y < f_{j:n}(1 - e^{-\alpha_{j:n}}) - \lambda_{j:n}\alpha_{j:n}, \\ 1 - \exp\left(-\alpha_{j:n} - \frac{y - f_{j:n}(1 - e^{-\alpha_{j:n}})}{\lambda_{j:n}}\right), & f_{j:n}(1 - e^{-\alpha_{j:n}}) - \lambda_{j:n}\alpha_{j:n} \\ & \leq y < f_{j:n}(1 - e^{-\alpha_{j:n}}), \\ f_{j:n}^{-1}(y), & f_{j:n}(1 - e^{-\alpha_{j:n}}) \leq y \\ & < f_{j:n}(1 - e^{-\beta_{j:n}}), \\ 1, & y \geq f_{j:n}(1 - e^{-\beta_{j:n}}), \end{cases} \tag{6.287}$$

uniquely determined up to the location and scale parameters μ and σ, respectively, with modified argument $x \mapsto y = \frac{x-\mu}{\sigma}B(j, n) + 1$, introduced for brevity.

Proof The crucial step of our reasoning consists in showing that the assumptions guarantee that the projection of $h_{j:n}(x) = f_{j:n}(1 - e^{-x}) - 1$ onto $\mathcal{C}_{\leq_c Z}$ has the form

$$P_{\leq_c Z} h_{j:n}(x) = \begin{cases} \lambda_{j:n}(x - \alpha_{j:n}) + f_{j:n}(1 - e^{-\alpha_{j:n}}) - 1, & 0 \leq x < \alpha_{j:n}, \\ f_{j:n}(1 - e^{-x}) - 1, & \alpha_{j:n} \leq x < \beta_{j:n}, \\ f_{j:n}(1 - e^{-\beta_{j:n}}) - 1, & x \geq \beta_{j:n}. \end{cases} \tag{6.288}$$

The tools are collected in Theorem 6.26.

Function (6.282) satisfies $M_{j:n}(0) < 0$, $M_{j:n}(c_{j:n}) > 0$, and increases in between (comp. the comments preceding Theorem 6.26). The first necessary condition for (6.288) is that the unique zero $\beta_{j:n}$ of (6.282) belongs to the interval $(b_{j:n}, c_{j:n})$ of concave increase of $h_{j:n}$. This obviously holds iff $M_{j:n}(b_{j:n}) < 0$. Point $\beta_{j:n}$ is the only admissible candidate for the change of the l-h-c type projection from $h_{j:n}$ itself to the constant.

The other condition is that $\mathcal{K}_{j:n}$ is nonempty, i.e., the interval $[b_{j:n}, \beta_{j:n})$ contains points α satisfying $K_{j:n}(\alpha) \geq 0$ and $L_{j:n}(\alpha) = 0$. The latter relation together with $M_{j:n}(\beta_{j:n}) = 0$ imply that the weighted integral of the l-h-c function glued at α and $\beta^*_{j:n}$ coincides with that of the original function $h_{j:n}$, which is a necessary condition for the projection (see (6.12)). Condition $K_{j:n}(\alpha) \geq 0$ implies that gluing $\lambda_{j:n}(\alpha)(x - \alpha) + h_{j:n}(\alpha)$ with $h_{j:n}(x)$ at $\alpha \in (b_{j:n}, \beta_{j:n})$ produces a concave function in a vicinity of α. If there were more points in $\mathcal{K}_{j:n}$, the projection is constructed with the use of the smallest one. Since all the necessary conditions are deduced from the assumptions of Theorem 6.32, the projection is actually equal to (6.288).

Due to (6.222), the upper bound on the expectation of standardized jth order statistic coincides with the norm of projection $P_{\preceq_c Z} h_{j:n}$. Since $P_{\preceq_c Z} h_{j:n} \not\equiv 0$, the bound is sharp. We present here slightly simpler analytic form of the norm based on the identity $B^2_{j:n} = ||P_{\preceq_c Z} h_{j:n}||^2_Z = ||P_{\preceq_c Z} f_{j:n} Z||^2_Z - 1$. It follows from the obvious relations $P_{\preceq_c Z} h_{j:n} = P_{\preceq_c Z}(f_{j:n} \circ Z - 1) = P_{\preceq_c Z} f_{j:n} \circ Z - 1$, and

$$
\begin{aligned}
||P_{\preceq_c Z} f_{j:n} \circ Z - 1||^2 = ||P_{\preceq_c Z} f_{j:n} \circ Z||^2 - 2\int_0^\infty P_{\preceq_c Z} f_{j:n} \circ Z(x) e^{-x} dx \\
+ \int_0^\infty e^{-x} dx = ||P_{\preceq_c Z} f_{j:n} \circ Z||^2 - 1
\end{aligned}
$$

valid due to

$$
\int_0^\infty P_{\preceq_c Z} f_{j:n} \circ Z(x) e^{-x} dx = \int_0^\infty f_{j:n} \circ Z(x) e^{-x} dx = \int_0^1 f_{j:n}(x) dx = 1
$$

(cf. (6.12)). Using

$$
\frac{F^{-1} \circ Z(x) - \mu}{\sigma} = \frac{P_{\preceq_c Z} \left(f_{j:n} \circ Z\right)(x) - 1}{B(j,n)}, \qquad 0 < x < \infty,
$$

and performing some tedious calculations, we obtain the explicit form (6.287) of the parent IFR distribution function attaining the bound. □

Up to linear transformations of the argument, the extreme distribution (6.287) is composed of three parts: the exponential one on the left, the inverse of some increasing part of the density function $f_{j:n}$, and the jump of height $e^{-\beta_{j:n}}$ at the right-end point. Note that the latter two are identical with the right part of the distribution function (4.84) attaining the general bound of Theorem 4.8. We carried out a number of numerical verifications of the assumptions of Theorem 6.32 for moderate n and all $3 \leq j \leq n-1$. It seems that the theorem provides the bound for the case $j = 3$, $n = 4$ only. The precise value of the bound and description of the IFR distribution attaining it is presented in the following example.

Example The sharp bound

$$\frac{\mathbb{E}X_{3:4} - \mu}{\sigma} \le 0.50522$$

is attained by the distribution function

$$F(x) = \begin{cases} 0, & \frac{x-\mu}{\sigma} < -1.73065, \\ 1 - \exp\left(-0.20544\frac{x-\mu}{\sigma} - 0.45773\right), & -1.73065 \le \frac{x-\mu}{\sigma} < 0.60205, \\ f_{3:4}^{-1}\left(0.50522\frac{x-\mu}{\sigma} + 1\right), & 0.60205 \le \frac{x-\mu}{\sigma} < 0.75138, \\ 1, & \frac{x-\mu}{\sigma} \ge 0.75138. \end{cases}$$

The exponential part on the left has probability 0.44089. The jump on the right has value 0.53753. The contribution of the inverse cubic function between them amounts to 0.02160 only.

Our conjecture is this example provides the only application of Theorem 6.32. For large n, the inflection point $b_{j:n}$ lies close to the maximal argument $c_{j:n}$, and consequently $h_{j:n}(b_{j:n})$ is only slightly less than the maximum $h_{j:n}(c_{j:n})$. However, by definition,

$$h_{j:n}(\beta_{j:n}) = \frac{\int_{\beta_{j:n}}^{\infty} h_{j:n}(x)e^{-x}dx}{\int_{\beta_{j:n}}^{\infty} e^{-x}dx}$$

cannot be too large, because $h_{j:n}(x)$ for large arguments x is essentially less than $h_{j:n}(c_{j:n})$. This implies that $\beta_{j:n} < b_{j:n}$, which violates condition $M_{j:n}(b_{j:n}) < 0$. Even for small n, when the relation holds, there is not enough space in the interval $(b_{j:n}, \beta_{j:n})$ for any points α satisfying $L_{j:n}(\alpha) = 0$ together with $K_{j:n}(\alpha) \ge 0$.

It seems that with the exception of the case presented in the above example, the bounds on the expectations of order statistics from IFR populations are determined with the use of the l-c-type projection. These are presented in Theorem 6.33 below. However, one should be aware of the fact that for given j and n, first the assumptions of Theorem 6.32 should be verified and the l-h-c projection excluded before one uses the formulas of Theorem 6.33. If the assumptions for the l-h-c shape of the projection do not hold, we define

$$A_{j:n}(\alpha) = (1 - 2\alpha e^{-\alpha} - e^{-2\alpha})[1 - e^{-\alpha} - F_{j:n}(1 - e^{-\alpha})] + e^{-\alpha}(1 - e^{-\alpha} - \alpha)\left[\sum_{k=1}^{j} \frac{F_{k:n}(1 - e^{-\alpha})}{n - k + 1} + e^{-\alpha} - 1\right], \tag{6.289}$$

$$\gamma_{j:n}(\alpha) = [1 - F_{j:n}(1 - e^{-\alpha})]e^{\alpha} - 1, \tag{6.290}$$

$$\lambda_{j:n}(\alpha) = \frac{\gamma_{j:n}(\alpha)}{e^{-\alpha} - 1 + \alpha}, \tag{6.291}$$

$$B_{j:n}^2(\alpha) = (\alpha^2 + 1)\lambda_{j:n}^2(\alpha) - [\lambda_{j:n}(\alpha) + \gamma_{j:n}(\alpha)]^2. \tag{6.292}$$

Theorem 6.33 (IFR Distributions) *Suppose that either* $M_{j:n}(b_{j:n}) \geq 0$ *or* $\mathcal{K}_{j:n} = \emptyset$ *for some fixed* $3 \leq j \leq n-1 \geq 3$. *Then set* $\mathcal{A}_{j:n} = \{\alpha \geq \beta_{j:n} : A_{j:n}(\alpha) = 0,\ \gamma_{j:n}(\alpha) > 0\}$ *is nonempty, and*

$$\frac{\mathbb{E}X_{j:n} - \mu}{\sigma} \leq B(j,n) = B_{j:n}(\alpha_{j:n}), \tag{6.293}$$

where $\alpha_{j:n} = \arg\max_{\alpha \in \mathcal{A}_{j:n}} B^2_{j:n}(\alpha)$. *The equality in* (6.293) *holds for the distribution function*

$$F(y) = \begin{cases} 0, & y < 0, \\ 1 - e^{-y}, & 0 \leq y < \alpha_{j:n}, \\ 1, & y \geq \alpha_{j:n} \end{cases} \tag{6.294}$$

for $y = y(x) = \frac{x-\mu}{\sigma\lambda_{j:n}} B(j,n) - \frac{\gamma_{j:n}}{\lambda_{j:n}} + \alpha_{j:n}$ *with* $\gamma_{j:n} = \gamma_{j:n}(\alpha_{j:n})$ *and* $\lambda_{j:n} = \lambda_{j:n}(\alpha_{j:n})$.

Proof Owing to the assumptions, the projection of (6.276) has an l-c form. Precisely, by (6.245), (6.278) and (6.280), we have

$$\begin{aligned} P_\alpha h_{j:n}(x) &= [1 - e^{-\alpha} - F_{j:n}(1 - e^{-\alpha})]e^{\alpha}\left[\frac{(x-\alpha)\mathbf{1}_{(0,\alpha)}(x)}{e^{-\alpha} - 1 + \alpha} + 1\right] \\ &= \lambda_{j:n}(\alpha)(x-\alpha)\mathbf{1}_{(0,\alpha)}(x) + \gamma_{j:n}(\alpha) \end{aligned}$$

(cf. (6.290) and (6.291)) for some $\alpha \geq \beta_{j:n}$ satisfying

$$\frac{1 - e^{-\alpha} - F_{j:n}(1 - e^{-\alpha})}{e^{-\alpha}} = \frac{-\left[\sum_{k=1}^{j} \frac{F_{k:n}(1 - e^{-\alpha})}{n-k+1} - \alpha\right](1 - e^{-\alpha} - \alpha)}{\alpha^2 - 2\alpha + 2 - 2e^{-\alpha} - (1 - \alpha - e^{-\alpha})^2} > 0$$

(cf. (6.247), (6.279), (6.281)). Observe that the last formula can be rewritten as $A_{j:n}(\alpha) = 0$ with $\gamma_{j:n}(\alpha) > 0$. This means that parameter α determining the projection should be chosen from set $\mathcal{A}_{j:n}$. Its emptiness contradicts the existence of projection. Its cardinality is also restricted, because function (6.289) cannot have infinitely many zeros.

For selecting the proper element of the set, we compare respective squared norms:

$$\begin{aligned} &||P_\alpha h_{j:n}||^2_Z \\ &= \lambda^2_{j:n}(\alpha)\int_0^\alpha (x-\alpha)^2 e^{-x}dx + 2\lambda_{j:n}(\alpha)\gamma_{j:n}(\alpha)\int_0^\alpha (x-\alpha)e^{-x}dx + \gamma^2_{j:n}(\alpha) \\ &= \lambda^2_{j:n}(\alpha)[\alpha^2 - 2(e^{-\alpha} - 1 + \alpha)] + \frac{2\gamma^2_{j:n}(\alpha)}{e^{-\alpha} - 1 + \alpha}(1 - e^{-\alpha} - \alpha) + \gamma^2_{j:n}(\alpha) \\ &= \lambda^2_{j:n}(\alpha)\alpha^2 - 2\lambda_{j:n}(\alpha)\gamma_{j:n}(\alpha) - \gamma^2_{j:n}(\alpha) \\ &= (\alpha^2 + 1)\lambda^2_{j:n}(\alpha) - [\lambda_{j:n}(\alpha) + \gamma_{j:n}(\alpha)]^2. \end{aligned}$$

There is a unique $\alpha_{j:n} \in \mathcal{A}_{j:n}$ minimizing $||P_\alpha h_{j:n}||_Z^2$, and this defines the unique non-zero projection $P_{\preceq_c Z} h_{j:n}$. By (6.222), the sharp upper mean-variance bound for the expectation of the jth order statistic is

$$||P_{\alpha_{j:n}} h_{j:n}||_Z = B(j,n) = [(\alpha_{j:n}^2 + 1)\lambda_{j:n}^2 - (\lambda_{j:n} + \gamma_{j:n})^2]^{1/2}.$$

The distribution function attaining the bound is characterized by the relation

$$\frac{F^{-1}(1-e^{-x}) - \mu}{\sigma} = \frac{\lambda_{j:n}(x - \alpha_{j:n})\mathbf{1}_{(0,\alpha)}(x) + \gamma_{j:n}}{B(j,n)}, \qquad 0 < x < \infty,$$

which determines (6.294). □

The bounds in (6.293) are attained by the right truncated and linearly transformed exponential random variables (cf. (6.294)), with the jumps of sizes $e^{-\alpha_{j:n}}$ on the right. Precisely, $X_i \overset{d}{=} \frac{\lambda_{j:n}\sigma}{B(j,n)}\left[\min\{Z_i - \alpha_{j:n}, 0\} + \frac{\gamma_{j:n}}{\lambda_{j:n}}\right] + \mu$ for Z_i, $i = 1, \ldots, n$, being standard exponential. The transformation is defined in this complicated way just to fulfil the first two moment conditions. The convex order transform is invariant under the location and scale modifications. This means that every exponential distribution truncated on the right at the level $1 - e^{-\alpha_{j:n}}$ attains the bound for the expected jth order statistic standardized with respective mean and standard deviation.

Numerical studies show that each set $\mathcal{A}_{j:n}$ contains only one element, and this is used in the construction of the projection and calculation of the bound. We cannot prove it formally, though. We are able to do it for the analogous hypothesis in the analysis of increasing density distributions. In that case, the counterparts of (6.282)–(6.285) and (6.289)–(6.291) are expressed as linear combinations of Bernstein polynomials. Then we could apply the respective variation diminishing property of Bernstein polynomials for evaluating the numbers of their zeros and extremes. The property is also useful in the analysis of the DFR case when the Bernstein polynomials of transformed argument $\alpha \mapsto 1 - e^{-\alpha}$ are studied. The method is not applicable here, because we consider functions composed of polynomials of argument $1 - e^{-\alpha}$ combined with small powers of α itself. Accordingly, specifying particular functions $h(x) = f_{j:n}(1 - e^{-x}) - 1$ does not allow us to obtain results stronger than those concluded directly from Theorem 6.26, being stated for general h satisfying assumptions (A3).

6.3.5 Positive Upper Bounds for Spacings from IFR Family

We now establish sharp upper bounds on the standardized expectations of spacings $\frac{\mathbb{E}(X_{j+1:n} - X_{j:n})}{\sigma}$, $1 \le j \le n-1$, coming from iid samples with IFR distributions. For this purpose, we project functions

$$h_{j,j+1:n}(x) = f_{j,j+1:n} \circ Z(x) = f_{j+1:n}(1-e^{-x}) - f_{j:n}(1-e^{-x}) \qquad (6.295)$$

onto the convex cone (6.277). We first take into account the cases $2 \le j \le n-2$, for which functions (6.295) satisfy assumptions (A3). Indeed, they decrease from 0 at 0 to their negative global minima at $a_{j,j+1:n} = \ln \frac{n}{n-j+\sqrt{\frac{j(n-j)}{n-1}}}$, then increase to the positive global maxima at $c_{j,j+1:n} = \ln \frac{n}{n-j-\sqrt{\frac{j(n-j)}{n-1}}}$, and finally decrease to zero at $d = +\infty$. In the intervals $(a_{j,j+1:n}, c_{j,j+1:n})$, where they increase, they are first convex and then concave. The tangency points cannot be written down explicitly. These are unique $b_{j,j+1:n} \in (a_{j,j+1:n}, c_{j,j+1:n})$ solving the equations

$$\frac{n(n-1)^2}{(n-j)(n-j-1)} \ln^3 x + \frac{(n-1)(3n^2-3n-3jn+4j)}{(n-j-1)(n-j)} \ln^2 x$$
$$- \frac{-3n^2+3jn-5j+6n-3}{n-j-1} \ln x + n - j - 1 = 0.$$

The bounds and justifications of their sharpness are presented in Theorems 6.34 and 6.35. Their statements are similar to those of Theorems 6.32 and 6.33, respectively. Their proofs are almost identical with those of their counterparts, and therefore we omit them. The only differences are in using modifications of functions (6.282)–(6.285) and (6.289)–(6.292). Noting the identity $h_{j,j+1:n} = h_{j+1:n} - h_{j:n}$ and linearity of operators (6.240)–(6.243) acting on functions h, we define

$$M_{j,j+1:n}(\beta) = M_{j+1:n}(\beta) - M_{j:n}(\beta) = -(n-j+1)B_{j-1,n}(1-e^{-\beta}) + (n-j-1)B_{j,n}(1-e^{-\beta}), \tag{6.296}$$

$$\lambda_{j,j+1:n}(\alpha) = \lambda_{j+1:n}(\alpha) - \lambda_{j:n}(\alpha)$$
$$= \frac{1}{\alpha^2-\alpha+2-2e^{-\alpha}} \left\{ \frac{1}{n-j} \sum_{k=j+1}^{n} B_{k,n}(1-e^{-\alpha}) - n(1-e^{-\alpha}-\alpha)[B_{j,n-1}(1-e^{-\alpha}) - B_{j-1,n-1}(1-e^{-\alpha})] \right\},$$

$$K_{j,j+1:n}(\alpha) = K_{j+1:n}(\alpha) - K_{j:n}(\alpha)$$
$$= \lambda_{j,j+1:n}(\alpha) + n\left[(n-j+1)B_{j-2,n-1}(1-e^{-\alpha}) + (n-j-1)B_{j,n-1}(1-e^{-\alpha}) - 2(n-j)B_{j-1,n-1}(1-e^{-\alpha})\right],$$

$$L_{j,j+1:n}(\alpha) = L_{j+1:n}(\alpha) - L_{j:n}(\alpha) = (j-1)B_{j,n}(1-e^{-\alpha}) - (j+1)B_{j+1,n}(1-e^{-\alpha}) - \lambda_{j,j+1:n}(\alpha)(1-e^{-\alpha}-\alpha).$$

Theorem 6.34 (IFR Distributions) *Suppose that $M_{j,j+1:n}(b_{j,j+1:n}) < 0$ so that the unique zero $a_{j,j+1:n} < \beta_{j,j+1:n} < c_{j,j+1:n}$ of* (6.296) *belongs to the interval $(b_{j,j+1:n}, c_{j,j+1:n})$. Also, suppose that*

$$\mathcal{K}_{j,j+1:n} = \{b_{j,j+1:n} < \alpha < \beta_{j,j+1:n} : K_{j,j+1:n}(\alpha) \geq 0, \; L_{j,j+1:n}(\alpha) = 0\}$$

is nonempty. Let $\alpha_{j,j+1:n}$ denote the smallest (possibly unique) element of $\mathcal{K}_{j,j+1:n}$, $f_{j,j+1:n} = f_{j+1:n} - f_{j:n}$, and $\lambda_{j,j+1:n} = \lambda_{j,j+1:n}(\alpha_{j,j+1:n})$. Then

$$\mathbb{E}\frac{X_{j+1:n} - X_{j:n}}{\sigma} \leq B = B(j, j+1, n),$$

where

$$\begin{aligned}
B^2 = B^2(j, j+1, n) &= f^2_{j,j+1:n}(1 - e^{-\alpha_{j,j+1:n}})(1 - e^{-\alpha_{j,j+1:n}}) \\
&+ 2\lambda_{j,j+1:n} f_{j,j+1:n}(1 - e^{-\alpha_{j,j+1:n}})(1 - e^{-\alpha_{j,j+1:n}} - \alpha_{j,j+1:n}) \\
&+ \lambda^2_{j,j+1:n}(\alpha^2_{j,j+1:n} - 2\alpha_{j,j+1:n} + 2 - 2e^{-\alpha_{j,j+1:n}}) \\
&+ f^2_{j,j+1:n}(1 - e^{-\beta_{j,j+1:n}})e^{-\beta_{j,j+1:n}} \\
&+ \frac{n\binom{2j}{j}\binom{2n-2j-2}{n-j-1}}{\binom{2n-1}{n}}\left[F_{2j+1:2n-1}(1-e^{-\beta_{j,j+1:n}}) - F_{2j+1:2n-1}(1-e^{-\alpha_{j,j+1:n}})\right] \\
&- \frac{2n\binom{2j-1}{j-1}\binom{2n-2j-1}{n-j}}{\binom{2n-1}{n}}\left[F_{2j:2n-1}(1-e^{-\beta_{j,j+1:n}}) - F_{2j:2n-1}(1-e^{-\alpha_{j,j+1:n}})\right] \\
&+ \frac{n\binom{2j-2}{j-1}\binom{2n-2j}{n-j}}{\binom{2n-1}{n}}\left[F_{2j-1:2n-1}(1-e^{-\beta_{j,j+1:n}}) - F_{2j-1:2n-1}(1-e^{-\alpha_{j,j+1:n}})\right].
\end{aligned}$$

The bound is attained by

$$F(y) = \begin{cases} 0, & y < a_1, \\ 1 - \exp\left(-\alpha_{j,j+1:n} - \frac{y - f_{j,j+1:n}(1-e^{-\alpha_{j,j+1:n}})}{\lambda_{j,j+1:n}}\right), & a_1 \leq y < a_2, \\ f^{-1}_{j,j+1:n}(y), & a_2 \leq y < a_3, \\ 1, & y \geq a_3, \end{cases}$$

uniquely determined up to the location and scale parameters μ and σ, respectively, with modified argument $x \mapsto y = \frac{x-\mu}{\sigma}B$, and abbreviations $a_1 = f_{j,j+1:n}(1 - e^{-\alpha_{j,j+1:n}}) - \lambda_{j,j+1:n}\alpha_{j,j+1:n}$, $a_2 = f_{j,j+1:n}(1 - e^{-\alpha_{j,j+1:n}})$, and $a_3 = f_{j,j+1:n}(1 - e^{-\beta_{j,j+1:n}})$.

Define now

$$A_{j,j+1:n}(\alpha) = A_{j+1:n}(\alpha) - A_{j:n}(\alpha) = nB_{j,n-1}(1-e^{-\alpha}) \times (1-e^{-2\alpha} - 2\alpha e^{-\alpha}) + (1-e^{-\alpha} - \alpha)F_{j+1:n}(1-e^{-\alpha}),$$

$$\gamma_{j,j+1:n}(\alpha) = \gamma_{j+1:n}(\alpha) - \gamma_{j:n}(\alpha) = \frac{n}{n-j}B_{j,n-1}(1-e^{-\alpha}),$$

$$\lambda_{j,j+1:n}(\alpha) = \lambda_{j+1:n}(\alpha) - \lambda_{j:n}(\alpha) = \frac{\gamma_{j,j+1:n}(\alpha)}{e^{-\alpha} - 1 + \alpha},$$

$$B^2_{j,j+1:n}(\alpha) = (\alpha+1)^2\lambda^2_{j,j+1:n}(\alpha) - [\lambda_{j,j+1:n}(\alpha) + \gamma_{j,j+1:n}(\alpha)]^2.$$

Theorem 6.35 (IFR Distributions) *Suppose that either* $M_{j,j+1:n}(\beta_{j,j+1:n}) \geq 0$ *or* $\mathcal{K}_{j,j+1:n} = \emptyset$ *for some fixed* $2 \leq j \leq n-2 \geq 2$. *Then, the set*

$$\mathcal{A}_{j,j+1:n} = \{\alpha \geq \beta_{j,j+1:n} : A_{j,j+1:n}(\alpha) = 0,\ \gamma_{j,j+1:n}(\alpha) > 0\}$$

is nonempty, and

$$\frac{\mathbb{E}(X_{j+1:n} - X_{j:n})}{\sigma} \leq B = B(j, j+1, n) = B_{j,j+1:n}(\alpha_{j,j+1:n})$$

for $\alpha_{j,j+1:n} = \arg\max_{\alpha \in \mathcal{A}_{j,j+1:n}} B^2_{j,j+1:n}(\alpha)$. *The equality holds for the distribution function*

$$F(y) = \begin{cases} 0, & y < 0, \\ 1 - e^{-y}, & 0 \leq y < \alpha_{j,j+1:n}, \\ 1, & y \geq \alpha_{j,j+1:n}, \end{cases}$$

for $y = y(x) = \frac{x-\mu}{\sigma\lambda_{j,j+1:n}}B - \frac{\gamma_{j,j+1:n}}{\lambda_{j,j+1:n}} + \alpha_{j,j+1:n}$ *with* $\gamma_{j,j+1:n} = \gamma_{j,j+1:n}(\alpha_{j,j+1:n})$ *and* $\lambda_{j,j+1:n} = \lambda_{j,j+1:n}(\alpha_{j,j+1:n})$.

We suspect that the assumptions of Theorem 6.34 never hold. The claim is verified for small n. For large n, the increase parts of $h_{j,j+1:n}$ are very steep, and only minor highly-located fragments are concave. It is very unlikely that their small pieces become parts of projections.

We finally discuss the cases of extreme spacings $X_{2:n} - X_{1:n}$ and $X_{n:n} - X_{n-1:n}$ coming from iid IFR samples. The solution in the first case is immediate. This is described in Theorem 6.31 which asserts that the optimal upper bounds for the first spacings in the family of general iid samples is attained by the parent convex distribution. Since convexity of the distribution function implies the same property of the corresponding cumulative failure rate, the general bound is attained by a member of the closure of IFR family as well. For establishing an analogous bound for the last spacing, we need more elaborate arguments.

In the first step, we project the functions

$$h_{n-1,n:n}(x) = f_{n:n}(1-e^{-x}) - f_{n-1:n}(1-e^{-x})$$
$$= n(1-e^{-x})^{n-2}(1-ne^{-x}),\ x>0,$$

onto (6.277). Starting from the origin, they decrease to the global minimum at $a_{n-1,n:n} = \ln\frac{n}{2}$, then convexly increase to the tangency point at $b_{n-1,n:n} = \ln\frac{n}{2-\sqrt{2\frac{n-2}{n-1}}}$, and eventually concavely increase to n at $+\infty$. The functions do not fulfil Assumptions (A3). Below we change them slightly and present a corresponding modification of Theorem 6.26. We say that Assumptions (A3') hold if (A3) are modified so that $c = d = +\infty$ and $\sup_{x>0} h(x) = \lim_{x\to\infty} h(x) > 0$.

Theorem 6.36 *Under Assumptions (A3'), with notation* (6.241)–(6.243), *the set* $\mathcal{K}' = \{\alpha > b :\ K(\alpha) \geq 0,\ L(\alpha) = 0\}$ *is nonempty, and for* $\alpha_* = \inf\{\alpha \in \mathcal{K}'\}$ *yields*

$$P_{\preceq_c W} h(x) = \begin{cases} h(\alpha_*) + \lambda_*(\alpha_*)(x-\alpha_*), & 0 \leq x < \alpha_*, \\ h(x), & x \geq \alpha_*. \end{cases}$$

In view of Theorem 6.26, the statement looks natural. A formal proof is postponed to the Appendix. Fixing $h = h_{n-1,n:n}$, we determine the following functions useful in description of the bounds:

$$\lambda_{n-1,n:n}(\alpha) = \frac{e^{-2\alpha}(1-n^2) + e^{-\alpha}[n-2+n^2(1-\alpha)] + 1 - n(1-\alpha)}{\alpha^2 - 2\alpha + 2 - 2e^{-\alpha}}$$
$$\times (1-e^{-\alpha})^{n-2},$$
$$K_{n-1,n:n}(\alpha) = \lambda_{n-1,n:n}(\alpha) - n(n-1)(1-e^{-\alpha})^{n-3}e^{-\alpha}(2-ne^{-\alpha}),$$
$$L_{n-1,n:n}(\alpha) = \frac{e^{-2\alpha}(1-n^2) + e^{-\alpha}[n-2+n^2(1-\alpha)] + 1 - n(1-\alpha)}{\alpha^2 + 2 - 2\alpha - 2e^{-\alpha}}$$
$$\times (1-e^{-\alpha}-\alpha) + n(1-e^{-\alpha})[1-e^{-\alpha}(n-1)], \qquad \alpha > 0.$$

Theorem 6.37 (IFR Distributions) *The set*

$$\mathcal{K}_{n-1,n:n} = \{\alpha \geq b_{n-1,n:n} :\ K_{n-1,n:n}(\alpha) \geq 0,\ L_{n-1,n:n}(\alpha) = 0\}$$

is nonempty and

$$\mathbb{E}\frac{X_{n:n} - X_{n-1:n}}{\sigma} \leq B = B(n-1,n,n),$$

where

$$B^2 = B^2(n-1,n,n) = \lambda^2_{n-1,n:n}(\alpha^2_{n-1,n:n} - 2\alpha_{n-1,n:n} + 2 - 2e^{-\alpha_{n-1,n:n}})$$

$$+ 2\lambda_{n-1,n:n}h_{n-1,n:n}(\alpha_{n-1,n:n})(1 - e^{-\alpha_{n-1,n:n}} - \alpha_{n-1,n:n})$$

$$+ h^2_{n-1,n:n}(\alpha_{n-1,n:n})(1 - e^{-\alpha_{n-1,n:n}}) + \frac{n^2(n-1)}{2n-1}\left[1 - \left(1 - e^{-\alpha_{n-1,n:n}}\right)^{2n-1}\right]$$

$$- n^3\left(1 - e^{-\alpha_{n-1,n:n}}\right)^{2n-2} e^{-\alpha_{n-1,n:n}}$$

$$- n^2(n-1)^2\left(1 - e^{-\alpha_{n-1,n:n}}\right)^{2n-3} e^{-2\alpha_{n-1,n:n}},$$

$\alpha_{n-1,n:n} = \inf \mathcal{K}_{n-1,n:n}$, *and* $\lambda_{n-1,n:n} = \lambda_{n-1,n:n}(\alpha_{n-1,n:n})$. *With the notation* $y = \frac{x-\mu}{\sigma}B$, $a_1 = f_{n-1,n:n}(1 - e^{-\alpha_{n-1,n:n}}) - \lambda_{n-1,n:n}\alpha_{n-1,n:n}$, *and* $a_2 = f_{n-1,n:n}(1 - e^{-\alpha_{n-1,n:n}})$, *the inequality becomes equality for*

$$F(y) = \begin{cases} 0, & y < a_1, \\ 1 - \exp\left(-\alpha_{n-1,n:n} - \frac{y - f_{n-1,n:n}(1-e^{-\alpha_{n-1,n:n}})}{\lambda_{n-1,n:n}}\right), & a_1 \le y < a_2, \\ f^{-1}_{n-1,n:n}(y), & a_2 \le y < n, \\ 1, & y \ge n. \end{cases}$$

The proof involves direct application of Theorem 6.36 to the function $h(x) = f_{n-1,n:n} \circ Z(x) = f_{n:n}(1 - e^{-x}) - f_{n-1:n}(1 - e^{-x})$, $x > 0$.

6.3.6 *Positive Upper Bounds for Order Statistics from IRFR Family*

Now we consider order statistics from the IRFR samples. For establishing the bounds, we use the projection method applying Theorem 6.26 for specific $W(x) = V(x) = e^x$, $h(x) = f_{j:n}(e^x) - 1$, $x < 0$. The particular counterparts of (6.240)–(6.243) have the forms

$$M_{j:n}(\beta) = f_{j:n}(e^\beta)(1 - e^\beta) + F_{j:n}(e^\beta) - 1$$

$$= -\sum_{i=0}^{j-2} B_{i,n}(e^\beta) + (n-j)B_{j-1,n}(e^\beta), \tag{6.297}$$

$$\lambda_{j:n}(\alpha) = \frac{1}{2e^\alpha}\left(e^\alpha f_{j:n}(e^\alpha) - \sum_{i=j}^{n}\frac{1}{i}F_{i:n}(e^\alpha)\right)$$

$$= \frac{n}{2}B_{j-1,n-1}(e^x) - \sum_{i=j}^{n}\left(\sum_{k=j}^{i}\frac{1}{k}\right)\frac{n}{2i}B_{i-1,n-1}(e^x), \tag{6.298}$$

$$
\begin{aligned}
K_{j:n}(\alpha) &= j f_{j+1:n}(\mathrm{e}^{\alpha}) - \left(j - \frac{3}{2}\right) f_{j:n}(\mathrm{e}^{\alpha}) - \frac{1}{2\mathrm{e}^{\alpha}} \sum_{i=j}^{n} \frac{1}{i} F_{i:n}(\mathrm{e}^{\alpha}) \\
&= -n\left(j - \frac{3}{2}\right) B_{j-1,n-1}(e^{x}) + nj B_{j,n-1}(e^{x}) \\
&\quad - \sum_{i=j}^{n} \left(\sum_{k=j}^{i} \frac{1}{k}\right) \frac{n}{2i} B_{i-1,n-1}(e^{x}),
\end{aligned}
\tag{6.299}
$$

$$
\begin{aligned}
L_{j:n}(\alpha) &= F_{j:n}(\mathrm{e}^{\alpha}) - \frac{1}{2}\left(\sum_{i=j}^{n} \frac{1}{i} F_{i:n}(\mathrm{e}^{\alpha}) + \mathrm{e}^{\alpha} f_{j:n}(\mathrm{e}^{\alpha})\right) \\
&= -\frac{n(j-1)^2}{2j(n+1)} B_{j,n}(e^{x})) + \sum_{i=j+1}^{n} \frac{n}{n+1}\left(1 - \frac{1}{2}\sum_{r=j}^{i} \frac{1}{r}\right) B_{i,n}(e^{x})).
\end{aligned}
\tag{6.300}
$$

The variation diminishing property of Bernstein polynomials allows us to deduce the following properties of the above functions. The first one (6.297) is first negative and then positive. Its zero is necessarily less than $c = v_0$ (see (6.32) of Lemma 6.4). For $\beta > v_0$, it has to be positive because it represents the difference of $f_{j:n}$ at e^{β} and its integral over $(e^{\beta}, 1)$. Then the function is decreasing over the interval, and the integral is less than the value of function at the left end-point of the integral. Function (6.298) is first positive and then negative, whereas (6.299) is either negative on the whole $\mathbb{R}_-$ or it has one interval of positivity separated from $-\infty$ and 0. The last one (6.300) has two zeros at most. If j is relatively small with respect to n so that $\sum_{r=j}^{n} \frac{1}{r} > 2$, then it is first negative and then positive. Otherwise, it has either 0 or 2 zeros.

Theorem 6.38 (IRFR Distributions) *We consider iid random variables $X_1, \ldots, X_n$ with an IRFR marginal distribution function, and $2 \le j \le n-1$. Suppose that the zero $-\infty < \beta_* < c = v_0$ of* (6.297) *belongs to $(b, c) = (v_1, v_0)$ (see* (6.32) *and* (6.33)*), and there exists a unique $b = v_1 < \alpha_* < \beta_*$ satisfying $K_{j:n}(\alpha) \ge 0$ and $L_{j:n}(\alpha) = 0$. Then*

$$
\mathbb{E}\frac{X_{j:n} - \mu}{\sigma} \le B = B(j, n),
\tag{6.301}
$$

where

$$B^2 = \frac{1}{2e^{\alpha_*}}\left[e^{2\alpha_*} f_{j:n}^2(e^{\alpha_*}) + \left(\sum_{i=j}^{n} \frac{1}{i} F_{i:n}(e^{\alpha_*})\right)^2\right]$$
$$+n\frac{\binom{2j-2}{j-1}\binom{2n-2j}{n-j}}{\binom{2n-1}{n}}\left[F_{2j-1:2n-1}(e^{\beta_*}) - F_{2j-1:2n-1}(e^{\alpha_*})\right]$$
$$+f_{j:n}^2(e^{\beta_*})(1-e^{\beta_*}) - 1. \tag{6.302}$$

The equality in (6.301) *holds for the distribution function*

$$F(x) = \begin{cases} \exp\left(\frac{1}{\lambda_{j:n}(\alpha_*)}\left(\frac{x-\mu}{\sigma}B + 1 - f_{j:n}(e^{\alpha_*})\right) + \alpha_*\right), & \frac{x-\mu}{\sigma} < \frac{f_{j:n}(e^{\alpha_*})-1}{B}, \\ f_{j:n}^{-1}\left(\frac{x-\mu}{\sigma}B + 1\right), & \frac{f_{j:n}(e^{\alpha_*})-1}{B} \le \frac{x-\mu}{\sigma} < \frac{f_{j:n}(e^{\beta_*})-1}{B}, \\ 1, & \frac{x-\mu}{\sigma} \ge \frac{f_{j:n}(e^{\beta_*})-1}{B}. \end{cases} \tag{6.303}$$

Otherwise, if the above conditions are not satisfied, the set of points $\beta_* \le \alpha < 0$ *being the solutions to the equation*

$$e^{\alpha} - F_{j:n}(e^{\alpha}) = \frac{1-e^{\alpha}}{2-e^{\alpha}}\left(e^{\alpha} - \sum_{i=j}^{n}\frac{1}{i}F_{i:n}(e^{\alpha})\right)$$

say $\mathcal{A}_{j:n}$, *is nonempty, and for* $\alpha_* = \arg\max_{\alpha \in \mathcal{A}_{j:n}} ||P_\alpha h_{j:n}||_V^2$, *with*

$$||P_\alpha h_{j:n}||_V^2 = \frac{2-e^{\alpha}}{e^{\alpha}}\left(\frac{e^{\alpha} - F_{j:n}(e^{\alpha})}{1-e^{\alpha}}\right)^2, \tag{6.304}$$

we have the bound

$$\mathbb{E}\frac{X_{j:n}-\mu}{\sigma} \le B = B(j,n) = \frac{e^{\alpha_*} - F_{j:n}(e^{\alpha_*})}{1-e^{\alpha_*}}\sqrt{\frac{2-e^{\alpha_*}}{e^{\alpha_*}}}. \tag{6.305}$$

The equality is attained for the distribution function

$$F(x) = \begin{cases} \exp\left(\left(\frac{x-\mu}{\sigma}\sqrt{\frac{(1-e^{\alpha_*})(2-e^{\alpha_*})}{e^{\alpha_*}-F_{j:n}(e^{\alpha_*})}} - 1\right) e^{\alpha_*} + \alpha_*\right), & \frac{x-\mu}{\sigma} < \sqrt{\frac{e^{\alpha_*}-F_{j:n}(e^{\alpha_*})}{(1-e^{\alpha_*})(2-e^{\alpha_*})}}, \\ 1, & \frac{x-\mu}{\sigma} \geq \sqrt{\frac{e^{\alpha_*}-F_{j:n}(e^{\alpha_*})}{(1-e^{\alpha_*})(2-e^{\alpha_*})}}. \end{cases} \tag{6.306}$$

Proof Assume first that $h_{j:n}(x) = f_{j:n}(e^x) - 1$, $x < 0$, satisfies the conditions of the first statement of Theorem 6.26, i.e., $\beta_* > v_1$ and $\mathcal{K}_{j:n} = \{v_1 < \alpha < \beta_* : K_{j:n}(\alpha) \geq 0, L_{j:n}(\alpha) = 0\}$ is nonempty. In fact, it consists of only one point. It follows from Proposition 6.1, because (6.299) is positive on a single interval.

The particular form of the projection (6.244) in our case is represented as

$$P_{\preceq V} h_{j:n}(x) = \begin{cases} h_{j:n}(\alpha_*) + \lambda_{j:n}(\alpha_*)(x - \alpha_*), & -\infty \leq x < \alpha_*, \\ h_{j:n}(x), & \alpha_* \leq x < \beta_*, \\ h_{j:n}(\beta_*), & \beta_* \leq x < 0. \end{cases} \tag{6.307}$$

Since

$$\begin{aligned} \int_{\alpha}^{\beta} f_{j:n}^2(e^x)e^x dx &= \int_{e^\alpha}^{e^\beta} n^2\binom{n-1}{j-1}^2 x^{2j-2}(1-x)^{2n-2j} dx \\ &= \frac{n^2\binom{n-1}{j-1}^2}{(2n-1)\binom{2n-2}{2j-2}} \int_{e_\alpha}^{e^\beta} f_{2j-1:2n-1}(x)dx \\ &= n\frac{\binom{2j-2}{j-1}\binom{2n-2j}{n-j}}{\binom{2n-1}{n}} \left[F_{2j-1:2n-1}(e^\beta) - F_{2j-1:2n-1}(e^\alpha)\right], \end{aligned}$$

we obtain the squared norm of (6.307), which is equivalent to

$$\begin{aligned} B^2 = {} & f_{j:n}^2(e^{\alpha_*})e^{\alpha_*} - 2\lambda_{j:n}(\alpha_*)e^{\alpha_*} f_{j:n}(e^{\alpha_*}) + 2\lambda_{j:n}^2(\alpha_*)e^{\alpha_*} \\ & + n\frac{\binom{2j-2}{j-1}\binom{2n-2j}{n-j}}{\binom{2n-1}{n}} \left[F_{2j-1:2n-1}(e^{\beta_*}) - F_{2j-1:2n-1}(e^{\alpha_*})\right] \\ & + f_{j:n}^2(e^{\beta_*})(1 - e^{\beta_*}) - 1. \end{aligned}$$

After plugging (6.298) into the above formula, we obtain (6.302).

Otherwise, if the above conditions do not hold, the projection of $h_{j:n}$ has a simpler form

$$P_\alpha h_{j:n}(x) = \frac{e^\alpha - F_{j:n}(e^\alpha)}{1 - e^\alpha}\left(\frac{x - \alpha}{e^\alpha}\mathbf{1}_{(-\infty,\alpha)}(x) + 1\right) \tag{6.308}$$

for some $\beta_* \leq \alpha < 0$. Its squared norm equals to (6.304). The projection of $h_{j:n}$ has the greatest value of the norm among all functions of the form (6.308). Therefore the maximal norm coincides with the desired bound.

The equality condition for both inequalities (6.301) and (6.305) follow from the relation

$$\frac{F^{-1}(e^x) - \mu}{\sigma} = \frac{P_{\preceq V} h_{j:n}(x)}{\|P_{\preceq V} h_{j:n}\|_V}, \tag{6.309}$$

which applied to appropriately adjusted (6.244) with (6.302), and (6.245) with (6.305) results in distribution functions given by (6.303) and (6.306), respectively. □

We applied the above result for numerical evaluations of expectations of order statistics for various sample sizes n and order statistics ranks $2 \leq j \leq n-1$. It never happened that the bound was calculated with the use of (6.302). The reason for this is that the intervals of concave increase of $h_{j:n}$ belong to very short left neighborhood of the function maximum, right to β_*. We believe that for all $2 \leq j \leq n-1$, the bounds are described by (6.305), but we cannot prove it formally. Therefore, the whole procedure of bound determination should consist of two steps. In the first one, we check if the conditions of the first statement of Theorem 6.38 hold. If so, the square root of (6.302) is used for the evaluation. If not, we take (6.305). In the case of sample maximum, the bound and its proof is substantially simpler.

Theorem 6.39 *If F has the increasing reversed failure rate, then we have the following optimal bound for the maximum of an iid sample:*

$$\mathbb{E}\frac{X_{n:n} - \mu}{\sigma} \leq 1 - \frac{1}{n}. \tag{6.310}$$

The equality is attained for the location-scale family of negative exponential distributions

$$F(x) = \begin{cases} \exp\left(\frac{x-\mu}{\sigma} - 1\right), & x < \mu + \sigma, \\ 1, & x \geq \mu + \sigma. \end{cases} \tag{6.311}$$

Proof We already noted that the function $h_{n:n}(x) = f_{j:n}(e^x) - 1$ is strictly increasing and convex on $\mathbb{R}_-$. This means that its projection onto (6.223) with $W = V$ and $(a, d) = \mathbb{R}_-$ is a linear and increasing function, say $P_{\preceq_c V} h_{n:n}(x) = ax + b$ for some $a > 0$ and $b \in \mathbb{R}$. Indeed, the graphs of every pair of strictly increasing convex function and nondecreasing concave function with the same integrals over $\mathbb{R}_-$ with the weight e^x have exactly two common points. The increasing linear function passing through these points runs between these convex and concave functions and it is a better approximation of the convex function than the original concave one. The parameters a, b of the optimal linear approximation can be derived by minimizing the following squared distance:

$$\int_{-\infty}^{0} [ne^{(n-1)x} - 1 - ax - b]^2 e^x \, dx = \int_{-\infty}^{0} [ne^{(n-1)x} - ax - c]^2 e^x \, dx$$

$$= n^2 \int_{-\infty}^{0} e^{(2n-1)x} \, dx + a^2 \int_{-\infty}^{0} x^2 e^x \, dx + c^2 \int_{-\infty}^{0} e^x \, dx$$

$$+ 2ac \int_{-\infty}^{0} xe^x \, dx - 2an \int_{-\infty}^{0} xe^{nx} \, dx - 2cn \int_{-\infty}^{0} e^{nx} \, dx$$

$$= \frac{n^2}{2n-1} + 2a^2 + c^2 - 2ac + \frac{2a}{n} - 2c$$

for $c = b + 1$. This is a quadratic convex function in each coordinate, and it is maximized by the pair $a = 1 - \frac{1}{n} > 0$ and $c = 2 - \frac{1}{n}$, i.e., for $a = b = 1 - \frac{1}{n}$. We have

$$||P_{\preceq V} h_{n:n}||_V^2 = \left(1 - \frac{1}{n}\right)^2 \int_{-\infty}^{0} (x+1)^2 e^x \, dx = \left(1 - \frac{1}{n}\right)^2,$$

and so $||P_{\preceq V} h_{n:n}||_V = 1 - \frac{1}{n}$ provides the optimal bound in (6.310). The marginal distribution for which the equality is attained is determined by the equation

$$\frac{F^{-1}(e^x) - \mu}{\sigma} = x + 1,$$

which is a particular form of (6.309) with $j = n$. The explicit form of the distribution function is presented in (6.311). □

6.3.7 *Bounds on Spacings from DRFR and IRFR Families*

The bounds on the expectations of spacings based on iid samples with monotone reversed failure rate distributions can be concluded from the analogous results for the spacings based on monotone classic failure rate families.

Corollary 6.4 *The sharp upper bounds on the expectations of jth spacings* $\mathbb{E}\frac{X_{j+1:n} - X_{j:n}}{\sigma}$, $j = 1, \ldots, n-1$, *based on iid samples from DRFR (IRFR)* *populations are identical to the sharp upper bounds for the expectations of $(n-j)$th spacings $\mathbb{E}\frac{X_{n-j+1:n}-X_{n-j:n}}{\sigma}$, $j = 1, \ldots, n-1$, from the IFR (DFR, respectively) family. If the latter is attained by a location-scale family $F\left(\frac{x-\mu}{\sigma}\right)$, $\mu \in \mathbb{R}$, $\sigma > 0$, for some F, then the former is attained by the family $1 - F\left(-\frac{x-\mu}{\sigma}-\right)$, $\mu \in \mathbb{R}$, $\sigma > 0$.*

Proof We consider in detail the upper bounds of jth spacings from the DRFR family. The bounds are obviously positive and they are equal to the projections of

$$f_{j,j+1:n} \circ V(x) = f_{j+1:n}(e^x) - f_{j:n}(e^x), \qquad x < 0, \; j = 1, \ldots, n-1, \tag{6.312}$$

onto the convex cone

$$\mathcal{C}_{\succeq_c V} = \{g \in L^2(\mathbb{R}_-, e^x dx) : g \text{ nondecreasing and convex}\}. \tag{6.313}$$

Note that

$$f_{j+1:n}(e^x) - f_{j:n}(e^x) = -[f_{n-j+1:n}(1-e^x) - f_{n-j:n}(1-e^x)], \qquad x < 0. \tag{6.314}$$

Looking for the projection $P_{\succeq_c V} f_{j,j+1:n} \circ V$, we minimize the squared distance between $f_{j,j+1:n} \circ V$ and $g \in \mathcal{C}_{\succeq_c V}$. This has the form

$$\begin{aligned} &\int_{-\infty}^{0} \{-[f_{n-j+1:n}(1-e^x) - f_{n-j:n}(1-e^x)] - g(x)\}^2 e^x \, dx \\ &= \int_{0}^{\infty} \{[f_{n-j+1:n}(1-e^{-x}) - f_{n-j:n}(1-e^{-x})] - [-g(-x)]\}^2 e^{-x} \, dx. \end{aligned} \tag{6.315}$$

We see that this problem is equivalent to finding the minimal $L^2(\mathbb{R}_+, e^{-x}dx)$-distance between $f_{n-j,n-j+1:n} \circ Z(x)$ and functions $-g(-x)$, $x > 0$, where $g(x)$, $x < 0$, are nondecreasing and convex. Under the assumptions, the functions $-g(-x)$ form the family of nondecreasing and concave functions in $L^2(\mathbb{R}_+, e^{-x}dx)$. Accordingly, the original problem is equivalent to the problem of finding the projection $P_{\preceq_c Z} f_{n-j,n-j+1:n} \circ Z$ of $f_{n-j,n-j+1:n} \circ Z$ onto $\mathcal{C}_{\preceq_c Z}$. Observe further that

$$\begin{aligned} \| P_{\preceq_c Z} f_{n-j,n-j+1:n} \circ Z \|_Z^2 &= \int_0^\infty P^2_{\preceq_c Z} f_{n-j,n-j+1:n} \circ Z(x) e^{-x} \, dx \\ &= \int_{-\infty}^{0} [-P_{\preceq_c Z} f_{n-j,n-j+1:n} \circ Z(-x)]^2 e^x \, dx \\ &= \| P_{\succeq_c V} f_{j,j+1:n} \circ V \|_V^2, \end{aligned} \tag{6.316}$$

which means that the sharp upper bound for $\mathbb{E}\frac{X_{j+1:n}-X_{j:n}}{\sigma}$ for $X_1, \ldots X_n$ being iid DRFR is identical with the sharp upper bound for $\mathbb{E}\frac{X_{n-j+1:n}-X_{n-j:n}}{\sigma}$ for iid $X_1, \ldots X_n$ from the IFR class. The respective attainability conditions are

$$\frac{F^{-1}(e^x)-\mu}{\sigma} = \frac{P_{\succeq_c V} f_{j,j+1:n} \circ V(x)}{\|P_{\succeq_c V} f_{j,j+1:n} \circ V\|_V}, \tag{6.317}$$

$$\frac{G^{-1}(1-e^{-x})-\mu}{\sigma} = \frac{P_{\preceq_c Z} f_{n-j,n-j+1:n} \circ Z(x)}{\|P_{\succeq_c V} f_{j,j+1:n} \circ V\|_V} = \frac{-P_{\succeq_c V} f_{j,j+1:n} \circ V(-x)}{\|P_{\succeq_c V} f_{j,j+1:n} \circ V\|_V}, \tag{6.318}$$

which immediately gives the relation $F(x) = 1 - G(-x-)$.

The proof of equivalence among the bounds for spacings from the IFRF and DFR families as well as the respective attainability conditions is analogous. □

The bounds for the first and last spacings from the DRFR samples are deduced from Theorem 6.37 and Corollary 4.4 (see also Theorem 6.31), respectively. The bounds for the remaining spacings are based on Theorem 6.35. The bounds for the IRFR families can be derived from Theorem 6.14 which describes the bounds for spacings in the DFR case.

The arguments of the above proof allow us to notice that the sharp upper bounds on any L-statistic $\sum_{i=1}^{n} c_i X_{i:n}$ with $\sum_{i=1}^{n} c_i = 0$ based on an iid DRFR (IRFR) sample is identical to the sharp upper bound on $\sum_{i=1}^{n} c_{n+1-i} X_{i:n}$ based on an IFR (DFR, respectively) sample. However, the precise values of the latter bounds are not known except on the special case of spacings.

6.3.8 Nonpositive Upper Bounds for Sample Minima

In contrast with the DD, DFR and DRFR populations, the nonpositive upper bounds on the differences between the expectations of order statistics and population means in the ID, IFR and IRFR cases are possible only for the sample minima.

Theorem 6.40 (ID, IFR, and IRFR Distributions) *Let $X_1, \ldots, X_n$ be iid random variables with a finite pth absolute central moment σ_p^p for some $p > 1$, and marginal distributions belonging to either of families ID, IFR and IRFR. Then, the following trivial bound is sharp:*

$$\frac{\mathbb{E}X_{1:n}-\mu}{\sigma_p} \le 0. \tag{6.319}$$

The equality is attained in the limit by the family of IRFR distributions being the mixtures of linearly transformed negative exponential distributions and atoms at the right end-points of their supports with probabilities e^{α} and $1-e^{\alpha}$, $\alpha < 0$, as $\alpha \to -\infty$.

Proof The bound is obviously trivial, and therefore it suffices to prove its optimality. Define the family of distribution functions

$$F_\alpha(x) = \begin{cases} \exp(e^\alpha(x-1)+\alpha), & x < 1, \\ 1, & x \geq 1, \end{cases} \qquad \alpha < 0. \tag{6.320}$$

The respective quantile functions composed with the exponential functions

$$F_\alpha^{-1}(e^x) = \begin{cases} e^{-\alpha}(x-\alpha)+1, & -\infty < x \leq \alpha, \\ 1, & \alpha \leq x < 0, \end{cases} \qquad \alpha < 0, \tag{6.321}$$

are concave which means that actually all the F_α, $\alpha < 0$, have the IRFR property. Simple calculations show that

$$\mathbb{E}_\alpha X_1 = \int_{-\infty}^{0} F_\alpha^{-1}(e^x)e^x\,dx = e^{-\alpha}\int_{-\infty}^{\alpha}(x-\alpha)e^x\,dx + \int_{-\infty}^{0} e^x\,dx = 0 \tag{6.322}$$

and

$$\begin{aligned}
\mathbb{E}_\alpha X_{1:n} &= \int_{-\infty}^{0} F_\alpha^{-1}(e^x) f_{1:n}(e^x)e^x\,dx \\
&= ne^{-\alpha}\int_{-\infty}^{\alpha}(x-\alpha)(1-e^x)^{n-1}e^x\,dx + n\int_{-\infty}^{0}(1-e^x)^{n-1}e^x\,dx \\
&= ne^{-\alpha}\sum_{i=0}^{n-1}\binom{n-1}{i}(-1)^{n-1-i}\int_{-\infty}^{\alpha}(x-\alpha)e^{(n-i)x}\,dx + 1 \\
&= \sum_{i=0}^{n-1}\binom{n}{i}(-1)^{n-i}e^{(n-1-i)\alpha} + 1,
\end{aligned} \tag{6.323}$$

because the integrals in the penultimate line amount to $-\frac{e^{(n-i)\alpha}}{n-i}$, $i = 0, \ldots, n-1$. We easily notice that

$$\lim_{\alpha\to-\infty} \mathbb{E}_\alpha X_{1:n} = \lim_{\alpha\to-\infty} \mathbb{E}_\alpha(X_{1:n} - X_1) = -n+1, \tag{6.324}$$

On the other hand,

$$\begin{aligned}\mathbb{E}_\alpha|X_1|^p &= \int_{-\infty}^{0} |F_\alpha^{-1}(e^x)|^p e^x\,dx \geq \int_{-\infty}^{0} (-\min\{F_\alpha^{-1}(e^x),0\})^p e^x\,dx \\ &= \int_{-\infty}^{\alpha-e^\alpha} [-e^\alpha(x-\alpha)-1]^p e^x\,dx \\ &= \exp((1-p)\alpha-e^\alpha)\int_0^\infty y^p e^{-y}\,dy \\ &= \exp((1-p)\alpha-e^\alpha)\Gamma(p+1), \end{aligned} \tag{6.325}$$

where the passage in the middle line is based on the change of variable $y = x - \alpha - e^\alpha$. Therefore

$$(\mathbb{E}_\alpha|X_1|^p)^{\frac{1}{p}} \geq \left(\exp((1-p)\alpha-e^\alpha)\Gamma(p+1)\right)^{\frac{1}{p}}, \tag{6.326}$$

which tends to $+\infty$ as $\alpha \to -\infty$. Combining this with (6.324), we obtain

$$\lim_{\alpha\to-\infty} \frac{\mathbb{E}_\alpha(X_{1:n}-X_1)}{(\mathbb{E}_\alpha|X_1|^p)^{\frac{1}{p}}} = 0. \tag{6.327}$$

Obviously one gets the same limit if we linearly transform (6.320) so to obtain the distribution with arbitrarily fixed real μ and positive σ_p. Since (6.320) are the IRFR distributions, they are ID and IFR as well. This completes the proof of optimality of the zero bounds in all three classes of distributions. □

The nontrivial sharp bounds in σ_1 units presented below come from Rychlik and Szymkowiak (2021).

Theorem 6.41 (ID Distributions) *Suppose that iid random variables $X_1, \ldots, X_n$ have an ID parent distribution with a finite mean.*

If $n = 2$, then the bound

$$\frac{\mathbb{E}X_{1:2}-\mu}{\sigma_1} \leq -\frac{1}{2} \tag{6.328}$$

is sharp, and it is attained in the limit by the mixtures of uniform distributions on

$$\left[\mu - \frac{2\sigma_1}{\alpha(2-\alpha)}, \mu + \frac{2\sigma_1}{(2-\alpha)^2}\right]$$

with the atoms at $\mu + \frac{2\sigma_1}{(2-\alpha)^2}$ when the probabilities of the atoms tend to 1.

For $n \geq 3$, we have

$$\frac{\mathbb{E}X_{1:n} - \mu}{\sigma_1} \leq -2\frac{n-1}{n+1}. \tag{6.329}$$

If $n = 3$, then the equality is attained by the mixture distributions described above for any fixed $\alpha \in (0, 1]$. For ≥ 4, the equality in (6.329) is attained by the uniform distribution on $[\mu - 2\sigma_1, \mu + 2\sigma_1]$ only.

Bound (6.328) is identical to that of Corollary 4.15 for arbitrary parent distributions. Bounds (6.329) coincide with the optimal evaluations (6.213) for symmetric unimodal distributions.

Proof Referring to the norm maximization method, we restrict ourselves to the quantile functions of distributions whose convex combinations constitute all the quantile functions which are concave. This extreme class contains the simple piecewise linear functions

$$F_\alpha^{-1}(x) = \begin{cases} \frac{x}{\alpha}, & 0 < x \leq \alpha, \\ 1, & \alpha \leq x < 1, \end{cases} \qquad 0 < \alpha < 1. \tag{6.330}$$

We easily check that for iid $Y_1, \ldots, Y_n$ with the above parent distributions, we have

$$\mathbb{E}_\alpha Y_1 = \int_0^\infty [1 - F_\alpha(x)]dx = \int_0^1 (1 - \alpha x)dx = 1 - \frac{\alpha}{2},$$

$$\mathbb{E}_\alpha Y_{1:n} = \int_0^\infty [1 - F_\alpha(x)]^n dx = \int_0^1 (1 - \alpha x)^n dx = \frac{1 - (1-\alpha)^{n+1}}{(n+1)\alpha}$$

and

$$N_1(\alpha) = \mathbb{E}_\alpha |Y_1 - \mathbb{E}_\alpha Y_1| = \alpha \int_0^1 \left| x - 1 + \frac{\alpha}{2} \right| dx + (1-\alpha)\frac{\alpha}{2} = \frac{1}{4}\alpha(2-\alpha)^2. \tag{6.331}$$

Consequently,

$$\frac{\mathbb{E}_\alpha Y_{1:n} - \mathbb{E}Y_1}{\mathbb{E}_\alpha |Y_1 - \mathbb{E}_\alpha Y_1|} = \frac{4}{n+1} \frac{1 - (n+1)\alpha + \frac{n+1}{2}\alpha^2 - (1-\alpha)^{n+1}}{\alpha^2(2-\alpha)^2} = B_n(\alpha), \tag{6.332}$$

say. The upper bound we look for amounts to the maximal value of the RHS of (6.332) with respect to $0 < \alpha \leq 1$.

We immediately check that $B_3(\alpha) = -1$, $0 < \alpha \leq 1$. This means that for $n = 3$, the upper bound is equal to -1, and it is attained by the arbitrary mixtures of standard uniform distributions with the atoms at 1. For other n, we use the following result.

Define functions

$$L_n(x) = 2x^{n+1} - (n+1)x^2 + n - 1, \qquad 0 < x < 1,\ n \geq 2,$$

with the derivatives

$$L_n'(x) = 2(n+1)(x^{n-1} - 1).$$

We consider the ratios $\frac{L_n(x)}{L_{n-1}(x)}$ for $n \geq 3$. Their derivatives satisfy

$$\begin{aligned}
2xL_{n-1}^2(x)\frac{d}{dx}\frac{L_n(x)}{L_{n-1}(x)} &= (n+1)(x^{n-1}-1)[2x^n - nx^2 + n - 2] \\
&- n(x^{n-2}-1)[2x^{n+1} - (n+1)x^2 + n - 1] \\
&= 2x^{2n-1} - n(n-1)x^{n+1} + (n-2)(n+1)x^n \\
&+ (n-2)(n+1)x^{n-1} - n(n-1)x^{n-2} + 2 \\
&= 2x^n(x^{n-1}-1) - 2(x^{n-1}-1) - n(n-1)(x^n - x^{n-2})(x-1) \\
&= 2(1-x^{n-1})(1-x^n) + n(n-1)x^n(1-x)^2(1+x) \\
&= (1-x)^2\left[2\sum_{i=0}^{n-2} x_i \sum_{i=0}^{n-1} x_i - n(n-1)(x^{n-2} + x^{n-1})\right].
\end{aligned}$$

The expression inside the brackets can be rewritten as

$$\begin{aligned}
&2\sum_{i=0}^{n-2}(i+1)(x^i + x^{2n-3-i}) - n(n-1)(x^{n-2} + x^{n-1}) \\
&= 2\sum_{i=0}^{n-2}(i+1)(x^i + x^{2n-3-i} - x^{n-2} - x^{n-1}) \\
&= 2\sum_{i=0}^{n-2}(i+1)x^i(1 - x^{n-1-i})(1 - x^{n-2-i}) > 0.
\end{aligned}$$

We conclude every fraction $\frac{L_n(x)}{L_{n-1}(x)}$, $n \geq 3$, is increasing $(0, 1)$, and as a consequence, all $\frac{L_n(1-x)}{L_{n-1}(1-x)}$, $n \geq 3$, are decreasing there.

Note that

$$B_n(\alpha) = -\frac{4}{n+1}\frac{L_n(1-\alpha)}{N_1(\alpha)}.$$

Obviously $N_1(\alpha) > 0$. The same applies to all functions L_n, because

$$-\frac{L_n(1-\alpha)}{2(n+1)\alpha} = \mathbb{E}Y_{1:n} - \mathbb{E}Y_1,$$

and the RHS is negative for all nondegenerate iid random variables. Also $\frac{L_3(1-\alpha)}{N_1(\alpha)} = 1$ for all α. For $n = 2$, function

$$B_2(\alpha) = -\frac{4}{3}\frac{L_3(1-\alpha)}{N_1(\alpha)}\frac{L_2(1-\alpha)}{L_3(1-\alpha)} = -\frac{4}{3}\frac{L_2(1-\alpha)}{L_3(1-\alpha)}$$

is decreasing, because the latter factor is positive increasing. Therefore

$$\sup_{0<\alpha\leq 1} B_2(\alpha) = \lim_{\alpha\to 0} B_2(\alpha) = -\frac{4}{3}\lim_{\alpha\to 0}\frac{-\frac{3}{2}\alpha+\alpha^3}{\alpha^2(2-\alpha)^2} = -\frac{1}{2}$$

(see (6.332)). This means that the bound for $n = 2$ is equal to $-\frac{1}{2}$, and it is attained in the limit by the convex combinations of the uniform distribution on $[0, 1]$ and atom at 1 as the contribution of the continuous part decreases to 0.

For $n \geq 4$, we have

$$B_n(\alpha) = -\frac{4}{n+1}\frac{L_3(1-\alpha)}{N_1(\alpha)}\prod_{i=4}^{n}\frac{L_i(1-\alpha)}{L_{i-1}(1-\alpha)} = -\frac{4}{n+1}\prod_{i=4}^{n}\frac{L_i(1-\alpha)}{L_{i-1}(1-\alpha)}.$$

The function is increasing because each factor of the product is positive decreasing, and attains its maximum $-2\frac{n-1}{n+1}$ at 1. This is the bound value and this is attained by the standard uniform distribution.

The attainability conditions can be easily generalized to the families of distributions with arbitrary $\mu \in \mathbb{R}$ and $\sigma_1 > 0$ if we replace the distributions with quantile functions (6.330) by ones of mixtures of uniform distributions on the interval $\left[\mu - \frac{2\sigma_1}{\alpha(2-\alpha)}, \mu + \frac{2\sigma_1}{(2-\alpha)^2}\right]$ with the atoms at $\mu + \frac{2\sigma_1}{(2-\alpha)^2}$. □

We finally present negative bounds on the expectations of sample minima coming from iid IFR samples.

Theorem 6.42 (IFR Distributions) *For the minimum of n iid random variables with a common distribution that has the convex cumulative failure fate and finite mean μ and first absolute central moment σ_1, the inequality*

$$\frac{\mathbb{E}X_{1:n}-\mu}{\sigma_1} \leq -\frac{n-1}{n}\frac{e}{2}, \tag{6.333}$$

holds. It becomes equality in the limit for the mixtures of exponential distributions with location parameter (left-end support point) $\mu - \sigma_1\frac{1-e^{-\alpha}}{N(\alpha)}$ and intensity α

truncated right at $\mu + \sigma_1 \frac{e^{-\alpha}}{N(\alpha)}$, *and atom at* $\mu + \sigma_1 \frac{e^{-\alpha}}{N(\alpha)}$ *with probabilities* $1 - e^{-\alpha}$ *and* $e^{-\alpha}$, *respectively, when* $\alpha \to \infty$, *where*

$$N(\alpha) = \frac{1}{\alpha}[2\exp(-e^{-\alpha} - 1) - e^{-\alpha} - e^{-2\alpha} - \alpha e^{-\alpha}]. \tag{6.334}$$

Proof Recalling the arguments of the first part of the proof of Theorem 6.41, we restrict our attention to the parent distribution functions F_α, $\alpha > 0$, satisfying

$$F_\alpha^{-1}(1 - e^{-x}) = \begin{cases} \frac{x}{\alpha}, & 0 < x \le \alpha, \\ 1, & x \ge \alpha. \end{cases}$$

The explicit forms of the distribution functions are

$$F_\alpha(x) = \begin{cases} 0, & x \le 0, \\ 1 - \exp(-\alpha x), & 0 \le x < 1, \\ 1, & x \ge 1, \end{cases} \tag{6.335}$$

which means that they have absolutely continuous parts with the density functions $f_\alpha(x) = \alpha e^{-\alpha x}$ on $[0, 1]$, and jumps of heights $e^{-\alpha}$ at 1.

If $Y_1, \ldots, Y_n$ are iid with the distribution function F_α, then

$$\mathbb{E}_\alpha Y_1 = \int_0^1 e^{-\alpha x}\, dx = \frac{1 - e^{-\alpha}}{\alpha},$$

$$\mathbb{E}_\alpha Y_{1:n} = \int_0^1 e^{-n\alpha x}\, dx = \frac{1 - e^{-n\alpha}}{n\alpha}$$

and

$$\mathbb{E}_\alpha(Y_{1:n} - Y_1) = \frac{1}{n\alpha}(1 - n + ne^{-\alpha} - e^{-n\alpha}).$$

We also determine

$$\mathbb{E}_\alpha|Y_1 - \mathbb{E}_\alpha Y_1| = \alpha \int_0^1 \left| x - \frac{1 - e^{-\alpha}}{\alpha} \right| e^{-x} dx + \left(1 - \frac{1 - e^{-\alpha}}{\alpha}\right) e^{-\alpha} = N(\alpha)$$

(see (6.334)). Accordingly,

$$\frac{\mathbb{E}_\alpha(Y_{1:n} - Y_1)}{\mathbb{E}_\alpha|Y_1 - \mathbb{E}_\alpha Y_1|} = \frac{1}{n} \frac{1 - n + ne^{-\alpha} - e^{-n\alpha}}{2\exp(-e^{-\alpha} - 1) - e^{-\alpha} - e^{-2\alpha} - \alpha e^{-\alpha}} = B_n(\alpha), \tag{6.336}$$

say. Our goal is to maximize each $B_n(\alpha)$, $n \ge 2$, with respect to $0 < \alpha < \infty$.

Fix first $n = 2$ and change the variable for $x = 1 - e^{-\alpha} \in (0, 1)$ in (6.336). Then we obtain

$$2B_2(-\ln(1-x)) = \frac{-x^2}{2e^{-x} - 2 + 3x - x^2 + (1-x)\ln(1-x)} = \frac{-x^2}{D(x)}.$$

We develop the denominator into the power series

$$\begin{aligned} D(x) &= 2\sum_{i=0}^{\infty} \frac{(-1)^i}{i!} x^i - 2 + 3x - x^2 - (1-x)\sum_{i=1}^{\infty} \frac{1}{i} x^i \\ &= \frac{x^2}{2} + \sum_{i=3}^{\infty} \left[\frac{2(-1)^i}{i!} - \frac{1}{i} + \frac{1}{i-1} \right] x^i. \end{aligned}$$

Therefore,

$$g_2(x) = \frac{-1}{2B_2(-\ln(1-x))} = \frac{D(x)}{x^2} = \frac{1}{2} + \sum_{i=1}^{\infty} \frac{1}{(i+1)(i+2)} \left[1 + \frac{2(-1)^i}{i!} \right] x^i. \tag{6.337}$$

The derivative has the expansion

$$g_2'(x) = \sum_{i=0}^{\infty} \frac{i+1}{(i+2)(i+3)} \left[1 - \frac{2(-1)^i}{(i+1)!} \right] x^i = \sum_{i=0}^{\infty} a_i x^i, \tag{6.338}$$

where $a_0 = -\frac{1}{6} < 0$, $a_1 = \frac{1}{3} > 0$, and $a_i > 0$, $i = 2, 3, \ldots$, as well, because $\left| \frac{2(-1)^i}{(i+1)!} \right| < 1$ for all $i \geq 2$. This means that $g_2'(0) < 0$ and g_2' increases on $(0, 1)$, where it is well defined. It follows that g_2 either always decreases on $(0, 1)$ or it first decreases and then increases there. Anyway, the global maximum is attained at either of the end-points. Consequently,

$$B_2(\alpha) = \frac{-1}{2g_2(1 - e^{-\alpha})}$$

attains its supremum either at 0 or at $+\infty$, because functions $(0, \infty) \ni \alpha \mapsto x = 1 - e^{-\alpha} \in (0, 1)$ and $(0, 1) \ni x \mapsto \frac{-1}{2x} \in \left(-\infty, -\frac{1}{2}\right)$ are increasing. We easily check that

$$\lim_{\alpha \to \infty} B_2(\alpha) = -\frac{e}{4} > \lim_{\alpha \to 0} B_2(\alpha) = -1.$$

Using the transformation $x = 1 - e^{-\alpha}$ for $n = 3$, we get

$$B_3(-\ln(1-x)) = \frac{2}{3}(3-x)B_2(x) = -\frac{1}{3}\frac{3-x}{g_2(x)}.$$

We verify variability of $g_3(x) = \frac{g_2(x)}{3-x}$ analyzing the sign of

$$g_3'(x)(3-x)^2 = g_2'(x)(3-x) + g_2(x).$$

Applying (6.337) and (6.338), by elementary calculations, we obtain the Taylor expansion of

$$g_3'(x)(3-x)^2 = \sum_{i=1}^{\infty}(b_i + c_i)x^i,$$

where

$$b_i = \frac{3(i+1)}{(i+2)(i+3)} - \frac{i+2}{(i+3)(i+4)} + \frac{1}{(i+1)(i+2)} = \frac{2i^3 + 14i^2 + 26i + 20}{(i+1)(i+2)(i+3)(i+4)}$$

and

$$c_i = \frac{2(-1)^i}{(i+4)!}[(i+3)(i+4) - 3(i+1)(i+4) - i - 2] = \frac{2(-1)^{i+1}}{(i+4)!}(2i^2 + 9i + 2)$$

change the sign. We easily check that

$$\frac{|c_i|}{b_i} = \frac{2i^2 + 9i + 2}{i^3 + 7i^3 + 13i + 10}\frac{1}{i!} < \frac{7i^3 + 13i + 10}{i^3 + 7i^3 + 13i + 10}\frac{1}{i!} \le 1,$$

which implies that $b_i + c_i > 0$, $i = 1, 2, \ldots$, and so the derivative of g_3 is positive. Accordingly,

$$B_3(\alpha) = -\frac{1}{3}\frac{1}{g_3(1-e^{-\alpha})}$$

is increasing.

We apply the claim for establishing the same conclusion for $n \ge 4$. Consider the functions

$$M_n(x) = x^n - nx + n - 1, \qquad 0 < x < 1,\ n \ge 3,$$

with the derivatives

$$M'_n(x) = n(x^{n-1} - 1),$$

and analyze variability of $\frac{M_{n+1}(x)}{M_n(x)}$, $n \geq 3$. We have

$$\begin{aligned}
M_n^2(x)\,\frac{d}{dx}\,\frac{M_{n+1}(x)}{M_n(x)} &= M'_{n+1}(x)M_n(x) - M_{n+1}(x)M'_n(x) \\
&= (n+1)(x^n - 1)[x^n - nx + n - 1] \\
&- n(x^{n-1} - 1)[x^{n+1} - (n+1)x + n] \\
&= x^{2n} - n^2x^{n+1} + 2(n^2 - 1)x^n - n^2x^{n-1} + 1 \\
&= x^{2n} - 2x^n + 1 - n^2x^{n-1}(x^2 - 2x + 1) \\
&= (1 - x^n)^2 - n^2x^{n-1}(1 - x)^2 \\
&= (1 - x)^2\left[\left(\sum_{i=0}^{n-1} x^i\right)^2 - nx^{n-1}\right]. \qquad (6.339)
\end{aligned}$$

Since

$$\left(\sum_{i=0}^{n-1} x^i\right)^2 = \sum_{i=0}^{n-2}(i + 1)(x^i + x^{2n-2-i}) + nx^{n-1}$$

and

$$n^2x^{n-1} = \sum_{i=0}^{n-2} 2(i + 1)x^{n-1} + nx^{n-1},$$

we can rewrite the expression inside the brackets of (6.339) as

$$\begin{aligned}
\left(\sum_{i=0}^{n-1} x^i\right)^2 - nx^{n-1} &= \sum_{i=0}^{n-2}(i + 1)(x^i - 2x^{n-1} + x^{2n-2-i}) \\
&= \sum_{i=0}^{n-2}(i + 1)x^i(1 - x^{n-1-i})^2 > 0.
\end{aligned}$$

This shows that the ratios $\frac{M_{n+1}(x)}{M_n(x)}$, $0 < x < 1$, are increasing for all $n \geq 3$, and $\frac{M_{n+1}(e^{-\alpha})}{M_n(e^{-\alpha})}$, $n \geq 3$, are decreasing with respect to $\alpha > 0$. Moreover,

$$M_n(e^{-\alpha}) = e^{-n\alpha} - ne^{-\alpha} + n - 1 = -n\alpha\mathbb{E}_\alpha(Y_{1:n} - Y_1) > 0.$$

We finally observe that

$$B_n(\alpha) = -\frac{1}{n}\frac{M_n(e^{-\alpha})}{D(1-e^{-\alpha})} = \frac{3}{n}\frac{M_n(e^{-\alpha})}{M_3(e^{-\alpha})}B_3(\alpha)$$
$$= \frac{3}{n}B_3(\alpha)\prod_{i=3}^{n-1}\frac{M_{i+1}(e^{-\alpha})}{M_i(e^{-\alpha})}, \qquad n \geq 4,$$

are negative increasing, because so is $B_3(\alpha)$, and each $\frac{M_{i+1}(e^{-\alpha})}{M_i(e^{-\alpha})}$, $i = 3, \ldots, n-1$, is positive decreasing.

Summing up, for every $n \geq 2$, we conclude that

$$\sup_{\alpha>0} B_n(\alpha) = \lim_{\alpha\to\infty} B_n(\alpha) = \lim_{x\to 0}\frac{1-n+nx-x^n}{2e^{-x-1}-x-x^2+x\ln x} = -\frac{n-1}{2n}e.$$

Distribution functions (6.335) which attain the bound in the limit represent the mixtures of the exponential distributions with intensity α truncated right to the interval $[0, 1]$ with probability $1-e^{-\alpha}$, and atom at 1 with probability $e^{-\alpha}$. Making a standard location-scale transformation, we obtain the distributions with arbitrarily chosen real μ and positive σ_1 which attain the bound in the limit. They are precisely described in the statement of Theorem 6.42. □

There is no known analog of Theorems 6.41 and 6.42 in the IRFR case.

6.4 DDA, DFRA, IDA and IFRA Distributions

6.4.1 General L-Statistics

We first consider bounds on expected L-statistics from samples with distribution functions F which are in the star order relation $F \succeq_* W$ with a given distribution function W. The distributions are characterized by the starshapedness property of the compositions $F^{-1} \circ W$. If F has finite mean μ and variance σ^2, then the compositions belong to the Hilbert space $L^2((a, d), W(dx))$ of square integrable with respect to W functions over the support (a, d) of W. Existence of finite variance of W implies that linear functions belong to $L^2((a, d), W(dx))$.

Functions $F^{-1} \circ W$ compose a convex cone

$$\mathcal{C}_{\succeq_* W} = \{g \in L^2((a, d), W(dx)) : g \text{ is nondecreasing and starshaped}\} \quad (6.340)$$

in the Hilbert space $L^2((a, d), W(dx))$ and the functions centered about μ are additionally orthogonal to constants. Using arguments of Sect. 6.1 (see Theorem 6.1), we can represent sharp positive bounds on the expectations of arbitrary L-statistics from the family of distributions satisfying $F \succeq_* W$ by means of projections of relevant functions onto (6.340). Let $P_{\succeq_* W}$ denote the operator of projection onto (6.340).

Theorem 6.43 *If $X_1, \ldots, X_n$ are iid with cdf $F \succeq_* W$ and with finite variance, then for arbitrary $\tilde{c} = (c_1, \ldots, c_n) \in \mathbb{R}^n$,*

$$\mathbb{E} \sum_{i=1}^{n} c_i \frac{X_{i:n} - \mu}{\sigma} \leq \sqrt{\|P_{\succeq_* W}(f_{\tilde{c}:n} \circ W)\|_W^2 - \left(\sum_{i=1}^{n} c_i\right)^2} \tag{6.341}$$

and if the RHS is positive, then an equality is attained iff

$$\frac{F^{-1} \circ W(x) - \mu}{\sigma} = \frac{P_{\succeq_* W}(f_{\tilde{c}:n} \circ W)(x) - \sum_{i=1}^{n} c_i}{\sqrt{\|P_{\succeq_* W}(f_{\tilde{c}:n} \circ W)\|_W^2 - (\sum_{i=1}^{n} c_i)^2}}. \tag{6.342}$$

Similar general results can be stated for samples with $F \preceq_* W$. Let

$$\mathcal{C}_{\preceq_* W} = \{g \in L^2((a,d), W(dx)) : g \text{ is nondecreasing and antistarshaped}\}, \tag{6.343}$$

and $P_{\preceq_* W}$ denote the respective projection operator. Then for all $F \preceq_* W$, the statements of Theorem 6.43 hold with $P_{\succeq_* W}$ replaced by $P_{\preceq_* W}$. Also, the counterpart of (6.342) is valid.

Note that the convex cones (6.340) and (6.343) contain the respective cones generated by the convex order. Therefore, bounds (6.341) are not less than (6.6) and equivalently (6.17), and they are strictly greater unless the respective projection $P_{\succeq_* W}(f_{\tilde{c}:n} \circ W) \in \mathcal{C}_{\succeq_c W}$. Families of distributions determined by the star order are of special interest in two cases $W = U$ and Z. In the former one, $F \succeq_* U$ means that F has an interval support with a finite left-end support point a_F, say, possibly an atom at a_F and a density function f which is nonincreasing on average which means that the function

$$x \mapsto \frac{1}{x - a_F} \int_{a_F}^{x} f(t)dt = \frac{F(x) - F(a_F)}{x - a_F} \tag{6.344}$$

is nonincreasing. Relation $F \preceq_* U$ implies that $F(a_F) = 0$ and (6.344) is nondecreasing on its support interval, and F may have an atom at the finite right end of the support. Intuitively, $F \succeq_* U$ and $F \preceq_* U$ means that the concentration of probability mass has a tendency to decrease and increase, respectively, although the density function is not necessarily monotone in its whole domain. Formally, the respective order relations mean that the distribution function is antistarshaped and starshaped, respectively. Moreover, repeating the arguments from the beginning of Sect. 6.1, we can claim that if the sharp bound in (6.341) is attained by some discontinuous $F_{\succeq_c} W$ ($F_{\preceq_c} W$) with $W = U$, then there exist sequences of absolutely continuous distributions for which (6.344) strictly decrease (increase), respectively, and tend weakly to F which means that their standardized expectations of L-statistics tend to the sharp bound. Therefore, for simplicity, we say that the distribution has either the decreasing or increasing density on average property (DDA and IDA, for short) if (6.344) is nonincreasing and nondecreasing, respectively. If $F \succeq_* Z$ and $F \preceq_* Z$ which means that

$$\bar{\lambda}_F(x) = \frac{1}{x - a_F} \int_{a_F}^{x} \frac{f(t)}{1 - F(t)} dt = \frac{\Lambda_F(x) - \Lambda_F(a_F)}{x - a_F} \tag{6.345}$$

is nonincreasing and nondecreasing, respectively, we say that F has decreasing and increasing failure rate on average (DFRA and IFRA, respectively). Formally, nonincrease and nondecrease of (6.345) mean that the cumulative hazard rate is antistarshaped and starshaped, respectively. The notion of monotone failure rate on average is well established in reliability theory.

6.4.2 *Positive Upper Bounds for Order Statistics and Spacings from DDA and DFRA Families*

Below we determine the bounds for simple order statistics based on the decreasing density and failure rate on average distributions, established in Rychlik (2002). Then we present and analogous results for spacings which are due to Danielak and Rychlik (2003b). The original results for the single order statistics are simplified here by the use of Lemma 6.1 presented in Sect. 6.1. For the problem of single order statistics from DDA and DFRA samples, we first mention some implications of Theorem 6.3 and 6.4. In the DD case with $j > \frac{n+1}{2}$, the DD bounds defined in (6.54)–(6.56) are positive, and so are the DDA bounds. These could not be smaller than the former ones, because they concern a wider class of distributions. For $j \le \frac{n+1}{2}$, we do not have positive bounds for DD distributions, because the uniform distributions belong to the class and $\mathbb{E}U_{j:n} \le \mu$. However, there are possible positive bounds for the DDA distributions then. Similarly, for the DFRA distributions under the condition $S(j, n) > 1$, the bounds are necessarily positive, but this is also possible when $S(j, n) < 1$.

In following we consider functions $h = f_{j:n} \circ W - 1$, $W = U, Z$, which belong to the family of functions which are bounded, twice differentiable, satisfy $\int_0^d h(x)W(dx) = 0$ and have the following shapes: they are strictly decreasing on $(0, a)$, strictly convex increasing on (a, b), strictly concave increasing on (b, c) and strictly decreasing on (c, d) for some $0 \le a \le b < c \le d \le \infty$. Our aim is to determine the projections of the above functions onto the cone (6.340). The above conditions coincide with the assumptions (A2) of Sect. 6.1 with the extension $0 \le a \le b$. So, we call them (A2'). Admitting $a > 0$ allows us to include spacings to our studies. Taking $a = b = 0$, we cover the cases of second order statistics $X_{2:n}$ and first spacing $X_{2:n} - X_{1:n}$

The projection of function satisfying (A2') onto the convex cone of nondecreasing starshaped function is described with use of the following functions:

$$K(\alpha, \beta) = \int_\beta^d \left[h(\alpha) + \frac{h(\beta) - h(\alpha)}{\beta} x - h(x) \right] x \, W(dx), \tag{6.346}$$

$$L(\alpha, \beta) = \int_0^\alpha [h(\alpha) - h(x)] W(dx) + \int_\beta^d \left[h(\alpha) + \frac{h(\beta) - h(\alpha)}{\beta} x - h(x) \right] W(dx) \tag{6.347}$$

for $a \le \alpha \le \beta \le c,$, $\beta \ne 0$, and

$$K(\alpha) = K(\alpha, \alpha) = \int_\alpha^d [h(\alpha) - h(x)] x \, W(dx), \tag{6.348}$$

$$L(\alpha) = L(\alpha, \alpha) = \int_0^d [h(\alpha) - h(x)] W(dx) = h(\alpha) \tag{6.349}$$

for $a \le \alpha \le c$. Obviously (6.349) is strictly increasing on $[a, c]$ and $L(a) < 0 < L(c)$.

Theorem 6.44 *Assume* (A2'). *If either* $K(0) \ge 0$ *or* $K(0) < 0$ *and*

$$K(\alpha_L) = -\int_{\alpha_L}^d h(x) x \, W(dx) \ge 0, \tag{6.350}$$

for α_L *being the unique zero of* h *(and* L*) in* (a, c), *then* $P_{\succeq_* W} h = 0$.

Otherwise, there exists a unique pair $a \le \alpha_* < c$, $\alpha_* < \beta_* < \bar{\beta}(\alpha_*)$ *solving the equations*

$$K(\alpha, \beta) = 0, \tag{6.351}$$

$$L(\alpha, \beta) = 0, \tag{6.352}$$

such that

$$P_{\succeq_* W} h(x) = \begin{cases} h(\alpha_*), & 0 \le x \le \alpha_*, \\ h(x), & \alpha_* \le x \le \beta_*, \\ h(\alpha_*) + \frac{h(\beta_*) - h(\alpha_*)}{\beta_*} x, & \beta_* \le x < d. \end{cases} \tag{6.353}$$

The projection function is first possibly constant, then coincides with the original function, and finally linear. The linear part has the minimal slope that admits to preserve the starshapedness of the projection. The proof of Theorem 6.44 is rather long and it is postponed to the Appendix (see Theorem A.5). It is presented in Rychlik (2002). Now we apply Theorem 6.44 for describing the bounds on order statistics from DDA and DFRA populations. The results presented below are based on projections of the form (6.353).

Theorem 6.45 (DDA Distributions) *Assume that $X_1, \ldots, X_n$ are iid random variables with antistarshaped distribution function F, finite mean and variance, and take $3 \le j \le n-1$. Let $\alpha_0 = \alpha_0(j,n)$ denote the first of two solutions to the equation $f_{j:n}(x) = 1$ in the interval $[0,1]$. If*

$$2j[1 - F_{j+1:n+1}(\alpha_0)] \le (n+1)(1-\alpha_0^2), \tag{6.354}$$

then

$$\mathbb{E}\frac{X_{j:n}-\mu}{\sigma} \le 0. \tag{6.355}$$

Otherwise, there exists a unique pair $0 < \alpha_* < \alpha_0$ and $\alpha_* < \beta_* < \frac{j-1}{n-1}$ solving the equations

$$\frac{1-\beta^3}{3\beta} f_{j:n}(\beta) - \frac{(1-\beta)^2(2+\beta)}{6\beta} f_{j:n}(\alpha) - \frac{j}{n+1}[1 - F_{j+1:n+1}(\beta)] = 0, \tag{6.356}$$

$$\frac{1-\beta^2}{2\beta} f_{j:n}(\beta) + \left[\alpha - \frac{(1-\beta)^2}{2\beta}\right] f_{j:n}(\alpha) - 1 + F_{j:n}(\beta) - F_{j:n}(\alpha) = 0. \tag{6.357}$$

As a consequence, we have the inequality

$$\mathbb{E}\frac{X_{j:n}-\mu}{\sigma} \le B, \tag{6.358}$$

with

$$\begin{aligned} B^2 = {} & \left[\alpha_* + \frac{(1-\beta_*)^3}{3\beta_*^2}\right] f_{j:n}^2(\alpha_*) + \frac{1-\beta_*^3}{3\beta_*^2} f_{j:n}^2(\beta_*) \\ & + \frac{(n!)^2}{(2n-1)!}\binom{2j-2}{j-1}\binom{2n-2j}{n-j}[F_{2j-1:2n-1}(\beta_*) - F_{2j-1:2n-1}(\alpha_*)] \\ & - \frac{(1-\beta_*)^2(2+\beta_*)}{3\beta_*^2} f_{j:n}(\alpha_*) f_{j:n}(\beta_*) - 1. \end{aligned} \tag{6.359}$$

The equality in (6.358) *holds iff the parent distribution function has the form*

$$F(x) = \begin{cases} 0, & \frac{x-\mu}{\sigma} < a_1, \\ f_{j:n}^{-1}\left(\frac{x-\mu}{\sigma}B + 1\right), & a_1 \le \frac{x-\mu}{\sigma} < a_2, \\ \frac{\beta_*}{f_{j:n}(\beta_*)-f_{j:n}(\alpha_*)}\left[\frac{x-\mu}{\sigma}B + 1 - f_{j:n}(\alpha_*)\right], & a_2 \le \frac{x-\mu}{\sigma} < a_3, \\ 1, & \frac{x-\mu}{\sigma} > a_3, \end{cases} \tag{6.360}$$

for $a_1 = \frac{f_{j:n}(\alpha_*)-1}{B}$, $a_2 = \frac{f_{j:n}(\beta_*)-1}{B}$ *and* $a_3 = \frac{f_{j:n}(\beta_*)-(1-\beta_*)f_{j:n}(\alpha_*)-\beta_*}{\beta_* B}$.

Obviously, relation (6.355) holds for the first and second order statistics as well. Note that condition (6.354) is more restrictive than $j \le \frac{n+1}{2}$ which assures inequality (6.355) in the decreasing density family. It follows from the fact that the DDA class of distributions contains the DD family, and the upper bounds of Theorem 6.45 cannot be less than their counterparts from Theorem 6.3.

Theorem 6.46 (DFRA Distributions) *Let* $X_1, \ldots, X_n$ *be iid with antistarshaped cumulative failure rate, finite mean* μ *and variance* σ^2. *Take* $3 \le j \le n$. *If for* $\alpha_0 = \alpha_0(j, n)$ *defined in Theorem* 6.45 *the following inequality*

$$(1+\alpha_0)e^{-\alpha_0} \ge \sum_{k=1}^{j} \frac{F_{n+1-k:n}(e^{-\alpha_0})}{n+1-k} + \alpha_0 F_{n+1-j:n}(e^{-\alpha_0}) \tag{6.361}$$

holds, then we have the inequality (6.355).

Otherwise, there exists a unique pair $0 < \alpha_* < -\ln(1-\alpha_0)$, $\alpha_* < \beta_* < -\ln\frac{n-j}{n-1}$ *solving the equations*

$$\frac{(\beta+1)^2+1}{\beta e^{\beta}} f_{j:n} \circ Z(\beta) - \frac{\beta+2}{\beta e^{\beta}} f_{j:n} \circ Z(\alpha)$$
$$- \sum_{i=1}^{j} \frac{1-F_{i:n} \circ Z(\beta)}{n+1-i} - \beta[1 - F_{j:n} \circ Z(\beta)] = 0, \tag{6.362}$$

$$\frac{\beta+1}{\beta e^{\beta}} f_{j:n}\circ Z(\beta) + \left[1 - \frac{1}{e^{\alpha}} - \frac{1}{\beta e^{\beta}}\right] f_{j:n}\circ Z(\alpha) - 1 + F_{j:n}\circ Z(\beta) - F_{j:n}\circ Z(\alpha) = 0, \tag{6.363}$$

and

$$\mathbb{E}\frac{X_{j:n}-\mu}{\sigma} \le B \tag{6.364}$$

for

$$
\begin{aligned}
B^2 &= \left[1 - \frac{1}{e^{\alpha_*}} + \frac{2}{\beta_*^2 e^{\beta_*}}\right] f_{j:n}^2 \circ Z(\alpha_*) + \frac{2 + 2\beta_* + \beta_*^2}{\beta_*^2 e^{\beta_*}} f_{j:n}^2 \circ Z(\beta_*) \\
&+ \frac{(n!)^2}{(2n-1)!}\binom{2j-2}{j-1}\binom{2n-2j}{n-j}[F_{2j-1:2n-1} \circ Z(\beta_*) - F_{2j-1:2n-1} \circ Z(\alpha_*)] \\
&- 2\frac{\beta_* + 2}{\beta_*^2 e^{\beta_*}} f_{j:n} \circ Z(\alpha_*) f_{j:n} \circ Z(\beta_*) - 1. \qquad (6.365)
\end{aligned}
$$

Inequality (6.364) *becomes the equality if*

$$
F(x) = \begin{cases} 0, & \frac{x-\mu}{\sigma} < a_4, \\ f_{j:n}^{-1}\left(\frac{x-\mu}{\sigma}B + 1\right), & a_4 \le \frac{x-\mu}{\delta} < a_5, \\ 1 - \exp\left(-\beta_* \frac{\frac{x-\mu}{\sigma}B + 1 - f_{j:n}(Z(\alpha_*))}{f_{j:n}(Z(\beta_*)) - f_{j:n}(Z(\alpha_*))}\right), & \frac{x-\mu}{\sigma} \ge a_5, \end{cases} \qquad (6.366)
$$

for $a_4 = \frac{f_{j:n} \circ Z(\alpha_*) - 1}{B}$ *and* $a_5 = \frac{f_{j:n} \circ Z(\beta_*) - 1}{B}$.

The proofs of Theorems 6.45 and 6.46 involves direct applications of Theorem 6.44 for particular functions $h = f_{j:n} - 1$ and $h = f_{j:n} \circ Z - 1$. Danielak (2003) presented generalizations of the above results to the trimmed means. Their structures are similar, but obviously the formulae are more sophisticated. We do not present them here.

In the case of spacings, the general projection problem has a similar solution. However, constant projections and zero bounds are not included here. Setting $h = f_{j,j+1:n} \circ W = f_{j+1:n} \circ W - f_{j:n} \circ W$, $W = U, Z$, we conclude the following statements.

Theorem 6.47 (DDA Distributions) *Assume that* $X_1, \ldots, X_n$ *are iid with a distribution function* $F \succeq_* U$ *and finite variance, and* $1 \le j \le n-2$. *Then,*

$$
\mathbb{E}\frac{X_{j+1:n} - X_{j:n}}{\sigma} \le B \qquad (6.367)
$$

for

$$
\begin{aligned}
B^2 = &\left(\alpha_* + \frac{1-\beta_*^3}{3\beta_*^2}\right) f_{j,j+1:n}^2(\alpha_*) + \frac{1-\beta_*^3}{3\beta_*^2} f_{j,j+1:n}^2(\beta_*) \\
&- \frac{\beta_*^3 - 3\beta_* + 2}{3\beta_*^2} f_{j,j+1:n}(\alpha_*) f_{j,j+1:n}(\beta_*) \\
&+ \frac{n}{\binom{2n-1}{n}} \left\{ \binom{2j}{j}\binom{2(n-j-1)}{n-j-1} [F_{2j+1:2n-1}(\beta_*) - F_{2j+1:2n-1}(\alpha_*)] \right. \\
&- 2\binom{2j-1}{j}\binom{2n-2j-1}{n-j} [F_{2j:2n-1}(\beta_*) - F_{2j:2n-1}(\alpha_*)] \\
&\left. + \binom{2j-2}{j-1}\binom{2n-2j}{n-j} [F_{2j-1:2n-1}(\beta_*) - F_{2j-1:2n-1}(\alpha_*)] \right\} \qquad (6.368)
\end{aligned}
$$

where $0 \le \alpha_* < \theta = \frac{j}{n}$, $\alpha_* < \beta_* < \bar{\beta}(\alpha_*)$ *are uniquely determined by the system of equations*

$$
\begin{aligned}
&2(1-\beta^3) f_{j,j+1:n}(\beta) + (1-\beta)^2(2+\beta) f_{j,j+1:n}(\alpha) \\
&= \frac{6\beta}{n+1}[1 + jF_{j+1:n+1}(\beta) - (j+1)F_{j+2:n+1}(\beta)] \qquad (6.369) \\
&[2\alpha\beta - (1-\beta)^2] f_{j,j+1:n}(\alpha) + (1-\beta^2) f_{j,j+1:n}(\beta) \\
&= 2\beta[F_{j:n}(\beta) - F_{j+1:n}(\beta) + F_{j:n}(\alpha) - F_{j+1:n}(\alpha)]. \qquad (6.370)
\end{aligned}
$$

The equality in (6.367) *occurs when*

$$
F(x) = \begin{cases} 0, & \frac{x-\mu}{\sigma} < a_6, \\ f_{j,j+1:n}^{-1}\left(B\frac{x-\mu}{\sigma}\right), & a_6 \le \frac{x-\mu}{\sigma} < a_7, \\ \frac{\beta_*[B\frac{x-\mu}{\sigma} - f_{j,j+1:n}(\alpha_*)]}{f_{j,j+1:n}(\beta_*) - f_{j,j+1:n}(\alpha_*)}, & a_7 \le \frac{x-\mu}{\sigma} < a_8, \\ 1, & \frac{x-\mu}{\delta} \ge a_8, \end{cases} \qquad (6.371)
$$

for $a_6 = \frac{f_{j,j+1:n}(\alpha_*)}{B}$, $a_7 = \frac{f_{j,j+1:n}(\beta_*)}{B}$ *and* $a_8 = \frac{f_{j,j+1:n}(\beta_*) - (1-\beta_*) f_{j,j+1:n}(\alpha_*)}{\beta_* B}$.

Theorem 6.48 (DFRA Distributions) *If* $X_1, \ldots, X_n$ *are iid with a distribution function* $F \succeq_* Z$ *and finite variance, and* $1 \le j \le n-1$, *then*

$$
\mathbb{E}\frac{X_{j+1:n} - X_{j:n}}{\sigma} \le B \qquad (6.372)
$$

for

$$
\begin{aligned}
B^2 &= \left[1 - \frac{1}{e^{\alpha_*}} + \frac{2}{\beta_*^2 e^{\beta_*}}\right] f_{j,j+1:n}^2 \circ Z(\alpha_*) + \frac{(\beta_*+1)^2+1}{\beta_*^2 e^{\beta_*}} f_{j,j+1:n}^2 \circ Z(\beta_*) \\
&- 2\frac{\beta_*+2}{\beta_*^2 e^{\beta_*}} f_{j,j+1:n} \circ Z(\alpha_*) f_{j,j+1:n} \circ Z(\beta_*) + \frac{n}{\binom{2n-1}{n}} \\
&\times \left\{ \binom{2j}{j}\binom{2(n-j-1)}{n-j-1} [F_{2j+1:2n-1} \circ Z(\beta_*) - F_{2j+1:2n-1} \circ Z(\alpha_*)] \right. \\
&- 2\binom{2j-1}{j}\binom{2n-2j-1}{n-j} [F_{2j:2n-1} \circ Z(\beta_*) - F_{2j:2n-1} \circ Z(\alpha_*)] \\
&\left. + \binom{2j-2}{j-1}\binom{2n-2j}{n-j} [F_{2j-1:2n-1} \circ Z(\beta_*) - F_{2j-1:2n-1} \circ Z(\alpha_*)] \right\},
\end{aligned} \tag{6.373}
$$

where $0 \le \alpha_* < -\ln\left(1 - \frac{j}{n}\right)$, $\alpha_* < \beta_* < \bar{\beta}(\alpha_*)$ *are uniquely defined by the equations*

$$
\begin{aligned}
&\frac{(\beta+1)^2+1}{\beta e^{\beta}} f_{j,j+1:n} \circ Z(\beta) + \frac{\beta+2}{\beta e^{\beta}} f_{j,j+1:n} \circ Z(\alpha) \\
&= \frac{1}{n-j} F_{n-j:n}(e^{-\beta}) + \beta[F_{n-j:n}(e^{-\beta}) - F_{n+1-j:n}(e^{-\beta})],
\end{aligned} \tag{6.374}
$$

$$
\begin{aligned}
&\left[1 - \frac{1}{e^{\alpha}} + \frac{1}{\beta e^{\beta}}\right] f_{j,j+1:n} \circ Z(\alpha) + \frac{\beta+1}{\beta e^{\beta}} f_{j,j+1:n} \circ Z(\beta) \\
&= F_{j:n} \circ Z(\beta) - F_{j+1:n} \circ Z(\beta) - F_{j:n} \circ Z(\alpha) + F_{j+1:n} \circ Z(\alpha).
\end{aligned} \tag{6.375}
$$

The distribution function providing the equality in (6.372) has the form

$$
F(x) = \begin{cases} 0, & \frac{x-\mu}{\sigma} < a_9, \\ f_{j,j+1:n}^{-1}\left(B\frac{x-\mu}{\sigma}\right), & a_9 \le \frac{x-\mu}{\sigma} < a_{10}, \\ 1 - \exp\left(-\frac{\beta_*[B\frac{x-\mu}{\sigma} - f_{j,j+1:n} \circ Z(\alpha_*)]}{f_{j,j+1:n} \circ Z(\beta_*) - f_{j,j+1:n} \circ Z(\alpha_*)}\right), & \frac{x-\mu}{\sigma} \ge a_{10}, \end{cases} \tag{6.376}
$$

for $a_9 = \frac{f_{j,j+1:n} \circ Z(\alpha_*)}{B}$ *and* $a_{10} = \frac{f_{j,j+1:n} \circ Z(\beta_*)}{B}$.

The case of last spacings from the DDA populations did not get included here. For $W = U$ and $n = 2$ we have $h(x) = 4x - 2$. Then $h \in \mathcal{C}_{\succeq_c U} \subset \mathcal{C}_{\succeq_* U} \subset \mathcal{C}^{\nearrow}$, and we have the well-known Plackett bound for the range of sample size $n = 2$ that is attained by the uniform distribution. For the remaining cases $W = U$, $j = n - 1 \ge 2$, the general bounds derived by means of Moriguti and Schwarz inequalities are attained by a distribution satisfying $F \succeq_* U$. We finally point out that the results of

Sects. 4.7 and 4.9 for spacings can be extended to the differences $X_{j+2:n} - X_{j:n}$, because $h = f_{j,j+2:n}$ and $h = f_{j,j+2:n} \circ Z$ satisfy conditions (A2) as well. This is not true for the differences of more distant order statistics, because respective h have more than one inflection point in (a, c).

6.4.3 Nonpositive Upper Bounds for Order Statistics from DDA and DFRA Families

Now we present nonpositive upper bounds on the deviations of the expectations of the small rank order statistics from the population means when the parent distributions belong to the DDA and DFRA families. We first concentrate on the DDA case. For all $1 \le j \le n-1$ and $p \ge 1$, we consider two functions

$$K_{j:n}(\alpha) = \frac{j}{n+1}[1 - F_{j+1:n+1}(\alpha)] - \frac{1-\alpha^2}{2}, \tag{6.377}$$

$$M_p(\alpha) = \left[\frac{2(p+1)\alpha(1-\alpha)^p + (1+\alpha^2)^{p+1} - (\alpha^2+2\alpha-1)|\alpha^2+2\alpha-1|^p}{2^{p+1}(p+1)}\right]^{\frac{1}{p}} \tag{6.378}$$

on the interval $(0, 1)$. For $2 \le j \le \frac{n}{2}$, let $a_{j:n}$ denote the smaller of two roots of equation $f_{j:n}(x) = 1$ in $(0, 1)$.

Theorem 6.49 (DDA Distributions) *Let $X_1, \ldots, X_n$ be idd with a distribution succeeding the uniform one in the star ordering and finite moment of order $p > 1$. If either $j = 1$ or $K_{j:n}(a_{j:n}) < 0$ for $2 \le j \le \frac{n}{2}$, then the bound*

$$\frac{\mathbb{E}_F X_{j:n} - \mu}{\sigma_p} \le 0 \tag{6.379}$$

is sharp, and it is attained in the limit by the family of mixtures of atoms at $\mu - \frac{\sigma_p}{2M_p(\alpha)}(1-\alpha^2)$ and uniform distributions on the intervals

$$\left[\mu + \frac{\sigma_p}{2M_p(\alpha)}(2\alpha - 1 + \alpha^2),\ \mu + \frac{\sigma_p}{2M_p(\alpha)}(1+\alpha^2)\right]$$

with weights α and $1-\alpha$, respectively, as $\alpha \to 1$.

Proof Obviously $\mathbb{E}X_{1:n} \le \mu$ holds with no assumptions. Theorem 6.45 asserts that this inequality is valid for $j \ge 2$ and DDA populations when $K_{j:n}(a_{j:n}) < 0$ holds (see (6.354)). Note that the condition is stronger than $j < \frac{n+1}{2}$. For $j \ge \frac{n+1}{2}$, there exist DD distributions for which $\mathbb{E}X_{j:n} \ge \mu$ which are the members of the DDA family as well. Therefore $K_{j:n}(a_{j:n}) \ge 0$ then by Theorem 6.45 again. Accordingly, it suffices to show that under the conditions of Theorem 6.49 inequality (6.379) is optimal, and is attained in the limit by the family of distributions described there. The respective quantile functions satisfy

$$\frac{F_\alpha^{-1}(x)-\mu}{\sigma_p} = \frac{x\,\mathbf{1}_{[\alpha,1)}(x) - \frac{1-\alpha^2}{2}}{M_p(\alpha)}, \qquad 0 \le x < 1. \tag{6.380}$$

Therefore,

$$\frac{F_\alpha^{-1}(x)-\mu+\frac{\sigma_p}{2M_p(\alpha)}(1-\alpha^2)}{x} = \frac{\sigma_p}{M_p(\alpha)}\mathbf{1}_{[\alpha,1)}(x), \qquad 0 \le x < 1,$$

are non-decreasing, and so F_α, $0 \le \alpha < 1$, are DDA. Since

$$\int_0^1 [F_\alpha^{-1}(x)-\mu]dx = \frac{\sigma_p}{M_p(\alpha)}\left[\int_0^1 x\mathbf{1}_{[\alpha,1)}(x)\,dx - \frac{1-\alpha^2}{2}\right] = 0,$$

we have $\mathbb{E}_\alpha X_1 = \mu$. Now we calculate

$$\int_0^1 \left| x\mathbf{1}_{[\alpha,1)}(x) - \frac{1-\alpha^2}{2}\right|^p dx = \int_0^\alpha \left(\frac{1-\alpha^2}{2}\right)^p dx$$
$$+ \begin{cases} \int_\alpha^{\frac{1-\alpha^2}{2}} \left(\frac{1-\alpha^2}{2} - x\right)^p dx + \int_{\frac{1-\alpha^2}{2}}^1 \left(x - \frac{1-\alpha^2}{2}\right)^p dx, & 0 \le \alpha \le \sqrt{2}-1, \\ \int_\alpha^1 \left(x - \frac{1-\alpha^2}{2}\right)^p dx, & \sqrt{2}-1 \le \alpha < 1, \end{cases}$$
$$= \frac{(1-\alpha^2)^p\alpha}{2^p} + \begin{cases} \frac{(1-\alpha^2-2\alpha)^{p+1}+(1+\alpha^2)^{p+1}}{2^{p+1}(p+1)}, & 0 \le \alpha \le \sqrt{2}-1, \\ \frac{(1+\alpha^2)^{p+1}-(1-\alpha^2-2\alpha)^{p+1}}{2^{p+1}(p+1)}, & \sqrt{2}-1 \le \alpha < 1, \end{cases} = M_p^p(\alpha).$$

Therefore,

$$\begin{aligned} \mathbb{E}_\alpha |X_1-\mu|^p &= \int_0^1 |F_\alpha^{-1}(x)-\mu|^p dx \\ &= \frac{\sigma_p^p}{M_p^p(\alpha)} \int_0^1 \left| x\mathbf{1}_{[\alpha,1)}(x) - \frac{1-\alpha^2}{2}\right|^p dx = \sigma_p^p, \end{aligned}$$

as claimed. Furthermore,

$$\frac{\mathbb{E}_\alpha X_{j:n} - \mu}{\sigma_p} = \int_0^1 \frac{F_\alpha^{-1}(x) - \mu}{\sigma_p} f_{j:n}(x)dx = \int_0^1 \frac{x\mathbf{1}_{[\alpha,1)}(x) - \frac{1-\alpha^2}{2}}{M_p(\alpha)} f_{j:n}(x)dx$$
$$= \frac{1}{M_p(\alpha)}\left[\int_\alpha^1 x f_{j:n}(x)dx - \frac{1-\alpha^2}{2}\right]$$
$$= \frac{1}{M_p(\alpha)}\left[\frac{j}{n+1}\int_\alpha^1 f_{j+1:n+1}(x)dx - \frac{1-\alpha^2}{2}\right] = \frac{K_{j:n}(\alpha)}{M_p(\alpha)}$$

(cf. (6.377) and (6.378)). Note that

$$K_{j:n}(\alpha) = \frac{j}{n+1}\sum_{i=0}^{j}\binom{n+1}{i}\alpha^i(1-\alpha)^{n+1-i} - \frac{1-\alpha^2}{2}$$
$$= (1-\alpha)\left[\frac{j}{n+1}\sum_{i=0}^{j}\binom{n+1}{i}\alpha^i(1-\alpha)^{n-i} - \frac{1+\alpha}{2}\right],$$

and the expression inside the square brackets tends to -1, as $\alpha \to 1$. Indeed, the sum does not converge there only for $j = n$. However, this case is excluded from our study. So we have $K_{j:n}(\alpha) \sim -(1-\alpha)$ as $\alpha \to 1$.

For $\alpha \geq \sqrt{2} - 1$, we have

$$M_p^p(\alpha) = \frac{2(p+1)\alpha(1-\alpha)^p + (1+\alpha^2)^{p+1} - (\alpha^2+2\alpha-1)^{p+1}}{2^{p+1}(p+1)}.$$

Applying the Taylor expansions, we obtain

$$(1+\alpha^2)^{p+1} = 2^{p+1} - (p+1)2^{p+1}(1-\alpha) + \mathcal{O}((1-\alpha)^2),$$
$$(\alpha^2+2\alpha-1)^{p+1} = 2^{p+1} - (p+1)2^{p+2}(1-\alpha) + \mathcal{O}((1-\alpha)^2).$$

Hence $M_p^p(\alpha) = (1-\alpha) + \mathcal{O}((1-\alpha)^{\min\{2,p\}})$, and so

$$\frac{K_{j:n}(\alpha)}{M_p(\alpha)} \sim -(1-\alpha)^{1-1/p} \to 0 \quad \text{as } \alpha \to 1.$$

which is the desired statement. □

Observe that for $p = 1$, formula (6.378) takes on the form

$$M_1(\alpha) = \frac{1}{8}\left[4\alpha(1-\alpha^2) + (1+\alpha^2)^2 - \mathrm{sgn}(\alpha^2+2\alpha-1)(\alpha^2+2\alpha-1)^2\right]$$
$$= \begin{cases} \frac{(\alpha^2+1)^2}{4}, & 0 \le \alpha \le \sqrt{2}-1, \\ \alpha(1-\alpha^2), & \sqrt{2}-1 \le \alpha < 1. \end{cases} \tag{6.381}$$

In the case $p = 1$, we have the following result.

Theorem 6.50 (DDA Distributions) *If the assumptions of Theorem* 6.49 *are satisfied with* $p = 1$, *then*

$$\frac{\mathbb{E}_F X_{j:n} - \mu}{\sigma_1} \le \sup_{0 \le \alpha < 1} \frac{K_{j:n}(\alpha)}{M_1(\alpha)} \tag{6.382}$$

(see (6.377) *and* (6.381)*). If the supremum in* (6.382) *is attained at some* $0 \le \alpha = \alpha_{j:n} < 1$, *then the bound is attained by the mixture of an atom at* $\mu - \frac{\sigma_1}{2M_1(\alpha)}(1-\alpha^2)$ *and the uniform distribution on* $\left[\mu + \frac{\sigma_1}{2M_1(\alpha)}(2\alpha - 1 + \alpha^2),\ \mu + \frac{\sigma_1}{2M_1(\alpha)}(1+\alpha^2)\right]$ *with weights* α *and* $1-\alpha$, *respectively. If*

$$\sup_{0 \le \alpha < 1} \frac{K_{j:n}(\alpha)}{M_1(\alpha)} = \lim_{\alpha \to 1} \frac{K_{j:n}(\alpha)}{M_1(\alpha)} = -\frac{1}{2},$$

then the bound is attained in the limit by the sequences of mixtures described above as $\alpha \to 1$.

The proof is based on the norm maximization method. The method allows us to verify that the maximization of the ratio $\mathbb{E}\frac{X_{j:n}-\mu}{\sigma}$ over the family of iid samples with a parent DDA distribution can be reduced to maximizing over the parametric family described by (6.380) with $p = 1$. This involves maximizing the fraction $\frac{K_{j:n}(\alpha)}{M_1(\alpha)}$ over the set of parameters $0 \le \alpha < 1$. We also easily check that the fraction tends to $-\frac{1}{2}$ as α tends to 1.

A more precise description of the result of Theorem 6.50 is due to Goroncy and Rychlik (2018) and it is presented below. For this purpose, we introduce several auxiliary notions. Take

$$\underline{M}_1(\alpha) = \frac{1}{4}(1+\alpha^2)^2, \tag{6.383}$$

$$\overline{M}_1(\alpha) = \alpha(1-\alpha^2), \tag{6.384}$$

$$\underline{B}_{j:n}(\alpha) = -\frac{K_{j:n}(\alpha)}{\underline{M}_1(\alpha)}, \tag{6.385}$$

$$\overline{B}_{j:n}(\alpha) = -\frac{K_{j:n}(\alpha)}{\overline{M}_1(\alpha)}. \tag{6.386}$$

We notice that

$$M_1(\alpha) = \begin{cases} \underline{M}_1(\alpha), \ 0 < \alpha \leq \sqrt{2} - 1, \\ \overline{M}_1(\alpha), \ \sqrt{2} - 1 \leq \alpha < 1, \end{cases} \tag{6.387}$$

which means that the negatives of the minimal values of

$$B_{j:n}(\alpha) = -\frac{K_{j:n}(\alpha)}{M_1(\alpha)} = \begin{cases} \underline{B}_{j:n}(\alpha), \ 0 < \alpha \leq \sqrt{2} - 1, \\ \overline{B}_{j:n}(\alpha), \ \sqrt{2} - 1 \leq \alpha < 1. \end{cases} \tag{6.388}$$

provide the bounds of Theorem 6.50. We study variability properties of (6.385) and (6.386) separately by analyzing the sign changes of the polynomial functions

$$\underline{L}(\alpha) = \frac{4\alpha^2 \underline{B}'(\alpha)}{1+\alpha^2} = \frac{4[K_{j:n}(\alpha)\underline{M}_1'(\alpha) - K_{j:n}'(\alpha)\underline{M}_1(\alpha)]}{1+\alpha^2}, \tag{6.389}$$

$$\overline{L}(\alpha) = \alpha^2 \overline{B}'(\alpha) = K_{j:n}(\alpha)\overline{M}_1(\alpha) - K_{j:n}'(\alpha)\overline{M}_1(\alpha), \tag{6.390}$$

which have the signs identical with those of derivatives $\underline{B}_{j:n}(\alpha)$ and $\overline{B}_{j:n}(\alpha)$, respectively, with use of the VDP property of the Bernstein polynomials. The following can be verified. If $j = 1$, then (6.389) is first positive, and then negative in (0, 1). If $2 \leq j \leq \frac{n}{2}$, then (6.389) is consecutively negative, positive and negative in (0, 1). For $1 \leq j \leq \frac{n}{2}$, the function (6.390) is negative for all $0 < \alpha < 1$ if j is small, and it is first negative, then positive, and finally negative in (0, 1) if j is large. Let $0 < \alpha_1 = \alpha_1(j, n) < \alpha_2 = \alpha_2(j, n) < 1$ stand for the only zeros of (6.389) for $j \geq 2$. Similarly, we denote zeros of (6.390) by $0 < \beta_1 = \beta_1(j, n) < \beta_2 = \beta_2(j, n) < 1$ (assuming that they exist). Using this notation, we formulate the refinement of Theorem 6.50 as follows.

Theorem 6.51 (DDA Distributions) *Assume that conditions of Theorem* 6.49 *with* $p = 1$ *hold.*

(i) If $j = 1$, *then*

$$\frac{\mathbb{E}X_{1:n} - \mu}{\sigma_1} \leq -\overline{B}_{1:n}(1-).$$

Suppose now that $j \geq 2$ *and* $K_{j:n}(a_{j:n}) > 0$.

(ii) If $\alpha_1 = \alpha_1(j, n) < \sqrt{2} - 1$ *and either* (6.390) *has no zeros in* (0, 1) *or* $\beta_1 = \beta_1(j, n) \leq \sqrt{2} - 1$, *then*

$$\frac{\mathbb{E}X_{j:n} - \mu}{\sigma_1} \leq -\min\{\underline{B}_{j:n}(\alpha_1), \overline{B}_{j:n}(1-)\}.$$

(iii) If $\alpha_1 = \alpha_1(j,n) < \sqrt{2}-1 < \beta_1 = \beta_1(j,n)$*, then*

$$\frac{\mathbb{E}X_{j:n}-\mu}{\sigma_1} \le -\min\{\underline{B}_{j:n}(\alpha_1), \overline{B}_{j:n}(\beta_1), \overline{B}_{j:n}(1-)\}.$$

(iv) If $\alpha_1(j,n), \beta_1(j,n) > \sqrt{2}-1$*, then*

$$\frac{\mathbb{E}X_{j:n}-\mu}{\sigma_1} \le -\min\{\overline{B}_{j:n}(\beta_1), \overline{B}_{j:n}(1-)\}.$$

We write $\overline{B}_{j:n}(1-)$ instead of its explicit value $\frac{1}{2}$ in order to point out how the possible bound is attained (see Theorem 6.50). Numerical examples show that for $j \ge 2$, cases (i) and (ii) appear most often. For small j, the bounds amount to $-\overline{B}(1-) = -\frac{1}{2}$, and they are strictly greater for j close to $\frac{n}{2}$. The proof of the theorem, especially of sign properties of polynomials (6.389) and (6.390), is long and is based on tedious calculations. We have not included it here. The interested readers are advised to study the paper by Goroncy and Rychlik (2018).

Analogs of Theorems 6.49 and 6.50 for the DFRA distributions are presented below. We define

$$L_{j:n}(x) = \sum_{i=1}^{j} \frac{1-F_{i:n}(x)}{n+1-i} - [1-F_{j:n}(x)]\ln(1-x) - (1-x)[1-\ln(1-x)] \tag{6.391}$$

and

$$N_p^p(\alpha) = (\alpha+1)^p e^{-p\alpha}(1-e^{-\alpha}) + \exp(-(\alpha+1)e^{-\alpha}) \times \begin{cases} \left[\int_0^{(\alpha+1)e^{-\alpha}-\alpha} x^p e^x dx + \Gamma(p+1)\right], & \alpha \le \alpha_*, \\ \int_{\alpha-(\alpha+1)e^{-\alpha}}^{\infty} x^p e^{-x} dx, & \alpha \ge \alpha_*, \end{cases} \tag{6.392}$$

with $\alpha_* \approx 0.80647$ uniquely determined by the equation $\alpha e^{\alpha} = \alpha + 1$.

Theorem 6.52 (DFRA Distributions) *Let* $X_1, \ldots, X_n$ *be iid random variables with a parent distribution function* F *whose support starts at a finite point* a_F*, and the ratio* $\frac{-\ln(1-F(x))}{x-a_F}$ *is nonincreasing for* $x > a_F$*. We also assume each* X_i *has mean* μ *and finite central absolute moment* σ_p^p *of some order* $p > 1$*. If either* $j = 1$ *or* $L_{j:n}(a_{j:n}) < 0$*, where* $a_{j:n}$ *is the smaller solution to the equation* $f_{j:n}(x) = 1$*, then the evaluation*

$$\frac{\mathbb{E}_F X_{j:n}-\mu}{\sigma_p} \le 0 \tag{6.393}$$

is sharp, and the equality is attained in the limit by the family of mixtures of the atom at $\mu - \frac{\sigma_p(1-\beta)[1-\ln(1-\beta)]}{N_p(-\ln(1-\beta))}$ *with probability* β*, and the exponential distribution with location* $\mu + \frac{\sigma_p[-(1-\beta)-\ln(1-\beta)]}{N_p(-\ln(1-\beta))}$ *and scale* $\frac{\sigma_p}{N_p(-\ln(1-\beta))}$ *with probability* $1-\beta$*, as* $\beta \to 1$.

Since the DFRA family is contained in the DDA family, the condition $L_{j:n}(a_{j:n}) < 0$ for nonpositivity of the difference $\mathbb{E}X_{j:n} - \mu$ in the DFRA case is less stringent than $K_{j:n}(a_{j:n}) < 0$ (see (6.377)) assuring the same in the DDA class.

Proof Theorem 6.46 guarantees nonpositivity of the bound under the assumptions of Theorem 6.52. Indeed, condition $L_{j:n}(a_{j:n}) < 0$, equivalent to (6.351), implies that (6.393) holds. Therefore, it only suffices to check the attainability conditions for the distributions presented in the statement. For the family of mixture distributions, the quantile functions satisfy

$$\frac{F_\beta^{-1}(y)-\mu}{\sigma_p} = \frac{-\ln(1-y)\mathbf{1}_{[\beta,1)}(y) - (1-\beta)[1-\ln(1-\beta)]}{N_p(-\ln(1-\beta))}.$$

Substituting $x = -\ln(1-y)$ and setting $\alpha = -\ln(1-\beta)$, we obtain

$$\frac{F_\beta^{-1}\left(1-e^{-x}\right)-\mu}{\sigma_p} = \frac{x\mathbf{1}_{[\alpha,\infty)}(x) - (\alpha+1)e^{-\alpha}}{N_p(\alpha)}, \tag{6.394}$$

which shows that all F_β, $0 \le \beta < 1$, are DFRA distribution functions. Since

$$\begin{aligned}\mathbb{E}_\beta X_1 - \mu &= \int_0^1 [F_\beta^{-1}(y)-\mu]dy = \int_0^\infty [F_\beta^{-1}(1-e^{-x})-\mu]e^{-x}dx \\ &= \frac{\sigma_p}{N_p(\alpha)}\int_0^\infty [x\mathbf{1}_{[\alpha,\infty)}(x) - (\alpha+1)e^{-\alpha}]e^{-x}dx = 0,\end{aligned}$$

and

$$\frac{N_p^p(\alpha)}{\sigma_p^p}\mathbb{E}_\beta|X_1-\mu|^p = \frac{N_p^p(\alpha)}{\sigma_p^p}\int_0^1 |F_\beta^{-1}(y)-\mu|^p dy$$

$$= \int_0^\infty |x\mathbf{1}_{[\alpha,\infty)}(x) - (\alpha+1)e^{-\alpha}|^p e^{-x}dx$$

$$= \int_0^\alpha (\alpha+1)^p e^{-p\alpha}e^{-x}dx + \exp(-(\alpha+1)e^{-\alpha})\int_{\alpha-(\alpha+1)e^{-\alpha}}^\infty |x|^p e^{-x}dx$$

$$= (\alpha+1)^p e^{-p\alpha}(1-e^{-\alpha}) + \exp(-(\alpha+1)e^{-\alpha})$$

$$\times \begin{cases} \int_{\alpha-(\alpha+1)e^{-\alpha}}^0 (-x)^p e^{-x}dx + \int_0^\infty x^p e^{-x}dx, & \alpha \le \alpha_*, \\ \int_{\alpha-(\alpha+1)e^{-\alpha}}^\infty x^p e^{-x}dx, & \alpha \ge \alpha_*, \end{cases} = N_p^p(\alpha),$$

distribution functions F_β, $0 \le \beta < 1$, satisfy both the moment conditions. Furthermore, we calculate

$$\begin{aligned}
&\frac{N_p(\alpha)}{\sigma_p}(\mathbb{E}_\beta X_{j:n} - \mu) = \frac{N_p(\alpha)}{\sigma_p}\int_0^1 [F_\beta^{-1}(y) - \mu] f_{j:n}(y)dy \\
&= \int_0^\infty [x\mathbf{1}_{[\alpha,\infty)}(x) - (\alpha+1)e^{-\alpha}] f_{j:n}(1-e^{-x})e^{-x}dx \\
&= \int_\alpha^\infty x f_{j:n}(1-e^{-x})e^{-x}dx - (\alpha+1)e^{-\alpha} \\
&= \alpha[1 - F_{j:n}(1-e^{-\alpha})] + \int_\alpha^\infty [1 - F_{j:n}(1-e^{-x})]dx - (\alpha+1)e^{-\alpha} \\
&= \sum_{i=0}^{j-1}\binom{n}{i}\int_\alpha^\infty (1-e^{-x})^i e^{-(n-1-i)x}e^{-x}dx + \alpha[1-F_{j:n}(1-e^{-\alpha})] - (\alpha+1)e^{-\alpha} \\
&= \sum_{i=0}^{j-1}\frac{1}{n-i}\int_{1-e^{-\alpha}}^1 f_{i+1:n}(y)dy + \alpha[1 - F_{j:n}(1-e^{-\alpha})] - (\alpha+1)e^{-\alpha} \\
&= \sum_{i=1}^{j}\frac{1 - F_{i:n}(1-e^{-x})}{n+1-i} + \alpha[1 - F_{j:n}(1-e^{-\alpha})] - (\alpha+1)e^{-\alpha} \\
&= L_{j:n}(1-e^{-\alpha}) \le 0.
\end{aligned} \tag{6.395}$$

Accordingly, it remains to prove that for $p > 1$

$$\lim_{\beta\to 1}\frac{\mathbb{E}_\beta X_{j:n} - \mu}{\sigma_p} = \lim_{\alpha\to\infty}\frac{L_{j:n}(1-e^{-\alpha})}{N_p(\alpha)} = 0.$$

We observe that

$$\lim_{\alpha\to\infty}\frac{L_{j:n}(1-e^{-\alpha})}{\alpha e^{-\alpha}} = -1,$$

because

$$\begin{aligned}
\lim_{\alpha\to\infty}\sum_{i=1}^{j}\frac{1 - F_{i:n}(1-e^{-x})}{(n+1-i)e^{-(n+1-j)\alpha}} &= \binom{n}{j-1}\frac{1}{n+1-j}, \\
\lim_{\alpha\to\infty}\frac{\alpha[1 - F_{j:n}(1-e^{-\alpha})]}{\alpha e^{-(n+1-j)\alpha}} &= \binom{n}{j-1}, \\
\lim_{\alpha\to\infty}\frac{(\alpha+1)e^{-\alpha}}{\alpha e^{-\alpha}} &= 1.
\end{aligned}$$

Also,

$$\lim_{\alpha\to\infty}\frac{N_p^p(\alpha)}{\alpha^p e^{-\alpha}}=1,$$

because

$$\lim_{\alpha\to\infty}\frac{(\alpha+1)^p e^{-p\alpha}(1-e^{-\alpha})}{\alpha^p e^{-p\alpha}}=1,$$

$$\lim_{a\to\infty}\frac{\int_a^\infty x^p e^{-x}dx}{a^p e^{-a}}=\lim_{a\to\infty}\frac{-a^p e^{-a}}{pa^{p-1}e^{-a}-a^p e^{-a}}$$

$$=\lim_{a\to\infty}\frac{a}{a-p}=1,$$

$$\lim_{\alpha\to\infty}\frac{\exp(-(\alpha+1)e^{-\alpha})\int_{\alpha-(\alpha+1)e^{-\alpha}}^\infty x^p e^{-x}dx}{\alpha^p e^{-\alpha}}=\lim_{\alpha\to\infty}\frac{(\alpha-(\alpha+1)e^{-\alpha})^p e^{-\alpha}}{\alpha^p e^{-\alpha}}=1.$$

Therefore,

$$\lim_{\alpha\to\infty}\frac{L_{j:n}(1-e^{-\alpha})}{N_p(\alpha)}=\lim_{\alpha\to\infty}\frac{\alpha e^{-\alpha}}{\alpha e^{-\alpha/p}}=0,$$

which completes the proof. □

A counterpart of Theorem 6.50 in the DFRA case is presented below. Formulating it, we use the modification of function (6.392) with $p=1$ of the following form:

$$Q_1(\beta)=N_1(-\ln(1-\beta))=\begin{cases}2\left(\frac{1-\beta}{e}\right)^{1-\beta}, & 0\le\beta\le\beta_*,\\ 2\beta(1-\beta)[1-\ln(1-\beta)], & \beta_*\le\beta<1,\end{cases}\tag{6.396}$$

with $\beta_*=1-e^{-\alpha_*}\approx 0.553567$ being the solution to the equation $-\ln(1-\beta)=\frac{1}{\beta}-1$.

Theorem 6.53 (DFRA Distributions) *Under the assumptions of Theorem 6.52 with $p=1$, we have*

$$\frac{\mathbb{E}_F X_{j:n}-\mu}{\sigma_1}\le\sup_{0\le\beta<1}\frac{L_{j:n}(\beta)}{Q_1(\beta)},\tag{6.397}$$

where the numerator and denominator of the RHS are defined in (6.391) and (6.396), respectively. If the supremum of the RHS of (6.397) is attained at some $0\le\beta<1$, then the bound is attained by the mixture of the pole at $\mu-$

$\frac{\sigma_1(1-\beta)[1-\ln(1-\beta)]}{Q_1(\beta))}$ *with probability* β, *and the exponential distribution with location* $\mu + \frac{\sigma_1[-(1-\beta)-\ln(1-\beta)]}{Q_1(\beta)}$ *and scale* $\frac{\sigma_1}{Q_1(\beta)}$ *with probability* $1-\beta$. *Otherwise, i.e., when*

$$\sup_{0\le\beta<1}\frac{L_{j:n}(\beta)}{Q_1(\beta)} = \lim_{\beta\to 1}\frac{L_{j:n}(\beta)}{Q_1(\beta)} = -\frac{1}{2},$$

then the bound is attained in the limit by the sequences of the mixture distributions as $\beta \to 1$.

The norm maximization method is used for proving the above theorem. It allows us to conclude that the maximal value of $\mathbb{E}\frac{X_{j:n}-\mu}{\sigma_1}$, when $X_1, \dots, X_n$ are iid with a DFRA marginal distribution, is attained at an element of the family defined in (6.394) with $p = 1$ for which

$$\mathbb{E}_\alpha \frac{X_{j:n}-\mu}{\sigma_1} = \frac{L_{j:n}(1-e^{-\alpha})}{Q_1(1-e^{-\alpha})}.$$

The details of reasoning are left to the readers. Since $\frac{K_{j:n}(\alpha)}{M_1(\alpha)}$ and $\frac{L_{j:n}(\beta)}{Q_1(\beta)}$ tend to $-\frac{1}{2}$ as α and β tend to 1, respectively, the bounds in Theorems 6.50 and 6.53 have to belong to the interval $\left[-\frac{1}{2}, 0\right)$. Numerical evaluations show that the bounds amount to $-\frac{1}{2}$ for small j, and become strictly greater when j increases. Note that $-\frac{1}{2}$ is the sharp bound for the sample minima from general iid samples. We do not have an analog of Theorem 6.51 in the DFRA case.

6.4.4 Upper Bounds for L-Statistics from IDA and IFRA Families

We finally show that it is impossible to improve the general Moriguti-Schwarz bounds for any L-statistic based on iid samples from distributions with increasing densities and failure rates on the average.

Theorem 6.54 *Suppose that a distribution function* W *has a bounded density* w *on its support* (a, d) *with* $a > -\infty$. *Then for every* $\bar{c} \in \mathbb{R}^n$, *there exists a sequence of distribution functions* $F_k \preceq_* W$, $k = 1, 2, \dots$, *with a given mean* μ *and variance* σ^2 *such that*

$$\mathbb{E}\sum_{i=1}^n c_i \frac{X_{i:n}(k)-\mu}{\sigma} \to \left[\int_0^1 (P^{\nearrow} f_{\bar{c}:n})^2(x)dx - \left(\sum_{i=1}^n c_i\right)^2\right]^{1/2} \quad (6.398)$$

as $k \to \infty$, with $X_1(k), \ldots, X_n(k)$, $k = 1, 2, \ldots$, denoting iid random variables with common distribution function F_k, and $P^{\nearrow} f_{\bar{c}:n}$ standing for the projection of

$$f_{\bar{c}:n}(x) = \sum_{i=1}^{x} c_i f_{i:n}(x) \tag{6.399}$$

onto the convex cone $C^{\nearrow}$ of nondecreasing functions in $L^2((0,1), dx)$ which is established by the greatest convex minorant method..

Proof The RHS of (6.398) is the best general upper bound for the expectation of the standardized L-statistic with coefficients $c_1, \ldots, c_n$. Also,

$$\bar{h} = \left(P^{\nearrow} f_{\bar{c}:n} - \sum_{i=1}^{n} c_i \mathbf{1} \right) \circ W = P^{\nearrow}(f_{\bar{c}:n} \circ W) - \sum_{i=1}^{n} c_i \mathbf{1} = P^{\nearrow} h \tag{6.400}$$

is (the right-continuous version of) the projection of $h = f_{\bar{c}:n} \circ W - \sum_{i=1}^{n} c_i \mathbf{1}$ onto the analogous convex cone in $L^2((a,d), W(dx))$ and

$$\|\bar{h}\|_W = \left[\int_a^d \bar{h}^2(x) W(dx) \right]^{1/2} = \sqrt{\|P^{\nearrow} f_{\bar{c}:n}\|_U^2 - \left(\sum_{i=1}^{n} c_i \right)^2}. \tag{6.401}$$

Note that

$$\sup_{a \le x < d} \bar{h}(x) = \sup_{0 \le x \le 1} \left| P^{\nearrow} f_{\bar{c}:n}(x) - \sum_{i=1}^{n} c_i \right| \le \sup_{0 \le x \le 1} \left| f_{\bar{c}:n}(x) - \sum_{i=1}^{n} c_i \right| < \infty, \tag{6.402}$$

because $P^{\nearrow} f_{\bar{c}:n}(x)$ is either just equal to $f_{\bar{c}:n}(x)$ or to the mean of $f_{\bar{c}:n}$ over an interval containing x. We also have

$$\sup_{a \le x < d} |\bar{h}'(x)|(x-a) \le \sup_{0 \le x \le 1} \left| f'_{\bar{c}:n}(x) \right| \sup_{a \le x < d} w(x)(x-a) < \infty. \tag{6.403}$$

Indeed, the former factor of the RHS is finite, and we show that so is the latter. This is obvious if d is finite. Otherwise,

$$\lim_{x \to \infty} w(x)(x-a) = 0 \tag{6.404}$$

because W has a finite expectation. There exists $a < d_0 < d = \infty$, such that

$$w(x)(x-a) < 1 \quad \text{for all} \quad x > d_0 \tag{6.405}$$

and

$$\sup_{a \le x < d} w(x)(x-a) \le \max\left\{(d_0 - a) \sup_{a \le x < d_0} w(x), 1\right\} < \infty. \tag{6.406}$$

Therefore

$$c_0 = \sup_{a \le x < d} |\bar{h}(x)| + \sup_{a \le x < d} |h'(x)|(x-a) < \infty. \tag{6.407}$$

Define

$$s_0(x) = \frac{\bar{h}(x) + c_0}{x - a}, \qquad a < x < d. \tag{6.408}$$

This is a nonnegative nonincreasing function, because

$$s_0(x) \ge \frac{1}{x-a} \sup_{a \le x < d} |h'(x)|(x-a) \ge 0 \tag{6.409}$$

and

$$s_0'(x) = \frac{\bar{h}'(x)(x-a) - \bar{h}(x) - c_0}{(x-a)^2} \le 0. \tag{6.410}$$

Put now

$$\bar{h}_k(x) = \begin{cases} s_0\left(a + \frac{1}{k}\right)(x-a) - c_0, & a \le x < a + \frac{1}{k}, \\ \bar{h}(x), & a + \frac{1}{k} \le x < d, \end{cases} \tag{6.411}$$

for $k = 1, 2, \ldots$. Each $\bar{h}_k(x)$ is continuous at $x = a + \frac{1}{k}$, and nondecreasing by (6.409) and the definition of $\bar{h}$. Moreover, by (6.410) and

$$\frac{\bar{h}_k(x) - \bar{h}_k(a)}{x - a} = s_0\left(\max\left\{a + \frac{1}{k}, x\right\}\right), \tag{6.412}$$

functions (6.411) are antistarshaped.

For every $x \in \left[a, a + \frac{1}{k}\right)$, we have

$$\begin{aligned} |\bar{h}_k(x) - \bar{h}(x)| &= \left|k(x-a)\bar{h}\left(a + \frac{1}{k}\right) - \bar{h}(x) - [1 - k(x-a)]c_0\right| \\ &\le \left|k(x-a)\bar{h}\left(a + \frac{1}{k}\right) - \bar{h}(x)\right| + [1 - k(x-a)]c_0 \\ &\le c_0 - \bar{h}(a) < \infty \end{aligned} \tag{6.413}$$

as $k \to \infty$, because $0 < 1 - k(x-a) < 1$, $\bar{h}$ is either constant zero or negative on some neighborhood of a, and $k(x-a)\bar{h}\left(a+\frac{1}{k}\right)$ lies between $\bar{h}(a)$ and 0 for sufficiently large k. Therefore

$$\begin{aligned}\|\bar{h}_k(x) - \bar{h}(x)\|_W^2 &= \int_a^{a+\frac{1}{k}} |\bar{h}_k(x) - \bar{h}(x)|^2 W(dx) \\ &\leq [c_0 - \bar{h}(a)]^2 W\left(a + \frac{1}{k}\right) \to 0. \end{aligned} \tag{6.414}$$

This implies that

$$\int_a^d \frac{\bar{h}_k(x)}{\|\bar{h}_k\|_W} g(x) W(dx) \to \int_a^d \frac{\bar{h}(x)}{\|\bar{h}\|_W} g(x) W(dx) \tag{6.415}$$

for every $g \in L^2([a,d), W(dx))$. Define distribution functions F_k, $k = 1, 2, \ldots,$ by the relations

$$\frac{F_k^{-1} \circ W(x) - \mu}{\sigma} = \frac{\bar{h}_k(x)}{\|\bar{h}_k\|_W}. \tag{6.416}$$

Then F_k have the common mean μ and variance σ^2 and satisfy $F_k \preceq_* W$. Furthermore, by the consecutive application of (6.416), (6.414), (6.400), and (6.401), we obtain

$$\begin{aligned}\mathbb{E}\sum_{i=1}^n c_i \frac{X_{i:n}(k) - \mu}{\sigma} &= \int_0^1 \frac{F_k^{-1}(x) - \mu}{\sigma}\left[f_{\bar{c}:n}(x) - \sum_{i=1}^n c_i\right] dx \\ &= \int_a^d \frac{F_k^{-1} \circ W(x) - \mu}{\sigma}\left[f_{\bar{c}:n} \circ W(x) - \sum_{i=1}^n c_i\right] W(dx) \\ &= \int_a^d \frac{\bar{h}_k(x)}{\|\bar{h}_k\|_W} h(x) W(dx) \\ &\to \int_a^d \frac{\bar{h}(x)}{\|\bar{h}\|_W} h(x) W(dx) = \|\bar{h}\|_W \\ &= \left[\int_0^1 (P^{\nearrow} f_{\bar{c}:n})^2(x)dx - \left(\sum_{i=1}^n c_i\right)^2\right]^{1/2}. \end{aligned} \tag{6.417}$$

The penultimate equality follows from the fact that $\bar{h}$ is the projection of h onto the convex cone of nondecreasing functions in $L^2((a,d), W(dx))$ (cf. (2.142) in Theorem 2.25). $\square$

Note that we do not have a similar claim for the IRFR family.

6.5 Numerical Evaluations

Here we present some numerical evaluations of the differences $\mathbb{E}X_{j:n} - \mu$ between the expectations of order statistics and population means when the iid samples come from restricted families of distributions. In all the cases we assume that the sample size is $n = 20$. Tables 6.1 and 6.2 contain the differences measured in the scale units being the population standard deviations.

In the first one, we consider the parent distributions with the increasing reversed failure rate, increasing density and failure rate functions, and increasing density and failure rate on the average. The common feature of these families is that the respective bounds are positive for all order statistics except for the minimal one, for which all the bounds amount to 0 (see Theorems 6.40 and 6.54).

The elements of the second column (IRFR case) come from Theorem 6.38 except for the last entry which is based on Theorem 6.39. The bounds there are attained by the mixtures of linear transformations of negative exponential distributions and atoms at the right ends of their supports. The contributions of the negative exponential parts increase with increase of j from 0.003690 at $j = 2$ to 1 for $j = n = 20$. For example, for $j = 10$ it amounts to 0.410680.

The numbers appearing in the third column and representing the ID distributions are calculated with use of Corollary 6.3 ($j = 2$), Theorem 6.24 ($j = n$) and Theorem 6.28 (the remaining order statistics). For $j = 2$, the bound is identical to the bound for arbitrary distributions. In the other cases, the bounds are attained by the mixtures of uniform distributions with the atoms at their right ends of supports. The weights of continuous parts increase from 0.022098 for $j = 3$ through 0.365499 for $j = 10$ till 1 for $j = 20$. This obviously means that the bound for the sample maximum is attained by the uniform distribution.

The bound for the second order statistic from the IFR population is identical with that for the ID and general cases. The bound for the sample maximum comes from Theorem 6.25. It is slightly greater than the standardized expectation of the maximum from the exponential sample. The reason is that this is attained by a distribution which is exponential on the left and coinciding with $f_{n:n}^{-1}(x) = \left(\frac{x}{n}\right)^{\frac{1}{n-1}}$ up to a linear transformation on the right. However, the mass of the exponential part approximately equals to 0.999741. The bounds for the non-extreme order statistics were calculated with the use of Theorem 6.33 after the preliminary exclusion of conditions of Theorem 6.32. They are attained by mixtures of right-truncated exponential distributions with the degenerate ones concentrated at the truncation points. As j increases, so does the contribution of the exponential part in the distribution attaining the bound. For $j = 3$, 10 and 19, it amounts to 0.022065, 0.359428 and 0.948991, respectively.

As Theorem 6.54 asserts, the bounds for the increasing density and failure rate on the average families coincide with the bounds for the iid samples with arbitrary parent distributions (A, for short). So, the values in the last column of Table 6.1 are identical with those of the last column of Table 5.2. These bounds are attained by the distribution functions with the shapes proportional to $f_{j:n}^{-1}$ on some intervals, and jumps at the right end of the intervals. The contributions of the continuous parts increase from 0.002770 at $j = 2$ to 1 at $j = 20$ (for example, it amounts to 0.319495 for $j = 10$). Note that the great contribution of the continuous strictly increasing part results in increase of the bound, because the Schwarz inequality on the respective subinterval of (0, 1) can be applied to $f_{j:n}$ itself rather than to its projection onto nondecreasing function which gives smaller values in the Schwarz inequality. Obviously, these bounds are attained in the limit by the elements of the IDA and IFRA families.

The entries in Table 6.1 obviously increase in rows when j increases. They also increase in the consecutive columns. This is implied by the relations IRFR $\subset$ ID $\subset$ IFR $\subset$ A. The bounds for the greater families cannot take smaller values. The values for ID, IFR, and A families with $j = 2$ are identical, because the general bounds for $X_{2:n}$ are attained by decreasing density distributions. As the order statistics ranks j increase, then the differences among the bounds for different families increase. Observe that the bounds for the sample maxima in the IRFR and ID families are bounded for all sample sizes n, and that they increase to 1 and $\sqrt{3} \approx 1.732051$, respectively, as $n \to \infty$. Analogous bounds for the IFR and arbitrary parent distributions tend to ∞ as $n \to \infty$ at the rates $\ln n$ and $\sqrt{\frac{n}{2}}$, respectively.

Table 6.2 contains the analogous bounds for the dual families of life distributions considered in Table 6.1. Precisely, they are the bounds for the expectations of order statistics $X_{j:20}$, $j = 9, \ldots, 20$, based on iid samples of size $n = 20$ coming from the populations with decreasing reversed failure rate, density function, failure

Table 6.1 Positive sharp upper bounds for $\mathbb{E}\frac{X_{j:n}-\mu}{\sigma_p}$, $p = 2$, for the iid samples coming from restricted families of distributions

j	IRFR	ID	IFR	IDA = IFRA = A
2	0.028359	0.030013	0.030013	0.030013
3	0.091927	0.100941	0.100964	0.101396
4	0.151831	0.170169	0.170271	0.172095
5	0.206700	0.235159	0.235481	0.239363
6	0.257964	0.297254	0.297993	0.304444
7	0.306815	0.357810	0.359235	0.368713
8	0.354128	0.417956	0.420430	0.433409
9	0.400561	0.478672	0.482688	0.499702
10	0.446634	0.540875	0.547117	0.568805
11	0.492776	0.605507	0.614938	0.642094
12	0.539362	0.673600	0.687517	0.721266
13	0.586723	0.746371	0.767057	0.808559
14	0.635159	0.825322	0.855909	0.907144
15	0.684924	0.912387	0.958148	1.021830
16	0.736202	1.010127	1.080240	1.160531
17	0.789025	1.121936	1.233895	1.337737
18	0.843096	1.251971	1.443891	1.584500
19	0.897407	1.403037	1.778727	1.988132
20	0.950000	1.567094	2.601216	3.042435

Table 6.2 Positive sharp upper bounds for $\mathbb{E}\frac{X_{j:n}-\mu}{\sigma_p}$, $p = 2$, for the iid samples coming from restricted families of distributions

j	DRFR	DD	DFR	DDA	DFRA	S	SU
9	0.120128	0.000000	0.000000	0.000000	0.000000	0.000000	0.000000
10	0.232484	0.000000	0.000000	0.025248	0.000000	0.000000	0.000000
11	0.339243	0.082479	0.000000	0.194421	0.000000	0.177728	0.082479
12	0.446271	0.247436	0.000000	0.360239	0.000000	0.501542	0.247436
13	0.558282	0.412393	0.004883	0.524533	0.088425	0.746864	0.412393
14	0.680172	0.577350	0.147740	0.690446	0.241882	0.904477	0.577350
15	0.818261	0.745356	0.314406	0.863098	0.417740	1.001511	0.742307
16	0.982159	0.930950	0.514406	1.050955	0.625337	1.082239	0.907265
17	1.188710	1.154720	0.764406	1.269121	0.882085	1.180946	1.072222
18	1.472832	1.453452	1.097885	1.549017	1.225604	1.324883	1.247290
19	1.928704	1.921739	1.623579	1.977308	1.760977	1.573638	1.531842
20	3.042435	3.042435	2.984174	3.042435	3.030057	2.264554	2.264554

rate, and decreasing density and failure rate on the average. For $j = 1, \dots, 8$, all these bounds, expressed in the standard deviation units, are equal to 0. In the last two columns, we have added the bounds for the populations with symmetric and symmetric unimodal distributions. They are not used for representing random lifetimes (except for possibly uniform distributions), but they also have zero bounds for order statistics with small ranks.

The evaluations for the DRFR families are concluded from Theorem 6.7. They are attained by the mixtures of two types of continuous distributions supported on neighboring intervals. The first part has the shape of increasing part of $f_{j:n}^{-1}$, and the other is negative exponential. The weights of the first part increases from 0.000089 at $j = 9$ to 1 for the sample maximum. Consequently, the bound for the sample maximum is identical with the Gumbel and Hartley-David bound.

The bounds for the DD and DFR distributions come from Theorems 6.3 and 6.4, respectively. The distributions attaining the DD bounds for $j = 11, 12, 13$ and 14 are uniform. For greater j, some contributions of continuous distributions related to $f_{j:n}^{-1}$ appear left to the uniform part. These contributions increase from 0.142857 at $j = 15$ till 0.742869 at $j = 19$. For the sample maximum the uniform part completely disappears, and we obtain the bound which is optimal for all parent F. The distributions which attain the bounds in the DFR case have similar structures to those that attain the bounds for the DD family. The positive bounds for $j = 13, \dots 17$ are attained by the exponential distributions. For $j = 18, 19$ and 20, the contributions of distributions based on $f_{j:n}^{-1}$ of sizes 0.093115, 0.449986 and 0.822339 appear on the left. Notice that as a, consequence the bound for the sample maximum is essentially less than in the general case.

The results in the DDA case are based on Theorem 6.45, and those for the DFRA families are deduced from Theorem 6.46. In the first case, the distributions attaining the bounds consist of three mutually neighboring parts: an atom on the left, a continuous component proportional to $f_{j:n}^{-1}$ in the middle, and a uniform part on the right. As j increases, the contributions of the atom and central component increase, and that of the uniform part decreases. The exemplary respective weights amount to 0.296166, 0.001398 and 0.702436 for $j = 10$, and 0.572162, 0.328298 and 0.099540 for $j = 19$. The bounds for $j = 20$ coincide with these of the general case. For the DFRA distributions, the structure of the marginal distributions attaining the bounds are similar with the only difference being that the uniform component is replaced by the exponential one. The tendency of weight changes of consecutive ingredients is similar to that in the DDA case. The probabilities of the atom, middle part proportional to $f_{j:n}^{-1}$ and the right exponential one are equal to 0.440850, 0.003530 and 0.555620, respectively, for $j = 13$, and 0.675001, 0.299198 and 0.025801, respectively, for $j = 20$.

The bounds for the symmetric case come from Corollary 4.22 (see also (4.325) for $j = n$). Their refinements for the symmetric unimodal distributions are based on Theorem 6.20. In the symmetric case, the distributions attaining the bounds have a continuous symmetric part being a linear transformation of $(f_{j:n} - f_{n+1-j:n})^{-1}$ on some interval, and the atoms of identical sizes at the interval ends. The weights of atoms decrease from 0.473684 for $j = 11$ till 0 for $j = 20$. This means that the bound for the sample maximum is attained by a continuous distribution. For the symmetric unimodal family, the bounds for $j = 11, \ldots, 17$ are attained by the uniform distributions. In the remaining cases, the distributions attaining the bounds have the structures similar to those attaining the bounds in the symmetric case. The only difference is that these distributions have identical contributions of continuous uniform distributions on the left and right instead of the atoms. These contributions decrease and attain the zero level for $j = 20$. Consequently, the bound for the sample maximum in the SU case is identical with that for the symmetric distributions.

The relations DFR $\subset$ DD $\subset$ DRFR, DFR $\subset$ DD $\subset$ DDA, DFR $\subset$ DFRA $\subset$ DDA, and SU $\subset$ S imply relations among the respective bound values in the columns of Table 6.2. A nontrivial observation is that the bounds in the DD family outrank those in the DFRA case. This obviously implies analogous inequalities between the DRFR and DFRA families.

It is certainly interesting to compare the bounds among the dual classes: IRFR with DRFR, ID with DD, and IFR with DFR. The latter amount to 0 for small j, and they are evidently smaller than the positive ones for the IRFR, ID and IFR families. It changes as j increases. For $j = 20$, the bounds for the DRFR, DD and DFR classes are close to the general ones, which is not true for their duals. It means that the bounds for the DRFR, DD and DFR families increase faster in j than their counterparts. These relations are not true for the families with monotone density and failure rate on the average. The bounds for the IDA and IFRA distributions coincide with the general bounds, and they are clearly greater for all indices j than their counterparts in the DDA and DFRA families.

It is hard to compare the bounds for the symmetric and symmetric unimodal distributions with the other bounds treated in Table 6.2. For instance, the bounds in the symmetric case are greater than all their counterparts in the life distributions presented in the table for $j = 12, \ldots, 16$, and less for $j = 19$ and 20. The bounds for the sample maxima in the S and SU cases increase to ∞ at the rate $\frac{\sqrt{n}}{2}$, whereas the rates of increase in the other families are greater by $\sqrt{2}$.

Although the mean-standard deviation bounds for small order statistics coming from the families represented in Table 6.2 amount to 0, the respective bounds expressed in the mean absolute deviations $\sigma_1 = \mathbb{E}|X_1 - \mathbb{E}X_1|$ can be nontrivial and strictly negative. They are presented in Table 6.3. There are some empty entries at the ends of some columns. In these cases, the bounds expressed in σ_1 units are certainly positive, but their values are not known.

The bounds for the DD and DFR families, given in the second and third columns, respectively, are based on Theorems 6.17 and 6.19, respectively. In the DD case, the bounds for $j = 1, \ldots, 7$ equal to $-\frac{1}{2}$ are attained in the limit by the uniform distributions with an atom at the left support end-point, as the size of the atom tends to 1. For $j = 8, 9, 10$, the uniform distributions attain the bounds. A similar property hold for the DFR family. For $j = 1, \ldots 9$, the bounds are attained by mixtures of a pole on the left, and the exponential distribution just right to it, when the contribution of the exponential part vanishes. The negative bounds for the DD and DFR families have the forms $\max\left\{-\frac{1}{2}, 2\sqrt{3}\left(\frac{j}{n-1} - \frac{1}{2}\right)\right\}$ and $\max\left\{-\frac{1}{2}, \frac{e}{2}(S(j,n) - 1)\right\}$, respectively. Note that $2\sqrt{3}\left(\frac{j}{n-1} - \frac{1}{2}\right)$ and $\frac{e}{2}(S(j,n)-1)$ are the values of $\mathbb{E}\frac{X_{j:n}-\mu}{\sigma_1}$ in the standard uniform and standard exponential cases, respectively. There are no known corresponding bounds for the DRFR family. We can suspect that they can be obtained in a similar way as for the DD and DFR classes. If so, they have the forms $\max\left\{-\frac{1}{2}, \mathbb{E}\frac{X_{j:n}-\mu}{\sigma_1}\right\} = \max\left\{-\frac{1}{2}, \frac{e}{2}(1 - S(n+1-j, n)\right\}$, with $X_{j:n}$ being the jth order statistic from the standard negative exponential distribution. Then we could obtain the bounds -0.5 for $j = 1, \ldots, 5$, and -0.427322, -0.200799 and -0.006636 for $j = 6, 7$ and 8, respectively. Note that the bounds for $\mathbb{E}\frac{X_{1:n}-\mu}{\sigma_1}$ based on the ID and IFR samples are equal to -1.809524 and -1.291184 (cf. Theorems 6.41 and 6.42), respectively. The analogous bound in the IRFR case is not known until now.

The numerical values of bounds for the DDA and DFRA classes are calculated with the use of Theorems 6.51 and 6.53, respectively. The conditions of attainability of the bound $-\frac{1}{2}$ are similar to those in the DD and DFR cases, respectively. The negative bounds greater than $-\frac{1}{2}$ in the DDA case are attained by the mixtures of atoms and uniform distributions supported in some distance right to the atoms. The probabilities of the atoms for $j = 8$ and 9 are equal to 0.225906 and 0.261855, respectively. For $j = 10$, 11 and 12, the bounds for the DFRA family are attained by similar combinations with the uniform distributions replaced by exponential ones. The corresponding weights of atoms are 0.315736, 0.357856 and 0.399111, respectively.

The last two columns, corresponding to the symmetric and symmetric unimodal parents, are based on Corollary 4.24 and Theorem 6.22, respectively, for $j = 1$, and Corollary 4.26 and Theorem 6.21, respectively, for $j = 2, \ldots, 10$. It is immediately noticed that they significantly differ from their neighbors on the left. The bounds for the sample minima are substantially smaller than their counterparts for the DD, DFR, DDA and DFRA families. For $2 \leq j \leq 9$, the bounds amount to 0, and are substantially greater. The negative bounds for the sample minima are attained by the two-point symmetric distributions in the symmetric case, and by continuous uniform distributions for the symmetric unimodal family. The zero bounds in the symmetric case are attained by three-point distributions as the weight of the central point increases to 1. In the symmetric unimodal case, the limiting distributions are uniform distributions with the atom at their medians, and the height of atom increasing to 1.

Relations DFR $\subset$ DD $\subset$ DDA and DFR $\subset$ DFRA $\subset$ DDA imply respective relations among the bounds for the order statistics coming from the families studied in Table 6.3. A nontrivial observation is that the bounds in the DD case are greater than the corresponding bounds for the DFRA family. The same relations among the bounds for these classes are seen in Table 6.2.

The bounds for the symmetric and symmetric unimodal distributions behave quite differently. They are significantly less in the sample minimum case than the respective bounds for the other classes considered in Table 6.3. For the S and SU distributions with $j = 2, \ldots, 10$, the bounds are equal to 0, and they are obviously greater than the negative bounds for the other families.

Table 6.3 Nonpositive sharp upper bounds for $\mathbb{E}\frac{X_{j:n}-\mu}{\sigma_p}$, $p = 1$, for the iid samples coming from restricted families of distributions

j	DD	DFR	DDA	DFRA	S	SU
1	−0.500000	−0.500000	−0.500000	−0.500000	−0.999998	−1.809524
2	−0.500000	−0.500000	−0.500000	−0.500000	0.000000	0.000000
3	−0.500000	−0.500000	−0.500000	−0.500000	0.000000	0.000000
4	−0.500000	−0.500000	−0.500000	−0.500000	0.000000	0.000000
5	−0.500000	−0.500000	−0.500000	−0.500000	0.000000	0.000000
6	−0.500000	−0.500000	−0.500000	−0.500000	0.000000	0.000000
7	−0.500000	−0.500000	−0.500000	−0.500000	0.000000	0.000000
8	−0.476190	−0.500000	−0.381335	−0.500000	0.000000	0.000000
9	−0.285714	−0.500000	−0.172557	−0.500000	0.000000	0.000000
10	−0.095238	−0.450186		−0.380926	0.000000	0.000000
11		−0.314272		−0.229016		
12		−0.163257		−0.063855		

6.6 Quantile Inequalities

6.6.1 *Single Order Statistics*

The quantile inequalities are mainly based on the convex transform order of the parent distributions and the Jensen inequality (see Theorem 2.43). The first result of this type was established by Blom (1958).

Theorem 6.55 (ID and DD Distributions) *Assume that $X_1, \ldots, X_n$ are iid with a convex (concave, respectively) parent distribution function F such that $\mathbb{E}X_{j:n}$ exists. Then*

$$\mathbb{E}X_{j:n} \le (\ge \; resp.) F^{-1}\left(\frac{j}{n+1}\right). \tag{6.418}$$

Under either of the assumptions, the equality is attained in (6.418) *if $X_1, \ldots, X_n$ are uniformly distributed on some interval.*

Proof The assumption means that F^{-1} is concave (convex, respectively). Let $U_1, \ldots, U_n$ denote an iid sample from the standard uniform distribution. Then, by the Jensen inequality,

$$\begin{aligned}\mathbb{E}X_{j:n} &= \int_0^1 F^{-1}(x) f_{j:n}(x)dx = \int_0^1 F^{-1}(x) F_{j:n}(dx) \\ &\le F^{-1}\left(\int_0^1 x F_{j:n}(dx)\right) = F^{-1}(\mathbb{E}U_{j:n}) = F^{-1}\left(\frac{j}{n+1}\right),\end{aligned}$$

and the inequality is reversed when F^{-1} is convex and F is concave. Under the assumption that F is in a convex transform relation with the standard uniform distribution function U, the equality appears in (6.418) when F is a location-scale transformation of U. □

The following extension is due to van Zwet (1964).

Theorem 6.56 *Let $X_1, \ldots, X_n$, and $Y_1, \ldots, Y_n$ be iid with parent distribution functions F and W, respectively, such that $\mathbb{E}X_{j:n}$ and $\mathbb{E}Y_{j:n}$ exist. If $F \preceq_c W$, then*

$$\mathbb{E}X_{j:n} \le F^{-1} \circ W(\mathbb{E}Y_{j:n}). \tag{6.419}$$

If $F \succeq_c W$, then the inequality is reversed. Moreover, in both the cases, the equality in (6.419) *holds when F is a location-scale transformation of W.*

Proof If $F \preceq_c W$, then $F^{-1} \circ W$ is concave. Therefore,

$$\begin{aligned}
\mathbb{E}X_{j:n} &= \int_0^1 F^{-1}(x) f_{j:n}(x)dx = \int_{\text{supp } W} F^{-1} \circ W(x) F_{j:n} \circ W(dx) \\
&= F^{-1} \circ W \left(\int_{\text{supp } W} x F_{j:n} \circ W(dx) \right) = F^{-1} \circ W(\mathbb{E}Y_{j:n}),
\end{aligned}$$

because $F_{j:n} \circ W$ is the distribution function of a probability measure on supp W. Relation $F \succeq_c W$ implies the reversed inequality. The equality conditions are also obvious. □

Below we present several conclusions from Theorem 6.56. For brevity, we assume that $F \preceq_c W$ for some specific W. The inequalities under the reversed ordering are reversed. The equality conditions are obvious as well.

Corollary 6.5 (IFR Distributions) *If $X_1, \ldots, X_n$ are iid, and their marginal distribution function has convex cumulative failure rate $-\ln(1-F)$, then*

$$\mathbb{E}X_{j:n} \le F^{-1}\left(1 - \exp\left(-\sum_{k=1}^{j} \frac{1}{n+1-k}\right)\right). \tag{6.420}$$

The integral approximations

$$-\ln\frac{n+1-j}{n+1} = \int_{n+1-j}^{n+1} \frac{dx}{x} < \sum_{k=1}^{j} \frac{1}{n+1-k} < \int_{n+\frac{1}{2}-j}^{n+\frac{1}{2}} \frac{dx}{x} = -\ln\frac{n+\frac{1}{2}-j}{n+\frac{1}{2}} \tag{6.421}$$

allow us to simplify (6.420) by the evaluations $\mathrm{E}X_{j:n} < F^{-1}\left(\frac{j}{n-\frac{1}{2}}\right)$ for the IFR variables, and write $\mathrm{E}X_{j:n} > F^{-1}\left(\frac{j}{n+1}\right)$ for the DFR ones.

Corollary 6.6 (α-IGFR Distributions) *Let $W_\alpha(x) = 1 - (1-\alpha x)^{1/\alpha}$ be the generalized Pareto distribution function with parameter $\alpha \neq 0$. If $X_1, \ldots X_n$ are iid with a baseline distribution function such that $\mathbb{E}X_{j:n}$ exists for some $j < n+1+\alpha$, and $F \preceq_c W_\alpha$, then*

$$\mathbb{E}X_{j:n} \le F^{-1}\left(1 - \left[\frac{B(j, n+1+\alpha-j)}{B(j, n+1-j)}\right]^{1/\alpha}\right). \tag{6.422}$$

Corollary 6.7 (α-IGRFR Distributions) *Let $V_\alpha(x) = (1+\alpha x)^{1/\alpha}$ denote the negative generalized Pareto distribution function with parameter $\alpha \neq 0$. If $X_1, \ldots X_n$ are iid with a baseline distribution function such that $\mathbb{E}X_{j:n}$ exists for some $j > -\alpha$, and $F \preceq_c V_\alpha$, then*

$$\mathbb{E}X_{j:n} \le F^{-1}\left(\left[\frac{B(j+\alpha, n+1-j)}{B(j, n+1-j)}\right]^{1/\alpha}\right). \tag{6.423}$$

The order assumptions of Corollaries 6.6 and 6.7 mean that $\frac{1-[1-F(x)]^\alpha}{\alpha}$ and $\frac{F^\alpha(x)-1}{\alpha}$ are convex, respectively. If F has a density f, then the assumptions can be expressed as increase of the α- generalized failure rate $[1-F(x)]^{\alpha-1}f(x)$ and α- generalized reversed failure rate $F^{\alpha-1}(x)f(x)$, respectively. If $\alpha = -1$, the inequalities (6.422) and (6.423) take on the forms

$$\mathbb{E}X_{j:n} \le F^{-1}\left(\frac{j}{n}\right), \qquad j \le n-1, \tag{6.424}$$

$$\mathbb{E}X_{j:n} \le F^{-1}\left(\frac{j-1}{n}\right), \qquad j \ge 1, \tag{6.425}$$

respectively. For $\alpha = 1$, in both the cases we obtain the statement of Theorem 6.55.

For completeness, we include the quantile bounds for the order statistics from the iid samples whose marginal distribution functions have increasing standard reversed failure rates. These distribution functions precede the negative exponential distribution function $V_0(x) = V(x) = e^x$, $x < 0$, in the convex transform order. The order statistics from iid negative exponential samples have the expectations

$$\mathbb{E}Y_{j:n} = -\sum_{k=j}^{n}\frac{1}{k}.$$

Corollary 6.8 (IRFR Distributions) *Let $X_1, \ldots, X_n$ be iid with a distribution function F such that* ln F *is convex. Then,*

$$\mathbb{E}X_{j:n} \le F^{-1}\left(\exp\left(-\sum_{k=j}^{n}\frac{1}{k}\right)\right) < F^{-1}\left(\frac{j-\frac{1}{2}}{n+\frac{1}{2}}\right). \tag{6.426}$$

The latter inequality follows from an integral approximation of harmonic sequence (see (6.421)).

Example We claim that for $X_1, \ldots, X_n$ being iid standard normal, the following holds:

$$\Phi^{-1}\left(\frac{j-1}{n}\right) \le \mathbb{E}X_{j:n} \le \Phi^{-1}\left(\frac{j}{n}\right), \qquad j = 1, \ldots, n. \tag{6.427}$$

To this end, it suffices to verify the Φ is (-1)-IGFR and (-1)-DGRFR (comp. (6.424) and (6.425)). The first property is true, because

$$\frac{d}{dx}\frac{\varphi(x)}{[1-\Phi(x)]^2} = \frac{\varphi(x)}{[1-\Phi(x)]^3}\{2\varphi(x) - x[1-\Phi(x)]\} > 0.$$

The inequality is obvious for nonpositive arguments. For $x > 0$, it is concluded from the relations

$$x[1-\Phi(x)] = \int_x^\infty x\varphi(t)dt < \int_x^\infty t\varphi(t)dt = \varphi(x).$$

This implies that $-x\Phi(x) < \varphi(x)$, $x < 0$ as well. Therefore

$$\frac{d}{dx}\frac{\varphi(x)}{\Phi^2(x)} = -\frac{\varphi(x)}{\Phi^3(x)}[2\varphi(x) + x\Phi(x)]] < 0,$$

which completes the proof of (6.427). Analogous inequalities hold for arbitrary normal distributions: it suffices to multiply the quantiles and expectations in (6.427) by arbitrary $\sigma > 0$, and add $\mu \in \mathbb{R}$ to them.

The example was presented by van Zwet (1964, pp. 57–58). He also determined similar inequalities for the order statistics for the iid samples with gamma and beta distributions. They depend on the shape parameters of the parent distributions, though.

6.6.2 Convex Combinations of Order Statistics

Rychlik (1998) extended the results of Theorem 6.56 to convex combinations of order statistics as follows.

Theorem 6.57 *Suppose that $\tilde{c} = (c_1, \ldots, c_n) \in \mathbb{R}^n$ satisfies $c_i \ge 0$, $i = 1, \ldots, n$, and $\sum_{i=1}^n c_i = 1$. Let $1 \le p \le q \le n$ be the numbers of the first and last nonzero element of $\tilde{c}$. Assume further that $X_1, \ldots, X_n$ and $Y_1, \ldots, Y_n$ are iid with parent distribution functions F and W, respectively, such that the expectations of the pth and qth order statistics are finite. If moreover, $F \preceq_c W$, then*

$$\mathbb{E}\sum_{i=1}^n c_i X_{i:n} \le F^{-1} \circ W\left(\mathbb{E}\sum_{i=1}^n c_i Y_{i:n}\right). \tag{6.428}$$

Proof We have

$$\begin{aligned}
\mathbb{E}\sum_{i=1}^{n} c_i X_{i:n} &= \int_0^1 F^{-1}(x) \sum_{i=1}^{n} c_i f_{i:n}(x)\, dx \\
&= \int_{\text{supp } W} F^{-1} \circ W(x) \sum_{i=1}^{n} c_i f_{i:n} \circ W(x)\, W(dx) \\
&= \int_{\text{supp } W} F^{-1} \circ W(x)\, F_{\bar{c}:n} \circ W(dx).
\end{aligned}$$

We can apply the Jensen inequality, because $F^{-1} \circ W$ is concave, and $F_{\bar{c}:n} \circ W = \sum_{i=1}^{n} c_i F_{i:n} \circ W$ is the distribution function of a probability measure on the support of W. Therefore

$$\begin{aligned}
\mathbb{E}\sum_{i=1}^{n} c_i X_{i:n} &\le F^{-1} \circ W \left(\int_{\text{supp } W} x\, F_{\bar{c}:n} \circ W(dx) \right) \\
&= F^{-1} \circ W \left(\int_{\text{supp } W} W^{-1}(x) \sum_{i=1}^{n} c_i f_{i:n}(x)\, dx \right) \\
&= F^{-1} \circ W \left(\mathbb{E}\sum_{i=1}^{n} c_i Y_{i:n} \right).
\end{aligned}$$

□

Theorem 6.57 has been proved by Arnold and Balakrishnan (1989, Section 3.4) under an extra assumption that $X_1, \ldots, X_n$ are nonnegative. It follows from the above proof that the assumptions of Theorem 6.57 can be weakened. Namely, it suffices to assume that $F_{\bar{c}:n} \circ W$ is a probability distribution function which is equivalent to the conditions $\sum_{i=1}^{n} c_i = 1$ with

$$\sum_{i=1}^{n} c_i f_{i:n}(x) = n \sum_{i=1}^{n} c_i B_{i-1,n-1}(x) \ge 0, \qquad 0 < x < 1. \tag{6.429}$$

The assumption of nonnegativity of all c_i is not necessary here. We obtain a more general result if we apply the Steffensen-Jensen inequality (see Theorem 2.45).

Theorem 6.58 *Under the assumptions of Theorem* 6.57*, with the condition of nonnegativity of* c_i*,* $i = 1, \ldots, n$*, replaced by the relations*

$$0 \le \sum_{i=1}^{n} c_i F_{i:n}(x) = \sum_{i=1}^{n} i c_i B_{i,n}(x) \le \sum_{i=1}^{n} c_i = 1, \qquad 0 < x < 1, \tag{6.430}$$

the inequality (6.428) *holds.*

The proof is similar to that of Theorem 6.57. Under the assumption (6.430), $\sum_{i=1}^{n} c_i F_{i:n} \circ W$ is a function of bounded variation on supp W, and attains its extreme values 0 and 1 at the left and right ends of the support. Therefore the Steffensen-Jensen inequality can be used in order to get (6.428). Note that (6.430), in contrast to (6.429), does not require that $\sum_{i=1}^{n} c_i F_{i:n}(x)$ is nondecreasing. The weak points of the assumptions (6.429) and (6.430) are the facts that they are not so easy to verify as nonnegativity of coefficients c_i, $i = 1, \ldots, n$.

Particular cases of Theorem 6.58 are presented below.

Corollary 6.9 (ID Distributions) *Under the assumption of Theorem* 6.58 *with* $W = U$ *being the standard uniform distribution, the inequality*

$$\mathbb{E}\sum_{i=1}^{n} c_i X_{i:n} \leq F^{-1}\left(\frac{1}{n+1}\sum_{i=1}^{n} i c_i\right)$$

holds.

Corollary 6.10 (IFR Distributions) *Under the assumption of Theorem* 6.58 *with* W *being the exponential distribution, we have*

$$\mathbb{E}\sum_{i=1}^{n} c_i X_{i:n} \leq F^{-1}\left(1 - \exp\left(-\sum_{j=1}^{n}\frac{1}{j}\sum_{i=n+1-j}^{n} c_i\right)\right). \tag{6.431}$$

Applying (6.421), we obtain a nicer non-sharp bound

$$\mathbb{E}\sum_{i=1}^{n} c_i X_{i:n} < F^{-1}\left(1 - \frac{1}{n+\frac{1}{2}}\prod_{i=1}^{n}\left(n+\frac{1}{2}-j\right)^{c_i}\right).$$

For the DFR parent distributions, we analogously derive the reversed inequality in (6.431) and

$$\mathbb{E}\sum_{i=1}^{n} c_i X_{i:n} > F^{-1}\left(1 - \frac{1}{n+1}\prod_{i=1}^{n}(n+1-j)^{c_i}\right).$$

More generally, the statement of Theorem 6.58 for α-IGFR parent distributions, $\alpha \neq 0$, has the form

$$\mathbb{E}\sum_{i=1}^{n} c_i X_{i:n} \leq F^{-1}\left(1 - \left[\sum_{i=1}^{n} c_i \frac{B(i, n+1-i+\alpha)}{B(i, n+1-i)}\right]^{1/\alpha}\right).$$

In the α-IGRFR cases, we have

$$\mathbb{E}\sum_{i=1}^{n} c_i X_{i:n} \le F^{-1}\left(\left[\sum_{i=1}^{n} c_i \frac{B(i+\alpha, n+1-i)}{B(i, n+1-i)}\right]^{1/\alpha}\right)$$

when $\alpha \neq 0$, and

$$\mathbb{E}\sum_{i=1}^{n} c_i X_{i:n} \le F^{-1}\left(\exp\left(-\sum_{k=1}^{n}\frac{1}{k}\sum_{i=1}^{k} c_i\right)\right)$$

for the parent distribution with the standard reversed hazard rate (cf. (6.422), (6.423), and (6.426)).

Barlow and Proschan (1966) noticed that the Jensen inequality can be applied for determining the quantile bounds for positive linear combinations of order statistics when $\sum_{i=1}^{n} c_i < 1$ as well, if we additionally assume that both F and W are life distributions. If fact, we can use the same arguments applying the Steffensen-Jensen inequality when we consider nonnegative X_i and Y_i and relax the assumption (6.430) by replacing the last equality by the inequality $\sum_{i=1}^{n} c_i \le 1$. Indeed, then we have

$$\begin{aligned}\mathbb{E}\sum_{i=1}^{n} c_i X_{i:n} &\le \sum_{i=1}^{n} c_i\, F^{-1}\circ W\left(\frac{\mathbb{E}\sum_{i=1}^{n} c_i Y_{i:n}}{\sum_{i=1}^{n} c_i}\right)\\ &= \sum_{i=1}^{n} c_i\, F^{-1}\circ W\left(\frac{\mathbb{E}\sum_{i=1}^{n} c_i Y_{i:n}}{\sum_{i=1}^{n} c_i}\right) + \left(1-\sum_{i=1}^{n} c_i\right) F^{-1}\circ W(0)\\ &\le F^{-1}\circ W\left(\mathbb{E}\sum_{i=1}^{n} c_i X_{i:n}\right).\end{aligned}$$

6.6.3 *Symmetric Distributions*

A similar trick can be used when we consider symmetrically distributed random variables $X_1, \ldots, X_n$ and $Y_1, \ldots Y_n$. We assume that the composition $F^{-1} \circ W$ is concave (convex) only right to the symmetry center of the support of W, and consequently, it is convex (concave, respectively) on the left. Such a relation defines the so-called s-ordering among symmetric distributions introduced by van Zwet (1964). For simplicity, we assume that they are symmetric about 0.

Theorem 6.59 *Let $X_1, \ldots, X_n$ and $Y_1, \ldots Y_n$ be independent, and identically symmetrically distributed about 0 with respective distribution functions F and W. We also assume that the composition $F^{-1} \circ W$ is concave on the nonnegative part of the support of W, and $\mathbb{E}X_{j:n}$ and $\mathbb{E}Y_{j:n}$ are finite for some $j > \frac{n+1}{2}$. Then,*

$$\mathbb{E}X_{j:n} \leq F^{-1} \circ W\left(\mathbb{E}Y_{j:n}\right). \tag{6.432}$$

Proof Let $0 < d \leq \infty$ denote the right end-point of the support of W. We have

$$\mathbb{E}X_{j:n} = \mathbb{E}\, F^{-1} \circ W(Y_{j:n}) = \int_{-d}^{d} F^{-1} \circ W(x)\, F_{j:n} \circ W(dx).$$

Since $F^{-1} \circ W(-x) = -F^{-1} \circ W(x)$ and $F_{j:n} \circ W(-x) = F_{j:n}(1 - W(x)) = F_{n+1-j:n} \circ W(x)$, under the change of variables $y = 1 - x$, we obtain

$$\int_{-d}^{0} F^{-1} \circ W(x)\, F_{j:n} \circ W(dx) = -\int_{0}^{d} F^{-1} \circ W(y)\, F_{n+1-j:n} \circ W(dy),$$

and as a consequence,

$$\mathbb{E}X_{j:n} = \int_{0}^{d} F^{-1} \circ W(x)\, F^{s}_{j:n} \circ W(dx), \tag{6.433}$$

where $F^{s}_{j:n}(x) = F_{j:n}(x) - F_{n+1-j:n}(x)$. Note that

$$\begin{aligned} -1 < -F_{n+1-j:n}\left(\frac{1}{2}\right) &< F_{j:n}\left(\frac{1}{2}\right) - F_{n+1-j:n}\left(\frac{1}{2}\right) \\ &= F^{s}_{j:n} \circ W(0) = -\sum_{k=n+1-j}^{j-1} \binom{n}{k} \frac{1}{2^n} < 0 \end{aligned}$$

and $F^{s}_{j:n} \circ W(d) = 0$. Moreover, $F^{s}_{j:n} \circ W(x)$ is nondecreasing for $0 < x < d$, because so is W, which ranges over $\left(\frac{1}{2}, 1\right)$ there, and

$$\begin{aligned} \frac{d}{dx} F^{s}_{j:n}(x) &= f_{j:n}(x) - f_{n+1-j:n}(x) \\ &= n\binom{n-1}{j-1} x^{n-j}(1-x)^{n-j}[x^{2j-n-1} - (1-x)^{2j-n-1}] > 0 \end{aligned}$$

for $\frac{1}{2} < x < 1$, because $2j - n - 1 > 0$. This means that $F^s_{j:n} \circ W$ is the distribution function of a positive measure on $[0, d]$ with total mass

$$C_{j:n} = -F^s_{j:n} \circ W(0) = F_{n+1-j:n}\left(\frac{1}{2}\right) - F_{j:n}\left(\frac{1}{2}\right) \in (0, 1).$$

Therefore, using twice the Jensen inequality, we finally obtain

$$\begin{aligned}
\mathbb{E}X_{j:n} &\leq C_{j:n}\, F^{-1} \circ W\left(\int_0^d \frac{x\, F^s_{j:n} \circ W(dx)}{C_{j:n}}\right) \\
&= C_{j:n}\, F^{-1} \circ W\left(\frac{\mathbb{E}Y_{j:n}}{C_{j:n}}\right) \\
&= C_{j:n}\, F^{-1} \circ W\left(\frac{\mathbb{E}Y_{j:n}}{C_{j:n}}\right) + \left[1 - C_{j:n}\right] F^{-1} \circ W(0) \\
&\leq F^{-1} \circ W(\mathbb{E}Y_{j:n}),
\end{aligned}$$

where the passage from the first line to the second one is justified by (6.433). □

Under the assumptions of Theorem 6.59, the reversed inequalities hold for the order statistics with indices $j < \frac{n+1}{2}$. Indeed, since $\mathbb{E}X_{j:n} = -\mathbb{E}X_{n+1-j:n}$ and $\mathbb{E}Y_{j:n} = -\mathbb{E}Y_{n+1-j:n}$, by symmetry of the parent distributions, formula (6.432) takes on the form

$$-\mathbb{E}X_{j:n} \leq F^{-1} \circ W(-\mathbb{E}Y_{j:n}) = -F^{-1} \circ W(\mathbb{E}Y_{j:n}).$$

If $F^{-1} \circ W$ is convex on the positive part of the support of W, then slightly modifying the proof of Theorem 6.59 we deduce that (6.432) holds for $j < \frac{n+1}{2}$, and its reversed version is valid for $j > \frac{n+1}{2}$. Clearly, if n is odd and $j = \frac{n+1}{2}$, we get the equality

$$0 = \mathbb{E}X_{\frac{n+1}{2}:n} = F^{-1} \circ W(0) = F^{-1} \circ W(-\mathbb{E}Y_{\frac{n+1}{2}:n}).$$

This is actually true for all symmetric distributions, and no extra shape conditions are needed here.

If X_i and Y_i, $i = 1, \ldots, n$, have distribution functions F and W symmetric about some μ and ν, not necessarily equal to zero, then (6.432) holds if $F^{-1} \circ W$ is concave on the interval joining the symmetry center ν and right end of the support of W. To see this, it suffices to apply (6.432) for $X_i^0 = X_i - \mu$ with quantile function $F^{-1} - \mu$ and $Y_i^0 = Y_i - \nu$ with distribution function $W(\cdot + \nu)$, and express the formula in terms of the original random variables and their distributions. Analogs of (6.432) can be derived similarly. The above considerations allow us to formulate quantile bounds for sample quasi-ranges as follows.

Corollary 6.11 *Under the assumptions of Theorem* 6.59,

$$\mathbb{E}(X_{j:n} - X_{n+1-j:n}) \leq 2\, F^{-1} \circ W(\mathbb{E}X_{j:n}), \qquad j > \frac{n+1}{2}. \tag{6.434}$$

If $F^{-1} \circ W$ is first concave, and then convex, then inequality (6.434) is reversed.

Corollary 6.12 (SU Distributions) *If $X_1, \ldots, X_n$ are iid with a symmetric unimodal distribution function F, then*

$$\mathbb{E}X_{j:n} \leq F^{-1}\left(\frac{j}{n+1}\right), \qquad j < \frac{n+1}{2}, \tag{6.435}$$

and

$$\mathbb{E}X_{j:n} \geq F^{-1}\left(\frac{j}{n+1}\right), \qquad j > \frac{n+1}{2}. \tag{6.436}$$

If F is symmetric U-shaped, then the inequalities in (6.435) *and* (6.436) *are reversed.*

The proof of Theorem 6.59 was essentially provided in Nagaraja (1981). Corollary 6.12 was derived by Ali and Chan (1965). Combining Corollaries 6.11 and 6.12, we obtain

$$\mathbb{E}(X_{j:n} - X_{n+1-j:n}) \geq 2\, F^{-1}\left(\frac{j}{n+1}\right), \qquad j > \frac{n+1}{2}, \tag{6.437}$$

when the parent distribution is symmetric unimodal, and the reversed inequality if it is symmetric U-shaped.

Example (van Zwet (1964)) Let $W(x) = \frac{1}{1+e^{-x}}$, $x \in \mathbb{R}$, denote the standard logistic distribution function. It has the quantile function $W^{-1}(x) = \ln \frac{x}{1-x}$. We show that its composition $W^{-1} \circ \Phi$ with the standard normal distribution function Φ is convex on the positive half-axis. We have

$$\frac{d^2}{dx^2} W^{-1} \circ \Phi(x) = \frac{d^2}{dx^2}\{\ln \Phi(x) - \ln[1 - \Phi(x)]\} = \frac{\varphi(x)}{\Phi^2(x)[1 - \Phi(x)]^2} A(x),$$

where

$$A(x) = [2\Phi(x) - 1]\varphi(x) - x\Phi(x)[1 - \Phi(x)],$$

and we check that $A(x) > 0$ for $x > 0$. To this end, we calculate

$$A'(x) = 2\varphi^2(x) - \Phi(x)[1 - \Phi(x)],$$
$$A''(x) = \varphi(x)B(x),$$

where $B(x) = 2\Phi(x) - 1 - 2x\varphi(x)$ and $B'(x) = 2x^2\varphi(x) > 0$. Since $B(0) = 0$, $B(x) > 0$ for all $x > 0$, and so is $A''(x)$. Relations $A'(0) = \frac{1}{\pi} - \frac{1}{4} > 0$ and $A(0) = 0$ imply that $A'(x) > 0$ and $A(x) > 0$ for $x > 0$ as well, which gives the desired claim.

The expectations of order statistics from the iid standard logistic samples satisfy

$$\mathbb{E}Y_{j:n} = \int_0^1 W^{-1}(x) f_{j:n}(x)dx = \int_0^1 \ln x f_{j:n}(x)dx - \int_0^1 \ln(1-x) f_{j:n}(x)dx.$$

The negative of the last integral is just the expectation of the jth order statistic $\mathbb{E}Z_{j:n}$ from the standard exponential sample. The former can be rewritten as

$$\begin{aligned}\int_0^1 \ln x f_{j:n}(x)dx &= \int_0^1 \ln(1-x) f_{j:n}(1-x)dx \\ &= \int_0^1 \ln(1-x) f_{n+1-j:n}(x)dx = -\mathbb{E}Z_{n+1-j:n}.\end{aligned}$$

Therefore,

$$\mathbb{E}Y_{j:n} = \begin{cases} -\sum_{k=j+1}^{n+1-j} \frac{1}{n+1-k}, & j < \frac{n+1}{2}, \\ 0, & j = \frac{n+1}{2}, \\ \sum_{k=n+2-j}^{j} \frac{1}{n+1-k}, & j < \frac{n+1}{2}, \end{cases} = \begin{cases} -\sum_{k=j}^{n-j} \frac{1}{k}, & j < \frac{n+1}{2}, \\ 0, & j = \frac{n+1}{2}, \\ \sum_{k=n+1-j}^{j-1} \frac{1}{k}, & j < \frac{n+1}{2}. \end{cases}$$

Combining Theorem 6.59 with the inequality $\sum_{k=n+1-j}^{j-1} \frac{1}{k} < \ln \frac{j-\frac{1}{2}}{n-j+\frac{1}{2}}$ (cf. (6.421)), we obtain the following upper bounds for the expectations of order statistics from the iid standard normal samples:

$$\begin{aligned}\mathbb{E}X_{j:n} &\le \Phi^{-1} \circ W(\mathbb{E}Y_{j:n}) = \Phi^{-1}\left(\left[1 + \exp\left(-\sum_{k=n+1-j}^{j-1} \frac{1}{k}\right)\right]^{-1}\right) \\ &< \Phi^{-1}\left(\frac{j-\frac{1}{2}}{n}\right), \qquad j > \frac{n+1}{2},\end{aligned}$$

which are essentially sharper than the upper bounds in (6.427). Similarly, we can improve the lower bounds of (6.427) by

$$\mathbb{E}X_{j:n} \ge \Phi^{-1}\left(\frac{j-\frac{1}{2}}{n}\right), \qquad j < \frac{n+1}{2}.$$

Theorem 6.59 can be generalized to linear combinations of order statistics as follows.

Theorem 6.60 *Take $c_i \in \mathbb{R}$, $i = p, \ldots, q$, $\frac{n+1}{2} < p < q \leq n$, such that $c_p c_q \neq 0$, and*

$$-1 \leq -C_{\tilde{c}:n} = F^s_{\tilde{c}:n}\left(\frac{1}{2}\right) \leq F^s_{\tilde{c}:n}(x) \leq 0, \qquad \frac{1}{2} \leq x \leq 1. \tag{6.438}$$

where

$$F^s_{\tilde{c}:n}(x) = \sum_{i=p}^{q} c_i[F_{i:n}(x) - F_{n+1-i:n}(x)] = -\sum_{i=p}^{q} c_i \sum_{k=n+1-i}^{i-1} B_{k,n}(x). \tag{6.439}$$

Assume that $X_1, \ldots, X_n$ and $Y_1, \ldots Y_n$ are independent, and identically symmetrically distributed with respective distribution functions F and W such that $\mathbb{E}X_{q:n}$ and $\mathbb{E}Y_{q:n}$ are finite, and the composition $F^{-1} \circ W$ is concave between the center and right end-point of the support of W. Then,

$$\mathbb{E}\sum_{i=p}^{q} c_i X_{i:n} \leq F^{-1} \circ W\left(\mathbb{E}\sum_{i=p}^{q} c_i Y_{i:n}\right). \tag{6.440}$$

Assumption (6.438) is satisfied if all $c_p, \ldots, c_q$ are nonnegative, and their sum does not exceed 1 (see the proof of Theorem 6.59).

Proof For simplicity of calculations, we assume that all the random variables are symmetric about 0. We start with the observation that our assumptions guarantee finiteness of the expectations appearing in (6.440). Furthermore,

$$\mathbb{E}\sum_{i=p}^{q} c_i X_{i:n} = \int_0^d F^{-1} \circ W(x)\, F_{\tilde{c}:n} \circ W(dx)$$

(comp. (6.433)), where d denotes the right-end of the support of W. Note that by (6.433) $F_{\tilde{c}:n} \circ W$ is the distribution function of a possibly signed measure on $[0, d]$ with the total mass $0 < C_{\tilde{c}:n} \leq 1$. It follows that the distribution function $\frac{F_{\tilde{c}:n} \circ W(x)}{C_{\tilde{c}:n}}$ with the function $F^{-1} \circ W(x)$, $0 \leq x \leq d$, satisfy the assumptions for the reversed inequality (2.278) in Theorem 2.45. Hence,

$$\begin{aligned}
\mathbb{E}\sum_{i=p}^{q} c_i X_{i:n} &= C_{\tilde{c}:n} \int_0^d F^{-1} \circ W(x) \frac{F_{\tilde{c}:n} \circ W(dx)}{C_{\tilde{c}:n}} \\
&= C_{\tilde{c}:n} F^{-1} \circ W\left(\int_0^d x \frac{F_{\tilde{c}:n} \circ W(dx)}{C_{\tilde{c}:n}}\right)
\end{aligned}$$

$$= C_{\tilde{c}:n}\, F^{-1} \circ W\left(\frac{\mathbb{E}\sum_{i=p}^{q} c_i Y_{i:n}}{C_{\tilde{c}:n}}\right) + (1 - C_{\tilde{c}:n})\, F^{-1} \circ W(0)$$

$$\leq F^{-1} \circ W\left(\mathbb{E}\sum_{i=p}^{q} c_i Y_{i:n}\right)$$

□

A counterpart of Corollary 6.11 for the linear combinations of order statistics is as follows.

Corollary 6.13 *Let $c_p, \ldots, c_q$, $X_1, \ldots, X_n$, and $Y_1, \ldots, Y_n$ satisfy the assumptions of Theorem* 6.60. *Define $\tilde{c} = (c_1, \ldots, c_n) \in \mathbb{R}^n$ as follows:*

$$c_i = \begin{cases} c_i, & i = p, \ldots, q, \\ -c_{n+1-i}, & i = n+1-q, \ldots, n+1-p, \\ 0, & otherwise. \end{cases}$$

Then

$$\mathbb{E}\sum_{i=1}^{n} c_i X_{i:n} \leq 2\, F^{-1} \circ W\left(\mathbb{E} \sum_{i > \frac{n+1}{2}} c_i Y_{i:n}\right).$$

The corollary applies to the symmetric L-estimates $\sum_{i=1}^{n} c_i X_{i:n}$ of scale with antisymmetric coefficients $c_i = -c_{n+1-i}$, $i = 1, \ldots, n$.

Example For the mean absolute deviation from the mean

$$\text{MAD}_n(\tilde{X}) = \sum_{i=1}^{n} c_i X_{i:n} = \frac{1}{n} \sum_{i \geq \frac{n}{2}+1} X_{i:n} - \frac{1}{n} \sum_{i \leq \frac{n}{2}} X_{i:n}$$

(see (3.33)), all c_i, $i > \frac{n+1}{2}$, are positive, and they sum up to $\frac{1}{2}$ if n is even, and $\frac{n-1}{2n}$ if n is odd. In both the cases, the sum is less than 1, which means that the vector of coefficients in the L-statistic representation satisfies the assumptions of Corollary 6.13 (see the comment following Theorem 6.59). In particular, for $W(x) = x + \frac{1}{2}$, $-\frac{1}{2} < x < \frac{1}{2}$, and F being symmetric U-shaped, we have

$$\mathbb{E}\text{MAD}_n(\tilde{X}) \leq 2F^{-1}\left(\mathbb{E} \sum_{i > \frac{n+1}{2}} \frac{1}{n} Y_{i:n} + \frac{1}{2}\right) = \begin{cases} 2\, F^{-1}\left(\frac{3n+2}{4(n+1)}\right), & \text{if } n \text{ is even}, \\ 2\, F^{-1}\left(\frac{3n-1}{4n}\right), & \text{if } n \text{ is odd}. \end{cases}$$

If F is symmetric unimodal, the inequality is reversed.

Example For the Gini mean differences

$$\mathrm{GMD}_n(\tilde{X}) = \frac{2}{n(n+1)}\sum_{i=1}^{n}(2i-n-1)X_{i:n}$$

(cf. (3.34)), we significantly reduce calculations once we notice that

$$\mathbb{E}\mathrm{GMD}_n(\tilde{X}) = \frac{2}{n(n+1)}\sum_{1\le i<j\le n}\mathbb{E}|X_i - X_j| = \mathbb{E}|X_1 - X_2| = \mathbb{E}(X_{2:2} - X_{1:2}).$$

Therefore, $\mathbb{E}\mathrm{GMD}_n(\tilde{X}) \le 2\,F^{-1}\circ W(\mathbb{E}Y_{2:2})$ if $F^{-1}\circ W$ is symmetric, and first convex and then concave. In particular, $\mathbb{E}\mathrm{GMD}_n(\tilde{X}) \le (\ge) 2\,F^{-1}\left(\frac{2}{3}\right)$ if F is symmetric and U-shaped (unimodal, respectively).

Application of the Steffensen-Jensen inequality to the quantile bounds for the general L-statistics based on iid symmetric populations is described in the following theorem.

Theorem 6.61 *Let $c_1, \ldots, c_n$ satisfy*

$$-1 \le F^s_{\tilde{c}:n}\left(\frac{1}{2}\right) \le F^s_{\tilde{c}:n}(x) \le 0, \qquad \frac{1}{2} \le x \le 1,$$

where $F^s_{\tilde{c}:n}(x) = \sum_{i=1}^n c_i[F_{i:n}(x) - F_{n+1-i:n}(x)]$. If $X_1, \ldots, X_n$ and $Y_1, \ldots, Y_n$ are iid with symmetric distribution functions F and W, respectively, and $F^{-1}\circ W$ is concave right to the symmetry center of W, then

$$\mathbb{E}\sum_{i=1}^{n} c_i X_{i:n} \le F^{-1}\circ W\left(\mathbb{E}\sum_{i=1}^{n} c_i X_{i:n}\right) \tag{6.441}$$

under the condition that the above expectations exist.

Obviously, the inequality is reversed if $F^{-1}\circ W$ is convex on the upper half of the support of W. Under the simplification that F and W are symmetric about 0, the proof is based on the representation

$$\begin{aligned}\mathbb{E}\sum_{i=1}^{n} c_i X_{i:n} &= \int_{\mathrm{supp}\,W} F^{-1}\circ W(x)\, F_{\tilde{c}:n}\circ W(dx)\\ &= \int_{\mathrm{supp}\,W\cap\mathbb{R}_+} F^{-1}\circ W(x)\, F^s_{\tilde{c}:n}\circ W(dx),\end{aligned}$$

and the observation that $F^s_{\tilde{c}:n}$ vanishes at 1. Function $F^s_{\tilde{c}:n}$ has a simpler representation

$$F^s_{\tilde{c}:n}(x) = \sum_{i=1}^{n} \left(\sum_{k=1}^{i} c_k \right) [B_{i,n}(x) + B_{n+1-i,n}(x)] - \sum_{i=1}^{n} c_i .$$

Lawrence (1975, Corollary 4.5) stated that the necessary and sufficient condition for the validity of inequality (6.441) for all F and W satisfying the assumptions of Theorem 6.61 is

$$-1 \le F^s_{\tilde{c}:n}(x) \le 0, \qquad \frac{1}{2} \le x \le 1.$$

The conclusion was based on conditions assuring general inequalities concerning integrals of convex-concave (and concave-convex) functions with respect to signed measures.

Appendix A

A.1 Projections onto Convex Functions

We consider the Hilbert space $L^2((a,d), W(dx))$, where $(a,d) \subset \mathbb{R}$ is an interval, and W is a distribution function of a probability measure on (a,d) with a density function w with respect to the Lebesgue measure. Define the convex cone

$$\mathcal{C}_{\succeq_c W} = \{g \in L^2((a,d), W(dx)) : g \text{ is nondecreasing, convex}\} \tag{A.1}$$

contained in the space. Our purpose is to determine the projection $P_{\succeq_c W} h$ of a function $h \in L^2((a,d), W(dx))$ with some fixed properties onto the cone. The functions W and h satisfy the following assumptions:

(A1) For some $-\infty \le a \le b < c \le d \le \infty$ and distribution function W with a positive probability density w on (a,d) and finite second moment, a function $h : [a,d) \to \mathbb{R}$ is nonnegative, bounded, twice differentiable, satisfies $h(a) = 0$,

$$\int_a^d h(x)w(x)dx = 1, \tag{A.2}$$

and it is strictly increasing and strictly convex on (a,b), strictly increasing and strictly concave on (b,c) and strictly decreasing on (c,d).

We admit that intervals (a,b) and (c,d) could be empty.

N. Balakrishnan, T. Rychlik, *Sharp Inequalities for Ordered Random Variables in Statistics and Reliability*, Frontiers in Probability and the Statistical Sciences,
https://doi.org/10.1007/978-3-031-61347-0

We first determine the shape of the projection in the following Lemma.

Lemma A.1 *Let $\mathcal{C}_* \subset \mathcal{C}_{\succeq_c W}$ be the class of functions satisfying either*

$$g_*(x) = \begin{cases} h(x), & a \le x \le \beta, \\ h(\beta) + \lambda(x - \beta), & \beta \le x < d, \end{cases} \tag{A.3}$$

for some $\lambda \ge h'(\beta)$ and $a \le \beta < b$, or

$$g_*(x) = h(\beta) + \lambda(x - \beta) \tag{A.4}$$

for $\lambda \ge 0$ and $a \le \beta < c$. Then for every $g \in \mathcal{C}_{\succeq_c W}$, there exists $g_* \in \mathcal{C}_*$ such that

$$\|g_* - h\|_W \le \|g - h\|_W. \tag{A.5}$$

Note that under the assumption $\int_a^d x^2 W(dx) < \infty$, both (A.3) and (A.4) belong to $L^2((a, d), W(dx))$.

Proof Suppose that $g \in \mathcal{C}_{\succeq_c W}$. Taking its translation $g_\alpha(x) = g(x) + \alpha$ such that

$$\int_a^d g_\alpha(x) w(x) dx = \int_a^d h(x) w(x) dx, \tag{A.6}$$

by Lemma 6.1(ii), we obtain

$$\|g_\alpha - h\|_W \le \|g - h\|_W. \tag{A.7}$$

Since $g_+ \in \mathcal{C}_{\succeq_c W}$ for $g_+(x) = \max\{g(x), 0\}$, and $h \ge 0$, we also have

$$\|g_+ - h\|_W \le \|g - h\|_W. \tag{A.8}$$

Therefore, we confine ourselves to functions $g \in \mathcal{C}_{\succeq_c W}$ which are nonnegative and satisfy $g < h$ and $g > h$ on some nondegenerate intervals. This implies in particular that $0 \le g(a) < h(c)$ and $g(x) < h(x)$ for some $x < c$. Our purpose is to construct a modification $g_* \in \mathcal{C}_*$ of $g \in \mathcal{C}_{\succeq_c W}$ such that for each $x \in [a, d)$

$$\text{either} \quad g(x) \le g_*(x) \le h(x) \quad \text{or} \quad h(x) \le g_*(x) \le g(x). \tag{A.9}$$

The particular constructions depend on some properties of g, but all of them have either of forms (A.3) and (A.4).

We first look at g and h at a right neighborhood of the inflection point b. Two cases are possible:

$$g(b+) > h(b+), \tag{A.10}$$

$$g(b+) < h(b+). \tag{A.11}$$

In the first case there are two further possibilities:

$$\text{(i)}\ \ g(x) \geq h(x) \qquad \text{for all } x \in (b, d), \tag{A.12}$$

$$\text{(ii)}\ \ g(x) < h(x) \qquad \text{for some } x \in (b, d). \tag{A.13}$$

(i) The former implies that either $g(x) > h(x)$ for $b < x < d$ or $g(\delta) = h(\delta)$ for a single point $\delta \in [b, c)$ and $g(x) < h(x)$ on some interval contained in (a, b). Then there exists a line ℓ_1 that separates the convex bodies above the graph of g and below the graph of h in (b, c). The line runs below convex g everywhere in (a, d). It is also above h in (b, d). Put

$$\tau = \max\{a \leq x \leq b \quad g(x) = h(x)\}. \tag{A.14}$$

Then ℓ_1 passes through h at some $\gamma \in [\tau, b]$. We easily see that

$$g_*(x) = \begin{cases} h(x), & \text{if } x \leq \gamma, \\ \ell_1(x), & \text{if } x \geq \gamma, \end{cases} \tag{A.15}$$

is convex increasing and lies closer to h than g.
(ii) In this case, Eq. (A.13) holds iff $x \in (\delta, \Delta)$ for some $b < \delta < c$ and $\delta < \Delta \leq d$. If $\Delta < d$, then

$$g(x) \begin{cases} < h(x), & \text{if } \delta < x < \Delta, \\ = h(x), & \text{if } x = \Delta, \\ > h(x), & \text{if } x > \Delta. \end{cases} \tag{A.16}$$

If $\Delta = d$, then $d < \infty$ and

$$g(x) \leq h(x), \qquad \delta < x < d. \tag{A.17}$$

In both the cases, define ℓ_2 as the line passing through $(\delta, h(\delta))$ and $(\Delta, h(\Delta))$. This lies above g in (δ, Δ) and below g elsewhere. We also have $\ell_2 < h$ in (δ, Δ), $\ell_2 > h$ in (Δ, d) (if nonempty), and $\ell_2 > h$ in some left neighborhood of δ containing b. If ℓ_2 crosses h at some $\gamma \in (a, b)$, then

$$g_*(x) = \begin{cases} h(x), & \text{if } a \leq x \leq \gamma, \\ \ell_2(x), & \text{if } \gamma \leq x < d, \end{cases} \tag{A.18}$$

is convex increasing and satisfies (A.9). If not, we simply put $g_*(x) = \ell_2(x)$.

Suppose now that (A.11) holds. Since $g(a) \geq h(a) = 0$, point (A.14) is well-defined, and we have

$$g(\tau+) < h(\tau+) \tag{A.19}$$

as well. Moreover, there is at most a single point $\Delta > b$ such that $g(\Delta) = h(\Delta)$. If $g < h$ in (τ, d), we put $\Delta = d$ (in case $\Delta = d = +\infty$, function g is constant and belongs to $\mathcal{C}_*$).

Define the family of linear functions

$$\ell_y(x) = h(y) + h'(y)(x - y), \qquad a \le y \le b, \tag{A.20}$$

that are tangent to h at y. Function

$$y \mapsto \ell_y(\Delta), \qquad a \le y \le b, \tag{A.21}$$

is continuously increasing. Under notation (A.14), two cases are possible:

$$\text{(iii)} \qquad \ell_\tau(\Delta) \le h(\Delta), \tag{A.22}$$

$$\text{(iv)} \qquad \ell_\tau(\Delta) > h(\Delta). \tag{A.23}$$

In case (iii), we have

$$\ell_\tau(\Delta) \le h(\Delta) < \ell_b(\Delta). \tag{A.24}$$

Consequently, there exists $\delta \in [\tau, b)$ such that $\ell_\delta(\Delta) = h(\Delta)$.

Function $\ell_\delta(x)$ has the following properties:

$$g(x) \le \ell_\delta(x) \le h(x), \qquad \text{if } x \in [\delta, \Delta), \tag{A.25}$$

$$h(x) \le \ell_\delta(x) \le g(x), \qquad \text{if } x \in [\Delta, d), \tag{A.26}$$

and $\ell'_\delta(x) = h'(\delta)$. Therefore,

$$g_*(x) = \begin{cases} h(x), & a \le x \le \delta, \\ \ell_\delta(x), & \delta \le x < d, \end{cases} \tag{A.27}$$

is convex increasing, satisfies (A.3) and lies closer to h than g.

(iv) If $\ell_\tau(\Delta > h(\Delta) \ge g(\Delta)$, define ℓ_4 as the straight line that runs through $(\tau, g(\tau))$ and $(\Delta, g(\Delta))$. It crosses ℓ_τ at τ and has a smaller slope. Then

$$\ell_4(x) \begin{cases} \le g(x), & \text{if } x \in [a, \tau) \cup [\Delta, d), \\ \ge g(x), & \text{if } x \in [\tau, \Delta), \end{cases} \tag{A.28}$$

and

$$\ell_4(x) \begin{cases} \le h(x), & \text{if } x \in [\tau, \Delta], \\ \ge h(x), & \text{if } x \in [\tau, \Delta). \end{cases} \tag{A.29}$$

If $\tau = 0$, then ℓ_4 lies between g and h everywhere and we put $g_* = \ell_4$. Otherwise, since $h(\tau) = \ell_4(\tau)$ and $h'(\tau) = \ell'_\tau > \ell'_4$, we also have $\ell_4(x) > h(x)$ in some left neighborhood of τ. If this holds for all $a < x < \tau$, we simply replace g by $g_* = \ell_4$. Otherwise we take

$$g_*(x) = \begin{cases} h(x), & \text{if } a \le x \le \delta, \\ \ell_4(x), & \text{if } \delta \le x < d, \end{cases} \tag{A.30}$$

for

$$\delta = \sup\{0 < x < \tau : \ell_4(x) = h(x)\}. \tag{A.31}$$

Then we have

$$h(x) \le \ell_4(x) \le g(x), \qquad \delta \le x \le \tau, \tag{A.32}$$

which shows that (A.9) holds. □

We proceed to presenting the analytic forms of the projection $P_{\succeq_c W} h$ for h satisfying (A1). To this end, we apply some auxiliary notions:

$$\lambda_*(\beta) = \frac{\int_\beta^d (x-\beta)[h(x)-h(\beta)]W(dx)}{\int_\beta^d (x-\beta)^2 W(dx)}, \tag{A.33}$$

$$K(\beta) = \lambda_*(\beta) - h'(\beta), \tag{A.34}$$

$$L(\beta) = \int_\beta^d h(x)W(dx) - \int_\beta^d [\lambda_*(\beta)(x-\beta) + h(\beta)]W(dx) \tag{A.35}$$

for $a \le \beta \le b$, and

$$\mathcal{K} = \{a < \beta \le b : K(\beta) \ge 0 \text{ and } L(\beta) = 0\}, \tag{A.36}$$

$$\lambda_* = \frac{\int_a^d (x-a)h(x)W(dx) - \int_a^d (x-a)W(dx)}{\int_a^d (x-a)^2 W(dx) - [\int_a^d (x-a)W(dx)]^2}, \tag{A.37}$$

$$\gamma_* = \frac{\int_a^d (x-a)^2 W(dx) - \int_a^d (x-a)h(x)W(dx) \int_a^d (x-c)W(dx)}{\int_a^d (x-a)^2 W(dx) - [\int_a^d (x-a)W(dx)]^2}. \tag{A.38}$$

Now we are in a position to formulate the following theorem.

Theorem A.1 (See Theorem 6.2) *If $\mathcal{K}$ is nonempty and $\beta_* = \sup \mathcal{K}$, then*

$$P_{\succeq_c W} h(x) = \begin{cases} h(x), & a \le x \le \beta_*, \\ \lambda_*(\beta_*)(x - \beta_*) + h(\beta_*), & \beta_* \le x < d. \end{cases} \tag{A.39}$$

If $\mathcal{K} = \emptyset$, and $a > -\infty$, and $\lambda_* > 0$, then

$$P_{\succeq_c W} h(x) = \lambda_*(x - a) + \gamma_*. \tag{A.40}$$

If $\mathcal{K} = \emptyset$, and either $a = -\infty$ or $\lambda_* \le 0$, then

$$P_{\succeq_c W} h(x) = 1. \tag{A.41}$$

Equation (A.33) presents the slope of the best approximation of h on $[\beta, d)$ among all the linear functions running through $(\beta, h(\beta))$. By (A.39), projection of form (A.3) has to have such an optimal slope. However, function (A.39) belongs to the cone of nondecreasing convex functions iff $K(\beta) \ge 0$. By Lemma 6.1, $L(\beta) = 0$ is the necessary condition for (A.39) to be the projection. Theorem A.1 asserts that if for some $a \le \beta \le b$ there exist functions (A.3) with optimal slope (A.33) that satisfy the necessary conditions $K(\beta) \ge 0$ and $L(\beta) = 0$, then the projection is that of the greatest such β. Otherwise the projection is linear. Equations (A.37) and (A.38) represent the slope and intercept, respectively, of the best linear approximation of h in $L^2([a, d), w(dx))$. The linear approximation may be the projection onto $\mathcal{C}_{\succeq_c W}$ if the slope is nonnegative and the approximation function is nonnegative on (a, d). This is impossible if $a = -\infty$. Therefore, the projection coincides with the optimal constant if either (A.37) is nonpositive or $a = -\infty$.

Proof of Theorem A.1 We first look for the projection in form (A.3), trying to establish optimal parameters λ and β. Keeping $\beta \in [a, b]$ fixed, we minimize

$$D(\lambda, \beta) = \|h - g\|_W^2 = \int_\beta^d [h(x) - h(\beta) - \lambda(x - \beta)]^2 W(dx) \tag{A.42}$$

with respect to $\lambda \ge h'(\beta)$. This is a quadratic function with the minimum at

$$\lambda_{**}(\beta) = \max\{h'(\beta), \lambda_*(\beta)\}, \tag{A.43}$$

where $\lambda_*(\beta)$ defined in (A.33) is the global minimum point.

Suppose that $h'(\beta) > \lambda_*(\beta)$ on some interval. Then

$$\begin{aligned} \frac{dD(\beta, h'(\beta))}{d\beta} &= -2h''(\beta) \int_\beta^d [h(x) - h'(\beta)(x - \beta) - h(\beta)](x - \beta) W(dx) \\ &\ge 0, \end{aligned} \tag{A.44}$$

because $h''(\beta) \geq 0$ and

$$\int_{\beta}^{d} [h(x) - h'(\beta)(x-\beta)h(\beta)](x-\beta)W(dx)$$
$$\leq \int_{\beta}^{d} [h(x) - \lambda_*(\beta)(x-\beta) - h(\beta)](x-\beta)W(dx) = 0 \tag{A.45}$$

by assumption and definition of $\lambda_*(\beta)$. Therefore, decreasing β we also decrease $D(\beta, \lambda_{**}(\beta)) = D(\beta, h'(\beta))$ until we arrive to β with $h'(\beta) = \lambda_*(\beta)$ or $\beta = a$. Summing up, we have proven that either $P_{\succeq_c W} h$ has the form

$$g_{*\beta}(x) = \begin{cases} h(x), & a \leq x \leq \beta, \\ h(\beta) + \lambda_*(\beta)(x-\beta), & \beta \leq x < d, \end{cases} \tag{A.46}$$

for some $a \leq \beta \leq b$ such that $K(\beta) \geq 0$ or it is linear.

Further, we have

$$\frac{dD(\beta, \lambda_*(\beta))}{d\beta}$$
$$= 2\int_{\beta}^{d} [h(x) - h(\beta) - \lambda_*(\beta)(x-\beta)][\lambda_*(\beta) - h'(\beta) - \lambda_*'(\beta)(x-\beta)]W(dx)$$
$$= 2K(\beta)L(\beta) - 2\lambda_*'(\beta)\int_{\beta}^{d} [h(x) - h(\beta) - \lambda_*(\beta)(x-\beta)](x-\beta)W(dx)$$
$$= 2K(\beta)L(\beta). \tag{A.47}$$

In fact, by Lemma 6.1, $L(\beta) = 0$ is necessary for (A.46) to be the projection. Therefore it suffices to consider $\beta \in \mathcal{K}$. One can check that $L(\beta) = 0$ is also the necessary and sufficient condition for $h(\beta) + \lambda_*(\beta)(x-\beta)$ to be the best linear approximation of h in $L^2([\beta, d), W(dx))$. Define

$$g_{**\beta}(x) = \begin{cases} h(x), & a \leq x \leq \beta, \\ \ell_\beta(x), & \beta \leq x < d, \end{cases} \qquad a \leq \beta \leq b, \tag{A.48}$$

where ℓ_β denotes the projection of h onto the linear subspace of linear functions on $L^2([\beta, d), W(dx))$. Since for $\beta_1 < \beta_2$

$$\int_{\beta_1}^{d} [h(x) - \ell_{\beta_1}(x)]^2 W(dx) \geq \int_{\beta_2}^{d} [h(x) - \ell_{\beta_1}(x)]^2 W(dx)$$
$$\geq \int_{\beta_2}^{d} [h(x) - \ell_{\beta_2}(x)]^2 W(dx), \tag{A.49}$$

function $\beta \to \|h - g_{**\beta}\|_W$ is nonincreasing. Consequently, if $\beta \in \mathcal{K}$, then (A.46) and (A.48) coincide, and

$$\|h - g_{**\beta}\|_W \le \|h - \ell_0\|_W \le \|h - \ell_0^+\|_W, \tag{A.50}$$

where ℓ_0 and ℓ_0^+ denote the best linear and linear nondecreasing approximations of h in $L^2([a, d), W(dx))$, respectively. Also, if $\beta_1, \beta_2 \in \mathcal{K}$ and $\beta_1 > \beta_2$, then

$$\|h - g_{**\beta_1}\|_W \le \|h - g_{**\beta_2}\|_W. \tag{A.51}$$

We have thus concluded the following. If (A.36) is nonempty, then $P_{\succeq_c W} h$ has the form (A.46) with parameter β being the maximal element of (A.36). Otherwise, it is linear.

In this case, if the projection onto the subspace of linear functions in nondecreasing, then it belongs to $\mathcal{C}_{\succeq_c W}$ and obviously becomes the projection. This has the form (A.40) with parameters (A.37) and (A.38). If it decreases, then we solve the linear approximation problem under the constraint to nonnegative slope parameters. We easily check that the solution is constant. This completes the proof. □

We can also deduce the following useful statement.

Proposition A.1 *Suppose that the set of zeros of* (A.35) *does not contain intervals and*

$$\mathcal{K}^+ = \{a < \beta < b : K(\beta) > 0\} \tag{A.52}$$

is nonempty. Then each subinterval of (A.52) *contains one zero at most, and* (A.35) *changes the sign from* $-$ *to* $+$*, and* $D(\beta, \lambda_*(\beta))$ *has a local minimum there. In particular, if* (A.52) *is a single interval and* (A.35) *has a zero there, then this is unique and provides the global minimum of* $D(\beta, \lambda_*(\beta))$*.*

Proof Differentiating the integral of the second line equation in Eq. (A.45) and taking into account that it vanishes, we obtain

$$-\int_\beta^d h(x)W(dx) - \lambda_*'(\beta)\int_\beta^d (x-\beta)^2 W(dx) + 2\lambda_*(\beta)\int_\beta^d (x-\beta)W(dx)$$
$$- h'(\beta)\int_\beta^d (x-\beta)W(dx) + h(\beta)\int_\beta^d W(dx) = 0,$$

which is equivalent to

$$L(\beta) = \int_\beta^d [K(\beta) - \lambda_*'(\beta)(x-\beta)](x-\beta)W(dx). \tag{A.53}$$

By definition, we also have

$$L'(\beta) = \int_{\beta}^{d} [\lambda_*(\beta) - h(\beta) - \lambda'(\beta)(x - \beta)]W(dx). \tag{A.54}$$

Suppose that $L(\beta) = 0$ at some $\beta \in \mathcal{K}^+$. Then, by (A.47), (A.54) and (A.53),

$$\begin{aligned}
\frac{d^2 D(\beta, \lambda_*(\beta))}{d\beta^2} &= 2K(\beta)L'(\beta) + 2K'(\beta)L(\beta) = 2K(\beta)L'(\beta) - 2\lambda_*'(\beta)L(\beta) \\
&= 2K(\beta) \int_{\beta}^{d} [K(\beta) - \lambda_*'(\beta)(x - \beta)]W(dx) \\
&- 2\lambda_*'(\beta) \int_{\beta}^{d} [K(\beta) - \lambda_*'(\beta)(x - \beta)](x - \beta)W(dx) \\
&= 2 \int_{\beta}^{d} [K(\beta) - \lambda_*'(\beta(x - \beta)]^2 W(dx) \geq 0.
\end{aligned} \tag{A.55}$$

Accordingly, every zero of (A.35) provides a local minimum of $D(\beta, \lambda_*(\beta))$, and it is impossible to have a different minimum and any maximum in any subinterval of (A.52). □

Now we consider functions satisfying the following assumptions:

(A2) For some $0 < a < b < c \leq d \leq +\infty$ and a distribution function W with a positive density w on $[0, d)$ and finite variance, function $h : [0, d) \to \mathbb{R}$ is bounded, twice differentiable, and satisfies $h(0) = 0$, $\lim_{x \to d} h(x) \geq 0$,

$$\int_{0}^{d} h(x)W(dx) = 0, \tag{A.56}$$

and it is decreasing on $(0, a)$ and (c, d) (if the latter is not empty), strictly convex increasing on (a, b), and strictly concave increasing on (b, c).

Conditions (A2) imply that h has a unique zero, say θ, in $(0, d)$ located between a and c, but the mutual relation between θ and the inflection point b is not determined.

Our purpose is to determine the projection of the function satisfying (A2) onto the convex cone

$$\mathcal{C}_{\geq_c W} = \{g \in L^2((0, d), W(dx)) : g \text{ is nondecreasing, convex}\}. \tag{A.57}$$

We first describe the shapes of the candidates for the projection as follows.

Lemma A.2 *Let $\mathcal{C}^* \subset \mathcal{C}_{\succeq_c W}$ be the class of functions of the form*

$$g^*(x) = \begin{cases} h(\alpha), & 0 \le x < \alpha, \\ h(x), & \alpha \le x < \beta, \\ \lambda(x-\beta) + h(\beta), & \beta \le x < d, \end{cases} \tag{A.58}$$

for some $a \le \alpha < \beta \le b$, $\lambda \ge h'(\beta)$, or

$$g^*(x) = \begin{cases} \gamma, & 0 \le x < \beta, \\ \lambda(x-\beta) + \gamma, & \beta \le x < d, \end{cases} \tag{A.59}$$

for $\lambda \ge 0$, $\gamma \in \mathbb{R}$. Then, for any $g \in \mathcal{C}_{\succeq_c W}$, there exists a function $g^* \in \mathcal{C}^*$ such that

$$\left\| h - g^* \right\|_W \le \| h - g \|_W .$$

Proof We show that for any $g \in \mathcal{C}_{\succeq_c W}$ we can find function $g^* \in \mathcal{C}^*$, which is less distant from h than g. Our proof starts with the observation that it suffices to consider functions g satisfying $g(0) < 0$. Monotonicity of g and the fact that g integrates to 0 imply that $g(0) < 0 < \lim_{x \to d} g(x)$ or $g(x) = 0$ for $x \in [0, d)$. We exclude the latter case since there exists a function that vanishes in $[0, \theta]$, and is linear increasing in (θ, d) is a better approximation of h than constant 0 (see Gajek & Rychlik, 1998). As $\max\{g, h(a)\}$ is nondecreasing convex and is less distant from h than g, it suffices to restrict our attention to functions g satisfying $0 > g(0) \ge h(a)$. Since $h - g$ is continuous, the set $\{x \in [0, d) : h(x) = g(x)\}$ is closed. It follows that there exist at most countably many closed intervals (possibly degenerate) where $h = g$. Note that it suffices to consider these g for which the set $\{h = g\}$ contains at most one nondegenerate interval. Conversely, suppose that there are at least two such intervals. They must be subsets of $[a, b]$, because g is nondecreasing convex. If $h = g$ in some $[\alpha_1, \beta_1] \cup [\alpha_2, \beta_2]$, $\beta_1 < \alpha_2$ and $h \ne g$ in (β_1, α_2), then

$$\tilde{g}(x) = \begin{cases} h(x), & x \in (\beta_1, \alpha_2), \\ g(x), & x \notin (\beta_1, \alpha_2), \end{cases}$$

is nondecreasing convex and $\|h - g\| \ge \|h - \tilde{g}\|$, which gives a contradiction. Now we need to consider two cases:

(I) the set $\{g = h\}$ contains a nondegenerate interval,
(II) the set $\{g = h\}$ does not contain any nondegenerate interval.

(I) Suppose first that $h = g$ on some $[\alpha, \beta] \subset [a, b]$, $\alpha < \beta$. We are to show that there exists function $\tilde{g}$ of the form (A.58), less distant from h than g. Take an arbitrary $\xi \in [\alpha, \beta]$ and denote by h_1 the closest to $h_{|[0,\xi]}$ nondecreasing function taking value $h(\xi)$ at ξ, and by h_2 the closest to $h_{|[\xi,d)}$ nondecreasing convex function such that $h_2(\xi) = h(\xi)$. We are now in position to show that h_1 is either constant equal to $h(\xi)$, or constant equal to $h(\eta)$, $a \le \eta < \xi$, on $[0, \eta]$ and equal to h on $[\eta, \xi]$, and h_2 is continuous and equal to h on $[\xi, \nu]$ and increasing linear on $[\nu, d)$ for some $\nu \in [\xi, d]$. Note that h_1 is convex, and so it is the best approximation of $h_{|[0,\xi]}$ in the class of nondecreasing convex functions. Furthermore, function

$$\tilde{g}(x) = \begin{cases} h_1(x), & x \in [0, \xi], \\ h_2(x), & x \in (\xi, d), \end{cases}$$

is nondecreasing and convex, satisfies (A.58) and is less distant from h than g.

Now our goal is to find the closest to $h_{|[0,\xi]}$ nondecreasing function, $P_1 h$, say. Applying the modification of the Moriguti method of obtaining greatest convex minorants, we observe that either $P_1 h(x) = \zeta$ and $\zeta > h(\xi)$, or $P_1 h$ is constant equal to $h(\eta)$, $a \le \eta < \xi$, on $[0, \eta]$, and equal to h on $[\eta, \xi]$. Only in the latter case the projection has the required form. We proceed to show that if the former holds, then the closest to $h_{|[0,\xi]}$ nondecreasing function taking value $h(\xi)$ at ξ is constant equal to $h(\xi)$ on $[0, \xi]$. Lemma 6.1 yields

$$\int_0^\xi h(x)\, W(dx) = \int_0^\xi P_1 h(x)\, W(dx) = \zeta \int_0^\xi W(dx).$$

Any nondecreasing function g such that $g(\xi) = h(\xi)$ can be represented as

$$g(x) = g_0(x) + h(\xi) - g_0(\xi),$$

where g_0 is nondecreasing and

$$\int_0^\xi g_0(x)\, W(dx) = \int_0^\xi h(x)\, W(dx) = \zeta \int_0^\xi W(dx),$$

and $g_0(0) \le \zeta \le g_0(\xi)$. Therefore

$$\begin{aligned} \|h - g\|_W^2 &= \|h - g_0 - [h(\xi) - g_0(\xi)]\,\mathbf{1}\|_W^2 \\ &= \|h - g_0\|_W^2 - 2\,[h(\xi) - g_0(\xi)]\,\langle h - g_0, \mathbf{1}\rangle \\ &\quad + [h(\xi) - g_0(\xi)]^2\,\langle \mathbf{1}, \mathbf{1}\rangle. \end{aligned}$$

Combining $\|P_1 h - h\|_W \le \|g_0 - h\|_W$, $\langle h - g_0, \mathbf{1}\rangle = 0$ and $\langle h - P_1 h, \mathbf{1}\rangle = 0$ with $h(\xi) \le \zeta \le g_0(\xi)$, we deduce that

$$\begin{aligned} \|h - g\|_W^2 &\ge \|h - P_1 h\|_W^2 + [h(\xi) - P_1 h(\xi)]^2\,\langle \mathbf{1}, \mathbf{1}\rangle \\ &= \|h - P_1 h - [h(\xi) - P_1 h(\xi)]\,\mathbf{1}\|_W^2 = \|h - h(\xi)\,\mathbf{1}\|_W^2. \end{aligned}$$

It follows that the closest to $h_{|[0,\xi]}$ nondecreasing function taking at ξ value $h(\xi)$ is either constant equal to $h(\xi)$ on $[0,\xi]$ or constant $h(\eta)$, $\eta < \xi$, on $[0,\eta]$ and equal to h on $[\eta,\xi]$.

It remains to observe that function h_2 is of the form described above. This has been proved in Lemma 1 of Gajek and Rychlik (1998) where the projections of functions satisfying (A1) onto the cones of nondecreasing, convex and starting from the same point as the original functions were constructed (cf. Theorem 6.8).

(II) Assume now that the set $\{g = h\}$ does not contain any interval. We prove that there exists function $\tilde{g}$ either of the form (A.58) or (A.59), such that $\|h - \tilde{g}\|_W \leq \|h - g\|_W$.

Suppose that there are some $a \leq \alpha < \beta \leq b$, such that $g(\alpha) = h(\alpha)$, $g(\beta) = h(\beta)$ and $g(x) < h(x)$ for $x \in (\alpha, \beta)$. Set

$$g_1(x) = \begin{cases} h(x) & x \in (\alpha, \beta), \\ g(x) & x \notin (\alpha, \beta). \end{cases}$$

Obviously, $\|h - g_1\|_W \leq \|h - g\|_W$ and g_1 is nondecreasing. If $g(\alpha-) < h(\alpha-)$, then $g'(\alpha) = h'(\alpha)$. If $g(\alpha-) > h(\alpha-)$, then $g'(\alpha) < h'(\alpha)$. Similarly, inequalities $g(\beta+) < h(\beta+)$ and $g(\beta+) > h(\beta+)$ imply $g'(\beta) = h'(\beta)$ and $g'(\beta) > h'(\beta)$, respectively. These imply that g_1 has nondecreasing derivative. Thus g_1 is convex. Note that the set $\{g_1 = h\}$ contains an interval, so we deal with the case (I) and can find the approximation of the form (A.58). The above shows that the projection cannot have two separate points in the interval (a, b) of convex increase of h where $P_{\preceq_c W} h = h$. We exclude such functions from our further considerations.

(i) Suppose now that $g(a+) \leq h(a+)$. Our claim is that there exists function $\tilde{g}$ of the form (A.58) which is less distant from h than g. As we assumed that $g(a) \geq g(0) \geq h(a)$, it suffices to consider the case $g(a) = h(a)$. If $g(a+) = h(a+)$, which means that $g = h$ on some interval, then we proceed as in the case (I).

Assume then that $g(a+) < h(a+)$. If there exists $\beta \in (a, b]$, such that $g(\beta) = h(\beta)$, then g can be replaced by function equal to h on some interval, what leads to the case (I). Otherwise, two more subcases are possible: either g crosses h at one point $\delta \in (b, d)$, or g runs beneath h in the whole (a, d). But then $d < \infty$ and we set $\delta = d$.

Let l_t denote tangent to h at t. The slopes of l_t continuously increase in $[a, b)$ and so do values $l_t(\delta)$ ranging from $l_a(\delta) < h(\delta)$ to $l_b(\delta) > h(\delta)$. It follows that there exists $\phi \in (a, b)$ such that $l_\phi(\delta) = h(\delta)$. Set

$$\tilde{g}(x) = \begin{cases} h(a), & 0 \leq x < a, \\ h(x), & a \leq x < \phi, \\ l_\phi(x), & \phi \leq x < d. \end{cases}$$

Obviously $\tilde{g}$ is of the form (A.58), it is nondecreasing and convex, and is less distant to h than g.

(ii) Consider the case $g(a+) > h(a+)$. Since $g(0) < 0$, it follows that there exists $\alpha \in [0, a]$ such that $h(\alpha) = g(\alpha)$. Clearly, there also exists $\alpha' \in [a, b]$ such that $h(\alpha') = h(\alpha)$. Note that the constant $h(\alpha)$ is less distant from h than g on $[0, \alpha']$.

Our next objective is to show that if $g(a+) > h(a+)$ and the set of degenerated intervals $\{g = h\} \cap (a, b]$ contains at least two points, then either of two cases holds:

(1) there exists function $\tilde{g}$ of the form (A.58) which is less distant from h than g,
(2) there are only two such points $\beta < \phi$, and then $g > h$ on (β, ϕ) and $g(\phi+) < h(\phi+)$.

If $h(x) = g(x)$ for $x \in \{\beta, \phi\}$, $\beta < \phi$ and $g < h$ on (β, ϕ), then, as we have already shown, g can be replaced by some function of the form (A.58). If $h(x) = g(x)$ for $x \in \{\beta, \phi\}$, $h(x) \neq g(x)$ for $x \in (a, \beta)$ and $g(\beta-) > h(\beta-)$, $g > h$ on (β, ϕ) and $g(\phi+) > h(\phi+)$, then function

$$g_1(x) = \begin{cases} h(x), & x \in (\beta, \phi), \\ g(x), & x \notin (\beta, \phi), \end{cases}$$

is less distant from h than g, belongs to the cone and can be replaced by function of the form (A.58), because the set $\{g_1 = h\}$ contains an interval. Therefore, inequality $g(a+) > h(a+)$ combined with the assumption that $\{g = h\}$ does not contain an interval, imply four cases:

(i) $\{g = h\} \cap (a, b) = \emptyset$, i.e. $g > h$ on (a, b),
(ii) $\{g = h\} \cap (a, b) = \{\phi\}$ and $g(\phi+) > h(\phi+)$,
(iii) $\{g = h\} \cap (a, b) = \{\phi\}$ and $g(\phi+) < h(\phi+)$,
(iv) $\{g = h\} \cap (a, b) = \{\beta, \phi\}$ and $g > h$ on (β, ϕ) and $g(\phi+) < h(\phi+)$.

If either (i) or (ii) hold, then there are two possibilities:

(a) $g \geq h$ on the whole (b, d) (g and h may be tangent at some $\delta \in (b, c)$),
(b) there exist $\delta \in (b, c)$ and $\Delta \in (\delta, d)$ such that $g > h$ on (b, δ) and (Δ, d) and $g < h$ on (δ, Δ).

If (a) holds, then there exists nondecreasing linear function l, such that $h \leq l \leq g$ on (b, c), where g is convex and h concave. Moreover, $h \leq l \leq g$ on (c, d), and $l \leq g$ on $(0, b)$. In case (i), we have

$$l(0) \leq g(0) \leq g(\alpha) = h(\alpha) = h(\alpha'),$$

so l crosses $h_1 = \max\{h(\alpha), h\}$ at some α''. Function

$$g_2(x) = \begin{cases} h(\alpha), & 0 \leq x \leq \alpha'', \\ l(x), & \alpha'' < x < d, \end{cases}$$

belongs to the family of nondecreasing and convex functions, and is of the form (A.58) if only $\alpha'' > \alpha'$. Otherwise it is of the form (A.59). The constant equal to $h(\alpha)$ is less distant from h than g on $(0, \alpha'')$ and $h \leq l \leq g$ holds on (α'', d). If case (ii) holds, then $g(\phi) = h(\phi)$. It follows that l crosses h at $\alpha'' \geq \phi$. Hence, g_2 is of the form (A.58) and is less distant from h than g.

Assume now that (b) holds. Take the line l_2 passing through $(\delta, h(\delta))$ and $(\Delta, h(\Delta))$. Then, $l_2(x) \leq g(x)$ for $x \in [0, \delta]$, $g(x) \leq l_2(x) \leq h(x)$ for $x \in [\delta, \Delta]$ and $h(x) \leq l_2(x) \leq g(x)$ for $x > \Delta$ and $h(\delta-) < l_2(\delta-)$. Applying the same arguments as in (a) we conclude that l_2 crosses $h_1 = \max\{h(\alpha), h\}$ at some α''. Therefore, we improve the approximation replacing g by

$$g_3(x) = \begin{cases} h(\alpha), & 0 \leq x \leq \alpha'', \\ l_2(x), & \alpha'' < x < d, \end{cases}$$

which is either of the form (A.58) or (A.59).

It remains to consider cases (iii) and (iv). Then, function g runs beneath h on $[\phi, b]$ where h is convex and either $g < h$ on (b, d) or g crosses h at a unique $\Delta \in (b, d)$. If the former holds, then we set $\Delta = d < \infty$. Let l_3 be the line passing through $(\phi, h(\phi))$ and $(\Delta, h(\Delta))$, with $h(\Delta) \geq g(\Delta)$. Then, $l_3(x) \leq g(x)$ for $x \in [0, \phi]$, $h(\phi-) < l_3(\phi-)$ and $g(x) \leq l_3(x) \leq h(x)$ for $x \in [\phi, \Delta]$ and $h(x) \leq l_3(x) \leq g(x)$ for $x > \Delta$. Since $l_3(x) \leq g(x)$ for $x < \phi$ and $g(0) < h(0)$, we see that $l_3(x) = h(x)$ for some $x > \phi$. Moreover, l_3 crosses $h_1 = \max\{h(\alpha), h\}$ at some $\alpha'' \in (\alpha, \phi]$. Let

$$g_4(x) = \begin{cases} h(\alpha), & 0 \leq x \leq \alpha'', \\ l_3(x), & \alpha'' < x < d. \end{cases}$$

If (iv) holds, then g_4 is of the form (A.58). If (iii) holds, then g_4 is either of the form (A.58) or (A.59). Obviously, $\|g_4 - h\|_W \leq \|g - h\|_W$. This completes the proof. □

The next Lemma is also useful in establishing the projections of functions satisfying (A2) onto (A.57).

Lemma A.3 *Let* $\bar{h} : (0, d] \longmapsto \mathbb{R}$ *be given by*

$$\bar{h}(\beta) = \frac{\int_0^\beta h(x)\, W(dx)}{\int_0^\beta W(dx)}.$$

Then

(i) $\bar{h}(d) = 0$ *and* $\bar{h}(\beta) < 0$ *for any* $\beta \in (0, d)$,
(ii) there exists a unique $\alpha_* \in (a, \theta)$ *such that*

$$\bar{h}(\alpha_*) = h(\alpha_*), \tag{A.60}$$

function $\bar{h}$ is decreasing and $\bar{h} > h$ in $(0, \alpha_*]$, and $\bar{h}$ is increasing and $\bar{h} < h$ in (α_*, d).

Proof

(i) From (A2), we have $\int_0^d h(x) W(dx) = 0$ and clearly $\bar{h}(d) = 0$. Since $h(\beta) \le 0$ for any $\beta \le \theta$, we see that $\int_0^\beta h(x) W(dx) < 0$, which implies $\bar{h}(\beta) < 0$ for $\beta \le \theta$. Likewise, $h(\beta) \ge 0$ for $\beta > \theta$, which gives $0 < \int_\beta^d h(x) W(dx) = -\int_0^\beta h(x) W(x)$. Thus $\bar{h}(\beta) < 0$ for $\beta > \theta$.

(ii) We have

$$\bar{h}'(\beta) = \frac{h(\beta) w(\beta) \int_0^\beta W(dx) - w(\beta) \int_0^\beta h(x) W(dx)}{\left[\int_0^\beta W(dx)\right]^2} = \frac{w(\beta)\left[h(\beta) - \bar{h}(\beta)\right]}{\int_0^\beta W(dx)}.$$

It suffices to show that $\bar{h} > h$ in $[0, \alpha_*]$ and $\bar{h} < h$ in (α_*, d) for some $\alpha_* \in (a, \theta)$. Function W is strictly increasing and $W(d) = 1$. With the notation

$$H_W(\gamma) = \int_0^\gamma h \circ W^{-1}(x)\, dx$$

for $\gamma \in [0, 1]$, we have

$$H_W(W(\beta)) = \int_0^{W(\beta)} h \circ W^{-1}(x)\, dx = \int_0^\beta h(x) W(dx). \tag{A.61}$$

Note that $H_W'(\gamma) = h \circ W^{-1}(\gamma)$ and $H_W'(W(\beta)) = h(\beta)$. Therefore $H_W'(0) = 0$, $H_W'(\gamma)$ decreases for $\gamma \in (0, W(a))$ and $\gamma \in (W(c), 1)$, increases for $\gamma \in (W(a), W(c))$, and $H_W'(W(\theta)) = 0$. It follows that $H_W(\gamma)$ is concave decreasing for $\gamma \in (0, W(a))$, convex decreasing for $\gamma \in (W(a), W(\theta))$, convex increasing for $\gamma \in (W(\theta), W(c))$, concave increasing for $\gamma \in (W(c), 1)$ and attains its local minimum at $\gamma = W(\theta)$. Moreover, $H_W(0) = 0$ and

$$H_W(1) = \int_0^1 h \circ W^{-1}(x)\, dx = \int_0^d h(x) W(dx) = 0.$$

According to the definition of W and (A.61), we have $\bar{h}(\beta) = \frac{H_W(W(\beta))}{W(\beta)}$. It follows that $\bar{h}(\beta)$ is the slope of the linear function l_β passing through $(0, 0)$ and $(W(\beta), H_W(W(\beta)))$. By concavity of H_W in $[0, W(a)]$, l_β lies beneath H_W for $W(\beta) \in (0, W(a))$, i.e., for $\beta \in [0, a]$. Every line tangent to H_W at point

$W(\beta) \in (0, W(a))$ lies over H_W in the considered interval. As $H'_W(W(\beta)) = h(\beta)$, we see that $\bar{h}(\beta) > h(\beta)$ for $\beta \in [0, a]$.

It follows from the convexity of H_W in $[W(a), W(\theta)]$ that there is a unique $\alpha_* \in [a, \theta]$ such that the difference $\bar{h} - h$ changes its sign from plus to minus. More precisely, α_* is such the argument such that the line passing through (0, 0) and $(W(\alpha_*), H_W(W(\alpha_*)))$ is tangent to H_W at $W(\alpha_*)$. Moreover, for $\beta \in (\theta, d)$, we have $\bar{h}(\beta) < 0 < h(\beta)$.

□

Lemma A.4 *Assume that* α_* *satisfies* (A.60). *Then for any* $\beta \in (0, c]$, *the function of the form*

$$g(x) = \begin{cases} \bar{h}(\beta), & 0 \le \beta \le \alpha_* \\ h(\max\{x, \alpha_*\}), & \alpha_* < \beta \le c, \end{cases} \qquad 0 < x < \beta, \tag{A.62}$$

is the projection of $h_{|[0,\beta]}$ *onto the convex cone of nondecreasing functions in the Hilbert space* $L^2((0, \beta), W(dx))$.

Proof Rychlik (2001, Example 3, pp. 14–16) presented the solution to the problem of projecting functions onto the convex cone of nondecreasing functions in spaces of the type L^2 with nonuniform weight function. In particular, the projection of $h_{|[0,\beta)} \in L^2([0, \beta), W(dx))$ is function $\left(\bar{H}_W^{\beta}\right)' \circ W$, where $\bar{H}_W^{\beta}$ is the greatest convex minorant of H_W on $[0, \beta)$. According to properties of H_W and superposition $H_W \circ W$, presented above, we conclude that $\left(\bar{H}_W^{\beta}\right)' \circ W(x)$ has the form (A.62) which completes the proof. □

Clearly, for any $\beta \le \max\{b, \alpha_*\}$, function (A.62) is the projection of $h_{|[0,\beta]}$ on the cone of nondecreasing and convex functions in $L^2([0, \beta), W(dx))$.

For constructing the projection, we use the following functions:

$$\lambda_*(\beta) = \frac{\int_\beta^d (x-\beta)[h(x) - h(\beta)]W(dx)}{\int_\beta^d (x-\beta)^2 W(dx)}, \tag{A.63}$$

$$K(\beta) = \lambda_*(\beta) - h'(\beta), \tag{A.64}$$

$$L(\beta) = \int_\beta^d h(x)W(dx) - \int_\beta^d [\lambda_*(\beta)(x-\beta) + h(\beta)]W(dx) \tag{A.65}$$

for $a \le \beta \le b$ (comp. (A.33)–(A.35)).

Theorem A.2 (See Theorem 6.12) *Assume that h satisfies the conditions* (A2) *and the respective* α_* *satisfies* (A.60). *If the set*

$$\mathcal{K} = \{\beta \in (\alpha_*, b) : K(\beta) \ge 0 \text{ and } L(\beta) = 0\}$$

is not empty, then

$$P_{\succeq_c w}h(x) = \begin{cases} h(\alpha_*), & 0 \le x \le \alpha_* \\ h(x), & \alpha_* < x \le \beta_* \\ h(\beta) + \lambda_*(\beta_*)(x - \beta_*), & \beta_* < x < d, \end{cases} \tag{A.66}$$

for $\beta_* = \sup\{\beta \in \mathcal{K}\}$ *and* λ_* *defined in* (A.63). *Otherwise,*

$$P_{\succeq_c w}h(x) = -\bar{h}(\beta)\left[\frac{(x-\beta)\mathbf{1}_{[\beta,d)}(x)}{\int_\beta^d (x-\beta)\, W(dx)} - 1\right], \tag{A.67}$$

for the greatest $0 < \beta \le \alpha_*$ *satisfying*

$$\begin{aligned} &\int_\beta^d h(x)\, W(dx)\left[\int_\beta^d (x-\beta)^2\, W(dx) - \left(\int_\beta^d (x-\beta)\, W(dx)\right)^2\right] \\ &= \int_0^\beta W(dx) \int_\beta^d (x-\beta)\, W(dx) \int_\beta^d (x-\beta)\, h(x)\, W(dx). \end{aligned} \tag{A.68}$$

Proof Suppose first that assumptions of the first claim are satisfied. Take an arbitrary $\xi \in (\alpha_*, \beta)$ and g defined by (A.66). By Lemma A.4, function $g_{|[0,\xi]}$ is the projection of $h_{|[0,\xi]}$ onto the cone of nondecreasing functions in $L^2([0,\xi), W(dx))$. Theorem A.1 implies that $g_{|[\xi,d]}$ is the projection of $h_{|[\xi,d]}$ onto the convex cone of nondecreasing convex functions in the Hilbert space $L^2([\xi,d), W(dx))$. Therefore, for any function f belonging to (A.57), we have

$$\int_0^d [f(x) - h(x)]^2\, W(dx) \ge \int_0^d [g(x) - h(x)]^2\, W(dx)$$

because the relations hold for the integrals over both intervals $[0,\xi]$ and $[\xi, d]$. It follows that g is the projection of h, because it is less distant from h than any other function f from the cone.

Suppose now that $P_{\succeq_c w}h$ is of the form

$$P_{\succeq_c w}h(x) = \begin{cases} h(\alpha), & 0 \le x \le \alpha, \\ h(x), & \alpha < x \le \beta, \\ h(\beta) + \lambda(x - \beta), & \beta < x < d, \end{cases} \tag{A.69}$$

but respective parameters α, β and/or λ do not satisfy assumptions of the theorem. We show that we can find better approximations of h than g, which gives a contradiction. First, assume that $\alpha \in (a, \beta)$ does not satisfy (A.60). Consider functions g of the form (A.66) restricted to $[0, \beta]$ with β fixed and $\alpha \in [a, \beta]$. Let

$$D(\alpha) = \int_0^{\beta} [g(x) - h(x)]^2 \, W(dx).$$

Then

$$D'(\alpha) = 2h'(\alpha)\left[h(\alpha) - \bar{h}(\alpha)\right] \int_0^{\alpha} W(dx).$$

If $D'(\alpha) < 0$, then $D(\alpha)$ decreases if α increases. If $D'(\alpha) > 0$, then $D(\alpha)$ decreases if α decreases. In both cases, approximation of h can be improved, which provides the contradiction.

Suppose now that g is of the form (A.69), α satisfies condition (A.60), but $\lambda \neq \lambda_*(\beta)$. Set

$$\begin{aligned} D(\beta, \lambda) &= \int_{\alpha}^{d} [g(x) - h(x)]^2 \, W(dx) \\ &= \int_{\beta}^{d} [h(\beta) + \lambda(x - \beta) - h(x)]^2 \, W(dx). \end{aligned}$$

We fix β and look for $\lambda \geq h'(\beta)$, for which $D(\beta, \lambda)$ is minimized. $D(\beta, \lambda)$ is a quadratic function of λ and attains its global minimum at $\lambda_*(\beta)$ defined in (A.57). Hence, we minimize $D(\beta, \lambda)$ taking $\lambda_{**}(\beta) = \max\left\{h'(\beta), \lambda_*(\beta)\right\}$. In this way, we can exclude all $\lambda \neq \lambda_{**}(\beta)$. Furthermore, if $\lambda_{**}(\beta) = h'(\beta) \neq \lambda_*(\beta)$, then we improve the approximation by decreasing β (cf. the proof of Theorem A.1, see also Gajek and Rychlik (1998, pp. 170–171)). Accordingly, we can also exclude $\lambda = \lambda(\beta) \neq \lambda_*(\beta)$.

Suppose then that g is of the form (A.69), $\alpha = \alpha_*$ satisfies (A.60) and $\lambda = \lambda_*(\beta)$. If $K(\beta) < 0$, then g is not convex. If $L(\beta) \neq 0$, then the necessary condition

$$\int_0^d P_{\succeq_c W} h(x) \, W(dx) = \int_0^d h(x) \, W(dx) \tag{A.70}$$

fails, and g is not a projection. Therefore, conditions (A.60), $K(\beta) \geq 0$ and $L(\beta) = 0$ are necessary for function g of the form (A.69) with $\lambda = \lambda_*(\beta)$ to be the projection of h onto (A.57). If there exist $\alpha < \beta_1 < \beta_2$ satisfying these conditions, then (A.69) with β_2 and $\lambda_*(\beta_2)$ approximates h better on $[\alpha, d)$ than (A.69) with β_1 and $\lambda_*(\beta_1)$. Indeed, this is a consequence of the fact that $h(\beta) + \lambda_*(\beta)(x - \beta)$ is the best linear approximation of $h(x)$ on the interval (β, d) when $L(\beta) = 0$ holds. In view of that, we see that (A.69) with β_2 and $\lambda_*(\beta_2)$ better approximates h on the both intervals $(0, \beta_2)$ and (β_2, d) than (A.69) with β_1 and $\lambda_*(\beta_1)$. Therefore, we take the greatest β_* satisfying above conditions.

Summing up, we have proved that assumptions of the first part of the proposition are necessary and sufficient. If they are not satisfied, then by Lemma A.2, the projection is of the form (A.59). We show that function satisfying assumptions of the second claim is the best approximation of h of this form.

Consider the function

$$\begin{aligned} D(\gamma,\lambda,\beta) &= \|g-h\|_W^2 \\ &= \int_0^\beta [h(x)-\gamma]^2\, W(dx) + \int_\beta^d [h(x)-\lambda(x-\beta)-\gamma]^2\, W(dx)\,. \end{aligned}$$

Denoting $\tilde{g}_{\lambda,\beta}(x) = h(x) - \lambda(x-\beta)\mathbf{1}_{[\beta,d)}(x)$, we can write

$$D(\gamma,\lambda,\beta) = \int_0^d \left[\tilde{g}_{\lambda,\beta}(x)-\gamma\right]^2 W(dx)\,.$$

Function $D(\gamma,\lambda,\beta)$ with λ and β fixed is quadratic convex and attains its minimum at

$$\gamma_* = \gamma_*(\lambda,\beta) = \int_0^d \tilde{g}_{\lambda,\beta}(x)\, W(dx) = -\lambda\int_\beta^d (x-\beta)\, W(dx)\,. \tag{A.71}$$

Set $D(\lambda,\beta) = D(\gamma_*,\lambda,\beta)$ and

$$\tilde{g}_\beta(x) = (x-\beta)\mathbf{1}_{[\beta,d)}(x) - \int_\beta^d (x-\beta)\, W(dx)\,.$$

Then

$$D(\lambda,\beta) = \int_0^d \left[h(x)-\lambda\tilde{g}_\beta(x)\right]^2 W(dx)$$

for any fixed $\beta \in (0,d)$ is minimized at

$$\lambda_* = \lambda_*(\beta) = \frac{\int_0^d h(x)\,\tilde{g}_\beta(x)\, W(dx)}{\int_0^d \tilde{g}_\beta^2(x)\, W(dx)} = \frac{Q(\beta)}{R(\beta)},$$

where

$$\begin{aligned} Q(\beta) &= \int_\beta^d (x-\beta)\,h(x)\, W(dx)\,, \\ R(\beta) &= \int_\beta^d (x-\beta)^2\, W(dx) - \left[\int_\beta^d (x-\beta)\, W(dx)\right]^2. \end{aligned}$$

We have

$$Q'(\beta) = -\int_\beta^d h(x) W(dx),$$

$$R'(\beta) = -2\int_\beta^d (x-\beta) W(dx) \int_0^\beta W(dx).$$

Note that $Q(d) = 0$, and $Q'(\beta) < 0$, and the same holds for R. It follows that $Q(\beta)$, $R(\beta)$ and $\lambda_*(\beta)$ are positive for any $\beta \in [0, d)$. Therefore, for arbitrary fixed β function of the form (A.59), with optimal parameters γ_* and λ_*, is nondecreasing and convex.

Let $D(\beta)$ stand for $D(\lambda_*(\beta), \beta)$. Then,

$$D(\beta) = \int_0^d h^2(x) W(dx) - 2\frac{Q(\beta)}{R(\beta)} \int_0^d h(x) \tilde{g}_\beta(x) W(dx) + \left(\frac{Q(\beta)}{R(\beta)}\right)^2 \int_0^d \tilde{g}_\beta^2(x) W(dx) = \int_0^d h^2(x) W(dx) - \frac{Q^2(\beta)}{R(\beta)}$$

and

$$D'(\beta) = 2\frac{Q(\beta)}{R(\beta)}\left[\lambda_*(\beta)\frac{R'(\beta)}{2} - Q'(\beta)\right].$$

Since $D'(\beta)$ is continuous and $D(\beta)$ does not attain its minimum for $\beta = 0$ or $\beta = d$, we see that condition $D'(\beta) = 0$ is necessary for $\beta \in (0, d)$ to be optimal. Thus we need

$$\lambda_*(\beta) = 2\frac{Q'(\beta)}{R'(\beta)} = -\frac{\bar{h}(\beta)}{\int_\beta^d (x-\beta) W(dx)}. \tag{A.72}$$

Substituting (A.72) into (A.71), we conclude that β is to satisfy

$$\gamma_*(\beta) = \frac{\int_0^\beta h(x) W(dx)}{\int_0^\beta W(dx)} = \bar{h}(\beta).$$

Plugging optimal $\gamma_*(\beta)$ and $\lambda_*(\beta)$ into (A.59) we obtain (A.67). Function g is the best constant approximation of h in $[0, \beta]$ because

$$\int_0^\beta g(x) W(dx) = \int_0^\beta h(x) W(dx),$$

and it is the best linear approximation in $[\beta, d]$ because it is of the form $\lambda_*(\beta)(x-\beta) + \gamma_*$. It remains to find $\beta \in (0, d)$ satisfying $D'(\beta) = 0$, for which function of the form (A.67) is the projection of h onto (A.57). We state that function of the form (A.67) is the projection if the following conditions are satisfied:

(i) constant approximation in $[0, \beta]$ is optimal in this interval,
(ii) point $(\beta, \gamma(\beta))$ lies on the curve $\left(\beta, \bar{h}(\beta)\right)$,
(iii) the linear part ℓ, say, is optimal linear approximation of h on $[\beta, d)$.

Assume that $D'(\beta) = 0$ for some $\beta > \alpha$. Then either g runs beneath h on (β, c), or g and h have at least one common point on (β, c). In both cases, we can find a better approximation of the form (A.66).

If conditions (i)–(iii) hold for $\beta = \alpha$, then the constant and linear part are optimal nondecreasing approximations of h on $[0, \alpha]$ and (α, d), respectively, and they together define the desired projection of h on $[0, d)$.

Assume that $\ell(\alpha) < h(\alpha) = \bar{h}(\alpha)$ for $\beta = \alpha$, where $\ell(\alpha)$ denotes the value of the optimal linear approximation of h on $[\alpha, d)$ at α. Then g can be replaced by function of the form (A.66), passing through $(\alpha, h(\alpha))$. So we have a contradiction. If function of the form (A.67) with $\beta = \alpha$ is not a projection, then the best linear approximation of h on (α, d) is the best nondecreasing and convex approximation on this interval and $\ell(\alpha) > h(\alpha)$. Now we decrease β starting from $\beta = \alpha$. Linear functions are still the best nondecreasing convex approximations of h on (β, d), values $\ell(\beta)$ change continuously until they reach the level $\bar{h}(\beta)$, earlier than $h(\beta)$, because $\bar{h}(\beta) > h(\beta)$ for $\beta < \alpha$. Then, function that is constant equal to $\bar{h}(\beta)$ on $[0, \beta]$ and equal to ℓ on (β, d) is the projection of h onto (A.57), because $\bar{h}(\beta)$ and ℓ are projections of h on $[0, \beta)$ and (β, d), respectively. □

It can be shown that the projection $P^{\nearrow}h$ of h onto the family of nondecreasing functions $\mathcal{C}^{\nearrow}$ in $L^2([0, d), W(dx))$ has two constant parts, one negative on the left hand side, and another positive on the right (provided that $c < d$) and coincides with h in the middle part. The first constant part is just $h(\alpha_*) = \bar{h}(\alpha_=)$, $0 \leq x \leq \alpha_*$. Splitting h into two parts $h_{|[0,\xi)}$ and $h_{|[\xi,d)}$ for some $\alpha_* < \xi < b$, we see that (A.66) for $x \leq \xi$ is the projection of $h_{|[0,\xi)}$ onto $\mathcal{C}^{\nearrow} \subset L^2([0, \xi), W(dx))$. This is convex so that it is also the projection onto the family of nondecreasing convex functions on $[0, \xi)$. By Theorem A.1, we conclude that (A.66) reduced to $[\xi, d)$ is also the projection of $h_{|[\xi,d)}$ onto the cone of nondecreasing convex functions on $[\xi, d)$. This justifies the first claim of Theorem A.2.

The broken line has the following properties: the constant part is a best nondecreasing approximation of h on $[0, \beta_*)$ and the linear one is the straight line closest to h on $[\beta_*, d)$. The partition point is chosen so as to make the composition continuous.

Another point of view on Theorem A.2 is also instructive. The nondecreasing function that is the least distant from h has two flat parts at the ends of the domain. The convexity requirement excludes the right one. This is replaced by the least convex function which is linear. This linear part has to absorb some regions where the nondecreasing projection is equal to the original h. As we move left, the condition on the slope of the line becomes less restrictive. In some cases, the whole original part and some of the lower constant one have to be absorbed.

A.2 Projections onto Concave Functions

We first describe the shape conditions of the projection which is applicable for establishing the bounds on the expectation of sample maxima in the iid IFR samples.

Lemma A.5 (See Lemma 6.7) *Let $h \in L^2(\mathbb{R}_+, W(dx))$ be bounded, differentiable, strictly increasing, first strictly convex and then concave, with the inflection point $b > 0$, say, where W is a probability distribution function supported on $\mathbb{R}_+$. Then, for every*

$$g \in \mathcal{C}_{\preceq_c W} = \{g \in L^2(\mathbb{R}_+, W(dx)) : g - nondecreasing,\ concave\}, \tag{A.73}$$

there exists $g_* \in \mathcal{C}_{\preceq_c W}$ of the form

$$g_*(x) = g_*(c, \alpha; x) = \begin{cases} \alpha(x - c) + h(c), & x \le c, \\ h(x), & x \ge c, \end{cases} \tag{A.74}$$

with $c \ge b$ and $\alpha \ge h'(c)$ such that $||h - g_*||_W \le ||h - g||_W$.

Proof For simplicity, we assume that $h(0) = 0$. We start by showing that the candidates $g \in \mathcal{C}_{\preceq_c Z}$ for the projection should satisfy $g(0) \le 0$ and $g(x) \le h(x)$ for sufficiently large arguments x. We prove the first property by contradiction. If $g(0) > 0 = h(0)$, by continuity we have $g(0) < g(\varepsilon) > h(\varepsilon)$ for some $0 < \varepsilon < b$. Then

$$\tilde{g}(x) = \begin{cases} \frac{g(\varepsilon)}{\varepsilon} x, & x < \varepsilon, \\ g(x), & x \ge \varepsilon, \end{cases}$$

is clearly increasing, and concave, because

$$\tilde{g}'(x) = \frac{g(\varepsilon)}{\varepsilon} \ge \frac{g(\varepsilon) - g(0)}{\varepsilon} \ge g'(\varepsilon+), \qquad 0 < x < \varepsilon,$$

and runs between h and g on $(0, \varepsilon)$, because

$$h(x) \le \frac{h(\varepsilon)}{\varepsilon} x < \frac{g(\varepsilon)}{\varepsilon} x \le \frac{g(\varepsilon) - g(0)}{\varepsilon} x + g(0) \le g(x)$$

there. This shows that we can restrict ourselves to the candidates g satisfying $g(0) \le h(0) = 0$.

If relation $g(x) \le h(x)$ does not hold for all x sufficiently large, there are two possibilities:

(i) The set $\{x \ge b: \ g(x) \le h(x)\}$ is non-empty, and consists of compact, possibly degenerate intervals. Take the first of them, $[\alpha, \beta]$, say. Then $g(\beta) = h(\beta)$ and $g'(\beta-) \ge h'(\beta)$, and so

$$\tilde{g}(x) = \begin{cases} g(x), \ x < \beta, \\ h(x), \ x \ge \beta, \end{cases}$$

is non-decreasing and concave, and evidently better approximates h than the original g.

(ii) We have $g(x) > h(x)$ for all $x \ge b$, and $g(0) \le 0$, as we have proved earlier. Consider the slopes of the lines tangent to the graph of h: they first increase to a maximal value at b, and then decrease to 0 at ∞. The values of the respective tangent lines at zero continuously decrease from 0 at the origin to the minimal negative value at b, and then increase to $0 < h(\infty) < \infty$. The unique tangency point $x_t > b$ determining the line $\ell_t(x) = h'(x_t)x$ passing thorough the origin satisfies $\ell_t \ge h$ on the whole positive half-axis. Since $\ell_t(0) = 0 \ge g(0)$, and $\ell_t(x_t) = h(x_t) < g(x_t)$, there exists $0 \le x_s < x_t$ such that $\ell_t(x_s) = g(x_s)$, $\ell_t(x) < g(x)$, $x_s < x < x_t$, and $\ell_t' \le g_t'(x_s-)$. Therefore

$$\tilde{g}(x) = \begin{cases} g(x), \ \ x < x_s, \\ \ell_t(x), \ x_s < x \le x_t, \\ h(x), \ \ x \ge x_t, \end{cases}$$

is concave non-decreasing and lies closer to h than g. This completes the proof of the second claim.

So we can concentrate on $g \in \mathcal{C}_{\le_c Z}$ such that $g(0) \le 0$ and $g(x) \le h(x)$ for all large arguments. If g crosses h twice in the convexity region $[0, b]$ at some $0 \le x_1 < x_2 \le b$, we can take the straight line ℓ passing through $(x_i, h(x_i))$, $i = 1, 2$. Then $g \le \ell \le h$ on $[0, x_1] \cup [x_2, b]$ and $h \le \ell \le g$ on $[x_1, x_2]$. For all $x > b$, inequality $g(x) \le \ell(x)$ holds. We also have $\ell < h$ at some right vicinity of b, but ultimately it must cross h at some $x_0 > b$ where $\ell' \ge h'(x_0)$, because $\lim_{x\to\infty} h(x) < \lim_{x\to\infty} \ell(x) = \infty$. So we can replace g by

$$\tilde{g}(x) = \begin{cases} \ell(x), \ x < x_0, \\ h(x), \ x \ge x_0. \end{cases} \tag{A.75}$$

If $g \le h$ on $[0, b]$ (i.e. the curves do not meet there or have a single common point), then there is a line ℓ separating the convex sets $\{(x, y): \ 0 \le x \le b, \ y \le g(x)\}$ and $\{(x, y): \ 0 \le x \le b, \ y \ge h(x)\}$. The line ℓ runs above g on the whole $\mathbb{R}_+$, and crosses h at some $x_0 > b$. So, the original g can be replaced by (A.75).

It remains to take into account the case that g crosses h at some $0 \leq x_1 \leq b$ and then runs above h till some $x_2 > b$. We consider two subcases. The first is that the line ℓ passing through the crossing points lies above h (and below g) in (x_1, x_2). We also have $g \leq \ell \leq h$ on $(0, x_1)$. For $x > x_2$ relation $g \leq \ell$ holds. Since $h_n(x_2-) < \ell(x_2-)$, we have $h_n'(x_2) \geq \ell'$. If the derivatives coincide, we can take (A.75) with $x_0 = x_2$, which is both concave and less distant from h than g. If $h'(x_2) > \ell'$, we obtain $h > \ell \geq g$ on some interval (x_2, x_0) so that $h(x_0) = \ell(x_0)$ and $h'(x_0) \leq \ell'$. Therefore, the approximation improvement may involve in taking the function of the form (A.75).

The other subcase is that ℓ crosses once h in (x_1, x_2) at some x_3 so that we have $h < \ell < g$ on (x_1, x_3) and $\ell < h < g$ on (x_3, x_2). We modify the line in the following way: we increase the slope keeping its constant value $h(x_1) = g(x_1)$ at x_1. Then the next meeting points x_3 and x_2 are replaced by ones belonging to (x_3, x_2) and gradually approaching one another. We stop the modification at the moment that these points coincide and we obtain the straight line $\tilde{\ell}$ which is tangent to h at some $x_0 \in (x_3, x_2)$. We observe that $g \leq \tilde{\ell} \leq h$ on $(0, x_1)$, $h \leq \tilde{\ell} \leq g$ on (x_1, x_0) and $\tilde{\ell}' = h'(x_0)$. Hence

$$\tilde{g}(x) = \begin{cases} \tilde{\ell}(x), & x < x_0, \\ h(x), & x \geq x_0, \end{cases}$$

belongs to $\mathcal{C}_{\preceq_c Z}$ and lies closer to h than g. □

Now we remind the following assumptions (A3):

(A3) Suppose that W is an absolutely continuous probability measure on (a_0, d) with a finite second moment, and h is a bounded, twice differentiable function on (a_0, d) such that

$$\int_{a_0}^{d} h(x) W(dx) = 0. \tag{A.76}$$

We further assume that h is strictly decreasing on (a_0, a), strictly convex increasing on (a, b), strictly concave increasing on (b, c) with $h(c) > 0 \geq h(0)$, and strictly decreasing on (c, d) with $h(d) = h(a_0)$ for some $-\infty \leq a_0 \leq a < b < c < d \leq \infty$.

Our purpose is to determine the projections of functions satisfying (A3) onto the convex cone

$$\mathcal{C}_{\preceq_c W} = \{g \in L^2((a_0, d), W(dx)) : g \text{ is nondecreasing, concave}\}. \tag{A.77}$$

These are used for determining the sharp bounds on the expectations of nonextreme order statistics and respective spacings based on iid samples with the marginal distributions belonging to the ID, IFR and IRFR families. We first describe the possible shapes of projections.

Lemma A.6 *Let h satisfy* (A3) *and let $\mathcal{C}_* \subset \mathcal{C}_{\preceq_c W}$ be the class of functions described by either of the formulae*

$$g_*(x) = \begin{cases} \lambda(x-\alpha) + h(\alpha), & a_0 \le x < \alpha, \\ h(x), & \alpha \le x < \beta, \\ h(\beta), & \beta \le x < d, \end{cases} \tag{A.78}$$

for $b \le \alpha < \beta \le c$, $\lambda \ge h'(\alpha)$, and

$$g_*(x) = \begin{cases} \lambda(x-\alpha) + \gamma, & a_0 \le x < \alpha, \\ \gamma, & \alpha \le x < d, \end{cases} \tag{A.79}$$

for $\lambda > 0$ and $a_0 < \alpha < d$. Then for every function $g \in \mathcal{C}_{\preceq_c W}$, there exists a function $g_* \in \mathcal{C}_*$ such that

$$||h - g_*||_W \le ||h - g||_W. \tag{A.80}$$

In shorts, we write that the candidate g_* for projection has either form l-h-c (first linear, then equal to h and finally constant) or $l - c$ (first linear and then constant) if it satisfies (A.78) and (A.79), respectively.

Proof Below we frequently use the following convention. We denote a preliminary candidate for projection by g. Then we construct its correction g_* which lies closer to h in the sense it runs between g and h on the whole domain $(a_0\ d)$. We aim at concluding that all the possible corrections have either of forms (A.78) and (A.79).

Since we have $\int_{a_0}^d P_{\preceq_c W} h(x) W(dx) = 0$ (see (A.76) and Lemma 6.1), we note that the constant candidate for the projection has to be equal to 0. From now on we focus on nonconstant candidates. Referring again to $\int_{a_0}^d P_{\preceq_c W} h(x) W(dx) = 0$, we notice that we can focus on functions $g \in \mathcal{C}_{\preceq_c W}$ which satisfy

$$g(a_0) \le 0 \le g(d).$$

We can also impose the following inequalities:

$$g(a_0) \le h(a_0), \tag{A.81}$$

$$g(d) \le h(c). \tag{A.82}$$

The first inequality is obvious if $a_0 = -\infty$, because every nondecreasing nonconstant and concave function tends to $-\infty$ as the arguments tend to $-\infty$. Otherwise, suppose that $g(a_0) > h(a_0)$ and so $g(\varepsilon) > h(\varepsilon)$ for some $a_0 < \varepsilon < b$. Therefore

$$g(x) \ge g(a_0) + \frac{g(\varepsilon) - g(a_0)}{\varepsilon - a_0}(x - a_0) > h(a_0) + \frac{g(\varepsilon) - h(a_0)}{\varepsilon - a_0}(x - a_0) > h(x)$$

for all $a_0 < x < \varepsilon$, and the function

$$g_*(x) = \begin{cases} h(a_0) + \frac{g(\varepsilon)-h(a_0)}{\varepsilon-a_0}(x - a_0), & a_0 \le x \le \varepsilon, \\ g(x), & \varepsilon \le x < d, \end{cases}$$

satisfies $g_*(a_0) \le h(a_0)$, (A.80), and belongs to $\mathcal{C}_{\preceq_c W}$, because $g_*'(\varepsilon-) \ge g'(\varepsilon-)$. The last relation means that $g_*'(x) \ge g'(x)$ for all x in some left neighborhood of ε. For establishing (A.82), it suffices to note that $g_* = \min\{g, h(c)\} \in \mathcal{C}_{\preceq_c W}$ lies closer to h than the original g.

Now we analyze the local behavior of the candidates for projection in the intervals $(a_0, b]$, (b, c) and $[c, d)$. The simplest situation is in the last interval: either g crosses h at one point $\gamma \in [c, d)$, say, from below, or $g \le h$ on the whole $[c, d)$. In the latter case, we put $\gamma = d$ for convenience, assuming tacitly that γ depends on g. Observe that when $\beta \in [b, c]$ is such that $h(\beta) = h(\gamma)$, then the horizontal line $h(\gamma)$ lies between g and h on $[\beta, d)$. Below we often use this constant in constructions of modifications g_* of original concave increasing functions g, gluing it with a piece of original h on the left. The following relations on $(a_0, b]$ are possible:

(i) $g \le h$ (i.e. either $g < h$ on $(a_0, b]$ or g and h have a single tangency point there), and then $g(b+) \le h(b+)$,
(ii) g and h cross each other twice, and then $g(b+) < h(b+)$,
(iii) g and h cross each other once, and then $g(b+) > h(b+)$.

A more sophisticated analysis is needed in the central interval (b, c). For any $g \in \mathcal{C}_{\preceq_c W}$, the set $A = \{g = h\} \cap (b, c)$ is a sum of closed, possibly degenerate, non-overlapping intervals $I_i = [\alpha_i, \beta_i]$, $(\alpha_i \le \beta_i)$ (the only possible exceptions are the ones starting at b, and that ending at c). We show that we can restrict our attention to functions g for which

(a) there can be at most one non-degenerated interval in A,
(b) the set $\{g > h\} \cap (b, c)$ is either empty or it is a single interval (b, α_1), $b < \alpha_1 < c$, being a right neighborhood of the inflection point b of h,
(c) the set $\{g < h\} \cap (b, c)$ is either empty or a single interval or the sum of two intervals such that one of them is a left neighborhood of the maximum point c of h.

Condition (a) follows immediately from the fact that g with two non-degenerate intervals $[\alpha_j, \beta_j]$, $[\alpha_k, \beta_k]$, $\alpha_j < \beta_j < \alpha_k < \beta_k$, can be replaced by g_* which amounts to h on (α_j, β_k), and g elsewhere. Clearly, the modification approximates h better than the original g.

Now we prove condition (b). Assume that $I_k = \{\alpha_k\}$, $I_{k+1} = \{\alpha_{k+1}\}$ and $g > h$ on (α_k, α_{k+1}). Then we have the following possibilities with respective implications:

- $g(\alpha_k-) < h(\alpha_k-)$, and then $g'(\alpha_k-) \ge h'(\alpha_k)$,
- $g(\alpha_k-) > h(\alpha_k-)$, and then $g'(\alpha_k-) = h'(\alpha_k)$,

- $g(\alpha_{k+1}+) < h(\alpha_{k+1}+)$, and then $g'(\alpha_{k+1}+) \le h'(\alpha_{k+1})$,
- $g(\alpha_{k+1}+) > h(\alpha_{k+1}+)$, and then $g'(\alpha_{k+1}+) = h'(\alpha_{k+1})$.

Therefore,

$$g'(\alpha_k-) \ge h'(\alpha_k) > h'(\alpha_{k+1}) \ge g'(\alpha_{k+1}+),$$

and we can replace g with h on the interval (α_k, α_{k+1}). If either of I_k and I_{k+1} is non-degenerate, our reasoning even simplifies. We also exclude the case $(b, c) \subset \{g > h\}$ by considering $g \in \mathcal{C}_{\preceq_c W}$ such that $g(x) > h(x)$, $b < x < c$. It follows that it crosses h once in $(a_0, b]$ at some δ. Take the lines ℓ_x, $b \le x \le c$, tangent to h at x. Function ℓ_c is horizontal and runs above h, whereas $\ell_b < h$ on (a_0, b). The slopes ℓ'_x decrease, and the lines gradually move up on $(a_0, b]$, as x increases. There is ℓ_{x_0} such that $\ell_{x_0}(\delta) = g(\delta) = h(\delta)$. It runs between g and h on $(a_0, x_0]$. Therefore g_* equal to ℓ_{x_0}, h, and $h(c)$ on $(a_0, x_0]$, $[x_0, c]$, and $[c, d]$, respectively, better approximates h than g.

Finally we verify (c). By contradiction, assume that there exist $I_k = \{\alpha_k\}$, $I_{k+1} = \{\alpha_{k+1}\}$ for some $k \ge 2$ so that we have $g(\alpha_k-) < h(\alpha_k-)$, $g(x) < h(x)$, $\alpha_k < x < \alpha_{k+1}$, and $g(\alpha_{k+1}-) < h(\alpha_{k+1}+)$. Hence $g'(\alpha_k-) \ge h'(\alpha_k) > h'(\alpha_{k+1}) \ge g(\alpha_{k+1}+)$, and we can replace g by h on the interval (α_k, α_{k+1}). The same construction can be performed if either I_k or I_{k+1} is non-degenerate. Summing up, the following mutual relations of g and h on (b, c) are possible:

- $A = \{g = h\} \cap (b, c) = \emptyset$ and $g < h$ on (b, c),
- $A = I_1 = [\alpha_1, \beta_1]$, where either $g < h$ or $g > h$ on (b, α_1), I_1 is either degenerate or non-degenerate, and $g < h$ on (β_1, c) if $\beta_1 < c$,
- $A = I_1 \cup I_2$, where $g > h$ on (b, α_1), I_1 is degenerate, $g < h$ on (α_1, α_2), I_2 is either a point or a non-degenerate interval, and $g < h$ on (β_2, c) if $\beta_2 < c$.

Having established properties of candidates g for $P_{\preceq_c W}h$ on various parts of (a_0, d), we are in a position to deduce the final shapes described in the statement of the lemma. We perform different construction procedures dependent on the mutual relation of g and h in the right vicinity of b.

Case I $g(b+) < h(b+)$ This implies that either g and h meet each other twice in $(a_0, b]$, or $g \le h$ on the whole interval. We focus first on the former subcase. Let ℓ denote the straight line passing through both the crossing points. It separates g and h on $(a_0, b]$. We also have $g \le \ell$ on $[b, d]$. Function ℓ crosses h either at some $\alpha \in (b, c)$ below the level $h(\gamma)$ (this is either the level of the meeting point of g and h in (c, d) or $h(d)$ otherwise), or further on, or nowhere in $[b, d)$. If $g < h$ on (b, c), and the first possibility happens, we take g_* equal to ℓ on $(a_0, \alpha]$, then to h on $[\alpha, \beta]$, where $\beta \in (\alpha, c]$ satisfies $h(\beta) = h(\gamma)$, and ultimately to the constant $h(\gamma)$. Otherwise, we can simply take $g_* = \min\{\ell, h(\gamma)\}$. If $A = [\alpha_1, \beta_1]$, a point or a proper interval, then ℓ hits h at some $b < \alpha \le \alpha_1$ from below. Then we can define g_* as ℓ on $(a_0, \alpha]$, then h on $[\alpha, \beta_1]$ and further on till β satisfying $h(\beta) = h(\gamma)$, and finally $h(\gamma)$. The construction lies between g and h on (a_0, d). In the subcase $g \le h$ on $(a_0, b]$, we can take any line ℓ separating the convex bodies

$\{(x, y) : a \le x \le b,\ y \le g(x)\}$ and $\{(x, y) : a \le x \le b,\ y \ge h(x)\}$. It satisfies $g(x) \le \ell(x) \le h(x)$ for all $a_0 < x \le b$. One can easily check that it suffices to replace the line passing through the two crossing points by the separating one and repeat all the arguments of the previous subcase. This would complete the proof in Case I.

Case II $g(b+) = h(b+)$ It can occur only if $g(x) < h(x)$ for $a_0 < x < b$. Then the line $\ell(x) = h'(b)(x-b)+h(b)$ separates g and h on $(a_0, b]$, and it can be concavely extended with h to the right. So the construction: g_* amounting to ℓ for $x \le b$, then h, and finally constant $h(\gamma)$ for $x \ge \beta \in (b, c)$ which satisfies $h(\beta) = h(\gamma)$, improves the approximation of h by original g.

Case III $g(b+) > h(b+)$ It is possible only if g crosses h once in $(a_0, b]$ at some δ, and implies that g and h meet either once or twice in (b, c). Let α_0 stand for the first meeting point, and ℓ_0 denote the line running through the points $(\delta, h(\delta))$ and $(\alpha_0, h(\alpha_0))$. Then $\ell_0(x) \le g(x)$ iff $\delta \le x \le \alpha_0$. Assume also first that $\ell_0(x) \ge h(x)$ for $\delta \le x \le \alpha_0$. Then $\ell_0' \le h'(\alpha_0)$. If $\ell_0' = h'(\alpha_0)$, it suffices to glue g_* from ℓ_0 on $(a_0, \alpha_0]$, h on $[\alpha_0, \beta]$, and $h(\beta) = h(\gamma)$ on $[\beta, d)$. Otherwise we have $\ell_0(\alpha_0+) < h(\alpha_0+)$. If ℓ_0 hits again h at some $\alpha < \beta$, where $b < \beta \le c$ satisfies $h(\beta) = h(\gamma)$, then we can compose g_* from ℓ_0 on $(a_0, \alpha]$, h on $[\alpha, \beta]$ and $h(\gamma)$ on $[\beta, d)$. Otherwise, it suffices to take $g_* = \min\{\ell_0, h(\gamma)\}$.

Suppose now that $\ell_0 \ge h$ does not hold true on the whole interval $[\delta, \alpha_0]$. We notice that $\ell_0 > h$ on (δ, α_0') and $\ell_0 < h$ on (α_0', α_0) for some $\delta < \alpha_0' < \alpha_0$. Indeed, $\ell_0(\delta+) > h(\delta+)$, and the relation cannot change on the interval where h decreases (if it overlaps with (δ, α_0)). The sign of $\ell_0 - h$ can change from plus to minus in $[a, b] \cap (\delta, \alpha_0)$, a region of convex increase of h, but it forces that the sign change is impossible in $(b, c) \cap (\delta, \alpha_0)$. Similarly, existence of zero for the difference $\ell_0 - h$ in the concave increase region $[b, c] \cap (\delta, \alpha_0)$ of h makes it impossible in $(a, b) \cap (\delta, \alpha_0)$. Consider now a procedure of constructing straight lines ℓ_x passing through $(\delta, h(\delta))$ with increasing slopes ℓ_x' greater than ℓ_0'. If the difference $\ell_x' - \ell_0' > 0$ is sufficiently small, then ℓ_x meets h at some $\alpha_0' < \alpha_x' < \alpha_x < \alpha_0$. If we gradually increase the slopes, the distances $\alpha_x - \alpha_x'$ would decrease. We terminate the procedure once the crossing points α_x' and α_x fuse into a single one α, say, and the respective line ℓ, say, becomes tangent to h at α. Note that α is a concavity point of h, because the tangent line ℓ runs above h in a vicinity of α. We have $g(x) \le \ell(x) \le h(x)$ for $a_0 \le x \le \delta$, the reversed inequalities for $\delta \le x \le \alpha$, and $h'(\alpha) = \ell'$. This allows us to repeat the first construction of g_* applied for the case $g(b+) > h(b+)$. We take first ℓ on $(a_0, \alpha]$, then h on $[\alpha, \beta]$, and ultimately $h(\gamma)$.

We finally show that assumption $h(a_0) \le 0$ excludes linear forms of projection. Suppose that g is linear increasing. Then $g(d) > 0 \ge h(d)$ which means that there exists $c < \gamma < d$ such that $g(\gamma) = h(\gamma)$. Consequently, $g_* = \min\{l, h(\gamma)\}$ is a better approximation of h than g. When $h(a_0) < 0$, the only possible constant candidate $g = 0$ for projection violates assumption $g(a_0) \le h(a_0)$. If $h(a_0) = 0$, then h has a unique zero $\theta \in (a_0, d)$ where it changes its sign from $-$ to $+$. It is formally deduced from further considerations, but is intuitively clear that the slope $\lambda_*(\theta)$ of the best approximation $l_*(x) = \lambda_*(\theta)(x - \theta)$ of h on $(a_0, \theta]$ among linear functions vanishing at θ has a strictly positive slope. Therefore, $g_* = \min\{l_*, 0\}$ has a smaller L^2-distance to h than the zero function. If $h(a_0) > 0$, linear functions have to remain among the candidates g_* for projection. □

We are now able to present the precise forms of the projections of functions satisfying (A3). Before we formulate the results, we introduce some auxiliary functions. Let

$$M(\beta) = h(\beta)[1 - W(\beta)] - \int_\beta^d h(x)W(dx), \qquad a_0 \le \beta \le d. \tag{A.83}$$

We see that $M(a_0) = h(a_0)$, and $M(a) < M(d) = 0 < M(c)$. Since $M'(\beta) = h'(\beta)[1 - W(\beta)]$, function (A.83) is decreasing on (a_0, a), increasing on (a, c), and decreasing on (c, d). Let β_* denote the unique zero of (A.83) in (a, c). Put further

$$\lambda_*(\alpha) = \frac{\int_{a_0}^{\alpha} (x - \alpha)[h(x) - h(\alpha)]W(dx)}{\int_{a_0}^{\alpha} (x - \alpha)^2 W(dx)}, \tag{A.84}$$

$$K(\alpha) = \lambda_*(\alpha) - h'(\alpha), \tag{A.85}$$

$$L(\alpha) = \int_{a_0}^{\alpha} [h(x) - h(\alpha) - \lambda_*(\alpha)(x - \alpha)]W(dx), \quad a_0 \le \alpha < d. \tag{A.86}$$

Theorem A.3 (See Theorem 6.26) *Suppose that the zero $a < \beta_* < c$ of* (A.83) *belongs to (b, c), and the set $\mathcal{K} = \{\alpha \in (b, \beta_*) : \ K(\alpha) \ge 0, \ L(\alpha) = 0\}$ is nonempty, and $\alpha_* = \inf\{\alpha \in \mathcal{K}\}$. Then,*

$$P_{\preceq_c W} h(x) = \begin{cases} h(\alpha_*) + \lambda_*(\alpha_*)(x - \alpha_*), & a_0 \le x < \alpha_* \\ h(x), & \alpha_* \le x < \beta_*, \\ h(\beta_*), & \beta_* \le x < d, \end{cases} \tag{A.87}$$

is the projection of h on (A.77). *Otherwise, we have*

$$P_{\preceq_c W} h(x) = \frac{\int_{\alpha_*}^d h(x)W(dx)}{1 - W(\alpha_*)} \left[\frac{(x - \alpha_*)\mathbf{1}_{(a_0, \alpha_*)}(x)}{-\int_{a_0}^{\alpha_*} (x - \alpha_*)W(dx)} + 1 \right] \tag{A.88}$$

for α_* maximizing

$$\|P_{\preceq_c W} h\|_W^2 = \frac{\left[\int_\alpha^d h(x)W(dx)\right]^2}{[1-W(\alpha)]^2\left[\int_{a_0}^\alpha (x-\alpha)W(dx)\right]^2}$$
$$\times \left[\int_{a_0}^\alpha (x-\alpha)^2 W(dx) - \left(\int_{a_0}^\alpha (x-\alpha)W(dx)\right)^2\right] \tag{A.89}$$

among $\alpha \geq \beta_*$ satisfying

$$\frac{\int_\alpha^d h(x)W(dx)}{1-W(\alpha)} = -\frac{\int_{a_0}^\alpha (x-\alpha)h(x)W(dx)\int_{a_0}^\alpha (x-\alpha)W(dx)}{\int_{a_0}^\alpha (x-\alpha)^2 W(dx) - \left(\int_{a_0}^\alpha (x-\alpha)W(dx)\right)^2} > 0. \tag{A.90}$$

The parameters α solving (A.90) in the latter case are not uniquely determined. We show in the proof that the equality in (A.90) implies that the increasing and constant parts of (A.88) are the best linear and constant approximations of h on the intervals $(a_0, \alpha]$ and $[\alpha, d)$, respectively. If several α's satisfy (A.90), we should just calculate the $L^2((a_0, d), W(dx))$-norms of the respective functions (A.88), and choose one which provides the maximal norm value. Note that the functions g of the form (A.88) with parameters solving (A.90) satisfy $\langle g, g-h\rangle = 0$, which implies that $\|g-h\|_W^2 = \|h\|_W^2 - \|g\|_W^2$. As a consequence, one maximizing the norm provides the smallest distance from the projected function, i.e., it is actually the projection. For particular choices of h, representing the expectation functionals for order statistics and spacings, we will be able to prove that condition (A.90) determines the projection uniquely.

Proof Suppose that a candidate g_* for the projection has form l-h-c (see (A.87)). We try to choose parameter $\alpha < \beta < c$ so as to minimize the distance of g_* to h on $[\alpha, d)$. Observe that the constant part on the right does not spoil the monotonicity and concavity conditions on g_*. The squared distance

$$D_1(\beta) = \int_\beta^d [h(x)-h(\beta)]^2 W(dx)$$

does not depend on α and has the derivatives

$$D_1'(\beta) = 2h'(\beta)M(\beta),$$
$$D_1''(\beta) = 2h''(\beta)M(\beta) + 2[h'(\beta)]^2[1-W(\beta)].$$

Since $h'(\beta) > 0$ for $\beta \in (a, c)$, the unique zero β_* of M is the only possible local extreme point. Observe that

$$D_1''(\beta_*) = 2[h'(\beta_*)]^2[1 - W(\beta_*)] > 0$$

implies that it is the minimum point of D_1. Since $M(c) > 0$, we also have $D_1'(c-) = 2h'(c-)M(c-) > 0$ as well. This implies that either D_1 is strictly increasing on the whole $[\alpha, c)$ (which means that a better candidate for projection has form l-c with the change point α) or $\alpha < \beta_* < c$ is the unique point where the projection l-h-c may change its form from h to the constant $h(\beta_*)$. Note that $M(b) < 0$ is a necessary condition for $b < \beta_* < c$, and existence of projection of the form l-h-c. Moreover, in this case the conditions

$$\int_{a_0}^{d} [P_{\preceq_c W} h(x) - h(x)]W(dx) = 0,$$

$$\int_{\beta_*}^{d} [h(\beta_*) - h(x)]W(dx)dx = 0,$$

imply that necessarily

$$\int_{a_0}^{\alpha} [h(x) - h(\alpha) - \lambda(x - \alpha)]W(x) = 0. \tag{A.91}$$

Having chosen the optimal $\beta = \beta_*$, we further try to calculate optimal parameters α and λ of the linear part by minimizing the squared distance

$$D_2(\alpha, \lambda) = \int_{a_0}^{\alpha} [\lambda(x - \alpha) + h(\alpha) - h(x)]^2 W(dx), \quad b \leq \alpha < \beta_*,\ \lambda \geq h'(\alpha),$$

of the l-h part of the projection. For a fixed α, this is a convex quadratic function of λ which has a global minimum at (A.84). Note that for the optimal slope $\lambda_*(\alpha)$, we have

$$\int_{a_0}^{\alpha} [\lambda_*(\alpha)(x - \alpha) + h(\alpha) - h(x)](x - \alpha)W(dx) = 0. \tag{A.92}$$

The constrained minimum under condition $\lambda \geq h'(\alpha)$ amounts to the maximum of $\lambda_*(\alpha)$ and $h'(\alpha)$.

Below we show that for the optimal parameter α_* we have $K(\alpha_*) = \lambda_*(\alpha_*) - h'(\alpha_*) \geq 0$. To this end, we first note that the set $\{K(\alpha) < 0,\ b < \alpha < \beta\}$ is open, i.e., it consists of a number of disjoint open intervals. We can also check that

$$K(b) < 0 < K(c). \tag{A.93}$$

The first inequality follows from the fact that the line $h'(b)(x-b)+h(b)$, tangent to h at its inflection point b, runs below the graph of h on the whole $(a_0, b]$, whereas the optimal linear approximation $\lambda_*(b)(x-b)+h(b)$ satisfies (A.92), i.e. crosses h somewhere in $(a_0, b]$, and runs above tangent line there. The latter holds, because $h'(c)=0$, and $\lambda_*(c)>0$. This inequality holds because $h(c)$ is the maximal value of h, and only an increasing line $\lambda_*(c)(x-c)+h(c)$ may cross h in (a_0, b). Now we calculate

$$\frac{dD_2(\alpha, h'(\alpha))}{d\alpha} = \frac{d}{d\alpha}\left[\int_{a_0}^{\alpha}[h'(\alpha)(x-\alpha)+h(\alpha)-h(x)]^2W(dx)\right]$$
$$= 2h''(\alpha)\int_{a_0}^{\alpha}[h'(\alpha)(x-\alpha)+h(\alpha)-h(x))](x-\alpha)W(dx).$$

If $K(\alpha)<0$, the last integral is strictly greater than the RHS of (A.92) which is zero. Since $h''(\alpha)<0$ on (b, β_*), we conclude the implication

$$K(\alpha)<0 \quad \Rightarrow \quad \frac{dD_2(\alpha, h'(\alpha))}{d\alpha}<0, \qquad b<\alpha<\beta_*.$$

Accordingly, if α belongs to an interval where $K(\alpha)<0$, we can decrease the distance to h if we move α to the right. Consequencently, we arrive either to a point where $K(\alpha)=0$ or to β_*. Note that in the latter case we obtain an l-c function.

Accordingly, we can focus on the intervals where $K(\alpha)\geq 0$, and examine $D_2(\alpha, \lambda_*(\alpha))$. Using (A.92), we obtain

$$\frac{dD_2(\alpha, \lambda_*(\alpha))}{d\alpha} = \frac{d}{d\alpha}\left[\int_{a_0}^{\alpha}[\lambda_*(\alpha)(x-\alpha)+h(\alpha)-h(x)]^2W(dx)\right]$$
$$= 2[h'(\alpha)-\lambda_*(\alpha)]\int_{a_0}^{\alpha}[\lambda_*(\alpha)(x-\alpha)+h(\alpha)-h(x)]W(dx)$$
$$= 2K(\alpha)L(\alpha). \tag{A.94}$$

Note that vanishing of $L(\alpha)$ is necessary condition for projection (cf. (A.91)). We also show the implication

$$L(\alpha)=0 \quad \Rightarrow \quad \frac{d^2D_2(\alpha, \lambda_*(\alpha))}{d\alpha^2}>0. \tag{A.95}$$

Differentiating the expressions of both sides of (A.92) with respect to α, we obtain the identity

$$L(\alpha) = \int_0^\alpha [\lambda_*'(\alpha)(x-\alpha) - \lambda_*(\alpha) + h'(\alpha)](x-\alpha)W(dx). \tag{A.96}$$

Therefore, for α satisfying $L(\alpha) = 0$, we obtain

$$\begin{aligned}
&\frac{d^2 D_2(\alpha, \lambda_*(\alpha))}{d\alpha^2} = 2K(\alpha)L'(\alpha) \\
&= 2\int_{a_0}^{\alpha} \left\{[h'(\alpha) - \lambda_*(\alpha)]^2 + [h'(\alpha) - \lambda_*(\alpha)]\lambda_*'(\alpha)(x-\alpha)\right\} W(dx) \\
&= 2\int_{a_0}^{\alpha} [h'(\alpha) - \lambda_*(\alpha) + \lambda_*'(\alpha)(x-\alpha)]^2 W(dx) \\
&- 2\int_{a_0}^{\alpha} \left\{[h'(\alpha) - \lambda_*(\alpha)]\lambda_*'(\alpha)(x-\alpha) + (\lambda_*'(\alpha))^2(x-\alpha)^2\right\} W(dx) \\
&> -2\lambda_*'(\alpha)\int_{a_0}^{\alpha} \left[h'(\alpha) - \lambda_*(\alpha) + \lambda_*'(\alpha)(x-\alpha)\right](x-\alpha)W(dx) \\
&= -2\lambda_*'(\alpha)L(\alpha) = 0.
\end{aligned}$$

The penultimate equality follows from (A.96). Due to (A.95), relation $L(\alpha) = 0$ implies a local minimum of $D_2(\alpha, \lambda_*(\alpha))$ at α. In every interval of $\{K(\alpha) > 0\}$ there is at most one such point, because two of them have to be separated by a local maximum, which is impossible in view of (A.94)–(A.96). We finally note that relations $K(\alpha) > 0$ and $L(\alpha) = 0$ imply that L changes its sign from plus to minus at α.

Summing up, we have proved that parameters α_*, λ_* with the unique β_* determine the projection if $L(\alpha_*) = 0$, $K(\alpha_*) \geq 0$, and $\lambda_* = \lambda_*(\alpha_*)$ (which provides (A.92)) for some $b \leq \alpha_* < \beta_*$. It is easy to check that relations (A.91) and (A.92) imply that $\lambda_*(\alpha_*)(x-\alpha_*) + h(\alpha_*)$ is the best linear approximation of h on the interval (a_0, α_*). Suppose now that two parameters $b \leq \alpha_1^* < \alpha_2^* < \beta_*$ satisfy the necessary conditions for projection. We show that the l-h-c function with parameters $(\alpha_1^*, \lambda_*(\alpha_1^*), \beta_*)$ better approximates h than one with parameters $(\alpha_2^*, \lambda_*(\alpha_2^*), \beta_*)$. Indeed, $\lambda_*(\alpha_1^*)(x-\alpha_1^*) + h(\alpha_1^*)$ lies closer to h on (a_0, α_1^*) than $\lambda_*(\alpha_2^*)(x-\alpha_2^*) + h(\alpha_2^*)$, because it is optimal linear there. Obviously h is closer to itself than $\lambda_*(\alpha_2^*)(x-\alpha_2^*) + h(\alpha_2^*)$ on $[\alpha_1^*, \alpha_2^*]$ as well. In conclusion, among possibly non-unique points $b \leq \alpha < \beta_*$ satisfying $L(\alpha) = 0$ and $K(\alpha) \geq 0$, only the minimal one determines the projection.

Suppose now that the assumptions of the first claim are not satisfied. Then the projection of h is first linear and then constant, as given in (A.79). Its squared distance from the original h is described by the function

$$D_3(\alpha, \lambda, \gamma) = ||g_* - h||_W^2 = \int_{a_0}^{\alpha} [h(x) - \lambda(x-\alpha) - \gamma]^2 W(dx) + \int_{\alpha}^{d} [h(x) - \gamma]^2 W(dx).$$

Simple calculations show that for every fixed α and λ, it is minimized by

$$\gamma_* = \gamma_*(\alpha, \lambda) = -\lambda \int_{a_0}^{\alpha} (x - \alpha) W(dx). \tag{A.97}$$

Consider now

$$\begin{aligned} D_3(\alpha, \lambda, \gamma_*(\alpha, \lambda)) = \lambda^2 &\left\{ \int_{a_0}^{\alpha} (x-\alpha)^2 W(dx) - \left[\int_{a_0}^{\alpha} (x - \alpha) W(dx) \right]^2 \right\} \\ &- 2\lambda \int_{a_0}^{\alpha} (x - \alpha) h(x) W(dx) + \int_{a_0}^{d} h^2(x) W(dx). \end{aligned}$$

For fixed α, it is minimized by

$$\lambda_* = \lambda_*(\alpha) = \frac{\int_{a_0}^{\alpha} (x - \alpha) h(x) W(dx)}{\int_{a_0}^{\alpha} (x-\alpha)^2 W(dx) - \left[\int_{a_0}^{\alpha} (x - \alpha) W(dx) \right]^2} = \frac{A(\alpha)}{B(\alpha)}, \tag{A.98}$$

say. Note that we have

$$\int_{a_0}^{\alpha} [\lambda_*(x - \alpha) + \gamma_* - h(x)](x - \alpha) W(dx) = 0. \tag{A.99}$$

The sign of (A.98) depends on the numerator $A(\alpha)$, because the denominator is positive by the Schwarz inequality. We further have

$$D_3(\alpha, \lambda_*(\alpha), \gamma_*(\alpha, \lambda_*(\alpha))) = \int_{a_0}^{d} h^2(x) W(dx) - \frac{A^2(\alpha)}{B(\alpha)}, \tag{A.100}$$

and so

$$\frac{dD_3(\alpha, \lambda_*(\alpha), \gamma_*(\alpha, \lambda_*(\alpha)))}{d\alpha} = \frac{A(\alpha)}{B^2(\alpha)} C(\alpha)$$
$$= \lambda_*(\alpha) B'(\alpha) \left[\lambda_*(\alpha) - 2\frac{A'(\alpha)}{B'(\alpha)} \right], \qquad (A.101)$$

where

$$C(\alpha) = A(\alpha)B'(\alpha) - 2A'(\alpha)B(\alpha),$$
$$A'(\alpha) = -\int_{a_0}^{\alpha} h(x)W(dx) = \int_{\alpha}^{d} h(x)W(dx),$$
$$B'(\alpha) = -2[1 - W(\alpha)] \int_{a_0}^{\alpha} (x - \alpha)W(dx) > 0.$$

We now recall conclusions of Lemma A.6. Firstly, the optimal $\lambda_*(\alpha_*) > 0$, because the projection cannot amount to a constant. Function C should change its sign at some $\alpha \in (0, 1)$ from minus to plus, because otherwise (A.100) would be minimized either by a constant or by an increasing linear function which also contradicts Lemma A.6.

Summing up, the projection is a broken l-c line, with the break at some α, where $C(\alpha) = 0$, and consequently

$$\lambda_*(\alpha) = \frac{-\int_{\alpha}^{d} h(x)W(dx)}{[1 - W(\alpha)] \int_{a_0}^{\alpha} (x - \alpha)W(dx)},$$

$$\gamma_*(\alpha, \lambda_*(\alpha)) = \frac{\int_{\alpha}^{d} h(x)W(dx)}{1 - W(\alpha)}.$$

The last formula implies that γ_* is the best constant approximation of h on $[\alpha, d)$. We can also check that for $\lambda_* = \lambda_*(\alpha)$ defined in (A.98) and $\gamma_* = \gamma_*(\alpha, \lambda_*(\alpha))$ defined in (A.97) with $\lambda_* = \lambda_*(\alpha)$, we have

$$\int_{a_0}^{\alpha} [\lambda_*(x - \alpha) + \gamma_* - h(x)]W(dx) = 0.$$

Therefore this relation and (A.99) imply that $\lambda_*(\alpha)(x - \alpha) + \gamma_*$ is the best linear approximation of h on (a_0, α). So the glued function $\lambda_*(\alpha)(x - \alpha)\mathbf{1}_{(a_0,\alpha)}(x) + \gamma_*$

with α being a zero of C is a good candidate for the projection of h onto (A.77). However, we cannot guarantee that in general condition $C(\alpha) = 0$ defines a unique breaking point α. If there are several ones sharing the property, one should simply consider all the resulting l-c functions and choose one which minimizes the distance to the original function h (or simply maximizes the norm). We can immediately exclude each point $\alpha > \beta_*$, because such an approximation can be easily improved in the left vicinity of β_* with the use of simple geometric arguments. □

Below we change slightly assumptions (A3), and present a respective modification of Theorem A.3. This is used for evaluating the expectation of the last spacing $X_{n:n} - X_{n-1:n}$ based on an iid sample from an IFR population. We say that Assumptions (A3') hold if (A3) are modified so that $c = d = +\infty$ and $\sup_{x>0} h(x) = \lim_{x\to\infty} h(x) > 0$.

Theorem A.4 (See Theorem 6.36) *Under Assumptions* (A3'), *with notation* (A.84)–(A.86), *the set* $\mathcal{K}' = \{\alpha > b : \ K(\alpha) \geq 0, \ L(\alpha) = 0\}$ *is nonempty, and for* $\alpha_* = \inf\{\alpha \in \mathcal{K}'\}$, *we have*

$$P_{\preceq_c W} h(x) = \begin{cases} h(\alpha_*) + \lambda_*(\alpha_*)(x - \alpha_*), & a_0 \leq x < \alpha_*, \\ h(x), & x \geq \alpha_*. \end{cases}$$

Proof *(Outline)* Mimicking the proof of Theorem A.3, we can show that the projection belongs to the family

$$Ph_{\alpha,\lambda}(x) = \begin{cases} \lambda(x - \alpha) + h(\alpha), & a_0 \leq x < \alpha, \\ h(x), & x \geq \alpha, \end{cases}$$

with $\alpha \geq b$, $\lambda \geq h'(\alpha)$. The only difference between Assumption (A3') and (A3) are that now function h does not have the decreasing part. The arguments in both the cases are identical, though. Especially, they are based on the fact that every concave nondecreasing function crosses at most two times the strictly convex increasing part of h, and so does if the convex increasing part is preceded by a strictly decreasing one.

We further note that for fixed $\alpha \geq b$,

$$||h - Ph_{\alpha,\lambda}||_W^2 = \int_{a_0}^{\alpha} [\lambda(x - \alpha) + h(\alpha) - h(x)]^2 W(dx)$$

is a quadratic convex function of argument λ. It is globally minimized at $\lambda_*(\alpha)$ defined in (A.84), and the constrained optimal slope is $\max\{\lambda_*(\alpha), h'(\alpha)\}$.

Now we claim that $K(\alpha) > 0$ for all sufficiently large α. Indeed, the linear functions $h'(\alpha)(x-\alpha)+h(\alpha)$ tend to the constant $\lim_{x\to\infty} h(x)$ as $\alpha \to \infty$. Hence, $h'(\alpha)(x - \alpha) + h(\alpha) \geq h(x)$ for all $x > a_0$ if α is sufficiently large. However, this does not hold for $\lambda_*(\alpha)(x - \alpha) + h(\alpha)$, because this is the best approximation of $h_{|(a_0,\alpha)}$ by functions $\lambda(x-\alpha)+h(\alpha)$, $\lambda \in \mathbb{R}$. This implies that $\lambda_*(\alpha)(x-\alpha)+h(\alpha)$ and $h(x)$ cross each other in (a_0, α), and so $\lambda_*(\alpha) > h'(\alpha)$, as required.

Due to continuity of functions λ_* and h', the set $\{\alpha > b : K(\alpha) < 0\}$ is a possibly empty sum of open intervals. Repeating arguments of the proof of Theorem A.3, we conclude that

$$K(\alpha) < 0 \quad \Rightarrow \quad \frac{d}{d\alpha}||h - Ph_{\alpha,h'(\alpha)}||_W^2 < 0.$$

This means that for every α such that $\lambda_*(\alpha) < h'(\alpha)$, we can decrease the distance $||Ph_{\alpha,h'(\alpha)} - h||_W$ moving α to the right as long as the inequality holds. Consequently, we can restrict ourselves to the family $\{Ph_{\alpha,\lambda_*(\alpha)} : \alpha \geq b,\ K(\alpha) \geq 0\}$.

Condition $L(\alpha) = 0$, being equivalent to

$$\int_{a_0}^{d} Ph_{\alpha,\lambda_*(\alpha)}(x)W(dx) = \int_{a_0}^{d} h(x)w(x)dx,$$

is a necessary condition for $Ph_{\alpha,\lambda_*(\alpha)} = P_{\preceq_c W}h$. Accordingly, we reduce the set of candidates for projection to $\{Ph_{c,\lambda_*(\alpha)} : \alpha \in \mathcal{K}'\}$. The set is nonempty, because the projection exists, and it is of the form $Ph_{\alpha,\lambda_*(\alpha)}$. If $\alpha_1 < \alpha_2$ for some $\alpha_1, \alpha_2 \in \mathcal{K}'$, the former provides a better approximation of h. It follows from the fact that $Ph_{\alpha,\lambda_*(\alpha)}$ with $L(\alpha) = 0$ is the projection of $h_{|(a_0,\alpha)}$ onto the subspace of linear functions. Hence, $\lambda_*(\alpha_1)(x-\alpha_1)+h(\alpha_1)$ lies closer to h on (a_0, α_1) than $\lambda_*(\alpha_2)(x-\alpha_2)+h(\alpha_2)$. The same clearly holds if we compare $h(x)$ itself with $\lambda_*(\alpha_2)(x-\alpha_2)+h(\alpha_2)$ in (α_1, α_2). □

A.3 Projections onto Starshaped Functions

Here we consider the functions $h \in L^2((0, d), W(dx))$ satisfying conditions (A2') of Sect. 6.4, i.e.,

(A2') For some $0 \leq a \leq b < c \leq d \leq +\infty$ and a distribution function W with a positive density w on $(0, d)$ and finite second moment, function $h \in L^2((0, d), W(dx)$ is bounded, twice differentiable, and satisfies $h(0) \geq 0$, $\lim_{x \to d} h(x) \leq 0$,

$$\int_0^d h(x)w(x)dx = 0, \tag{A.102}$$

and it is decreasing on $(0, a)$ and (c, d) (if they are not empty), strictly convex increasing on (a, b) (if it is not empty), and strictly concave increasing on (b, c).

Our purpose is to determine the projections $P_{\succeq * W}h$ of fixed h onto the convex cone

$$\mathcal{C}_{\succeq_* W} = \{g \in L^2((0,d), W(dx)) \colon g \text{ is nondecreasing and starshaped}\}. \tag{A.103}$$

We claim that $P_{\succeq_* W}h$ has the representation

$$g(x) = g_{\alpha\beta}(x) = \begin{cases} h(\alpha), & 0 \le x \le \alpha, \\ h(x), & \alpha \le x \le \beta, \\ h(\alpha) + \frac{h(\beta)-h(\alpha)}{\beta}x, & \beta \le x < d, \end{cases} \tag{A.104}$$

for some $\alpha \in [0, c)$ and $0 \ne \beta \in [\alpha, c)$.

Function (A.104) is first constant, then identical with h, and finally linear. Parameters α and β are chosen so that (A.104) is continuous nondecreasing, and the extension of the linear part to the left runs through $(0, h(\alpha))$. If $0 \le \alpha = \beta < c$, then (A.104) is constant. Observe that (A.104) starts from a level that belongs to the range of h, and this is a necessary condition for projection. Indeed, relation $g(0) \ge h(c) = \max\{h(x) : x \in [0, d)\}$ with its nondecrease would contradict the requirement

$$\int_0^d g(x)\, W(dx) = \int_0^d h(x)\, W(dx) \tag{A.105}$$

necessary for g being the projection. Also, $g \in \mathcal{C}_{\succeq_* W}$ satisfying $g(0) < h(a) = \min\{h(x) : x \in [0, d)\}$ can be replaced by $g_+(x) = \max\{g(x), h(a)\}$, which is also nondecreasing and starshaped, and uniformly better approximates h than g in the whole domain $[0, d)$. In order to make (A.104) starshaped, we should add some further restrictions on the parameters. They are concluded from the following lemma.

Lemma A.7 *Suppose that h satisfies* (A2'). *For $\alpha \in [a, c)$, define the functions*

$$s_\alpha(\beta) = \frac{h(\beta) - h(\alpha)}{\beta}, \qquad a \ne \beta \in [\alpha, d). \tag{A.106}$$

Then for each α, there exists a unique $\bar{\beta} = \bar{\beta}(\alpha) \in [\max\{\alpha, b\}, c)$ solving the equation

$$s_\alpha(\beta) = h'(\beta), \tag{A.107}$$

and such that (A.106) *is increasing on $(\alpha, \bar{\beta})$ and decreasing on $(\bar{\beta}, c)$, where the former interval may be empty.*

Proof Suppose first that $0 = a = b = \alpha$ which means that h is concave increasing on $(0, c)$. Then for every $0 < \beta < c$, the linear function

$$\ell_{0,\beta}(x) = \frac{h(\beta) - h(0)}{\beta} x + h(0) = s_0(\beta)x + h(0) \tag{A.108}$$

crosses h at 0 and β. Since h is strictly concave on $(0, c)$, all the points $(\gamma, h(\gamma))$, $\beta < \gamma < c$ run below $\ell_{0,\beta}$. Since the respective lines $\ell_{0\gamma}$ start from $(0, h(0))$, we have $s_0(\gamma) < s_0(\beta)$ for all $0 < \beta < \gamma < c$. This means that $s_0(\beta)$ are decreasing in $(0, c)$.

Suppose now that $a \le \alpha \le b$ and construct the function

$$\tilde{h}(\beta) = \begin{cases} h(\alpha),\ 0 \le \beta \le \alpha, \\ h(\beta),\ \alpha \le \beta \le b, \end{cases} \tag{A.109}$$

which is convex on $[0, b]$ and strictly convex on $[\alpha, b]$. It is strictly starshaped on $[\alpha, b]$ as a consequence, and so $s_\alpha(\beta) = \frac{h(\beta)-h(\alpha)}{\beta}$ is increasing on (α, b).

Thus, only the case of $b < \beta < c$ remains. For $\beta = b$, function $\ell_{\alpha,b}$ (see (A.108)) crosses h from top to bottom, because b belongs to the interval of convex increase of h. For $\beta = c$, the difference $\ell_{\alpha,c} - h$ changes the sign from minus to plus, because $\ell_{\alpha,c}$ is increasing and $h'(c) = 0$. Consider the function $\beta \mapsto \ell_{\alpha,\beta}(\beta) - h(\beta)$, $b \le \beta \le c$, with the derivative $s_\alpha(\beta) - h'(\beta)$. We have $s_\alpha(b) - h'(b) < 0 < s_\alpha(c) - h'(c)$. By continuity of the derivative, there exists $b < \bar{\beta} = \bar{\beta}(\alpha) < c$ such that $s_\alpha(\bar{\beta}) = h'(\bar{\beta}) > 0$, which also implies that $h(\bar{\beta}) > h(\alpha)$ and $\bar{\beta} > \alpha$. The line $\ell_{\alpha,\bar{\beta}}(\alpha)$ is tangent to h at $\bar{\beta}$. Since h is strictly concave in (b, c), all points of the graph of h are situated below the graph of $\ell_{\alpha,\bar{\beta}}(\alpha)$, and so $s_\alpha(\beta) \le s_\alpha(\bar{\beta})$ for all $\beta \in [\alpha, c) \subset [b, c)$.

We also note that

$$s_\alpha(\bar{\beta}) > s_\alpha(c) > s_\alpha(x), \qquad c < x < d. \tag{A.110}$$

The proof is thus complete. □

Now we are in a position to state the following lemma.

Lemma A.8 *Assume that $h \in L^2((0, d), W(dx))$ satisfies* (A2'). *Then for every $g \in C_{\ge_* W}$, there exists $g_* \in C_{\ge_* W}$ of the form* (A.104) *for some $a \le \alpha < c$ and $\alpha \le \beta \le \bar{\beta}(\alpha)$, $\beta > 0$, such that*

$$||h - g_*||_W \le ||h - g||_W. \tag{A.111}$$

Proof Suppose first that a nondecreasing starshaped function g satisfies $h(a) \leq g(0) \leq h(0)$. Since h is decreasing in $(0, a)$, there exists $0 \leq \alpha_0 \leq a$ such that $g(\alpha_0-) \leq h(\alpha_0) \leq g(\alpha_0+)$ and the function

$$g_0(x) = \begin{cases} h(\alpha_0), & 0 < x \leq \alpha_0, \\ g(x), & \alpha_0 < x < d, \end{cases}$$

lies closer to h than the original g. From now on, we focus on functions $g \in \mathcal{C}_{\geq_* W}$ such that $g(0) = h(\alpha)$ for some $\alpha \in [a, c)$ and constant on $(0, \alpha_0)$ if $h(\alpha) = h(\alpha_0) < h(0)$.

We assume that a given nondecreasing starshaped function g satisfies $g(0) = h(\alpha)$ for some $a \leq \alpha < c$. Suppose first that $g(\bar{\beta}) \geq h(\bar{\beta})$ for $\bar{\beta} = \bar{\beta}(\alpha)$ defined by (A.107). Then define g_* as (A.104) with $\beta = \bar{\beta}(\alpha)$. This is starshaped. Moreover,

$$h(x) \leq h(\alpha) = g_*(x) \leq g(x), \qquad 0 \leq x \leq \alpha. \tag{A.112}$$

Since $g_*(x) = h(x)$, for $\alpha \leq x \leq \bar{\beta}(\alpha)$, g_* is obviously the best approximation of h there. Finally, for $x > \bar{\beta}(\alpha)$, we have

$$h(x) \leq \frac{h(\bar{\beta}) - h(\alpha)}{\bar{\beta}} x + h(\alpha) = g_*(x) \leq g(x). \tag{A.113}$$

The straight line defined in (A.113) is tangent to h at $\bar{\beta}$. Thus the first inequality holds by concavity of $h(x)$ in $(\bar{\beta}, c)$ and its decrease in (c, d). The second one is a consequence of the fact that the line is the smallest starshaped function in $[\bar{\beta}, d)$ which starts at $A = (0, h(\alpha))$ and passes either through or above $\bar{B} = (\bar{\beta}, h(\bar{\beta}))$. Clearly, this construction cannot provide the projection because $h \neq g^* \geq h$ which contradicts the necessary condition (A.105).

Suppose now that $g(\bar{\beta}) < h(\bar{\beta})$ and define

$$\Delta = \inf\{x \geq \bar{\beta} : g(x) \geq h(x)\}, \tag{A.114}$$

putting $\Delta = d$ if $g(x) < h(x)$ for all $x > \bar{\beta}$. The latter case occurs if either $g(x) = \text{const} = h(\alpha)$ and so satisfies (A.104) or $d < \infty$. Define the line

$$\ell_{AD}(x) = \frac{h(\Delta) - h(\alpha)}{\Delta} x + h(\alpha) \tag{A.115}$$

that runs through $A = (0, h(\alpha))$ and $D = (\Delta, h(\Delta))$. First, we observe that

$$h(x) < \ell_{AD}(x), \qquad \Delta < x < d, \tag{A.116}$$

provided that $(\Delta, d) \neq \emptyset$, because h is concave in (Δ, c) (if nonempty) and decreasing in $(\max\{\Delta, c\}, d)$. We also have

$$\ell_{AD}(x) < h(x), \qquad \bar{\beta} < x < \Delta. \tag{A.117}$$

Indeed, if $(c, \Delta) \neq \emptyset$, then $\ell_{AD}(x) - h(x)$ is decreasing there and becomes zero at Δ. If $\Delta \in (\bar{\beta}, c)$, then $\ell_{AD}(x) - h(x)$ is concave there, and we obtain

$$\ell_{AD}(\bar{\beta}) \le \ell_{A\bar{B}}(\bar{\beta}) = h(\bar{\beta}) \tag{A.118}$$

by (A.110).

Now we show that there is one point in $(a, \bar{\beta})$ where $h(x)$ and $\ell_{AD}(x)$ coincide. Such a point exists, because $h(a) \le \ell_{AD}(a) = h(\alpha)$, $\ell_{AD}(d) \ge h(d)$, (A.118) holds, and both h and ℓ_{AD} are continuous. Put

$$\delta = \sup\{x \in [a, \beta) : h(x) = \ell_{AD}(x)\}. \tag{A.119}$$

Evidently, $\ell_{AD}(x) - h(x)$ is positive and negative at the left and right neighborhoods of δ, respectively. Assume first that $\delta \ge b$. Since $\ell_{AD}(x) - h(x)$ changes the sign from plus to minus at δ, and is strictly convex in (b, δ), it is positive in $[b, \delta)$. It is also strictly concave in (a, b), and nonnegative at the ends, and so positive in (a, b) as well. If $a < \delta < b$, then $\ell_{AD}(x) - h(x) > 0$ for all $x \in (a, \delta)$, because this is strictly concave there and nonnegative at a and δ. We have thus proved that there is a unique $\delta \in (a, \bar{\beta})$ satisfying $h(\delta) = \ell_{AD}(\delta)$.

We have established that

$$h(x) \le \ell_{AD}(x), \qquad x \in (a, \delta], \tag{A.120}$$

$$h(x) \ge \ell_{AD}(x), \qquad x \in [\delta, \bar{\beta}]. \tag{A.121}$$

We also have

$$g(x) \le \ell_{AD}(x), \qquad x \in (a, \Delta], \tag{A.122}$$

$$\ell_{AD}(x) \le g(x), \qquad x \in [\Delta, d). \tag{A.123}$$

because $\ell_{AD}(a) = g(a) = h(\alpha)$, $g(\Delta-) \le \ell_{AD}(\Delta) = h(\Delta) \le g(\Delta+)$, $g(x)$ is starshaped and $\ell_{AD}(x)$ is linear. We can write

$$\ell_{AD}(x) = h(\delta) + \frac{h(\delta) - h(\alpha)}{\delta - a}(x - a) \tag{A.124}$$

and define

$$g_*(x) = \begin{cases} h(\alpha), & 0 \le x \le \alpha, \\ h(x), & \alpha \le x \le \delta, \\ h(\alpha) + \frac{h(\delta)-h(\alpha)}{\delta}x, & \delta \le x < d. \end{cases}$$

We easily check that this function is starshaped. Moreover, it lies closer to h than the original g. Indeed, if $h(\alpha) < h(0)$, then $g_*(x) = g(x)$ for $0 < x \le \alpha_0$, where $0 < \alpha_0 \le a$ is defined by $h(\alpha_0) = h(\alpha)$. We also have $h(x) \le g_*(x) \le g(x)$ for $x \in [\alpha_0, \alpha]$. If $h(\alpha) \ge h(0)$, then $h(x) \le g_*(x) \le g(x)$ for all $x \in (0, \alpha)$. In $[\alpha, \delta]$, we simply get $g_* = h$ which is the optimal approximation. We further obtain $g(x) \le g_*(x) \le h(x)$ in the interval $[\delta, \Delta]$. We finally observe that $h(\alpha) + \frac{h(\delta)-h(\alpha)}{\delta}x$ is the smallest starshaped function passing through $A = (0, h(\alpha))$ and $D = (\Delta, h(\Delta))$. This completes the proof. □

We now determine parameters α_*, β_* of the representation (A.104) of the projection $P_{\succeq_* W}h$ of h satisfying assumptions of Lemma A.8. Set first

$$K(\alpha, \beta) = \int_\beta^d \left[h(\alpha) + \frac{h(\beta) - h(\alpha)}{\beta}x - h(x)\right] x \; W(dx), \tag{A.125}$$

$$L(\alpha, \beta) = \int_0^\alpha [h(\alpha) - h(x)] W(dx) + \int_\beta^d \left[h(\alpha) + \frac{h(\beta) - h(\alpha)}{\beta}x - h(x)\right] W(dx), \tag{A.126}$$

for $a \le \alpha \le \beta \le c$, $\beta \ne 0$, and

$$K(\alpha) = K(\alpha, \alpha) = \int_\alpha^d [h(\alpha) - h(x)]x \; W(dx), \tag{A.127}$$

$$L(\alpha) = L(\alpha, \alpha) = \int_0^d [h(\alpha) - h(x)] W(dx) = h(\alpha), \tag{A.128}$$

for $a \le \alpha \le c$. Note that (A.128) is strictly increasing on $[a, c]$ and $L(a) < 0 < L(c)$.

Theorem A.5 (See Theorem 6.44) *Assume* (A2'). *If either $K(0) \ge 0$ or $K(0) < 0$ and*

$$K(\alpha_L) = -\int_{\alpha_L}^d h(x)x \; W(dx) \ge 0, \tag{A.129}$$

for α_L being the unique zero of h (and L) in (a, c), then $P_{\succeq_* W}h = \mathbf{0}$.

Otherwise, there exists a unique pair $a \le \alpha_* < c$, $\alpha_* < \beta_* < \bar{\beta}(\alpha_*)$ solving the equations

$$K(\alpha, \beta) = 0, \tag{A.130}$$

$$L(\alpha, \beta) = 0, \tag{A.131}$$

such that

$$P_{\geq_* W} h(x) = \begin{cases} h(\alpha_*), & 0 \leq x \leq \alpha_*, \\ h(x), & \alpha_* \leq x \leq \beta_*, \\ h(\alpha_*) + \frac{h(\beta_*)-h(\alpha_*)}{\beta_*} x, & \beta_* \leq x < d. \end{cases} \tag{A.132}$$

Proof Owing to Lemma A.8, it suffices to minimize

$$D(\alpha, \beta) = \int_a^d [h(x) - g_{\alpha\beta}(x)]^2 W(dx) = \int_a^\alpha [h(x) - h(\alpha)]^2 W(dx)$$
$$+ \int_\beta^d \left[h(x) - h(\alpha) - \frac{h(\beta) - h(\alpha)}{\beta} x \right]^2 W(dx). \tag{A.133}$$

with respect to $a \leq \alpha < c$ and $\alpha \leq \beta < \bar{\beta}(\alpha)$, $\beta \neq 0$. Keeping α fixed, we calculate

$$\frac{\partial D(\alpha, \beta)}{\partial \beta} = 2 \frac{h'(\beta) - \ell'_{AB}}{\beta} K(\alpha, \beta), \tag{A.134}$$

where ℓ_{AB} denotes the line running through $A = (0, h(\alpha))$ and $B = (\beta, h(\beta))$, and $K(\alpha, \beta)$ is defined in (A.125). The fraction is positive, because $h(x) - \ell_{AB}(x)$ changes the sign from minus to plus at β. Therefore, the sign of (A.134) coincides with the sign of (A.125). We also see that

$$\frac{\partial K(\alpha, \beta)}{\partial \beta} = \frac{h'(\beta) - \ell'_{AB}}{\beta} \int_0^d x^2 W(dx) > 0. \tag{A.135}$$

Moreover, $K(\alpha, \bar{\beta}(\alpha)) > 0$, because the integrand in (A.125) is positive then. Indeed, the line $h(\alpha) + \frac{h(\bar{\beta})-h(\alpha)}{\bar{\beta}} x$ is tangent to h at $\bar{\beta}$ and runs above h in $[\bar{\beta}, d)$.

Now consider

$$K(\alpha) = \lim_{\beta \to \alpha} K(\alpha, \beta). \tag{A.136}$$

We note $K(c) > 0$, and

$$K'(\alpha) = h'(\alpha) \int_\alpha^d x W(dx) > 0, \qquad 0 < \alpha < c. \tag{A.137}$$

Suppose that $K(0) \geq 0$ so that $K(\alpha)$ is a positive function. Due to (A.134), function (A.133) is minimized for $\alpha = \beta \to 1$, which means that the projection (A.132) is a constant function. Since $\int_0^d h(x) W(dx) = 0$, the optimal constant amounts to 0.

Assume now that $K(0) < 0$. This implies that (A.127) is increasing and has a unique zero at some $\alpha_K \in (a, c)$. Accordingly, we arrive at the conclusions

$$D(\alpha) = \inf_{\alpha \le \beta \le \bar{\beta}(\alpha)} D(\alpha, \beta) = D(\alpha, \beta_{**}(\alpha)) = \begin{cases} D(\alpha, \beta_*(\alpha)), & \text{if } \alpha \le \alpha_K, \\ D(\alpha, \alpha), & \text{if } \alpha \ge \alpha_K, \end{cases} \tag{A.138}$$

where $\beta_*(\alpha)$ is the unique solution to (A.130) for a given $\alpha \le \alpha_K$. Now we aim at minimizing (A.138) with respect to α. By Lemma 6.1, a necessary condition to that is

$$L(\alpha, \beta_{**}(\alpha)) = 0. \tag{A.139}$$

Since the projection problem has a solution, there exists at least one α satisfying (A.139). Below we show that the LHS of (A.139) is strictly increasing, and so its unique zero α_* minimizes (A.138) and solves the projection problem.

If $\alpha \le \alpha_K$, using the differentiation rule for implicit functions, we obtain

$$\begin{aligned} \beta_*'(\alpha) &= -\frac{\partial K(\alpha, \beta_*(\alpha))}{\partial \alpha} \Big/ \frac{\partial K(\alpha, \beta_*(\alpha))}{\partial \beta} \\ &= \frac{h'(\alpha) \int_{\beta_*(\alpha)}^d [x - \beta_*(\alpha)] x \, W(dx)}{[h'(\beta_*(\alpha)) - \ell'_{AB_*}] \int_{\beta_*(\alpha)}^d x^2 \, W(dx)}. \end{aligned} \tag{A.140}$$

This is positive because so are all the factors in the numerator and denominator. This implies that $\beta_*(\alpha)$ increases with respect to α till $\beta_*(\alpha_K) = \alpha_K$. By (A.140), for $\alpha \le \alpha_K$, we have

$$\begin{aligned} \frac{dL(\alpha, \beta_*(\alpha))}{d\alpha} &== \frac{\partial L(\alpha, \beta_*(\alpha))}{\partial \alpha} + \beta_*'(\alpha) \frac{\partial L(\alpha, \beta_*(\alpha))}{\partial \beta} \\ &= h'(\alpha) \left[\int_0^\alpha W(dx) + \int_{\beta_*(\alpha)}^d W(dx) - \frac{\int_{\beta_*(\alpha)}^d x \, W(dx)}{\beta_*(\alpha)} \right] \\ &\quad + \beta_*'(\alpha)[h'(\beta_*(\alpha)) - \ell'_{AB_*(\alpha)}] \frac{\int_{\beta_*(\alpha)}^d x \, W(dx)}{\beta_*(\alpha)} \\ &= \frac{h'(\alpha)}{\int_{\beta_*(\alpha)}^d x^2 W(dx)} \left\{ \int_a^\alpha W(dx) \int_{\beta_*}^d x^2 W(dx) \right. \\ &\quad \left. + \int_{\beta_*(\alpha)}^d W(dx) \int_{\beta_*(\alpha)}^d x^2 W(dx) - \left[\int_{\beta_*(\alpha)}^d x W(dx) \right]^2 \right\}. \end{aligned} \tag{A.141}$$

Due to the Schwarz inequality, the expression in the last line is positive All the factors in the preceding line of (A.141) are positive as well. Hence, $L(\alpha\ \beta_{**}(\alpha))$ is increasing for $\alpha \leq \alpha_k$. The same conclusion holds for $\alpha \geq \alpha_k$ because then

$$L(\alpha, \beta_{**}(\alpha)) = L(\alpha) = h(\alpha), \tag{A.142}$$

and we obtain the desired claim.

If $\alpha_k \leq \alpha_L$ which means $K(\alpha_L) \geq 0$ (see (A.129)), then the optimal parameters are identical and amount to α_L. Thus the projection is constant and equal to $\mathbf{0}$. Otherwise, the system of equations (A.130) and (A.131) uniquely determines parameters $\alpha_*, \beta_*(\alpha_*)$ of the projection of the form (A.132). This completes the proof. □

References

Ali, M. M., & Chan, L. K. (1965). Some bounds for expected values of order statistics. *Annals of Mathematical Statistics, 36*, 1055–1057.

Arnold, B. C. (1985). p-Norm bounds on the expectation of the maximum of possibly dependent sample. *Journal of Multivariate Analysis, 17*, 316–332.

Arnold, B. C., & Balakrishnan, N. (1989). *Relations, bounds and approximations for order statistics*. Lecture notes in statistics (Vol. 53). New York: Springer-Verlag.

Arnold, B. C., Balakrishnan, N., & Nagaraja, H. N. (1992). *A first course in order statistics*. New York: Wiley.

Balakrishnan, A. V. (1981). *Applied functional analysis* (2nd ed.). New York: Springer-Verlag.

Balakrishnan, N. (1984). Algorithm AS 200. Approximating the sum of squares of normal scores. *Journal of the Royal Statistical Society, Series C, 33*, 242–245.

Balakrishnan, N. (1990). Improving the Hartley-David-Gumbel bound for the mean of extreme order statistics. *Statistics & Probability Letters, 9*, 291–294.

Balakrishnan, N. (1993). A simple application of binomial-negative binomial relationship in the derivation of sharp bounds for moments of order statistics based on greatest convex minorants. *Statistics & Probability Letters, 18*, 301–305.

Balakrishnan, N., Charalambides, C., & Papadatos, N. (2003). Bounds on the expectation of order statistic from a finite population. *Journal of Statistical Planning and Inference, 113*, 569–588.

Balakrishnan, N., & Joshi, P. C. (1985). Bounds for the mean of second largest order statistic in large samples. *Statistics, 16*, 457–454.

Balakrishnan, N., & Rao, C. R. (Eds.). (1998a). *Handbook of statistics 16: Order statistics: Theory and methods*. Amsterdam: North-Holland.

Balakrishnan, N., & Rao, C. R. (Eds.). (1998b). *Handbook of statistics 17: Order statistics: Applications*. Amsterdam: North-Holland.

Balakrishnan, N., & Rychlik, T. (2006). Evaluating expectations of L-statistics by the Steffensen inequality. *Metrika, 63*, 371–384.

Balakrishnan, N., & Samanta, R. (2023). Improved imputation-type bounds for moments of order statistics (submitted for publication).

Barlow, R. E., Bartholomew, D. J., Bremner, J. M., & Brunk, H. D. (1972). *Statistical inference under order restrictions. The theory and application of isotonic regression*. New York: Wiley.

Barlow, R. E., & Proschan, F. (1966). Inequalities for linear combinations of order statistics from restricted families. *Annals of Mathematical Statistics, 37*, 1574–1592.

Barnett, V. D. (1966). Order statistics estimators of the location of the Cauchy distribution. *Journal of the American Statistical Association, 61*, 1205–1218. Correction, *63*, 383–385.

N. Balakrishnan, T. Rychlik, *Sharp Inequalities for Ordered Random Variables in Statistics and Reliability*, Frontiers in Probability and the Statistical Sciences,
https://doi.org/10.1007/978-3-031-61347-0

Beesack, P. R. (1973). On bounds for the range of ordered variates. *Publication of the Electrotechnical Faculty of Belgrade University, Mathematics and Physics Series, 428*, 93–96.

Beesack, P. R. (1976). On bounds for the range of ordered variates II. *Aequationes Mathematicae, 14*, 293–301.

Billingsley, P. (1986). *Probability and measure* (2nd ed.). New York: Wiley.

Blom, G. (1958). *Statistical estimates and transformed beta-variables*. Uppsala: Almqvist and Wiksell.

Borenius, G. (1966). On the limit distribution of an extreme value in a sample from a normal distribution. *Skandinavisk Aktuarietidskrift, 1965*, 1–15.

Boyd, A. V. (1971). Bounds for order statistics. *Publications of the Electrotechnical Faculty of Belgrade University Mathematics and Physics Series, 365*, 31–32.

Bruckner, A. M., & Ostrow, E. (1962). Some function classes related to the class of convex functions. *Pacific Journal of Mathematics, 12*, 1203–1215.

Čiginas, A., & Pumputis, D. (2014). A note on the upper bound to variance of the sample extreme from a finite population. *Communications in Statistics—Theory and Methods, 43*, 1793–1799.

Danielak, K. (2003). Sharp upper mean-variance bounds for trimmed means from restricted families. *Statistics, 37*, 305–324.

Danielak, K., & Rychlik, T. (2003a). Exact bounds for the bias of trimmed means. *Australian & New Zealand Journal of Statistics, 45*, 83–96.

Danielak, K., & Rychlik, T. (2003b). Sharp bounds for expectations of spacings from DDA and DFRA families. *Statistics & Probability Letters, 65*, 303–316.

Danielak, K., & Rychlik, T. (2004). Sharp bounds for expectations of spacings from decreasing density and failure rate families. *Applied Mathematics (Warsaw), 31*, 369–395.

David, H. A. (1988). General bounds and inequalities in order statistics. *Communications in Statistics-Theory Methods, 17*, 2119–2134.

David, H. A., & Nagaraja, H. N. (2003). *Order statistics* (3rd ed.). Hoboken, New Jersey: Wiley.

Descartes, R. (1954). *The geometry of Rene descartes with a facsimile of the First Edition*. New York: Dover.

Deutsch, F. (2001). *Best approximation in inner product spaces*. New York: Springer-Verlag.

Fahmy, S., & Proschan, F. (1981). Bounds on differences of order statistics. *American Statistician, 35*, 46–47.

Fréchet, M. (1907). Sur les ensembles de fonctions et les opérations linéaires. *Comptes rendus de l'Académie des Sciences, 144*, 1414–1416 (in French).

Gajek, L., & Okolewski, A. (1997). Steffensen-type inequalities for order and record statistics. *Annales Universitatis Mariae Curie-Sklodowska, Section A, 51*, 41–59.

Gajek, L., & Okolewski, A. (2000a). Sharp bounds on moments of generalized order statistics. *Metrika, 52*, 27–43.

Gajek, L., & Okolewski, A. (2000b). Inequalities for generalized order statistics from some restricted family of distributions. *Communications in Statistics-Theory Methods, 29*, 2427–2438.

Gajek, L., & Okolewski, A. (2001). Improved Steffensen type bounds on expectations of record statistics. *Statistics & Probability Letters, 55*, 205–212.

Gajek, L., & Rychlik, T. (1998). Projection method for moment bounds on order statistics from restricted families. II. Independent case. *Journal of Multivariate Analysis, 64*, 156–182.

Goroncy, A. (2009). Lower bounds on positive L-statistics. *Communications in Statistics-Theory Methods, 38*, 1989–2002.

Goroncy, A., & Rychlik, T. (2006). How deviant can you be? The complete solution. *Mathematical Inequalities & Applications, 9*, 633–647.

Goroncy, A., & Rychlik, T. (2006b). Lower bounds on expectations of positive L-statistics based on samples drawn with replacement. *Statistics, 40*, 389–408.

Goroncy, A., & Rychlik, T. (2008). Lower bounds on expectations of positive L-statistics from without replacement models. *Journal of Statistical Planning and Inference, 138*, 3647–3659.

Goroncy, A., & Rychlik, T. (2009). Lower bounds on spacings based on with-replacement model. *Communications in Statistics-Theory Methods, 38*, 1982–1988.

Goroncy, A., & Rychlik, T. (2015). Optimal bounds on expectations of order statistics and spacings from nonparametric families of distributions generated by convex transform order. *Metrika, 78*, 175–204.

Goroncy, A., & Rychlik, T. (2016). Evaluations of expected order statistics and spacings based on the IFR distributions. *Metrika, 79*, 635–657.

Goroncy, A., & Rychlik, T. (2018). Refined solution to upper bound problem for the expectations of order statistics from decreasing density on the average distributions. *Communications in Statistics-Theory Methods, 47*, 4029–4041.

Goroncy, A., & Rychlik, T. (2024). Bounds on the expectations of order statistics for the monotone reversed failure rate distributions. *Metrika* (to appear). http://dx.doi.org/10.1007/s00184-024-00968-y

Groeneveld, R. A. (1982). Best bounds for order statistics and their expectations in range and mean units with applications. *Communications in Statistics-Theory Methods, 11*, 1809–1815.

Gumbel, E. L. (1954). The maxima of the mean largest value and of the range. *Annals of Mathematical Statistics, 25*, 76–84.

Hardy, G., Littlewood, J. E., & Pólya, G. (1934). *Inequalities*. Cambridge: Cambridge University Press.

Harter, H. L. (1961). Expected values of normal order statistics. *Biometrika 48*, 151–165. Correction, **48**, 476.

Harter, H. L., & Balakrishnan, N. (1996). *CRC handbook of tables for the use of order statistics in estimation*. Boca Raton: CRC Press.

Hartley, H. O., & David, H. A. (1954). Universal bounds for the mean range and extreme observation. *Annals of Mathematical Statistics, 25*, 85–99.

Hawkins, D. M. (1971). On the bounds of the range of order statistics. *Journal of the American Statistical Association, 66*, 644–645.

Hoeffding, W. (1940). Masstabinvariante Korrelations-Theorie. *Schriften Mathematics Institute of University of Berlin, 5*, 181–233.

Irwin, J. O. (1925). The further theory of Francis Galton's individual difference problem. *Biometrika, 17*, 100–128.

Jasiński, K., & Rychlik, T. (2013). Maximum variance of order statistics from symmetric populations revisited. *Statistics, 47*, 422–438.

Jasiński, K., & Rychlik, T. (2016). Inequalities for variances of order statistics originating from urn models. *Journal of Applied Probability, 53*, 162–173.

Jones, M. C., & Balakrishnan, N. (2002). How are moments and moments of spacings related to distribution function. *Journal of Statistical Planning and Inference, 103*, 377–390.

Joshi, P. C. (1969). Bounds and approximations for the moments of order statistics. *Journal of the American Statistical Association, 64*, 1617–1624.

Joshi, P. C., & Balakrishnan, N. (1983). Bounds for the moments of extreme order statistics for large samples. *Mathematisch Operationsforschung und Statistik, 14*, 387–396.

Karlin, S., & Studden, W. J. (1966). *Tchebycheff systems: With applications in analysis and statistics*. New York: Wiley.

Komornik, V. (2006). Another short proof of Descartes's rule of signs. *American Mathematical Monthly, 113*, 829–830.

Kozyra, P. M. (2017). *Sharp bounds on the moments of linear combinations of order statistics and kth records*. Ph.D. thesis. Warsaw: Inst. Math., Polish Acad. Sci.

Kozyra, P. M., & Rychlik, T. (2017a). Sharp bounds on the expectations of L-statistics expressed in the Gini mean differences units. *Communications in Statistics-Theory Methods, 46*, 2921–2941.

Kozyra, P. M., & Rychlik, T. (2017b). Lower and upper bounds on the variances of spacings. *Annals of the Institute of Mathematical Statistics, 69*, 417–428.

Krein, M., & Milman, D. (1940). On extreme points of regular convex sets. *Studia Mathematica, 9*, 133–138.

Lai, T. L., & Robbins, H. (1976). Maximally dependent random variables. *Proceedings of the National Academy of Sciences of USA, 73*, 286–288.

Lai, T. L., & Robbins, H. (1978). A class of dependent random variables and their maxima. *Zeitschrift für Wahrscheinlichkeitstheorie und verwandte Gebiete, 42*, 89–111.

Lawrence, M. J. (1975). Inequalities for s-ordered distributions. *Annals of Statistics, 3*, 413–428.

Lehmann, E. L. (1966). Some concepts of dependence. *Annals of Mathematical Statistics, 37*, 1137–1153.

Lin, G., & Huang, J. (1989). Variances of sample medians. *Statistics & Probability Letters, 8*, 143–146.

Lindenstrauss, J., & Tzafriri, L. (1977). *Classical Banach spaces. I. Sequence spaces*. Berlin: Springer-Verlag.

Lindenstrauss, J., & Tzafriri, L. (1979). *Classical Banach spaces. II. Function spaces*. Berlin: Springer-Verlag.

Loh, W.-Y. (1984). Bounds on AREs for restricted classes of distributions defined via tail-orderings. *Annals of Statistics, 12*, 685–701.

López-Blázquez, F. (1998). Discrete distributions with maximum expected value of the maximum. *Journal of Statistical Planning and Inference, 70*, 201–207.

López-Blázquez, F. (2000). Bounds for the expected value of spacings from discrete distributions. *Journal of Statistical Planning and Inference, 84*, 1–9.

López-Blázquez, F., & Rychlik, T. (2008). Sharp upper bounds for the expected values of non-extreme order statistics from discrete distributions. *Journal of Statistical Planning and Inference, 138*, 3635–3646.

Ludwig, O. (1973). Differenzen der Erwatungswerte von Ranggrössen in kleinen Stichproben. In: B. Bereanu, M. Iosifescu, T. Postelnicu, & P. Tăutu (Eds.), *Proc. 4th Conf. on Probab. Theory*, Editura Acad., Braşov (pp. 299–303).

Mallows, C. L., & Richter, D. (1969). Inequalities of Chebyshev type involving conditional expectations. *Annals of Mathematical Statistics, 40*, 1922–1932.

Mărgăritescu, E., & Nicolae, T. (1990). Best linear bounds for order statistics in terms of the mean and the generalized range. *Studii şi Cercetări de Matematică, 42*, 41–45.

Marshall, A. W., Olkin, I., & Arnold, B. C. (2011). *Inequalities: Theory of majorization and its applications*. New York: Springer.

Marshall, A. W., & Proschan, F. (1970). Mean life of series and parallel systems. *Journal of Applied Probability, 7*, 165–170.

Mathai, A. M. (1975). Bounds for moments through a general orthogonal expansion in a pre-Hilbert space I. *Canadian Journal of Statistics, 3*, 13–34.

Mitrinović, D. S. (1970). *Analytic inequalities*. Berlin: Spring-Verlag.

Mitrinović, D. S., Pečarić, J. E., & Fink, A. M. (1991). *Inequalities involving functions and their integrals and derivatives*. Dordrecht: Kluwer Academic Publishers Group.

Moriguti, S. (1951). Extremal properties of extreme value distributions. *Annals of Mathematical Statistics, 22*, 523–536.

Moriguti, S. (1953). A modification of the Schwarz's inequality with applications to distributions. *Annals of Mathematical Statistics, 24*, 107–113.

Nagaraja, H. N. (1981). Some finite sample results for the selection differential. *Annals of Mathematical Statistics, 33*, 437–448.

Nair, K. R. (1948). The distribution of the extreme deviate from the sample mean and its studentized form. *Biometrika, 35*, 118–144.

Olkin, I. (1992). A matrix formulation on how deviant can an observation be. *American Statistician, 46*, 205–209.

Papadatos, N. (1995). Maximum variance of order statistics. *Annals of the Institute of Statistical Mathematics, 47*, 185–193.

Papadatos, N. (1997). A note on maximum variance of order statistics from symmetric populations. *Annals of the Institute of Statistical Mathematics, 49*, 117–121.

Papadatos, N. (2021). Optimal moment inequalities for order statistics from nonnegative random variables. *Probability in the Engineering and Informational Sciences, 35*, 316–330.

Papadatos, N., & Rychlik, T. (2004). Bounds on expectations of L-statistics from without replacement samples. *Journal of Statistical Planning and Inference, 124*, 317–336.

Parrish, R. S. (1992). Computing variances and covariances of normal order statistics. *Communications in Statistics–Simulation and Computation, 21*, 71–101.

Pearson, E. S., & Chandra Sekar, C. (1936). The efficiency of statistical tools and a criterion for the rejection of outlying observations. *Biometrika, 28*, 308–319.

Pearson, K. (1902). Note on Francis Galton's problem. *Biometrika, 1*, 390–399.

Pečarić, J. E., Proschan, F., & Tong, Y. L. (1992). *Convex functions, partial orderings, and statistical applications*. Boston: Academic Press.

Plackett, R. L. (1947). Limits of the ratio of mean range to standard deviation. *Biometrika, 34*, 120–122.

Rao, M. M., & Ren, Z. D. (1991). *Theory of Orlicz spaces*. New York: Marcel Dekker.

Riesz, F. (1907). Sur une espece de géométrie analytique des systemes de fonctions sommables. *Comptes rendus de l'Académie des Sciences, 144*, 1409–1411 (in French).

Riesz, F. (1909). Untersuchungen über Systeme integrierbarer Funktionen *Mathematische Annalen, 69*, 449–497 (in German).

Robertson, T., Wright, F. T., & Dykstra, R. L. (1988). *Order restricted statistical inference*. New York: Wiley.

Rockafellar, R. T. (1970). *Convex analysis*. Princeton Mathematical Series (Vol. 28). Princeton: Princeton University Press.

Ruben, H. (1954). On the moments of order statistics in samples from normal populations. *Biometrika, 41*, 200–227.

Rudin, W. (1991). *Functional analysis* (2nd ed.). New York: McGraw-Hill.

Rychlik, T. (1992). Sharp inequalities for linear combinations of elements of monotone sequences. *Bulletin of the Polish Academy of Sciences. Mathematics, 40*, 247–254.

Rychlik, T. (1993). Sharp bounds on L-estimates and their expectations for dependent samples. *Communications in Statistics-Theory Methods, 22*, 1053–1068. Corrigendum in *Communications in Statistics-Theory Methods, 23* (1994), 305–306.

Rychlik, T. (1998). Bounds for expectations of L-estimates. In: *Handbook of statistics 16–Order statistics: Theory & methods* (Chapter 6, pp. 105–145). Amsterdam: North-Holland.

Rychlik, T. (2001). *Projecting statistical functionals*. Lecture notes in statistics (Vol. 160). New York: Springer-Verlag.

Rychlik, T. (2002). Optimal mean-variance bounds on order statistics from families determined by star ordering. *Applied Mathematics (Warsaw), 29*, 15–32.

Rychlik, T. (2004). Optimal bounds on L-statistics based on samples drawn with replacement from finite populations. *Statistics, 38*, 391–412.

Rychlik, T. (2009a). Bounds on expectations of small order statistics from decreasing density populations. *Metrika, 70*, 369–381.

Rychlik, T. (2009b). Non-positive upper bounds on expectations of low rank order statistics from DFR populations. *Statistics, 43*, 53–63.

Rychlik, T. (2009c). Tight evaluations for expectations of small order statistics from symmetric and symmetric unimodal populations. *Statistics & Probability Letters, 79*, 1488–1493.

Rychlik, T. (2014a). Non-positive upper bounds on expectations of small order statistics from DDA and DFRA populations. *Metrika, 77*, 539–557.

Rychlik, T. (2014b). Maximal expectations of extreme order statistics from increasing density and failure rate populations. *Communications in Statistics-Theory and Methods, 43*, 2199–2213.

Rychlik, T. (2021). Bounds on the expectations of L-statistics from iid symmetric populations in various scale units. *Mathematical Methods of Statistics, 30*, 47–62.

Rychlik, T. (2022). Bounds on the expectations of L-statistics based on iid life distributions. *Mathematical Methods of Statistics, 31* (2022), 43–56.

Rychlik, T., & Szymkowiak, M. (2021). Bounds on the lifetime expectations of series systems with IFR component lifetimes. *Entropy, 23*, Paper no. 385, 14 pp.

Samuelson, P. A. (1968). How deviant can you be? *Journal of the American Statistical Association, 63*, 1522–1525.

Sansone, G. (1959). *Orthogonal functions*. New York: Dover Publications.
Saw, J. G., & Chow, B. (1966). The curve through the expected values of ordered variates and the sum of squares of normal scores. *Biometrika, 53*, 252–255.
Scott, J. M. C. (1936). Appendix to paper by Pearson and Chandra Sekar. *Biometrika, 28*, 319–320.
Sen, P. K. (1959). On moments of the sample quantiles. *Calcutta Statistical Associate Bulletin, 9*, 1–19.
Segner, I. A. (1728). Dissertatio epistolica, qua regulam Harriotti..., University of Jena.
Segner, I. A. (1756). Démonstration de la règle de Descartes. *Histoire de l'académie de Berlin*, 292–299.
Sen, P. K. (1959). On the moments of the sample quantiles. *Calcutta Statistical Association Bulletin, 9*, 1–19.
Shaked, M., & Shanthikumar, J. (2007). *Stochastic orders*. New York: Springer.
Steinhaus, H. (1919). Additive und stetige Funktionaloperationen. *Mathematische Zeitschrift, 5*, 186–221 (in German).
Stoops, G., & Barr, D. (1971). Moments of certain Cauchy order statistics. *American Statistician, 25*, 51.
Sugiura, N. (1962). On the orthogonal inverse expansion with an application to the moments of order statistics. *Osaka Mathematical Journal, 14*, 253–263.
Sugiura, N. (1964). The bivariate orthogonal inverse expansion and the moments of order statistics. *Osaka Journal of Mathematics, 1*, 45–59.
Szegö, G. (1959). *Orthogonal polynomials*. Providence: American Mathematical Society.
Szymkowiak, M. (2020). *Lifetime analysis by aging intensity functions*. Studies in systems, decision and control (Vol. 192). Cham: Springer.
Takahasi, K., & Futatsuya, M. (1998). Dependence between order statistics in samples from finite population and its application to ranked set sampling. *Annals of the Institute of Statistical Mathematics, 50*, 49–70.
Thomson, G. W. (1955). Bounds on the ratio of range to standard deviation. *Biometrika, 42*, 268–269.
Tietjen, G. L., Kahaner, D. L., & Beckman, R. J. (1977). Variances and covariances of the normal order statistics for sample sizes 2 to 50. *Selected Tables in Mathematical Statistics, 5*, 1–74.
Tolstov, G. P. (1962). *Fourier series*. New York: Dover Publications.
van Zwet, W. R. (1964). *Convex transformations of random variables*. Mathematical Centre tracts (Vol. 7). Amsterdam: Mathematisch Centrum.
Yamauti, Z. (Ed.). (1972). *Statistical tables and formulas with computer applications*. Tokyo: Japanese Standard Association.
Yang, H. J. (1982). On the variances of median and some order statistics. *Bulletin of the Institute of Mathematics Academia Sinica, 10*, 197–204.

SPRINGER NATURE

GPSR Compliance

The European Union's (EU) General Product Safety Regulation (GPSR) is a set of rules that requires consumer products to be safe and our obligations to ensure this.

If you have any concerns about our products, you can contact us on ProductSafety@springernature.com

In case Publisher is established outside the EU, the EU authorized representative is:

Springer Nature Customer Service Center GmbH
Europaplatz 3
69115 Heidelberg, Germany

The manufacturer's authorised representative in the EU is Springer Nature Customer Service Centre GmbH, Europaplatz 3, 69115 Heidelberg, Germany. If you have any concerns regarding our products, please contact ProductSafety@springernature.com

Printed and bound by CPI Group (UK) Ltd, Croydon, CR0 4YY
01/10/2026
02225897-0020